R.P. NASH

THEORY OF PLATES AND SHELLS

ENGINEERING SOCIETIES MONOGRAPHS

Bakhmeteff: *Hydraulics of Open Channels*
Bleich: *Buckling Strength of Metal Structures*
Crandall: *Engineering Analysis*
Elevatorski: *Hydraulic Energy Dissipators*
Leontovich: *Frames and Arches*
Nadai: *Theory of Flow and Fracture of Solids*
Timoshenko and Gere: *Theory of Elastic Stability*
Timoshenko and Goodier: *Theory of Elasticity*
Timoshenko and Woinowsky-Krieger: *Theory of Plates and Shells*

Five national engineering societies, the American Society of Civil Engineers, the American Institute of Mining, Metallurgical, and Petroleum Engineers, the American Society of Mechanical Engineers, the American Institute of Electrical Engineers, and the American Institute of Chemical Engineers, have an arrangement with the McGraw-Hill Book Company, Inc., for the production of a series of selected books adjudged to possess usefulness for engineers and industry.

The purposes of this arrangement are: to provide monographs of high technical quality within the field of engineering; to rescue from obscurity important technical manuscripts which might not be published commercially because of two limited sale without special introduction; to develop manuscripts to fill gaps in existing literature; to collect into one volume scattered information of especial timeliness on a given subject.

The societies assume no responsibility for any statements made in these books. Each book before publication has, however, been examined by one or more representatives of the societies competent to express an opinion on the merits of the manuscript.

Ralph H. Phelps, CHAIRMAN
Engineering Societies Library
New York

ENGINEERING SOCIETIES MONOGRAPHS COMMITTEE

A. S. C. E.
    Howard T. Critchlow
    H. Alden Foster

A. I. M. E.
    Nathaniel Arbiter
    John F. Elliott

A. S. M. E.
    Calvin S. Cronan
    John de S. Coutinho

A. I. E. E.
    F. Malcolm Farmer
    Royal W. Sorensen

A. I. Ch. E.
    Joseph F. Skelly
    Charles E. Reed

THEORY OF PLATES AND SHELLS

S. TIMOSHENKO

Professor Emeritus of Engineering Mechanics
Stanford University

S. WOINOWSKY-KRIEGER

Professor of Engineering Mechanics
Laval University

SECOND EDITION

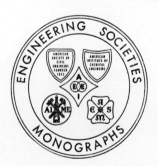

McGRAW-HILL BOOK COMPANY, INC.

New York   Toronto   London

1959

THEORY OF PLATES AND SHELLS

Copyright © 1959 by the McGraw-Hill Book Company, Inc. Copyright, 1940, by the United Engineering Trustees, Inc. Printed in the United States of America. All rights reserved. This book, or parts thereof, may not be reproduced in any form without permission of the publishers. *Library of Congress Catalog Card Number* 58-59675

III

64779

THE MAPLE PRESS COMPANY, YORK, PA.

PREFACE

Since the publication of the first edition of this book, the application of the theory of plates and shells in practice has widened considerably, and some new methods have been introduced into the theory. To take these facts into consideration, we have had to make many changes and additions. The principal additions are (1) an article on deflection of plates due to transverse shear, (2) an article on stress concentrations around a circular hole in a bent plate, (3) a chapter on bending of plates resting on an elastic foundation, (4) a chapter on bending of anisotropic plates, and (5) a chapter reviewing certain special and approximate methods used in plate analysis. We have also expanded the chapter on large deflections of plates, adding several new cases of plates of variable thickness and some numerical tables facilitating plate analysis.

In the part of the book dealing with the theory of shells, we limited ourselves to the addition of the stress-function method in the membrane theory of shells and some minor additions in the flexural theory of shells.

The theory of shells has been developing rapidly in recent years, and several new books have appeared in this field. Since it was not feasible for us to discuss these new developments in detail, we have merely referred to the new bibliography, in which persons specially interested in this field will find the necessary information.

S. Timoshenko
S. Woinowsky-Krieger

CONTENTS

Preface v

Notation xiii

Introduction 1

Chapter 1. Bending of Long Rectangular Plates to a Cylindrical Surface . . . 4

1. Differential Equation for Cylindrical Bending of Plates 4
2. Cylindrical Bending of Uniformly Loaded Rectangular Plates with Simply Supported Edges 6
3. Cylindrical Bending of Uniformly Loaded Rectangular Plates with Built-in Edges 13
4. Cylindrical Bending of Uniformly Loaded Rectangular Plates with Elastically Built-in Edges 17
5. The Effect on Stresses and Deflections of Small Displacements of Longitudinal Edges in the Plane of the Plate 20
6. An Approximate Method of Calculating the Parameter u 24
7. Long Uniformly Loaded Rectangular Plates Having a Small Initial Cylindrical Curvature 27
8. Cylindrical Bending of a Plate on an Elastic Foundation 30

Chapter 2. Pure Bending of Plates 33

9. Slope and Curvature of Slightly Bent Plates 33
10. Relations between Bending Moments and Curvature in Pure Bending of Plates 37
11. Particular Cases of Pure Bending 42
12. Strain Energy in Pure Bending of Plates 46
13. Limitations on the Application of the Derived Formulas 47
14. Thermal Stresses in Plates with Clamped Edges 49

Chapter 3. Symmetrical Bending of Circular Plates 51

15. Differential Equation for Symmetrical Bending of Laterally Loaded Circular Plates 51
16. Uniformly Loaded Circular Plates 54
17. Circular Plate with a Circular Hole at the Center 58
18. Circular Plate Concentrically Loaded 63
19. Circular Plate Loaded at the Center 67
20. Corrections to the Elementary Theory of Symmetrical Bending of Circular Plates 72

Chapter 4. Small Deflections of Laterally Loaded Plates 79

21. The Differential Equation of the Deflection Surface 79

22. Boundary Conditions. 83
23. Alternate Method of Derivation of the Boundary Conditions 88
24. Reduction of the Problem of Bending of a Plate to That of Deflection of a Membrane 92
25. Effect of Elastic Constants on the Magnitude of Bending Moments . . 97
26. Exact Theory of Plates 98

Chapter 5. Simply Supported Rectangular Plates 105

27. Simply Supported Rectangular Plates under Sinusoidal Load 105
28. Navier Solution for Simply Supported Rectangular Plates 108
29. Further Applications of the Navier Solution 111
30. Alternate Solution for Simply Supported and Uniformly Loaded Rectangular Plates 113
31. Simply Supported Rectangular Plates under Hydrostatic Pressure . . 124
32. Simply Supported Rectangular Plate under a Load in the Form of a Triangular Prism 130
33. Partially Loaded Simply Supported Rectangular Plate 135
34. Concentrated Load on a Simply Supported Rectangular Plate . . . 141
35. Bending Moments in a Simply Supported Rectangular Plate with a Concentrated Load 143
36. Rectangular Plates of Infinite Length with Simply Supported Edges . . 149
37. Bending Moments in Simply Supported Rectangular Plates under a Load Uniformly Distributed over the Area of a Rectangle. 158
38. Thermal Stresses in Simply Supported Rectangular Plates 162
39. The Effect of Transverse Shear Deformation on the Bending of Thin Plates 165
40. Rectangular Plates of Variable Thickness 173

Chapter 6. Rectangular Plates with Various Edge Conditions 180

41. Bending of Rectangular Plates by Moments Distributed along the Edges . 180
42. Rectangular Plates with Two Opposite Edges Simply Supported and the Other Two Edges Clamped 185
43. Rectangular Plates with Three Edges Simply Supported and One Edge Built In 192
44. Rectangular Plates with All Edges Built In 197
45. Rectangular Plates with One Edge or Two Adjacent Edges Simply Supported and the Other Edges Built In 205
46. Rectangular Plates with Two Opposite Edges Simply Supported, the Third Edge Free, and the Fourth Edge Built In or Simply Supported . . . 208
47. Rectangular Plates with Three Edges Built In and the Fourth Edge Free . 211
48. Rectangular Plates with Two Opposite Edges Simply Supported and the Other Two Edges Free or Supported Elastically 214
49. Rectangular Plates Having Four Edges Supported Elastically or Resting on Corner Points with All Edges Free 218
50. Semi-infinite Rectangular Plates under Uniform Pressure 221
51. Semi-infinite Rectangular Plates under Concentrated Loads 225

Chapter 7. Continuous Rectangular Plates. 229

52. Simply Supported Continuous Plates 229
53. Approximate Design of Continuous Plates with Equal Spans 236
54. Bending of Plates Supported by Rows of Equidistant Columns (Flat Slabs) 245
55. Flat Slab Having Nine Panels and Slab with Two Edges Free 253
56. Effect of a Rigid Connection with Column on Moments of the Flat Slab . 257

Chapter 8. Plates on Elastic Foundation 259

57. Bending Symmetrical with Respect to a Center 259
58. Application of Bessel Functions to the Problem of the Circular Plate . . 265
59. Rectangular and Continuous Plates on Elastic Foundation . . . 269
60. Plate Carrying Rows of Equidistant Columns 276
61. Bending of Plates Resting on a Semi-infinite Elastic Solid 278

Chapter 9. Plates of Various Shapes 282

62. Equations of Bending of Plates in Polar Coordinates 282
63. Circular Plates under a Linearly Varying Load 285
64. Circular Plates under a Concentrated Load 290
65. Circular Plates Supported at Several Points along the Boundary . . . 293
66. Plates in the Form of a Sector 295
67. Circular Plates of Nonuniform Thickness 298
68. Annular Plates with Linearly Varying Thickness 303
69. Circular Plates with Linearly Varying Thickness 305
70. Nonlinear Problems in Bending of Circular Plates 308
71. Elliptical Plates 310
72. Triangular Plates 313
73. Skewed Plates 318
74. Stress Distribution around Holes 319

Chapter 10. Special and Approximate Methods in Theory of Plates . . . 325

75. Singularities in Bending of Plates 325
76. The Use of Influence Surfaces in the Design of Plates 328
77. Influence Functions and Characteristic Functions 334
78. The Use of Infinite Integrals and Transforms 336
79. Complex Variable Method 340
80. Application of the Strain Energy Method in Calculating Deflections . . 342
81. Alternative Procedure in Applying the Strain Energy Method 347
82. Various Approximate Methods 348
83. Application of Finite Differences Equations to the Bending of Simply Supported Plates 351
84. Experimental Methods 362

Chapter 11. Bending of Anisotropic Plates 364

85. Differential Equation of the Bent Plate 364
86. Determination of Rigidities in Various Specific Cases 366
87. Application of the Theory to the Calculation of Gridworks 369
88. Bending of Rectangular Plates 371
89. Bending of Circular and Elliptic Plates 376

Chapter 12. Bending of Plates under the Combined Action of Lateral Loads and Forces in the Middle Plane of the Plate 378

90. Differential Equation of the Deflection Surface 378
91. Rectangular Plate with Simply Supported Edges under the Combined Action of Uniform Lateral Load and Uniform Tension 380
92. Application of the Energy Method 382
93. Simply Supported Rectangular Plates under the Combined Action of Lateral Loads and of Forces in the Middle Plane of the Plate 387
94. Circular Plates under Combined Action of Lateral Load and Tension or Compression 391
95. Bending of Plates with a Small Initial Curvature 393

Chapter 13. Large Deflections of Plates 396

96. Bending of Circular Plates by Moments Uniformly Distributed along the Edge 396
97. Approximate Formulas for Uniformly Loaded Circular Plates with Large Deflections 400
98. Exact Solution for a Uniformly Loaded Circular Plate with a Clamped Edge 404
99. A Simply Supported Circular Plate under Uniform Load 408
100. Circular Plates Loaded at the Center 412
101. General Equations for Large Deflections of Plates 415
102. Large Deflections of Uniformly Loaded Rectangular Plates . . . 421
103. Large Deflections of Rectangular Plates with Simply Supported Edges . 425

Chapter 14. Deformation of Shells without Bending 429

104. Definitions and Notation 429
105. Shells in the Form of a Surface of Revolution and Loaded Symmetrically with Respect to Their Axis 433
106. Particular Cases of Shells in the Form of Surfaces of Revolution . . 436
107. Shells of Constant Strength 442
108. Displacements in Symmetrically Loaded Shells Having the Form of a Surface of Revolution 445
109. Shells in the Form of a Surface of Revolution under Unsymmetrical Loading 447
110. Stresses Produced by Wind Pressure 449
111. Spherical Shell Supported at Isolated Points 453
112. Membrane Theory of Cylindrical Shells 457
113. The Use of a Stress Function in Calculating Membrane Forces of Shells . 461

Chapter 15. General Theory of Cylindrical Shells 466

114. A Circular Cylindrical Shell Loaded Symmetrically with Respect to Its Axis 466
115. Particular Cases of Symmetrical Deformation of Circular Cylindrical Shells 471
116. Pressure Vessels 481
117. Cylindrical Tanks with Uniform Wall Thickness 485
118. Cylindrical Tanks with Nonuniform Wall Thickness. 488
119. Thermal Stresses in Cylindrical Shells 497
120. Inextensional Deformation of a Circular Cylindrical Shell 501
121. General Case of Deformation of a Cylindrical Shell 507
122. Cylindrical Shells with Supported Edges 514
123. Deflection of a Portion of a Cylindrical Shell 516
124. An Approximate Investigation of the Bending of Cylindrical Shells . 519
125. The Use of a Strain and Stress Function 522
126. Stress Analysis of Cylindrical Roof Shells 524

Chapter 16. Shells Having the Form of a Surface of Revolution and Loaded Symmetrically with Respect to Their Axis 533

127. Equations of Equilibrium 533
128. Reduction of the Equations of Equilibrium to Two Differential Equations of the Second Order 537
129. Spherical Shell of Constant Thickness 540

130. Approximate Methods of Analyzing Stresses in Spherical Shells . . . 547
131. Spherical Shells with an Edge Ring 555
132. Symmetrical Bending of Shallow Spherical Shells 558
133. Conical Shells 562
134. General Case of Shells Having the Form of a Surface of Revolution . . 566

Name Index 569
Subject Index 575

NOTATION

x, y, z Rectangular coordinates
r, θ Polar coordinates
r_x, r_y Radii of curvature of the middle surface of a plate in xz and yz planes, respectively
h Thickness of a plate or a shell
q Intensity of a continuously distributed load
p Pressure
P Single load
γ Weight per unit volume
$\sigma_x, \sigma_y, \sigma_z$ Normal components of stress parallel to x, y, and z axes
σ_n Normal component of stress parallel to n direction
σ_r Radial stress in polar coordinates
σ_t, σ_θ Tangential stress in polar coordinates
τ Shearing stress
$\tau_{xy}, \tau_{xz}, \tau_{yz}$ Shearing stress components in rectangular coordinates
u, v, w Components of displacements
ϵ Unit elongation
$\epsilon_x, \epsilon_y, \epsilon_z$ Unit elongations in x, y, and z directions
ϵ_r Radial unit elongation in polar coordinates
$\epsilon_t, \epsilon_\theta$ Tangential unit elongation in polar coordinates
$\epsilon_\varphi, \epsilon_\theta$ Unit elongations of a shell in meridional direction and in the direction of parallel circle, respectively
$\gamma_{xy}, \gamma_{xz}, \gamma_{yz}$ Shearing strain components in rectangular coordinates
$\gamma_{r\theta}$ Shearing strain in polar coordinates
E Modulus of elasticity in tension and compression
G Modulus of elasticity in shear
ν Poisson's ratio
V Strain energy
D Flexural rigidity of a plate or shell
M_x, M_y Bending moments per unit length of sections of a plate perpendicular to x and y axes, respectively
M_{xy} Twisting moment per unit length of section of a plate perpendicular to x axis
M_n, M_{nt} Bending and twisting moments per unit length of a section of a plate perpendicular to n direction
Q_x, Q_y Shearing forces parallel to z axis per unit length of sections of a plate perpendicular to x and y axes, respectively
Q_n Shearing force parallel to z axis per unit length of section of a plate perpendicular to n direction
N_x, N_y Normal forces per unit length of sections of a plate perpendicular to x and y directions, respectively

N_{xy}	Shearing force in direction of y axis per unit length of section of a plate perpendicular to x axis
M_r, M_t, M_{rt}	Radial, tangential, and twisting moments when using polar coordinates
Q_r, Q_t	Radial and tangential shearing forces
N_r, N_t	Normal forces per unit length in radial and tangential directions
r_1, r_2	Radii of curvature of a shell in the form of a surface of revolution in meridional plane and in the normal plane perpendicular to meridian, respectively
$\chi_\varphi, \chi_\theta$	Changes of curvature of a shell in meridional plane and in the plane perpendicular to meridian, respectively
$\chi_{\theta\varphi}$	Twist of a shell
X, Y, Z	Components of the intensity of the external load on a shell, parallel to x, y, and z axes, respectively
$N_\varphi, N_\theta, N_{\varphi\theta}$	Membrane forces per unit length of principal normal sections of a shell
M_θ, M_φ	Bending moments in a shell per unit length of meridional section and a section perpendicular to meridian, respectively
χ_x, χ_φ	Changes of curvature of a cylindrical shell in axial plane and in a plane perpendicular to the axis, respectively
$N_\varphi, N_x, N_{x\varphi}$	Membrane forces per unit length of axial section and a section perpendicular to the axis of a cylindrical shell
M_φ, M_x	Bending moments per unit length of axial section and a section perpendicular to the axis of a cylindrical shell, respectively
$M_{x\varphi}$	Twisting moment per unit length of an axial section of a cylindrical shell
Q_φ, Q_x	Shearing forces parallel to z axis per unit length of an axial section and a section perpendicular to the axis of a cylindrical shell, respectively
log	Natural logarithm
$\log_{10}$, Log	Common logarithm

INTRODUCTION

The bending properties of a plate depend greatly on its thickness as compared with its other dimensions. In the following discussion, we shall distinguish between three kinds of plates: (1) thin plates with small deflections, (2) thin plates with large deflections, (3) thick plates.

Thin Plates with Small Deflection. If deflections w of a plate are small in comparison with its thickness h, a very satisfactory approximate theory of bending of the plate by lateral loads can be developed by making the following assumptions:

1. There is no deformation in the middle plane of the plate. This plane remains *neutral* during bending.
2. Points of the plate lying initially on a normal-to-the-middle plane of the plate remain on the normal-to-the-middle surface of the plate after bending.
3. The normal stresses in the direction transverse to the plate can be disregarded.

Using these assumptions, all stress components can be expressed by deflection w of the plate, which is a function of the two coordinates in the plane of the plate. This function has to satisfy a linear partial differential equation, which, together with the boundary conditions, completely defines w. Thus the solution of this equation gives all necessary information for calculating stresses at any point of the plate.

The second assumption is equivalent to the disregard of the effect of shear forces on the deflection of plates. This assumption is usually satisfactory, but in some cases (for example, in the case of holes in a plate) the effect of shear becomes important and some corrections in the theory of thin plates should be introduced (see Art. 39).

If, in addition to lateral loads, there are external forces acting in the middle plane of the plate, the first assumption does not hold any more, and it is necessary to take into consideration the effect on bending of the plate of the stresses acting in the middle plane of the plate. This can be done by introducing some additional terms into the above-mentioned differential equation of plates (see Art. 90).

Thin Plates with Large Deflection. The first assumption is completely satisfied only if a plate is bent into a developable surface. In other cases bending of a plate is accompanied by strain in the middle plane, but calculations show that the corresponding stresses in the middle plane are negligible if the deflections of the plate are small in comparison with its thickness. If the deflections are not small, these supplementary stresses must be taken into consideration in deriving the differential equation of plates. In this way we obtain nonlinear equations and the solution of the problem becomes much more complicated (see Art. 96). In the case of large deflections we have also to distinguish between immovable edges and edges free to move in the plane of the plate, which may have a considerable bearing upon the magnitude of deflections and stresses of the plate (see Arts. 99, 100). Owing to the curvature of the deformed middle plane of the plate, the supplementary tensile stresses, which predominate, act in opposition to the given lateral load; thus, the given load is now transmitted partly by the flexural rigidity and partly by a membrane action of the plate. Consequently, very thin plates with negligible resistance to bending behave as membranes, except perhaps for a narrow edge zone where bending may occur because of the boundary conditions imposed on the plate.

The case of a plate bent into a developable, in particular into a cylindrical, surface should be considered as an exception. The deflections of such a plate may be of the order of its thickness without necessarily producing membrane stresses and without affecting the linear character of the theory of bending. Membrane stresses would, however, arise in such a plate if its edges are immovable in its plane and the deflections are sufficiently large (see Art. 2). Therefore, in "plates with small deflection" membrane forces caused by edges immovable in the plane of the plate can be practically disregarded.

Thick Plates. The approximate theories of thin plates, discussed above, become unreliable in the case of plates of considerable thickness, especially in the case of highly concentrated loads. In such a case the thick-plate theory should be applied. This theory considers the problem of plates as a three-dimensional problem of elasticity. The stress analysis becomes, consequently, more involved and, up to now, the problem is completely solved only for a few particular cases. Using this analysis, the necessary corrections to the thin-plate theory at the points of application of concentrated loads can be introduced.

The main suppositions of the theory of thin plates also form the basis for the usual theory of thin shells. There exists, however, a substantial difference in the behavior of plates and shells under the action of external loading. The static equilibrium of a plate element under a lateral load is only possible by action of bending and twisting moments, usually accompanied by shearing forces, while a shell, in general, is able to transmit the surface load by "membrane" stresses which act parallel to the tangential plane at a given point of the middle surface and are distributed uniformly over the thickness of the shell. This property of shells makes them, as a rule, a much more rigid and a more economical structure than a plate would be under the same conditions.

In principle, the membrane forces are independent of bending and are wholly defined by the conditions of static equilibrium. The methods of determination of these forces represent the so-called "membrane theory of shells." However, the reactive forces and deformation obtained by the use of the membrane theory at the shell's boundary usually become incompatible with the actual boundary conditions. To remove this discrepancy the bending of the shell in the edge zone has to be considered, which may affect slightly the magnitude of initially calculated membrane forces. This bending, however, usually has a very localized[1] character and may be calculated on the basis of the same assumptions which were used in the case of small deflections of thin plates. But there are problems, especially those concerning the elastic stability of shells, in which the assumption of small deflections should be discontinued and the "large-deflection theory" should be used.

If the thickness of a shell is comparable to the radii of curvature, or if we consider stresses near the concentrated forces, a more rigorous theory, similar to the thick-plate theory, should be applied.

[1] There are some kinds of shells, especially those with a negative Gaussian curvature, which provide us with a lot of exceptions. In the case of developable surfaces such as cylinders or cones, large deflection without strain of the middle surface is possible, and, in some cases, membrane stresses can be neglected and consideration of the bending stresses alone may be sufficient.

CHAPTER 1

BENDING OF LONG RECTANGULAR PLATES TO A CYLINDRICAL SURFACE

1. Differential Equation for Cylindrical Bending of Plates. We shall begin the theory of bending of plates with the simple problem of the bending of a long rectangular plate that is subjected to a transverse load that does not vary along the length of the plate. The deflected surface of a portion of such a plate at a considerable distance from the ends[1] can be assumed cylindrical, with the axis of the cylinder parallel to the length of the plate. We can therefore restrict ourselves to the investigation of the bending of an elemental strip cut from the plate by two planes perpendicular to the length of the plate and a unit distance (say 1 in.) apart. The deflection of this strip is given by a differential equation which is similar to the deflection equation of a bent beam.

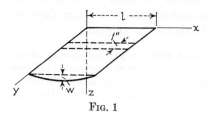

FIG. 1

To obtain the equation for the deflection, we consider a plate of uniform thickness, equal to h, and take the xy plane as the middle plane of the plate before loading, i.e., as the plane midway between the faces of the plate. Let the y axis coincide with one of the longitudinal edges of the plate and let the positive direction of the z axis be downward, as shown in Fig. 1. Then if the width of the plate is denoted by l, the elemental strip may be considered as a bar of rectangular cross section which has a length of l and a depth of h. In calculating the bending stresses in such a bar we assume, as in the ordinary theory of beams, that cross sections of the bar remain plane during bending, so that they undergo only a rotation with respect to their neutral axes. If no normal forces are applied to the end sections of the bar, the neutral surface of the bar coincides with the middle surface of the plate, and the unit elongation of a fiber parallel to the x axis is proportional to its distance z

[1] The relation between the length and the width of a plate in order that the maximum stress may approximate that in an infinitely long plate is discussed later; see pp. 118 and 125.

from the middle surface. The curvature of the deflection curve can be taken equal to $-d^2w/dx^2$, where w, the deflection of the bar in the z direction, is assumed to be small compared with the length of the bar l. The unit elongation ϵ_x of a fiber at a distance z from the middle surface (Fig. 2) is then $-z\, d^2w/dx^2$.

Making use of Hooke's law, the unit elongations ϵ_x and ϵ_y in terms of the normal stresses σ_x and σ_y acting on the element shown shaded in Fig. 2a are

$$\epsilon_x = \frac{\sigma_x}{E} - \frac{\nu \sigma_y}{E}$$
$$\epsilon_y = \frac{\sigma_y}{E} - \frac{\nu \sigma_x}{E} = 0 \qquad (1)$$

where E is the modulus of elasticity of the material and ν is Poisson's ratio. The lateral strain in the y direction must be zero in order to maintain continuity in the plate during bending, from which it follows by the second of the equations (1) that $\sigma_y = \nu \sigma_x$. Substituting this value in the first of the equations (1), we obtain

$$\epsilon_x = \frac{(1 - \nu^2)\sigma_x}{E}$$

and
$$\sigma_x = \frac{E \epsilon_x}{1 - \nu^2} = -\frac{Ez}{1 - \nu^2} \frac{d^2w}{dx^2} \qquad (2)$$

If the plate is submitted to the action of tensile or compressive forces acting in the x direction and uniformly distributed along the longitudinal sides of the plate, the corresponding direct stress must be added to the stress (2) due to bending.

Having the expression for bending stress σ_x, we obtain by integration the bending moment in the elemental strip:

$$M = \int_{-h/2}^{h/2} \sigma_x z\, dz = -\int_{-h/2}^{h/2} \frac{Ez^2}{1-\nu^2} \frac{d^2w}{dx^2} dz = -\frac{Eh^3}{12(1-\nu^2)} \frac{d^2w}{dx^2}$$

Introducing the notation

$$\frac{Eh^3}{12(1-\nu^2)} = D \qquad (3)$$

we represent the equation for the deflection curve of the elemental strip in the following form:

$$D \frac{d^2w}{dx^2} = -M \qquad (4)$$

in which the quantity D, taking the place of the quantity EI in the case of beams, is called the *flexural rigidity* of the plate. It is seen that the calculation of deflections of the plate reduces to the integration of Eq. (4), which has the same form as the differential equation for deflection of beams. If there is only a lateral load acting on the plate and the edges are free to approach each other as deflection occurs, the expression for the bending moment M can be readily derived, and the deflection curve is then obtained by integrating Eq. (4). In practice the problem is more complicated, since the plate is usually attached to the boundary and its edges are not free to move. Such a method of support sets up tensile reactions along the edges as soon as deflection takes place. These reactions depend on the magnitude of the deflection and affect the magnitude of the bending moment M entering in Eq. (4). The problem reduces to the investigation of bending of an elemental strip submitted to the action of a lateral load and also an axial force which depends on the deflection of the strip.[1] In the following we consider this problem for the particular case of uniform load acting on a plate and for various conditions along the edges.

2. Cylindrical Bending of Uniformly Loaded Rectangular Plates with Simply Supported Edges. Let us consider a uniformly loaded long rectangular plate with longitudinal edges which are free to rotate but cannot move toward each other during bending. An elemental strip cut out

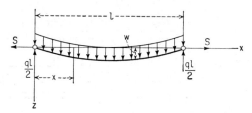

Fig. 3

from this plate, as shown in Fig. 1, is in the condition of a uniformly loaded bar submitted to the action of an axial force S (Fig. 3). The magnitude of S is such as to prevent the ends of the bar from moving along the x axis. Denoting by q the intensity of the uniform load, the bending moment at any cross section of the strip is

$$M = \frac{ql}{2}x - \frac{qx^2}{2} - Sw$$

[1] In such a form the problem was first discussed by I. G. Boobnov; see the English translation of his work in *Trans. Inst. Naval Architects*, vol. 44, p. 15, 1902, and his "Theory of Structure of Ships," vol. 2, p. 545, St. Petersburg, 1914. See also the paper by Stewart Way presented at the National Meeting of Applied Mechanics, ASME, New Haven, Conn., June, 1932; from this paper are taken the curves used in Arts. 2 and 3.

Substituting in Eq. (4), we obtain

$$\frac{d^2w}{dx^2} - \frac{Sw}{D} = -\frac{qlx}{2D} + \frac{qx^2}{2D} \tag{a}$$

Introducing the notation

$$\frac{S}{D}\frac{l^2}{4} = u^2 \tag{5}$$

the general solution of Eq. (a) can be written in the following form:

$$w = C_1 \sinh \frac{2ux}{l} + C_2 \cosh \frac{2ux}{l} + \frac{ql^3 x}{8u^2 D} - \frac{ql^2 x^2}{8u^2 D} - \frac{ql^4}{16u^4 D} \tag{b}$$

The constants of integration C_1 and C_2 will be determined from the conditions at the ends. Since the deflections of the strip at the ends are zero, we have

$$w = 0 \quad \text{for } x = 0 \text{ and } x = l \tag{c}$$

Substituting for w its expression (b), we obtain from these two conditions

$$C_1 = \frac{ql^4}{16u^4 D} \frac{1 - \cosh 2u}{\sinh 2u} \qquad C_2 = \frac{ql^4}{16u^4 D}$$

and the expression (b) for the deflection w becomes

$$w = \frac{ql^4}{16u^4 D}\left(\frac{1 - \cosh 2u}{\sinh 2u} \sinh \frac{2ux}{l} + \cosh \frac{2ux}{l} - 1\right) + \frac{ql^3 x}{8u^2 D} - \frac{ql^2 x^2}{8u^2 D}$$

Substituting

$$\cosh 2u = \cosh^2 u + \sinh^2 u \qquad \sinh 2u = 2 \sinh u \cosh u$$
$$\cosh^2 u = 1 + \sinh^2 u$$

we can represent this expression in a simpler form:

$$w = \frac{ql^4}{16u^4 D}\left(\frac{-\sinh u \sinh \frac{2ux}{l} + \cosh u \cosh \frac{2ux}{l}}{\cosh u} - 1\right) + \frac{ql^2 x}{8u^2 D}(l - x)$$

or

$$w = \frac{ql^4}{16u^4 D}\left[\frac{\cosh u \left(1 - \frac{2x}{l}\right)}{\cosh u} - 1\right] + \frac{ql^2 x}{8u^2 D}(l - x) \tag{6}$$

Thus, deflections of the elemental strip depend upon the quantity u, which, as we see from Eq. (5), is a function of the axial force S. This force can be determined from the condition that the ends of the strip (Fig. 3) do not move along the x axis. Hence the extension of the strip produced by the forces S is equal to the difference between the length of the arc along the deflection curve and the chord length l. This difference for small deflections can be represented by the formula[1]

$$\lambda = \frac{1}{2} \int_0^l \left(\frac{dw}{dx}\right)^2 dx \qquad (7)$$

In calculating the extension of the strip produced by the forces S, we assume that the lateral strain of the strip in the y direction is prevented and use Eq. (2). Then

$$\lambda = \frac{S(1-\nu^2)l}{hE} = \frac{1}{2} \int_0^l \left(\frac{dw}{dx}\right)^2 dx \qquad (d)$$

Substituting expression (6) for w and performing the integration, we obtain the following equation for calculating S:

$$\frac{S(1-\nu^2)l}{hE} = \frac{q^2 l^7}{D^2}\left(\frac{5}{256}\frac{\tanh u}{u^7} + \frac{1}{256}\frac{\tanh^2 u}{u^6} - \frac{5}{256 u^6} + \frac{1}{384 u^4}\right)$$

or substituting $S = 4u^2 D/l^2$, from Eq. (5), and the expression for D, from Eq. (3), we finally obtain the equation

$$\frac{E^2 h^8}{(1-\nu^2)^2 q^2 l^8} = \frac{135}{16}\frac{\tanh u}{u^9} + \frac{27}{16}\frac{\tanh^2 u}{u^8} - \frac{135}{16 u^8} + \frac{9}{8 u^6} \qquad (8)$$

For a given material, a given ratio h/l, and a given load q the left-hand side of this equation can be readily calculated, and the value of u satisfying the equation can be found by a trial-and-error method. To simplify this solution, the curves shown in Fig. 4 can be used. The abscissas of these curves represent the values of u and the ordinates represent the quantities $\log_{10}(10^4 \sqrt{U_0})$, where U_0 denotes the numerical value of the right-hand side of Eq. (8). $\sqrt{U_0}$ is used because it is more easily calculated from the plate constants and the load; and the factor 10^4 is introduced to make the logarithms positive. In each particular case we begin by calculating the square root of the left-hand side of Eq. (8), equal to $Eh^4/(1-\nu^2)ql^4$, which gives $\sqrt{U_0}$. The quantity $\log_{10}(10^4 \sqrt{U_0})$ then gives the ordinate which must be used in Fig. 4, and the corresponding value of u can be readily obtained from the curve. Having u, we obtain the value of the axial force S from Eq. (5).

In calculating stresses we observe that the total stress at any cross section of the strip consists of a bending stress proportional to the bending moment and a tensile stress of magnitude S/h which is constant along the length of the strip. The maximum stress occurs at the middle of the strip, where the bending moment is a maximum. From the differential equation (4) the maximum bending moment is

$$M_{\max} = -D\left(\frac{d^2 w}{dx^2}\right)_{x=l/2}$$

[1] See Timoshenko, "Strength of Materials," part I, 3d ed., p. 178, 1955.

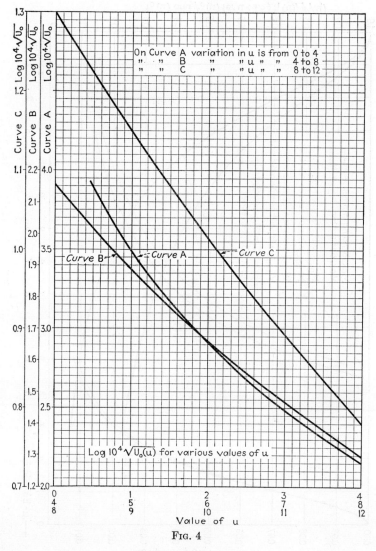

Fig. 4

Substituting expression (6) for w, we obtain

$$M_{\max} = \frac{ql^2}{8} \psi_0(u) \tag{9}$$

where

$$\psi_0 = \frac{1 - \operatorname{sech} u}{\dfrac{u^2}{2}} \tag{e}$$

The values of ψ_0 are given by curves in Fig. 5. It is seen that these values diminish rapidly with increase of u, and for large u the maximum

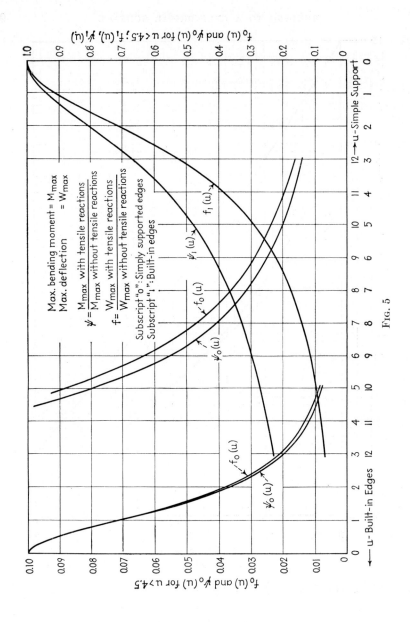

Fig. 5

bending moment is several times smaller than the moment $ql^2/8$ which would be obtained if there were no tensile reactions at the ends of the strip.

The direct tensile stress σ_1 and the maximum bending stress σ_2 are now readily expressed in terms of u, q, and the plate constants as follows:

$$\sigma_1 = \frac{S}{h} = \frac{4u^2 D}{hl^2} = \frac{Eu^2}{3(1-\nu^2)}\left(\frac{h}{l}\right)^2 \qquad (10)$$

$$\sigma_2 = \frac{6}{h^2} M_{max} = \frac{3}{4} q \left(\frac{l}{h}\right)^2 \psi_0 \qquad (11)$$

The maximum stress in the plate is then

$$\sigma_{max} = \sigma_1 + \sigma_2$$

To show how the curves in Figs. 4 and 5 can be used in calculating maximum stresses, let us take a numerical example and assume that a long rectangular steel plate 50 in. wide and $\frac{1}{2}$ in. thick carries a uniformly distributed load $q = 20$ psi. We start with a computation of $\sqrt{U_0}$:

$$\sqrt{U_0} = \frac{E}{(1-\nu^2)q}\left(\frac{h}{l}\right)^4 = \frac{30 \cdot 10^6}{(1-0.3^2)20}\frac{1}{10^8} = 0.01648$$

Then, from tables,

$$\log_{10}(10^4 \sqrt{U_0}) = 2.217$$

From the curve A in Fig. 4 we find $u = 3.795$, and from Fig. 5 we obtain $\psi_0 = 0.1329$.

Now, computing stresses by using Eqs. (10) and (11), we find

$$\sigma_1 = \frac{30 \cdot 10^6 \cdot 3.795^2}{3(1-0.3^2)}\frac{1}{10^4} = 15{,}830 \text{ psi}$$
$$\sigma_2 = \tfrac{3}{4} \cdot 20 \cdot 10^4 \cdot 0.1329 = 19{,}930 \text{ psi}$$
$$\sigma_{max} = \sigma_1 + \sigma_2 = 35{,}760 \text{ psi}$$

In calculating the maximum deflection we substitute $x = l/2$ in Eq. (6) of the deflection curve. In this manner we obtain

$$w_{max} = \frac{5ql^4}{384D} f_0(u) \qquad (12)$$

where

$$f_0(u) = \frac{\operatorname{sech} u - 1 + \dfrac{u^2}{2}}{\dfrac{5u^4}{24}}$$

To simplify calculations, values of $f_0(u)$ are given by the curve in Fig. 5. If there were no tensile reactions at the ends of the strip, the maximum deflection would be $5ql^4/384D$. The effect of the tensile reactions is given by the factor $f_0(u)$, which diminishes rapidly with increasing u.

Using Fig. 5 in the numerical example previously discussed, we find that for $u = 3.795$ the value of $f_0(u)$ is 0.145. Substituting this value in Eq. (12), we obtain

$$w_{\max} = 4.74 \cdot 0.145 = 0.688 \text{ in.}$$

It is seen from Eq. (8) that the tensile parameter u depends, for a given material of the plate, upon the intensity of the load q and the

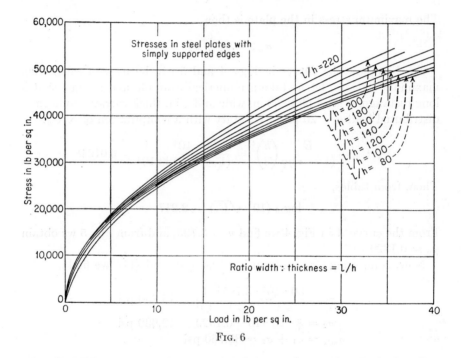

Fig. 6

ratio l/h of width to thickness of the plate. From Eqs. (10) and (11) we see that the stresses σ_1 and σ_2 are also functions of u, q, and l/h. Therefore, the maximum stress in the plate depends only on the load q and the ratio l/h. This means that we can plot a set of curves giving maximum stress in terms of q, each curve in the set corresponding to a particular value of l/h. Such curves are given in Fig. 6. It is seen that because of the presence of tensile forces S, which increase with the load, the maximum stress is not proportional to the load q; and for large values of q this stress does not vary much with the thickness of the plate. By taking the curve marked $l/h = 100$ and assuming $q = 20$ psi, we obtain from the curve the value $\sigma_{\max}$ calculated before in the numerical example.

3. Cylindrical Bending of Uniformly Loaded Rectangular Plates with Built-in Edges.

We assume that the longitudinal edges of the plate are fixed in such a manner that they cannot rotate. Taking an elemental strip of unit width in the same manner as before (Fig. 1) and denoting by M_0 the bending moment per unit length acting on the longitudinal edges of the plate, the forces acting on the strip will be as shown in Fig. 7. The bending moment at any cross section of the strip is

$$M = \frac{ql}{2}x - \frac{qx^2}{2} - Sw + M_0$$

Substituting this expression in Eq. (4), we obtain

$$\frac{d^2w}{dx^2} - \frac{S}{D}w = -\frac{qlx}{2D} + \frac{qx^2}{2D} - \frac{M_0}{D} \qquad (a)$$

The general solution of this equation, using notation (5), will be represented in the following form:

$$w = C_1 \sinh \frac{2ux}{l} + C_2 \cosh \frac{2ux}{l} + \frac{ql^3 x}{8u^2 D} - \frac{ql^2 x^2}{8u^2 D} - \frac{ql^4}{16u^4 D} + \frac{M_0 l^2}{4u^2 D} \qquad (b)$$

Observing that the deflection curve is symmetrical with respect to the middle of the strip, we determine the constants of integration C_1, C_2, and

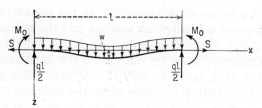

Fig. 7

the moment M_0 from the following three conditions:

$$\begin{aligned}\frac{dw}{dx} &= 0 \quad \text{for } x = 0 \text{ and } x = \frac{l}{2} \\ w &= 0 \quad \text{for } x = 0\end{aligned} \qquad (c)$$

Substituting expression (b) for w, we obtain from these conditions

$$C_1 = -\frac{ql^4}{16u^3 D} \qquad C_2 = \frac{ql^4}{16u^3 D} \coth u$$

$$M_0 = \frac{ql^2}{4u^2} - \frac{ql^2}{4u} \coth u = -\frac{ql^2}{12}\psi_1(u) \qquad (13)$$

where

$$\psi_1(u) = \frac{3(u - \tanh u)}{u^2 \tanh u}$$

The deflection w is therefore given by the expression

$$w = -\frac{ql^4}{16u^3 D}\sinh\frac{2ux}{l} + \frac{ql^4}{16u^3 D}\coth u \cosh\frac{2ux}{l}$$
$$+ \frac{ql^3 x}{8u^2 D} - \frac{ql^2 x^2}{8u^2 D} - \frac{ql^4}{16u^3 D}\coth u$$

This can be further simplified and finally put in the following form:

$$w = \frac{ql^4}{16u^3 D \tanh u}\left\{\frac{\cosh\left[u\left(1 - \frac{2x}{l}\right)\right]}{\cosh u} - 1\right\} + \frac{ql^2(l - x)x}{8u^2 D} \quad (14)$$

For calculating the parameter u we proceed as in the previous article and use Eq. (d) of that article. Substituting in it expression (14) for w and performing the integration, we obtain

$$\frac{S(1 - \nu^2)l}{hE} = \frac{q^2 l^7}{D^2}\left(-\frac{3}{256 u^5 \tanh u} - \frac{1}{256 u^4 \sinh^2 u} + \frac{1}{64 u^6} + \frac{1}{384 u^4}\right)$$

Substituting S from Eq. (5) and expression (3) for D, the equation for calculating u finally becomes

$$\frac{E^2 h^8}{(1 - \nu^2)^2 q^2 l^8} = -\frac{81}{16 u^7 \tanh u} - \frac{27}{16 u^6 \sinh^2 u} + \frac{27}{4 u^8} + \frac{9}{8 u^6} \quad (15)$$

To simplify the solution of this equation we use the curve in Fig. 8, in which the parameter u is taken as abscissa and the ordinates are equal to $\log_{10}(10^4 \sqrt{U_1})$, where U_1 denotes the right-hand side of Eq. (15). For any given plate we begin by calculating the square root of the left-hand side of Eq. (15), equal to $Eh^4/[(1 - \nu^2)ql^4]$, which gives us $\sqrt{U_1}$. The quantity $\log_{10}(10^4 \sqrt{U_1})$ then gives the ordinate of the curve in Fig. 8, and the corresponding abscissa gives the required value of u.

Having u, we can begin calculating the maximum stresses in the plate. The total stress at any point of a cross section of the strip consists of the constant tensile stress σ_1 and the bending stress. The maximum bending stress σ_2 will act at the built-in edges where the bending moment is the largest. Using Eq. (10) to calculate σ_1 and Eq. (13) to calculate the bending moment M_0, we obtain

$$\sigma_1 = \frac{Eu^2}{3(1 - \nu^2)}\left(\frac{h}{l}\right)^2 \quad (16)$$

$$\sigma_2 = -\frac{6M_0}{h^2} = \frac{q}{2}\left(\frac{l}{h}\right)^2 \psi_1(u) \quad (17)$$

$$\sigma_{\max} = \sigma_1 + \sigma_2$$

To simplify the calculation of the stress σ_2, the values of the function $\psi_1(u)$ are given by a curve in Fig. 5.

The maximum deflection is at the middle of the strip and is obtained by

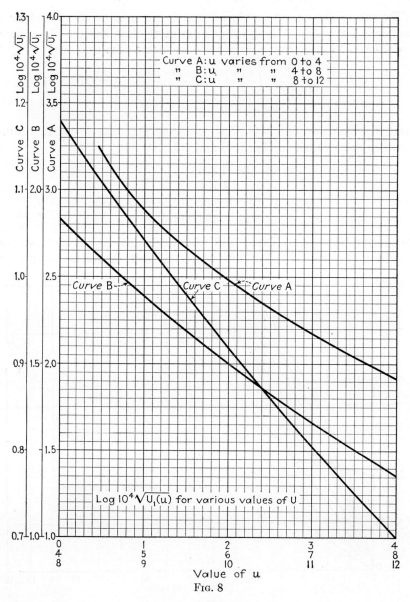

Fig. 8

substituting $x = l/2$ in Eq. (14), from which

$$w_{\max} = \frac{ql^4}{384D} f_1(u) \qquad (18)$$

where
$$f_1(u) = \frac{24}{u^4}\left(\frac{u^2}{2} + \frac{u}{\sinh u} - \frac{u}{\tanh u}\right)$$

The function $f_1(u)$ is also given by a curve in Fig. 5.

The use of the curves in Figs. 5 and 8 will now be illustrated by a numerical example. A long rectangular steel plate has the dimensions $l = 50$ in., $h = \frac{1}{2}$ in., and $q = 10$ psi. In such a case we have

$$\sqrt{U_1} = \frac{E}{(1-\nu^2)q}\left(\frac{h}{l}\right)^4 = \frac{30 \cdot 10^6}{(1-0.3^2)10 \cdot 10^4} = 0.032966$$

$$\log_{10} 10^4 \sqrt{U_1} = 2.5181$$

From Fig. 8 we now find $u = 1.894$; and from Fig. 5, $\psi_1 = 0.8212$. Substituting these values in Eqs. (16) and (17), we find

$$\sigma_1 = \frac{30 \cdot 10^6 \cdot 1.894^2}{3(1-0.3^2)10^4} = 3{,}940 \text{ psi}$$

$$\sigma_2 = \tfrac{1}{2} \cdot 10 \cdot 10^4 \cdot 0.8212 = 41{,}060 \text{ psi}$$

$$\sigma_{\max} = \sigma_1 + \sigma_2 = 45{,}000 \text{ psi}$$

Comparing these stress values with the maximum stresses obtained for a plate of the same size, but with twice the load, on the assumption of

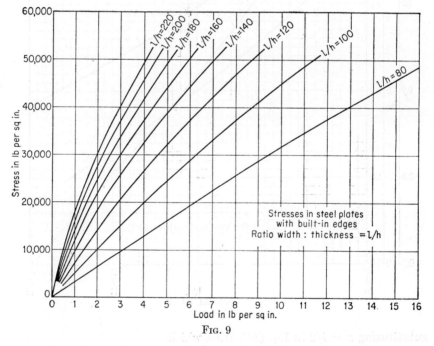

Fig. 9

simply supported edges (see page 11), it can be concluded that, owing to clamping of the edges, the direct tensile stress decreases considerably, whereas the maximum bending stress increases several times, so that finally the maximum total stress in the case of clamped edges becomes larger than in the case of simply supported edges.

Proceeding as in the previous article it can be shown that the maximum stress in a plate depends only on the load q and the ratio l/h, and we can plot a set of curves giving maximum stress in terms of q, each curve in the set corresponding to a particular value of l/h. Such curves are given in Fig. 9. It is seen that for small values of the intensity of the load q, when the effect of the axial force on the deflections of the strip is small, the maximum stress increases approximately in the same ratio as q increases. But for larger values of q the relation between the load and the maximum stress becomes nonlinear.

In conclusion, we give in Table 1 the numerical values of all the functions plotted in Figs. 4, 5, and 8. This table can be used instead of the curves in calculating maximum stresses and maximum deflections of long, uniformly loaded rectangular plates.

4. Cylindrical Bending of Uniformly Loaded Rectangular Plates with Elastically Built-in Edges. Let us assume that when bending occurs, the longitudinal edges of the plate rotate through an angle proportional to the bending moment at the edges. In such a case the forces acting on an elemental strip will again be of the type shown in Fig. 7, and we shall obtain expression (b) of the previous article for the deflections w. However, the conditions at the edges, from which the constants of integration and the moment M_0 are determined, are different; viz., the slope of the deflection curve at the ends of the strip is no longer zero but is proportional to the magnitude of the moment M_0, and we have

$$\left(\frac{dw}{dx}\right)_{x=0} = -\beta M_0 \qquad (a)$$

where β is a factor depending on the rigidity of restraint along the edges. If this restraint is very flexible, the quantity β is large, and the conditions at the edges approach those of simply supported edges. If the restraint is very rigid, the quantity β becomes small, and the edge conditions approach those of absolutely built-in edges. The remaining two end conditions are the same as in the previous article. Thus we have

$$\left(\frac{dw}{dx}\right)_{x=0} = -\beta M_0 \qquad \left(\frac{dw}{dx}\right)_{x=l/2} = 0 \qquad (b)$$
$$(w)_{x=0} = 0$$

Using these conditions, we find both the constants of integration and the magnitude of M_0 in expression (b) of the previous article. Owing to flexibility of the boundary, the end moments M_0 will be smaller than those given by Eq. (13) for absolutely built-in edges, and the final result can be put in the form

$$M_0 = -\gamma \frac{ql^2}{12} \psi_1(u) \qquad (19)$$

TABLE 1

u	$\log_{10} 10^4 \sqrt{U_0}$	$\log_{10} 10^4 \sqrt{U_1}$	$\log_{10} 10^4 \sqrt{U_2}$	$f_0(u)$	$f_1(u)$	$\psi_0(u)$	$\psi_1(u)$	u
0	∞	∞	∞	1.000	1.000	1.000	1.000	0
0.5	3.889	3.217	3.801	0.908	0.976	0.905	0.984	0.5
	406	331	425					
1.0	3.483	2.886	3.376	0.711	0.909	0.704	0.939	1.0
	310	223	336					
1.5	3.173	2.663	3.040	0.532	0.817	0.511	0.876	1.5
	262	182	292					
2.0	2.911	2.481	2.748	0.380	0.715	0.367	0.806	2.0
	227	161	257					
2.5	2.684	2.320	2.491	0.281	0.617	0.268	0.736	2.5
	198	146	228					
3.0	2.486	2.174	2.263	0.213	0.529	0.200	0.672	3.0
	175	134	202					
3.5	2.311	2.040	2.061	0.166	0.453	0.153	0.614	3.5
	156	124	180					
4.0	2.155	1.916	1.881	0.132	0.388	0.120	0.563	4.0
	141	115	163					
4.5	2.014	1.801	1.718	0.107	0.335	0.097	0.519	4.5
	128	107	148					
5.0	1.886	1.694	1.570	0.088	0.291	0.079	0.480	5.0
	118	100	135					
5.5	1.768	1.594	1.435	0.074	0.254	0.066	0.446	5.5
	108	93	124					
6.0	1.660	1.501	1.311	0.063	0.223	0.055	0.417	6.0
	100	88	115					
6.5	1.560	1.413	1.196	0.054	0.197	0.047	0.391	6.5
	93	82	107					
7.0	1.467	1.331	1.089	0.047	0.175	0.041	0.367	7.0
	87	78	100					
7.5	1.380	1.253	0.989	0.041	0.156	0.036	0.347	7.5
	82	74	94					
8.0	1.298	1.179	0.895	0.036	0.141	0.031	0.328	8.0
	77	70	89					
8.5	1.221	1.109	0.806	0.032	0.127	0.028	0.311	8.5
	73	67	83					
9.0	1.148	1.042	0.723	0.029	0.115	0.025	0.296	9.0
	69	63	80					
9.5	1.079	0.979	0.643	0.026	0.105	0.022	0.283	9.5
	65	61	75					
10.0	1.014	0.918	0.568	0.024	0.096	0.020	0.270	10.0
	63	58	72					
10.5	0.951	0.860	0.496	0.021	0.088	0.018	0.259	10.5
	59	55	69					
11.0	0.892	0.805	0.427	0.020	0.081	0.017	0.248	11.0
	57	54	65					
11.5	0.835	0.751	0.362	0.018	0.075	0.015	0.238	11.5
	55	51	63					
12.0	0.780	0.700	0.299	0.016	0.069	0.014	0.229	12.0

where γ is a numerical factor smaller than unity and given by the formula

$$\gamma = \frac{\tanh u}{\frac{2\beta}{l} Du + \tanh u}$$

It is seen that the magnitude of the moments M_0 at the edges depends upon the magnitude of the coefficient β defining the rigidity of the restraint. When β is very small, the coefficient γ approaches unity, and the moment M_0 approaches the value (13) calculated for absolutely built-in edges. When β is very large, the coefficient γ and the moment M_0 become small, and the edge conditions approach those of simply supported edges.

The deflection curve in the case under consideration can be represented in the following form:

$$w = \frac{ql^4}{16u^4 D} \frac{\tanh u - \gamma(\tanh u - u)}{\tanh u} \left\{ \frac{\cosh\left[u\left(1 - \frac{2x}{l}\right)\right]}{\cosh u} - 1 \right\} + \frac{ql^2}{8u^2 D} x(l - x) \quad (20)$$

For $\gamma = 1$ this expression reduces to expression (14) for deflections of a plate with absolutely built-in edges. For $\gamma = 0$ we obtain expression (6) for a plate with simply supported edges.

In calculating the tensile parameter u we proceed as in the previous cases and determine the tensile force S from the condition that the extension of the elemental strip is equal to the difference between the length of the arc along the deflection curve and the chord length l. Hence

$$\frac{S(1 - \nu^2)l}{hE} = \frac{1}{2} \int_0^l \left(\frac{dw}{dx}\right)^2 dx$$

Substituting expression (20) in this equation and performing the integration, we obtain

$$\frac{E^2 h^8}{(1 - \nu^2)^2 q^2 l^8} = (1 - \gamma) U_0 + \gamma U_1 - \gamma(1 - \gamma) U_2 \quad (21)$$

where U_0 and U_1 denote the right-hand sides of Eqs. (8) and (15), respectively, and

$$U_2 = \frac{27}{16} \frac{(u - \tanh u)^2}{u^9 \tanh^2 u} (u \tanh^2 u - u + \tanh u)$$

The values of $\log_{10}(10^4 \sqrt{U_2})$ are given in Table 1. By using this table, Eq. (21) can be readily solved by the trial-and-error method. For any particular plate we first calculate the left-hand side of the equation and, by using the curves in Figs. 4 and 8, determine the values of the parameter u (1) for simply supported edges and (2) for absolutely built-in edges. Naturally u for elastically built-in edges must have a value intermediate between these two. Assuming one such value for u, we calculate U_0, U_1, and U_2 by using Table 1 and determine the value of the right-hand side of Eq. (21). Generally this value will be different from the value of the left-hand side calculated previously, and a new trial calculation with a new assumed value for u must be made. Two such trial calculations will usually be sufficient to determine by interpolation the value of u satisfying Eq. (21). As soon as the parameter u is determined, the bending moments M_0 at the ends may be calculated from Eq. (19). We can also calculate the moment at the middle of the strip and find the maximum stress. This stress will occur at the ends or at the middle, depending on the degree of rigidity of the constraints at the edges.

5. The Effect on Stresses and Deflections of Small Displacements of Longitudinal Edges in the Plane of the Plate. It was assumed in the previous discussion that, during bending, the longitudinal edges of the plate have no displacement in the plane of the plate. On the basis of this assumption the tensile force S was calculated in each particular case. Let us assume now that the edges of the plate undergo a displacement toward each other specified by Δ. Owing to this displacement the extension of the elemental strip will be diminished by the same amount, and the equation for calculating the tensile force S becomes

$$\frac{Sl(1-\nu^2)}{hE} = \frac{1}{2}\int_0^l \left(\frac{dw}{dx}\right)^2 dx - \Delta \qquad (a)$$

At the same time Eqs. (6), (14), and (20) for the deflection curve hold true regardless of the magnitude of the tensile force S. They may be differentiated and substituted under the integral sign in Eq. (a). After evaluating this integral and substituting $S = 4u^2D/l^2$, we obtain for simply supported edges

$$\frac{E^2h^8}{q^2(1-\nu^2)^2l^8} \frac{u^2 + \dfrac{3l\Delta}{h^2}}{u^2} = U_0 \qquad (22)$$

and for built-in edges

$$\frac{E^2h^8}{q^2(1-\nu^2)^2l^8} \frac{u^2 + \dfrac{3l\Delta}{h^2}}{u^2} = U_1 \qquad (23)$$

If Δ is made zero, Eqs. (22) and (23) reduce to Eqs. (8) and (15), obtained previously for immovable edges.

The simplest case is obtained by placing compression bars between the longitudinal sides of the boundary to prevent free motion of one edge of the plate toward the other during bending. Tensile forces S in the plate produce contraction of these bars, which results in a displacement Δ proportional to S.* If k is the factor of proportionality depending on the elasticity and cross-sectional area of the bars, we obtain

$$S = k\Delta$$

or, substituting $S = 4u^2D/l^2$, we obtain

$$\Delta = \frac{1}{k}\frac{Eu^2h^3}{3l^2(1-\nu^2)}$$

and

$$\frac{u^2 + \dfrac{3l\Delta}{h^2}}{u^2} = 1 + \frac{Eh}{kl(1-\nu^2)}$$

Thus the second factor on the left-hand side of Eqs. (22) and (23) is a constant that can be readily calculated if the dimensions and the elastic properties of the structure are known. Having the magnitude of this factor, the solution of Eqs. (22) and (23) can be accomplished in exactly the same manner as used for immovable edges.

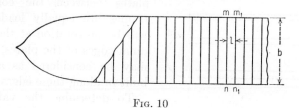

Fig. 10

In the general case the second factor on the left-hand side of Eqs. (22) and (23) may depend on the magnitude of the load acting on the structure, and the determination of the parameter u can be accomplished only by the trial-and-error method. This procedure will now be illustrated by an example that is encountered in analyzing stresses in the hull of a ship when it meets a wave. The bottom plates in the hull of a ship are subjected to a uniformly distributed water pressure and also to forces in the plane of the plates due to bending of the hull as a beam. Let b be the width of the ship at a cross section mn (Fig. 10) and l be the frame spacing at the bottom. When the hollow of a wave is amidships (Fig. 11b), the buoyancy is decreased there and increased at the ends. The effect of this change on the structure is that a sagging bending moment is produced and the normal distance l between the frames at the bottom is increased by a certain amount. To calculate this displacement accurately we must consider not only the action of the bending moment M on the hull but also the effect on this bending of a certain change in

---

* The edge support is assumed to be such that Δ is uniform along the edges.

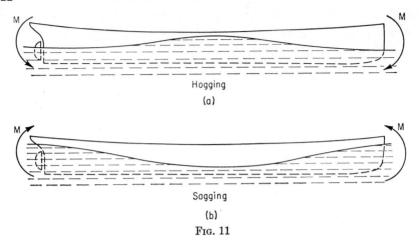

Fig. 11

tensile forces S distributed along the edges mn and m_1n_1 of the bottom plate mnm_1n_1 (Fig. 10), which will be considered as a long rectangular plate uniformly loaded by water pressure. Owing to the fact that the plates between the consecutive frames are equally loaded, there will be no rotation at the longitudinal edges of the plates, and they may be considered as absolutely built in along these edges.

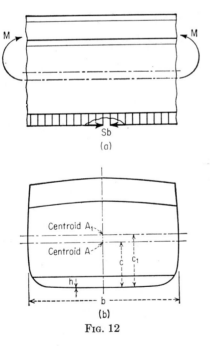

Fig. 12

To determine the value of Δ, which denotes, as before, the displacement of the edge mn toward the edge m_1n_1 in Fig. 10 and which is produced by the hull bending moment M and the tensile reactions S per unit length along the edges mn and m_1n_1 of the bottom plate, let us imagine that the plate mnm_1n_1 is removed and replaced by uniformly distributed forces S so that the total force along mn and m_1n_1 is Sb (Fig. 12a). We can then say that the displacement Δ of one frame relative to another is due to the bending moment M and to the eccentric load Sb applied to the hull without bottom plating.

If A, I, and c are the cross-sectional area, the centroidal moment of inertia, and the distance from the bottom plate to the neutral axis of the complete hull section, and if A_1, I_1, and c_1 are the corresponding quantities for the hull section without bottom plates, the latter set of quantities can be derived from the former by the relations

$$A_1 = A - bh$$
$$c_1 = \frac{Ac}{A_1} \qquad (b)$$
$$I_1 = I - bhc^2 - A_1(c_1 - c)^2$$

The relative displacement Δ_1 produced by the eccentrically applied forces Sb is

$$\Delta_1 = \frac{l(1 - \nu^2)}{E}\left(\frac{Sb}{A_1} + \frac{Sbc_1^2}{I_1}\right)$$

in which the factor $1 - \nu^2$ must be introduced if one neglects the lateral strain. The displacement due to the bending moment M is

$$\Delta_2 = -\frac{Mc_1 l}{EI_1}$$

Hence the total displacement is

$$\Delta = \Delta_1 + \Delta_2 = \frac{l(1 - \nu^2)}{E}\left[\frac{Sb}{A_1} + \frac{Sbc_1}{I_1} - \frac{Mc_1}{I_1(1 - \nu^2)}\right] \qquad (c)$$

Substituting in this expression

$$S = \frac{4u^2 D}{l^2} = \frac{Eu^2 h^3}{3l^2(1 - \nu^2)}$$

we finally obtain

$$\Delta = \frac{u^2 h^3}{3l}\left(\frac{b}{A_1} + \frac{bc_1^2}{I_1}\right) - \frac{Mlc_1}{EI_1} \qquad (d)$$

This quantity must be substituted in Eq. (23) for determining the tensile parameter u.

Let us apply this theory to a numerical example. Assume $b = 54$ ft, $I = 1{,}668$ ft^4, $A = 13.5$ ft^2, $c = 12.87$ ft, $h = 0.75$ in. $= 0.0625$ ft, $l = 45$ in. $= 3.75$ ft, $q = 10$ psi, $M = 123{,}500$ ft-tons. From Eqs. (b) we obtain

$$A_1 = 13.5 - 0.0625 \cdot 54 = 10.125 \text{ ft}^2$$
$$c_1 = \frac{13.5 \cdot 12.87}{10.125} = 17.16 \text{ ft}$$
$$I_1 = 1{,}668 - 559.0 - 10.125(17.16 - 12.87)^2 = 922.7 \text{ ft}^4$$

Substituting these values in expression (d), we calculate Δ and finally obtain

$$\frac{3\Delta l}{h^2} = 1.410u^2 - 11.48$$

Equation (23) then becomes

$$\frac{E^2 h^8}{q^2(1-\nu^2)^2 l^8} \cdot \frac{u^2 + 1.410 u^2 - 11.48}{u^2} = U_1$$

or

$$\frac{1.552 E h^4}{q(1-\nu^2) l^4} \sqrt{\frac{u^2 - 4.763}{u^2}} = \sqrt{U_1}$$

Substituting numerical values and taking logarithms of both sides, we obtain

$$3.597 + \log_{10} \sqrt{\frac{u^2 - 4.763}{u^2}} = \log_{10}(10^4 \sqrt{U_1})$$

Using the curve in Fig. 8, this equation can be readily solved by the trial-and-error method, and we obtain $u = 2.187$ and, from Fig. 5, $\psi_1(u) = 0.780$. The maximum stress is now calculated by using Eqs. (16) and (17), from which

$$\sigma_1 = \frac{30 \cdot 10^6 \cdot 4.783}{3 \cdot 0.91 \cdot 60^2} = 14{,}600 \text{ psi}$$

$$\sigma_2 = \tfrac{1}{2} \cdot 10 \cdot 60^2 \cdot 0.780 = 14{,}040 \text{ psi}$$

$$\sigma_{\max} = \sigma_1 + \sigma_2 = 28{,}640 \text{ psi}$$

If the bending stress in the plate due to water pressure were neglected and if the bottom plate stress were calculated from the formula $\sigma = Mc/I$, we would arrive at a figure of only 13,240 psi.

6. An Approximate Method of Calculating the Parameter u. In calculating the parameter u for plates in which the longitudinal edges do not move in the plane of the plate, we used the equation

$$\frac{Sl(1-\nu^2)}{hE} = \frac{1}{2} \int_0^l \left(\frac{dw}{dx}\right)^2 dx \qquad (a)$$

which states that the extension of an elemental strip produced by the forces S is equal to the difference between the length of the arc along the deflection curve of the strip and the chord length l. In the particular cases considered in the previous articles, exact expressions for the deflections w were derived, and numerical tables and curves for the right-hand side of Eq. (a) were given. When such tables are not at hand, the solution of the equation becomes complicated, and to simplify the problem recourse should be had to an approximate method. From the discussion of bending of beams it is known[1] that, in the case of simply supported ends with all lateral loads acting in the same direction, the deflection curve of an elemental strip produced by a combination of a lateral load and an axial tensile force S (Fig. 3) can be represented with sufficient

[1] See Timoshenko, "Strength of Materials," part II, 3d ed., p. 52, 1956.

accuracy by the equation

$$w = \frac{w_0}{1+\alpha} \sin \frac{\pi x}{l} \qquad (b)$$

in which w_0 denotes the deflection at the middle of the strip produced by the lateral load alone, and the quantity α is given by the equation

$$\alpha = \frac{S}{S_{cr}} = \frac{Sl^2}{\pi^2 D} \qquad (c)$$

Thus, α represents the ratio of the axial force S to the Euler critical load for the elemental strip.

Substituting expression (b) in Eq. (a) and integrating, we obtain

$$\frac{Sl(1-\nu^2)}{hE} = \frac{\pi^2 w_0^2}{4l(1+\alpha)^2}$$

Now, using notation (c) and substituting for D its expression (3), we finally obtain

$$\alpha(1+\alpha)^2 = \frac{3w_0^2}{h^2} \qquad (24)$$

From this equation the quantity α can be calculated in each particular case, and the parameter u is now determined from the equation

$$u^2 = \frac{S}{D}\frac{l^2}{4} = \frac{\pi^2 \alpha}{4} \qquad (d)$$

To show the application of the approximate Eq. (24) let us take a numerical example. A long rectangular steel plate with simply supported edges and of dimensions $l = 50$ in. and $h = \frac{1}{2}$ in. is loaded with a uniformly distributed load $q = 20$ psi. In such a case

$$w_0 = \frac{5}{384}\frac{ql^4}{D}$$

and, after substituting numerical values, Eq. (24) becomes

$$\alpha(1+\alpha)^2 = 269.56$$

The solution of the equation can be simplified by letting

$$1 + \alpha = x \qquad (e)$$

Then
$$x^3 - x^2 = 269.56$$

i.e., the quantity x is such that the difference between its cube and its square has a known value. Thus x can be readily determined from a slide rule or a suitable table, and we find in our case

$$x = 6.8109 \quad \text{and} \quad \alpha = 5.8109$$

Then, from Eq. (d)
$$u = 3.7865$$
and from the formula (e) (see page 9)
$$\psi_0 = 0.13316$$

For calculating direct stress and maximum bending stress we use Eqs. (10) and (11). In this way we find

$$\sigma_1 = 15{,}759 \text{ psi}$$
$$\sigma_2 = 19{,}974 \text{ psi}$$
$$\sigma_{\max} = \sigma_1 + \sigma_2 = 35{,}733 \text{ psi}$$

The calculations made in Art. 2 (page 11) give, for this example,

$$\sigma_{\max} = 35{,}760 \text{ psi}$$

Thus the accuracy of the approximate Eq. (24) is in this case very high. In general, this accuracy depends on the magnitude of u. The error increases with increase of u. Calculations show that for $u = 1.44$ the error in the maximum stress is only 0.065 of 1 per cent and that for $u = 12.29$, which corresponds to very flexible plates, it is about 0.30 of 1 per cent. These values of u will cover the range ordinarily encountered in practice, and we conclude that Eq. (24) can be used with sufficient accuracy in all practical cases of uniformly loaded plates with simply supported edges.

It can also be used when the load is not uniformly distributed, as in the case of a hydrostatic pressure nonuniformly distributed along the elemental strip. If the longitudinal force is found by using the approximate Eq. (24), the deflections may be obtained from Eq. (b), and the bending moment at any cross section may be found as the algebraic sum of the moment produced by the lateral load and the moment due to the longitudinal force.[1]

In the case of built-in edges the approximate expression for the deflection curve of an elemental strip can be taken in the form

$$w = \frac{w_0}{1 + \alpha/4} \frac{1}{2}\left(1 - \cos \frac{2\pi x}{l}\right) \qquad (f)$$

in which w_0 is the deflection of the built-in beam under the lateral load acting alone and α has the same meaning as before. Substituting this expression in Eq. (a) and integrating, we obtain for determining α the equation

---

[1] More accurate values for the deflections and for the bending moments can be obtained by substituting the approximate value of the longitudinal force in Eq. (4) and integrating this equation, which gives Eqs. (12) and (9).

$$\alpha\left(1 + \frac{\alpha}{4}\right)^2 = \frac{3w_0^2}{h^2} \qquad (25)$$

which can be solved in each particular case by the method suggested for solving Eq. (24).

When α is found, the parameter u is determined from Eq. (d); the maximum stress can be calculated by using Eqs. (16) and (17); and the maximum deflection, by using Eq. (18).

If, during bending, one edge moves toward the other by an amount Δ, the equation

$$\frac{Sl(1-\nu^2)}{hE} = \frac{1}{2}\int_0^l \left(\frac{dw}{dx}\right)^2 dx - \Delta \qquad (g)$$

must be used instead of Eq. (a). Substituting expression (b) in this equation, we obtain for determining α in the case of simply supported edges the equation

$$\alpha(1+\alpha)^2 \frac{\alpha + 12\dfrac{\Delta l}{\pi^2 h^2}}{\alpha} = \frac{3w_0^2}{h^2} \qquad (26)$$

In the case of built-in edges we use expression (f). Then for determining α we obtain

$$\alpha\left(1+\frac{\alpha}{4}\right)^2 \frac{\alpha + 12\dfrac{\Delta l}{\pi^2 h^2}}{\alpha} = \frac{3w_0^2}{h^2} \qquad (27)$$

If the dimensions of the plate and the load q are given, and the displacement Δ is known, Eqs. (26) and (27) can both be readily solved in the same manner as before. If the displacement Δ is proportional to the tensile force S, the second factor on the left-hand sides of Eqs. (26) and (27) is a constant and can be determined as explained in the previous article (see page 21). Thus again the equations can be readily solved.

7. Long Uniformly Loaded Rectangular Plates Having a Small Initial Cylindrical Curvature. It is seen from the discussions in Arts. 2 and 3 that the tensile forces S contribute to the strength of the plates by counteracting the bending produced by lateral load. This action increases with an increase in deflection. A further reduction of maximum stress can be accomplished by giving a suitable initial curvature to a plate. The effect on stresses and deflections of such an initial curvature can be investigated[1] by using the approximate method developed in the previous article.

Let us consider the case of a long rectangular plate with simply supported edges (Fig. 13), the initial curvature of which is given by the equation

---

[1] See S. Timoshenko's paper in "Festschrift zum siebzigsten Geburtstage August Föppls," p. 74, Berlin, 1923.

$$w_1 = \delta \sin \frac{\pi x}{l} \tag{a}$$

If tensile forces S are applied to the edges of the plate, the initial deflections (a) will be reduced in the ratio $1/(1 + \alpha)$, where α has the same meaning as in the previous article[1] (page 25). The lateral load in combination with the forces S will produce deflections that can be expressed approximately by Eq. (b) of the previous article. Thus the total deflection of the plate, indicated in Fig. 13 by the dashed line, is

$$w = \frac{\delta}{1 + \alpha} \sin \frac{\pi x}{l} + \frac{w_0}{1 + \alpha} \sin \frac{\pi x}{l} = \frac{\delta + w_0}{1 + \alpha} \sin \frac{\pi x}{l} \tag{b}$$

Assuming that the longitudinal edges of the plate do not move in the plane of the plate, the tensile force S is found from the condition that the extension of the elemental strip produced by the forces S is equal to

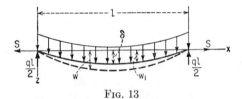

FIG. 13

the difference between the length of the arc along the deflection curve of the elemental strip and the initial length of the strip. This difference, in the case of small deflections, is given by the equation

$$\lambda = \frac{1}{2} \int_0^l \left(\frac{dw}{dx}\right)^2 dx - \frac{1}{2} \int_0^l \left(\frac{dw_1}{dx}\right)^2 dx \tag{c}$$

Substituting expressions (a) and (b) for w and w_1 and integrating, we obtain

$$\lambda = \frac{\pi^2}{4l} \left[\left(\frac{\delta + w_0}{1 + \alpha}\right)^2 - \delta^2\right]$$

Putting λ equal to the extension of the strip $Sl(1 - \nu^2)/hE$, we finally obtain

$$\alpha(1 + \alpha)^2 = \frac{3(\delta + w_0)^2}{h^2} - \frac{3\delta^2(1 + \alpha)^2}{h^2} \tag{28}$$

If we take $\delta = 0$, this equation reduces to Eq. (24) for a plate without initial curvature.

To show the effect of the initial curvature on the maximum stress in a plate, let us apply Eq. (28) to a numerical example. Assume that a steel plate having $l = 45$ in. and $h = \frac{3}{8}$ in. is submitted to the action of

---

[1] See Timoshenko, "Strength of Materials," part II, 3d ed., p. 52, 1956.

a uniformly distributed load $q = 10$ psi. If there is no initial deflection, $\delta = 0$ and Eq. (28) becomes

$$\alpha(1 + \alpha)^2 = 290$$

from which

$$\alpha = 5.97 \quad \text{and} \quad u = \frac{\pi}{2}\sqrt{\alpha} = 3.83$$

From Eq. (10) we then obtain

$$\sigma_1 = 11{,}300 \text{ psi}$$

and from Eq. (11)

$$\sigma_2 = 14{,}200 \text{ psi}$$

The maximum stress in the plate is

$$\sigma_{\max} = \sigma_1 + \sigma_2 = 25{,}500 \text{ psi}$$

Let us now assume that there is an initial deflection in the plate such that $\delta = h = \frac{3}{8}$ in. In such a case Eq. (28) gives

$$\alpha(1 + \alpha)^2 = 351.6 - 3(1 + \alpha)^2$$

Letting

$$1 + \alpha = x$$

we obtain

$$x^3 + 2x^2 = 351.6$$

from which

$$x = 6.45 \qquad \alpha = 5.45 \qquad u = \frac{\pi}{2}\sqrt{\alpha} = 3.67$$

The tensile stress, from Eq. (10), is

$$\sigma_1 = 10{,}200 \text{ psi}$$

In calculating the bending stress we must consider only the change in deflections

$$w - w_1 = \frac{w_0}{1 + \alpha}\sin\frac{\pi x}{l} - \frac{\alpha\delta}{1+\alpha}\sin\frac{\pi x}{l} \qquad (d)$$

The maximum bending stress, corresponding to the first term on the right-hand side of Eq. (d), is the same as for a flat plate with $u = 3.67$. From Table 1 we find $\psi_0 = 0.142$ and from Eq. (11)

$$\sigma_2' = 15{,}300 \text{ psi}$$

The bending moment corresponding to the second term in Eq. (d) is

$$-D\frac{d^2}{dx^2}\left(-\frac{\alpha\delta}{1+\alpha}\sin\frac{\pi x}{l}\right) = -\frac{\alpha\pi^2\,\delta D}{(1+\alpha)l^2}\sin\frac{\pi x}{l}$$

This moment has a negative sign, and a corresponding maximum stress of

$$\sigma_2'' = \frac{6}{h^2} \frac{\alpha \pi^2 \, \delta D}{(1+\alpha)l^2} = 9{,}500 \text{ psi}$$

must be subtracted from the bending stress σ_2' calculated above. Hence the maximum stress for the plate with the initial deflection is

$$\sigma_{\max} = 10{,}200 + 15{,}300 - 9{,}500 = 16{,}000 \text{ psi}$$

Comparison of this result with that obtained for the plane plate shows that the effect of the initial curvature is to reduce the maximum stress from 25,500 to 16,000 psi. This result is obtained assuming the initial deflection equal to the thickness of the plate. By increasing the initial deflection, the maximum stress can be reduced still further.

8. Cylindrical Bending of a Plate on an Elastic Foundation. Let us consider the problem of bending of a long uniformly loaded rectangular plate supported over the entire surface by an elastic foundation and rigidly supported along the edges (Fig. 14).

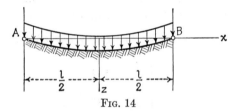

Fig. 14

Cutting out from the plate an elemental strip, as before, we may consider it as a beam on an elastic foundation. Assuming that the reaction of the foundation at any point is proportional to the deflection w at that point, and using Eq. (4), we obtain by double differentiation of that equation[1]

$$D \frac{d^4w}{dx^4} = q - kw \quad (29)$$

where q is the intensity of the load acting on the plate and k is the reaction of the foundation per unit area for a deflection equal to unity. Introducing the notation

$$\beta = \frac{l}{2} \sqrt{\frac{k}{4D}} \quad (30)$$

the general solution of Eq. (29) can be written as follows:

$$w = \frac{q}{k} + C_1 \sin \frac{2\beta x}{l} \sinh \frac{2\beta x}{l} + C_2 \sin \frac{2\beta x}{l} \cosh \frac{2\beta x}{l} + C_3 \cos \frac{2\beta x}{l} \sinh \frac{2\beta x}{l}$$
$$+ C_4 \cos \frac{2\beta x}{l} \cosh \frac{2\beta x}{l} \quad (a)$$

The four constants of integration must now be determined from the conditions at the ends of the strip. In the case under consideration the deflection is symmetrical with respect to the middle of the strip. Thus, taking the coordinate axes as shown in Fig.

[1] *Ibid.*, p. 21.

14, we conclude[1] that $C_2 = C_3 = 0$. The constants C_1 and C_4 are found from the conditions that the deflection and the bending moment of the strip are zero at the end ($x = l/2$). Hence

$$(w)_{x=l/2} = 0$$

$$\left(\frac{d^2w}{dx^2}\right)_{x=l/2} = 0 \qquad (b)$$

Substituting expression (a) for w and observing that $C_2 = C_3 = 0$, we obtain

$$\frac{q}{k} + C_1 \sin \beta \sinh \beta + C_4 \cos \beta \cosh \beta = 0 \qquad (c)$$

$$C_1 \cos \beta \cosh \beta - C_4 \sin \beta \sinh \beta = 0$$

from which we find

$$C_1 = -\frac{q}{k} \frac{\sin \beta \sinh \beta}{\sin^2 \beta \sinh^2 \beta + \cos^2 \beta \cosh^2 \beta} = -\frac{q}{k} \frac{2 \sin \beta \sinh \beta}{\cos 2\beta + \cosh 2\beta}$$

$$C_4 = -\frac{q}{k} \frac{\cos \beta \cosh \beta}{\sin^2 \beta \sinh^2 \beta + \cos^2 \beta \cosh^2 \beta} = -\frac{q}{k} \frac{2 \cos \beta \cosh \beta}{\cos 2\beta + \cosh 2\beta}$$

Substituting these values of the constants in expression (a) and using Eq. (30), we finally represent the deflection of the strip by the equation

$$w = \frac{ql^4}{64D\beta^4}\left(1 - \frac{2 \sin \beta \sinh \beta}{\cos 2\beta + \cosh 2\beta} \sin \frac{2\beta x}{l} \sinh \frac{2\beta x}{l} \right.$$
$$\left. - \frac{2 \cos \beta \cosh \beta}{\cos 2\beta + \cosh 2\beta} \cos \frac{2\beta x}{l} \cosh \frac{2\beta x}{l}\right) \qquad (d)$$

The deflection at the middle is obtained by substituting $x = 0$, which gives

$$(w)_{x=0} = \frac{5ql^4}{384D} \varphi(\beta) \qquad (31)$$

where

$$\varphi(\beta) = \frac{6}{5\beta^4}\left(1 - \frac{2 \cos \beta \cosh \beta}{\cos 2\beta + \cosh 2\beta}\right)$$

To obtain the angles of rotation of the edges of the plate, we differentiate expression (d) with respect to x and put $x = -l/2$. In this way we obtain

$$\left(\frac{dw}{dx}\right)_{x=-l/2} = \frac{ql^3}{24D} \varphi_1(\beta) \qquad (32)$$

where

$$\varphi_1(\beta) = \frac{3}{4\beta^3} \frac{\sinh 2\beta - \sin 2\beta}{\cosh 2\beta + \cos 2\beta}$$

The bending moment at any cross section of the strip is obtained from the equation

$$M = -D\frac{d^2w}{dx^2}$$

Substituting expression (d) for w, we find for the middle of the strip

$$(M)_{x=0} = \frac{ql^2}{8} \varphi_2(\beta) \qquad (33)$$

where

$$\varphi_2(\beta) = \frac{2}{\beta^2} \frac{\sinh \beta \sin \beta}{\cosh 2\beta + \cos 2\beta}$$

[1] It is seen that the terms with coefficients C_2 and C_3 change sign when x is replaced by $-x$.

To simplify the calculation of deflections and stresses, numerical values of functions φ, φ_1, and φ_2 are given in Table 2. For small values of β, that is, for a yielding foundation, the functions φ and φ_2 do not differ greatly from unity. Thus the maximum deflection and bending stresses are close to those for a simply supported strip without an elastic foundation. With an increase in β, the effect of the foundation becomes more and more important.

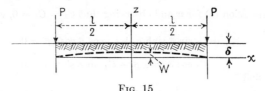

Fig. 15

Conditions similar to those represented in Fig. 14 are obtained if a long rectangular plate of width l is pressed into an elastic foundation by loads uniformly distributed along the edges and of the amount P per unit length (Fig. 15). The plate will be

TABLE 2

β	φ	φ_1	φ_2	β	φ	φ_1	φ_2
0.1	1.000	1.000	1.000	1.6	0.186	0.200	0.164
0.2	0.999	0.999	0.999	1.7	0.151	0.166	0.129
0.3	0.996	0.995	0.995	1.8	0.124	0.138	0.101
0.4	0.984	0.983	0.983	1.9	0.102	0.116	0.079
0.5	0.961	0.961	0.959	2.0	0.084	0.099	0.062
0.6	0.921	0.923	0.919	2.2	0.058	0.072	0.037
0.7	0.863	0.866	0.859	2.4	0.042	0.055	0.021
0.8	0.787	0.791	0.781	2.6	0.029	0.043	0.011
0.9	0.698	0.702	0.689	2.8	0.022	0.034	0.005
1.0	0.602	0.609	0.591	3.0	0.016	0.028	0.002
1.1	0.508	0.517	0.494	3.2	0.012	0.023	0.000
1.2	0.421	0.431	0.405	3.4	0.010	0.019	−0.001
1.3	0.345	0.357	0.327	3.6	0.007	0.016	−0.002
1.4	0.281	0.294	0.262	3.8	0.006	0.014	−0.002
1.5	0.228	0.242	0.208	4.0	0.005	0.012	−0.002

pressed into the elastic foundation and bent, as shown by the dashed line. If δ denotes the deflection at the edges of the plate, the reaction of the foundation at any point is

$$k(\delta - w) = k\delta - kw$$

where w is given by Eq. (d) with $q = k\delta$. The magnitude of δ is then obtained from the condition that the load is balanced by the reaction of the foundation. Hence

$$P = \frac{k\delta l}{2} - k \int_0^{l/2} w\, dx$$

Plates on elastic foundation with other conditions at the longitudinal edges can also be discussed in a similar manner.

CHAPTER 2

PURE BENDING OF PLATES

9. Slope and Curvature of Slightly Bent Plates. In discussing small deflections of a plate we take the *middle plane* of the plate, before bending occurs, as the xy plane. During bending, the particles that were in the xy plane undergo small displacements w perpendicular to the xy plane and form the *middle surface* of the plate. These displacements of the middle surface are called *deflections* of a plate in our further discussion. Taking a normal section of the plate parallel to the xz plane (Fig. 16a), we find that the slope of the middle surface in the x direction is $i_x = \partial w/\partial x$. In the same manner the slope in the y direction is $i_y = \partial w/\partial y$. Taking now any direction an in the xy plane (Fig. 16b) making an angle α with the x axis, we find that the difference in the deflections of the two adjacent points a and a_1 in the an direction is

$$dw = \frac{\partial w}{\partial x}\,dx + \frac{\partial w}{\partial y}\,dy$$

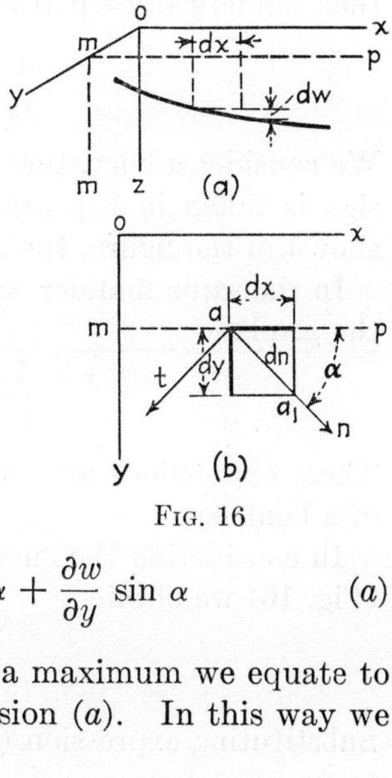

and that the corresponding slope is

Fig. 16

$$\frac{\partial w}{\partial n} = \frac{\partial w}{\partial x}\frac{dx}{dn} + \frac{\partial w}{\partial y}\frac{dy}{dn} = \frac{\partial w}{\partial x}\cos\alpha + \frac{\partial w}{\partial y}\sin\alpha \qquad (a)$$

To find the direction α_1 for which the slope is a maximum we equate to zero the derivative with respect to α of expression (a). In this way we obtain

$$\tan\alpha_1 = \frac{\partial w}{\partial y}\bigg/\frac{\partial w}{\partial x} \qquad (b)$$

Substituting the corresponding values of $\sin\alpha_1$ and $\cos\alpha_1$ in (a), we obtain for the maximum slope the expression

$$\left(\frac{\partial w}{\partial n}\right)_{\max} = \sqrt{\left(\frac{\partial w}{\partial x}\right)^2 + \left(\frac{\partial w}{\partial y}\right)^2} \qquad (c)$$

By setting expression (a) equal to zero we obtain the direction for which the slope of the surface is zero. The corresponding angle α_2 is determined from the equation

$$\tan \alpha_2 = -\frac{\partial w}{\partial x} \bigg/ \frac{\partial w}{\partial y} \qquad (d)$$

From Eqs. (b) and (d) we conclude that

$$\tan \alpha_1 \tan \alpha_2 = -1$$

which shows that the directions of zero slope and of maximum slope are perpendicular to each other.

In determining the curvature of the middle surface of the plate we observe that the deflections of the plate are very small. In such a case the slope of the surface in any direction can be taken equal to the angle that the tangent to the surface in that direction makes with the xy plane, and the square of the slope may be neglected compared to unity. The curvature of the surface in a plane parallel to the xz plane (Fig. 16) is then numerically equal to

$$\frac{1}{r_x} = -\frac{\partial}{\partial x}\left(\frac{\partial w}{\partial x}\right) = -\frac{\partial^2 w}{\partial x^2} \qquad (e)$$

We consider a curvature positive if it is convex downward. The minus sign is taken in Eq. (e), since for the deflection convex downward, as shown in the figure, the second derivative $\partial^2 w/\partial x^2$ is negative.

In the same manner we obtain for the curvature in a plane parallel to the yz plane

$$\frac{1}{r_y} = -\frac{\partial}{\partial y}\left(\frac{\partial w}{\partial y}\right) = -\frac{\partial^2 w}{\partial y^2} \qquad (f)$$

These expressions are similar to those used in discussing the curvature of a bent beam.

In considering the curvature of the middle surface in any direction an (Fig. 16) we obtain

$$\frac{1}{r_n} = -\frac{\partial}{\partial n}\left(\frac{\partial w}{\partial n}\right)$$

Substituting expression (a) for $\partial w/\partial n$ and observing that

$$\frac{\partial}{\partial n} = \frac{\partial}{\partial x}\cos\alpha + \frac{\partial}{\partial y}\sin\alpha$$

we find

$$\begin{aligned}
\frac{1}{r_n} &= -\left(\frac{\partial}{\partial x}\cos\alpha + \frac{\partial}{\partial y}\sin\alpha\right)\left(\frac{\partial w}{\partial x}\cos\alpha + \frac{\partial w}{\partial y}\sin\alpha\right) \\
&= -\left(\frac{\partial^2 w}{\partial x^2}\cos^2\alpha + 2\frac{\partial^2 w}{\partial x\,\partial y}\sin\alpha\cos\alpha + \frac{\partial^2 w}{\partial y^2}\sin^2\alpha\right) \\
&= \frac{1}{r_x}\cos^2\alpha - \frac{1}{r_{xy}}\sin 2\alpha + \frac{1}{r_y}\sin^2\alpha \qquad (g)
\end{aligned}$$

It is seen that the curvature in any direction n at a point of the middle surface can be calculated if we know at that point the curvatures

$$\frac{1}{r_x} = -\frac{\partial^2 w}{\partial x^2} \qquad \frac{1}{r_y} = -\frac{\partial^2 w}{\partial y^2}$$

and the quantity

$$\frac{1}{r_{xy}} = \frac{\partial^2 w}{\partial x\, \partial y} \tag{h}$$

which is called the *twist of the surface* with respect to the x and y axes.

If instead of the direction an (Fig. 16b) we take the direction at perpendicular to an, the curvature in this new direction will be obtained from expression (g) by substituting $\pi/2 + \alpha$ for α. Thus we obtain

$$\frac{1}{r_t} = \frac{1}{r_x}\sin^2\alpha + \frac{1}{r_{xy}}\sin 2\alpha + \frac{1}{r_y}\cos^2\alpha \tag{i}$$

Adding expressions (g) and (i), we find

$$\frac{1}{r_n} + \frac{1}{r_t} = \frac{1}{r_x} + \frac{1}{r_y} \tag{34}$$

which shows that at any point of the middle surface the sum of the curvatures in two perpendicular directions such as n and t is independent of the angle α. This sum is usually called the *average curvature* of the surface at a point.

The twist of the surface at a with respect to the an and at directions is

$$\frac{1}{r_{nt}} = \frac{d}{dt}\left(\frac{dw}{dn}\right)$$

In calculating the derivative with respect to t, we observe that the direction at is perpendicular to an. Thus we obtain the required derivative by substituting $\pi/2 + \alpha$ for α in Eq. (a). In this manner we find

$$\begin{aligned}\frac{1}{r_{nt}} &= \left(\frac{\partial}{\partial x}\cos\alpha + \frac{\partial}{\partial y}\sin\alpha\right)\left(-\frac{\partial w}{\partial x}\sin\alpha + \frac{\partial w}{\partial y}\cos\alpha\right) \\ &= \frac{1}{2}\sin 2\alpha \left(-\frac{\partial^2 w}{\partial x^2} + \frac{\partial^2 w}{\partial y^2}\right) + \cos 2\alpha\,\frac{\partial^2 w}{\partial x\,\partial y} \\ &= \frac{1}{2}\sin 2\alpha \left(\frac{1}{r_x} - \frac{1}{r_y}\right) + \cos 2\alpha\,\frac{1}{r_{xy}}\end{aligned} \tag{j}$$

In our further discussion we shall be interested in finding in terms of α the directions in which the curvature of the surface is a maximum or a minimum and in finding the corresponding values of the curvature. We obtain the necessary equation for determining α by equating the derivative of expression (g) with respect to α to zero, which gives

$$\frac{1}{r_x}\sin 2\alpha + \frac{2}{r_{xy}}\cos 2\alpha - \frac{1}{r_y}\sin 2\alpha = 0 \tag{k}$$

whence

$$\tan 2\alpha = -\frac{\dfrac{2}{r_{xy}}}{\dfrac{1}{r_x} - \dfrac{1}{r_y}} \tag{35}$$

From this equation we find two values of α, differing by $\pi/2$. Substituting these in Eq. (g) we find two values of $1/r_n$, one representing the maximum and the other the minimum curvature at a point a of the surface. These two curvatures are called the *principal curvatures* of the surface; and the corresponding planes naz and taz, the *principal planes of curvature*.

Observing that the left-hand side of Eq. (k) is equal to the doubled value of expression (j), we conclude that, if the directions an and at (Fig. 16) are in the principal planes, the corresponding twist $1/r_{nt}$ is equal to zero.

We can use a circle, similar to Mohr's circle representing combined stresses, to show how the curvature and the twist of a surface vary with the angle α.* To simplify the discussion we assume that the coordinate planes xz and yz are taken parallel to the principal planes of curvature at the point a. Then

$$\frac{1}{r_{xy}} = 0$$

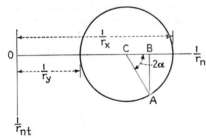

Fig. 17

and we obtain from Eqs. (g) and (j) for any angle α

$$\frac{1}{r_n} = \frac{1}{r_x}\cos^2\alpha + \frac{1}{r_y}\sin^2\alpha$$
$$\frac{1}{r_{nt}} = \frac{1}{2}\left(\frac{1}{r_x} - \frac{1}{r_y}\right)\sin 2\alpha \tag{36}$$

Taking the curvatures as abscissas and the twists as ordinates and constructing a circle on the diameter $1/r_x - 1/r_y$, as shown in Fig. 17, we see that the point A defined by the angle 2α has the abscissa

$$\overline{OB} = \overline{OC} + \overline{CB} = \frac{1}{2}\left(\frac{1}{r_x} + \frac{1}{r_y}\right) + \frac{1}{2}\left(\frac{1}{r_x} - \frac{1}{r_y}\right)\cos 2\alpha$$
$$= \frac{1}{r_x}\cos^2\alpha + \frac{1}{r_y}\sin^2\alpha$$

and the ordinate

$$\overline{AB} = \frac{1}{2}\left(\frac{1}{r_x} - \frac{1}{r_y}\right)\sin 2\alpha$$

Comparing these results with formulas (36), we conclude that the coordi-

---
* See S. Timoshenko, "Strength of Materials," part I, 3d ed., p. 40, 1955.

nates of the point A define the curvature and the twist of the surface for any value of the angle α. It is seen that the maximum twist, represented by the radius of the circle, takes place when $\alpha = \pi/4$, i.e., when we take two perpendicular directions bisecting the angles between the principal planes.

In our example the curvature in any direction is positive; hence the surface is bent convex downward. If the curvatures $1/r_x$ and $1/r_y$ are both negative, the curvature in any direction is also negative, and we have a bending of the plate convex upward. Surfaces in which the curvatures in all planes have like signs are called *synclastic*. Sometimes we shall deal with surfaces in which the two principal curvatures have opposite signs. A saddle is a good example. Such surfaces are called *anticlastic*. The circle in Fig. 18 represents a particular case of such surfaces when

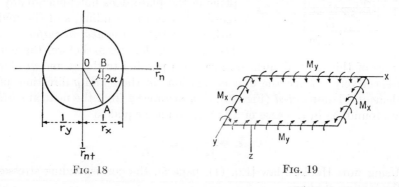

Fig. 18   Fig. 19

$1/r_y = -1/r_x$. It is seen that in this case the curvature becomes zero for $\alpha = \pi/4$ and for $\alpha = 3\pi/4$, and the twist becomes equal to $\pm 1/r_x$.

10. Relations between Bending Moments and Curvature in Pure Bending of Plates. In the case of pure bending of prismatic bars a rigorous solution for stress distribution is obtained by assuming that cross sections of the bar remain plane during bending and rotate only with respect to their neutral axes so as to be always normal to the deflection curve. Combination of such bending in two perpendicular directions brings us to pure bending of plates. Let us begin with pure bending of a rectangular plate by moments that are uniformly distributed along the edges of the plate, as shown in Fig. 19. We take the xy plane to coincide with the middle plane of the plate before deflection and the x and y axes along the edges of the plate as shown. The z axis, which is then perpendicular to the middle plane, is taken positive downward. We denote by M_x the bending moment per unit length acting on the edges parallel to the y axis and by M_y the moment per unit length acting on the edges parallel to the x axis. These moments we consider positive when they are directed as shown in the figure, i.e., when they produce compression in the upper surface of the plate and tension in the lower. The thickness of the plate we denote, as before, by h and consider it small in comparison with other dimensions.

Let us consider an element cut out of the plate by two pairs of planes parallel to the xz and yz planes, as shown in Fig. 20. Since the case shown in Fig. 19 represents the combination of two uniform bendings, the stress conditions are identical in all elements, as shown in Fig. 20, and we have a uniform bending of the plate. Assuming that during bending of the plate the lateral sides of the element remain plane and rotate about the neutral axes nn so as to remain normal to the deflected middle surface of the plate, it can be concluded that the middle plane of the plate does not undergo any extension during this bending, and the middle surface is therefore the *neutral surface*.[1] Let

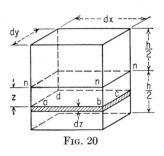

Fig. 20

$1/r_x$ and $1/r_y$ denote, as before, the curvatures of this neutral surface in sections parallel to the xz and yz planes, respectively. Then the unit elongations in the x and y directions of an elemental lamina $abcd$ (Fig. 20), at a distance z from the neutral surface, are found, as in the case of a beam, and are equal to

$$\epsilon_x = \frac{z}{r_x} \qquad \epsilon_y = \frac{z}{r_y} \qquad (a)$$

Using now Hooke's law [Eq. (1), page 5], the corresponding stresses in the lamina $abcd$ are

$$\sigma_x = \frac{Ez}{1 - \nu^2}\left(\frac{1}{r_x} + \nu \frac{1}{r_y}\right)$$
$$\sigma_y = \frac{Ez}{1 - \nu^2}\left(\frac{1}{r_y} + \nu \frac{1}{r_x}\right) \qquad (b)$$

These stresses are proportional to the distance z of the lamina $abcd$ from the neutral surface and depend on the magnitude of the curvatures of the bent plate.

The normal stresses distributed over the lateral sides of the element in Fig. 20 can be reduced to couples, the magnitudes of which per unit length evidently must be equal to the external moments M_x and M_y. In this way we obtain the equations

$$\int_{-h/2}^{h/2} \sigma_x z \, dy \, dz = M_x \, dy$$
$$\int_{-h/2}^{h/2} \sigma_y z \, dx \, dz = M_y \, dx \qquad (c)$$

---

[1] It will be shown in Art. 13 that this conclusion is accurate enough if the deflections of the plate are small in comparison with the thickness h.

PURE BENDING OF PLATES

Substituting expressions (b) for σ_x and σ_y, we obtain

$$M_x = D\left(\frac{1}{r_x} + \nu \frac{1}{r_y}\right) = -D\left(\frac{\partial^2 w}{\partial x^2} + \nu \frac{\partial^2 w}{\partial y^2}\right) \quad (37)$$

$$M_y = D\left(\frac{1}{r_y} + \nu \frac{1}{r_x}\right) = -D\left(\frac{\partial^2 w}{\partial y^2} + \nu \frac{\partial^2 w}{\partial x^2}\right) \quad (38)$$

where D is the flexural rigidity of the plate defined by Eq. (3), and w denotes small deflections of the plate in the z direction.

Let us now consider the stresses acting on a section of the lamina $abcd$ parallel to the z axis and inclined to the x and y axes. If acd (Fig. 21) represents a portion of the lamina cut by such a section, the stress acting on the side ac can be found by means of the equations of statics. Resolving this stress into a normal component σ_n and a shearing component τ_{nt},

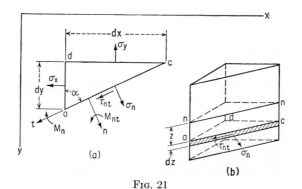

Fig. 21

the magnitudes of these components are obtained by projecting the forces acting on the element acd on the n and t directions respectively, which gives the known equations

$$\begin{aligned}\sigma_n &= \sigma_x \cos^2 \alpha + \sigma_y \sin^2 \alpha \\ \tau_{nt} &= \tfrac{1}{2}(\sigma_y - \sigma_x) \sin 2\alpha\end{aligned} \quad (d)$$

in which α is the angle between the normal n and the x axis or between the direction t and the y axis (Fig. 21a). The angle is considered positive if measured in a clockwise direction.

Considering all laminas, such as acd in Fig. 21b, over the thickness of the plate, the normal stresses σ_n give the bending moment acting on the section ac of the plate, the magnitude of which per unit length along ac is

$$M_n = \int_{-h/2}^{h/2} \sigma_n z \, dz = M_x \cos^2 \alpha + M_y \sin^2 \alpha \quad (39)$$

The shearing stresses τ_{nt} give the twisting moment acting on the section

ac of the plate, the magnitude of which per unit length of ac is

$$M_{nt} = -\int_{-h/2}^{h/2} \tau_{nt} z \, dz = \tfrac{1}{2} \sin 2\alpha (M_x - M_y) \qquad (40)$$

The signs of M_n and M_{nt} are chosen in such a manner that the positive values of these moments are represented by vectors in the positive directions of n and t (Fig. 21a) if the rule of the right-hand screw is used. When α is zero or π, Eq. (39) gives $M_n = M_x$. For $\alpha = \pi/2$ or $3\pi/2$, we obtain $M_n = M_y$. The moments M_{nt} become zero for these values of α. Thus we obtain the conditions shown in Fig. 19.

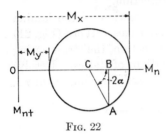

Fig. 22

Equations (39) and (40) are similar to Eqs. (36), and by using them the bending and twisting moments can be readily calculated for any value of α. We can also use the graphical method for the same purpose and find the values of M_n and M_{nt} from Mohr's circle, which can be constructed as shown in the previous article by taking M_n as abscissa and M_{nt} as ordinate. The diameter of the circle will be equal to $M_x - M_y$, as shown in Fig. 22. Then the coordinates $\overline{OB}$ and $\overline{AB}$ of a point A, defined by the angle 2α, give the moments M_n and M_{nt} respectively.

Let us now represent M_n and M_{nt} as functions of the curvatures and twist of the middle surface of the plate. Substituting in Eq. (39) for M_x and M_y their expressions (37) and (38), we find

$$M_n = D\left(\frac{1}{r_x}\cos^2\alpha + \frac{1}{r_y}\sin^2\alpha\right) + \nu D\left(\frac{1}{r_x}\sin^2\alpha + \frac{1}{r_y}\cos^2\alpha\right)$$

Using the first of the equations (36) of the previous article, we conclude that the expressions in parentheses represent the curvatures of the middle surface in the n and t directions respectively. Hence

$$M_n = D\left(\frac{1}{r_n} + \nu \frac{1}{r_t}\right) = -D\left(\frac{\partial^2 w}{\partial n^2} + \nu \frac{\partial^2 w}{\partial t^2}\right) \qquad (41)$$

To obtain the corresponding expression for the twisting moment M_{nt}, let us consider the distortion of a thin lamina $abcd$ with the sides ab and ad parallel to the n and t directions and at a distance z from the middle plane (Fig. 23). During bending of the plate the points a, b, c, and d undergo small displacements. The components of the displacement of the point a in the n and t directions we denote by u and v respectively. Then the displacement of the adjacent point d in the n direction is $u + (\partial u/\partial t)\, dt$, and the displacement of the point b in the t direction is $v + (\partial v/\partial n)\, dn$. Owing to these displacements, we obtain for the shearing strain

$$\gamma_{nt} = \frac{\partial u}{\partial t} + \frac{\partial v}{\partial n} \qquad (e)$$

The corresponding shearing stress is

$$\tau_{nt} = G\left(\frac{\partial u}{\partial t} + \frac{\partial v}{\partial n}\right) \qquad (f)$$

From Fig. 23b, representing the section of the middle surface made by the normal plane through the n axis, it may be seen that the angle of rotation in the counterclockwise direction of an element pq, which initially was perpendicular to the xy plane, about an axis perpendicular to the nz plane is equal to $-\partial w/\partial n$. Owing to this rotation a point of the

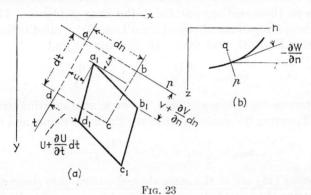

Fig. 23

element at a distance z from the neutral surface has a displacement in the n direction equal to

$$u = -z\frac{\partial w}{\partial n}$$

Considering the normal section through the t axis, it can be shown that the same point has a displacement in the t direction equal to

$$v = -z\frac{\partial w}{\partial t}$$

Substituting these values of the displacements u and v in expression (f), we find

$$\tau_{nt} = -2Gz\frac{\partial^2 w}{\partial n\,\partial t} \qquad (42)$$

and expression (40) for the twisting moment becomes

$$M_{nt} = -\int_{-h/2}^{h/2}\tau_{nt}z\,dz = \frac{Gh^3}{6}\frac{\partial^2 w}{\partial n\,\partial t} = D(1-\nu)\frac{\partial^2 w}{\partial n\,\partial t} \qquad (43)$$

It is seen that the twisting moment for the given perpendicular directions n and t is proportional to the twist of the middle surface corresponding to those directions. When the n and t directions coincide with the x and y axes, there are only bending moments M_x and M_y acting on the sections perpendicular to those axes (Fig. 19). Hence the corresponding twist is zero, and the curvatures $1/r_x$ and $1/r_y$ are the principal curvatures of the middle surface of the plate. They can readily be calculated from Eqs. (37) and (38) if the bending moments M_x and M_y are given. The curvature in any other direction, defined by an angle α, can then be calculated by using the first of the equations (36), or it can be taken from Fig. 17.

Regarding the stresses in a plate undergoing pure bending, it can be concluded from the first of the equations (d) that the maximum normal stress acts on those sections parallel to the xz or yz planes. The magnitudes of these stresses are obtained from Eqs. (b) by substituting $z = h/2$ and by using Eqs. (37) and (38). In this way we find

$$(\sigma_x)_{\max} = \frac{6M_x}{h^2} \qquad (\sigma_y)_{\max} = \frac{6M_y}{h^2} \qquad (44)$$

If these stresses are of opposite sign, the maximum shearing stress acts in the plane bisecting the angle between the xz and yz planes and is equal to

$$\tau_{\max} = \frac{1}{2}(\sigma_x - \sigma_y) = \frac{3(M_x - M_y)}{h^2} \qquad (45)$$

If the stresses (44) are of the same sign, the maximum shear acts in the plane bisecting the angle between the xy and xz planes or in that bisecting the angle between the xy and yz planes and is equal to $\frac{1}{2}(\sigma_y)_{\max}$ or $\frac{1}{2}(\sigma_x)_{\max}$, depending on which of the two principal stresses $(\sigma_y)_{\max}$ or $(\sigma_x)_{\max}$ is greater.

11. Particular Cases of Pure Bending. In the discussion of the previous article we started with the case of a rectangular plate with uniformly distributed bending moments acting along the edges. To obtain a general case of pure bending of a plate, let us imagine that a portion of any shape is cut out from the plate considered above (Fig. 19) by a cylindrical or prismatic surface perpendicular to the plate. The conditions of bending of this portion will remain unchanged provided that bending and twisting moments that satisfy Eqs. (39) and (40) are distributed along the boundary of the isolated portion of the plate. Thus we arrive at the case of pure bending of a plate of any shape, and we conclude that pure bending is always produced if along the edges of the plate bending moments M_n and twisting moments M_{nt} are distributed in the manner given by Eqs. (39) and (40).

Let us take, as a first example, the particular case in which

$$M_x = M_y = M$$

It can be concluded, from Eqs. (39) and (40), that in this case, for a plate of any shape, the bending moments are uniformly distributed along the entire boundary and the twisting moments vanish. From Eqs. (37) and (38) we conclude that

$$\frac{1}{r_x} = \frac{1}{r_y} = \frac{M}{D(1+\nu)} \qquad (46)$$

i.e., the plate in this case is bent to a spherical surface the curvature of which is given by Eq. (46).

In the general case, when M_x is different from M_y, we put

$$M_x = M_1 \quad \text{and} \quad M_y = M_2$$

Then, from Eqs. (37) and (38), we find

$$\begin{aligned}\frac{\partial^2 w}{\partial x^2} &= -\frac{M_1 - \nu M_2}{D(1 - \nu^2)} \\ \frac{\partial^2 w}{\partial y^2} &= -\frac{M_2 - \nu M_1}{D(1 - \nu^2)}\end{aligned} \qquad (a)$$

and in addition

$$\frac{\partial^2 w}{\partial x\, \partial y} = 0 \qquad (b)$$

Integrating these equations, we find

$$w = -\frac{M_1 - \nu M_2}{2D(1 - \nu^2)} x^2 - \frac{M_2 - \nu M_1}{2D(1 - \nu^2)} y^2 + C_1 x + C_2 y + C_3 \qquad (c)$$

where C_1, C_2, and C_3 are constants of integration. These constants define the plane from which the deflections w are measured. If this plane is taken tangent to the middle surface of the plate at the origin, the constants of integration must be equal to zero, and the deflection surface is given by the equation

$$w = -\frac{M_1 - \nu M_2}{2D(1 - \nu^2)} x^2 - \frac{M_2 - \nu M_1}{2D(1 - \nu^2)} y^2 \qquad (d)$$

In the particular case where $M_1 = M_2 = M$, we get from Eq. (d)

$$w = -\frac{M(x^2 + y^2)}{2D(1 + \nu)} \qquad (e)$$

i.e., a paraboloid of revolution instead of the spherical surface given by Eq. (46). The inconsistency of these results arises merely from the use of the approximate expressions $\partial^2 w/\partial x^2$ and $\partial^2 w/\partial y^2$ for the curvatures $1/r_x$ and $1/r_y$ in deriving Eq. (e). These second derivatives of the deflections, rather than the exact expressions for the curvatures, will be used also in all further considerations, in accordance with the assumptions made in Art. 9. This procedure greatly simplifies the fundamental equations of the theory of plates.

Returning now to Eq. (d), let us put $M_2 = -M_1$. In this case the principal curvatures, from Eqs. (a), are

$$\frac{1}{r_x} = -\frac{1}{r_y} = -\frac{\partial^2 w}{\partial x^2} = \frac{M_1}{D(1-\nu)} \qquad (f)$$

and we obtain an anticlastic surface the equation of which is

$$w = -\frac{M_1}{2D(1-\nu)}(x^2 - y^2) \qquad (g)$$

Straight lines parallel to the x axis become, after bending, parabolic curves convex downward (Fig. 24), whereas straight lines in the y direction become parabolas convex upward. Along the lines bisecting the angles between the x and y axes we have $x = y$, or $x = -y$; thus deflections along these lines, as seen from Eq. (g), are zero. All lines parallel to these bisecting lines before bending remain straight during bending, rotating only by some angle. A rectangle $abcd$ bounded by such lines will be twisted as shown in Fig. 24. Imagine normal sections of the plate along lines ab, bc, cd, and ad. From Eqs. (39) and (40) we conclude that bending moments along these sections are zero and that twisting moments along sections ad and bc are equal to M_1 and along sections ab and cd are equal to $-M_1$. Thus the portion $abcd$ of the plate is in the condition of a plate undergoing pure bending produced by twisting moments uniformly distributed along the edges (Fig. 25a). These twisting moments are formed by the horizontal shearing stresses continuously distributed over the edge [Eq. (40)]. This horizontal stress distribution can be replaced by vertical shearing forces which produce the same effect as the actual distribution of stresses. To show this, let the edge ab be divided into infinitely narrow rectangles, such as $mnpq$ in Fig. 25b. If Δ is the small width of the rectangle, the corresponding twisting couple is $M_1\Delta$ and can be formed by two vertical forces equal to M_1 acting along the vertical sides of the rectangle. This replacement of the distributed horizontal forces by a statically equivalent system of two vertical forces cannot cause any sensible disturbance in the plate, except within a distance comparable with the thickness of the plate,[1] which is assumed small. Proceeding in the same manner with all the rectangles, we find that all forces M_1 acting along the vertical sides of the rectangles balance one another and only two forces M_1 at the corners a and d are left. Making

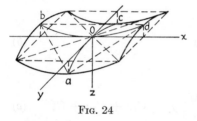

Fig. 24

---

[1] This follows from *Saint Venant's principle;* see S. Timoshenko and J. N. Goodier, "Theory of Elasticity," 2d ed., p. 33, 1951.

the same transformation along the other edges of the plate, we conclude that bending of the plate to the anticlastic surface shown in Fig. 25a can be produced by forces concentrated at the corners[1] (Fig. 25c). Such an experiment is comparatively simple to perform, and was used for the experimental verification of the theory of bending of plates discussed above.[2] In these experiments the deflections of the plate along the line *bod* (Fig. 24) were measured and were found to be in very satisfactory agreement with the theoretical results obtained from Eq. (*g*). Some discrepancies were found only near the edges, and they were more pro-

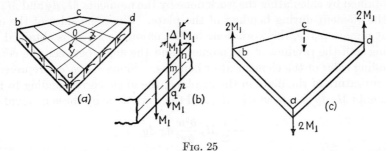

FIG. 25

nounced in the case of comparatively thick plates, as would be expected from the foregoing discussion of the transformation of twisting couples along the edges.

As a last example let us consider the bending of a plate (Fig. 19) to a cylindrical surface having its generating line parallel to the *y* axis. In such a case $\partial^2 w/\partial y^2 = 0$, and we find, from Eqs. (37) and (38),

$$M_x = -D\frac{\partial^2 w}{\partial x^2} \qquad M_y = -\nu D\frac{\partial^2 w}{\partial x^2} \qquad (h)$$

It is seen that to produce bending of the plate to a cylindrical surface we must apply not only the moments M_x but also the moments M_y. Without these latter moments the plate will be bent to an anticlastic surface.[3] The first of equations (*h*) has already been used in Chap. 1 in discussing the bending of long rectangular plates to a cylindrical surface. Although in that discussion we had a bending of plates by lateral loads and there were not only bending stresses but also vertical shearing stresses

---

[1] This transformation of the force system acting along the edges was first suggested by Lord Kelvin and P. G. Tait; see "Treatise on Natural Philosophy," vol. 1, part 2, p. 203, 1883.

[2] Such experiments were made by A. Nádai, *Forschungsarb.*, vols. 170, 171, Berlin, 1915; see also his book "Elastische Platten," p. 42, Berlin, 1925.

[3] We always assume very small deflections or else bending to a developable surface. The case of bending to a nondevelopable surface when the deflections are not small will be discussed later; see p. 47.

acting on sections perpendicular to the x axis, it can be concluded from a comparison with the usual beam theory that the effect of the shearing forces is negligible in the case of thin plates, and the equations developed for the case of pure bending can be used with sufficient accuracy for lateral loading.

12. Strain Energy in Pure Bending of Plates. If a plate is bent by uniformly distributed bending moments M_x and M_y (Fig. 19) so that the xz and yz planes are the principal planes of the deflection surface of the plate, the strain energy stored in an element, such as shown in Fig. 20, is obtained by calculating the work done by the moments $M_x\,dy$ and $M_y\,dx$ on the element during bending of the plate. Since the sides of the element remain plane, the work done by the moments $M_x\,dy$ is obtained by taking half the product of the moment and the angle between the corresponding sides of the element after bending. Since $-\partial^2 w/\partial x^2$ represents the curvature of the plate in the xz plane, the angle corresponding to the moments $M_x\,dy$ is $-(\partial^2 w/\partial x^2)\,dx$, and the work done by these moments is

$$-\frac{1}{2} M_x \frac{\partial^2 w}{\partial x^2}\,dx\,dy$$

An analogous expression is also obtained for the work produced by the moments $M_y\,dx$. Then the total work, equal to the strain energy of the element, is

$$dV = -\frac{1}{2}\left(M_x \frac{\partial^2 w}{\partial x^2} + M_y \frac{\partial^2 w}{\partial y^2}\right) dx\,dy$$

Substituting for the moments their expressions (37) and (38), the strain energy of the elements is represented in the following form:

$$dV = \frac{1}{2} D \left[\left(\frac{\partial^2 w}{\partial x^2}\right)^2 + \left(\frac{\partial^2 w}{\partial y^2}\right)^2 + 2\nu \frac{\partial^2 w}{\partial x^2}\frac{\partial^2 w}{\partial y^2}\right] dx\,dy \qquad (a)$$

Since in the case of pure bending the curvature is constant over the entire surface of the plate, the total strain energy of the plate will be obtained if we substitute the area A of the plate for the elementary area $dx\,dy$ in expression (a). Then

$$V = \frac{1}{2} DA \left[\left(\frac{\partial^2 w}{\partial x^2}\right)^2 + \left(\frac{\partial^2 w}{\partial y^2}\right)^2 + 2\nu \frac{\partial^2 w}{\partial x^2}\frac{\partial^2 w}{\partial y^2}\right] \qquad (47)$$

If the directions x and y do not coincide with the principal planes of curvature, there will act on the sides of the element (Fig. 20) not only the bending moments $M_x\,dy$ and $M_y\,dx$ but also the twisting moments $M_{xy}\,dy$ and $M_{yx}\,dx$. The strain energy due to bending moments is represented by expression (a). In deriving the expression for the strain energy due to twisting moments $M_{xy}\,dy$ we observe that the corresponding angle of twist is equal to the rate of change of the slope $\partial w/\partial y$, as x varies, multiplied with dx; hence the strain energy due to $M_{xy}\,dy$ is

$$\frac{1}{2} M_{xy} \frac{\partial^2 w}{\partial x\, \partial y}\, dx\, dy$$

which, applying Eq. (43), becomes

$$\frac{1}{2} D(1 - \nu) \left(\frac{\partial^2 w}{\partial x\, \partial y}\right)^2 dx\, dy$$

The same amount of energy will also be produced by the couples $M_{yx}\,dx$, so that the strain energy due to both twisting couples is

$$D(1 - \nu) \left(\frac{\partial^2 w}{\partial x\, \partial y}\right)^2 dx\, dy \qquad (b)$$

Since the twist does not affect the work produced by the bending moments, the total strain energy of an element of the plate is obtained by adding together the energy of bending (a) and the energy of twist (b). Thus we obtain

$$dV = \frac{1}{2} D \left[\left(\frac{\partial^2 w}{\partial x^2}\right)^2 + \left(\frac{\partial^2 w}{\partial y^2}\right)^2 + 2\nu \frac{\partial^2 w}{\partial x^2} \frac{\partial^2 w}{\partial y^2}\right] dx\, dy \\ + D(1 - \nu) \left(\frac{\partial^2 w}{\partial x\, \partial y}\right)^2 dx\, dy$$

or

$$dV = \frac{1}{2} D \left\{\left(\frac{\partial^2 w}{\partial x^2} + \frac{\partial^2 w}{\partial y^2}\right)^2 - 2(1 - \nu) \left[\frac{\partial^2 w}{\partial x^2} \frac{\partial^2 w}{\partial y^2} - \left(\frac{\partial^2 w}{\partial x\, \partial y}\right)^2\right]\right\} dx\, dy \qquad (48)$$

The strain energy of the entire plate is now obtained by substituting the area A of the plate for the elemental area $dx\, dy$. Expression (48) will be used later in more complicated cases of bending of plates.

13. Limitations on the Application of the Derived Formulas. In discussing stress distribution in the case of pure bending (Art. 10) it was assumed that the middle surface is the neutral surface of the plate. This condition can be rigorously satisfied only if the middle surface of the bent plate is a *developable surface*. Considering, for instance, pure bending of a plate to a cylindrical surface, the only limitation on the application of the theory will be the requirement that the thickness of the plate be small in comparison with the radius of curvature. In the problems of bending of plates to a cylindrical surface by lateral loading, discussed in the previous chapter, it is required that deflections be small in comparison with the width of the plate, since only under this condition will the approximate expression used for the curvature be accurate enough.

If a plate is bent to a nondevelopable surface, the middle surface undergoes some stretching during bending, and the theory of pure bending developed previously will be accurate enough only if the stresses corresponding to this stretching of the middle surface are small in comparison with the maximum bending stresses given by Eqs. (44) or, what is equivalent, if the strain in the middle surface is small in comparison with the maximum bending strain $h/2r_{\min}$. This requirement puts an additional limitation on deflections of a plate, *viz.*, that the deflections w of the plate must be small in comparison with its thickness h.

To show this, let us consider the bending of a circular plate by bending couples M uniformly distributed along the edge. The deflection surface, for small deflections, is spherical with radius r as defined by Eq. (46). Let AOB (Fig. 26) represent a diametral section of the bent circular plate, a its outer radius before bending, and δ the deflection at the middle. We assume at first that there is no stretching of the middle surface of the plate in the radial direction. In such a case the arc OB must be equal to the initial outer radius a of the plate. The angle φ and the radius b of the plate after bending are then given by the following equations:

$$\varphi = \frac{a}{r} \qquad b = r \sin \varphi$$

It is seen that the assumed bending of the plate implies a compressive strain of the middle surface in the circumferential direction. The magnitude of this strain at the edge of the plate is

$$\epsilon = \frac{a - b}{a} = \frac{r\varphi - r \sin \varphi}{r\varphi} \qquad (a)$$

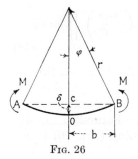

Fig. 26

For small deflections we can take

$$\sin \varphi = \varphi - \frac{\varphi^3}{6}$$

which, substituted in Eq. (a), gives

$$\epsilon = \frac{\varphi^2}{6} \qquad (b)$$

To represent this strain as a function of the maximum deflection δ, we observe that

$$\delta = r(1 - \cos \varphi) \approx \frac{r\varphi^2}{2}$$

Hence
$$\varphi^2 = \frac{2\delta}{r}$$

Substituting in Eq. (b), we obtain

$$\epsilon = \frac{\delta}{3r} \qquad (49)$$

This represents an upper limit for the circumferential strain at the edge of the plate. It was obtained by assuming that the radial strain is zero. Under actual conditions there is some radial strain, and the circumferential compression is somewhat smaller[1] than that given by Eq. (49).

From this discussion it follows that the equations obtained in Art. 10, on the assumption that the middle surface of the bent plate is its neutral surface, are accurate provided the strain given by expression (49) is small in comparison with the maximum bending strain $h/2r$, or, what is equivalent, if the deflection δ is small in comparison with the thickness h of the plate. A similar conclusion can also be obtained in the more general case of pure bending of a plate when the two principal curvatures are not equal.[2] Generalizing these conclusions we can state that the equations of Art. 10 can always be applied with sufficient accuracy if the deflections of a plate from its initial plane or from a true developable surface are small in comparison with the thickness of the plate.

14. Thermal Stresses in Plates with Clamped Edges. Equation (46) for the bending of a plate to a spherical surface can be used in calculating thermal stresses in a plate for certain cases of nonuniform heating. Assume that the variation of the temperature through the thickness of the plate follows a linear law and that the temperature does not vary in planes parallel to the surfaces of the plate. In such a case, by measuring the temperature with respect to that of the middle surface, it can be concluded that temperature expansions and contractions are proportional to the distance from the middle surface. Thus we have exactly the same condition as in the pure bending of a plate to a spherical surface. If the edges of the nonuniformly heated plate are entirely free, the plate will bend to a spherical surface.[3] Let α be the coefficient of linear expansion of the material of the plate, and let t denote the difference in temperature of the upper and lower faces of the plate. The difference between the maximum thermal expansion and the expansion at the middle surface is $\alpha t/2$, and the curvature resulting from the nonuniform heating can be found from the equation

$$\frac{\alpha t}{2} = \frac{h}{2r} \qquad (a)$$

from which

$$\frac{1}{r} = \frac{\alpha t}{h} \qquad (50)$$

This bending of the plate does not produce any stresses, provided the

---

[1] This question is discussed later; see Art. 96.

[2] See Kelvin and Tait, *op. cit.*, vol. 1, part 2, p. 172.

[3] It is assumed that deflections are small in comparison with the thickness of the plate.

edges are free and deflections are small in comparison with the thickness of the plate.

Assume now that the middle plane of the plate is free to expand but that the edges are clamped so that they cannot rotate. In such a case the nonuniform heating will produce bending moments uniformly distributed along the edges of the plate. The magnitude of these moments is such as to eliminate the curvature produced by the nonuniform heating [Eq. (50)], since only in this way can the condition at the clamped edge be satisfied. Using Eq. (46) for the curvature produced by the bending moments, we find for determining the magnitude M of the moment per unit length of the boundary the equation[1]

$$\frac{M}{D(1 + \nu)} = \frac{\alpha t}{h}$$

from which

$$M = \frac{\alpha t D(1 + \nu)}{h} \quad (b)$$

The corresponding maximum stress can be found from Eqs. (44) and is equal to

$$\sigma_{max} = \frac{6M}{h^2} = \frac{6\alpha t D(1 + \nu)}{h^3}$$

Substituting for D its expression (3), we finally obtain

$$\sigma_{max} = \frac{\alpha t E}{2(1 - \nu)} \quad (51)$$

It is seen that the stress is proportional to the coefficient of thermal expansion α, to the temperature difference t between the two faces of the plate, and to the modulus of elasticity E. The thickness h of the plate does not enter into formula (51); but since the difference t of temperatures usually increases in proportion to the thickness of the plate, it can be concluded that greater thermal stresses are to be expected in thick plates than in thin ones.

[1] The effect of pure bending upon the curvature of the entire plate thus is equivalent but opposite in sign to the effect of the temperature gradient. Now, if the plate remains, in the end, perfectly plane, the conditions of a built-in edge are evidently satisfied along any given boundary. Also, since in our case the bending moments are equal everywhere and in any direction, the clamping moments along that given boundary are always expressed by the same Eq. (b).

CHAPTER 3

SYMMETRICAL BENDING OF CIRCULAR PLATES

15. Differential Equation for Symmetrical Bending of Laterally Loaded Circular Plates.[1] If the load acting on a circular plate is symmetrically distributed about the axis perpendicular to the plate through its center, the deflection surface to which the middle plane of the plate is bent will also be symmetrical. In all points equally distant from the center of the plate the deflections will be the same, and it is sufficient to consider deflections in one diametral section through the axis of symmetry (Fig. 27). Let us take the origin of coordinates O at the center of the undeflected plate and denote by r the radial distances of points in the middle plane of the plate and by w their deflections in the downward direction. The maximum slope of the deflection surface at any point A is then equal to $-dw/dr$, and the curvature of the middle surface of the plate in the diametral section rz for small deflections is

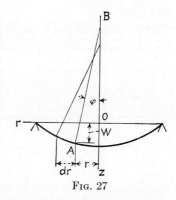

Fig. 27

$$\frac{1}{r_n} = -\frac{d^2w}{dr^2} = \frac{d\varphi}{dr} \qquad (a)$$

where φ is the small angle between the normal to the deflection surface at A and the axis of symmetry OB. From symmetry we conclude that $1/r_n$ is one of the principal curvatures of the deflection surface at A. The second principal curvature will be in the section through the normal AB and perpendicular to the rz plane. Observing that the normals, such as AB, for all points of the middle surface with radial distance r form a conical surface with apex B, we conclude that the length AB is the radius of the second principal curvature which we denote by r_t. Then, from the figure, we obtain

$$\frac{1}{r_t} = -\frac{1}{r}\frac{dw}{dr} = \frac{\varphi}{r} \qquad (b)$$

---

[1] The solution of these problems of bending of circular plates was given by Poisson; see "Memoirs of the Academy," vol. 8, Paris, 1829.

Having expressions (a) and (b) for the principal curvatures, we can obtain the corresponding values of the bending moments assuming that relations (37) and (38), derived for pure bending, also hold between these moments and the curvatures.[1] Using these relations, we obtain

$$M_r = -D\left(\frac{d^2w}{dr^2} + \frac{\nu}{r}\frac{dw}{dr}\right) = D\left(\frac{d\varphi}{dr} + \frac{\nu}{r}\varphi\right) \qquad (52)$$

$$M_t = -D\left(\frac{1}{r}\frac{dw}{dr} + \nu\frac{d^2w}{dr^2}\right) = D\left(\frac{\varphi}{r} + \nu\frac{d\varphi}{dr}\right) \qquad (53)$$

where, as before, M_r and M_t denote bending moments per unit length. The moment M_r acts along circumferential sections of the plate, such as the section made by the conical surface with the apex at B, and M_t acts along the diametral section rz of the plate.

Equations (52) and (53) contain only one variable, w or φ, which can be determined by considering the equilibrium of an element of the plate such as element $abcd$ in Fig. 28 cut out from the plate by two cylindrical sections ab and cd and by two diametral sections ad and bc. The couple acting on the side cd of the element is

$$M_r r \, d\theta \qquad (c)$$

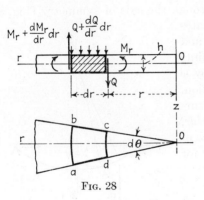

Fig. 28

The corresponding couple on the side ab is

$$\left(M_r + \frac{dM_r}{dr}dr\right)(r + dr)\,d\theta \qquad (d)$$

The couples on the sides ad and bc of the element are each $M_t\,dr$, and they give a resultant couple in the plane rOz equal to

$$M_t\,dr\,d\theta \qquad (e)$$

From symmetry it can be concluded that the shearing forces that may act on the element must vanish on diametral sections of the plate but that they are usually present on cylindrical sections such as sides cd and ab of the element. Denoting by Q the shearing force per unit length of

[1] The effect on deflections of shearing stresses acting on normal sections of the plate perpendicular to meridians, such as the section cut by the conical surface with the apex at B, is neglected here. Their effect is slight in the case of plates in which the thickness is small in comparison with the diameter. Further discussion of this subject will be given in Art. 20. The stresses perpendicular to the surface of the plate are also neglected, which is justifiable in all cases when the load is not highly concentrated (see p. 69).

the cylindrical section of radius r, the total shearing force acting on the side cd of the element is $Qr\,d\theta$, and the corresponding force on the side ab is

$$\left[Q + \left(\frac{dQ}{dr}\right)dr\right](r + dr)\,d\theta$$

Neglecting the small difference between the shearing forces on the two opposite sides of the element, we can state that these forces give a couple in the rz plane equal to

$$Qr\,d\theta\,dr \qquad (f)$$

Summing up the moments (c), (d), (e), and (f) with proper signs and neglecting the moment due to the external load on the element as a small quantity of higher order, we obtain the following equation of equilibrium of the element $abcd$:

$$\left(M_r + \frac{dM_r}{dr}dr\right)(r + dr)\,d\theta - M_r r\,d\theta - M_t\,dr\,d\theta + Qr\,d\theta\,dr = 0$$

from which we find, by neglecting a small quantity of higher order,

$$M_r + \frac{dM_r}{dr}r - M_t + Qr = 0 \qquad (g)$$

Substituting expressions (52) and (53) for M_r and M_t, Eq. (g) becomes

$$\frac{d^2\varphi}{dr^2} + \frac{1}{r}\frac{d\varphi}{dr} - \frac{\varphi}{r^2} = -\frac{Q}{D} \qquad (54)$$

or, in another form,

$$\frac{d^3w}{dr^3} + \frac{1}{r}\frac{d^2w}{dr^2} - \frac{1}{r^2}\frac{dw}{dr} = \frac{Q}{D} \qquad (55)$$

In any particular case of a symmetrically loaded circular plate the shearing force Q can easily be calculated by dividing the load distributed within the circle of radius r by $2\pi r$; then Eq. (54) or (55) can be used to determine the slope φ and the deflection w of the plate. The integration of these equations is simplified if we observe that they can be put in the following forms:

$$\frac{d}{dr}\left[\frac{1}{r}\frac{d}{dr}(r\varphi)\right] = -\frac{Q}{D} \qquad (56)$$

$$\frac{d}{dr}\left[\frac{1}{r}\frac{d}{dr}\left(r\frac{dw}{dr}\right)\right] = \frac{Q}{D} \qquad (57)$$

If Q is represented by a function of r, these equations can be integrated without any difficulty in each particular case.

Sometimes it is advantageous to represent the right-hand side of Eq. (57) as a function of the intensity q of the load distributed over the plate. For this purpose we multiply both sides of the equation by $2\pi r$. Then,

observing that
$$Q2\pi r = \int_0^r q 2\pi r \, dr$$

we obtain
$$r\frac{d}{dr}\left[\frac{1}{r}\frac{d}{dr}\left(r\frac{dw}{dr}\right)\right] = \frac{1}{D}\int_0^r qr \, dr$$

Differentiating both sides of this equation with respect to r and dividing by r, we finally obtain

$$\frac{1}{r}\frac{d}{dr}\left\{r\frac{d}{dr}\left[\frac{1}{r}\frac{d}{dr}\left(r\frac{dw}{dr}\right)\right]\right\} = \frac{q}{D} \qquad (58)$$

This equation can easily be integrated if the intensity of the load q is given as a function of r.

16. Uniformly Loaded Circular Plates. If a circular plate of radius a carries a load of intensity q uniformly distributed over the entire surface of the plate, the magnitude of the shearing force Q at a distance r from the center of the plate is determined from the equation

$$2\pi r Q = \pi r^2 q$$

from which
$$Q = \frac{qr}{2} \qquad (a)$$

Substituting in Eq. (57), we obtain

$$\frac{d}{dr}\left[\frac{1}{r}\frac{d}{dr}\left(r\frac{dw}{dr}\right)\right] = \frac{qr}{2D} \qquad (b)$$

By one integration we find

$$\frac{1}{r}\frac{d}{dr}\left(r\frac{dw}{dr}\right) = \frac{qr^2}{4D} + C_1 \qquad (c)$$

where C_1 is a constant of integration to be found later from the conditions at the center and at the edge of the plate. Multiplying both sides of Eq. (c) by r, and making the second integration, we find

$$r\frac{dw}{dr} = \frac{qr^4}{16D} + \frac{C_1 r^2}{2} + C_2$$

and
$$\frac{dw}{dr} = \frac{qr^3}{16D} + \frac{C_1 r}{2} + \frac{C_2}{r} \qquad (59)$$

The new integration then gives

$$w = \frac{qr^4}{64D} + \frac{C_1 r^2}{4} + C_2 \log\frac{r}{a} + C_3 \qquad (60)$$

Let us now calculate the constants of integration for various particular cases.

Circular Plate with Clamped Edges. In this case the slope of the deflection surface in the radial direction must be zero for $r = 0$ and $r = a$. Hence, from Eq. (59),

$$\left(\frac{qr^3}{16D} + \frac{C_1 r}{2} + \frac{C_2}{r}\right)_{r=0} = 0$$

$$\left(\frac{qr^3}{16D} + \frac{C_1 r}{2} + \frac{C_2}{r}\right)_{r=a} = 0$$

From the first of these equations we conclude that $C_2 = 0$. Substituting this in the second equation, we obtain

$$C_1 = -\frac{qa^2}{8D}$$

With these values of the constants, Eq. (59) gives the following expression for the slope:

$$\varphi = -\frac{dw}{dr} = \frac{qr}{16D}(a^2 - r^2) \tag{61}$$

Equation (60) gives

$$w = \frac{qr^4}{64D} - \frac{qa^2 r^2}{32D} + C_3 \tag{d}$$

At the edge of the plate the deflection is zero. Hence,

$$\frac{qa^4}{64D} - \frac{qa^4}{32D} + C_3 = 0$$

and we obtain

$$C_3 = \frac{qa^4}{64D}$$

Substituting in Eq. (d), we find

$$w = \frac{q}{64D}(a^2 - r^2)^2 \tag{62}$$

The maximum deflection is at the center of the plate and, from Eq. (62), is equal to

$$w_{max} = \frac{qa^4}{64D} \tag{e}$$

This deflection is equal to three-eighths of the deflection of a uniformly loaded strip with built-in ends having a flexural rigidity equal to D, a width of unity, and a length equal to the diameter of the plate.

Having expression (61) for the slope, we obtain now the bending moments M_r and M_t by using expressions (52) and (53), from which we find

$$M_r = \frac{q}{16}[a^2(1 + \nu) - r^2(3 + \nu)] \tag{63}$$

$$M_t = \frac{q}{16}[a^2(1 + \nu) - r^2(1 + 3\nu)] \tag{64}$$

Substituting $r = a$ in these expressions, we find for the bending moments at the boundary of the plate

$$(M_r)_{r=a} = -\frac{qa^2}{8} \qquad (M_t)_{r=a} = -\frac{\nu q a^2}{8} \tag{65}$$

At the center of the plate where $r = 0$,

$$M_r = M_t = \frac{qa^2}{16}(1 + \nu) \tag{66}$$

From expressions (65) and (66) it is seen that the maximum stress is at the boundary of the plate where

$$(\sigma_r)_{max} = -\frac{6M_r}{h^2} = \frac{3}{4}\frac{qa^2}{h^2} \tag{f}$$

The variation of stresses σ_r and σ_t at the lower face of the plate along the radius of the plate is shown in Fig. 29.

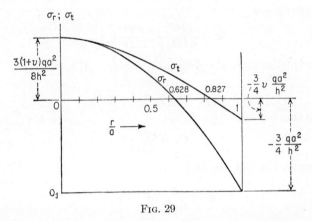

Fig. 29

Circular Plate with Supported Edges. In calculating deflections for this case we apply the method of superposition. It was shown that in the case of clamped edges there are negative bending moments $M_r = -qa^2/8$ acting along the edge (Fig. 30a). If this case is combined with that of pure bending shown in Fig. 30b, the bending moments M_r at the edge will be eliminated, and we obtain the bending of a plate supported at the edge. The deflection surface in the case of pure bending by the moments $qa^2/8$, from Eq. (46) or Eq. (e) on page 43, is

$$w = \frac{qa^2}{16D(1 + \nu)}(a^2 - r^2)$$

Adding this to the deflections (62) of the clamped plate, we find for the plate with a simply supported edge

SYMMETRICAL BENDING OF CIRCULAR PLATES

$$w = \frac{q(a^2 - r^2)}{64D} \left(\frac{5 + \nu}{1 + \nu} a^2 - r^2 \right) \qquad (67)$$

Substituting $r = 0$ in this expression we obtain the deflection of the plate at the center:

$$w_{\max} = \frac{(5 + \nu)qa^4}{64(1 + \nu)D} \qquad (68)$$

For $\nu = 0.3$ this deflection is about four times as great as that for the plate with clamped edge.

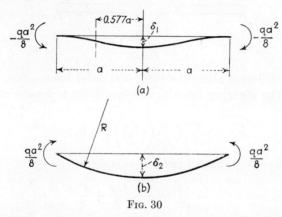

Fig. 30

In calculating bending moments in this case we must add the constant bending moment $qa^2/8$ to the moments (63) and (64) found above for the case of clamped edges. Hence in the case of supported edges

$$M_r = \frac{q}{16}(3 + \nu)(a^2 - r^2) \qquad (69)$$

$$M_t = \frac{q}{16}[a^2(3 + \nu) - r^2(1 + 3\nu)] \qquad (70)$$

The maximum bending moment is at the center of the plate where

$$M_r = M_t = \frac{3 + \nu}{16} qa^2$$

The corresponding maximum stress is

$$(\sigma_r)_{\max} = (\sigma_t)_{\max} = \frac{6M_r}{h^2} = \frac{3(3 + \nu)qa^2}{8h^2} \qquad (71)$$

To get the maximum stress at any distance r from the center we must add to the stress calculated for the plate with clamped edges the constant value

$$\frac{6}{h^2} \frac{qa^2}{8}$$

58   THEORY OF PLATES AND SHELLS

corresponding to the pure bending shown in Fig. 30b. The same stress is obtained also from Fig. 29 by measuring the ordinates from the horizontal axis through O_1. It may be seen that by clamping the edge a more favorable stress distribution in the plate is obtained.

17. Circular Plate with a Circular Hole at the Center. Let us begin with a discussion of the bending of a plate by the moments M_1 and M_2

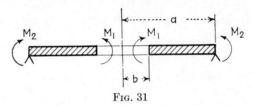

FIG. 31

uniformly distributed along the inner and outer boundaries, respectively (Fig. 31). The shearing force Q vanishes in such a case, and Eq. (57) becomes

$$\frac{d}{dr}\left[\frac{1}{r}\frac{d}{dr}\left(r\frac{dw}{dr}\right)\right] = 0$$

By integrating this equation twice we obtain

$$-\frac{dw}{dr} = \varphi = \frac{C_1 r}{2} + \frac{C_2}{r} \tag{a}$$

Integrating again, we find the deflection

$$w = -\frac{C_1 r^2}{4} - C_2 \log \frac{r}{a} + C_3 \tag{b}$$

The constants of integration are now to be determined from the conditions at the edges. Substituting expression (a) into Eq. (52), we find

$$M_r = D\left[\frac{C_1}{2} - \frac{C_2}{r^2} + \nu\left(\frac{C_1}{2} + \frac{C_2}{r^2}\right)\right] \tag{c}$$

This moment must be equal to M_1 for $r = b$ and equal to M_2 for $r = a$. Hence equations for determining constants C_1 and C_2 are

$$D\left[\frac{C_1}{2}(1+\nu) - \frac{C_2}{b^2}(1-\nu)\right] = M_1$$

$$D\left[\frac{C_1}{2}(1+\nu) - \frac{C_2}{a^2}(1-\nu)\right] = M_2$$

from which

$$C_1 = \frac{2(a^2 M_2 - b^2 M_1)}{(1+\nu)D(a^2 - b^2)} \qquad C_2 = \frac{a^2 b^2 (M_2 - M_1)}{(1-\nu)D(a^2 - b^2)} \tag{d}$$

To determine the constant C_3 in Eq. (b), the deflections at the edges of the plate must be considered. Assume, for example, that the plate in Fig. 31 is supported along the outer edge. Then $w = 0$ for $r = a$, and we find, from (b),

$$C_3 = \frac{C_1 a^2}{4} = \frac{a^2(a^2 M_2 - b^2 M_1)}{2(1+\nu)D(a^2 - b^2)}$$

In the particular case when $M_2 = 0$ we obtain

$$C_1 = -\frac{2b^2 M_1}{(1+\nu)D(a^2-b^2)} \qquad C_2 = -\frac{a^2 b^2 M_1}{(1-\nu)D(a^2-b^2)}$$

$$C_3 = -\frac{a^2 b^2 M_1}{2(1+\nu)D(a^2-b^2)}$$

and expressions (a) and (b) for the slope and the deflection become

$$\frac{dw}{dr} = \frac{a^2 b^2 M_1}{D(1-\nu)(a^2-b^2)}\left(\frac{1}{r} + \frac{1-\nu}{1+\nu}\frac{r}{a^2}\right) \tag{72}$$

$$w = -\frac{b^2 M_1}{2(1+\nu)D(a^2-b^2)}(a^2 - r^2) + \frac{a^2 b^2 M_1}{(1-\nu)D(a^2-b^2)}\log\frac{r}{a} \tag{73}$$

As a second example we consider the case of bending of a plate by shearing forces Q_0 uniformly distributed along the inner edge (Fig. 32). The shearing force per unit length of a circumference of radius r is

$$Q = \frac{Q_0 b}{r} = \frac{P}{2\pi r}$$

where $P = 2\pi b Q_0$ denotes the total load applied to the inner boundary of the plate. Substituting this in Eq. (57) and integrating, we obtain

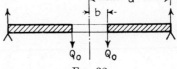

Fig. 32

$$\frac{dw}{dr} = \frac{Pr}{8\pi D}\left(2\log\frac{r}{a} - 1\right) - \frac{C_1 r}{2} - \frac{C_2}{r} \tag{e}$$

and

$$w = \frac{Pr^2}{8\pi D}\left(\log\frac{r}{a} - 1\right) - \frac{C_1 r^2}{4} - C_2 \log\frac{r}{a} + C_3 \tag{f}$$

The constants of integration will now be calculated from the boundary conditions. Assuming that the plate is simply supported along the outer edge, we have

$$(w)_{r=a} = 0 \qquad -D\left(\frac{d^2w}{dr^2} + \frac{\nu}{r}\frac{dw}{dr}\right)_{r=a} = 0 \tag{g}$$

For the inner edge of the plate we have

$$-D\left(\frac{d^2w}{dr^2} + \frac{\nu}{r}\frac{dw}{dr}\right)_{r=b} = 0 \tag{h}$$

Substituting expressions (e) and (f) in Eqs. (g) and (h), we find

$$C_1 = \frac{P}{4\pi D}\left(\frac{1-\nu}{1+\nu} - \frac{2b^2}{a^2-b^2}\log\frac{b}{a}\right)$$

$$C_2 = -\frac{(1+\nu)P}{(1-\nu)4\pi D}\frac{a^2 b^2}{a^2-b^2}\log\frac{b}{a} \qquad (i)$$

$$C_3 = \frac{Pa^2}{8\pi D}\left(1 + \frac{1}{2}\frac{1-\nu}{1+\nu} - \frac{b^2}{a^2-b^2}\log\frac{b}{a}\right)$$

With these values of the constants substituted in expressions (e) and (f), we find the slope and the deflection at any point of the plate shown in Fig. 32. For the slope at the inner edge, which will be needed in the further discussion, we obtain

$$\left(\frac{dw}{dr}\right)_{r=b} = \frac{Pb}{8\pi D}\left[2\log\frac{b}{a} - 1 - \frac{1-\nu}{1+\nu}\right.$$
$$\left. + \frac{2b^2}{a^2-b^2}\log\frac{b}{a}\left(1 + \frac{a^2}{b^2}\frac{1+\nu}{1-\nu}\right)\right] \qquad (j)$$

In the limiting case where b is infinitely small, $b^2 \log(b/a)$ approaches zero, and the constants of integration become

$$C_1 = \frac{1-\nu}{1+\nu}\frac{P}{4\pi D} \qquad C_2 = 0 \qquad C_3 = \frac{Pa^2}{8\pi D}\left(1 + \frac{1}{2}\frac{1-\nu}{1+\nu}\right)$$

Substituting these values in expression (f), we obtain

$$w = \frac{P}{8\pi D}\left[\frac{3+\nu}{2(1+\nu)}(a^2 - r^2) + r^2 \log\frac{r}{a}\right] \qquad (k)$$

This coincides with the deflection of a plate without a hole and loaded at the center [see Eq. (89), page 68]. Thus a very small hole at the center does not affect the deflection of the plate.

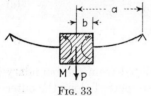

Fig. 33

Combining the loadings shown in Figs. 31 and 32, we can obtain the solution for the case of a plate built in along the inner edge and uniformly loaded along the outer edge (Fig. 33). Since the slope at the built-in edge is zero in this case, using expressions (72) and (j), we obtain the following equation for determining the bending moment M_1 at the built-in edge:

$$-\frac{a^2 b^2 M_1}{D(1-\nu)(a^2-b^2)}\left(\frac{1}{b} + \frac{1-\nu}{1+\nu}\frac{b}{a^2}\right) = \frac{Pb}{8\pi D}\left[2\log\frac{b}{a} - 1\right.$$
$$\left. - \frac{1-\nu}{1+\nu} + \frac{2b^2}{a^2-b^2}\log\frac{b}{a}\left(1 + \frac{a^2}{b^2}\frac{1+\nu}{1-\nu}\right)\right]$$

from which

$$M_1 = \frac{P}{4\pi\left[(1+\nu)\dfrac{a^2}{b^2} + 1 - \nu\right]}\left[(1-\nu)\left(\frac{a^2}{b^2} - 1\right) + 2(1+\nu)\frac{a^2}{b^2}\log\frac{a}{b}\right] \quad (74)$$

Having this expression for the moment M_1, we obtain the deflections of the plate by superposing expression (73) and expression (f), in which the constants of integration are given by expressions (i).

By using the same method of superposition we can obtain also the solution for the case shown in Fig. 34, in which the plate is supported along the outer edge and carries a uniformly distributed load. In this case we use the solution obtained in the previous article for the plate without a hole at the center. Considering the section of this plate cut by the cylindrical surface of radius b and perpendicular to the plate, we find that along this section there act a shearing force $Q = \pi q b^2/2\pi b = qb/2$ and a bending moment of the intensity [see Eq. (69)]

$$M_r = \frac{q}{16}(3+\nu)(a^2 - b^2)$$

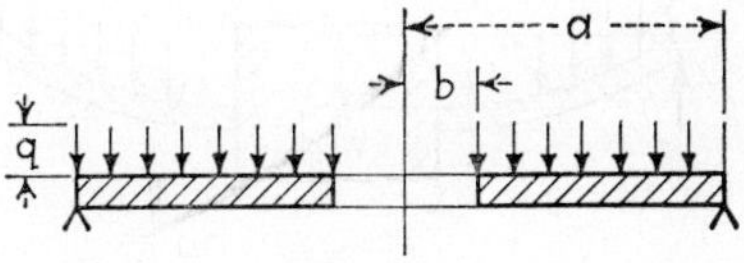

Fig. 34

Hence to obtain the stresses and deflections for the case shown in Fig. 34, we have to superpose on the stresses and deflections obtained for the plate without a hole the stresses and deflections produced by the bending moments and shearing forces shown in Fig. 35. These latter quantities are obtained from expressions (72), (73), (e), and (f), with due attention being given to the sign of the applied shears and moments.

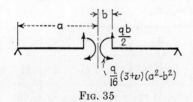

Fig. 35

Several cases of practical importance are represented in Fig. 36. In all these cases the maximum stress is given by a formula of the type

$$\sigma_{\max} = k\frac{qa^2}{h^2} \quad \text{or} \quad \sigma_{\max} = \frac{kP}{h^2} \quad (75)$$

depending on whether the applied load is uniformly distributed over the surface or concentrated along the edge. The numerical values of the factor k, calculated[1] for several values of the ratio a/b and for Poisson's ratio $\nu = 0.3$, are given in Table 3.

---

[1] The calculations for cases 1 to 8 inclusive were made by A. M. Wahl and G. Lobo, *Trans. ASME*, vol. 52, 1930. Further data concerning symmetrically loaded circular plates with and without a hole may be found in K. Beyer, "Die Statik im Stahlbetonbau," 2d ed., p. 652, Berlin, 1948.

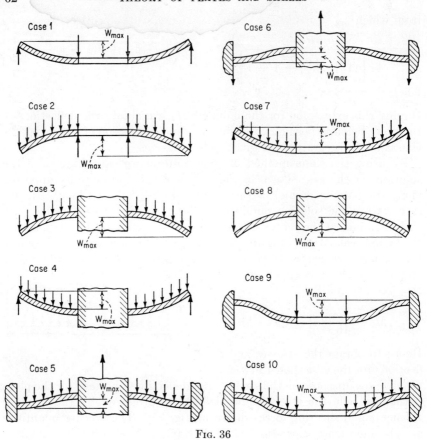

Fig. 36

Table 3. Coefficients k and k_1 in Eqs. (75) and (76) for the Ten Cases Shown in Fig. 36

$a/b =$	1.25		1.5		2		3		4		5	
Case	k	k_1	k	k_1	k	k_1	k	k_1	k	k_1	k	k_1
1	1.10	0.341	1.26	0.519	1.48	0.672	1.88	0.734	2.17	0.724	2.34	0.704
2	0.66	0.202	1.19	0.491	2.04	0.902	3.34	1.220	4.30	1.300	5.10	1.310
3	0.135	0.00231	0.410	0.0183	1.04	0.0938	2.15	0.293	2.99	0.448	3.69	0.564
4	0.122	0.00343	0.336	0.0313	0.74	0.1250	1.21	0.291	1.45	0.417	1.59	0.492
5	0.090	0.00077	0.273	0.0062	0.71	0.0329	1.54	0.110	2.23	0.179	2.80	0.234
6	0.115	0.00129	0.220	0.0064	0.405	0.0237	0.703	0.062	0.933	0.092	1.13	0.114
7	0.592	0.184	0.976	0.414	1.440	0.664	1.880	0.824	2.08	0.830	2.19	0.813
8	0.227	0.00510	0.428	0.0249	0.753	0.0877	1.205	0.209	1.514	0.293	1.745	0.350
9	0.194	0.00504	0.320	0.0242	0.454	0.0810	0.673	0.172	1.021	0.217	1.305	0.238
10	0.105	0.00199	0.259	0.0139	0.480	0.0575	0.657	0.130	0.710	0.162	0.730	0.175

The maximum deflections in the same cases are given by formulas of the type

$$w_{max} = k_1 \frac{qa^4}{Eh^3} \quad \text{or} \quad w_{max} = k_1 \frac{Pa^2}{Eh^3} \qquad (76)$$

The coefficients k_1 are also given in Table 3.

When the ratio a/b approaches unity, the values of the coefficients k and k_1 in Eqs. (75) and (76) can be obtained with sufficient accuracy by considering a radial strip as a beam with end conditions and loading as in the actual plate. The effect of the moments M_t on bending is then entirely neglected.

18. Circular Plate Concentrically Loaded. We begin with the case of a simply supported plate in which the load is uniformly distributed along a circle of radius b (Fig. 37a). Dividing the plate into two parts as shown in Fig. 37b and c, it may be seen that the inner portion of the plate is in the condition of pure bending produced by the uniformly distributed moments M_1 and that the outer part is bent by the moments M_1 and the shearing forces Q_1. Denoting by P the total load applied, we find that

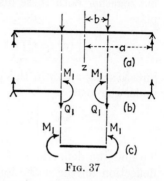

Fig. 37

$$Q_1 = \frac{P}{2\pi b} \qquad (a)$$

The magnitude of the moment M_1 is found from the condition of continuity along the circle $r = b$, from which it follows that both portions of the plate have, at that circle, the same slope. Using Eqs. (72) and (j) of the preceding article, we find the slope for the inner boundary of the outer portion of the plate equal to

$$\left(\frac{dw}{dr}\right)_{r=b} = \frac{a^2 b^2 M_1}{D(1-\nu)(a^2-b^2)} \left(\frac{1}{b} + \frac{1-\nu}{1+\nu}\frac{b}{a^2}\right)$$
$$+ \frac{Pb}{8\pi D}\left[2\log\frac{b}{a} - 1 - \frac{1-\nu}{1+\nu} \right.$$
$$\left. + \frac{2b^2}{a^2-b^2}\log\frac{b}{a}\left(1 + \frac{a^2}{b^2}\frac{1+\nu}{1-\nu}\right)\right] \qquad (b)$$

The inner portion of the plate is bent to a spherical surface, the curvature of which is given by expression (46). Therefore the corresponding slope at the boundary is

$$\left(\frac{dw}{dr}\right)_{r=b} = -\frac{M_1 b}{D(1+\nu)} \qquad (c)$$

Equating expressions (b) and (c), we obtain

$$M_1 = \frac{(1-\nu)P(a^2-b^2)}{8\pi a^2} - \frac{(1+\nu)P \log \frac{b}{a}}{4\pi} \quad (d)$$

Substituting this expression for M_1 in Eq. (73), we obtain deflections of the outer part of the plate due to the moments M_1. The deflections due to the forces Q_1 are obtained from Eq. (f) of the preceding article. Adding together both these deflections, we obtain for the outer part of the plate

$$w = \frac{P}{8\pi D}\left[(a^2-r^2)\left(1+\frac{1}{2}\frac{1-\nu}{1+\nu}\frac{a^2-b^2}{a^2}\right) + (b^2+r^2)\log\frac{r}{a}\right] \quad (77)$$

Substituting $r = b$ in this expression, we obtain the following deflection under the load:

$$(w)_{r=b} = \frac{P}{8\pi D}\left[(a^2-b^2)\left(1+\frac{1}{2}\frac{1-\nu}{1+\nu}\frac{a^2-b^2}{a^2}\right) + 2b^2\log\frac{b}{a}\right] \quad (e)$$

To find the deflections of the inner portion of the plate, we add to the deflection (e) the deflections due to pure bending of that portion of the plate. In this manner we obtain

$$\begin{aligned}w &= \frac{P}{8\pi D}\left[(a^2-b^2)\left(1+\frac{1}{2}\frac{1-\nu}{1+\nu}\frac{a^2-b^2}{a^2}\right) + 2b^2\log\frac{b}{a}\right]\\&\quad + \frac{b^2-r^2}{2D(1+\nu)}\left[\frac{(1-\nu)P(a^2-b^2)}{8\pi a^2} - \frac{(1+\nu)P\log\frac{b}{a}}{4\pi}\right]\\&= \frac{P}{8\pi D}\left[(b^2+r^2)\log\frac{b}{a} + r^2 - b^2 + (a^2-r^2)\frac{(3+\nu)a^2-(1-\nu)b^2}{2(1+\nu)a^2}\right]\\&= \frac{P}{8\pi D}\left[(b^2+r^2)\log\frac{b}{a} + (a^2-b^2)\frac{(3+\nu)a^2-(1-\nu)r^2}{2(1+\nu)a^2}\right] \quad (78)\end{aligned}$$

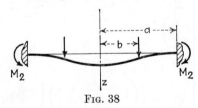

Fig. 38

If the outer edge of the plate is built in, the deflections of the plate are obtained by superposing on the deflections (77) and (78) the deflections produced by the bending moments M_2 uniformly distributed along the outer edge of the plate (Fig. 38) and of such a magnitude that the slope of the deflection surface at that edge is equal to zero. From expression (77) the slope at the edge of a simply supported plate is

$$\left(\frac{dw}{dr}\right)_{r=a} = -\frac{P}{4\pi D}\frac{1}{1+\nu}\frac{a^2-b^2}{a} \quad (f)$$

SYMMETRICAL BENDING OF CIRCULAR PLATES 65

The slope produced by the moments M_2 is

$$\left(\frac{dw}{dr}\right)_{r=a} = \frac{M_2 a}{D(1 + \nu)} \qquad (g)$$

Equating the sum of expressions (f) and (g) to zero, we obtain

$$M_2 = \frac{P}{4\pi} \frac{a^2 - b^2}{a^2}$$

Deflections produced by this moment are

$$w = \frac{M_2}{D(1 + \nu)} \frac{r^2 - a^2}{2} = \frac{P}{8\pi D(1 + \nu)} \frac{a^2 - b^2}{a^2} (r^2 - a^2) \qquad (h)$$

Adding these deflections to the deflections (77) and (78) we obtain for the outer portion of a plate with a built-in edge

$$w = \frac{P}{8\pi D}\left[(a^2 - r^2)\frac{a^2 + b^2}{2a^2} + (b^2 + r^2)\log\frac{r}{a}\right] \qquad (79)$$

and for the inner portion,

$$w = \frac{P}{8\pi D}\left[(b^2 + r^2)\log\frac{b}{a} + r^2 - b^2 + \frac{(a^2 - r^2)(a^2 + b^2)}{2a^2}\right]$$
$$= \frac{P}{8\pi D}\left[(b^2 + r^2)\log\frac{b}{a} + \frac{(a^2 + r^2)(a^2 - b^2)}{2a^2}\right] \qquad (80)$$

Having the deflections for the case of a load uniformly distributed along a concentric circle, any case of bending of a circular plate symmetrically loaded with respect to the center can be solved by using the method of superposition. Let us consider, for example, the case in which the load is uniformly distributed over the inner portion of the plate

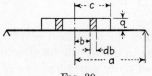

Fig. 39

bounded by a circle of radius c (Fig. 39). Expression (77) is used to obtain the deflection at any point of the unloaded portion of the plate $(a > r > c)$. The deflection produced by an elementary loading distributed over a ring surface of radius b and width db (see Fig. 39) is obtained by substituting $P = 2\pi bq\, db$ in that expression, where q is the intensity of the uniform load. Integrating the expression thus obtained with respect to b, we obtain the deflection

$$w = \frac{q}{4D} \int_0^c \left\{ (a^2 - r^2) \frac{3+\nu}{2(1+\nu)} + r^2 \log \frac{r}{a} \right.$$
$$\left. + b^2 \left[\log \frac{r}{a} - \frac{(1-\nu)(a^2 - r^2)}{2(1+\nu)a^2} \right] \right\} b\, db$$
$$= \frac{qc^2}{8D} \left[\frac{3+\nu}{2(1+\nu)} (a^2 - r^2) + r^2 \log \frac{r}{a} \right]$$
$$+ \frac{qc^4}{16D} \left[\log \frac{r}{a} - \frac{1-\nu}{2(1+\nu)} \frac{a^2 - r^2}{a^2} \right]$$

or, denoting the total load $\pi c^2 q$ by P,

$$w = \frac{P}{16\pi D} \left\{ \frac{3+\nu}{1+\nu} (a^2 - r^2) + 2r^2 \log \frac{r}{a} \right.$$
$$\left. + c^2 \left[\log \frac{r}{a} - \frac{1-\nu}{2(1+\nu)} \frac{a^2 - r^2}{a^2} \right] \right\} \quad (81)$$

Expression (78) is used to obtain the deflection at the center. Substituting $r = 0$ and $P = 2\pi bq\, db$ in this expression and integrating, we find

$$(w)_{r=0} = \frac{q}{4D} \int_0^c \left[b^2 \log \frac{b}{a} + \frac{a^2 - b^2}{2} \left(\frac{3+\nu}{1+\nu} \right) \right] b\, db$$
$$= \frac{P}{16\pi D} \left[\frac{3+\nu}{1+\nu} a^2 + c^2 \log \frac{c}{a} - \frac{7+3\nu}{4(1+\nu)} c^2 \right] \quad (82)$$

where $P = \pi c^2 q$.

The maximum bending moment is at the center and is found by using expression (d). Substituting $2\pi bq\, db$ for P in this expression and integrating, we find

$$M_{\max} = q \int_0^c \left(\frac{1-\nu}{4} \frac{a^2 - b^2}{a^2} - \frac{1+\nu}{2} \log \frac{b}{a} \right) b\, db$$
$$= \frac{P}{4\pi} \left[(1+\nu) \log \frac{a}{c} + 1 - \frac{(1-\nu)c^2}{4a^2} \right] \quad (83)$$

where, as before, P denotes the total load $\pi c^2 q$.*

Expression (81) is used to obtain the bending moments M_r and M_t at any point of the unloaded outer portion of the plate. Substituting this expression in the general formulas (52) and (53), we find

$$M_r = \frac{(1+\nu)P}{4\pi} \log \frac{a}{r} + \frac{(1-\nu)Pc^2}{16\pi} \left(\frac{1}{r^2} - \frac{1}{a^2} \right) \quad (84)$$

$$M_t = \frac{P}{4\pi} \left[(1+\nu) \log \frac{a}{r} + 1 - \nu \right] - \frac{(1-\nu)Pc^2}{16\pi} \left(\frac{1}{r^2} + \frac{1}{a^2} \right) \quad (85)$$

* This expression applies only when c is at least several times the thickness h. The case of a very small c is discussed in Art. 19.

The maximum values of these moments are obtained at the circle $r = c$, where

$$M_r = \frac{(1 + \nu)P}{4\pi} \log \frac{a}{c} + \frac{(1 - \nu)P(a^2 - c^2)}{16\pi a^2} \tag{86}$$

$$M_t = \frac{P}{4\pi}\left[(1 + \nu) \log \frac{a}{c} + 1 - \nu\right] - \frac{(1 - \nu)P(a^2 + c^2)}{16\pi a^2} \tag{87}$$

The same method of calculating deflections and moments can be used also for any kind of symmetrical loading of a circular plate.

The deflection at the center of the plate can easily be calculated also for any kind of unsymmetrical loading by using the following consideration.

Owing to the complete symmetry of the plate and of its boundary conditions, the deflection produced at its center by an isolated load P depends only on the magnitude of the load and on its radial distance from the center. This deflection remains unchanged if the load P is moved to another position provided the radial distance of the load from the center remains the same. The deflection remains unchanged also if the load P is replaced by several loads the sum of which is equal to P and the radial distances of which are the same as that of the load P. From this it follows that in calculating the deflection of the plate at the center we can replace an isolated load P by a load P uniformly distributed along a circle the radius of which is equal to the radial distance of the isolated load. For the load uniformly distributed along a circle of radius b the deflection at the center of a plate supported at the edges is given by Eq. (78) and is

$$(w)_{r=0} = \frac{P}{8\pi D}\left[\frac{3 + \nu}{2(1 + \nu)}(a^2 - b^2) - b^2 \log \frac{a}{b}\right] \tag{i}$$

This formula gives the deflection at the center of the plate produced by an isolated load P at a distance b from the center of the plate. Having this formula the deflection at the center for any other kind of loading can be obtained by using the method of superposition.[1] It should be noted that the deflections and stresses in a circular plate with or without a hole can be efficiently reduced by reinforcing the plate with either concentric[2] or radial ribs. In the latter case, however, the stress distribution is no longer symmetrical with respect to the center of the plate.

19. Circular Plate Loaded at the Center. The solution for a concentrated load acting at the center of the plate can be obtained from the

---

[1] This method of calculating deflections at the center of the plate was indicated by Saint Venant in his translation of the "Théorie de l'élasticité des corps solides," by Clebsch, p. 363, Paris, 1883. The result (*i*) can also be obtained by applying Maxwell's reciprocal theorem to the circular plate.

[2] This case is discussed by W. A. Nash, *J. Appl. Mechanics*, vol. 15, p. 25, 1948. See also C. B. Biezeno and R. Grammel, "Technische Dynamik," 2d ed., vol. 1, p. 497, 1953.

discussion of the preceding article by assuming that the radius c of the circle within which the load is distributed becomes infinitely small, whereas the total load P remains finite. Using this assumption, we find that the maximum deflection at the center of a simply supported plate, by Eq. (82), is

$$w_{\max} = \frac{(3+\nu)Pa^2}{16\pi(1+\nu)D} \tag{88}$$

The deflection at any point of the plate at a distance r from the center, by Eq. (81), is

$$w = \frac{P}{16\pi D}\left[\frac{3+\nu}{1+\nu}(a^2 - r^2) + 2r^2 \log\frac{r}{a}\right] \tag{89}$$

The bending moment for points with $r > c$ may be found by omitting the terms in Eqs. (84) and (85) which contain c^2. This gives

$$M_r = \frac{P}{4\pi}(1+\nu)\log\frac{a}{r} \tag{90}$$

$$M_t = \frac{P}{4\pi}\left[(1+\nu)\log\frac{a}{r} + 1 - \nu\right] \tag{91}$$

To obtain formulas for a circular plate with clamped edges we differentiate Eq. (89) and find for the slope at the boundary of a simply supported plate

$$-\left(\frac{dw}{dr}\right)_{r=a} = \frac{Pa}{4(1+\nu)\pi D} \tag{a}$$

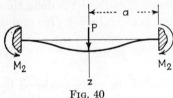

Fig. 40

The bending moments M_2 uniformly distributed along the clamped edge (Fig. 40) produce a bending of the plate to a spherical surface the radius of which is given by Eq. (46), and the corresponding slope at the boundary is

$$-\frac{M_2 a}{(1+\nu)D} \tag{b}$$

Using (a) and (b), the condition that the built-in edge does not rotate gives

$$(M_r)_{r=a} = M_2 = -\frac{P}{4\pi} \tag{c}$$

Deflections produced by moments M_2 by Eq. (h) of the preceding article are

$$\frac{P(r^2 - a^2)}{8\pi D(1+\nu)}$$

Superposing these deflections on the deflections of a simply supported plate in Eq. (89), we obtain the following expression for the deflections of a clamped plate loaded at the center:

$$w = \frac{Pr^2}{8\pi D} \log \frac{r}{a} + \frac{P}{16\pi D}(a^2 - r^2) \tag{92}$$

Adding Eq. (c) to Eqs. (90) and (91) for a simply supported plate, we obtain the following equations for the bending moment at any point not very close to the load:

$$M_r = \frac{P}{4\pi}\left[(1+\nu)\log\frac{a}{r} - 1\right] \tag{93}$$

$$M_t = \frac{P}{4\pi}\left[(1+\nu)\log\frac{a}{r} - \nu\right] \tag{94}$$

When r approaches zero, expressions (90), (91), (93), and (94) approach infinity and hence are not suitable for calculating the bending moments. Moreover, the assumptions that serve as the basis for the elementary theory of bending of circular plates do not hold near the point of application of a concentrated load. As the radius c of the circle over which P is distributed decreases, the intensity $P/\pi c^2$ of the pressure increases till it can no longer be neglected in comparison with the bending stresses as is done in the elementary theory. Shearing stresses which are also disregarded in the simple theory likewise increase without limit as c approaches zero, since the cylindrical surface $2\pi ch$ over which the total shear force P is distributed approaches zero.

Discarding the assumptions on which the elementary theory is based, we may obtain the stress distribution near the point of application of the load by considering that portion of the plate as a body all three dimensions of which are of the same order of magnitude. To do this imagine the central loaded portion separated from the rest of the plate by a cylindrical surface whose radius b is several times as large as the thickness h of the plate, as shown in Fig. 41. It may be assumed that the elementary theory of bending is accurate enough at a distance b from the point of application of the load P and that the corresponding stresses may be calculated by means of Eq. (90). The problem of stress distribution near the center of the plate is thus reduced to the problem of a symmetrical stress distribution in a circular cylinder of height h and radius b acted upon by a load P distributed over a small circle of radius c and by reactions along the lateral boundary.[1] The solution of this problem shows that the maximum compressive

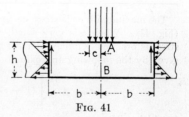

FIG. 41

---

[1] Several examples of symmetrical stress distribution are discussed in S. Timoshenko and J. N. Goodier, "Theory of Elasticity," 2d ed., p. 384, 1951. The case shown in Fig. 41 was studied by A. Nádai (see his book "Elastische Platten," p. 308) and also by S. Woinowsky-Krieger (see his paper in *Ingr.-Arch.*, vol. 4, p. 305, 1933). The results given here are from the latter paper.

stress at the center A of the upper face of the plate can be expressed by the following approximate formula:[1]

$$\sigma_r = \sigma_t = \sigma_1 - \frac{P}{\pi c^2}\left[\frac{1+2\nu}{2} - (1+\nu)\alpha\right] \quad (95)$$

in which σ_1 is the value of the compressive bending stress[2] obtained from the approximate theory, say, by using Eq. (83) for the case of a simply supported plate, and α is a numerical factor depending on $2c/h$, the ratio of the diameter of the loaded area to the

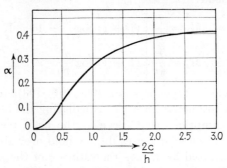

Fig. 42

thickness of the plate. Several values of this factor are given in Table 4. Its variation with the ratio $2c/h$ is shown also in Fig. 42. When c approaches zero, the stress calculated by Eq. (95) approaches infinity.

Table 4. Values of Factor α in Eq. (95)

$2c/h =$	0.10	0.25	0.50	0.75	1.00	1.50	2.00	2.50
$\alpha =$	0.0106	0.0466	0.1234	0.200	0.263	0.348	0.386	0.398

The maximum tensile stress occurs at B, the center of the lower surface of the plate (Fig. 41). When c is very small, i.e., for a strong load concentration, this tensile stress is practically independent of the ratio $2c/h$ and for a simply supported plate is given by the following approximate formula:[3]

$$\sigma_{max} = \frac{P}{h^2}\left[(1+\nu)\left(0.485 \log \frac{a}{h} + 0.52\right) + 0.48\right] \quad (96)$$

in which a is the outer radius.

To obtain the compressive stresses σ_r and σ_t at the center of the upper surface of a clamped plate, we must decrease the value of the compressive stress σ_1 in Eq. (95) by an amount equal to

$$\frac{P}{4\pi}\frac{6}{h^2} = \frac{3}{2}\frac{P}{\pi h^2} \quad (d)$$

[1] When c is very small, the compressive stress $P/\pi c^2$ becomes larger than the value of σ_{max} given by Eq. (95) (see Fig. 43).
[2] This quantity should be taken with negative sign in Eq. (95).
[3] See Woinowsky-Krieger, op. cit.

on account of the action of the moments $M_2 = -P/4\pi$. The maximum tensile stress at the center of the lower surface of a clamped plate for a strong concentration of the load ($c = 0$) is found by subtracting Eq. (d) from Eq. (96). This stress is

$$\sigma_{\max} = \frac{P}{h^2}(1 + \nu)\left(0.485 \log \frac{a}{h} + 0.52\right) \tag{97}$$

The stress distribution across a thick circular plate ($h/a = 0.4$) with built-in edges is shown in Fig. 43. These stresses are calculated for $c = 0.1a$ and $\nu = 0.3$. For this case the maximum compressive stress σ_z normal to the surface of the plate is larger than the maximum compressive stress in bending given by Eq. (95). The maximum

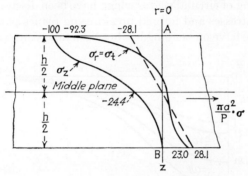

Fig. 43

tensile stress is smaller than the tensile stress given by the elementary theory of bending. The value of the latter across the thickness of the plate is shown in the figure by the dashed line. It was calculated from the equation for bending moment

$$M_{\max} = \frac{P}{4\pi}\left[(1 + \nu) \log \frac{a}{c} - \frac{(1 - \nu)c^2}{4a^2}\right] \tag{98}$$

obtained by adding the moment $M_2 = -P/4\pi$ to Eq. (83).

In determining the safe dimensions of a circular plate loaded at the center, we can usually limit our investigations to the calculation of the maximum tensile bending stresses at the bottom of the plate by means of Eqs. (96) and (97). Although the compressive stresses at the top of the plate may be many times as large as the tensile stresses at the bottom in the case of a strong concentration of the load, they do not represent a direct danger because of their highly localized character. The local yielding in the case of a ductile material will not affect the deformation of the plate in general if the tensile stresses at the bottom of the plate remain within safe limits. The compressive strength of a brittle material is usually many times greater than its tensile strength, so that a plate of such a material will also be safe if the tensile stress at the bottom is within the limit of safety.

The local disturbance produced by a concentrated load in the vicinity of its point of application must also be considered if we want an exact description of the deflection of the plate. This disturbance is mainly confined to a cylindrical region of radius several times h, and thus its effect on the total deflection becomes of practical importance when the thickness of the plate is not very small compared with its radius. As an illustration there are shown in Fig. 44 the deflections of circular plates with built-in edges and a central concentrated load for which the ratio of thickness to radius h/a

is 0.2, 0.4, and 0.6.[1] The deflection given by the elementary theory [Eq. (92)] is shown by the dashed line. It may be seen that the discrepancy between the elementary theory and the exact solution diminishes rapidly as the ratio h/a diminishes. In the next article we shall show that this discrepancy is due principally to the effect of shearing forces which are entirely neglected in the elementary theory.

20. Corrections to the Elementary Theory of Symmetrical Bending of Circular Plates. The relations (37) and (38) between bending moments and curvatures, which were derived for the case of pure bending, have been used as the basis for the solution of the various problems of symmetrical bending of circular plates which have been discussed. The effect that shearing stresses and normal pressures on planes parallel to the surface of the plate have on bending has not been taken into account. Hence

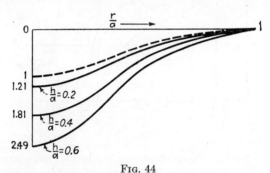

Fig. 44

only the solution for a plate bent to a spherical surface and the solution for the annular plate loaded with moments uniformly distributed along the inner and outer boundaries (Fig. 31) are rigorous. In all other cases discussed, the formulas obtained are approximate, and their accuracy depends on the ratio of the thickness of the plate to its outer radius. More accurate formulas may be obtained by considering in an approximate manner[2] the effect of shearing stresses and lateral pressures on deflections.

Let us consider first a circular plate without a hole supported along its edge and uniformly loaded. The shearing force Q per unit length of arc

---

[1] The curves in Fig. 44 are the results of the exact solution of Woinowsky-Krieger, *loc. cit.*

[2] A rigorous theory of plates was originated by Saint Venant in his translation of Clebsch's "Théorie de l'élasticité des corps solides," p. 337. A valuable criticism of this work is given in "History of the Theory of Elasticity," by I. Todhunter and K. Pearson, vol. 2, part 1, p. 217. Further development of the theory is due to J. H. Michell, *Proc. London Math. Soc.*, vol. 31, p. 100, 1900, and to A. E. H. Love, "Mathematical Theory of Elasticity," 4th ed., p. 465. A list of references on this subject is given by Woinowsky-Krieger, *op. cit.*, p. 203. Some examples of rigorous theory are given in Art. 26 (see p. 98).

along a circle of radius r is

$$Q = \tfrac{1}{2}qr$$

From the exact solution for plates whose thickness is not assumed to be small,[1] it is known that the shearing stresses τ_{rz} vary across the thickness of the plate according to the parabolic law in the same way as in beams of narrow rectangular cross section. Hence the maximum shearing stress is at the middle surface of the plate, and its magnitude is

$$(\tau_{rz})_{\max} = \frac{3}{2}\frac{qr}{2h} \qquad (a)$$

The corresponding shearing strain is

$$\frac{dw_1}{dr} = -\frac{3}{2}\frac{qr}{2Gh} \qquad (b)$$

where w_1 is the additional deflection of the middle surface of the plate due to the shearing stress. By integration the deflections produced by the shearing stresses are found to be

$$w_1 = \frac{3}{2}\frac{q}{4Gh}(a^2 - r^2) \qquad (c)$$

and at the center of the plate,

$$(w_1)_{\max} = \frac{3}{2}\frac{qa^2}{4Gh} \qquad (d)$$

The lateral pressure acting on the plate produces a positive curvature, convex downward, similar to that which occurs in a uniformly loaded beam.[2] The pressure q per unit area produces a radial elongation of $\nu q/E$ at the upper surface of the plate. At the middle surface of the plate this elongation is $\nu q/2E$, and at the bottom of the plate it is zero. Assuming a straight-line relation to hold, an approximate value of the radius of curvature R can be found from the equation

$$\frac{\nu q}{2E} = \frac{h}{2R}$$

from which

$$\frac{1}{2R} = \frac{\nu q}{2hE}$$

and the negative deflection is

$$w_2 = -\frac{1}{2R}(a^2 - r^2) = -\frac{\nu q}{2hE}(a^2 - r^2) \qquad (e)$$

---

[1] Timoshenko and Goodier, *op. cit.*, p. 351.
[2] See *ibid.*, p. 43.

Adding Eqs. (c) and (e) to Eq. (67), a more exact expression for deflection is found to be

$$w = \frac{q}{64D}(a^2 - r^2)\left(\frac{5+\nu}{1+\nu}a^2 - r^2\right) + \frac{qh^2}{8D}\frac{3+\nu}{6(1-\nu^2)}(a^2 - r^2)$$

At the center of the plate this becomes

$$w_{\max} = \frac{qa^4}{64D}\left(\frac{5+\nu}{1+\nu} + \frac{4}{3}\frac{3+\nu}{1-\nu^2}\frac{h^2}{a^2}\right) \qquad (f)$$

The second term in Eq. (f) represents the correction for shearing stresses and lateral pressure. This correction is seen to be small when the ratio of the thickness of the plate to its radius is small. The value of this correction given by the exact solution is[1]

$$\frac{qa^4}{64D}\frac{2}{5}\frac{8+\nu+\nu^2}{1-\nu^2}\frac{h^2}{a^2} \qquad (g)$$

For $\nu = 0.3$ the exact value is about 20 per cent less than that given by Eq. (f).

In a uniformly loaded circular plate with clamped edges the negative deflection w_2 due to pressure cannot occur, and hence only the deflection w_1 due to shear need be considered. Adding this deflection to Eq. (62), we obtain as a more accurate value of the deflection

$$w = \frac{q}{64D}\left[(a^2 - r^2)^2 + \frac{4h^2}{1-\nu}(a^2 - r^2)\right] \qquad (h)$$

It is interesting to note that this coincides with the exact solution.[2]

Consider next the deflections produced by shearing stresses in the annular plate loaded with shearing forces uniformly distributed along the inner edge of the plate as shown in Fig. 32. The maximum shearing stress at a distance r from the center is

$$(\tau_{rz})_{\max} = \frac{3}{2}\frac{P}{2\pi rh}$$

where P denotes the total shear load. The corresponding shear strain is[3]

$$\frac{dw_1}{dr} = -\frac{3}{2}\frac{P}{2\pi rhG} \qquad (i)$$

Integrating, we obtain for the deflection produced by shear

$$w_1 = \frac{3}{4}\frac{P}{\pi hG}\log\frac{a}{r} = \frac{Ph^2}{8\pi(1-\nu)D}\log\frac{a}{r} \qquad (j)$$

[1] See Love, op. cit., p. 481.
[2] See ibid., p. 485.
[3] If the plate has no hole, the right-hand side of Eq. (i) should be multiplied by a factor $(1-\nu)/(1+\nu)$, in accordance with the result (t) given below.

This deflection must be added to Eq. (k) on page 60 to get a more accurate value of the deflection of the plate shown in Fig. 32. When the radius b of the hole is very small, the expression for the total deflection becomes

$$w = \frac{P}{8\pi D}\left[\frac{3+\nu}{2(1+\nu)}(a^2 - r^2) + r^2 \log \frac{r}{a}\right] + \frac{Ph^2}{8\pi(1-\nu)D}\log\frac{a}{r} \quad (k)$$

The deflection at the edge of the hole is

$$w_{\max} = \frac{Pa^2}{8\pi D}\left[\frac{3+\nu}{2(1+\nu)} + \frac{1}{1-\nu}\frac{h^2}{a^2}\log\frac{a}{b}\right] \quad (l)$$

The second term in this expression represents the correction due to shear. It increases indefinitely as b approaches zero, as a consequence of our assumption that the load P is always finite. Thus when b approaches zero, the corresponding shearing stress and shearing strain become infinitely large.

The term in Eq. (l) which represents the correction for shear cannot be applied to a plate without a hole. The correction for a plate without a hole may be expected to be somewhat smaller because of the wedging effect produced by the concentrated load P applied at the center of the upper surface of the plate. Imagine that the central portion of the plate is removed by means of a cylindrical section of small radius b and that its action on the remainder of the plate is replaced by vertical shearing forces equivalent to P and by radial forces S representing the wedging effect of the load and

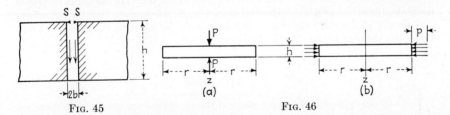

Fig. 45   Fig. 46

distributed along the upper edge of the hole as shown in Fig. 45. It is evident that the latter forces produce stretching of the middle surface of the plate together with some deflection of the plate in the upward direction. This indicates that we must decrease the correction term in expression (k) to make it apply to a plate without a hole. To get an idea of the magnitude of the radial forces S, let us consider the plate under the two loading conditions shown in Fig. 46. In the first case the plate is compressed by two equal and opposite forces P acting along the axis of symmetry z. In the second case the plate is subjected to uniform compression in its plane by a pressure p uniformly distributed over the cylindrical surface bounding the plate. As a result of lateral expansion these pressures produce an increase of the thickness of the plate by the amount

$$\Delta h = \frac{2\nu p}{E}h$$

We can now obtain from this expression the increase Δr in the radius r of the plate due to the action of the forces P (Fig. 46a) by applying the reciprocal theorem to the two conditions of loading shown in Fig. 46. This gives

$$P \Delta h = 2\pi r h p \Delta r$$

from which

$$\Delta r = \frac{P \Delta h}{2\pi r h p} = \frac{2\nu}{E} \frac{P}{2\pi r} \tag{m}$$

Let us compare this radial expansion with the radial expansion produced in a thick-walled cylinder by an internal pressure p_i. If the inner radius b of the cylinder is very small compared with the outer radius r, the increase in the outer radius by Lamé's formula[1] is

$$\Delta r = \frac{1+\nu}{E} \frac{p_i b^2}{r} \tag{n}$$

Comparing expressions (m) and (n), we conclude that the radial expansion which the forces P in Fig. 46a produce in the plate has the same magnitude as the radial expansion produced in a plate with a small cylindrical hole at the center (Fig. 45) by internal pressure p_i whose magnitude is given by the equation

$$\frac{2\nu P}{E 2\pi r} = \frac{1+\nu}{E} \frac{p_i b^2}{r}$$

From this we obtain

$$p_i = \frac{\nu P}{(1+\nu)\pi b^2} \tag{o}$$

Returning to the case of one concentrated force at the center of the upper surface of the plate, the action of which is illustrated by Fig. 45, we conclude that the force S per unit length of the circumference of the hole must be equal to the pressure $p_i h/2$. Using the value of p_i from Eq. (o), we obtain

$$S = \frac{\nu P h}{2(1+\nu)\pi b^2}$$

These forces applied in the upper plane of the plate produce upward deflections w_1, the magnitude of which is found by substituting

$$M_1 = \frac{Sh}{2} = \frac{\nu P h^2}{4(1+\nu)\pi b^2}$$

in Eq. (73) and neglecting b^2 in comparison with a^2. In this manner we obtain

$$w_1 = -\frac{\nu P h^2}{8\pi(1+\nu)^2 D} \frac{a^2 - r^2}{a^2} - \frac{\nu P h^2}{4(1-\nu^2)\pi D} \log \frac{a}{r} \tag{p}$$

Adding this to expression (k), we obtain the following more accurate formula for the deflection of a plate without a hole and carrying a load P concentrated at the center of the upper surface of the plate:

[1] See S. Timoshenko, "Strength of Materials," part II, 3d ed., p. 210, 1956.

$$w = \frac{P}{8\pi D}\left[\frac{3+\nu}{2(1+\nu)}(a^2-r^2) + r^2\log\frac{r}{a}\right] + \frac{Ph^2}{8\pi(1+\nu)D}\log\frac{a}{r}$$
$$- \frac{\nu Ph^2}{8\pi(1+\nu)^2 D}\frac{a^2-r^2}{a^2} \quad (q)$$

This equation can be used to calculate the deflection of all points of the plate that are not very close to the point of application of the load. When r is of the same order of magnitude as the thickness of the plate, Eq. (q) is no longer applicable; and to obtain a satisfactory solution the central portion of the plate must be considered, as explained in the preceding article. We can get an approximate value of the deflection of this central portion considered as a plate of small radius b by adding the deflection due to local disturbance in stress distribution near the point of application of the load to the deflection given by the elementary theory.[1] The deflection due to local disturbance near the center is affected very little by the conditions at the edge of the plate and hence can be evaluated approximately by means of the curves in Fig. 44. The dashed-line curve in this figure is obtained by using Eq. (92). The additional deflections due to local stress disturbance are equal to the differences between the ordinates of the full lines and those of the dashed line.

As an example, consider a plate the radius of the inner portion of which is $b = 5h$. The deflection of the inner portion calculated from Eq. (92) and taken as unity in Fig. 44 is

$$\delta_1 = \frac{Pb^2}{16\pi D} = \frac{P}{16\pi D}(5h)^2$$

Using the curve $h/a = 0.2$ in Fig. 44, the additional deflection due to local stress disturbance is

$$\delta_2 = 0.21\delta_1 = 0.21\frac{P}{16\pi D}(5h)^2 \quad (r)$$

If we consider a plate for which $b = 2.5h$ and use the curve for $h/a = 0.4$ in Fig. 44, we obtain

$$\delta_2 = 0.81\frac{P}{16\pi D}(2.5h)^2 \quad (s)$$

which differs only slightly from that given in expression (r) for $b = 5h$. It will be unsatisfactory to take b smaller than $2.5h$, since for smaller radii the edge condition of the thick plate becomes of importance and the curves in Fig. 44, calculated for a built-in edge, may not be accurate enough for our case.

Finally, to obtain the deflection of the plate under the load we calculate the deflection by means of Eq. (q), putting $r = 0$ in the first term and $r = b = 2.5h$ in both other terms. To this deflection we add the deflection of the central portion of the plate due to the shear forces as given by expression (s).

In the particular case of $\nu = 0.3$ the deflections of simply supported circular plates may also be obtained by a simple superposition of the curves plotted in Fig. 44,[*] with the deflection

$$\frac{P(a^2 - r^2)}{8\pi D(1+\nu)}$$

[1] In the case under consideration this deflection can be calculated by using the first term in expression (q) and substituting b for a.

[*] Figure 44 was calculated for $\nu = 0.3$.

due to the pure bending by radial moments $P/4\pi$ applied along the boundary of the plate.

It should be noted also that, for small values of the ratio r/a, the effect of the shearing force $P/2\pi r$ upon the deflection is represented mainly by the second term on the right-hand side of Eq. (q). To this term corresponds a slope

$$\frac{dw_1}{dr} = -\frac{3}{2}\frac{1-\nu}{1+\nu}\frac{P}{2\pi rhG} \tag{t}$$

Comparing this result with the expression (i), we conclude that the factor

$$k = \frac{3}{2}\frac{1-\nu}{1+\nu} \tag{u}$$

if introduced into Eq. (i) instead of $k = \frac{3}{2}$, would give a more accurate value of the deformation due to shear in the case of a plate without a hole.

All preceding considerations are applicable only to circular plates bent to a surface of revolution. A more general theory of bending taking into account the effect of the shear forces on the deformation of the plate will be given in Arts. 26 and 39.

CHAPTER 4

SMALL DEFLECTIONS OF LATERALLY LOADED PLATES

21. The Differential Equation of the Deflection Surface. We assume that the load acting on a plate is normal to its surface and that the deflections are small in comparison with the thickness of the plate (see Art. 13). At the boundary we assume that the edges of the plate are free to move in the plane of the plate; thus the reactive forces at the edges are normal to the plate. With these assumptions we can neglect any strain in the middle plane of the plate during bending. Taking, as

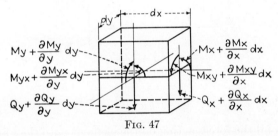

Fig. 47

before (see Art. 10), the coordinate axes x and y in the middle plane of the plate and the z axis perpendicular to that plane, let us consider an element cut out of the plate by two pairs of planes parallel to the xz and yz planes, as shown in Fig. 47. In addition to the bending moments M_x and M_y and the twisting moments M_{xy} which were considered in the pure bending of a plate (see Art. 10), there are vertical shearing forces[1] acting on the sides of the element. The magnitudes of these shearing forces per unit length parallel to the y and x axes we denote by Q_x and Q_y, respectively, so that

$$Q_x = \int_{-h/2}^{h/2} \tau_{xz}\, dz \qquad Q_y = \int_{-h/2}^{h/2} \tau_{yz}\, dz \qquad (a)$$

Since the moments and the shearing forces are functions of the coordinates x and y, we must, in discussing the conditions of equilibrium of the element, take into consideration the small changes of these quantities when the coordinates x and y change by the small quantities dx and dy.

[1] There will be no horizontal shearing forces and no forces normal to the sides of the element, since the strain of the middle plane of the plate is assumed negligible.

The middle plane of the element is represented in Fig. 48a and b, and the directions in which the moments and forces are taken as positive are indicated.

We must also consider the load distributed over the upper surface of the plate. The intensity of this load we denote by q, so that the load acting on the element[1] is $q\,dx\,dy$.

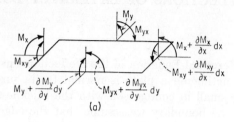

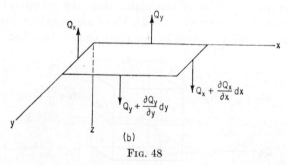

Fig. 48

Projecting all the forces acting on the element onto the z axis we obtain the following equation of equilibrium:

$$\frac{\partial Q_x}{\partial x}\,dx\,dy + \frac{\partial Q_y}{\partial y}\,dy\,dx + q\,dx\,dy = 0$$

from which

$$\frac{\partial Q_x}{\partial x} + \frac{\partial Q_y}{\partial y} + q = 0 \qquad (99)$$

Taking moments of all the forces acting on the element with respect to the x axis, we obtain the equation of equilibrium

$$\frac{\partial M_{xy}}{\partial x}\,dx\,dy - \frac{\partial M_y}{\partial y}\,dy\,dx + Q_y\,dx\,dy = 0 \qquad (b)$$

[1] Since the stress component σ_z is neglected, we actually are not able to apply the load on the upper or on the lower surface of the plate. Thus, every transverse single load considered in the thin-plate theory is merely a discontinuity in the magnitude of the shearing forces, which vary according to the parabolic law through the thickness of the plate. Likewise, the weight of the plate can be included in the load q without affecting the accuracy of the result. If the effect of the surface load becomes of special interest, thick-plate theory has to be used (see Art. 19).

The moment of the load q and the moment due to change in the force Q_y are neglected in this equation, since they are small quantities of a higher order than those retained. After simplification, Eq. (b) becomes

$$\frac{\partial M_{xy}}{\partial x} - \frac{\partial M_y}{\partial y} + Q_y = 0 \tag{c}$$

In the same manner, by taking moments with respect to the y axis, we obtain

$$\frac{\partial M_{yx}}{\partial y} + \frac{\partial M_x}{\partial x} - Q_x = 0 \tag{d}$$

Since there are no forces in the x and y directions and no moments with respect to the z axis, the three equations (99), (c), and (d) completely define the equilibrium of the element. Let us eliminate the shearing forces Q_x and Q_y from these equations by determining them from Eqs. (c) and (d) and substituting into Eq. (99). In this manner we obtain

$$\frac{\partial^2 M_x}{\partial x^2} + \frac{\partial^2 M_{yx}}{\partial x\, \partial y} + \frac{\partial^2 M_y}{\partial y^2} - \frac{\partial^2 M_{xy}}{\partial x\, \partial y} = -q \tag{e}$$

Observing that $M_{yx} = -M_{xy}$, by virtue of $\tau_{xy} = \tau_{yx}$, we finally represent the equation of equilibrium (e) in the following form:

$$\frac{\partial^2 M_x}{\partial x^2} + \frac{\partial^2 M_y}{\partial y^2} - 2\frac{\partial^2 M_{xy}}{\partial x\, \partial y} = -q \tag{100}$$

To represent this equation in terms of the deflections w of the plate, we make the assumption here that expressions (41) and (43), developed for the case of pure bending, can be used also in the case of laterally loaded plates. This assumption is equivalent to neglecting the effect on bending of the shearing forces Q_x and Q_y and the compressive stress σ_z produced by the load q. We have already used such an assumption in the previous chapter and have seen that the errors in deflections obtained in this way are small provided the thickness of the plate is small in comparison with the dimensions of the plate in its plane. An approximate theory of bending of thin elastic plates, taking into account the effect of shearing forces on the deformation, will be given in Art. 39, and several examples of exact solutions of bending problems of plates will be discussed in Art. 26.

Using x and y directions instead of n and t, which were used in Eqs. (41) and (43), we obtain

$$M_x = -D\left(\frac{\partial^2 w}{\partial x^2} + \nu \frac{\partial^2 w}{\partial y^2}\right) \qquad M_y = -D\left(\frac{\partial^2 w}{\partial y^2} + \nu \frac{\partial^2 w}{\partial x^2}\right) \tag{101}$$

$$M_{xy} = -M_{yx} = D(1-\nu)\frac{\partial^2 w}{\partial x\, \partial y} \tag{102}$$

Substituting these expressions in Eq. (100), we obtain[1]

$$\frac{\partial^4 w}{\partial x^4} + 2\frac{\partial^4 w}{\partial x^2 \partial y^2} + \frac{\partial^4 w}{\partial y^4} = \frac{q}{D} \qquad (103)$$

This latter equation can also be written in the symbolic form

$$\Delta\Delta w = \frac{q}{D} \qquad (104)$$

where
$$\Delta w = \frac{\partial^2 w}{\partial x^2} + \frac{\partial^2 w}{\partial y^2} \qquad (105)$$

It is seen that the problem of bending of plates by a lateral load q reduces to the integration of Eq. (103). If, for a particular case, a solution of this equation is found that satisfies the conditions at the boundaries of the plate, the bending and twisting moments can be calculated from Eqs. (101) and (102). The corresponding normal and shearing stresses are found from Eq. (44) and the expression

$$(\tau_{xy})_{\max} = \frac{6M_{xy}}{h^2}$$

Equations (c) and (d) are used to determine the shearing forces Q_x and Q_y, from which

$$Q_x = \frac{\partial M_{yx}}{\partial y} + \frac{\partial M_x}{\partial x} = -D\frac{\partial}{\partial x}\left(\frac{\partial^2 w}{\partial x^2} + \frac{\partial^2 w}{\partial y^2}\right) \qquad (106)$$

$$Q_y = \frac{\partial M_y}{\partial y} - \frac{\partial M_{xy}}{\partial x} = -D\frac{\partial}{\partial y}\left(\frac{\partial^2 w}{\partial x^2} + \frac{\partial^2 w}{\partial y^2}\right) \qquad (107)$$

or, using the symbolic form,

$$Q_x = -D\frac{\partial}{\partial x}(\Delta w) \qquad Q_y = -D\frac{\partial}{\partial y}(\Delta w) \qquad (108)$$

The shearing stresses τ_{xz} and τ_{yz} can now be determined by assuming that they are distributed across the thickness of the plate according to the parabolic law.[2] Then

$$(\tau_{xz})_{\max} = \frac{3}{2}\frac{Q_x}{h} \qquad (\tau_{yz})_{\max} = \frac{3}{2}\frac{Q_y}{h}$$

---

[1] This equation was obtained by Lagrange in 1811, when he was examining the memoir presented to the French Academy of Science by Sophie Germain. The history of the development of this equation is given in I. Todhunter and K. Pearson, "History of the Theory of Elasticity," vol. 1, pp. 147, 247, 348, and vol. 2, part 1, p. 263. See also the note by Saint Venant to Art. 73 on page 689 of the French translation of "Théorie de l'élasticité des corps solides," by Clebsch, Paris, 1883.

[2] It will be shown in Art. 26 that in certain cases this assumption is in agreement with the exact theory of bending of plates.

It is seen that the stresses in a plate can be calculated provided the deflection surface for a given load distribution and for given boundary conditions is determined by integration of Eq. (103).

22. Boundary Conditions. We begin the discussion of boundary conditions with the case of a rectangular plate and assume that the x and y axes are taken parallel to the sides of the plate.

Built-in Edge. If the edge of a plate is built in, the deflection along this edge is zero, and the tangent plane to the deflected middle surface along this edge coincides with the initial position of the middle plane of the plate. Assuming the built-in edge to be given by $x = a$, the boundary conditions are

$$(w)_{x=a} = 0 \qquad \left(\frac{\partial w}{\partial x}\right)_{x=a} = 0 \qquad (109)$$

Simply Supported Edge. If the edge $x = a$ of the plate is simply supported, the deflection w along this edge must be zero. At the same time this edge can rotate freely with respect to the edge line; *i.e.*, there are no bending moments M_x along this edge. This kind of support is represented in Fig. 49. The analytical expressions for the boundary conditions in this case are

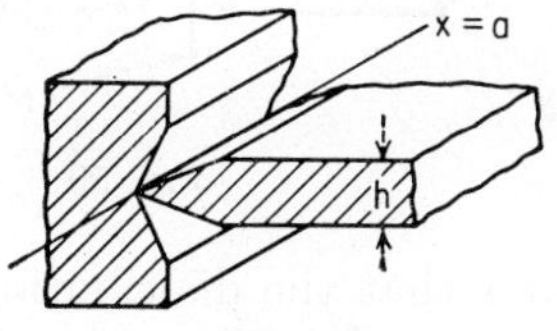

$$(w)_{x=a} = 0 \qquad \left(\frac{\partial^2 w}{\partial x^2} + \nu \frac{\partial^2 w}{\partial y^2}\right)_{x=a} = 0 \qquad (110)$$

FIG. 49

Observing that $\partial^2 w/\partial y^2$ must vanish together with w along the rectilinear edge $x = a$, we find that the second of the conditions (110) can be rewritten as $\partial^2 w/\partial x^2 = 0$ or also $\Delta w = 0$. Equations (110) are therefore equivalent to the equations

$$(w)_{x=a} = 0 \qquad (\Delta w)_{x=a} = 0 \qquad (111)$$

which do not involve Poisson's ratio ν.

Free Edge. If an edge of a plate, say the edge $x = a$ (Fig. 50), is entirely free, it is natural to assume that along this edge there are no bending and twisting moments and also no vertical shearing forces, *i.e.*, that

$$(M_x)_{x=a} = 0 \qquad (M_{xy})_{x=a} = 0 \qquad (Q_x)_{x=a} = 0$$

The boundary conditions for a free edge were expressed by Poisson[1] in this form. But later on, Kirchhoff[2] proved that three boundary conditions are too many and that two conditions are sufficient for the complete determination of the deflections w satisfying Eq. (103). He showed

---

[1] See the discussion of this subject in Todhunter and Pearson, *op. cit.*, vol. 1, p. 250, and in Saint Venant, *loc. cit.*

[2] See *J. Crelle*, vol. 40, p. 51, 1850.

also that the two requirements of Poisson dealing with the twisting moment M_{xy} and with the shearing force Q_x must be replaced by one boundary condition. The physical significance of this reduction in the number of boundary conditions has been explained by Kelvin and Tait.[1] These authors point out that the bending of a plate will not be changed if the horizontal forces giving the twisting couple $M_{xy}\,dy$ acting on an element of the length dy of the edge $x = a$ are replaced by two vertical forces of magnitude M_{xy} and dy apart, as shown in Fig. 50. Such a replacement does not change the magnitude of twisting moments and produces only local changes in the stress distribution at the edge of the plate, leaving the stress condition of the rest of the plate unchanged.

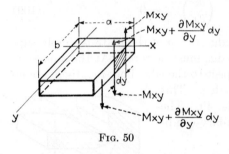

Fig. 50

We have already discussed a particular case of such a transformation of the boundary force system in considering pure bending of a plate to an anticlastic surface (see Art. 11). Proceeding with the foregoing replacement of twisting couples along the edge of the plate and considering two adjacent elements of the edge (Fig. 50), we find that the distribution of twisting moments M_{xy} is statically equivalent to a distribution of shearing forces of the intensity

$$Q'_x = -\left(\frac{\partial M_{xy}}{\partial y}\right)_{x=a}$$

Hence the joint requirement regarding twisting moment M_{xy} and shearing force Q_x along the free edge $x = a$ becomes

$$V_x = \left(Q_x - \frac{\partial M_{xy}}{\partial y}\right)_{x=a} = 0 \qquad (a)$$

Substituting for Q_x and M_{xy} their expressions (106) and (102), we finally obtain for a free edge $x = a$:

$$\left[\frac{\partial^3 w}{\partial x^3} + (2 - \nu)\frac{\partial^3 w}{\partial x\,\partial y^2}\right]_{x=a} = 0 \qquad (112)$$

The condition that bending moments along the free edge are zero requires

$$\left(\frac{\partial^2 w}{\partial x^2} + \nu\frac{\partial^2 w}{\partial y^2}\right)_{x=a} = 0 \qquad (113)$$

[1] See "Treatise of Natural Philosophy," vol. 1, part 2, p. 188, 1883. Independently the same question was explained by Boussinesq, *J. Math.*, ser. 2, vol. 16, pp. 125–274, 1871; ser. 3, vol. 5, pp. 329–344, Paris, 1879.

Equations (112) and (113) represent the two necessary boundary conditions along the free edge $x = a$ of the plate.

Transforming the twisting couples as explained in the foregoing discussion and as shown in Fig. 50, we obtain not only shearing forces Q'_x distributed along the edge $x = a$ but also two concentrated forces at the ends of that edge, as indicated in Fig. 51. The magnitudes of these forces are equal to the magnitudes of the twisting couple[1] M_{xy} at the corresponding corners of the plate. Making the analogous transformation of twisting couples M_{yx} along the edge $y = b$, we shall find that in this case again, in addition to the distributed shearing forces Q'_y, there will be concentrated forces M_{yx} at the corners. This indicates that a rectangular plate supported in some way along the edges and loaded laterally will usually produce not only reactions distributed along the boundary but also concentrated reactions at the corners.

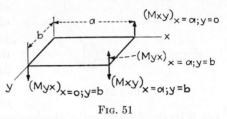

Fig. 51

Regarding the directions of these concentrated reactions, a conclusion can be drawn if the general shape of the deflection surface is known. Take, for example, a uniformly loaded square plate simply supported along the edges. The general shape of the deflection surface is indicated in Fig. 52a by dashed lines representing the section of the middle surface of the plate by planes parallel to the xz and yz coordinate planes. Considering these lines, it may be seen that near the corner A the derivative $\partial w/\partial x$, representing the slope of the deflection surface in the x direction, is negative and decreases numerically with increasing y. Hence $\partial^2 w/\partial x\, \partial y$ is positive at the corner A. From Eq. (102) we conclude that M_{xy} is positive and M_{yx} is negative at that corner. From this and from the directions of M_{xy} and M_{yx} in Fig. 48a it follows that both concentrated forces, indicated at the point $x = a$, $y = b$ in Fig. 51, have a downward direction. From symmetry we conclude also that the forces have the same magnitude and direction at all corners of the plate. Hence the conditions are as indicated in Fig. 52b, in which

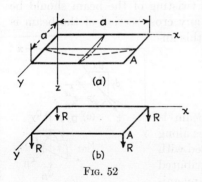

Fig. 52

$$R = 2(M_{xy})_{x=a,y=a} = 2D(1 - \nu)\left(\frac{\partial^2 w}{\partial x\, \partial y}\right)_{x=a,y=a}$$

---

[1] The couple M_{xy} is a moment per unit length and has the dimension of a force.

It can be seen that, when a square plate is uniformly loaded, the corners in general have a tendency to rise, and this is prevented by the concentrated reactions at the corners, as indicated in the figure.

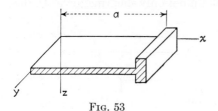

Fig. 53

Elastically Supported and Elastically Built-in Edge. If the edge $x = a$ of a rectangular plate is rigidly joined to a supporting beam (Fig. 53), the deflection along this edge is not zero and is equal to the deflection of the beam. Also, rotation of the edge is equal to the twisting of the beam. Let B be the flexural and C the torsional rigidity of the beam. The pressure in the z direction transmitted from the plate to the supporting beam, from Eq. (*a*), is

$$-V_x = -\left(Q_x - \frac{\partial M_{xy}}{\partial y}\right)_{x=a} = D\frac{\partial}{\partial x}\left[\frac{\partial^2 w}{\partial x^2} + (2-\nu)\frac{\partial^2 w}{\partial y^2}\right]_{x=a}$$

and the differential equation of the deflection curve of the beam is

$$B\left(\frac{\partial^4 w}{\partial y^4}\right)_{x=a} = D\frac{\partial}{\partial x}\left[\frac{\partial^2 w}{\partial x^2} + (2-\nu)\frac{\partial^2 w}{\partial y^2}\right]_{x=a} \tag{114}$$

This equation represents one of the two boundary conditions of the plate along the edge $x = a$.

To obtain the second condition, the twisting of the beam should be considered. The angle of rotation[1] of any cross section of the beam is $-(\partial w/\partial x)_{x=a}$, and the rate of change of this angle along the edge is

$$-\left(\frac{\partial^2 w}{\partial x\, \partial y}\right)_{x=a}$$

Hence the twisting moment in the beam is $-C(\partial^2 w/\partial x\, \partial y)_{x=a}$. This moment varies along the edge, since the plate, rigidly connected with the beam, transmits continuously distributed twisting moments to the beam. The magnitude of these applied moments per unit length is equal and opposite to the bending moments M_x in the plate. Hence, from a consideration of the rotational equilibrium of an element of the beam, we obtain

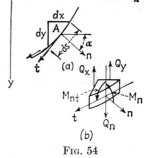

Fig. 54

$$-C\frac{\partial}{\partial y}\left(\frac{\partial^2 w}{\partial x\, \partial y}\right)_{x=a} = -(M_x)_{x=a}$$

---

[1] The right-hand-screw rule is used for the sign of the angle.

or, substituting for M_x its expression (101),

$$-C\frac{\partial}{\partial y}\left(\frac{\partial^2 w}{\partial x\,\partial y}\right)_{x=a} = D\left(\frac{\partial^2 w}{\partial x^2} + \nu\frac{\partial^2 w}{\partial y^2}\right)_{x=a} \quad (115)$$

This is the second boundary condition at the edge $x = a$ of the plate.

In the case of a plate with a curvilinear boundary (Fig. 54), we take at a point A of the edge the coordinate axes in the direction of the tangent t and the normal n as shown in the figure. The bending and twisting moments at that point are

$$M_n = \int_{-h/2}^{h/2} z\sigma_n\,dz \qquad M_{nt} = -\int_{-h/2}^{h/2} z\tau_{nt}\,dz \quad (b)$$

Using for the stress components σ_n and τ_{nt} the known expressions[1]

$$\sigma_n = \sigma_x \cos^2\alpha + \sigma_y \sin^2\alpha + 2\tau_{xy}\sin\alpha\cos\alpha$$
$$\tau_{nt} = \tau_{xy}(\cos^2\alpha - \sin^2\alpha) + (\sigma_y - \sigma_x)\sin\alpha\cos\alpha$$

we can represent expressions (b) in the following form:

$$\begin{aligned}M_n &= M_x \cos^2\alpha + M_y \sin^2\alpha - 2M_{xy}\sin\alpha\cos\alpha \\ M_{nt} &= M_{xy}(\cos^2\alpha - \sin^2\alpha) + (M_x - M_y)\sin\alpha\cos\alpha\end{aligned} \quad (c)$$

The shearing force Q_n at point A of the boundary will be found from the equation of equilibrium of an element of the plate shown in Fig. 54b, from which

$$Q_n\,ds = Q_x\,dy - Q_y\,dx$$
or
$$Q_n = Q_x \cos\alpha + Q_y \sin\alpha \quad (d)$$

Having expressions (c) and (d), the boundary condition in each particular case can be written without difficulty.

If the curvilinear edge of the plate is built in, we have for such an edge

$$w = 0 \qquad \frac{\partial w}{\partial n} = 0 \quad (e)$$

In the case of a simply supported edge we have

$$w = 0 \qquad M_n = 0 \quad (f)$$

Substituting for M_n its expression from the first of equations (c) and using Eqs. (101) and (102), we can represent the boundary conditions (f) in terms of w and its derivatives.

If the edge of a plate is free, the boundary conditions are

$$M_n = 0 \qquad V_n = Q_n - \frac{\partial M_{nt}}{\partial s} = 0 \quad (g)$$

---

[1] The x and y directions are not the principal directions as in the case of pure bending; hence the expressions for M_n and M_{nt} will be different from those given by Eqs. (39) and (40).

where the term $-\partial M_{nt}/\partial s$ is obtained in the manner shown in Fig. 50 and represents the portion of the edge reaction which is due to the distribution along the edge of the twisting moment M_{nt}. Substituting expressions (c) and (d) for M_n, M_{nt}, and Q_n and using Eqs. (101), (102), (106), and (107), we can represent boundary conditions (g) in the following form:

$$\nu \Delta w + (1 - \nu)\left(\cos^2 \alpha \frac{\partial^2 w}{\partial x^2} + \sin^2 \alpha \frac{\partial^2 w}{\partial y^2} + \sin 2\alpha \frac{\partial^2 w}{\partial x\, \partial y}\right) = 0$$

$$\cos \alpha \frac{\partial}{\partial x} \Delta w + \sin \alpha \frac{\partial}{\partial y} \Delta w + (1 - \nu) \frac{\partial}{\partial s}\left[\cos 2\alpha \frac{\partial^2 w}{\partial x\, \partial y}\right.$$
$$\left. + \frac{1}{2} \sin 2\alpha \left(\frac{\partial^2 w}{\partial y^2} - \frac{\partial^2 w}{\partial x^2}\right)\right] = 0 \quad (116)$$

where, as before,

$$\Delta w = \frac{\partial^2 w}{\partial x^2} + \frac{\partial^2 w}{\partial y^2}$$

Another method of derivation of these conditions will be shown in the next article.

23. Alternative Method of Derivation of the Boundary Conditions. The differential equation (104) of the deflection surface of a plate and the boundary conditions can be obtained by using the principle of virtual displacements together with the expression for the strain energy of a bent plate.[1] Since the effect of shearing stress on the deflections was entirely neglected in the derivation of Eq. (104), the corresponding expression for the strain energy will contain only terms depending on the action of bending and twisting moments as in the case of pure bending discussed in Art. 12. Using Eq. (48) we obtain for the strain energy in an infinitesimal element

$$dV = \frac{1}{2} D \left\{\left(\frac{\partial^2 w}{\partial x^2} + \frac{\partial^2 w}{\partial y^2}\right)^2 - 2(1 - \nu)\left[\frac{\partial^2 w}{\partial x^2} \frac{\partial^2 w}{\partial y^2} - \left(\frac{\partial^2 w}{\partial x\, \partial y}\right)^2\right]\right\} dx\, dy \quad (a)$$

The total strain energy of the plate is then obtained by integration as follows:

$$V = \frac{1}{2} D \iint \left\{\left(\frac{\partial^2 w}{\partial x^2} + \frac{\partial^2 w}{\partial y^2}\right)^2 - 2(1 - \nu)\left[\frac{\partial^2 w}{\partial x^2} \frac{\partial^2 w}{\partial y^2} - \left(\frac{\partial^2 w}{\partial x\, \partial y}\right)^2\right]\right\} dx\, dy \quad (117)$$

where the integration is extended over the entire surface of the plate.

Applying the principle of virtual displacements, we assume that an infinitely small variation δw of the deflections w of the plate is produced. Then the corresponding change in the strain energy of the plate must be equal to the work done by the external forces during the assumed virtual displacement. In calculating this work we must consider not only the lateral load q distributed over the surface of the plate but also the bending moments M_n and transverse forces $Q_n - (\partial M_{nt}/\partial s)$ distributed along the boundary of the plate. Hence the general equation, given by the principle of virtual displacements, is

[1] This is the method by which the boundary conditions were satisfactorily established for the first time; see G. Kirchhoff in *J. Crelle*, vol. 40, 1850, and also his Vorlesungen über Mathematische Physik, *Mechanik*, p. 450, 1877. Lord Kelvin took an interest in Kirchhoff's derivations and spoke with Helmholtz about them; see the biography of Kelvin by Sylvanus Thompson, vol. 1, p. 432.

SMALL DEFLECTIONS OF LATERALLY LOADED PLATES

$$\delta V = \iint q\, \delta w\, dx\, dy - \int M_n \frac{\partial\, \delta w}{\partial n} ds + \int \left(Q_n - \frac{\partial M_{nt}}{\partial s}\right) \delta w\, ds \qquad (b)$$

The first integral on the right-hand side of this equation represents the work of the lateral load during the displacement δw. The second, extended along the boundary of the plate, represents the work of the bending moments due to the rotation $\partial(\delta w)/\partial n$ of the edge of the plate. The minus sign follows from the directions chosen for M_n and the normal n indicated in Fig. 54. The third integral represents the work of the transverse forces applied along the edge of the plate.

In the calculation of the variation δV of the strain energy of the plate we use certain transformations which will be shown in detail for the first term of expression (117). The small variation of this term is

$$\delta \iint \left(\frac{\partial^2 w}{\partial x^2}\right)^2 dx\, dy = 2 \iint \frac{\partial^2 w}{\partial x^2} \frac{\partial^2\, \delta w}{\partial x^2} dx\, dy$$

$$= 2 \iint \left[\frac{\partial}{\partial x}\left(\frac{\partial^2 w}{\partial x^2} \frac{\partial\, \delta w}{\partial x}\right) - \frac{\partial^3 w}{\partial x^3} \frac{\partial\, \delta w}{\partial x}\right] dx\, dy$$

$$= 2 \iint \left[\frac{\partial}{\partial x}\left(\frac{\partial^2 w}{\partial x^2} \frac{\partial\, \delta w}{\partial x}\right) - \frac{\partial}{\partial x}\left(\frac{\partial^3 w}{\partial x^3} \delta w\right) + \frac{\partial^4 w}{\partial x^4} \delta w\right] dx\, dy \qquad (c)$$

In the first two terms after the last equality sign in expression (c) the double integration can be replaced by simple integrals if we remember that for any function F of x and y the following formulas hold:

$$\iint \frac{\partial F}{\partial x} dx\, dy = \int F \cos \alpha\, ds$$
$$\iint \frac{\partial F}{\partial y} dx\, dy = \int F \sin \alpha\, ds \qquad (d)$$

In these expressions the simple integrals are extended along the boundary, and α is the angle between the outer normal and the x axis, as shown in Fig. 54. Using the first of formulas (d), we can represent expression (c) as follows:

$$\delta \iint \left(\frac{\partial^2 w}{\partial x^2}\right)^2 dx\, dy = 2 \iint \frac{\partial^4 w}{\partial x^4} \delta w\, dx\, dy + 2 \int \left(\frac{\partial^2 w}{\partial x^2} \frac{\partial\, \delta w}{\partial x} - \frac{\partial^3 w}{\partial x^3} \delta w\right) \cos \alpha\, ds \qquad (e)$$

Advancing along the boundary in the direction shown in Fig. 54, we have

$$\frac{\partial\, \delta w}{\partial x} = \frac{\partial\, \delta w}{\partial n} \frac{dn}{dx} + \frac{\partial\, \delta w}{\partial s} \frac{ds}{dx} = \frac{\partial\, \delta w}{\partial n} \cos \alpha - \frac{\partial\, \delta w}{\partial s} \sin \alpha$$

With this transformation, expression (e) becomes

$$\delta \iint \left(\frac{\partial^2 w}{\partial x^2}\right)^2 dx\, dy = 2 \iint \frac{\partial^4 w}{\partial x^4} \delta w\, dx\, dy$$
$$+ 2 \int \frac{\partial^2 w}{\partial x^2} \left(\frac{\partial\, \delta w}{\partial n} \cos \alpha - \frac{\partial\, \delta w}{\partial s} \sin \alpha\right) \cos \alpha\, ds - 2 \int \frac{\partial^3 w}{\partial x^3} \delta w \cos \alpha\, ds \qquad (f)$$

Integrating by parts, we have

$$\int \frac{\partial^2 w}{\partial x^2} \sin \alpha \cos \alpha \frac{\partial\, \delta w}{\partial s} ds = \left|\frac{\partial^2 w}{\partial x^2} \sin \alpha \cos \alpha\, \delta w\right| - \int \frac{\partial}{\partial s}\left(\frac{\partial^2 w}{\partial x^2} \sin \alpha \cos \alpha\right) \delta w\, ds$$

The first term on the right-hand side of this expression is zero, since we are integrating along the closed boundary of the plate. Thus we obtain

$$\int \frac{\partial^2 w}{\partial x^2} \sin \alpha \cos \alpha \, \frac{\partial \, \delta w}{\partial s} \, ds = -\int \frac{\partial}{\partial s}\left(\frac{\partial^2 w}{\partial x^2} \sin \alpha \cos \alpha\right) \delta w \, ds$$

Substituting this result in Eq. (f), we finally obtain the variation of the first term in the expression for the strain energy in the following form:

$$\delta \iint \left(\frac{\partial^2 w}{\partial x^2}\right)^2 dx\, dy = 2 \iint \frac{\partial^4 w}{\partial x^4} \delta w \, dx\, dy + 2\int \frac{\partial^2 w}{\partial x^2} \cos^2 \alpha \, \frac{\partial \, \delta w}{\partial n} \, ds$$
$$+ 2 \int \left[\frac{\partial}{\partial s}\left(\frac{\partial^2 w}{\partial x^2} \sin \alpha \cos \alpha\right) - \frac{\partial^3 w}{\partial x^3} \cos \alpha\right] \delta w \, ds \quad (g)$$

Transforming in a similar manner the variations of the other terms of expression (117), we obtain

$$\delta \iint \left(\frac{\partial^2 w}{\partial y^2}\right)^2 dx\, dy = 2 \iint \frac{\partial^4 w}{\partial y^4} \delta w \, dx\, dy + 2\int \frac{\partial^2 w}{\partial y^2} \sin^2 \alpha \, \frac{\partial \, \delta w}{\partial n} \, ds$$
$$- 2 \int \left[\frac{\partial}{\partial s}\left(\frac{\partial^2 w}{\partial y^2} \sin \alpha \cos \alpha\right) + \frac{\partial^3 w}{\partial y^3} \sin \alpha\right] \delta w \, ds \quad (h)$$

$$\delta \iint \frac{\partial^2 w}{\partial x^2}\frac{\partial^2 w}{\partial y^2} dx\, dy = 2\iint \frac{\partial^4 w}{\partial x^2 \partial y^2} \delta w \, dx\, dy$$
$$+ \int \left(\frac{\partial^2 w}{\partial y^2} \cos^2 \alpha + \frac{\partial^2 w}{\partial x^2} \sin^2 \alpha\right) \frac{\partial \, \delta w}{\partial n} \, ds - \int \left\{\frac{\partial^3 w}{\partial x^2 \partial y} \sin \alpha + \frac{\partial^3 w}{\partial x \partial y^2} \cos \alpha \right.$$
$$\left. + \frac{\partial}{\partial s}\left[\left(\frac{\partial^2 w}{\partial x^2} - \frac{\partial^2 w}{\partial y^2}\right) \sin \alpha \cos \alpha\right]\right\} \delta w \, ds \quad (i)$$

$$\delta \iint \left(\frac{\partial^2 w}{\partial x\, \partial y}\right)^2 dx\, dy = 2 \iint \frac{\partial^4 w}{\partial x^2 \partial y^2} \delta w \, dx\, dy$$
$$+ 2 \int \frac{\partial^2 w}{\partial x\, \partial y} \sin \alpha \cos \alpha \, \frac{\partial \, \delta w}{\partial n} \, ds + \int \left\{\frac{\partial}{\partial s}\left[\frac{\partial^2 w}{\partial x\, \partial y}(\sin^2 \alpha - \cos^2 \alpha)\right] \right.$$
$$\left. - \frac{\partial^3 w}{\partial x\, \partial y^2} \cos \alpha - \frac{\partial^3 w}{\partial x^2\, \partial y} \sin \alpha\right\} \delta w \, ds \quad (j)$$

By using these formulas the variation of the potential energy will be represented in the following form:

$$\delta V = D \left(\iint \Delta \Delta w \, \delta w \, dx\, dy \right.$$
$$+ \int \left[(1 - \nu)\left(\frac{\partial^2 w}{\partial x^2} \cos^2 \alpha + 2\frac{\partial^2 w}{\partial x\, \partial y} \sin \alpha \cos \alpha + \frac{\partial^2 w}{\partial y^2} \sin^2 \alpha\right) + \nu \Delta w\right] \frac{\partial \, \delta w}{\partial n} \, ds$$
$$+ \int \left\{(1 - \nu) \frac{\partial}{\partial s}\left[\left(\frac{\partial^2 w}{\partial x^2} - \frac{\partial^2 w}{\partial y^2}\right) \sin \alpha \cos \alpha - \frac{\partial^2 w}{\partial x\, \partial y}(\cos^2 \alpha - \sin^2 \alpha)\right] \right.$$
$$\left. \left. - \left(\frac{\partial^3 w}{\partial x^3} + \frac{\partial^3 w}{\partial x\, \partial y^2}\right) \cos \alpha - \left(\frac{\partial^3 w}{\partial y^3} + \frac{\partial^3 w}{\partial x^2\, \partial y}\right) \sin \alpha \right\} \delta w \, ds \right) \quad (118)$$

Substituting this expression in Eq. (b) and remembering that δw and $\partial(\delta w)/\partial n$ are arbitrary small quantities satisfying the boundary conditions, we conclude that Eq. (b)

SMALL DEFLECTIONS OF LATERALLY LOADED PLATES

will be satisfied only if the following three equations are satisfied:

$$\iint (D\Delta\Delta w - q)\, \delta w\, dx\, dy = 0 \tag{k}$$

$$\int \left\{ D\left[(1-\nu)\left(\frac{\partial^2 w}{\partial x^2}\cos^2\alpha + 2\frac{\partial^2 w}{\partial x\, \partial y}\sin\alpha\cos\alpha + \frac{\partial^2 w}{\partial y^2}\sin^2\alpha\right) + \nu\Delta w\right] + M_n\right\} \frac{\partial\, \delta w}{\partial n}\, ds = 0 \tag{l}$$

$$\int \left(D\left\{(1-\nu)\frac{\partial}{\partial s}\left[\left(\frac{\partial^2 w}{\partial x^2} - \frac{\partial^2 w}{\partial y^2}\right)\sin\alpha\cos\alpha - \frac{\partial^2 w}{\partial x\, \partial y}(\cos^2\alpha - \sin^2\alpha)\right]\right.\right.$$
$$\left.\left. - \left(\frac{\partial^3 w}{\partial x^3} + \frac{\partial^3 w}{\partial x\, \partial y^2}\right)\cos\alpha - \left(\frac{\partial^3 w}{\partial y^3} + \frac{\partial^3 w}{\partial x^2\, \partial y}\right)\sin\alpha\right\} - \left(Q_n - \frac{\partial M_{nt}}{\partial s}\right)\right) \delta w\, ds = 0 \tag{m}$$

The first of these equations will be satisfied only if in every point of the middle surface of the plate we have

$$D\Delta\Delta w - q = 0$$

i.e., the differential equation (104) of the deflection surface of the plate. Equations (*l*) and (*m*) give the boundary conditions.

If the plate is built in along the edge, δw and $\partial(\delta w)/\partial n$ are zero along the edge; and Eqs. (*l*) and (*m*) are satisfied. In the case of a simply supported edge, $\delta w = 0$ and $M_n = 0$. Hence Eq. (*m*) is satisfied, and Eq. (*l*) will be satisfied if

$$(1-\nu)\left(\frac{\partial^2 w}{\partial x^2}\cos^2\alpha + 2\frac{\partial^2 w}{\partial x\, \partial y}\sin\alpha\cos\alpha + \frac{\partial^2 w}{\partial y^2}\sin^2\alpha\right) + \nu\Delta w = 0 \tag{n}$$

In the particular case of a rectilinear edge parallel to the y axis, $\alpha = 0$; and we obtain from Eq. (*n*)

$$\frac{\partial^2 w}{\partial x^2} + \nu\frac{\partial^2 w}{\partial y^2} = 0$$

as it should be for a simply supported edge.

If the edge of a plate is entirely free, the quantity δw and $\partial(\delta w)/\partial n$ in Eqs. (*l*) and (*m*) are arbitrary; furthermore, $M_n = 0$ and $Q_n - (\partial M_{nt}/\partial s) = 0$. Hence, from Eqs. (*l*) and (*m*), for a free edge we have

$$(1-\nu)\left(\frac{\partial^2 w}{\partial x^2}\cos^2\alpha + 2\frac{\partial^2 w}{\partial x\, \partial y}\sin\alpha\cos\alpha + \frac{\partial^2 w}{\partial y^2}\sin^2\alpha\right) + \nu\Delta w = 0$$

$$(1-\nu)\frac{\partial}{\partial s}\left[\left(\frac{\partial^2 w}{\partial x^2} - \frac{\partial^2 w}{\partial y^2}\right)\sin\alpha\cos\alpha - \frac{\partial^2 w}{\partial x\, \partial y}(\cos^2\alpha - \sin^2\alpha)\right]$$
$$- \left(\frac{\partial^3 w}{\partial x^3} + \frac{\partial^3 w}{\partial x\, \partial y^2}\right)\cos\alpha - \left(\frac{\partial^3 w}{\partial y^3} + \frac{\partial^3 w}{\partial x^2\, \partial y}\right)\sin\alpha = 0$$

These conditions are in agreement with Eqs. (116) which were obtained previously (see page 88). In the particular case of a free rectilinear edge parallel to the y axis, $\alpha = 0$, and we obtain

$$\frac{\partial^2 w}{\partial x^2} + \nu\frac{\partial^2 w}{\partial y^2} = 0$$

$$\frac{\partial^3 w}{\partial x^3} + (2-\nu)\frac{\partial^3 w}{\partial x\, \partial y^2} = 0$$

These equations coincide with Eqs. (112) and (113) obtained previously.

In the case when given moments M_n and transverse forces $Q_n - (\partial M_{nt}/\partial s)$ are distributed along the edge of a plate, the corresponding boundary conditions again can be easily obtained by using Eqs. (*l*) and (*m*).

24. Reduction of the Problem of Bending of a Plate to That of Deflection of a Membrane. There are cases in which it is advantageous to replace the differential equation (103) of the fourth order developed for a plate by two equations of the second order which represent the deflections of a membrane.[1] For this purpose we use form (104) of this equation:

$$\left(\frac{\partial^2}{\partial x^2} + \frac{\partial^2}{\partial y^2}\right)\left(\frac{\partial^2 w}{\partial x^2} + \frac{\partial^2 w}{\partial y^2}\right) = \frac{q}{D} \qquad (a)$$

and observe that by adding together the two expressions (101) for bending moments (see page 81) we have

$$M_x + M_y = -D(1 + \nu)\left(\frac{\partial^2 w}{\partial x^2} + \frac{\partial^2 w}{\partial y^2}\right) \qquad (b)$$

Introducing a new notation

$$M = \frac{M_x + M_y}{1 + \nu} = -D\left(\frac{\partial^2 w}{\partial x^2} + \frac{\partial^2 w}{\partial y^2}\right) \qquad (119)$$

the two Eqs. (*a*) and (*b*) can be represented in the following form:

$$\begin{aligned}\frac{\partial^2 M}{\partial x^2} + \frac{\partial^2 M}{\partial y^2} &= -q \\ \frac{\partial^2 w}{\partial x^2} + \frac{\partial^2 w}{\partial y^2} &= -\frac{M}{D}\end{aligned} \qquad (120)$$

Both these equations are of the same kind as that obtained for a uniformly stretched and laterally loaded membrane.[2]

The solution of these equations is very much simplified in the case of a simply supported plate of polygonal shape, in which case along each rectilinear portion of the boundary we have $\partial^2 w/\partial s^2 = 0$ since $w = 0$ at the boundary. Observing that $M_n = 0$ at a simply supported edge, we conclude also that $\partial^2 w/\partial n^2 = 0$ at the boundary. Hence we have [see Eq. (34)]

$$\frac{\partial^2 w}{\partial s^2} + \frac{\partial^2 w}{\partial n^2} = \frac{\partial^2 w}{\partial x^2} + \frac{\partial^2 w}{\partial y^2} = -\frac{M}{D} = 0 \qquad (c)$$

at the boundary in accordance with the second of the equations (111). It is seen that the solution of the plate problem reduces in this case to the integration of the two equations (120) in succession. We begin with

---

[1] This method of investigating the bending of plates was introduced by H. Marcus in his book "Die Theorie elastischer Gewebe," 2d ed., p. 12, Berlin, 1932.

[2] See S. Timoshenko and J. N. Goodier, "Theory of Elasticity," 2d ed., p. 269, 1951.

the first of these equations and find a solution satisfying the condition $M = 0$ at the boundary.[1] Substituting this solution in the second equation and integrating it, we find the deflections w. Both problems are of the same kind as the problem of the deflection of a uniformly stretched and laterally loaded membrane having zero deflection at the boundary. This latter problem is much simpler than the plate problem, and it can always be solved with sufficient accuracy by using an approximate method of integration such as Ritz's or the method of finite differences. Some examples of the application of these latter methods will be discussed later (see Arts. 80 and 83). Several applications of Ritz's method are given in discussing torsional problems.[2]

A simply supported plate of polygonal shape, bent by moments M_n uniformly distributed along the boundary, is another simple case of the application of Eqs. (120). Equations (120) in such a case become

$$\frac{\partial^2 M}{\partial x^2} + \frac{\partial^2 M}{\partial y^2} = 0$$
$$\frac{\partial^2 w}{\partial x^2} + \frac{\partial^2 w}{\partial y^2} = -\frac{M}{D}$$
(121)

Along a rectilinear edge we have again $\partial^2 w/\partial s^2 = 0$. Hence

$$M_n = -D\frac{\partial^2 w}{\partial n^2}$$

and we have at the boundary

$$\frac{\partial^2 w}{\partial x^2} + \frac{\partial^2 w}{\partial y^2} = \frac{\partial^2 w}{\partial n^2} = -\frac{M_n}{D} = -\frac{M}{D}$$

This boundary condition and the first of the equations (121) will be satisfied if we take for the quantity M the constant value $M = M_n$ at all points of the plate, which means that the sum of the bending moments M_x and M_y remains constant over the entire surface of the plate. The deflections of the plate will then be found from the second of the equations (121),[3] which becomes

$$\frac{\partial^2 w}{\partial x^2} + \frac{\partial^2 w}{\partial y^2} = -\frac{M_n}{D} \qquad (d)$$

It may be concluded from this that, in the case of bending of a simply supported polygonal plate by moments M_n uniformly distributed along the boundary, the deflection surface of the plate is the same as that of

---

[1] Note that if the plate is not of a polygonal shape, M generally does not vanish at the boundary when $M_n = 0$.
[2] See Timoshenko and Goodier, *op. cit.*, p. 280.
[3] This was shown first by S. Woinowsky-Krieger, *Ingr.-Arch.*, vol. 4, p. 254, 1933.

a uniformly stretched membrane with a uniformly distributed load. There are many cases for which the solutions of the membrane problem are known. These can be immediately applied in discussing the corresponding plate problems.

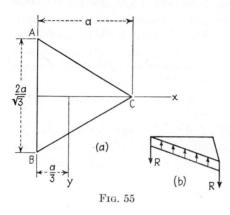

FIG. 55

Take, for example, a simply supported equilateral triangular plate (Fig. 55) bent by moments M_n uniformly distributed along the boundary. The deflection surface of the plate is the same as that of a uniformly stretched and uniformly loaded membrane. The latter can be easily obtained experimentally by stretching a soap film on the triangular boundary and loading it uniformly by air pressure.[1]

The analytical expression of the deflection surface is also comparatively simple in this case. We take the product of the left-hand sides of the equations of the three sides of the triangle:

$$\left(x + \frac{a}{3}\right)\left(\frac{x}{\sqrt{3}} + y - \frac{2a}{3\sqrt{3}}\right)\left(\frac{x}{\sqrt{3}} - y - \frac{2a}{3\sqrt{3}}\right)$$
$$= \frac{x^3 - 3y^2 x}{3} - \frac{a(x^2 + y^2)}{3} + \frac{4a^3}{3 \cdot 27}$$

This expression evidently becomes zero at the boundary. Hence the boundary condition $w = 0$ for the membrane is satisfied if we take for deflections the expression

$$w = N\left[\frac{x^3 - 3y^2 x}{3} - \frac{a(x^2 + y^2)}{3} + \frac{4a^3}{3 \cdot 27}\right] \quad (e)$$

where N is a constant factor the magnitude of which we choose in such a manner as to satisfy Eq. (d). In this way we obtain the required solution:

$$w = \frac{M_n}{4aD}\left[x^3 - 3y^2 x - a(x^2 + y^2) + \frac{4}{27}a^3\right] \quad (f)$$

Substituting $x = y = 0$ in this expression, we obtain the deflection at the centroid of the triangle

$$w_0 = \frac{M_n a^2}{27 D} \quad (g)$$

[1] Such experiments are used in solving torsional problems; see Timoshenko and Goodier, op. cit., p. 289.

The expressions for the bending and twisting moments, from Eqs. (101) and (102), are

$$M_x = \frac{M_n}{2}\left[1 + \nu - (1-\nu)\frac{3x}{a}\right]$$

$$M_y = \frac{M_n}{2}\left[1 + \nu + (1-\nu)\frac{3x}{a}\right] \qquad (h)$$

$$M_{xy} = -\frac{3(1-\nu)M_n y}{2a}$$

Shearing forces, from Eqs. (106) and (107), are

$$Q_x = Q_y = 0$$

Along the boundary, from Eq. (d) of Art. 22, the shearing force $Q_n = 0$, and the bending moment is equal to M_n. The twisting moment along the side BC (Fig. 55) from Eqs. (c) of Art. 22 is

$$M_{nt} = \frac{3(1-\nu)M_n}{4a}(y - \sqrt{3}\,x)$$

The vertical reactions acting on the plate along the side BC (Fig. 55) are

$$V_n = Q_n - \frac{\partial M_{nt}}{\partial s} = -\frac{3(1-\nu)}{2a}M_n \qquad (i)$$

From symmetry we conclude that the same uniformly distributed reactions also act along the two other sides of the plate. These forces are balanced by the concentrated reactions at the corners of the triangular plate, the magnitude of which can be found as explained on page 85 and is equal to

$$R = 2(M_{nt})_{x=\frac{2}{3}a,y=0} = (1-\nu)\sqrt{3}\,M_n \qquad (j)$$

The distribution of the reactive forces along the boundary is shown in Fig. 55b. The maximum bending stresses are at the corners and act on the planes bisecting the angles. The magnitude of the corresponding bending moment, from Eqs. (h), is

$$(M_y)_{\max} = (M_y)_{x=\frac{2}{3}a} = \frac{M_n(3-\nu)}{2} \qquad (k)$$

This method of determining the bending of simply supported polygonal plates by moments uniformly distributed along the boundary can be applied to the calculation of the thermal stresses produced in such plates by nonuniform heating. In discussing thermal stresses in clamped plates, it was shown in Art. 14 [Eq. (b)] that nonuniform heating produces uniformly distributed bending moments along the boundary of the plate which prevent any bending of the plate. The magnitude of these moments is[1]

$$M_n = \frac{\alpha t D(1 + \nu)}{h} \qquad (l)$$

To obtain thermal stresses in the case of a simply supported plate we need only to superpose on the stresses produced in pure bending by the moments (l) the stresses that are produced in a plate with simply supported edges by the bending moments $-\alpha t D(1 + \nu)/h$ uniformly distributed along the boundary. The solution of the latter problem, as was already explained, can be obtained without much difficulty in the case of a plate of polygonal shape.[2]

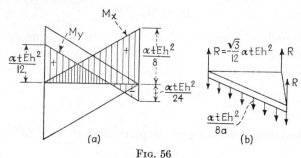

Fig. 56

Take again, as an example, the equilateral triangular plate. If the edges of the plate are clamped, the bending moments due to nonuniform heating are

$$M'_x = M'_y = \frac{\alpha t D(1 + \nu)}{h} \qquad (m)$$

To find the bending moments M_x and M_y for a simply supported plate we must superpose on the moments (m) the moments that will be obtained from Eqs. (h) by letting $M_n = -\alpha t D(1 + \nu)/h$. In this way we finally obtain

$$M_x = \frac{\alpha t D(1 + \nu)}{h} - \frac{\alpha t D(1 + \nu)}{2h}\left[1 + \nu - (1 - \nu)\frac{3x}{a}\right]$$

$$= \frac{\alpha t E h^2}{24}\left(1 + \frac{3x}{a}\right)$$

$$M_y = \frac{\alpha t D(1 + \nu)}{h} - \frac{\alpha t D(1 + \nu)}{2h}\left[1 + \nu + (1 - \nu)\frac{3x}{a}\right]$$

$$= \frac{\alpha t E h^2}{24}\left(1 - \frac{3x}{a}\right)$$

$$M_{xy} = \frac{1}{8}\frac{\alpha t E h^2 y}{a}$$

---

[1] It is assumed that the upper surface of the plate is kept at a higher temperature than the lower one and that the plate thus has the tendency to bend convexly upward.

[2] See dissertation by J. L. Maulbetsch, *J. Appl. Mechanics*, vol. 2, p. 141, 1935.

The reactive forces can now be obtained from Eqs. (i) and (j) by substitution of $M_n = -\alpha t D(1 + \nu)/h$. Hence we find

$$V_n = Q_n - \frac{\partial M_{nt}}{\partial s} = \frac{\alpha t E h^2}{8a} \qquad R = -\frac{\sqrt{3}\,\alpha t E h^2}{12}$$

The results obtained for moments and reactive forces due to nonuniform heating are represented in Fig. 56a and b, respectively.

25. Effect of Elastic Constants on the Magnitude of Bending Moments. It is seen from Eqs. (101) and (102) that the magnitude of the bending and twisting moments in a plate is considerably affected by the numerical value of Poisson's ratio ν. On the other hand, it can be easily shown that in the case of a transverse load the magnitude of the quantity Dw is independent of both constants E and ν if the plate is either simply supported at rectilinear edges or clamped along some edges, whether rectilinear or not.

Assuming such boundary conditions in any combination, let us consider the following problem. Some values of the bending moments M_x and M_y being given numerically for an assumed numerical value of ν, these moments must be computed for a new value, say ν', of the same elastic constant. Let M'_x and M'_y be the new values of the bending moments. Writing Eqs. (101) first for ν, then for ν', eliminating from them the curvatures $\partial^2 w/\partial x^2$ and $\partial^2 w/\partial y^2$, and solving the resulting equations for M'_x and M'_y, we obtain

$$M'_x = \frac{1}{1-\nu^2}[(1-\nu\nu')M_x + (\nu'-\nu)M_y]$$
$$M'_y = \frac{1}{1-\nu^2}[(1-\nu\nu')M_y + (\nu'-\nu)M_x] \qquad (122)$$

Thus M'_x and M'_y can be readily calculated if M_x and M_y are known.

If the constant ν is implied in some of the given boundary conditions, as in the case of a free edge [Eq. (112)], Eqs. (122) do not hold any more.

If the plate is elastically supported or elastically clamped, the moments also depend on the flexural rigidity D of the plate with respect to the stiffness of its restraint.

The thermal stresses, finally, are affected not only by all the above-mentioned factors, but also by the absolute value of the rigidity D of the plate.

Average values of ν for some materials are given in Table 5. The last value of the table varies widely, depending on the age of the concrete, on the type of aggregate, and on other factors.[1]

TABLE 5. AVERAGE VALUES OF POISSON'S RATIO ν

Material	ν
Steel	0.30
Aluminum	0.30
Glass	0.25
Concrete	0.15–0.25

[1] The German Code (DIN 4227) gives values of ν which approximately can be expressed by $\nu = \sqrt{f'_c}/350$, f'_c being the compressive strength of concrete at 28 days in pounds per square inch. See also J. C. Simmons, *Mag. of Concrete Research*, vol. 8, p. 39, 1956.

26. Exact Theory of Plates. The differential equation (103), which, together with the boundary conditions, defines the deflections of plates, was derived (see Art. 21) by neglecting the effect on bending of normal stresses σ_z and shearing stresses τ_{xz} and τ_{yz}. This means that in the derivation each thin layer of the plate parallel to the middle plane was considered to be in a state of plane stress in which only the stress components σ_x, σ_y, and τ_{xy} may be different from zero. One of the simplest cases of this kind is that of pure bending. The deflection surface in this case is a second-degree function in x and y [see Eq. (c), Art. 11] that satisfies Eq. (103). The stress components σ_x, σ_y, and τ_{xy} are proportional to z and independent of x and y.

There are other cases of bending in which a plane stress distribution takes place and Eq. (103) holds rigorously. Take, for example, a circular plate with a central circular hole bent by moments M_r uniformly distributed along the boundary of the hole (Fig. 57). Each thin layer of the plate cut out by two adjacent planes parallel to the middle plane is in the same stress condition as a thick-walled cylinder subjected to a uniform internal pressure or tension (Fig. 57b). The sum $\sigma_r + \sigma_t$ of the two principal stresses is constant in such a case,[1] and it can be concluded that the deformation of the layer in the z direction is also constant and does not interfere with the deformation of adjacent layers. Hence we have again a planar stress distribution, and Eq. (103) holds.

Let us discuss now the general question regarding the shape of the deflection surface of a plate when bending results in a planar stress distribution. To answer this question it is necessary to consider the three differential equations of equilibrium together with the six compatibility conditions. If body forces are neglected, these equations are[2]

$$\frac{\partial \sigma_x}{\partial x} + \frac{\partial \tau_{xy}}{\partial y} + \frac{\partial \tau_{xz}}{\partial z} = 0$$

$$\frac{\partial \sigma_y}{\partial y} + \frac{\partial \tau_{xy}}{\partial x} + \frac{\partial \tau_{yz}}{\partial z} = 0 \qquad (a)$$

$$\frac{\partial \sigma_z}{\partial z} + \frac{\partial \tau_{xz}}{\partial x} + \frac{\partial \tau_{yz}}{\partial y} = 0$$

$$\Delta_1 \sigma_x = -\frac{1}{1+\nu}\frac{\partial^2 \theta}{\partial x^2}$$

$$\Delta_1 \sigma_y = -\frac{1}{1+\nu}\frac{\partial^2 \theta}{\partial y^2} \qquad (b)$$

$$\Delta_1 \sigma_z = -\frac{1}{1+\nu}\frac{\partial^2 \theta}{\partial z^2}$$

$$\Delta_1 \tau_{xy} = -\frac{1}{1+\nu}\frac{\partial^2 \theta}{\partial x\, \partial y}$$

$$\Delta_1 \tau_{xz} = -\frac{1}{1+\nu}\frac{\partial^2 \theta}{\partial x\, \partial z} \qquad (c)$$

$$\Delta_1 \tau_{yz} = -\frac{1}{1+\nu}\frac{\partial^2 \theta}{\partial y\, \partial z}$$

in which

$$\theta = \sigma_x + \sigma_y + \sigma_z$$

and

$$\Delta_1 = \frac{\partial^2}{\partial x^2} + \frac{\partial^2}{\partial y^2} + \frac{\partial^2}{\partial z^2}$$

[1] See Timoshenko and Goodier, *op. cit.*, p. 60.
[2] See *ibid.*, pp. 229, 232.

Adding Eqs. (b), we find that

$$\frac{\partial^2 \theta}{\partial x^2} + \frac{\partial^2 \theta}{\partial y^2} + \frac{\partial^2 \theta}{\partial z^2} = \Delta_1 \theta = 0 \qquad (d)$$

i.e., the sum of the three normal stress components represents a harmonic function. In the case of a planar stress $\tau_{xz} = \tau_{yz} = \sigma_z = 0$, and it can be concluded from the last two of the equations (c) and the last of the equations (b) that $\partial \theta / \partial z$ must be a constant, say β. Hence the general expression for θ in the case of planar stress is

$$\theta = \theta_0 + \beta z \qquad (e)$$

where θ_0 is a plane harmonic function, i.e.,

$$\frac{\partial^2 \theta_0}{\partial x^2} + \frac{\partial^2 \theta_0}{\partial y^2} = \Delta \theta_0 = 0$$

We see that in the case of planar stress the function θ consists of two parts: θ_0 independent of z and βz proportional to z. The first part does not vary through the thickness of the plate. It depends on deformation of the plate in its own plane and can be omitted if we are interested only in bending of plates. Thus we can take in our further discussion

$$\theta = \beta z \qquad (f)$$

Equations of equilibrium (a) will be satisfied in the case of a planar stress distribution if we take

$$\sigma_x = \frac{\partial^2 \varphi}{\partial y^2} \qquad \sigma_y = \frac{\partial^2 \varphi}{\partial x^2} \qquad \tau_{xy} = -\frac{\partial^2 \varphi}{\partial x \, \partial y} \qquad (g)$$

where φ is the stress function. Let us consider now the general form of this function.

Substituting expressions (g) in Eq. (f), we obtain

$$\frac{\partial^2 \varphi}{\partial x^2} + \frac{\partial^2 \varphi}{\partial y^2} = \beta z \qquad (h)$$

Furthermore, from the first of the equations (b) we conclude that

$$\Delta_1 \frac{\partial^2 \varphi}{\partial y^2} = 0 \quad \text{or} \quad \frac{\partial^2}{\partial y^2} \Delta_1 \varphi = 0$$

which, by using Eq. (h), can be put in the following form:

$$\frac{\partial^2}{\partial y^2}\left(\frac{\partial^2 \varphi}{\partial z^2}\right) = 0 \qquad (i)$$

In the same manner, from the second and the third of the equations (b), we find

$$\frac{\partial^2}{\partial x^2}\left(\frac{\partial^2 \varphi}{\partial z^2}\right) = 0 \qquad \frac{\partial^2}{\partial x \, \partial y}\left(\frac{\partial^2 \varphi}{\partial z^2}\right) = 0 \qquad (j)$$

From Eqs. (i) and (j) it follows that $\partial^2 \varphi / \partial z^2$ is a linear function of x and y. This function may be taken to be zero without affecting the magnitudes of the stress components given by expressions (g). In such a case the general expression of the stress function is

$$\varphi = \varphi_0 + \varphi_1 z$$

where φ_0 is a plane harmonic function and φ_1 satisfies the equation

$$\frac{\partial^2 \varphi_1}{\partial x^2} + \frac{\partial^2 \varphi_1}{\partial y^2} = \beta \qquad (k)$$

Since we are not interested in the deformations of plates in their plane, we can omit φ_0 in our further discussion and take as a general expression for the stress function

$$\varphi = \varphi_1 z \qquad (l)$$

Substituting this in Eqs. (g), the stress components can now be calculated, and the displacements can be found from the equations

$$\frac{\partial u}{\partial x} = \frac{1}{E}(\sigma_x - \nu\sigma_y) \qquad \frac{\partial v}{\partial y} = \frac{1}{E}(\sigma_y - \nu\sigma_x) \qquad \frac{\partial w}{\partial z} = -\frac{\nu}{E}(\sigma_x + \sigma_y)$$

$$\frac{\partial u}{\partial y} + \frac{\partial v}{\partial x} = \frac{1}{G}\tau_{xy} \qquad \frac{\partial u}{\partial z} + \frac{\partial w}{\partial x} = 0 \qquad \frac{\partial v}{\partial z} + \frac{\partial w}{\partial y} = 0 \qquad (m)$$

For the displacements w perpendicular to the plate we obtain in this way[1]

$$w = -\frac{\beta}{2E}(x^2 + y^2 + \nu z^2) + \frac{1+\nu}{E}\varphi_1$$

and the deflection of the middle surface of the plate is

$$w = -\frac{\beta}{2E}(x^2 + y^2) + \frac{1+\nu}{E}\varphi_1 \qquad (n)$$

The corresponding stress components, from Eqs. (g) and (l), are

$$\sigma_x = z\frac{\partial^2 \varphi_1}{\partial y^2} \qquad \sigma_y = z\frac{\partial^2 \varphi_1}{\partial x^2} \qquad \tau_{xy} = -z\frac{\partial^2 \varphi_1}{\partial x\, \partial y}$$

and the bending and twisting moments are

$$M_x = \int_{-h/2}^{h/2} \sigma_x z\, dz = \frac{h^3}{12}\frac{\partial^2 \varphi_1}{\partial y^2} \qquad M_y = \int_{-h/2}^{h/2} \sigma_y z\, dz = \frac{h^3}{12}\frac{\partial^2 \varphi_1}{\partial x^2}$$

$$M_{xy} = -\int_{-h/2}^{h/2} \tau_{xy} z\, dz = \frac{h^3}{12}\frac{\partial^2 \varphi_1}{\partial x\, \partial y} \qquad (o)$$

For the curvatures and the twist of a plate, we find, from Eq. (n)

$$\frac{\partial^2 w}{\partial x^2} = -\frac{\beta}{E} + \frac{1+\nu}{E}\frac{\partial^2 \varphi_1}{\partial x^2} \qquad \frac{\partial^2 w}{\partial y^2} = -\frac{\beta}{E} + \frac{1+\nu}{E}\frac{\partial^2 \varphi_1}{\partial y^2} \qquad \frac{\partial^2 w}{\partial x\, \partial y} = \frac{1+\nu}{E}\frac{\partial^2 \varphi_1}{\partial x\, \partial y}$$

from which, by using Eqs. (k) and (o), we obtain

$$\frac{\partial^2 w}{\partial x^2} + \nu\frac{\partial^2 w}{\partial y^2} = -\frac{1-\nu^2}{E}\frac{\partial^2 \varphi_1}{\partial y^2} = -\frac{M_x}{D}$$

$$\frac{\partial^2 w}{\partial y^2} + \nu\frac{\partial^2 w}{\partial x^2} = -\frac{1-\nu^2}{E}\frac{\partial^2 \varphi_1}{\partial x^2} = -\frac{M_y}{D} \qquad (p)$$

$$\frac{\partial^2 w}{\partial x\, \partial y} = \frac{1+\nu}{E}\frac{\partial^2 \varphi_1}{\partial x\, \partial y} = \frac{M_{xy}}{(1-\nu)D}$$

[1] Several examples of calculating u, v, and w from Eqs. (m) are given in *ibid.*

From this analysis it may be concluded that, in the case of bending of plates resulting in a planar stress distribution, the deflections w [see Eq. (n)] rigorously satisfy Eq. (103) and also Eqs. (101) and (102) representing bending and twisting moments. If a solution of Eq. (k) is taken in the form of a function of the second degree in x and y, the deflection surface (n) is also of the second degree which represents the deflection for

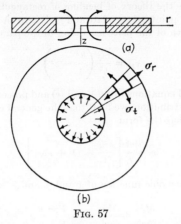

Fig. 57

pure bending. Generally we can conclude, from Eq. (k), that the deflection of the plate in the case of a planar stress distribution is the same as that of a uniformly stretched and uniformly loaded membrane. The plate shown in Fig. 57 represents a particular case of such bending, viz., the case for which the solution of Eq. (k), given in polar coordinates, is

$$\varphi_1 = Ar^2 + B \log r + C$$

where A, B, and C are constants that must be chosen so as to satisfy the boundary conditions.

Plates of a polygonal shape simply supported and bent by moments uniformly distributed along the boundary (see Art. 24) represent another example of bending in

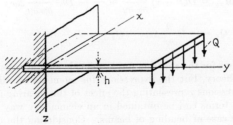

Fig. 58

which the deflection surface has a form satisfying Eq. (n), and Eqs. (101), (102), and (103) hold rigorously. In all these cases, as we may see from Eqs. (k) and (o), we have

$$M_x + M_y = \frac{h^3}{12}\left(\frac{\partial^2 \varphi_1}{\partial x^2} + \frac{\partial^2 \varphi_1}{\partial y^2}\right) = \frac{\beta h^3}{12}$$

i.e., the sum of the bending moments in two perpendicular directions remains constant over the entire plate.

Let us consider now the case in which bending of a plate results in a generalized planar stress distribution, i.e., one in which the normal stress component σ_z is zero at all points of the plate and the shearing stress components τ_{xz} and τ_{yz} are zero on the surfaces $z = \pm h/2$ of the plate. The deflection of a rectangular plate clamped along one edge and uniformly loaded along the opposite edge (Fig. 58) represents an example of such bending. From the theory of bending of rectangular beams we know that in this case $\sigma_z = 0$ at all points of the plate and τ_{xz} is zero on the surfaces of the plate and varies along the depth of the plate according to the parabolic law

$$\tau_{xz} = \frac{6Q}{h^3}\left(\frac{h^2}{4} - z^2\right)$$

Using again the general equations (a), (b), and (c) and proceeding as in the preceding case of a planar stress distribution, we find[1] that the general expression for the deflection surface in this case has the form

$$w = \frac{1}{E}\left[\frac{h^2 \varphi}{4} + (1+\nu)\varphi_1\right] \qquad (q)$$

in which φ is a planar harmonic function of x and y, and φ_1 satisfies the equation

$$\frac{\partial^2 \varphi_1}{\partial x^2} + \frac{\partial^2 \varphi_1}{\partial y^2} = -\frac{1-\nu}{1+\nu}\varphi$$

It can be concluded that in this case again the differential equation (103) holds with $q = 0$.

The equations for the bending and twisting moments and for the shearing forces in this case are

$$\begin{aligned}
M_x &= -D\left(\frac{\partial^2 w}{\partial x^2} + \nu\frac{\partial^2 w}{\partial y^2}\right) + \frac{8+\nu}{40}Dh^2\frac{\partial^2}{\partial y^2}\Delta w \\
M_y &= -D\left(\frac{\partial^2 w}{\partial y^2} + \nu\frac{\partial^2 w}{\partial x^2}\right) + \frac{8+\nu}{40}Dh^2\frac{\partial^2}{\partial x^2}\Delta w \\
M_{xy} &= D(1-\nu)\frac{\partial^2 w}{\partial x\,\partial y} + \frac{8+\nu}{40}Dh^2\frac{\partial^2}{\partial x\,\partial y}\Delta w \\
Q_x &= -D\frac{\partial}{\partial x}\Delta w \qquad Q_y = -D\frac{\partial}{\partial y}\Delta w
\end{aligned} \qquad (123)$$

Hence the expressions for the shearing forces coincide with expressions (108) given by the approximate theory, but the expressions for moments are different, the second terms of those expressions representing the effect of the shearing forces.

These correction terms can be obtained in an elementary way by using the same reasoning as in the case of bending of beams. Considering the curvature in the xz plane, we can state that the total curvature is produced by two factors, the bending moments M_x, M_y and the shearing force Q_x. The curvature produced by the bending

---

[1] The rigorous solution for this case was given by Saint Venant; see his translation of Clebsch's "Théorie de l'élasticité des corps solides," p. 337. A general discussion of the rigorous theory of bending of plates was given by J. H. Michell, *Proc. London Math. Soc.*, vol. 31, p. 100, 1900. See also A. E. H. Love, "The Mathematical Theory of Elasticity," p. 473, 1927. The results given in our further discussion are taken from the latter book.

moments is obtained by subtracting from the total curvature $-\partial^2 w/\partial x^2$ the portion $-\partial(kQ_x/hG)/\partial x$ produced by the shearing force.[1] Substituting

$$-\frac{\partial^2 w}{\partial x^2} + \frac{\partial\left(\frac{kQ_x}{hG}\right)}{\partial x}$$

and $-(\partial^2 w/\partial y^2) + \partial(kQ_y/hG)/\partial y$ for $-\partial^2 w/\partial x^2$ and $-\partial^2 w/\partial y^2$ in Eqs. (101) and using the last two equations of the system (123), we find for the bending moments the expressions

$$M_x = -D\left(\frac{\partial^2 w}{\partial x^2} + \nu\frac{\partial^2 w}{\partial y^2}\right) + \frac{kDh^2}{6}\frac{\partial^2}{\partial y^2}\Delta w$$

$$M_y = -D\left(\frac{\partial^2 w}{\partial y^2} + \nu\frac{\partial^2 w}{\partial x^2}\right) + \frac{kDh^2}{6}\frac{\partial^2}{\partial x^2}\Delta w$$

These equations coincide with the first two equations of the system (123) if we take

$$\frac{k}{6} = \frac{8+\nu}{40}$$

For $\nu = 0.3$ this gives $k = 1.245$.

From the theory of bending of beams we know that the correction due to the action of the shearing force is small and can be neglected if the depth h is small in comparison with the span of the beam. The same conclusion also holds in the case of plates.

The exact expressions for stress components are

$$\sigma_x = -\frac{Ez}{1-\nu^2}\left(\frac{\partial^2 w}{\partial x^2} + \nu\frac{\partial^2 w}{\partial y^2}\right) + \frac{E}{1-\nu^2}\left(\frac{h^2 z}{4} - \frac{2-\nu}{6}z^3\right)\frac{\partial^2}{\partial y^2}\Delta w$$

$$\sigma_y = -\frac{Ez}{1-\nu^2}\left(\frac{\partial^2 w}{\partial y^2} + \nu\frac{\partial^2 w}{\partial x^2}\right) + \frac{E}{1-\nu^2}\left(\frac{h^2 z}{4} - \frac{2-\nu}{6}z^3\right)\frac{\partial^2}{\partial x^2}\Delta w \quad (r)$$

$$\tau_{xy} = -\frac{Ez}{1+\nu}\frac{\partial^2 w}{\partial x\,\partial y} - \frac{E}{1-\nu^2}\left(\frac{h^2 z}{4} - \frac{2-\nu}{6}z^3\right)\frac{\partial^2}{\partial x\,\partial y}\Delta w$$

$$\tau_{xz} = -\frac{E(h^2 - 4z^2)}{8(1-\nu^2)}\frac{\partial}{\partial x}\Delta w \qquad \tau_{yz} = -\frac{E(h^2 - 4z^2)}{8(1-\nu^2)}\frac{\partial}{\partial y}\Delta w \qquad \sigma_z = 0$$

The second terms on the right-hand sides of the equations for σ_x, σ_y, and τ_{xy} are the corrections due to the effect of shearing forces on bending. It is seen that the stresses σ_x, σ_y, and τ_{xy} are no longer proportional to the distance z from the middle plane but contain a term proportional to z^3. Shearing stresses τ_{xz} and τ_{yz} vary according to the same parabolic law as for rectangular beams. In the case of a plane stress distribution, Δw is a constant, and formulas (r) coincide with those given by the approximate theory.

The problem of a uniformly loaded plate can also be treated rigorously in the same way. Thus it can be shown that the general expression for deflections in this case is obtained by adding to expression (q) the term

$$\frac{1}{64}\frac{q}{D}(x^2 + y^2)\left(x^2 + y^2 - \frac{2h^2}{1-\nu}\right) \qquad (s)$$

---

[1] k is a numerical factor that in the case of beams depends on the shape of the cross section.

which again satisfies Eq. (103) of the approximate theory. The equations for bending moments do not coincide with Eqs. (101) of the approximate theory but contain some additional correction terms. If the thickness of the plate is small in comparison with the other dimensions, these terms are small and can be neglected.

In all previous cases general solutions of plate bending problems were discussed without considering the boundary conditions. There are also rigorous solutions of several problems in which boundary conditions are considered.[1] These solutions indicate that, provided the plate can be considered "thin," the customary theory is accurate enough for practical purposes except (1) in the vicinity of a highly concentrated transverse load and (2) in narrow edge zones, especially near the corners of plates and around holes with a diameter of the order of magnitude of the plate thickness itself.

In the first of these two cases the stress components σ_z and the transverse shearing stresses must be considered equally important in their effect on the deformation of the plate. In obtaining the necessary correction to the stresses given by the approximate theory (see page 70) the boundary conditions can be eliminated from consideration. In such circumstances the thick-plate theory proves most convenient for the solution of the problem.

In the second case the effect of the stress components σ_z on the deformation becomes secondary as compared with the effect of the transverse shearing stresses τ_{xz} and τ_{yz}. Primarily taking into account this latter effect, several modified thin-plate theories have been developed recently (see Art. 39). These theories are better suited for the analysis of the stress distribution in the edge zone of the plates than the more rigorous thick-plate theory.

---

[1] In recent times the rigorous theory of plates has attracted the interest of engineers, and several important papers in this field have been published. We shall mention here the following: S. Woinowsky-Krieger, *Ingr.-Arch.*, vol. 4, pp. 203 and 305, 1933. B. Galerkin, *Compt. rend.*, vol. 190, p. 1047; vol. 193, p. 568; vol. 194, p. 1440. G. D. Birkhoff, *Phil. Mag.*, vol. 43, p. 953, 1922. C. A. Garabedian, *Trans. Am. Math. Soc.*, vol. 25, p. 343, 1923; *Compt. rend.*, vols. 178 (1924), 180 (1925), 186 (1928), 195 (1932). R. Archie Higdon and D. L. Holl, *Duke Math. J.*, vol. 3, p. 18, 1937. A. C. Stevenson, *Phil. Mag.*, ser. 7, vol. 33, p. 639, 1942; R. Ohlig, *Ingr.-Arch.*, vol. 13, p. 155, 1942; I. N. Sneddon, *Proc. Cambridge Phil. Soc.*, vol. 42, p. 260, 1946; L. Leibenson, "Works," vol. 1, p. 111, Moscow, 1951; H. Jung, *Z. angew. Math. Mech.*, vol. 32, p. 57, 1952; E. Koppe, *Z. angew. Math. Mech.*, vol. 37, p. 38, 1957. For thermal stresses see K. Marguerre, *Z. angew. Math. Mech.*, vol. 15, p. 369, 1935; and I. S. Sokolnikoff and E. S. Sokolnikoff, *Trans. Am. Math. Soc.*, vol. 45, p. 235, 1939.

CHAPTER 5

SIMPLY SUPPORTED RECTANGULAR PLATES

27. Simply Supported Rectangular Plates under Sinusoidal Load. Taking the coordinate axes as shown in Fig. 59, we assume that the load distributed over the surface of the plate is given by the expression

$$q = q_0 \sin \frac{\pi x}{a} \sin \frac{\pi y}{b} \qquad (a)$$

in which q_0 represents the intensity of the load at the center of the plate. The differential equation (103) for the deflection surface in this case becomes

$$\frac{\partial^4 w}{\partial x^4} + 2 \frac{\partial^4 w}{\partial x^2 \partial y^2} + \frac{\partial^4 w}{\partial y^4} = \frac{q_0}{D} \sin \frac{\pi x}{a} \sin \frac{\pi y}{b} \qquad (b)$$

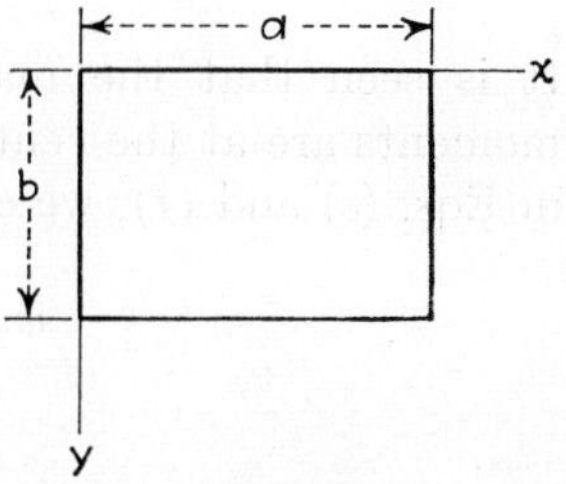

The boundary conditions for simply supported edges are

$$w = 0 \quad M_x = 0 \quad \text{for } x = 0 \text{ and } x = a$$
$$w = 0 \quad M_y = 0 \quad \text{for } y = 0 \text{ and } y = b$$

Fig. 59

Using expression (101) for bending moments and observing that, since $w = 0$ at the edges, $\partial^2 w/\partial x^2 = 0$ and $\partial^2 w/\partial y^2 = 0$ for the edges parallel to the x and y axes, respectively, we can represent the boundary conditions in the following form:

$$\begin{array}{llll} (1) \ w = 0 & (2) \ \dfrac{\partial^2 w}{\partial x^2} = 0 & \text{for } x = 0 \text{ and } x = a \\ (3) \ w = 0 & (4) \ \dfrac{\partial^2 w}{\partial y^2} = 0 & \text{for } y = 0 \text{ and } y = b \end{array} \qquad (c)$$

It may be seen that all boundary conditions are satisfied if we take for deflections the expression

$$w = C \sin \frac{\pi x}{a} \sin \frac{\pi y}{b} \qquad (d)$$

in which the constant C must be chosen so as to satisfy Eq. (b). Substituting expression (d) into Eq. (b), we find

$$\pi^4 \left(\frac{1}{a^2} + \frac{1}{b^2} \right)^2 C = \frac{q_0}{D}$$

and we conclude that the deflection surface satisfying Eq. (b) and boundary conditions (c) is

$$w = \frac{q_0}{\pi^4 D \left(\frac{1}{a^2} + \frac{1}{b^2}\right)^2} \sin \frac{\pi x}{a} \sin \frac{\pi y}{b} \quad (e)$$

Having this expression and using Eqs. (101) and (102), we find

$$M_x = \frac{q_0}{\pi^2 \left(\frac{1}{a^2} + \frac{1}{b^2}\right)^2} \left(\frac{1}{a^2} + \frac{\nu}{b^2}\right) \sin \frac{\pi x}{a} \sin \frac{\pi y}{b}$$

$$M_y = \frac{q_0}{\pi^2 \left(\frac{1}{a^2} + \frac{1}{b^2}\right)^2} \left(\frac{\nu}{a^2} + \frac{1}{b^2}\right) \sin \frac{\pi x}{a} \sin \frac{\pi y}{b} \quad (f)$$

$$M_{xy} = \frac{q_0(1-\nu)}{\pi^2 \left(\frac{1}{a^2} + \frac{1}{b^2}\right)^2 ab} \cos \frac{\pi x}{a} \cos \frac{\pi y}{b}$$

It is seen that the maximum deflection and the maximum bending moments are at the center of the plate. Substituting $x = a/2$, $y = b/2$ in Eqs. (e) and (f), we obtain

$$w_{\max} = \frac{q_0}{\pi^4 D \left(\frac{1}{a^2} + \frac{1}{b^2}\right)^2} \quad (124)$$

$$(M_x)_{\max} = \frac{q_0}{\pi^2 \left(\frac{1}{a^2} + \frac{1}{b^2}\right)^2} \left(\frac{1}{a^2} + \frac{\nu}{b^2}\right)$$

$$(M_y)_{\max} = \frac{q_0}{\pi^2 \left(\frac{1}{a^2} + \frac{1}{b^2}\right)^2} \left(\frac{\nu}{a^2} + \frac{1}{b^2}\right) \quad (125)$$

In the particular case of a square plate, $a = b$, and the foregoing formulas become

$$w_{\max} = \frac{q_0 a^4}{4\pi^4 D} \qquad (M_x)_{\max} = (M_y)_{\max} = \frac{(1+\nu)q_0 a^2}{4\pi^2} \quad (126)$$

We use Eqs. (106) and (107) to calculate the shearing forces and obtain

$$Q_x = \frac{q_0}{\pi a \left(\frac{1}{a^2} + \frac{1}{b^2}\right)} \cos \frac{\pi x}{a} \sin \frac{\pi y}{b}$$

$$Q_y = \frac{q_0}{\pi b \left(\frac{1}{a^2} + \frac{1}{b^2}\right)} \sin \frac{\pi x}{a} \cos \frac{\pi y}{b} \quad (g)$$

To find the reactive forces at the supported edges of the plate we proceed as was explained in Art. 22. For the edge $x = a$ we find

$$V_x = \left(Q_x - \frac{\partial M_{xy}}{\partial y}\right)_{x=a} = -\frac{q_0}{\pi a \left(\frac{1}{a^2} + \frac{1}{b^2}\right)^2}\left(\frac{1}{a^2} + \frac{2-\nu}{b^2}\right)\sin\frac{\pi y}{b} \quad (h)$$

In the same manner, for the edge $y = b$,

$$V_y = \left(Q_y - \frac{\partial M_{xy}}{\partial x}\right)_{y=b} = -\frac{q_0}{\pi b \left(\frac{1}{a^2} + \frac{1}{b^2}\right)^2}\left(\frac{1}{b^2} + \frac{2-\nu}{a^2}\right)\sin\frac{\pi x}{a} \quad (i)$$

Hence the pressure distribution follows a sinusoidal law. The minus sign indicates that the reactions on the plate act upward. From symmetry it may be concluded that formulas (h) and (i) also represent pressure distributions along the sides $x = 0$ and $y = 0$, respectively. The resultant of distributed pressures is

$$\frac{2q_0}{\pi \left(\frac{1}{a^2} + \frac{1}{b^2}\right)^2}\left[\frac{1}{a}\left(\frac{1}{a^2} + \frac{2-\nu}{b^2}\right)\int_0^b \sin\frac{\pi y}{b}\,dy\right.$$

$$\left. + \frac{1}{b}\left(\frac{1}{b^2} + \frac{2-\nu}{a^2}\right)\int_0^a \sin\frac{\pi x}{a}\,dx\right] = \frac{4q_0 ab}{\pi^2} + \frac{8q_0(1-\nu)}{\pi^2 ab\left(\frac{1}{a^2} + \frac{1}{b^2}\right)^2} \quad (j)$$

Observing that

$$\frac{4q_0 ab}{\pi^2} = \int_0^a \int_0^b q_0 \sin\frac{\pi x}{a} \sin\frac{\pi y}{b}\,dx\,dy \quad (k)$$

it can be concluded that the sum of the distributed reactions is larger than the total load on the plate given by expression (k). This result can be easily explained if we note that, proceeding as described in Art. 22, we obtain not only the distributed reactions but also reactions concentrated at the corners of the plate. These concentrated reactions are equal, from symmetry; and their magnitude, as may be seen from Fig. 51, is

$$R = 2(M_{xy})_{x=a,y=b} = \frac{2q_0(1-\nu)}{\pi^2 ab\left(\frac{1}{a^2} + \frac{1}{b^2}\right)^2} \quad (l)$$

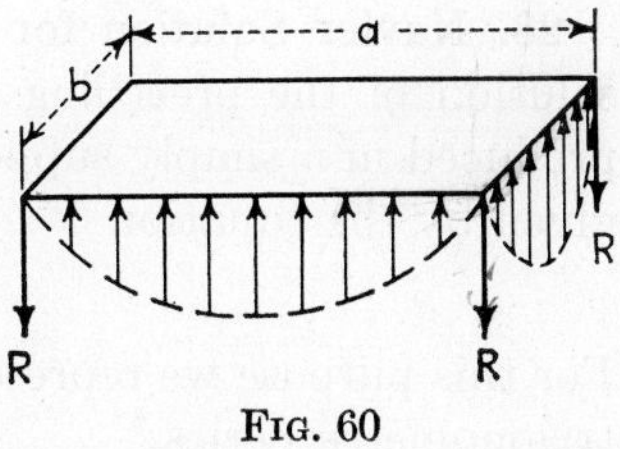

Fig. 60

The positive sign indicates that the reactions act downward. Their sum is exactly equal to the second term in expression (j). The distributed and the concentrated reactions which act on the plate and keep the load, defined by Eq. (a), in equilibrium are shown graphically in Fig. 60. It may be seen that the corners of the plate have a tendency to rise up under the action of the applied load and that the concentrated forces R must be applied to prevent this.

The maximum bending stress is at the center of the plate. Assuming that $a > b$, we find that at the center $M_y > M_x$. Hence the maximum bending stress is

$$(\sigma_y)_{\max} = \frac{6(M_y)_{\max}}{h^2} = \frac{6q_0}{\pi^2 h^2 \left(\dfrac{1}{a^2} + \dfrac{1}{b^2}\right)^2} \left(\frac{\nu}{a^2} + \frac{1}{b^2}\right)$$

The maximum shearing stress will be at the middle of the longer sides of the plate. Observing that the total transverse force $V_y = Q_y - \dfrac{\partial M_{xy}}{\partial x}$ is distributed along the thickness of the plate according to the parabolic law and using Eq. (i), we obtain

$$(\tau_{yz})_{\max} = \frac{3q_0}{2\pi bh \left(\dfrac{1}{a^2} + \dfrac{1}{b^2}\right)^2} \left(\frac{1}{b^2} + \frac{2-\nu}{a^2}\right)$$

If the sinusoidal load distribution is given by the equation

$$q = q_0 \sin \frac{m\pi x}{a} \sin \frac{n\pi y}{b} \qquad (m)$$

where m and n are integer numbers, we can proceed as before, and we shall obtain for the deflection surface the following expression:

$$w = \frac{q_0}{\pi^4 D \left(\dfrac{m^2}{a^2} + \dfrac{n^2}{b^2}\right)^2} \sin \frac{m\pi x}{a} \sin \frac{n\pi y}{b} \qquad (127)$$

from which the expressions for bending and twisting moments can be readily obtained by differentiation.

28. Navier Solution for Simply Supported Rectangular Plates. The solution of the preceding article can be used in calculating deflections produced in a simply supported rectangular plate by any kind of loading given by the equation

$$q = f(x,y) \qquad (a)$$

For this purpose we represent the function $f(x,y)$ in the form of a double trigonometric series:[1]

$$f(x,y) = \sum_{m=1}^{\infty} \sum_{n=1}^{\infty} a_{mn} \sin \frac{m\pi x}{a} \sin \frac{n\pi y}{b} \qquad (128)$$

[1] The first solution of the problem of bending of simply supported rectangular plates and the use for this purpose of double trigonometric series are due to Navier, who

To calculate any particular coefficient $a_{m'n'}$ of this series we multiply both sides of Eq. (128) by $\sin(n'\pi y/b)\, dy$ and integrate from 0 to b. Observing that

$$\int_0^b \sin\frac{n\pi y}{b}\sin\frac{n'\pi y}{b}\, dy = 0 \quad \text{when } n \neq n'$$

$$\int_0^b \sin\frac{n\pi y}{b}\sin\frac{n'\pi y}{b}\, dy = \frac{b}{2} \quad \text{when } n = n'$$

we find in this way

$$\int_0^b f(x,y)\sin\frac{n'\pi y}{b}\, dy = \frac{b}{2}\sum_{m=1}^{\infty} a_{mn'}\sin\frac{m\pi x}{a} \qquad (b)$$

Multiplying both sides of Eq. (b) by $\sin(m'\pi x/a)\, dx$ and integrating from 0 to a, we obtain

$$\int_0^a \int_0^b f(x,y)\sin\frac{m'\pi x}{a}\sin\frac{n'\pi y}{b}\, dx\, dy = \frac{ab}{4} a_{m'n'}$$

from which

$$a_{m'n'} = \frac{4}{ab}\int_0^a \int_0^b f(x,y)\sin\frac{m'\pi x}{a}\sin\frac{n'\pi y}{b}\, dx\, dy \qquad (129)$$

Performing the integration indicated in expression (129) for a given load distribution, *i.e.*, for a given $f(x,y)$, we find the coefficients of series (128) and represent in this way the given load as a sum of partial sinusoidal loadings. The deflection produced by each partial loading was discussed in the preceding article, and the total deflection will be obtained by summation of such terms as are given by Eq. (127). Hence we find

$$w = \frac{1}{\pi^4 D}\sum_{m=1}^{\infty}\sum_{n=1}^{\infty}\frac{a_{mn}}{\left(\frac{m^2}{a^2}+\frac{n^2}{b^2}\right)^2}\sin\frac{m\pi x}{a}\sin\frac{n\pi y}{b} \qquad (130)$$

Take the case of a load uniformly distributed over the entire surface of the plate as an example of the application of the general solution (130). In such a case

$$f(x,y) = q_0$$

where q_0 is the intensity of the uniformly distributed load. From formula (129) we obtain

$$a_{mn} = \frac{4q_0}{ab}\int_0^a \int_0^b \sin\frac{m\pi x}{a}\sin\frac{n\pi y}{b}\, dx\, dy = \frac{16q_0}{\pi^2 mn} \qquad (c)$$

---

presented a paper on this subject to the French Academy in 1820. The abstract of the paper was published in *Bull. soc. phil.-math.*, Paris, 1823. The manuscript is in the library of l'École des Ponts et Chaussées.

where m and n are odd integers. If m or n or both of them are even numbers, $a_{mn} = 0$. Substituting in Eq. (130), we find

$$w = \frac{16q_0}{\pi^6 D} \sum_{m=1}^{\infty} \sum_{n=1}^{\infty} \frac{\sin \frac{m\pi x}{a} \sin \frac{n\pi y}{b}}{mn \left(\frac{m^2}{a^2} + \frac{n^2}{b^2}\right)^2} \tag{131}$$

where $m = 1, 3, 5, \ldots$ and $n = 1, 3, 5, \ldots$.

In the case of a uniform load we have a deflection surface symmetrical with respect to the axes $x = a/2$, $y = b/2$; and quite naturally all terms with even numbers for m or n in series (131) vanish, since they are unsymmetrical with respect to the above-mentioned axes. The maximum deflection of the plate is at its center and is found by substituting $x = a/2$, $y = b/2$ in formula (131), giving

$$w_{\max} = \frac{16q_0}{\pi^6 D} \sum_{m=1}^{\infty} \sum_{n=1}^{\infty} \frac{(-1)^{\frac{m+n}{2}-1}}{mn \left(\frac{m^2}{a^2} + \frac{n^2}{b^2}\right)^2} \tag{132}$$

This is a rapidly converging series, and a satisfactory approximation is obtained by taking only the first term of the series, which, for example, in the case of a square plate gives

$$w_{\max} = \frac{4q_0 a^4}{\pi^6 D} = 0.00416 \frac{q_0 a^4}{D}$$

or, by substituting expression (3) for D and assuming $\nu = 0.3$,

$$w_{\max} = 0.0454 \frac{q_0 a^4}{E h^3}$$

This result is about $2\frac{1}{2}$ per cent in error (see Table 8).

From expression (132) it may be seen that the deflections of two plates that have the same thickness and the same value of the ratio a/b increase as the fourth power of the length of the sides.

The expressions for bending and twisting moments can be obtained from the general solution (131) by using Eqs. (101) and (102). The series obtained in this way are not so rapidly convergent as series (131), and in the further discussion (see Art. 30) another form of solution will be given, more suitable for numerical calculations. Since the moments are expressed by the second derivatives of series (131), their maximum values, if we keep q_0 and D the same, are proportional to the square of linear dimensions. Since the total load on the plate, equal to $q_0 ab$, is also proportional to the square of the linear dimensions, we conclude that, for two plates of equal thickness and of the same value of the ratio a/b, the maximum bending moments and hence the maximum stresses are equal if the total loads on the two plates are equal.[1]

29. Further Applications of the Navier Solution. From the discussion in the preceding article it is seen that the deflection of a simply supported rectangular plate (Fig. 59) can always be represented in the form of a double trigonometric series (130), the coefficients a_{mn} being given by Eq. (129).

Let us apply this result in the case of a single load P uniformly distributed over the area of the rectangle shown in Fig. 61. By virtue of Eq. (129) we have

$$a_{mn} = \frac{4P}{abuv} \int_{\xi-u/2}^{\xi+u/2} \int_{\eta-v/2}^{\eta+v/2} \sin \frac{m\pi x}{a} \sin \frac{n\pi y}{b} \, dx \, dy$$

or

$$a_{mn} = \frac{16P}{\pi^2 mnuv} \sin \frac{m\pi\xi}{a} \sin \frac{n\pi\eta}{b} \sin \frac{m\pi u}{2a} \sin \frac{n\pi v}{2b} \quad (a)$$

If, in particular, $\xi = a/2$, $\eta = b/2$, $u = a$, and $v = b$, Eq. (a) yields the expression (c) obtained in Art. 28 for the uniformly loaded plate.

Another case of practical interest is a single load concentrated at any given point $x = \xi$, $y = \eta$ of the plate. Using Eq. (a) and letting u and v tend to zero we arrive at the expression

$$a_{mn} = \frac{4P}{ab} \sin \frac{m\pi\xi}{a} \sin \frac{n\pi\eta}{b} \quad (b)$$

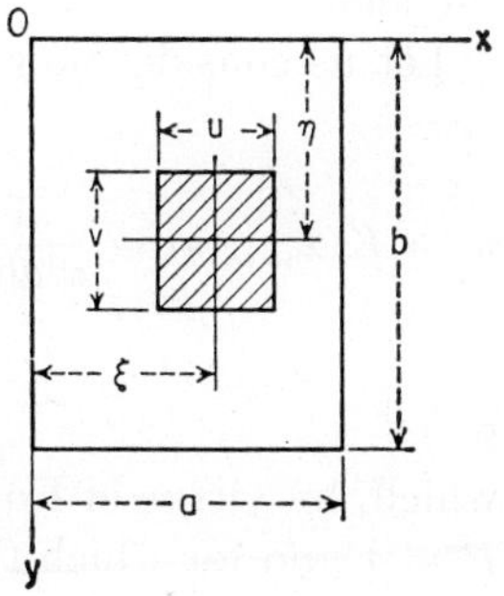

and, by Eq. (130), at the deflection

$$w = \frac{4P}{\pi^4 ab D} \sum_{m=1}^{\infty} \sum_{n=1}^{\infty} \frac{\sin \frac{m\pi\xi}{a} \sin \frac{n\pi\eta}{b}}{\left(\frac{m^2}{a^2} + \frac{n^2}{b^2}\right)^2} \sin \frac{m\pi x}{a} \sin \frac{n\pi y}{b} \quad (133)$$

Fig. 61

The series converges rapidly, and we can obtain the deflection at any point of the plate with sufficient accuracy by taking only the first few terms of the series. Let us, for example, calculate the deflection at the middle when the load is applied at the middle as well. Then we have $\xi = x = a/2$, $\eta = y = b/2$, and the series (133) yields

$$w_{\max} = \frac{4P}{\pi^4 ab D} \sum_{m=1}^{\infty} \sum_{n=1}^{\infty} \frac{1}{\left(\frac{m^2}{a^2} + \frac{n^2}{b^2}\right)^2} \quad (c)$$

---

[1] This conclusion was established by Mariotte in the paper "Traité du mouvement des eaux," published in 1686. See Mariotte's scientific papers, new ed., vol. 2, p. 467, 1740.

where $m = 1, 3, 5, \ldots$ and $n = 1, 3, 5, \ldots$. In the case of a square plate, expression (c) becomes

$$w_{\max} = \frac{4Pa^2}{\pi^4 D} \sum_{m=1}^{\infty} \sum_{n=1}^{\infty} \frac{1}{(m^2 + n^2)^2}$$

Taking the first four terms of the series we find that

$$w_{\max} = \frac{0.01121 Pa^2}{D}$$

which is about $3\frac{1}{2}$ per cent less than the correct value (see Table 23, page 143).

As for the series (128) representing the intensity of the concentrated load it is divergent at $x = \xi$, $y = \eta$, and so also are the series expressing the bending moments and shearing forces at the point of application of the load.

Let us consider now the expression

$$w = K(x,y,\xi,\eta) = \frac{4}{\pi^4 ab D} \sum_{m=1}^{\infty} \sum_{n=1}^{\infty} \frac{\sin \frac{m\pi\xi}{a} \sin \frac{n\pi\eta}{b} \sin \frac{m\pi x}{a} \sin \frac{n\pi y}{b}}{\left(\frac{m^2}{a^2} + \frac{n^2}{b^2}\right)^2} \quad (134)$$

which, by virtue of Eq. (132), represents the deflection due to a unit load $P = 1$ and for which the notation $K(x,y,\xi,\eta)$ is introduced for brevity.

Regarding x and y as the variables, $w = K(x,y,\xi,\eta)$ is the equation of the elastic surface of the plate submitted to a unit load at a fixed point $x = \xi$, $y = \eta$. Now considering ξ and η as variable, Eq. (134) defines the influence surface for the deflection of the plate at a fixed point x, y, the position of the traveling unit load being given by ξ and η. If, therefore, some load of intensity $f(\xi,\eta)$ distributed over an area A is given, the corresponding deflection at any point of the plate may easily be obtained. In fact, applying an elementary load $f(\xi,\eta) \, d\xi \, d\eta$ at $x = \xi$, $y = \eta$ and using the principle of superposition, we arrive at the deflection

$$w = \iint_A f(\xi,\eta) K(x,y,\xi,\eta) \, d\xi \, d\eta \quad (135)$$

the double integral being extended over the loaded area and $K(x,y,\xi,\eta)$ being given by Eq. (134).

The function $K(x,y,\xi,\eta)$ is sometimes called Green's function of the plate. When given as by Eq. (134), this function is associated with the boundary conditions of the simply supported rectangular plate. Many properties of Green's function, however, are independent of those restrictions. An example is the property of symmetry,

expressed by the relation
$$K(x,y,\xi,\eta) = K(\xi,\eta,x,y)$$
which follows from the well-known reciprocal theorem of Maxwell[1] and is easy to verify in the particular case of the function (134).

As the last example in the application of Navier's solution let us consider the case of as ingle load P uniformly distributed over the area of a circle with radius c and with center at $x = \xi$, $y = \eta$. Introducing polar coordinates ρ, θ with the origin at the center of the loaded area and replacing the elementary area $dx\,dy$ in Eq. (129) by the area $\rho\,d\rho\,d\theta$, we have, by this latter equation,

$$a_{mn} = \frac{4}{ab}\frac{P}{\pi c^2}\int_0^c\int_0^{2\pi}\sin\frac{m\pi(\xi+\rho\cos\theta)}{a}\sin\frac{n\pi(\eta+\rho\sin\theta)}{b}\rho\,d\rho\,d\theta \qquad (d)$$

Provided that the circle $\rho = c$ remains entirely inside the boundary of the plate the evaluation of the integral (d) gives the expression[2]

$$a_{mn} = \frac{8P}{abc\gamma_{mn}}J_1(\gamma_{mn}c)\sin\frac{m\pi\xi}{a}\sin\frac{n\pi\eta}{b} \qquad (e)$$

in which $\gamma_{mn} = \pi\sqrt{(m/a)^2 + (n/b)^2}$ and $J_1(\gamma_{mn}c)$ is the Bessel function of order one, with the argument $\gamma_{mn}c$. The required deflection now is obtainable by substitution of the expression (e) into Eq. (130).

It is seen that the form of the Navier solution remains simple even in relatively complex cases of load distribution. On the other hand, the double series of this solution are not convenient for numerical computation especially if higher derivatives of the function w are involved. So, another form of solution for the bending of the rectangular plate, more suitable for this purpose, will be discussed below.

30. Alternate Solution for Simply Supported and Uniformly Loaded Rectangular Plates. In discussing problems of bending of rectangular plates that have two opposite edges simply supported, M. Lévy[3] suggested taking the solution in the form of a series

$$w = \sum_{m=1}^{\infty} Y_m \sin\frac{m\pi x}{a} \qquad (136)$$

where Y_m is a function of y only. It is assumed that the sides $x = 0$ and $x = a$ (Fig. 62) are simply supported. Hence each term of series (136) satisfies the boundary conditions $w = 0$ and $\partial^2 w/\partial x^2 = 0$ on these two sides. It remains to determine Y_m in such a form as to satisfy the bound-

---

[1] See, for instance, S. Timoshenko and D. H. Young, "Theory of Structures," p. 250, 1945.

[2] See S. Woinowsky-Krieger, *Ingr.-Arch.*, vol. 3, p. 240, 1932.

[3] See *Compt. rend.*, vol. 129, pp. 535–539, 1899. The solution was applied to several particular cases of bending of rectangular plates by E. Estanave, "Thèses," Paris, 1900; in this paper the transformation of the double series of the Navier solution to the simple series of M. Lévy is shown.

ary conditions on the sides $y = \pm b/2$ and also the equation of the deflection surface

$$\frac{\partial^4 w}{\partial x^4} + 2 \frac{\partial^4 w}{\partial x^2 \, \partial y^2} + \frac{\partial^4 w}{\partial y^4} = \frac{q}{D} \qquad (a)$$

In applying this method to uniformly loaded and simply supported rectangular plates, a further simplification can be made by taking the solution of Eq. (a) in the form[1]

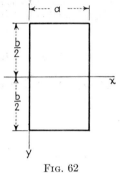

FIG. 62

$$w = w_1 + w_2 \qquad (b)$$

and letting

$$w_1 = \frac{q}{24D}(x^4 - 2ax^3 + a^3 x) \qquad (c)$$

i.e., w_1 represents the deflection of a uniformly loaded strip parallel to the x axis. It satisfies Eq. (a) and also the boundary conditions at the edges $x = 0$ and $x = a$.

The expression w_2 evidently has to satisfy the equation

$$\frac{\partial^4 w_2}{\partial x^4} + 2 \frac{\partial^4 w_2}{\partial x^2 \, \partial y^2} + \frac{\partial^4 w_2}{\partial y^4} = 0 \qquad (137)$$

and must be chosen in such a manner as to make the sum (b) satisfy all boundary conditions of the plate. Taking w_2 in the form of the series (136) in which, from symmetry, $m = 1, 3, 5, \ldots$ and substituting it into Eq. (137), we obtain

$$\sum_{m=1}^{\infty}\left(Y_m^{\mathrm{IV}} - 2\frac{m^2\pi^2}{a^2}Y_m'' + \frac{m^4\pi^4}{a^4}Y_m\right)\sin\frac{m\pi x}{a} = 0$$

This equation can be satisfied for all values of x only if the function Y_m satisfies the equation

$$Y_m^{\mathrm{IV}} - 2\frac{m^2\pi^2}{a^2}Y_m'' + \frac{m^4\pi^4}{a^4}Y_m = 0 \qquad (d)$$

The general integral of this equation can be taken in the form[2]

$$Y_m = \frac{qa^4}{D}\left(A_m \cosh\frac{m\pi y}{a} + B_m \frac{m\pi y}{a}\sinh\frac{m\pi y}{a} \right.$$
$$\left. + C_m \sinh\frac{m\pi y}{a} + D_m \frac{m\pi y}{a}\cosh\frac{m\pi y}{a}\right) \qquad (138)$$

[1] This form of solution was used by A. Nádai, *Forschungsarb.*, nos. 170 and 171, Berlin, 1915; see also his book "Elastische Platten," Berlin, 1925.

[2] A somewhat different form for Y_m, more convenient to satisfy some particular boundary conditions, has been suggested by P. F. Papkovitch, *Priklad. Mat. Mekh.*, vol. 5, 1941.

SIMPLY SUPPORTED RECTANGULAR PLATES

Observing that the deflection surface of the plate is symmetrical with respect to the x axis (Fig. 62), we keep in the expression (138) only even functions of y and let the integration constants $C_m = D_m = 0$.

The deflection surface (b) is then represented by the following expression:

$$w = \frac{q}{24D}(x^4 - 2ax^3 + a^3 x)$$
$$+ \frac{qa^4}{D}\sum_{m=1}^{\infty}\left(A_m \cosh\frac{m\pi y}{a} + B_m \frac{m\pi y}{a}\sinh\frac{m\pi y}{a}\right)\sin\frac{m\pi x}{a} \quad (e)$$

which satisfies Eq. (a) and also the boundary conditions at the sides $x = 0$ and $x = a$. It remains now to adjust the constants of integration A_m and B_m in such a manner as to satisfy the boundary conditions

$$w = 0 \qquad \frac{\partial^2 w}{\partial y^2} = 0 \quad (f)$$

on the sides $y = \pm b/2$. We begin by developing expression (c) in a trigonometric series, which gives[1]

$$\frac{q}{24D}(x^4 - 2ax^3 + a^3 x) = \frac{4qa^4}{\pi^5 D}\sum_{m=1}^{\infty}\frac{1}{m^5}\sin\frac{m\pi x}{a}$$

where $m = 1, 3, 5, \ldots$. The deflection surface (e) will now be represented in the form

$$w = \frac{qa^4}{D}\sum_{m=1}^{\infty}\left(\frac{4}{\pi^5 m^5} + A_m \cosh\frac{m\pi y}{a} + B_m \frac{m\pi y}{a}\sinh\frac{m\pi y}{a}\right)\sin\frac{m\pi x}{a} \quad (g)$$

where $m = 1, 3, 5, \ldots$. Substituting this expression in the boundary conditions (f) and using the notation

$$\frac{m\pi b}{2a} = \alpha_m \quad (h)$$

we obtain the following equations for determining the constants A_m and B_m:

$$\frac{4}{\pi^5 m^5} + A_m \cosh\alpha_m + \alpha_m B_m \sinh\alpha_m = 0$$
$$(A_m + 2B_m)\cosh\alpha_m + \alpha_m B_m \sinh\alpha_m = 0$$

from which

$$A_m = -\frac{2(\alpha_m \tanh\alpha_m + 2)}{\pi^5 m^5 \cosh\alpha_m} \qquad B_m = \frac{2}{\pi^5 m^5 \cosh\alpha_m} \quad (i)$$

---
[1] See S. Timoshenko, "Strength of Materials," 3d ed., part II, p. 50, 1956.

Substituting these values of the constants in Eq. (g), we obtain the deflection surface of the plate, satisfying Eq. (a) and the boundary conditions, in the following form:

$$w = \frac{4qa^4}{\pi^5 D} \sum_{m=1,3,5,\ldots}^{\infty} \frac{1}{m^5} \left(1 - \frac{\alpha_m \tanh \alpha_m + 2}{2 \cosh \alpha_m} \cosh \frac{2\alpha_m y}{b} \right.$$

$$\left. + \frac{\alpha_m}{2 \cosh \alpha_m} \frac{2y}{b} \sinh \frac{2\alpha_m y}{b} \right) \sin \frac{m\pi x}{a} \quad (139)$$

from which the deflection at any point can be calculated by using tables of hyperbolic functions.[1] The maximum deflection is obtained at the middle of the plate ($x = a/2$, $y = 0$), where

$$w_{\max} = \frac{4qa^4}{\pi^5 D} \sum_{m=1,3,5,\ldots}^{\infty} \frac{(-1)^{(m-1)/2}}{m^5} \left(1 - \frac{\alpha_m \tanh \alpha_m + 2}{2 \cosh \alpha_m}\right) \quad (j)$$

Disregarding the second term in the parentheses, this series represents the deflection of the middle of a uniformly loaded strip. Hence we can represent expression (j) in the following form:

$$w_{\max} = \frac{5}{384} \frac{qa^4}{D} - \frac{4qa^4}{\pi^5 D} \sum_{m=1,3,5,\ldots}^{\infty} \frac{(-1)^{(m-1)/2}}{m^5} \frac{\alpha_m \tanh \alpha_m + 2}{2 \cosh \alpha_m} \quad (140)$$

The series in this expression converges very rapidly,[2] and sufficient accuracy is obtained by taking only the first term. Taking a square plate as an example, we know from Eq. (h) that

$$\alpha_1 = \frac{\pi}{2} \qquad \alpha_3 = \frac{3\pi}{2} \qquad \cdots$$

and Eq. (140) gives

$$w_{\max} = \frac{5}{384} \frac{qa^4}{D} - \frac{4qa^4}{\pi^5 D} (0.68562 - 0.00025 + \cdots) = 0.00406 \frac{qa^4}{D}$$

It is seen that the second term of the series in parentheses is negligible

---

[1] See, for example, "Tables of Circular and Hyperbolic Sines and Cosines," 1939, and "Table of Circular and Hyperbolic Tangents and Cotangents," 1943, Columbia University Press, New York; also British Association for the Advancement of Science, "Mathematical Tables," 3d ed., vol. 1, Cambridge University Press, 1951; finally, F. Lösch, "Siebenstellige Tafeln der elementaren transzendenten Funktionen," Berlin, 1954.

[2] We assume that $b \geq a$, as in Fig. 62.

and that by taking only the first term the formula for deflection is obtained correct to three significant figures.

Making use of the formula (140), we can represent the maximum deflection of a plate in the form

$$w_{\max} = \alpha \frac{qa^4}{D} \qquad (141)$$

where α is a numerical factor depending on the ratio b/a of the sides of the plate. Values of α are given in Table 8 (page 120).

The bending moments M_x and M_y are calculated by means of expression (e). Substituting the algebraic portion of this expression in Eqs. (101), we find that

$$M'_x = \frac{qx(a-x)}{2} \qquad M'_y = \nu \frac{qx(a-x)}{2} \qquad (k)$$

The substitution of the series of expression (e) in the same equations gives

$$M''_x = (1-\nu)qa^2\pi^2 \sum_{m=1}^{\infty} m^2 \left[A_m \cosh \frac{m\pi y}{a} \right.$$
$$\left. + B_m \left(\frac{m\pi y}{a} \sinh \frac{m\pi y}{a} - \frac{2\nu}{1-\nu} \cosh \frac{m\pi y}{a} \right) \right] \sin \frac{m\pi x}{a} \qquad (l)$$

$$M''_y = -(1-\nu)qa^2\pi^2 \sum_{m=1}^{\infty} m^2 \left[A_m \cosh \frac{m\pi y}{a} \right.$$
$$\left. + B_m \left(\frac{m\pi y}{a} \sinh \frac{m\pi y}{a} + \frac{2}{1-\nu} \cosh \frac{m\pi y}{a} \right) \right] \sin \frac{m\pi x}{a}$$

The total bending moments are obtained by summation of expressions (k) and (l). Along the x axis the expression for the bending moments becomes

$$(M_x)_{y=0} = \frac{qx(a-x)}{2} - qa^2\pi^2 \sum_{m=1,3,5,\ldots}^{\infty} m^2[2\nu B_m - (1-\nu)A_m] \sin \frac{m\pi x}{a}$$

$$(M_y)_{y=0} = \nu \frac{qx(a-x)}{2} - qa^2\pi^2 \sum_{m=1,3,5,\ldots}^{\infty} m^2[2B_m + (1-\nu)A_m] \sin \frac{m\pi x}{a}$$

Both series converge rapidly and the moments can readily be computed and represented in the form

$$(M_x)_{y=0} = \beta' qa^2 \qquad (M_y)_{y=0} = \beta'_1 qa^2 \qquad (m)$$

The numerical values of the factors β' and β'_1 are given in Table 6.

The bending moments acting along the middle line $x = a/2$ can be computed in a similar manner and represented in the form

$$(M_x)_{x=a/2} = \beta'' q a^2 \qquad (M_y)_{x=a/2} = \beta_1'' q a^2 \qquad (n)$$

Values of β'' and β_1'' are given in Table 7.

The maximum values of these moments,

$$(M_x)_{\max} = \beta q a^2 \qquad (M_y)_{\max} = \beta_1 q a^2 \qquad (o)$$

are at the center of the plate ($x = a/2$, $y = 0$), and the corresponding factors β and β_1 are found in Table 8. The distribution of the moments in the particular case of a square plate is shown in Fig. 63.

TABLE 6. NUMERICAL FACTORS β' AND β_1' FOR BENDING MOMENTS OF SIMPLY SUPPORTED RECTANGULAR PLATES UNDER UNIFORM PRESSURE q
$\nu = 0.3$, $b \geq a$

b/a	$M_x = \beta' q a^2$, $y = 0$					$M_y = \beta_1' q a^2$, $y = 0$				
	$x = 0.1a$	$x = 0.2a$	$x = 0.3a$	$x = 0.4a$	$x = 0.5a$	$x = 0.1a$	$x = 0.2a$	$x = 0.3a$	$x = 0.4a$	$x = 0.5a$
1.0	0.0209	0.0343	0.0424	0.0466	0.0479	0.0168	0.0303	0.0400	0.0459	0.0479
1.1	0.0234	0.0389	0.0486	0.0541	0.0554	0.0172	0.0311	0.0412	0.0475	0.0493
1.2	0.0256	0.0432	0.0545	0.0607	0.0627	0.0174	0.0315	0.0417	0.0480	0.0501
1.3	0.0277	0.0472	0.0599	0.0671	0.0694	0.0175	0.0316	0.0419	0.0482	0.0503
1.4	0.0297	0.0509	0.0649	0.0730	0.0755	0.0175	0.0315	0.0418	0.0481	0.0502
1.5	0.0314	0.0544	0.0695	0.0783	0.0812	0.0173	0.0312	0.0415	0.0478	0.0498
1.6	0.0330	0.0572	0.0736	0.0831	0.0862	0.0171	0.0309	0.0411	0.0472	0.0492
1.7	0.0344	0.0599	0.0773	0.0874	0.0908	0.0169	0.0306	0.0405	0.0466	0.0486
1.8	0.0357	0.0623	0.0806	0.0913	0.0948	0.0167	0.0301	0.0399	0.0459	0.0479
1.9	0.0368	0.0644	0.0835	0.0948	0.0985	0.0165	0.0297	0.0393	0.0451	0.0471
2.0	0.0378	0.0663	0.0861	0.0978	0.1017	0.0162	0.0292	0.0387	0.0444	0.0464
2.5	0.0413	0.0729	0.0952	0.1085	0.1129	0.0152	0.0272	0.0359	0.0412	0.0430
3.0	0.0431	0.0763	0.1000	0.1142	0.1189	0.0145	0.0258	0.0340	0.0390	0.0406
4.0	0.0445	0.0791	0.1038	0.1185	0.1235	0.0138	0.0246	0.0322	0.0369	0.0384
∞	0.0450	0.0800	0.1050	0.1200	0.1250	0.0135	0.0240	0.0315	0.0360	0.0375

From Table 8 it is seen that, as the ratio b/a increases, the maximum deflection and the maximum moments of the plate rapidly approach the values calculated for a uniformly loaded strip or for a plate bent to a cylindrical surface obtained by making $b/a = \infty$. For $b/a = 3$ the difference between the deflection of the strip and the plate is about $6\frac{1}{2}$ per cent. For $b/a = 5$ this difference is less than $\frac{1}{2}$ per cent. The differences between the maximum bending moments for the same ratios of

TABLE 7. NUMERICAL FACTORS β'' AND β_1'' FOR BENDING MOMENTS OF SIMPLY SUPPORTED RECTANGULAR PLATES UNDER UNIFORM PRESSURE q
$\nu = 0.3, b \geq a$

b/a	$M_x = \beta'' q a^2, x = a/2$					$M_y = \beta_1'' q a^2, x = a/2$				
	$y = 0.4a$	$y = 0.3a$	$y = 0.2a$	$y = 0.1a$	$y = 0$	$y = 0.4a$	$y = 0.3a$	$y = 0.2a$	$y = 0.1a$	$y = 0$
1.0	0.0168	0.0303	0.0400	0.0459	0.0479	0.0209	0.0343	0.0424	0.0466	0.0479
1.1	0.0197	0.0353	0.0465	0.0532	0.0554	0.0225	0.0363	0.0442	0.0481	0.0493
1.2	0.0225	0.0401	0.0526	0.0600	0.0627	0.0239	0.0379	0.0454	0.0490	0.0501
1.3	0.0252	0.0447	0.0585	0.0667	0.0694	0.0252	0.0391	0.0462	0.0494	0.0503
1.4	0.0275	0.0491	0.0639	0.0727	0.0755	0.0263	0.0402	0.0468	0.0495	0.0502
1.5	0.0302	0.0532	0.0690	0.0781	0.0812	0.0275	0.0410	0.0470	0.0493	0.0498
1.6	0.0324	0.0571	0.0737	0.0832	0.0862	0.0288	0.0417	0.0471	0.0489	0.0492
1.7	0.0348	0.0607	0.0780	0.0877	0.0908	0.0295	0.0423	0.0470	0.0484	0.0486
1.8	0.0371	0.0641	0.0819	0.0917	0.0948	0.0304	0.0428	0.0469	0.0478	0.0479
1.9	0.0392	0.0673	0.0854	0.0953	0.0985	0.0314	0.0433	0.0467	0.0472	0.0471
2.0	0.0413	0.0703	0.0887	0.0986	0.1017	0.0322	0.0436	0.0464	0.0465	0.0464
2.5	0.0505	0.0828	0.1012	0.1102	0.1129	0.0360	0.0446	0.0447	0.0435	0.0430
3	0.0586	0.0923	0.1092	0.1168	0.1189	0.0389	0.0447	0.0431	0.0413	0.0406
4	0.0723	0.1054	0.1180	0.1224	0.1235	0.0426	0.0436	0.0406	0.0389	0.0384
∞	0.1250	0.1250	0.1250	0.1250	0.1250	0.0375	0.0375	0.0375	0.0375	0.0375

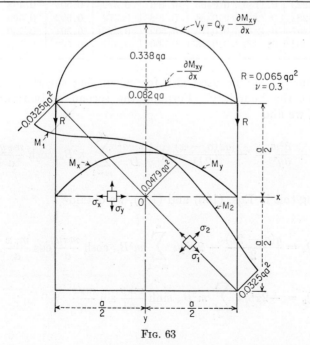

FIG. 63

b/a are 5 and $\frac{1}{3}$ per cent, respectively. It may be concluded from this comparison that for $b/a > 3$ the calculations for a plate can be replaced by those for a strip without substantial error.

TABLE 8. NUMERICAL FACTORS α, β, γ, δ, n FOR UNIFORMLY LOADED AND SIMPLY SUPPORTED RECTANGULAR PLATES
$\nu = 0.3$

b/a	$w_{\max}$ $= \alpha \dfrac{qa^4}{D}$	$(M_x)_{\max}$ $= \beta qa^2$	$(M_y)_{\max}$ $= \beta_1 qa^2$	$(Q_x)_{\max}$ $= \gamma qa$	$(Q_y)_{\max}$ $= \gamma_1 qa$	$(V_x)_{\max}$ $= \delta qa$	$(V_y)_{\max}$ $= \delta_1 qa$	R $= nqa^2$
	α	β	β_1	γ	γ_1	δ	δ_1	n
1.0	0.00406	0.0479	0.0479	0.338	0.338	0.420	0.420	0.065
1.1	0.00485	0.0554	0.0493	0.360	0.347	0.440	0.440	0.070
1.2	0.00564	0.0627	0.0501	0.380	0.353	0.455	0.453	0.074
1.3	0.00638	0.0694	0.0503	0.397	0.357	0.468	0.464	0.079
1.4	0.00705	0.0755	0.0502	0.411	0.361	0.478	0.471	0.083
1.5	0.00772	0.0812	0.0498	0.424	0.363	0.486	0.480	0.085
1.6	0.00830	0.0862	0.0492	0.435	0.365	0.491	0.485	0.086
1.7	0.00883	0.0908	0.0486	0.444	0.367	0.496	0.488	0.088
1.8	0.00931	0.0948	0.0479	0.452	0.368	0.499	0.491	0.090
1.9	0.00974	0.0985	0.0471	0.459	0.369	0.502	0.494	0.091
2.0	0.01013	0.1017	0.0464	0.465	0.370	0.503	0.496	0.092
3.0	0.01223	0.1189	0.0406	0.493	0.372	0.505	0.498	0.093
4.0	0.01282	0.1235	0.0384	0.498	0.372	0.502	0.500	0.094
5.0	0.01297	0.1246	0.0375	0.500	0.372	0.501	0.500	0.095
∞	0.01302	0.1250	0.0375	0.500	0.372	0.500	0.500	0.095

Expression (e) can be used also for calculating shearing forces and reactions at the boundary. Forming the second derivatives of this expression, we find

$$\Delta w = \frac{\partial^2 w}{\partial x^2} + \frac{\partial^2 w}{\partial y^2} = -\frac{qx(a-x)}{2D} + \frac{2\pi^2 qa^2}{D} \sum_{m=1}^{\infty} m^2 B_m \cosh \frac{m\pi y}{a} \sin \frac{m\pi x}{a}$$

Substituting this in Eqs. (106) and (107), we obtain

$$Q_x = \frac{q(a-2x)}{2} - 2\pi^3 qa \sum_{m=1}^{\infty} m^3 B_m \cosh \frac{m\pi y}{a} \cos \frac{m\pi x}{a}$$

$$Q_y = -2\pi^3 qa \sum_{m=1}^{\infty} m^3 B_m \sinh \frac{m\pi y}{a} \sin \frac{m\pi x}{a}$$

SIMPLY SUPPORTED RECTANGULAR PLATES

For the sides $x = 0$ and $y = -b/2$ we find

$$(Q_x)_{x=0} = \frac{qa}{2} - 2\pi^3 qa \sum_{m=1}^{\infty} m^3 B_m \cosh \frac{m\pi y}{a}$$

$$= \frac{qa}{2} - \frac{4qa}{\pi^2} \sum_{m=1,3,5,\ldots}^{\infty} \frac{\cosh \frac{m\pi y}{a}}{m^2 \cosh \alpha_m}$$

$$(Q_y)_{y=-b/2} = 2\pi^3 qa \sum_{m=1}^{\infty} m^3 B_m \sinh \alpha_m \sin \frac{m\pi x}{a}$$

$$= \frac{4qa}{\pi^2} \sum_{m=1,3,5,\ldots}^{\infty} \frac{\tanh \alpha_m}{m^2} \sin \frac{m\pi x}{a}$$

These shearing forces have their numerical maximum value at the middle of the sides, where

$$(Q_x)_{x=0,y=0} = \frac{qa}{2} - \frac{4qa}{\pi^2} \sum_{m=1,3,5,\ldots}^{\infty} \frac{1}{m^2 \cosh \alpha_m} = \gamma qa$$

$$(Q_y)_{x=a/2, y=-b/2} = \frac{4qa}{\pi^2} \sum_{m=1,3,5,\ldots}^{\infty} \frac{(-1)^{(m-1)/2}}{m^2} \tanh \alpha_m = \gamma_1 qa$$
(p)

The numerical factors γ and γ_1 are also given in Table 8.

The reactive forces along the side $x = 0$ are given by the expression

$$V_x = \left(Q_x - \frac{\partial M_{xy}}{\partial y} \right)_{x=0} = \frac{qa}{2} - \frac{4qa}{\pi^2} \sum_{m=1,3,5,\ldots}^{\infty} \frac{\cosh \frac{m\pi y}{a}}{m^2 \cosh \alpha_m}$$

$$+ \frac{2(1-\nu)qa}{\pi^2} \sum_{m=1,3,5,\ldots}^{\infty} \frac{1}{m^2 \cosh^2 \alpha_m}$$

$$\left(\alpha_m \sinh \alpha_m \cosh \frac{m\pi y}{a} - \frac{m\pi y}{a} \cosh \alpha_m \sinh \frac{m\pi y}{a} \right)$$

The maximum numerical value of this pressure is at the middle of the side ($y = 0$), at which point we find

$$(V_x)_{x=0,y=0} = qa \left[\frac{1}{2} - \frac{4}{\pi^2} \sum_{m=1,3,5,\ldots}^{\infty} \frac{1}{m^2 \cosh \alpha_m} \right.$$

$$\left. + \frac{2(1-\nu)}{\pi^2} \sum_{m=1,3,5,\ldots}^{\infty} \frac{\alpha_m \sinh \alpha_m}{m^2 \cosh^2 \alpha_m} \right] = \delta qa \quad (q)$$

where δ is a numerical factor depending on ν and on the ratio b/a, which can readily be obtained by summing up the rapidly converging series that occur in expression (q). Numerical values of δ and of δ_1, which corresponds to the middle of the sides parallel to the x axis, are given in Table 8. The distribution of the pressures (q) along the sides of a square plate is shown in Fig. 63. The portion of the pressures produced by the

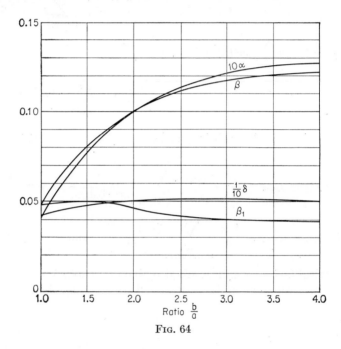

Fig. 64

twisting moments M_{xy} is also shown. These latter pressures are balanced by reactive forces concentrated at the corners of the plate. The magnitude of these forces is given by the expression

$$R = 2(M_{xy})_{x=a,y=b/2} = 2D(1-\nu)\left(\frac{\partial^2 w}{\partial x\, \partial y}\right)_{x=a,y=b/2}$$

$$= \frac{4(1-\nu)qa^2}{\pi^3} \sum_{m=1,3,5,\ldots}^{\infty} \frac{1}{m^3 \cosh \alpha_m}[(1 + \alpha_m \tanh \alpha_m)\sinh \alpha_m - \alpha_m \cosh \alpha_m] = nqa^2 \quad (r)$$

The forces are directed downward and prevent the corners of a plate from rising up during bending. The values of the coefficient n are given in the last column of Table 8.

The values of the factors α, β, β_1, δ as functions of the ratio b/a are represented by the curves in Fig. 64.

In the presence of the forces R, which act downward and are by no means small, anchorage must be provided at the corners of the plate if the plate is not solidly joined with the supporting beams.

In order to determine the moments arising at the corner let us consider the equilibrium of the element abc of the plate next to its corner (Fig. 65) and let us introduce, for the same purpose, new coordinates 1, 2 at an angle of 45° to the coordinates x, y in Fig. 59. We can then readily verify that the bending moments acting at the sides ab and cb of the element are $M_1 = -R/2$ and $M_2 = +R/2$, respectively, and that the corresponding twisting moments are zero. In fact, using Eq. (39), we obtain for the side ac, that is, for the element of the edge, given by $\alpha = -45°$, the bending moment

$$M_n = M_1 \cos^2 \alpha + M_2 \sin^2 \alpha = 0$$

in accordance with the boundary conditions of a simply supported plate. The magnitude of the twisting moment applied at the same edge element is obtained in like manner by means of Eq. (40). Putting $\alpha = -45°$ we have

$$M_{nt} = \frac{1}{2} \sin 2\alpha (M_1 - M_2) = \frac{R}{2}$$

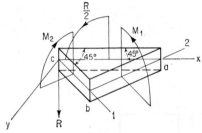

Fig. 65

according to Eq. (r). Thus, the portion of the plate in the vicinity of the corner is bent to an anticlastic surface, the moments $\pm R/2$ at the corner itself being of the same order of magnitude as the bending moments at the middle of the plate (see Table 8).

The clamping effect of the corners of a simply supported plate is plainly illustrated by the distribution of the bending moments M_1 and M_2 of a square plate (Fig. 63). If the corners of the rectangular plate are not properly secured against lifting, the clamping becomes ineffective and the bending moments in the center portion of the plate increase accordingly. The values of $(M_x)_{max}$ and $(M_y)_{max}$ given in Table 8 must then be multiplied by some factor $k > 1$. The approximate expression[1]

$$k = \frac{a^4 - \frac{5}{12}a^2b^2 + b^4}{a^4 - \frac{5}{6}a^2b^2 + b^4}$$

may be used for that purpose.

It should be noted that in the case of a polygonal plate with simply supported edges no single reactive forces arise at a corner point provided the angle between both adjacent sides of the plate is other than a right angle.[2]

Even in rectangular plates, however, no corner reactions are obtained if the transverse shear deformation is taken into account. In view of the strongly concentrated

---

[1] Recommended by the German Code for Reinforced Concrete (1943) and based on a simplified theory of thin plates due to H. Marcus; see his book "Die vereinfachte Berechnung biegsamer Platten," 2d ed., Berlin, 1925.

[2] For a simple proof see, for example, H. Marcus, "Die Theorie elastischer Gewebe," 2d ed., p. 46, Berlin, 1932.

reactive forces this shear deformation obviously is no longer negligible, and the customary thin-plate theory disregarding it completely must be replaced by a more exact theory. The latter, which will be discussed in Art. 39, actually leads to a distribution of reactive pressures which include no forces concentrated at the corners of the plate (see Fig. 81).

31. Simply Supported Rectangular Plates under Hydrostatic Pressure.

Assume that a simply supported rectangular plate is loaded as shown in Fig. 66. Proceeding as in the case of a uniformly distributed load, we take the deflection of the plate in the form[1]

$$w = w_1 + w_2 \qquad (a)$$

in which

$$w_1 = \frac{q_0}{360D}\left(\frac{3x^5}{a} - 10ax^3 + 7a^3x\right) = \frac{2q_0 a^4}{D\pi^5} \sum_{m=1,2,3,\ldots}^{\infty} \frac{(-1)^{m+1}}{m^5} \sin\frac{m\pi x}{a} \qquad (b)$$

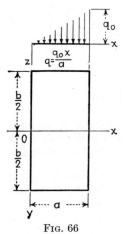

FIG. 66

represents the deflection of a strip under the triangular load. This expression satisfies the differential equation

$$\frac{\partial^4 w}{\partial x^4} + 2\frac{\partial^4 w}{\partial x^2\, \partial y^2} + \frac{\partial^4 w}{\partial y^4} = \frac{q}{D} = \frac{q_0 x}{aD} \qquad (c)$$

and the boundary conditions

$$w = 0 \qquad \frac{\partial^2 w}{\partial x^2} = 0 \qquad \text{for } x = 0 \text{ and } x = a$$

The part w_2 is taken in the form of a series

$$w_2 = \sum_{m=1}^{\infty} Y_m \sin\frac{m\pi x}{a} \qquad (d)$$

where the functions Y_m have the same form as in the preceding article, and $m = 1, 2, 3, \ldots$ Substituting expressions (b) and (d) into Eq. (a), we obtain

$$w = \frac{q_0 a^4}{D} \sum_{m=1}^{\infty} \left[\frac{2(-1)^{m+1}}{\pi^5 m^5} + A_m \cosh\frac{m\pi y}{a} + B_m \frac{m\pi y}{a}\sinh\frac{m\pi y}{a}\right]\sin\frac{m\pi x}{a} \qquad (e)$$

where the constants A_m and B_m are to be determined from the conditions

$$w = 0 \qquad \frac{\partial^2 w}{\partial y^2} = 0 \qquad \text{for } y = \pm\frac{b}{2}$$

[1] This problem was discussed by E. Estanave, *op. cit.* The numerical tables of deflections and moments were calculated by B. G. Galerkin, *Bull. Polytech. Inst.*, St. Petersburg, vols. 26 and 27, 1918.

From these conditions we find

$$\frac{2(-1)^{m+1}}{\pi^5 m^5} + A_m \cosh \alpha_m + B_m \alpha_m \sinh \alpha_m = 0$$

$$(2B_m + A_m) \cosh \alpha_m + B_m \alpha_m \sinh \alpha_m = 0$$

In these equations we use, as before, the notation

$$\alpha_m = \frac{m\pi b}{2a}$$

Solving them, we find

$$A_m = -\frac{(2 + \alpha_m \tanh \alpha_m)(-1)^{m+1}}{\pi^5 m^5 \cosh \alpha_m} \qquad B_m = \frac{(-1)^{m+1}}{\pi^5 m^5 \cosh \alpha_m} \qquad (f)$$

The deflection of the plate along the x axis is

$$(w)_{y=0} = \frac{q_0 a^4}{D} \sum_{m=1}^{\infty} \left[\frac{2(-1)^{m+1}}{\pi^5 m^5} + A_m \right] \sin \frac{m\pi x}{a}$$

For a square plate $a = b$, and we find

$$(w)_{y=0} = \frac{q_0 a^4}{D} \left(0.002055 \sin \frac{\pi x}{a} - 0.000177 \sin \frac{2\pi x}{a} \right.$$
$$\left. + 0.000025 \sin \frac{3\pi x}{a} - \cdots \right) \qquad (g)$$

The deflection at the center of the plate is

$$(w)_{x=a/2, y=0} = 0.00203 \frac{q_0 a^4}{D} \qquad (h)$$

which is one-half the deflection of a uniformly loaded plate (see page 116) as it should be. By equating the derivative of expression (g) to zero, we find that the maximum deflection is at the point $x = 0.557a$. This maximum deflection, which is $0.00206 \, q_0 a^4/D$, differs only very little from the deflection at the middle as given by formula (h). The point of maximum deflection approaches the center of the plate as the ratio b/a increases. For $b/a = \infty$, as for a strip [see expression (b)], the maximum deflection is at the point $x = 0.5193a$. When $b/a < 1$, the point of maximum deflection moves away from the center of the plate as the ratio b/a decreases. The deflections at several points along the x axis (Fig. 66) are given in Table 9. It is seen that, as the ratio b/a increases, the deflections approach the values calculated for a strip. For $b/a = 4$ the differences in these values are about $1\frac{1}{2}$ per cent. We can always calculate the deflection of a plate for which $b/a > 4$ with satisfactory accuracy by using formula (b) for the deflection of a strip under triangular load. The bending moments M_x and M_y are found by substituting

TABLE 9. NUMERICAL FACTOR α FOR DEFLECTIONS OF A SIMPLY SUPPORTED RECTANGULAR PLATE UNDER HYDROSTATIC PRESSURE $q = q_0 x/a$
$b > a$
$w = \alpha q_0 a^4 / D$, $y = 0$

b/a	$x = 0.25a$	$x = 0.50a$	$x = 0.60a$	$x = 0.75a$
1	0.00131	0.00203	0.00201	0.00162
1.1	0.00158	0.00243	0.00242	0.00192
1.2	0.00186	0.00282	0.00279	0.00221
1.3	0.00212	0.00319	0.00315	0.00248
1.4	0.00235	0.00353	0.00348	0.00273
1.5	0.00257	0.00386	0.00379	0.00296
1.6	0.00277	0.00415	0.00407	0.00317
1.7	0.00296	0.00441	0.00432	0.00335
1.8	0.00313	0.00465	0.00455	0.00353
1.9	0.00328	0.00487	0.00475	0.00368
2.0	0.00342	0.00506	0.00494	0.00382
3.0	0.00416	0.00612	0.00592	0.00456
4.0	0.00437	0.00641	0.00622	0.00477
5.0	0.00441	0.00648	0.00629	0.00483
∞	0.00443	0.00651	0.00632	0.00484

expression (e) for deflections in Eqs. (101). Along the x axis ($y = 0$) the expression for M_x becomes

$$(M_x)_{y=0} = q_0 a^2 \sum_{m=1}^{\infty} \frac{2(-1)^{m+1}}{\pi^3 m^3} \sin \frac{m\pi x}{a}$$
$$+ q_0 a^2 \pi^2 \sum_{m=1}^{\infty} m^2[(1-\nu)A_m - 2\nu B_m] \sin \frac{m\pi x}{a} \quad (i)$$

The first sum on the right-hand side of this expression represents the bending moment for a strip under the action of a triangular load and is equal to $(q_0/6)(ax - x^3/a)$. Using expressions (f) for the constants A_m and B_m in the second sum, we obtain

$$(M_x)_{y=0} = \frac{q_0(a^2 x - x^3)}{6a}$$
$$- \frac{q_0 a^2}{\pi^3} \sum_{m=1}^{\infty} \frac{(-1)^{m+1}}{m^3 \cosh \alpha_m} [2 + (1-\nu)\alpha_m \tanh \alpha_m] \sin \frac{m\pi x}{a} \quad (j)$$

The series thus obtained converges rapidly, and a sufficiently accurate value of M_x can be realized by taking only the first few terms. In this

TABLE 10. Numerical Factors β and β_1 for Bending Moments of Simply Supported Rectangular Plates under Hydrostatic Pressure $q = q_0x/a$
$\nu = 0.3, b > a$

b/a	$M_x = \beta a^2 q_0$, $y = 0$				$M_y = \beta_1 a^2 q_0$, $y = 0$			
	$x = 0.25a$	$x = 0.50a$	$x = 0.60a$	$x = 0.75a$	$x = 0.25a$	$x = 0.50a$	$x = 0.60a$	$x = 0.75a$
1.0	0.0132	0.0239	0.0264	0.0259	0.0149	0.0239	0.0245	0.0207
1.1	0.0156	0.0276	0.0302	0.0289	0.0155	0.0247	0.0251	0.0211
1.2	0.0179	0.0313	0.0338	0.0318	0.0158	0.0250	0.0254	0.0213
1.3	0.0200	0.0346	0.0371	0.0344	0.0160	0.0252	0.0255	0.0213
1.4	0.0221	0.0376	0.0402	0.0367	0.0160	0.0253	0.0254	0.0212
1.5	0.0239	0.0406	0.0429	0.0388	0.0159	0.0249	0.0252	0.0210
1.6	0.0256	0.0431	0.0454	0.0407	0.0158	0.0246	0.0249	0.0207
1.7	0.0272	0.0454	0.0476	0.0424	0.0155	0.0243	0.0246	0.0205
1.8	0.0286	0.0474	0.0496	0.0439	0.0153	0.0239	0.0242	0.0202
1.9	0.0298	0.0492	0.0513	0.0452	0.0150	0.0235	0.0238	0.0199
2.0	0.0309	0.0508	0.0529	0.0463	0.0148	0.0232	0.0234	0.0197
3.0	0.0369	0.0594	0.0611	0.0525	0.0128	0.0202	0.0207	0.0176
4.0	0.0385	0.0617	0.0632	0.0541	0.0120	0.0192	0.0196	0.0168
5.0	0.0389	0.0623	0.0638	0.0546	0.0118	0.0187	0.0193	0.0166
∞	0.0391	0.0625	0.0640	0.0547	0.0117	0.0187	0.0192	0.0165

way the bending moment at any point of the x axis can be represented by the equation

$$(M_x)_{y=0} = \beta q_0 a^2 \tag{k}$$

where β is a numerical factor depending on the abscissa x of the point. In a similar manner we get

$$(M_y)_{y=0} = \beta_1 q_0 a^2 \tag{l}$$

The numerical values of the factors β and β_1 in formulas (k) and (l) are given in Table 10. It is seen that for $b \gtreqless 4a$ the moments are very close to the values of the moments in a strip under a triangular load.

Equations (106) and (107) are used to calculate shearing forces. From the first of these equations, by using expression (j), we obtain for points on the x axis

$$(Q_x)_{y=0} = -D\frac{\partial}{\partial x}\left(\frac{\partial^2 w}{\partial x^2} + \frac{\partial^2 w}{\partial y^2}\right)_{y=0}$$
$$= \frac{q_0(a^2 - 3x^2)}{6a} - \frac{2q_0 a}{\pi^2} \sum_{m=1}^{\infty} \frac{(-1)^{m+1}}{m^2 \cosh \alpha_m} \cos \frac{m\pi x}{a}$$

The general expressions for shearing forces Q_x and Q_y are

$$Q_x = \frac{q_0(a^2 - 3x^2)}{6a} - \frac{2q_0 a}{\pi^2} \sum_{m=1}^{\infty} \frac{(-1)^{m+1} \cosh \frac{m\pi y}{a}}{m^2 \cosh \alpha_m} \cos \frac{m\pi x}{a} \qquad (m)$$

$$Q_y = -\frac{2q_0 a}{\pi^2} \sum_{m=1}^{\infty} \frac{(-1)^{m+1} \sinh \frac{m\pi y}{a}}{m^2 \cosh \alpha_m} \sin \frac{m\pi x}{a} \qquad (n)$$

The magnitude of the vertical reactions V_x and V_y along the boundary is obtained by combining the shearing forces with the derivatives of the twisting moments. Along the sides $x = 0$ and $x = a$ these reactions can be represented in the form

$$V_x = \left(Q_x - \frac{\partial M_{xy}}{\partial y} \right)_{x=0, x=a} = \pm \delta q_0 a \qquad (o)$$

Table 11. Numerical Factors δ and δ_1 for Reactions of Simply Supported Rectangular Plates under Hydrostatic Pressure $q = q_0 x/a$
$\nu = 0.3$, $b > a$

b/a	Reactions $\delta q_0 a$				Reactions $\delta_1 q_0 b$			
	$x = 0$		$x = a$		$y = \pm b/2$			
	$y = 0$	$y = 0.25b$	$y = 0$	$y = 0.25b$	$x = 0.25a$	$x = 0.50a$	$x = 0.60a$	$x = 0.75a$
1.0	0.126	0.098	0.294	0.256	0.115	0.210	0.234	0.239
1.1	0.136	0.107	0.304	0.267	0.110	0.199	0.221	0.224
1.2	0.144	0.114	0.312	0.276	0.105	0.189	0.208	0.209
1.3	0.150	0.121	0.318	0.284	0.100	0.178	0.196	0.196
1.4	0.155	0.126	0.323	0.292	0.095	0.169	0.185	0.184
1.5	0.159	0.132	0.327	0.297	0.090	0.160	0.175	0.174
1.6	0.162	0.136	0.330	0.302	0.086	0.151	0.166	0.164
1.7	0.164	0.140	0.332	0.306	0.082	0.144	0.157	0.155
1.8	0.166	0.143	0.333	0.310	0.078	0.136	0.149	0.147
1.9	0.167	0.146	0.334	0.313	0.074	0.130	0.142	0.140
2.0	0.168	0.149	0.335	0.316	0.071	0.124	0.135	0.134
3.0	0.169	0.163	0.336	0.331	0.048	0.083	0.091	0.089
4.0	0.168	0.167	0.334	0.334	0.036	0.063	0.068	0.067
5.0	0.167	0.167	0.334	0.335	0.029	0.050	0.055	0.054
∞	0.167	0.167	0.333	0.333				

and along the sides $y = \pm b/2$ in the form

$$V_y = \left(Q_y - \frac{\partial M_{xy}}{\partial x}\right)_{y=\pm b/2} = \mp \delta_1 q_0 b \tag{p}$$

in which δ and δ_1 are numerical factors depending on the ratio b/a and on the coordinates of the points taken on the boundary. Several values of these factors are given in Table 11.

The magnitude of concentrated forces that must be applied to prevent the corners of the plate rising up during bending can be found from the values of the twisting moments M_{xy} at the corners. Since the load is not symmetrical, the reactions R_1 at $x = 0$ and $y = \pm b/2$ are different from the reactions R_2 at $x = a$ and $y = \pm b/2$. These reactions can be represented in the following form:

$$R_1 = n_1 q_0 ab \qquad R_2 = n_2 q_0 ab \tag{q}$$

The values of the numerical factors n_1 and n_2 are given in Table 12.

TABLE 12. NUMERICAL FACTORS n_1 AND n_2 IN EQS. (q) FOR REACTIVE FORCES R_1 AND R_2 AT THE CORNERS OF SIMPLY SUPPORTED RECTANGULAR PLATES UNDER HYDROSTATIC PRESSURE $q = q_0 x/a$
$\nu = 0.3, b > a$

b/a	1.0	1.1	1.2	1.3	1.4	1.5	1.6	1.7	1.8	1.9	2.0	3.0	4.0	5.0
n_1	0.026	0.026	0.026	0.026	0.025	0.024	0.023	0.022	0.021	0.021	0.020	0.014	0.010	0.008
n_2	0.039	0.038	0.037	0.036	0.035	0.033	0.032	0.030	0.029	0.028	0.026	0.018	0.014	0.011

Since a uniform load q_0 is obtained by superposing the two triangular loads $q = q_0 x/a$ and $q_0(a - x)/a$, it can be concluded that for corresponding values of b/a the sum $n_1 + n_2$ of the factors given in Table 12 multiplied by b/a must equal the corresponding value of n, the last column in Table 8.

If the relative dimensions of the plate are such that a in Fig. 66 is greater than b, then more rapidly converging series will be obtained by representing w_1 and w_2 by the following expressions:

$$w_1 = \frac{q_0 x}{a} \frac{1}{384D} (16y^4 - 24b^2 y^2 + 5b^4) \tag{r}$$

$$w_2 = \sum_{m=1}^{\infty} X_{2m-1} \cos \frac{(2m-1)\pi y}{b} \tag{s}$$

The first of these expressions is the deflection of a narrow strip parallel to the y axis, supported at $y = \pm b/2$ and carrying a uniformly distributed

130   THEORY OF PLATES AND SHELLS

TABLE 13. NUMERICAL FACTORS α FOR DEFLECTIONS OF SIMPLY SUPPORTED RECTANGULAR PLATES UNDER HYDROSTATIC PRESSURE $q = q_0 x/a$
$b < a$
$w = \alpha q_0 b^4/D$, $y = 0$

a/b	$x = 0.25a$	$x = 0.50a$	$x = 0.60a$	$x = 0.75a$
∞	0.00325	0.00651	0.00781	0.00976
5	0.00325	0.00648	0.00778	0.00965
4	0.00325	0.00641	0.00751	0.00832
3	0.00321	0.00630	0.00692	0.00707
2	0.00288	0.00506	0.00542	0.00492
1.9	0.00281	0.00487	0.00518	0.00465
1.8	0.00270	0.00465	0.00491	0.00434
1.7	0.00261	0.00441	0.00463	0.00404
1.6	0.00249	0.00415	0.00432	0.00372
1.5	0.00234	0.00386	0.00399	0.00339
1.4	0.00218	0.00353	0.00363	0.00304
1.3	0.00199	0.00319	0.00325	0.00269
1.2	0.00179	0.00282	0.00286	0.00234
1.1	0.00153	0.00243	0.00245	0.00199
1.0	0.00131	0.00202	0.00201	0.00162

load of intensity $q_0 x/a$. This expression satisfies the differential equation (c) and also the boundary conditions $w = 0$ and $\partial^2 w/\partial y^2 = 0$ at $y = \pm b/2$. Expression (s) represents an infinite series each term of which also satisfies the conditions at the edges $y = \pm b/2$. The functions X_{2m-1} of x are chosen in such a manner that each of them satisfies the homogeneous equation (137) of the preceding article (see page 114) and so that expression (a) satisfies the boundary conditions at the edges $x = 0$ and $x = a$. Since the method of determining the functions X_{2m-1} is similar to that already used in determining the functions Y_m, we shall limit ourselves to giving only the final numerical results, which are represented by Tables 13, 14, 15, and 16. The notation in these tables is the same as in the foregoing tables for the hydrostatic pressure.

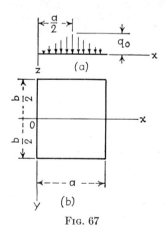

Fig. 67

32. Simply Supported Rectangular Plate under a Load in the Form of a Triangular Prism. Assume that the intensity of the load is represented

TABLE 14. NUMERICAL FACTORS β AND β_1 FOR BENDING MOMENTS IN SIMPLY SUPPORTED RECTANGULAR PLATES UNDER HYDROSTATIC PRESSURE $q = q_0 x/a$
$\nu = 0.3, b < a$

a/b	$M_x = \beta b^2 q_0, y = 0$				$M_y = \beta_1 b^2 q_0, y = 0$			
	$x = 0.25a$	$x = 0.50a$	$x = 0.60a$	$x = 0.75a$	$x = 0.25a$	$x = 0.50a$	$x = 0.60a$	$x = 0.75a$
∞	0.0094	0.0187	0.0225	0.0281	0.0312	0.0625	0.0750	0.0937
5.0	0.0094	0.0187	0.0230	0.0309	0.0312	0.0623	0.0742	0.0877
4.0	0.0094	0.0192	0.0237	0.0326	0.0312	0.0617	0.0727	0.0820
3.0	0.0096	0.0202	0.0256	0.0345	0.0309	0.0594	0.0678	0.0715
2.0	0.0108	0.0232	0.0285	0.0348	0.0284	0.0508	0.0554	0.0523
1.9	0.0111	0.0235	0.0288	0.0345	0.0278	0.0492	0.0533	0.0498
1.8	0.0115	0.0239	0.0291	0.0341	0.0269	0.0474	0.0509	0.0470
1.7	0.0117	0.0243	0.0293	0.0337	0.0261	0.0454	0.0485	0.0442
1.6	0.0120	0.0246	0.0294	0.0331	0.0251	0.0431	0.0457	0.0412
1.5	0.0123	0.0249	0.0294	0.0324	0.0239	0.0406	0.0428	0.0381
1.4	0.0126	0.0253	0.0292	0.0315	0.0225	0.0376	0.0396	0.0348
1.3	0.0129	0.0252	0.0290	0.0304	0.0209	0.0346	0.0360	0.0314
1.2	0.0131	0.0250	0.0284	0.0291	0.0192	0.0313	0.0323	0.0279
1.1	0.0134	0.0247	0.0276	0.0276	0.0169	0.0276	0.0285	0.0245
1.0	0.0132	0.0239	0.0264	0.0259	0.0149	0.0239	0.0245	0.0207

by an isosceles triangle as shown in Fig. 67a. The deflection surface can again be represented in the form

$$w = w_1 + w_2 \quad (a)$$

in which w_1 represents the deflection of a simply supported strip parallel to the x axis, and w_2 has the same form as in the preceding article [Eq. (d)]. To represent the deflection w_1 in the form of a trigonometric series we observe that the deflection produced by a concentrated force P applied at a distance ξ from the left end of a strip is[1]

$$\frac{2Pa^3}{D\pi^4} \sum_{m=1}^{\infty} \frac{1}{m^4} \sin \frac{m\pi\xi}{a} \sin \frac{m\pi x}{a} \quad (b)$$

Substituting $q\,d\xi$ for P and using $q = 2q_0\xi/a$ for $\xi < a/2$ and $q = 2q_0(a - \xi)/a$ for $\xi > a/2$, the deflection of the strip by an elemental load is obtained. The deflection produced by the total load on

[1] See Timoshenko, "Strength of Materials," part II, 3d ed., p. 49, 1956.

TABLE 15. Numerical Factors δ and δ_1 for Reactions in Simply Supported Rectangular Plates under Hydrostatic Pressure $q = q_0 x/a$
$\nu = 0.3,\ b < a$

a/b	Reactions $\delta q_0 a$				Reactions $\delta_1 q_0 b$			
	$x = 0$		$x = a$		$y = \pm b/2$			
	$y = 0$	$y = b/4$	$y = 0$	$y = b/4$	$x = 0.25a$	$x = 0.50a$	$x = 0.60a$	$x = 0.75a$
∞	.....	.....	.....	.....	0.125	0.250	0.300	0.375
5.0	0.008	0.006	0.092	0.076	0.125	0.250	0.301	0.379
4.0	0.013	0.010	0.112	0.093	0.125	0.251	0.301	0.377
3.0	0.023	0.018	0.143	0.119	0.125	0.252	0.304	0.368
2.0	0.050	0.038	0.197	0.166	0.127	0.251	0.296	0.337
1.9	0.055	0.041	0.205	0.172	0.127	0.251	0.294	0.331
1.8	0.060	0.045	0.213	0.179	0.128	0.249	0.291	0.325
1.7	0.066	0.050	0.221	0.187	0.127	0.248	0.288	0.318
1.6	0.073	0.055	0.230	0.195	0.127	0.245	0.284	0.311
1.5	0.080	0.060	0.240	0.204	0.127	0.243	0.279	0.302
1.4	0.088	0.067	0.250	0.213	0.126	0.239	0.273	0.292
1.3	0.097	0.074	0.260	0.223	0.124	0.234	0.266	0.281
1.2	0.106	0.081	0.271	0.233	0.122	0.227	0.257	0.269
1.1	0.116	0.090	0.282	0.244	0.120	0.220	0.247	0.255
1.0	0.126	0.098	0.294	0.256	0.115	0.210	0.234	0.239

the strip is now obtained by integration in the following form:

$$w_1 = \frac{4q_0 a^2}{D\pi^4} \sum_{m=1}^{\infty} \frac{1}{m^4} \sin \frac{m\pi x}{a} \left[\int_0^{a/2} \xi \sin \frac{m\pi \xi}{a} d\xi + \int_{a/2}^{a} (a - \xi) \sin \frac{m\pi \xi}{a} d\xi \right]$$

$$= \frac{8 q_0 a^4}{D\pi^6} \sum_{m=1,3,5,\ldots} \frac{(-1)^{(m-1)/2}}{m^6} \sin \frac{m\pi x}{a} \quad (c)$$

Substituting this in Eq. (a) and using Eq. (d) of the preceding article, we obtain

$$w = \frac{q_0 a^4}{D} \sum_{m=1,3,5,\ldots}^{\infty} \left[\frac{8(-1)^{(m-1)/2}}{\pi^6 m^6} + A_m \cosh \frac{m\pi y}{a} + B_m \frac{m\pi y}{a} \sinh \frac{m\pi y}{a} \right] \sin \frac{m\pi x}{a} \quad (d)$$

This expression satisfies Eq. (103) and also the boundary conditions at the edges $x = 0$ and $x = a$. The constants A_m and B_m can be found from the conditions along the edges $y = \pm b/2$, which are the same as in the preceding article and which give

$$\frac{8(-1)^{(m-1)/2}}{\pi^6 m^6} + A_m \cosh \alpha_m + B_m \alpha_m \sinh \alpha_m = 0$$
$$(2B_m + A_m) \cosh \alpha_m + B_m \alpha_m \sinh \alpha_m = 0 \qquad (e)$$

where, as before, we use the notation

$$\alpha_m = \frac{m\pi b}{2a}$$

Solving Eqs. (e), we find

$$A_m = -\frac{4(2 + \alpha_m \tanh \alpha_m)(-1)^{(m-1)/2}}{\pi^6 m^6 \cosh \alpha_m}$$
$$B_m = \frac{4(-1)^{(m-1)/2}}{\pi^6 m^6 \cosh \alpha_m} \qquad (f)$$

To obtain the deflection of the plate along the x axis we put $y = 0$ in

TABLE 16. NUMERICAL FACTORS n_1 AND n_2 IN EQS. (q) (ART. 31) FOR REACTIVE FORCES R_1 AND R_2 AT THE CORNERS OF SIMPLY SUPPORTED RECTANGULAR PLATES UNDER HYDROSTATIC PRESSURE $q = q_0 x/a$
$\nu = 0.3, b < a$

a/b	5	4	3	2	1.9	1.8	1.7	1.6	1.5	1.4	1.3	1.2	1.1	1.0
n_1	0.002	0.004	0.006	0.013	0.014	0.016	0.017	0.018	0.020	0.021	0.023	0.024	0.025	0.026
n_2	0.017	0.020	0.025	0.033	0.034	0.035	0.036	0.037	0.037	0.038	0.039	0.039	0.039	0.039

expression (d). Then

$$(w)_{y=0} = \frac{q_0 a^4}{D} \sum_{m=1,3,5,\ldots}^{\infty} \left[\frac{8(-1)^{(m-1)/2}}{\pi^6 m^6} + A_m \right] \sin \frac{m\pi x}{a}$$

The maximum deflection is at the center of the plate, where

$$w_{\max} = \frac{q_0 a^4}{D} \sum_{m=1,3,5,\ldots}^{\infty} \left[\frac{8}{\pi^6 m^6} + A_m (-1)^{(m-1)/2} \right]$$

It can be represented in the form

$$w_{\max} = \alpha \frac{q_0 a^4}{D}$$

in which α is a numerical factor depending on the magnitude of the ratio b/a. Several values of this factor are given in Table 17.[1]

[1] The tables are taken from the paper by Galerkin, *loc. cit.*

TABLE 17. NUMERICAL FACTORS α, β, γ, δ, n FOR SIMPLY SUPPORTED RECTANGULAR PLATES UNDER A LOAD IN FORM OF A TRIANGULAR PRISM
$\nu = 0.3$, $b > a$

b/a	$w_{max} = \alpha \dfrac{q_0 a^4}{D}$	$(M_x)_{max} = \beta q_0 a^2$	$(M_y)_{max} = \beta_1 q_0 a^2$	$(Q_x)_{max} = \gamma q_0 a$	$(Q_y)_{max} = \gamma_1 q_0 b$	$(V_x)_{max} = \delta q_0 a$	$(V_y)_{max} = \delta_1 q_0 b$	$R = n q_0 a b$
	α	β	β_1	γ	γ_1	δ	δ_1	n
1.0	0.00263	0.0340	0.0317	0.199	0.315	0.147	0.250	0.038
1.1	0.00314	0.0390	0.0326	0.212	0.297	0.161	0.232	0.038
1.2	0.00364	0.0436	0.0330	0.222	0.280	0.173	0.216	0.037
1.3	0.00411	0.0479	0.0332	0.230	0.265	0.184	0.202	0.036
1.4	0.00455	0.0518	0.0331	0.236	0.250	0.193	0.189	0.035
1.5	0.00496	0.0554	0.0329	0.241	0.236	0.202	0.178	0.034
1.6	0.00533	0.0586	0.0325	0.246	0.224	0.208	0 168	0.033
1.7	0.00567	0.0615	0.0321	0.247	0 212	0.214	0.158	0.031
1.8	0.00597	0.0641	0.0316	0.249	0.201	0.220	0.150	0.030
1.9	0.00625	0.0664	0.0311	0.251	0.191	0.224	0.142	0.029
2.0	0.00649	0.0685	0.0306	0.252	0.183	0.228	0.135	0.028
3.0	0.00783	0.0794	0.0270	0.253	0.122	0.245	0.090	0.019
∞	0.00833	0.0833	0.0250	0.250	....	0.250		

Using expression (d) and proceeding as in the preceding article, we can readily obtain the expressions for bending moments M_x and M_y. The maximum values of these moments in this case are evidently at the center of the plate and can be represented in the following form:

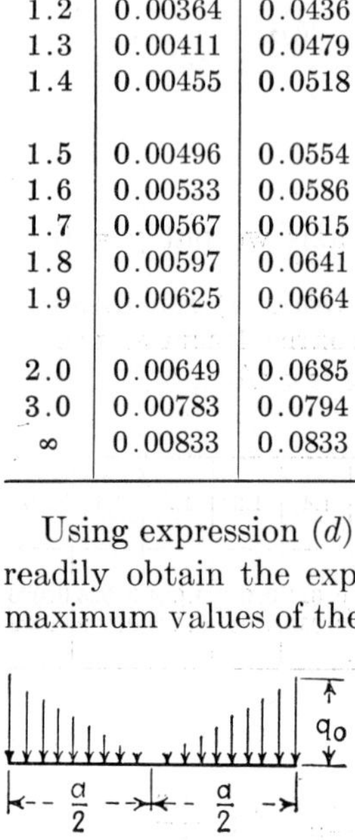

$$(M_x)_{max} = \beta q_0 a^2 \qquad (M_y)_{max} = \beta_1 q_0 a^2$$

FIG. 68

The values of the numerical factors β and β_1 are also given in Table 17. This table also gives numerical factors γ, γ_1, δ, δ_1, and n for calculating (1) shearing forces $(Q_x)_{max} = \gamma q_0 a$, $(Q_y)_{max} = \gamma_1 q_0 b$ at the middle of the sides $x = 0$ and $y = -b/2$ of the plate, (2) reactive forces

$$V_x = \left(Q_x - \frac{\partial M_{xy}}{\partial y} \right)_{max} = \delta q_0 a$$

$$V_y = \left(Q_y - \frac{\partial M_{xy}}{\partial x} \right)_{max} = \delta_1 q_0 b$$

at the same points, and (3) concentrated reactions $R = n q_0 a b$ at the corners of the plate which are acting downward and prevent the corners of the plate from rising. All these values are given for $b > a$. When $b < a$, a better convergency can be obtained by taking the portion w_1

TABLE 18. NUMERICAL FACTORS α, β, γ, δ, n FOR SIMPLY SUPPORTED
RECTANGULAR PLATES UNDER A LOAD IN FORM OF A TRIANGULAR PRISM
$\nu = 0.3, b < a$

a/b	$w_{max} = \alpha \dfrac{q_0 b^4}{D}$	$(M_x)_{max} = \beta q_0 b^2$	$(M_y)_{max} = \beta_1 q_0 b^2$	$(Q_x)_{max} = \gamma q_0 a$	$(Q_y)_{max} = \gamma_1 q_0 b$	$(V_x)_{max} = \delta q_0 a$	$(V_y)_{max} = \delta_1 q_0 b$	$R = n q_0 a b$
	α	β	β_1	γ	γ_1	δ	δ_1	n
∞	0.01302	0.0375	0.1250	.....	0.500	.....	0.500	
3.0	0.00868	0.0387	0.0922	0.045	0.442	0.027	0.410	0.010
2.0	0.00686	0.0392	0.0707	0.091	0.412	0.057	0.365	0.023
1.9	0.00656	0.0392	0.0681	0.098	0.407	0.062	0.358	0.024
1.8	0.00624	0.0391	0.0651	0.106	0.402	0.098	0.350	0.026
1.7	0.00588	0.0390	0.0609	0.115	0.396	0.074	0.342	0.028
1.6	0.00549	0.0388	0.0585	0.124	0.389	0.081	0.332	0.029
1.5	0.00508	0.0386	0.0548	0.135	0.381	0.090	0.322	0.031
1.4	0.00464	0.0382	0.0508	0.146	0.371	0.099	0.311	0.033
1.3	0.00418	0.0376	0.0464	0.158	0.360	0.109	0.298	0.035
1.2	0.00367	0.0368	0.0418	0.171	0.347	0.120	0.284	0.036
1.1	0.00316	0.0356	0.0369	0.185	0.332	0.133	0.268	0.037
1.0	0.00263	0.0340	0.0317	0.199	0.315	0.147	0.250	0.038

of the deflection of the plate in the form of the deflection of a strip parallel to the y direction. We omit the derivations and give only the numerical results assembled in Table 18.

Combining the load shown in Fig. 67a with the uniform load of intensity q_0, the load shown in Fig. 68 is obtained. Information regarding deflections and stresses in this latter case can be obtained by combining the data of Table 8 with those of Table 17 or 18.

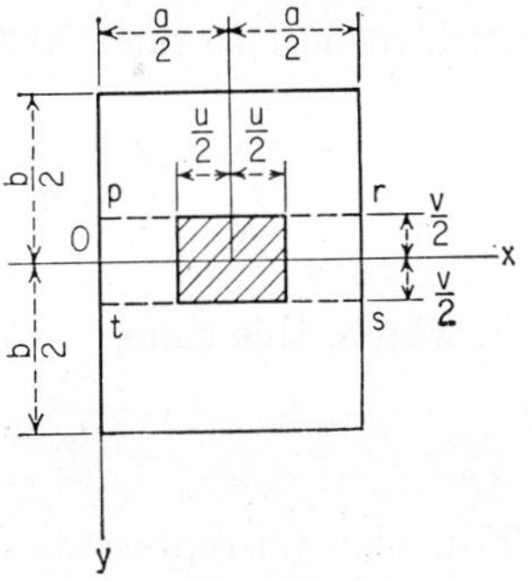

33. Partially Loaded Simply Supported Rectangular Plate. Let us consider a symmetrical case of bending in which a uniform load q is distributed over the shaded rectangle (Fig. 69) with the sides u and v.

We begin by developing the load in the series

FIG. 69

$$\frac{2}{a} \sum_{m=1}^{\infty} \sin \frac{m\pi x}{a} \int_{\frac{1}{2}(a-u)}^{\frac{1}{2}(a+u)} q \sin \frac{m\pi \xi}{a} d\xi$$

$$= \frac{4q}{\pi} \sum_{m=1,3,5,\ldots}^{\infty} \frac{(-1)^{(m-1)/2}}{m} \sin \frac{m\pi u}{2a} \sin \frac{m\pi x}{a} \quad (a)$$

which represents the load for the portion $prst$ of the plate. The corresponding deflection of this portion of the plate is governed by the differential equation (103), which becomes

$$\frac{\partial^4 w}{\partial x^4} + 2\frac{\partial^4 w}{\partial x^2 \partial y^2} + \frac{\partial^4 w}{\partial y^4} = \frac{4q}{\pi D} \sum_{m=1,3,5,\ldots}^{\infty} \frac{(-1)^{(m-1)/2}}{m} \sin\frac{m\pi u}{2a} \sin\frac{m\pi x}{a} \quad (b)$$

Let us again take the deflection in the form

$$w = w_1 + w_2 \quad (c)$$

where w_1 is a particular solution of Eq. (b), independent of the variable y, that is, satisfying the equation

$$\frac{\partial^4 w_1}{\partial x^4} = \frac{4q}{\pi D} \sum_{m=1,3,5\ldots}^{\infty} \frac{(-1)^{(m-1)/2}}{m} \sin\frac{m\pi u}{2a} \sin\frac{m\pi x}{a}$$

Integrating this latter equation with respect to x, we obtain

$$w_1 = \frac{4qa^4}{\pi^5 D} \sum_{m=1,3,5,\ldots}^{\infty} \frac{(-1)^{(m-1)/2}}{m^5} \sin\frac{m\pi u}{2a} \sin\frac{m\pi x}{a} \quad (d)$$

Then w_2 must be a solution of Eq. (137) (page 114). Choosing the form (136) for this solution and keeping in the expression (138) for Y_m only even functions of y, because of the symmetry of the deflection surface with respect to the x axis, we have, by Eq. (c),

$$w = \sum_{m=1,3,5,\ldots}^{\infty} \left(a_m + A_m \cosh\frac{m\pi y}{a} + B_m \frac{m\pi y}{a} \sinh\frac{m\pi y}{a} \right) \sin\frac{m\pi x}{a} \quad (e)$$

in which, this time,

$$a_m = \frac{4qa^4}{\pi^5 m^5 D} (-1)^{(m-1)/2} \sin\frac{m\pi u}{2a} \quad (f)$$

Equation (e) represents deflections of the portion $prst$ of the plate.

Considering now the unloaded portion of the plate below the line ts we can take the deflection surface in the form

$$w' = \sum_{m=1,3,5,\ldots}^{\infty} \left(A'_m \cosh\frac{m\pi y}{a} + B'_m \frac{m\pi y}{a} \sinh\frac{m\pi y}{a} \right.$$
$$\left. + C'_m \sinh\frac{m\pi y}{a} + D'_m \frac{m\pi y}{a} \cosh\frac{m\pi y}{a} \right) \sin\frac{m\pi x}{a} \quad (g)$$

It is now necessary to choose the constants $A_m, B_m, \ldots, D'_m$ in the

SIMPLY SUPPORTED RECTANGULAR PLATES

series (*e*) and (*g*) in such a manner as to satisfy the boundary conditions at $y = b/2$ and the continuity conditions along the line ts. To represent these conditions in a simpler form, let us introduce the notation

$$\alpha_m = \frac{m\pi b}{2a} \qquad \gamma_m = \frac{m\pi v}{4a} \qquad (h)$$

The geometric conditions along the line ts require that

$$w = w' \quad \text{and} \quad \frac{\partial w}{\partial y} = \frac{\partial w'}{\partial y} \quad \text{for } y = \frac{v}{2} \qquad (i)$$

Furthermore, since there are no concentrated forces applied along the line ts, the bending moments M_y and the shearing forces Q_y must be continuous along this line. Observing Eqs. (*i*) these latter conditions can be written down in the form

$$\frac{\partial^2 w}{\partial y^2} = \frac{\partial^2 w'}{\partial y^2} \quad \text{and} \quad \frac{\partial^3 w}{\partial y^3} = \frac{\partial^3 w'}{\partial y^3} \quad \text{for } y = \frac{v}{2} \qquad (j)$$

Substituting expressions (*e*) and (*g*) in Eqs. (*i*) and (*j*) and using notation (*h*), we can represent these equations in the following form:

$$\begin{aligned}
(A_m - A'_m) \cosh 2\gamma_m &+ (B_m - B'_m) 2\gamma_m \sinh 2\gamma_m \\
&- C'_m \sinh 2\gamma_m - D'_m 2\gamma_m \cosh 2\gamma_m + a_m = 0 \\
(A_m - A'_m) \sinh 2\gamma_m &+ (B_m - B'_m)(\sinh 2\gamma_m + 2\gamma_m \cosh 2\gamma_m) \\
&- C'_m \cosh 2\gamma_m - D'_m (\cosh 2\gamma_m + 2\gamma_m \sinh 2\gamma_m) = 0 \\
(A_m - A'_m) \cosh 2\gamma_m &+ (B_m - B'_m)(2 \cosh 2\gamma_m + 2\gamma_m \sinh 2\gamma_m) \\
&- C'_m \sinh 2\gamma_m - D'_m (2 \sinh 2\gamma_m + 2\gamma_m \cosh 2\gamma_m) = 0 \\
(A_m - A'_m) \sinh 2\gamma_m &+ (B_m - B'_m)(3 \sinh 2\gamma_m + 2\gamma_m \cosh 2\gamma_m) \\
&- C'_m \cosh 2\gamma_m - D'_m (3 \cosh 2\gamma_m + 2\gamma_m \sinh 2\gamma_m) = 0
\end{aligned} \qquad (k)$$

From these equations we find

$$\begin{aligned}
A_m - A'_m &= a_m(\gamma_m \sinh 2\gamma_m - \cosh 2\gamma_m) \\
B_m - B'_m &= \frac{a_m}{2} \cosh 2\gamma_m \\
C'_m &= a_m(\gamma_m \cosh 2\gamma_m - \sinh 2\gamma_m) \\
D'_m &= \frac{a_m}{2} \sinh 2\gamma_m
\end{aligned} \qquad (l)$$

To these four equations, containing six constants $A_m, \ldots, D'_m$, we add two equations representing the boundary conditions at the edge $y = b/2$. Substituting expression (*g*) in the conditions $w' = 0$, $\partial^2 w'/\partial y^2 = 0$ at $y = b/2$ we obtain

$$\begin{aligned}
A'_m \cosh \alpha_m + B'_m \alpha_m \sinh \alpha_m + C'_m \sinh \alpha_m + D'_m \alpha \cosh \alpha_m &= 0 \\
B'_m \cosh \alpha_m + D'_m \sinh \alpha_m &= 0
\end{aligned} \qquad (m)$$

Equations (m), together with Eqs. (l), yield the constants

$$A_m = -\frac{a_m}{\cosh \alpha_m}\left[\cosh(\alpha_m - 2\gamma_m) + \gamma_m \sinh(\alpha_m - 2\gamma_m) + \alpha_m \frac{\sinh 2\gamma_m}{2 \cosh \alpha_m}\right] \quad (n)$$

$$B_m = \frac{a_m}{2\cosh \alpha_m}\cosh(\alpha_m - 2\gamma_m)$$

Substituting these and expression (f) in Eq. (e), we obtain

$$w = \frac{4qa^4}{D\pi^5}\sum_{m=1,3,5,\ldots}^{\infty}\frac{(-1)^{(m-1)/2}}{m^5}\sin\frac{m\pi u}{2a}\left\{1 - \frac{\cosh\frac{m\pi y}{a}}{\cosh \alpha_m}\right.$$

$$\left[\cosh(\alpha_m - 2\gamma_m) + \gamma_m \sinh(\alpha_m - 2\gamma_m) + \alpha_m \frac{\sinh 2\gamma_m}{2\cosh \alpha_m}\right]$$

$$\left. + \frac{\cosh(\alpha_m - 2\gamma_m)}{2\cosh \alpha_m}\frac{m\pi y}{a}\sinh\frac{m\pi y}{a}\right\}\sin\frac{m\pi x}{a} \quad (142)$$

From this equation the deflection at any point of the loaded portion of the plate can be calculated.

In the particular case where $u = a$ and $v = b$ we have, from Eqs. (h), $\gamma_m = \alpha_m/2$. Expressions (n) become

$$A_m = -\frac{a_m}{\cosh \alpha_m}\left(1 + \frac{\alpha_m}{2}\tanh \alpha_m\right) \qquad B_m = \frac{a_m}{2\cosh \alpha_m}$$

and Eq. (142) coincides with Eq. (139) (page 116) derived for a uniformly loaded rectangular plate.

The maximum deflection of the plate is at the center and is obtained by substituting $y = 0$, $x = a/2$ in formula (142), which gives

$$w_{\max} = \frac{4qa^4}{D\pi^5}\sum_{m=1,3,5,\ldots}^{\infty}\frac{1}{m^5}\sin\frac{m\pi u}{2a}\left\{1 - \frac{1}{\cosh \alpha_m}\right.$$

$$\left.\left[\cosh(\alpha_m - 2\gamma_m) + \gamma_m \sinh(\alpha_m - 2\gamma_m) + \alpha_m \frac{\sinh 2\gamma_m}{2\cosh \alpha_m}\right]\right\} \quad (143)$$

As a particular example let us consider the case where $u = a$ and v is very small. This case represents a uniform distribution of load along the x axis. Considering γ_m as small in Eq. (143) and retaining only small terms of the first order, we obtain, using the notation $qv = q_0$,

$$w_{\max} = \frac{q_0 a^3}{D\pi^4}\sum_{m=1,3,5,\ldots}^{\infty}\frac{(-1)^{(m-1)/2}}{m^4}\left(\tanh \alpha_m - \frac{\alpha_m}{\cosh^2 \alpha_m}\right) \quad (144)$$

For a square plate this equation gives

$$w_{\max} = 0.00674 \frac{q_0 a^3}{D}$$

In the general case the maximum deflection can be represented in the form

$$w_{\max} = \alpha \frac{q_0 a^3}{D} \quad \text{for } a < b$$

$$= \alpha \frac{q_0 b^3}{D} \quad \text{for } a > b$$

Several values of the coefficient α are given in Table 19.

TABLE 19. DEFLECTIONS OF SIMPLY SUPPORTED RECTANGULAR PLATES UNIFORMLY LOADED ALONG THE AXIS OF SYMMETRY PARALLEL TO THE DIMENSION a
$w_{\max} = \alpha q_0 a^3 / D$

b/a	2	1.5	1.4	1.3	1.2	1.1	1.0
α	0.00987	0.00911	0.00882	0.00844	0.00799	0.00742	0.00674

a/b	1.1	1.2	1.3	1.4	1.5	2.0	∞
α	0.00802	0.00926	0.01042	0.01151	0.01251	0.01629	0.02083

Returning to the general case where v is not necessarily small and u may have any value, the expressions for the bending moments M_x and M_y can be derived by using Eq. (142). The maximum values of these

TABLE 20. COEFFICIENTS β FOR $(M_x)_{\max}$ IN SIMPLY SUPPORTED PARTIALLY LOADED SQUARE PLATES
$\nu = 0.3$

$u/a =$	0	0.1	0.2	0.3	0.4	0.5	0.6	0.7	0.8	0.9	1.0
v/a				Coefficients β in the expression $(M_x)_{\max} = \beta P$							
0	∞	0.321	0.251	0.209	0.180	0.158	0.141	0.125	0.112	0.102	0.092
0.1	0.378	0.284	0.232	0.197	0.170	0.150	0.134	0.120	0.108	0.098	0.088
0.2	0.308	0.254	0.214	0.184	0.161	0.142	0.127	0.114	0.103	0.093	0.084
0.3	0.262	0.225	0.195	0.168	0.151	0.134	0.120	0.108	0.098	0.088	0.080
0.4	0.232	0.203	0.179	0.158	0.141	0.126	0.113	0.102	0.092	0.084	0.076
0.5	0.208	0.185	0.164	0.146	0.131	0.116	0.106	0.096	0.087	0.079	0.071
0.6	0.188	0.168	0.150	0.135	0.121	0.109	0.099	0.090	0.081	0.074	0.067
0.7	0.170	0.153	0.137	0.124	0.112	0.101	0.091	0.083	0.076	0.069	0.062
0.8	0.155	0.140	0.126	0.114	0.103	0.094	0.085	0.077	0.070	0.063	0.057
0.9	0.141	0.127	0.115	0.104	0.094	0.086	0.078	0.070	0.064	0.058	0.053
1.0	0.127	0.115	0.105	0.095	0.086	0.078	0.071	0.064	0.058	0.053	0.048

TABLE 21. COEFFICIENTS β AND β_1 FOR $(M_x)_{\max}$ AND $(M_y)_{\max}$ IN PARTIALLY LOADED RECTANGULAR PLATES WITH $b = 1.4a$
$\nu = 0.3$

$u/a =$	0	0.2	0.4	0.6	0.8	1.0	0	0.2	0.4	0.6	0.8	1.0
v/a	Coefficient β in the expression $(M_x)_{\max} = \beta P$						Coefficient β_1 in the expression $(M_y)_{\max} = \beta_1 P$					
0	∞	0.276	0.208	0.163	0.134	0.110	∞	0.299	0.230	0.183	0.151	0.125
0.2	0.332	0.239	0.186	0.152	0.125	0.103	0.246	0.208	0.175	0.147	0.124	0.102
0.4	0.261	0.207	0.168	0.138	0.115	0.095	0.177	0.157	0.138	0.119	0.101	0.083
0.6	0.219	0.181	0.151	0.126	0.105	0.086	0.138	0.125	0.111	0.097	0.083	0.069
0.8	0.187	0.158	0.134	0.112	0.094	0.078	0.112	0.102	0.091	0.080	0.069	0.058
1.0	0.162	0.139	0.118	0.100	0.084	0.070	0.093	0.085	0.077	0.068	0.058	0.049
1.2	0.141	0.122	0.104	0.089	0.075	0.062	0.079	0.072	0.065	0.058	0.050	0.042
1.4	0.123	0.106	0.091	0.077	0.065	0.054	0.068	0.062	0.056	0.050	0.043	0.036

TABLE 22. COEFFICIENTS β AND β_1 FOR $(M_x)_{\max}$ AND $(M_y)_{\max}$ IN PARTIALLY LOADED RECTANGULAR PLATES WITH $b = 2a$
$\nu = 0.3$

$u/a =$	0	0.2	0.4	0.6	0.8	1.0	0	0.2	0.4	0.6	0.8	1.0
v/a	Coefficient β in expression $(M_x)_{\max} = \beta P$						Coefficient β_1 in expression $(M_y)_{\max} = \beta_1 P$					
0	∞	0.289	0.220	0.175	0.144	0.118	∞	0.294	0.225	0.179	0.148	0.122
0.2	0.347	0.252	0.199	0.163	0.135	0.111	0.242	0.203	0.170	0.143	0.120	0.099
0.4	0.275	0.221	0.181	0.150	0.125	0.103	0.172	0.152	0.133	0.114	0.097	0.081
0.6	0.233	0.195	0.164	0.138	0.115	0.095	0.133	0.120	0.106	0.093	0.079	0.066
0.8	0.203	0.174	0.148	0.126	0.106	0.088	0.107	0.097	0.087	0.076	0.065	0.054
1.0	0.179	0.155	0.134	0.115	0.097	0.080	0.089	0.081	0.073	0.064	0.055	0.046
1.2	0.161	0.141	0.122	0.105	0.089	0.074	0.074	0.068	0.061	0.054	0.046	0.039
1.4	0.144	0.127	0.111	0.096	0.081	0.068	0.064	0.058	0.052	0.046	0.040	0.033
1.6	0.130	0.115	0.101	0.087	0.074	0.062	0.056	0.051	0.046	0.040	0.035	0.029
1.8	0.118	0.104	0.091	0.079	0.067	0.056	0.049	0.045	0.041	0.036	0.031	0.026
2.0	0.107	0.094	0.083	0.072	0.061	0.051	0.044	0.041	0.037	0.032	0.028	0.023

moments occur at the center of the plate and can be represented by the formulas

$$(M_x)_{\max} = \beta uvq = \beta P \qquad (M_y)_{\max} = \beta_1 uvq = \beta_1 P$$

where $P = uvq$ is the total load. The values of the numerical factors β for a square plate and for various sizes of the loaded rectangle are given in Table 20. The coefficients β_1 can also be obtained from this table by interchanging the positions of the letters u and v.

SIMPLY SUPPORTED RECTANGULAR PLATES

The numerical factors β and β_1 for plates with the ratios $b = 1.4a$ and $b = 2a$ are given in Tables 21 and 22, respectively.[1]

34. Concentrated Load on a Simply Supported Rectangular Plate. Using Navier's method an expression in double-series form has been obtained in Art. 29 for the deflection of a plate carrying a single load P at some given point $x = \xi$, $y = \eta$ (Fig. 70). To obtain an equivalent solution in the form of a simple series we begin by representing the Navier solution (133) in the following manner:

$$w = \frac{4Pb^3}{\pi^4 a} \sum_{m=1}^{\infty} S_m \sin \frac{m\pi\xi}{a} \sin \frac{m\pi x}{a} \quad (a)$$

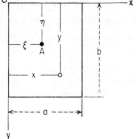

Fig. 70

the coefficient S_m being given by

$$S_m = \sum_{n=1}^{\infty} \frac{\sin \frac{n\pi\eta}{b} \sin \frac{n\pi y}{b}}{\left(\frac{m^2 b^2}{a^2} + n^2\right)^2} \quad (b)$$

Introducing the notation

$$S'_m = \sum_{n=1}^{\infty} \frac{\cos \frac{n\pi(y-\eta)}{b}}{\left(\frac{m^2 b^2}{a^2} + n^2\right)^2} \qquad S''_m = \sum_{n=1}^{\infty} \frac{\cos \frac{n\pi(\eta+y)}{b}}{\left(\frac{m^2 b^2}{a^2} + n^2\right)^2} \quad (c)$$

we can also represent expression (b) in the form

$$S_m = \tfrac{1}{2}(S'_m - S''_m) \quad (d)$$

To evaluate the sums (c) we use the known series

$$\sum_{n=1}^{\infty} \frac{\cos nz}{\alpha^2 + n^2} = -\frac{1}{2\alpha^2} + \frac{\pi}{2\alpha} \frac{\cosh \alpha(\pi - z)}{\sinh \pi\alpha} \quad (e)$$

which holds for $0 \leq z < 2\pi$ and which we regard, first of all, as a function $S(\alpha)$ of α. Differentiation of the left-hand side of Eq. (e) with respect to α gives

$$\frac{\partial S(\alpha)}{\partial \alpha} = -2\alpha \sum_{n=1}^{\infty} \frac{\cos nz}{(\alpha^2 + n^2)^2} \quad (f)$$

After differentiating also the right-hand side of Eq. (e) and substituting

---

[1] The values of M_x and M_y for various ratios a/b, u/a, and v/b are also given in the form of curves by G. Pigeaud, *Ann. ponts et chaussées*, 1929. See also Art. 37 of this book.

the result in Eq. (f), we conclude that

$$\sum_{n=1}^{\infty} \frac{\cos nz}{(\alpha^2 + n^2)^2} = -\frac{1}{2\alpha} \frac{\partial S(\alpha)}{\partial \alpha} = -\frac{1}{2\alpha^4} + \frac{\pi}{4\alpha^3} \frac{\cosh \alpha(\pi - z)}{\sinh \pi\alpha}$$
$$- \frac{\pi(\pi - z)}{4\alpha^2} - \frac{\sinh \alpha(\pi - z)}{\sinh \pi\alpha} + \frac{\pi^2}{4\alpha^2} \frac{\cosh \alpha(\pi - z) \cosh \pi\alpha}{\sinh^2 \pi\alpha} \quad (g)$$

Now, to obtain the values of the sums (c) we have to put, in Eq. (g), first $z = (\pi/b)(y - \eta)$, then $z = (\pi/b)(y + \eta)$ and, in addition, $\alpha = mb/a$. Using these values for substitution in Eqs. (d) and (a) we arrive, finally, at the following expression for the deflection of the plate:

$$w = \frac{Pa^2}{\pi^3 D} \sum_{m=1}^{\infty} \left(1 + \beta_m \coth \beta_m - \frac{\beta_m y_1}{b} \coth \frac{\beta_m y_1}{b} - \frac{\beta_m \eta}{b} \coth \frac{\beta_m \eta}{b}\right)$$
$$\frac{\sinh \frac{\beta_m \eta}{b} \sinh \frac{\beta_m y_1}{b} \sin \frac{m\pi \xi}{a} \sin \frac{m\pi x}{a}}{m^3 \sinh \beta_m} \quad (145)$$

in which

$$\beta_m = \frac{m\pi b}{a} \qquad y_1 = b - y \qquad \text{and} \qquad y \geq \eta$$

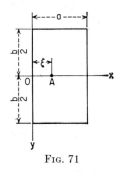

Fig. 71

In the case of $y < \eta$ the quantity y_1 must be replaced by y and the quantity η by $\eta_1 = b - \eta$, in using expression (145).

Let us consider more closely the particular case of a load P concentrated at a point A on the axis of symmetry of the plate, which may be used as the x axis (Fig. 71). With $\eta = b/2$ and the notation

$$\alpha_m = \frac{m\pi b}{2a} = \frac{\beta_m}{2} \quad (h)$$

the general expression (145) for the deflection of the plate becomes

$$w = \frac{Pa^2}{2\pi^3 D} \sum_{m=1}^{\infty} \left[(1 + \alpha_m \tanh \alpha_m) \sinh \frac{\alpha_m}{b}(b - 2y)\right.$$
$$\left. - \frac{\alpha_m}{b}(b - 2y) \cosh \frac{\alpha_m}{b}(b - 2y)\right] \frac{\sin \frac{m\pi \xi}{a} \sin \frac{m\pi x}{a}}{m^3 \cosh \alpha_m} \quad (146)$$

which is valid for $y \geq 0$, that is, below the x axis in Fig. 71. Putting, in particular, $y = 0$ we obtain the deflection of the plate along the x axis in the form

$$(w)_{y=0} = \frac{Pa^2}{2\pi^3 D} \sum_{m=1}^{\infty} \left(\tanh \alpha_m - \frac{\alpha_m}{\cosh^2 \alpha_m} \right) \frac{\sin \frac{m\pi\xi}{a} \sin \frac{m\pi x}{a}}{m^3} \qquad (i)$$

This series converges rapidly, and the first few terms give the deflections with sufficient accuracy. In the case of a load P applied at the center of the plate, the maximum deflection, which is at the center, is obtained by substituting $x = \xi = a/2$ in expression (i). In this way we arrive at the result

$$w_{\max} = \frac{Pa^2}{2\pi^3 D} \sum_{m=1}^{\infty} \frac{1}{m^3} \left(\tanh \alpha_m - \frac{\alpha_m}{\cosh^2 \alpha_m} \right) = \alpha \frac{Pa^2}{D} \qquad (147)$$

Values of the numerical factor α for various values of the ratio b/a are given in Table 23.

TABLE 23. FACTOR α FOR DEFLECTION (147) OF A CENTRALLY LOADED RECTANGULAR PLATE

$b/a =$	1.0	1.1	1.2	1.4	1.6	1.8	2.0	3.0	∞
$\alpha =$	0.01160	0.01265	0.01353	0.01484	0.01570	0.01620	0.01651	0.01690	0.01695

It is seen that the maximum deflection rapidly approaches that of an infinitely long plate[1] as the length of the plate increases. The comparison of the maximum deflection of a square plate with that of a centrally loaded circular plate inscribed in the square (see page 68) indicates that the deflection of the circular plate is larger than that of the corresponding square plate. This result may be attributed to the action of the reactive forces concentrated at the corners of the square plate which have the tendency to produce deflection of the plate convex upward.

The calculation of bending moments is discussed in Arts. 35 and 37.

35. Bending Moments in a Simply Supported Rectangular Plate with a Concentrated Load. To determine the bending moments along the central axis $y = 0$ of the plate loaded according to Fig. 71 we calculate the second derivatives of expression (146), which become

$$\left(\frac{\partial^2 w}{\partial x^2} \right)_{y=0} = -\frac{P}{2D\pi} \sum_{m=1}^{\infty} \frac{\sin \frac{m\pi\xi}{a}}{m} \left(\tanh \alpha_m - \frac{\alpha_m}{\cosh^2 \alpha_m} \right) \sin \frac{m\pi x}{a}$$

$$\left(\frac{\partial^2 w}{\partial y^2} \right)_{y=0} = -\frac{P}{2D\pi} \sum_{m=1}^{\infty} \frac{\sin \frac{m\pi\xi}{a}}{m} \left(\tanh \alpha_m + \frac{\alpha_m}{\cosh^2 \alpha_m} \right) \sin \frac{m\pi x}{a}$$

---

[1] The deflection of plates by a concentrated load was investigated experimentally by M. Bergsträsser; see *Forschungsarb.*, vol. 302, Berlin, 1928; see also the report of N. M. Newmark and H. A. Lepper, *Univ. Illinois Bull.*, vol. 36, no. 84, 1939.

Substituting these derivatives into expressions (101) for the bending moments, we obtain

$$(M_x)_{y=0} = \frac{P}{2\pi} \sum_{m=1}^{\infty} \frac{\sin \frac{m\pi\xi}{a}}{m} \left[(1+\nu) \tanh \alpha_m - \frac{(1-\nu)\alpha_m}{\cosh^2 \alpha_m} \right] \sin \frac{m\pi x}{a}$$

$$(M_y)_{y=0} = \frac{P}{2\pi} \sum_{m=1}^{\infty} \frac{\sin \frac{m\pi\xi}{a}}{m} \left[(1+\nu) \tanh \alpha_m + \frac{(1-\nu)\alpha_m}{\cosh^2 \alpha_m} \right] \sin \frac{m\pi x}{a} \quad (a)$$

When b is very large in comparison with a, we can put

$$\tanh \alpha_m \approx 1 \qquad \frac{\alpha_m}{\cosh^2 \alpha_m} \approx 0$$

Then $\quad (M_x)_{y=0} = (M_y)_{y=0} = \dfrac{(1+\nu)P}{2\pi} \sum\limits_{m=1}^{\infty} \dfrac{1}{m} \sin \dfrac{m\pi\xi}{a} \sin \dfrac{m\pi x}{a} \quad (b)$

This series does not converge rapidly enough for a satisfactory calculation of the moments in the vicinity of the point of application of the load P, so it is necessary to derive another form of representation of the moments near that point. From the discussion of bending of a circular plate by a force applied at the center (see Art. 19) we know that the shearing forces and bending moments become infinitely large at the point of application of the load. We have similar conditions also in the case of a rectangular plate. The stress distribution within a circle of small radius with its center at the point of application of the load is substantially the same as that near the center of a centrally loaded circular plate. The bending stress at a point within this circle may be considered as consisting of two parts: one is the same as that in the case of a centrally loaded circular plate of radius a, and the other represents the difference between the stresses in a circular and those in a rectangular plate. As the distance r between the point of application of the load and the point under consideration becomes smaller and smaller, the first part of the stresses varies as $\log (a/r)$ and becomes infinite at the center, whereas the second part, representing the effect of the difference in the boundary conditions of the two plates, remains continuous.

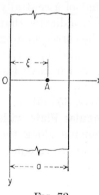

Fig. 72

To obtain the expressions for bending moments in the vicinity of the point of application of the load we begin with the simpler case of an infinitely long plate (Fig. 72). The deflection of such a plate can readily be derived from expression (146) by increasing the length of the side b, and consequently the quantity $\alpha_m = m\pi b/2a$, indefinitely, i.e., by putting

$$\tanh \alpha_m \approx 1 \qquad \cosh \alpha_m \approx \tfrac{1}{2} e^{\alpha_m}$$
$$\sinh \frac{\alpha_m}{b}(b-2y) \approx \cosh \frac{\alpha_m}{b}(b-2y) \approx \frac{1}{2} e^{(\alpha_m/b)(b-2y)}$$

Substituting this into Eq. (146) the required deflection of the simply supported strip carrying a concentrated load P at $x = \xi$, $y = 0$ becomes[1]

$$w = \frac{Pa^2}{2\pi^3 D} \sum_{m=1}^{\infty} \frac{1}{m^3} \sin \frac{m\pi \xi}{a} \sin \frac{m\pi x}{a} \left(1 + \frac{m\pi y}{a}\right) e^{-m\pi y/a} \qquad (148)$$

which holds for $y \geq 0$, that is, below the x axis (Fig. 72).

The corresponding expressions for the bending moments and the twisting moment are readily obtained by means of Eqs. (101) and (102). We have

$$M_x = \frac{P}{2\pi} \sum_{m=1}^{\infty} \frac{1}{m} \sin \frac{m\pi \xi}{a} \sin \frac{m\pi x}{a} \left[1 + \nu + (1-\nu)\frac{m\pi y}{a}\right] e^{-m\pi y/a}$$

$$M_y = \frac{P}{2\pi} \sum_{m=1}^{\infty} \frac{1}{m} \sin \frac{m\pi \xi}{a} \sin \frac{m\pi x}{a} \left[1 + \nu - (1-\nu)\frac{m\pi y}{a}\right] e^{-m\pi y/a} \qquad (149)$$

$$M_{xy} = -\frac{P}{2a} y(1-\nu) \sum_{m=1}^{\infty} \sin \frac{m\pi \xi}{a} \cos \frac{m\pi x}{a} e^{-m\pi y/a}$$

Once again using the quantity $M = (M_x + M_y)/(1 + \nu)$ introduced on page 92, we have

$$M = -D\left(\frac{\partial^2 w}{\partial x^2} + \frac{\partial^2 w}{\partial y^2}\right) = \frac{P}{\pi} \sum_{m=1}^{\infty} \frac{1}{m} \sin \frac{m\pi \xi}{a} \sin \frac{m\pi x}{a} e^{-m\pi y/a} \qquad (150)$$

The moments (149) can be expressed now in terms of the function M in the following simple manner:

$$M_x = \frac{1}{2}\left[(1+\nu)M - (1-\nu)y \frac{\partial M}{\partial y}\right]$$
$$M_y = \frac{1}{2}\left[(1+\nu)M + (1-\nu)y \frac{\partial M}{\partial y}\right] \qquad (151)$$
$$M_{xy} = -\frac{1}{2}(1-\nu)y \frac{\partial M}{\partial x}$$

---

[1] This important case of bending of a plate has been discussed in detail by A. Nádai; see his book "Elastische Platten," pp. 78–109, Berlin, 1925.

Summing up the series (150), we obtain the expression[1]

$$M = \frac{P}{4\pi} \log \frac{\cosh \frac{\pi y}{a} - \cos \frac{\pi(x+\xi)}{a}}{\cosh \frac{\pi y}{a} - \cos \frac{\pi(x-\xi)}{a}} \qquad (152)$$

and, using Eqs. (151), we are able now to represent the moments of the infinitely long plate in a closed form. Observing, furthermore, that $\Delta\Delta w = 0$ everywhere, except at the point $(x = \xi, y = 0)$ of the application of the load, we conclude that the function $M = -D\,\Delta w$ satisfies (except at the above-mentioned point) the equation $\Delta M = 0$. By virtue of the second of the equations (111) the boundary condition $M = 0$ along the edges $x = 0$ and $x = a$ is also satisfied by the function M.

For the points along the x axis Eqs. (151) yield $M_x = M_y$ and therefore

$$(M_x)_{y=0} = (M_y)_{y=0} = (M)_{y=0} \frac{1+\nu}{2} \qquad (c)$$

Using Eqs. (c) and Eq. (152) in the particular case of a load applied at the center axis of the strip, $\xi = a/2$, we obtain

$$(M_x)_{y=0} = (M_y)_{y=0} = \frac{P(1+\nu)}{8\pi} \log \frac{1 + \sin \frac{\pi x}{a}}{1 - \sin \frac{\pi x}{a}} \qquad (d)$$

a result which also can be obtained by summation of the series (b).

Now let us return to the calculation of bending moments for points which are close to the point of application of the load but not necessarily on the x axis. In this case the quantities $(x - \xi)$ and y are small and, using expression (152), we can put

$$\cos \frac{\pi(x-\xi)}{a} \approx 1 - \frac{\pi^2(x-\xi)^2}{2a^2} \qquad \cosh \frac{\pi y}{a} \approx 1 + \frac{\pi^2 y^2}{2a^2}$$

Thus we arrive at the result

$$M = \frac{P}{4\pi} \log \frac{1 - \cos \frac{2\pi\xi}{a}}{1 + \frac{\pi^2 y^2}{2a^2} - 1 + \frac{\pi^2(x-\xi)^2}{2a^2}}$$

$$= \frac{P}{4\pi} \log \left(\frac{2a \sin \frac{\pi\xi}{a}}{\pi r}\right)^2 = \frac{P}{2\pi} \log \frac{2a \sin \frac{\pi\xi}{a}}{\pi r} \qquad (153)$$

[1] See, for instance, W. Magnus and F. Oberhettinger, "Formeln und Sätze fur die speziellen Funktionen der mathematischen Physik," 2d ed., p. 214, Berlin, 1948.

in which
$$r = \sqrt{(x - \xi)^2 + y^2}$$
represents the distance of the point under consideration from the point of application of the load P. Now, using expression (153) for substitution in Eqs. (151) we obtain the following expressions, valid for points in the vicinity of the concentrated load:

$$M_x = \frac{1}{2}\left[(1 + \nu)\frac{P}{2\pi}\log\frac{2a\sin\frac{\pi\xi}{a}}{\pi r} + \frac{(1 - \nu)Py^2}{2\pi r^2}\right]$$

$$M_y = \frac{1}{2}\left[(1 + \nu)\frac{P}{2\pi}\log\frac{2a\sin\frac{\pi\xi}{a}}{\pi r} - \frac{(1 - \nu)Py^2}{2\pi r^2}\right] \quad (154)$$

It is interesting to compare this result with that for a centrally loaded, simply supported circular plate (see Art. 19). Taking a radius r under an angle α to the x axis, we find, from Eqs. (90) and (91), for a circular plate

$$M_x = M_n \cos^2 \alpha + M_t \sin^2 \alpha = \frac{P}{4\pi}(1 + \nu)\log\frac{a}{r} + (1 - \nu)\frac{P}{4\pi}\frac{y^2}{r^2}$$
$$M_y = M_n \sin^2 \alpha + M_t \cos^2 \alpha = \frac{P}{4\pi}(1 + \nu)\log\frac{a}{r} + (1 - \nu)\frac{P}{4\pi}\frac{x^2}{r^2} \quad (e)$$

The first terms of expressions (154) and (e) will coincide if we take the outer radius of the circular plate equal to

$$\frac{2a}{\pi}\sin\frac{\pi\xi}{a}$$

Under this condition the moments M_x are the same for both cases. The moment M_y for the long rectangular plate is obtained from that of the circular plate by subtraction of the constant quantity[1] $(1 - \nu)P/4\pi$. From this it can be concluded that in a long rectangular plate the stress distribution around the point of application of the load is obtained by superposing on the stresses of a centrally loaded circular plate with radius $(2a/\pi)\sin(\pi\xi/a)$ a simple bending produced by the moments $M_y = -(1 - \nu)P/4\pi$.

It may be assumed that the same relation between the moments of circular and long rectangular plates also holds in the case of a load P uniformly distributed over a circular area of small radius c. In such a case, for the center of a circular plate we obtain from Eq. (83), by neglecting the term containing c^2,

$$M_{\max} = \frac{P}{4\pi}\left[(1 + \nu)\log\frac{a}{c} + 1\right]$$

[1] We observe that $x^2 = r^2 - y^2$.

Hence at the center of the loaded circular area of a long rectangular plate we obtain from Eqs. (154)

$$M_x = \frac{P}{4\pi}\left[(1+\nu)\log\frac{2a\sin\frac{\pi\xi}{a}}{\pi c}+1\right]$$

$$M_y = \frac{P}{4\pi}\left[(1+\nu)\log\frac{2a\sin\frac{\pi\xi}{a}}{\pi c}+1\right]-\frac{(1-\nu)P}{4\pi}$$

(155)

From this comparison of a long rectangular plate with a circular plate it may be concluded that all information regarding the local stresses at the point of application of the load P, derived for a circular plate by using the thick-plate theory (see Art. 19), can also be applied in the case of a long rectangular plate.

When the plate is not very long, Eqs. (a) should be used instead of Eq. (b) in the calculation of the moments M_x and M_y along the x axis. Since $\tanh \alpha_m$ approaches unity rapidly and $\cosh \alpha_m$ becomes a large number when m increases, the differences between the sums of series (a) and the sum of series (b) can easily be calculated, and the moments M_x and M_y along the x axis and close to the point of application of the load can be represented in the following form:

$$M_x = \frac{(1+\nu)P}{2\pi}\sum_{m=1}^{\infty}\frac{1}{m}\sin\frac{m\pi\xi}{a}\sin\frac{m\pi x}{a}+\gamma_1\frac{P}{4\pi}$$

$$= \frac{P(1+\nu)}{4\pi}\log\frac{2a\sin\frac{\pi\xi}{a}}{\pi r}+\gamma_1\frac{P}{4\pi}$$

$$M_y = \frac{(1+\nu)P}{2\pi}\sum_{m=1}^{\infty}\frac{1}{m}\sin\frac{m\pi\xi}{a}\sin\frac{m\pi x}{a}+\gamma_2\frac{P}{4\pi}$$

$$= \frac{P(1+\nu)}{4\pi}\log\frac{2a\sin\frac{\pi\xi}{a}}{\pi r}+\gamma_2\frac{P}{4\pi}$$

(156)

in which γ_1 and γ_2 are numerical factors the magnitudes of which depend on the ratio b/a and the position of the load on the x axis. Several values of these factors for the case of central application of the load are given in Table 24.

Again the stress distribution near the point of application of the load is substantially the same as for a centrally loaded circular plate of radius $(2a/\pi)\sin(\pi\xi/a)$. To get the bending moments M_x and M_y near the load we have only to superpose on the moments of the

TABLE 24. FACTORS γ_1 AND γ_2 IN EQS. (156)

b/a	1.0	1.2	1.4	1.6	1.8	2.0	∞
γ_1	-0.565	-0.350	-0.211	-0.125	-0.073	-0.042	0
γ_2	$+0.135$	$+0.115$	$+0.085$	$+0.057$	$+0.037$	$+0.023$	0

circular plate the uniform bending by the moments $M'_x = \gamma_1 P/4\pi$ and $M'_y = -(1 - \nu - \gamma_2)P/4\pi$. Assuming that this conclusion holds also when the load P is uniformly distributed over a circle of a small radius c, we obtain for the center of the circle

$$M_x = \frac{P}{4\pi}\left[(1+\nu)\log\frac{2a\sin\frac{\pi\xi}{a}}{\pi c} + 1\right] + \frac{\gamma_1 P}{4\pi}$$

$$M_y = \frac{P}{4\pi}\left[(1+\nu)\log\frac{2a\sin\frac{\pi\xi}{a}}{\pi c} + 1\right] - (1-\nu-\gamma_2)\frac{P}{4\pi}$$

(157)

Just as in the case of a distributed load, reactive forces acting downward and considerable clamping moments are produced by concentrated loads at the corners of a rectangular plate. The corner reactions

$$R = nP \qquad (f)$$

due to a central load P are given in Table 25 by the numerical values of the factor n, whereas the clamping moments have the value of $-R/2$ (see page 85). The computation of the values of R has been carried out by a simple method which will be described in Art. 36.

TABLE 25. NUMERICAL FACTOR n FOR REACTIVE FORCES R AT THE CORNERS OF SIMPLY SUPPORTED RECTANGULAR PLATES UNDER CENTRAL LOAD
$\nu = 0.3$

$b/a =$	1.0	1.2	1.4	1.6	1.8	2.0	3.0	∞
$n =$	0.1219	0.1162	0.1034	0.0884	0.0735	0.0600	0.0180	0

The distribution of the bending moments and reactive pressures in the particular case of a square plate with a central load is shown in Fig. 73. The dashed portion of the curves holds for a uniform distribution of the load P over the shadowed circular area with a radius of $c = 0.05a$.

36. Rectangular Plates of Infinite Length with Simply Supported Edges. In our foregoing discussions infinitely long plates have been considered in several cases. The deflections and moments in such plates were usually obtained from the corresponding solutions for a finite plate by letting the length of the plate increase indefinitely. In some cases

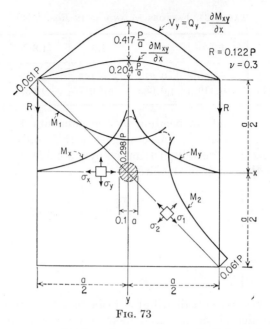

Fig. 73

it is advantageous to obtain solutions for an infinitely long plate first and combine them in such a way as to obtain the solution for a finite plate. Several examples of this method of solution will be given in this article. We begin with the case of an infinitely long plate of width a loaded along the x axis as shown in Fig. 74. Since the deflection surface is symmetrical with respect to the x axis, we need consider only the portion of the plate corresponding to positive values of y in our further discussion. Since the load is distributed only along the x axis, the deflection w of the plate satisfies the equation

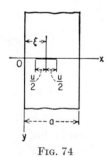

Fig. 74

$$\frac{\partial^4 w}{\partial x^4} + 2\frac{\partial^4 w}{\partial x^2\, \partial y^2} + \frac{\partial^4 w}{\partial y^4} = 0 \qquad (a)$$

We take the solution of this equation in the form

$$w = \sum_{m=1}^{\infty} Y_m \sin \frac{m\pi x}{a} \qquad (b)$$

which satisfies the boundary conditions along the simply supported longitudinal edges of the plate. To satisfy Eq. (a), functions Y_m must be chosen so as to satisfy the equation

$$Y_m^{IV} - 2\frac{m^2\pi^2}{a^2} Y_m'' + \frac{m^4\pi^4}{a^4} Y_m = 0$$

Taking the solution of this equation in the form

$$Y_m = A_m e^{m\pi y/a} + B_m \frac{m\pi y}{a} e^{m\pi y/a} + C_m e^{-m\pi y/a} + D_m \frac{m\pi y}{a} e^{-m\pi y/a} \quad (c)$$

and observing that the deflections and their derivatives approach zero at a great distance from the x axis, it may be concluded that the constants A_m and B_m should be taken equal to zero. Hence solution (b) can be represented as follows:

$$w = \sum_{m=1}^{\infty} \left(C_m + D_m \frac{m\pi y}{a} \right) e^{-m\pi y/a} \sin \frac{m\pi x}{a} \quad (d)$$

From the condition of symmetry we have

$$\left(\frac{\partial w}{\partial y} \right)_{y=0} = 0$$

This condition is satisfied by taking $C_m = D_m$ in expression (d). Then

$$w = \sum_{m=1}^{\infty} C_m \left(1 + \frac{m\pi y}{a} \right) e^{-m\pi y/a} \sin \frac{m\pi x}{a} \quad (e)$$

The constants C_m can be readily calculated in each particular case provided the load distribution along the x axis is given.

As an example, assume that the load is uniformly distributed along the entire width of the plate. The intensity of loading can then be represented by the following trigonometric series:

$$q = \frac{4}{\pi} q_0 \sum_{m=1,3,5,\ldots}^{\infty} \frac{1}{m} \sin \frac{m\pi x}{a}$$

in which q_0 is the load per unit length. Since the load is equally divided between the two halves of the plate, we see that

$$(Q_y)_{y=0} = -D \frac{\partial}{\partial y} \left(\frac{\partial^2 w}{\partial x^2} + \frac{\partial^2 w}{\partial y^2} \right)_{y=0} = -\frac{2}{\pi} q_0 \sum_{m=1,3,5\ldots}^{\infty} \frac{1}{m} \sin \frac{m\pi x}{a} \quad (f)$$

Substituting expression (e) for w, we obtain

$$\frac{2D\pi^3}{a^3} \sum_{m=1}^{\infty} C_m m^3 \sin \frac{m\pi x}{a} = \frac{2q_0}{\pi} \sum_{m=1,3,5,\ldots}^{\infty} \frac{1}{m} \sin \frac{m\pi x}{a}$$

from which

$$C_m = \frac{q_0 a^3}{D\pi^4 m^4} \quad \text{where } m = 1, 3, 5, \ldots$$

Hence
$$w = \frac{q_0 a^3}{\pi^4 D} \sum_{m=1,3,5,\ldots}^{\infty} \frac{1}{m^4} \left(1 + \frac{m\pi y}{a}\right) e^{-m\pi y/a} \sin \frac{m\pi x}{a} \qquad (g)$$

The deflection is a maximum at the center of the plate ($x = a/2, y = 0$), where

$$(w)_{\max} = \frac{q_0 a^3}{\pi^4 D} \sum_{m=1,3,5,\ldots}^{\infty} \frac{(-1)^{(m-1)/2}}{m^4} = \frac{5\pi q_0 a^3}{1{,}536 D} \qquad (h)$$

The same result can be obtained by setting $\tanh \alpha_m = 1$ and $\cosh \alpha_m = \infty$ in Eq. (144) (see page 138).

As another example of the application of solution (e), consider a load of length u uniformly distributed along a portion of the x axis (Fig. 74). Representing this load distribution by a trigonometric series, we obtain

$$q = \frac{4q_0}{\pi} \sum_{m=1}^{\infty} \frac{1}{m} \sin \frac{m\pi \xi}{a} \sin \frac{m\pi u}{2a} \sin \frac{m\pi x}{a}$$

where q_0 is the intensity of the load along the loaded portion of the x axis. The equation for determining the constants C_m, corresponding to Eq. (f), is

$$D \frac{\partial}{\partial y}\left(\frac{\partial^2 w}{\partial x^2} + \frac{\partial^2 w}{\partial y^2}\right)_{y=0} = \frac{2q_0}{\pi} \sum_{m=1}^{\infty} \frac{1}{m} \sin \frac{m\pi \xi}{a} \sin \frac{m\pi u}{2a} \sin \frac{m\pi x}{a}$$

Substituting expression (e) for w, we obtain

$$\frac{2D\pi^3}{a^3} \sum_{m=1}^{\infty} C_m m^3 \sin \frac{m\pi x}{a} = \frac{2q_0}{\pi} \sum_{m=1}^{\infty} \frac{1}{m} \sin \frac{m\pi \xi}{a} \sin \frac{m\pi u}{2a} \sin \frac{m\pi x}{a}$$

from which

$$C_m = \frac{q_0 a^3}{\pi^4 D m^4} \sin \frac{m\pi \xi}{a} \sin \frac{m\pi u}{2a}$$

Expression (e) for the deflections then becomes

$$w = \frac{q_0 a^3}{\pi^4 D} \sum_{m=1}^{\infty} \frac{1}{m^4} \sin \frac{m\pi \xi}{a} \sin \frac{m\pi u}{2a} \left(1 + \frac{m\pi y}{a}\right) e^{-m\pi y/a} \sin \frac{m\pi x}{a} \qquad (i)$$

The particular case of a concentrated force applied at a distance ξ from the origin is obtained by making the length u of the loaded portion of the x axis infinitely small. Substituting

SIMPLY SUPPORTED RECTANGULAR PLATES

$$q_0 u = P \quad \text{and} \quad \sin\frac{m\pi u}{a} \approx \frac{m\pi u}{a}$$

in Eq. (i), we obtain

$$w = \frac{Pa^2}{2\pi^3 D} \sum_{m=1}^{\infty} \frac{1}{m^3} \sin\frac{m\pi\xi}{a}\left(1 + \frac{m\pi y}{a}\right) e^{-m\pi y/a} \sin\frac{m\pi x}{a} \qquad (158)$$

an expression that coincides with expression (148) of the preceding article.

We can obtain various other cases of loading by integrating expression (i) for the deflection of a long plate under a load distributed along a portion u of the x axis. As an example, consider the case of a load of intensity q uniformly distributed over a rectangle with sides equal to u and v (shown shaded in Fig. 75). Taking an infinitesimal element of a load of magnitude $qu\,d\eta$ at a distance η from the x axis, the corresponding deflection produced by this load at points with $y > \eta$ is obtained by substituting $q\,d\eta$ for q_0 and $y - \eta$ for y in expression (i). The deflection produced by the entire load, at points for which $y \geqq v/2$, is now obtained by integration as follows:

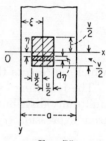

Fig. 75

$$w = \frac{qa^3}{\pi^4 D} \sum_{m=1}^{\infty} \frac{1}{m^4} \frac{m\pi\xi}{a} \sin\frac{m\pi u}{2a} \sin\frac{m\pi x}{a}$$
$$\int_{-v/2}^{v/2} \left[1 + \frac{m\pi(y-\eta)}{a}\right] e^{-\frac{m\pi(y-\eta)}{a}} d\eta$$

$$= \frac{qa^3}{\pi^4 D} \sum_{m=1}^{\infty} \frac{1}{m^4} \sin\frac{m\pi\xi}{a} \sin\frac{m\pi u}{2a} \sin\frac{m\pi x}{a}$$
$$\left[\left(\frac{2a}{m\pi} + y - \frac{v}{2}\right) e^{-\frac{m\pi(2y-v)}{2a}} - \left(\frac{2a}{m\pi} + y + \frac{v}{2}\right) e^{-\frac{m\pi(2y+v)}{2a}}\right] \qquad (j)$$

By a proper change of the limits of integration the deflection at points with $y < v/2$ can also be obtained. Let us consider the deflection along the x axis (Fig. 75). The deflection produced by the upper half of the load is obtained from expression (j) by substituting the quantity $v/4$ for y and for $v/2$. By doubling the result obtained in this way we also take into account the action of the lower half of the load and finally obtain

$$(w)_{y=0} = \frac{4qa^4}{\pi^5 D} \sum_{m=1}^{\infty} \frac{1}{m^5} \sin\frac{m\pi\xi}{a} \sin\frac{m\pi u}{2a} \sin\frac{m\pi x}{a}\left[1 - \left(1 + \frac{m\pi v}{4a}\right) e^{-m\pi v/2a}\right]$$
$$(k)$$

When $v = \infty$, the load, indicated in Fig. 75, is expanded along the entire length of the plate, and the deflection surface is cylindrical. The corresponding deflection, from expression (k), is

$$(w)_{y=0} = \frac{4qa^4}{\pi^5 D} \sum_{m=1}^{\infty} \frac{1}{m^5} \sin\frac{m\pi\xi}{a} \sin\frac{m\pi u}{2a} \sin\frac{m\pi x}{a} \quad (l)$$

Making $\xi = u/2 = a/2$ in this expression, we obtain

$$(w)_{y=0} = \frac{4qa^4}{\pi^5 D} \sum_{m=1,3,5,\ldots}^{\infty} \frac{1}{m^5} \sin\frac{m\pi x}{a}$$

which represents the deflection curve of a uniformly loaded strip.

The following expressions for bending moments produced by the load uniformly distributed along a portion u of the x axis are readily obtained from expression (i) for deflection w:

$$M_x = \frac{q_0 a}{\pi^2} \sum_{m=1}^{\infty} \frac{1}{m^2} \sin\frac{m\pi\xi}{a} \sin\frac{m\pi u}{2a} \sin\frac{m\pi x}{a}$$
$$\left[1 + \nu + (1-\nu)\frac{m\pi y}{a}\right] e^{-m\pi y/a} \quad (m)$$
$$M_y = \frac{q_0 a}{\pi^2} \sum_{m=1}^{\infty} \frac{1}{m^2} \sin\frac{m\pi\xi}{a} \sin\frac{m\pi u}{2a} \sin\frac{m\pi x}{a}$$
$$\left[1 + \nu - (1-\nu)\frac{m\pi y}{a} e^{-m\pi y/a}\right]$$

These moments have their maximum values on the x axis, where

$$(M_x)_{y=0} = (M_y)_{y=0}$$
$$= \frac{q_0 a(1+\nu)}{\pi^2} \sum_{m=1}^{\infty} \frac{1}{m^2} \sin\frac{m\pi\xi}{a} \sin\frac{m\pi u}{2a} \sin\frac{m\pi x}{a} \quad (n)$$

In the particular case when $\xi = u/2 = a/2$, that is, when the load is distributed along the entire width of the plate,

$$(M_x)_{y=0} = (M_y)_{y=0} = \frac{q_0 a(1+\nu)}{\pi^2} \sum_{m=1,3,5,\ldots}^{\infty} \frac{1}{m^2} \sin\frac{m\pi x}{a}$$

The maximum moment is at the center of the plate where

$$(M_x)_{\max} = (M_y)_{\max} = \frac{q_0 a(1+\nu)}{\pi^2} \sum_{m=1,3,5,\ldots}^{\infty} \frac{(-1)^{(m-1)/2}}{m^2} = 0.0928 q_0 a(1+\nu)$$

When u is very small, *i.e.*, in the case of a concentrated load, we put

$$\sin \frac{m\pi u}{2a} \approx \frac{m\pi u}{2a} \quad \text{and} \quad q_0 u = P$$

Then, from expression (n), we obtain

$$(M_x)_{y=0} = (M_y)_{y=0} = \frac{P(1+\nu)}{2\pi} \sum_{m=1}^{\infty} \frac{1}{m} \sin \frac{m\pi \xi}{a} \sin \frac{m\pi x}{a} \qquad (o)$$

which coincides with expression (b) of the preceding article and can be expressed also in a closed form (see page 146).

In the case of a load q uniformly distributed over the area of a rectangle (Fig. 75), the bending moments for the portion of the plate for which $y \geqq v/2$ are obtained by integration of expressions (m) as follows:

$$M_x = \frac{qa}{\pi^2} \sum_{m=1}^{\infty} \frac{1}{m^2} \sin \frac{m\pi \xi}{a} \sin \frac{m\pi u}{2a} \sin \frac{m\pi x}{a}$$

$$\int_{-v/2}^{+v/2} \left[1 + \nu + (1-\nu) \frac{m\pi(y-\eta)}{a} \right] e^{-\frac{m\pi(y-\eta)}{a}} d\eta$$

$$= \frac{qa}{\pi^2} \sum_{m=1}^{\infty} \frac{1}{m^2} \sin \frac{m\pi \xi}{a} \sin \frac{m\pi u}{2a} \sin \frac{m\pi x}{a}$$

$$\left\{ \left[\frac{2a}{m\pi} + (1-\nu)\left(y - \frac{v}{2}\right) \right] e^{-\frac{m\pi(2y-v)}{2a}} \right.$$

$$\left. - \left[\frac{2a}{m\pi} + (1-\nu)\left(y + \frac{v}{2}\right) \right] e^{-\frac{m\pi(2y+v)}{2a}} \right\} \qquad (159)$$

$$M_y = \frac{qa}{\pi^2} \sum_{m=1}^{\infty} \frac{1}{m^2} \sin \frac{m\pi \xi}{a} \sin \frac{m\pi u}{2a} \sin \frac{m\pi x}{a}$$

$$\left\{ \left[\frac{2\nu a}{m\pi} - (1-\nu)\left(y - \frac{v}{2}\right) \right] e^{-\frac{m\pi(2y-v)}{2a}} \right.$$

$$\left. - \left[\frac{2\nu a}{m\pi} - (1-\nu)\left(y + \frac{v}{2}\right) \right] e^{-\frac{m\pi(2y+v)}{2a}} \right\}$$

The moments for the portion of the plate for which $y < v/2$ can be calculated in a similar manner. To obtain the moments along the x axis, we have only to substitute $v/2$ for v and $v/4$ for y in formulas (159) and double the results thus obtained. Hence

$$(M_x)_{y=0} = \frac{4qa^2}{\pi^3} \sum_{m=1}^{\infty} \frac{1}{m^3} \sin \frac{m\pi\xi}{a} \sin \frac{m\pi u}{2a} \sin \frac{m\pi x}{a}$$
$$\left\{ 1 - \left[1 + (1-\nu) \frac{m\pi v}{4a} \right] e^{-\frac{m\pi v}{2a}} \right\}$$
$$(M_y)_{y=0} = \frac{4qa^2}{\pi^3} \sum_{m=1}^{\infty} \frac{1}{m^3} \sin \frac{m\pi\xi}{a} \sin \frac{m\pi u}{2a} \sin \frac{m\pi x}{a} \qquad (160)$$
$$\left\{ \nu - \left[\nu - (1-\nu) \frac{m\pi v}{4a} \right] e^{-\frac{m\pi v}{2a}} \right\}$$

If values of the moments at the center of the loaded rectangular area are required, the calculation may also be carried out by means of expressions (167), which will be given in Art. 37. When v is very small, Eqs. (160) coincide with Eq. (n) if we observe that qv must be replaced in such a case by q_0. When v is very large, we have the deflection of the plate to a cylindrical surface, and Eqs. (160) become

$$(M_x)_{y=0} = \frac{4qa^2}{\pi^3} \sum_{m=1}^{\infty} \frac{1}{m^3} \sin \frac{m\pi\xi}{a} \sin \frac{m\pi u}{2a} \sin \frac{m\pi x}{a}$$

$$(M_y)_{y=0} = \frac{4\nu qa^2}{\pi^3} \sum_{m=1}^{\infty} \frac{1}{m^3} \sin \frac{m\pi\xi}{a} \sin \frac{m\pi u}{2a} \sin \frac{m\pi x}{a}$$

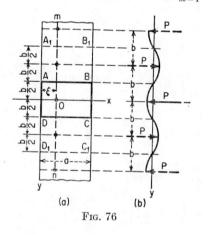

FIG. 76

The expressions for the deflections and bending moments in a plate of finite length can be obtained from the corresponding quantities in an infinitely long plate by using the *method of images*.[1] Let us begin with the case of a concentrated force P applied on the axis of symmetry x of the rectangular plate with sides a and b in Fig. 76a. If we now imagine the plate prolonged in both the positive and the negative y directions and loaded with a series of forces P applied along the line mn at a distance b from one another and in alternate directions, as shown in Fig. 76b, the deflections of such an infinitely

[1] This method was used by A. Nádai (see *Z. angew. Math. Mech.*, vol. 2, p. 1, 1922) and by M. T. Huber (see *Z. angew. Math. Mech.*, vol. 6, p. 228, 1926).

long plate are evidently equal to zero along the lines A_1B_1, AB, CD, C_1D_1, ... The bending moments along the same lines are also zero, and we may consider the given plate $ABCD$ as a portion of the infinitely long plate loaded as shown in Fig. 76b. Hence the deflection and the stresses produced in the given plate at the point of application O of the concentrated force can be calculated by using formulas derived for infinitely long plates. From Eq. (158) we find that the deflection produced at the x axis of the infinitely long plate by the load P applied at the point O is

$$w_1 = \frac{Pa^2}{2\pi^3 D} \sum_{m=1}^{\infty} \frac{1}{m^3} \sin \frac{m\pi\xi}{a} \sin \frac{m\pi x}{a}$$

The two adjacent forces P applied at the distances b from the point O (Fig. 76b) produce at the x axis the deflection

$$w_2 = -\frac{Pa^2}{\pi^3 D} \sum_{m=1}^{\infty} \frac{1}{m^3} \sin \frac{m\pi\xi}{a} (1 + 2\alpha_m) e^{-2\alpha_m} \sin \frac{m\pi x}{a}$$

in which, as before,

$$\alpha_m = \frac{m\pi b}{2a}$$

The forces P at the distance $2b$ from the point O produce at the x axis the deflection

$$w_3 = \frac{Pa^2}{\pi^3 D} \sum_{m=1}^{\infty} \frac{1}{m^3} \sin \frac{m\pi\xi}{a} (1 + 4\alpha_m) e^{-4\alpha_m} \sin \frac{m\pi x}{a}$$

and so on. The total deflection at the x axis will be given by the summation

$$w = w_1 + w_2 + w_3 + \cdots \quad (p)$$

Observing that

$$\tanh \alpha_m = \frac{1 - e^{-2\alpha_m}}{1 + e^{-2\alpha_m}} = 1 - 2e^{-2\alpha_m} + 2e^{-4\alpha_m} \cdots$$

$$\frac{1}{\cosh^2 \alpha_m} = \frac{4}{(e^{\alpha_m} + e^{-\alpha_m})^2} = \frac{4e^{-2\alpha_m}}{(1 + e^{-2\alpha_m})^2}$$
$$= 4e^{-2\alpha_m}(1 - 2e^{-2\alpha_m} + 3e^{-4\alpha_m} - 4e^{-6\alpha_m} + \cdots)$$

we can bring expression (p) into coincidence with expression (146) of Art. 34.

Let us apply the method of images to the calculation of the reactive force

$$R = -2M_{xy}$$

acting at the corner D of the rectangular plate $ABCD$ (Fig. 76) and produced by a load P at the center of this plate. Using Eqs. (151) and (152), we find that the general expression for the twisting moment of an infinitely long plate in the case of a single load becomes

$$M_{xy} = -\frac{1}{2}(1-\nu)y\frac{\partial M}{\partial x}$$

$$= \frac{P(1-\nu)y}{8a}\left[\frac{\sin\frac{\pi(x-\xi)}{a}}{\cosh\frac{\pi y}{a} - \cos\frac{\pi(x-\xi)}{a}} - \frac{\sin\frac{\pi(x+\xi)}{a}}{\cosh\frac{\pi y}{a} - \cos\frac{\pi(x+\xi)}{a}}\right] \quad (q)$$

Hence a load P concentrated at $x = \xi = a/2$, $y = 0$ produces at $x = 0$ the twisting moment

$$M_{xy} = -\frac{P(1-\nu)}{4a}\frac{y}{\cosh\frac{\pi y}{a}} \quad (r)$$

Now, putting $y = b/2, 3b/2, 5b/2, \ldots$ consecutively, we obtain the twisting moments produced by the loads $\pm P$ acting above the line DC. Taking the sum of these moments we obtain

$$M_{xy} = -\frac{Pb(1-\nu)}{8a}\left(\frac{1}{\cosh\frac{\pi b}{2a}} - \frac{3}{\cosh\frac{3\pi b}{2a}} + \frac{5}{\cosh\frac{5\pi b}{2a}} \cdots\right) \quad (s)$$

To take into account the loads acting below the line DC we have to double the effect (s) of loads acting above the line DC in order to obtain the effect of all given loads. Thus we arrive at the final result

$$M_{xy} = -\frac{Pb(1-\nu)}{4a}\sum_{m=1,3,5,\ldots}^{\infty}(-1)^{(m-1)/2}\frac{m}{\cosh\frac{m\pi b}{2a}} \quad (t)$$

As for the reactive force acting downward at the point D, and consequently at the other corners of the plate, it is equal to $R = -2M_{xy}$, M_{xy} being given by Eq. (t).

The method of images can be used also when the point of application of P is not on the axis of symmetry (Fig. 77a). The deflections and moments can be calculated by introducing a system of auxiliary forces as shown in the figure and using the formulas derived for an infinitely long plate. If the load is distributed over a rectangle, formulas (167), which will be given in Art. 37, can be used for calculating the bending moments produced by actual and auxiliary loads.

37. Bending Moments in Simply Supported Rectangular Plates under a Load Uniformly Distributed over the Area of a Rectangle. Let us consider once more the practically important case of the loading represented in Fig. 78. If we proceed as described in Art. 33, we find that for small values of u/a and v/b the series representing the bending moments at the center of the loaded area converge slowly and become unsuitable for numerical computation.

In order to derive more convenient formulas[1] in this case let us introduce, in extension of Eq. (119), the following notation:

[1] See S. Woinowsky-Krieger, *Ingr.-Arch.*, vol. 21, p. 331, 1953.

$$M = \frac{M_x + M_y}{1+\nu} = -D\left(\frac{\partial^2 w}{\partial x^2} + \frac{\partial^2 w}{\partial y^2}\right)$$

$$N = \frac{M_x - M_y}{1-\nu} = -D\left(\frac{\partial^2 w}{\partial x^2} - \frac{\partial^2 w}{\partial y^2}\right)$$

(161)

Hence
$$M_x = \tfrac{1}{2}(1+\nu)M + \tfrac{1}{2}(1-\nu)N$$
$$M_y = \tfrac{1}{2}(1+\nu)M - \tfrac{1}{2}(1-\nu)N$$

(162)

At first let us consider a *clamped circular plate* of a radius a_0 with a central load, distributed as shown in Fig. 78. The bending moments at the center of such a plate can be obtained by use of the Michell solution, for an eccentric single load. If u and v are small in

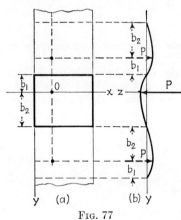

Fig. 77

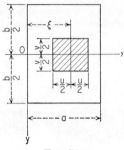

Fig. 78

comparison with a_0, the result, evaluated by due integration of expression (197) (p. 293), can be put in the form

$$M = \frac{P}{4\pi}\left(2 + 2\log\frac{2a_0}{d} - \varphi\right)$$
$$N = \frac{P}{4\pi}\psi$$

(a)

in which

$$\varphi = k\arctan\frac{1}{k} + \frac{1}{k}\arctan k$$
$$\psi = k\arctan\frac{1}{k} - \frac{1}{k}\arctan k$$
$$k = \frac{v}{u} \quad \text{and} \quad d = \sqrt{u^2 + v^2}$$

(163)

For a *simply supported circular plate* with the same radius a_0 as before, we have to add a term $P/4\pi$ to M_x and M_y (see p. 68), i.e., a term $P/2\pi(1+\nu)$ to M and nothing to N, so that these latter quantities become

$$M = \frac{P}{4\pi}\left(2 + 2\log\frac{2a_0}{d} - \varphi\right) + \frac{P}{2\pi(1+\nu)}$$
$$N = \frac{P}{4\pi}\psi$$

(b)

Finally, to obtain the corresponding expressions for an *infinite strip* (Fig. 75), we must assume $a_0 = 2a/\pi \sin(\pi\xi/a)$ and introduce an additional moment $M_y = -(1-\nu)P/4\pi$ (see p. 147). This latter operation changes the quantity M by $-(1-\nu)P/4\pi(1+\nu)$ and the quantity N by $+P/4\pi$. Introducing this in Eqs. (b) we arrive at the result

$$M = \frac{P}{4\pi}\left(2\log\frac{4a\sin\frac{\pi\xi}{a}}{\pi d} + 3 - \varphi\right)$$
$$N = \frac{P}{4\pi}(1+\psi)$$
(164)

The values of the factors φ and ψ, depending only on the ratio v/u, are given in Table 26.

Considering now the case of a *rectangular plate* (Fig. 78), we have only to take into account the effect of the auxiliary loads[1] $\pm P$ (Fig. 77) and to add this effect to the values (164) of M and N. The final result, in the case shown in Fig. 78, can then be put in the form

$$M = \frac{P}{4\pi}\left(2\log\frac{4a\sin\frac{\pi\xi}{a}}{\pi d} + \lambda - \varphi\right)$$
$$N = \frac{P}{4\pi}(\mu+\psi)$$
(165)

where φ, ψ, d are given by expressions (163) and Table 26, and

$$\lambda = 3 - 4\sum_{m=1}^{\infty}\frac{e^{-\alpha_m}}{\cosh\alpha_m}\sin^2\frac{m\pi\xi}{a}$$
$$\mu = 1 - \frac{2\pi b}{a}\sum_{m=1}^{\infty}\frac{1}{\cosh^2\alpha_m}\sin^2\frac{m\pi\xi}{a}$$
(166)

with $\alpha_m = m\pi b/2a$. The terms λ and μ, expressed by rapidly convergent series, are wholly independent of the dimensions u and v (and even the shape) of the loaded area. Their numerical values are given in Table 27.

From Eqs. (162) we obtain the expressions for the bending moments

$$M_x = \frac{P}{8\pi}\left[\left(2\log\frac{4a\sin\frac{\pi\xi}{a}}{\pi d} + \lambda - \varphi\right)(1+\nu) + (\mu+\psi)(1-\nu)\right]$$
$$M_y = \frac{P}{8\pi}\left[\left(2\log\frac{4a\sin\frac{\pi\xi}{a}}{\pi d} + \lambda - \varphi\right)(1+\nu) - (\mu+\psi)(1-\nu)\right]$$
(167)

acting at the center of the loaded area (Fig. 78). Expressions (165) and (167) are also applicable to the calculation of moments of a simply supported infinite strip as a particular case.

[1] It is permissible to regard them as concentrated provided u and v are small.

TABLE 26. VALUES OF THE FACTOR φ AND ψ DEFINED BY EQS. (163)
$k = v/u$

k	φ	ψ	k	φ	ψ	k	φ	ψ
0	1.000	−1.000	1.0	1.571	0.000	2.5	1.427	0.475
0.05	1.075	−0.923	1.1	1.569	0.054	3.0	1.382	0.549
0.1	1.144	−0.850	1.2	1.564	0.104	4.0	1.311	0.648
0.2	1.262	−0.712	1.3	1.556	0.148	5.0	1.262	0.712
0.3	1.355	−0.588	1.4	1.547	0.189	6.0	1.225	0.757
0.4	1.427	−0.475	1.5	1.537	0.227	7.0	1.197	0.789
0.5	1.481	−0.374	1.6	1.526	0.261	8.0	1.176	0.814
0.6	1.519	−0.282	1.7	1.515	0.293	9.0	1.158	0.834
0.7	1.545	−0.200	1.8	1.504	0.322	10	1.144	0.850
0.8	1.560	−0.127	1.9	1.492	0.349	20	1.075	0.923
0.9	1.568	−0.060	2.0	1.481	0.374	∞	1.000	1.000

TABLE 27. VALUES OF THE FACTORS λ AND μ (EQ. 166) FOR SIMPLY SUPPORTED RECTANGULAR PLATES

b/a	λ for $\xi/a =$					μ for $\xi/a =$				
	0.1	0.2	0.3	0.4	0.5	0.1	0.2	0.3	0.4	0.5
0.5	2.792	2.352	1.945	1.686	1.599	0.557	−0.179	−0.647	−0.852	−0.906
0.6	2.861	2.545	2.227	2.011	1.936	0.677	0.053	−0.439	−0.701	−0.779
0.7	2.904	2.677	2.433	2.259	2.198	0.758	0.240	−0.229	−0.514	−0.605
0.8	2.933	2.768	2.584	2.448	2.399	0.814	0.391	−0.031	−0.310	−0.404
0.9	2.952	2.832	2.694	2.591	2.553	0.856	0.456	0.148	−0.108	−0.198
1.0	2.966	2.879	2.776	2.698	2.669	0.887	0.611	0.304	0.080	0.000
1.2	2.982	2.936	2.880	2.836	2.820	0.931	0.756	0.551	0.393	0.335
1.4	2.990	2.966	2.936	2.912	2.903	0.958	0.849	0.719	0.616	0.578
1.6	2.995	2.982	2.966	2.953	2.948	0.975	0.908	0.828	0.764	0.740
1.8	2.997	2.990	2.982	2.975	2.972	0.985	0.945	0.897	0.858	0.843
2.0	2.999	2.995	2.990	2.987	2.985	0.991	0.968	0.939	0.915	0.906
3.0	3.000	3.000	3.000	2.999	2.999	0.999	0.998	0.996	0.995	0.994
∞	3.000	3.000	3.000	3.000	3.000	1.000	1.000	1.000	1.000	1.000

Extending the integration over circular, elliptic, and other areas, the corresponding expressions for M and N for these loadings are readily found. Taking, for instance, a circular loaded area (Fig. 79) we obtain for its center

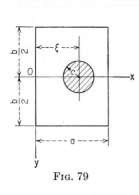

Fig. 79

$$M = \frac{P}{4\pi}\left(2\log\frac{2a\sin\frac{\pi\xi}{a}}{\pi c} + \lambda - 2\right) \quad (168)$$

$$N = \frac{P}{4\pi}\mu$$

these expressions being equivalent to the result (157). Comparing (168) with expressions (165) for $k = 1$, we may conclude that a circular and a square loaded area are equivalent with respect to the bending moments they produce at the center of the area, if

$$c = \frac{u}{\sqrt{2}}e^{\pi/4-1} = 0.57u \quad \text{or} \quad u = 0.88 \times 2c \quad (c)$$

It should be noted that, as the load becomes more and more concentrated, the accuracy of the approximate logarithmic formulas for the bending moments, such as given by Eqs. (157) and (167), increases while the convergence of the customary series representing these moments becomes slower. Numerical calculations[1] show also that the accuracy of those approximate formulas is entirely sufficient for practical purposes.

38. Thermal Stresses in Simply Supported Rectangular Plates. Let us assume that the upper surface of a rectangular plate is kept at a higher temperature than the lower surface so that the plate has a tendency to bend convexly upward because of nonuniform heating. Because of the constraint along the simply supported edges of the plate, which prevents the edges from leaving the plane of the supports, the nonuniform heating of the plate produces certain reactions along the boundary of the plate and certain bending stresses at a distance from the edges. The method described in Art. 24 will be used in calculating these stresses.[2] We assume first that the edges of the plate are clamped. In such a case the nonuniform heating produces uniformly distributed bending moments along the boundary whose magnitude is (see page 50)

$$M_n = \frac{\alpha t D(1 + \nu)}{h} \quad (a)$$

where t is the difference between the temperatures of the upper and the lower surfaces of the plate and α is the coefficient of thermal expansion.

---

[1] See S. Woinowsky-Krieger, *Ingr.-Arch.*, vol. 3, p. 340, 1932; and *Ingr.-Arch.*, vol. 21, pp. 336, 337, 1953.

[2] See paper by J. L. Maulbetsch, *J. Appl. Mechanics*, vol. 2, p. 141, 1935; see also E. Melan and H. Parkus, "Wärmespannungen infolge stationärer Temperaturfelder," Vienna, 1953, which includes a bibliography on thermal stresses. For stresses due to assemblage errors in plates, see W. Nowacki, *Bull. acad. polon. sci.*, vol. 4, p. 79, 1956.

To get the bending moments M_x and M_y for a simply supported plate (Fig. 62), we must superpose on the uniformly distributed moments given by Eq. (a) the moments that are produced in a simply supported rectangular plate by the moments $M'_n = -\alpha t D(1 + \nu)/h$ uniformly distributed along the edges. We shall use Eqs. (120) (see page 92) in discussing this latter problem. Since the curvature in the direction of an edge is zero in the case of simply supported edges, we have $M'_t = \nu M'_n$. Hence at the boundary

$$M = \frac{M_x + M_y}{1 + \nu} = \frac{M'_n + M'_t}{1 + \nu} = -\frac{\alpha t D(1 + \nu)}{h} \qquad (b)$$

Thus the first of equations (120) is satisfied by taking M constant along the entire plate and equal to its boundary value (b). Then the second of equations (120) gives

$$\frac{\partial^2 w}{\partial x^2} + \frac{\partial^2 w}{\partial y^2} = \frac{\alpha t(1 + \nu)}{h} \qquad (c)$$

Hence the deflection surface of the plate produced by nonuniform heating is the same as that of a uniformly stretched and uniformly loaded rectangular membrane and is obtained by finding the solution of Eq. (c) that satisfies the condition that $w = 0$ at the boundary.

Proceeding as before, we take the deflection surface of the plate in the form

$$w = w_1 + w_2 \qquad (d)$$

in which w_1 is the deflection of a perfectly flexible string loaded uniformly and stretched axially in such a way that the intensity of the load divided by the axial force is equal to $-\alpha t(1 + \nu)/h$. In such a case the deflection curve is a parabola which can be represented by a trigonometric series as follows:

$$w_1 = -\frac{\alpha t(1 + \nu)}{h} \frac{x(a - x)}{2}$$

$$= -\frac{\alpha t(1 + \nu)}{h} \frac{4a^2}{\pi^3} \sum_{m=1,3,5,\ldots}^{\infty} \frac{\sin \frac{m\pi x}{a}}{m^3} \qquad (e)$$

This expression satisfies Eq. (c). The deflection w_2, which must satisfy the equation

$$\frac{\partial^2 w_2}{\partial x^2} + \frac{\partial^2 w_2}{\partial y^2} = 0 \qquad (f)$$

can be taken in the form of the series

$$w_2 = \sum_{m=1,3,5,\ldots}^{\infty} Y_m \sin \frac{m\pi x}{a} \qquad (g)$$

in which Y_m is a function of y only. Substituting (g) in Eq. (f), we find

$$Y_m'' - \frac{m^2\pi^2}{a^2} Y_m = 0$$

Hence
$$Y_m = A_m \sinh \frac{m\pi y}{a} + B_m \cosh \frac{m\pi y}{a} \qquad (h)$$

From the symmetry of the deflection surface with respect to the x axis it may be concluded that Y_m must be an even function of y. Hence the constant A_m in the expression (h) must be taken equal to zero, and we finally obtain

$$w = w_1 + w_2 = \sum_{m=1,3,5,\ldots}^{\infty} \sin \frac{m\pi x}{a} \left[-\frac{\alpha t(1+\nu)}{h} \frac{4a^2}{\pi^3 m^3} + B_m \cosh \frac{m\pi y}{a} \right] \qquad (i)$$

This expression satisfies the boundary conditions $w = 0$ at the edges $x = 0$ and $x = a$. To satisfy the same condition at the edges $y = \pm b/2$, we must have

$$B_m \cosh \frac{m\pi b}{2a} - \frac{\alpha t(1+\nu)}{h} \frac{4a^2}{\pi^3 m^3} = 0$$

Substituting the value of B_m obtained from this equation in Eq. (i), we find that

$$w = -\frac{\alpha t(1+\nu)4a^2}{\pi^3 h} \sum_{m=1,3,5,\ldots}^{\infty} \frac{\sin \frac{m\pi x}{a}}{m^3} \left(1 - \frac{\cosh \frac{m\pi y}{a}}{\cosh \alpha_m} \right) \qquad (j)$$

in which, as before, $\alpha_m = m\pi b/2a$.

Having this expression for the deflections w, we can find the corresponding values of bending moments; and, combining them with the moments (a), we finally obtain

$$\begin{aligned} M_x &= \frac{\alpha t D(1+\nu)}{h} - D\left(\frac{\partial^2 w}{\partial x^2} + \nu \frac{\partial^2 w}{\partial y^2}\right) \\ &= \frac{4D\alpha t(1-\nu^2)}{\pi h} \sum_{m=1,3,5,\ldots}^{\infty} \frac{\sin \frac{m\pi x}{a} \cosh \frac{m\pi y}{a}}{m \cosh \alpha_m} \\ M_y &= \frac{\alpha t D(1+\nu)}{h} - D\left(\frac{\partial^2 w}{\partial y^2} + \nu \frac{\partial^2 w}{\partial x^2}\right) \\ &= \frac{\alpha t(1-\nu^2)D}{h} - \frac{4D\alpha t(1-\nu^2)}{\pi h} \sum_{m=1,3,5,\ldots}^{\infty} \frac{\sin \frac{m\pi x}{a} \cosh \frac{m\pi y}{a}}{m \cosh \alpha_m} \end{aligned} \qquad (k)$$

The sum of the series that appears in these expressions can be readily found if we put it in the following form:

$$\sum_{m=1,3,5,\ldots}^{\infty} \frac{\sin \frac{m\pi x}{a} \cosh \frac{m\pi y}{a}}{m \cosh \alpha_m}$$

$$= \sum_{m=1,3,5,\ldots}^{\infty} \left(\frac{\sin \frac{m\pi x}{a} \cosh \frac{m\pi y}{a}}{m \cosh \alpha_m} - \frac{e^{m\pi y/a} \sin \frac{m\pi x}{a}}{m e^{\alpha_m}} \right)$$

$$+ \sum_{m=1,3,5,\ldots}^{\infty} \frac{e^{m\pi y/a}}{m e^{\alpha_m}} \sin \frac{m\pi x}{a} \quad (l)$$

The first series on the right-hand side of this equation converges rapidly, since $\cosh (m\pi y/a)$ and $\cosh \alpha_m$ rapidly approach $e^{m\pi y/a}$ and e^{α_m} as m increases. The second series can be represented as follows.[1]

$$\sum_{m=1,3,5,\ldots}^{\infty} \frac{e^{m\pi y/a} \sin \frac{m\pi x}{a}}{m e^{\alpha_m}} = \frac{1}{2} \arctan \frac{\sin \frac{\pi x}{a}}{\sinh \left(\frac{\pi b}{2a} - \frac{\pi y}{a} \right)} \quad (m)$$

The bending moments M_x and M_y have their maximum values at the boundary. These values are

$$(M_x)_{y=\pm b/2} = (M_y)_{x=0,x=a} = \frac{\alpha t (1-\nu^2) D}{h} = Eh^2 \frac{\alpha t}{12} \quad (n)$$

It is seen that these moments are obtained by multiplying the value of M_n in formula (a) by $(1-\nu)$. The same conclusion is reached if we observe that the moments M'_n which were applied along the boundary produce in the perpendicular direction the moments

$$M'_t = \nu M'_n = -\nu \frac{\alpha t D (1+\nu)}{h}$$

which superposed on the moment (a) give the value (n).

39. The Effect of Transverse Shear Deformation on the Bending of Thin Plates. We have seen that the customary theory of thin elastic plates leads to a differential equation (103) of the fourth order for the

---

[1] See W. E. Byerly, "Elementary Treatise on Fourier Series and Spherical, Cylindrical and Ellipsoidal Harmonics," p. 100, Boston, 1893. The result can be easily obtained by using the known series

$$\frac{1}{2} \arctan \frac{2x \sin \varphi}{1-x^2} = x \sin \varphi + \frac{x^3}{3} \sin 3\varphi + \frac{x^5}{5} \sin 5\varphi + \cdots$$

deflection and, accordingly, to two boundary conditions which can and must be satisfied at each edge. For a plate of a finite thickness, however, it appears more natural to require the fulfillment of three boundary conditions than of two. The formal reason for the impossibility of satisfying more than two conditions by the customary theory has been the order of the basic equation of this theory; physically this reason lies in the fact that the distortion of the elements of the plate due to transverse forces such as Q (page 52), Q_x, and Q_y (page 79) has been neglected in establishing the relations between the stresses and the deflection of the plate. The disregard of the deformation due to the transverse stress component obviously is equivalent to the assumption of a shearing modulus $G_z = \infty$; proceeding in this way we replace the actual material of the plate, supposed to be isotropic, by a hypothetic material of no perfect isotropy. Owing to the assumption $G_z = \infty$ the plate does not respond to a rotation of some couple applied at the cylindrical surface of the plate, if the vector of the couple coincides with the normal to this surface. This enables us to identify the variation $\partial M_{xy}/\partial y$ of twisting couples due to horizontal shearing stresses and acting along an edge $x = a$ with the effect of vertical forces Q_x applied at the same edge, thus reducing the number of the edge conditions from three to two (page 83). The stress analysis of the elastic plates is greatly simplified by this reduction. On the other hand, in attributing some purely hypothetic properties to the material of the plate we cannot expect complete agreement of the theoretical stress distribution with the actual one. The inaccuracy of the customary thin-plate theory becomes of practical interest in the edge zones of plates and around holes that have a diameter which is not large in comparison with the thickness of the plate.

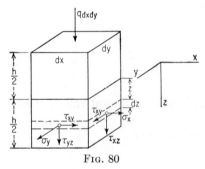

Fig. 80

The generalization of the customary theory with respect to the effect of shear deformation is substantially due to E. Reissner.[1]

Let us consider an element of the plate submitted to the external transversal load $q\,dx\,dy$ and to a system of stress components (Fig. 80). In accordance with E. Reissner's theory we assume a linear law for the distribution of the stress components σ_x, σ_y, and τ_{xy} through the thickness of the plate. By equations of equilibrium (a) on page 98 the distribu-

---

[1] See *J. Math. and Phys.*, vol. 23, p. 184, 1944; *J. Appl. Mechanics*, vol. 12, p. A-68, 1945; *Quart. Appl. Math.*, vol. 5, p. 55, 1947. For the history of this question going back to a controversy between M. Lévy and Boussinesq, see L. Bolle, *Bull. tech. Suisse romande*, October, 1947.

tion of the components τ_{xz} and τ_{yz} then follows a parabolic law. As for the stress component σ_z it is readily obtained from the third of equations of equilibrium (a) if one takes into account the conditions

$$(\sigma_z)_{z=-h/2} = -q \qquad (\sigma_z)_{z=h/2} = 0$$

at the upper and lower surface of the plate. We arrive, in this manner, at the following expressions for the stress components in terms of their resultants and the coordinate z:

$$\sigma_x = \frac{12M_x z}{h^3} \qquad \sigma_y = \frac{12M_y z}{h^3} \qquad \tau_{xy} = -\frac{12M_{xy} z}{h^3} \qquad (a)$$

$$\tau_{xz} = \frac{3Q_x}{2h}\left[1 - \left(\frac{2z}{h}\right)^2\right] \qquad \tau_{yz} = \frac{3Q_y}{2h}\left[1 - \left(\frac{2z}{h}\right)^2\right]$$

$$\sigma_z = -\frac{3q}{4}\left[\frac{2}{3} - \frac{2z}{h} + \frac{1}{3}\left(\frac{2z}{h}\right)^3\right] \qquad (b)$$

Except for Eq. (b) the foregoing system of equations coincides with the corresponding relations of the customary theory. In like manner we can rewrite the following conditions of equilibrium of the stress resultants (see pages 80, 81):

$$\frac{\partial Q_x}{\partial x} + \frac{\partial Q_y}{\partial y} + q = 0 \qquad (c)$$

$$\frac{\partial M_x}{\partial x} - \frac{\partial M_{xy}}{\partial y} - Q_x = 0$$
$$\frac{\partial M_y}{\partial y} - \frac{\partial M_{xy}}{\partial x} - Q_y = 0 \qquad (d)$$

Assuming an isotropic material and supposing the displacements u_0, v_0, w_0 of any point of the plate to be small as compared with its thickness h, we make use of the general stress-strain relations

$$\frac{\partial u_0}{\partial x} = \frac{1}{E}[\sigma_x - \nu(\sigma_y + \sigma_z)]$$
$$\frac{\partial v_0}{\partial y} = \frac{1}{E}[\sigma_y - \nu(\sigma_x + \sigma_z)]$$
$$\frac{\partial u_0}{\partial y} + \frac{\partial v_0}{\partial x} = \frac{1}{G}\tau_{xy} \qquad (e)$$
$$\frac{\partial u_0}{\partial z} + \frac{\partial w_0}{\partial x} = \frac{1}{G}\tau_{xz}$$
$$\frac{\partial v_0}{\partial z} + \frac{\partial w_0}{\partial y} = \frac{1}{G}\tau_{yz}$$

in which $G = E/2(1 + \nu)$. We do not use the sixth relation

$$\frac{\partial w_0}{\partial z} = \frac{1}{E}[\sigma_z - \nu(\sigma_x + \sigma_y)]$$

however, since this latter proves to be in contradiction with the assumed linear law for the distribution of the stresses σ_x, σ_y, τ_{xy}.

Next,[1] we introduce some average value w of the transverse displacement, taken over the thickness of the plate, as well as some average values φ_x and φ_y of the rotation of the sections $x = $ constant and $y = $ constant, respectively. We define these quantities by equating the work of the resultant couples on the average rotations and the work of the resultant forces on the average displacement to the work of the corresponding stresses on the actual displacements u_0, v_0, and w_0 in the same section; *i.e.*, we put

$$\int_{-h/2}^{h/2} \sigma_x u_0 \, dz = M_x \varphi_x \qquad -\int_{-h/2}^{h/2} \tau_{xy} v_0 \, dz = M_{xy} \varphi_y$$
$$\int_{-h/2}^{h/2} \sigma_y v_0 \, dz = M_y \varphi_y \qquad -\int_{-h/2}^{h/2} \tau_{xy} u_0 \, dz = M_{xy} \varphi_x \qquad (f)$$
$$\int_{-h/2}^{h/2} \tau_{xz} w_0 \, dz = Q_x w \qquad \int_{-h/2}^{h/2} \tau_{yz} w_0 \, dz = Q_y w$$

Now, substituting expressions (a) for the stresses in Eqs. (f), we arrive at the following relations between the average and the actual displacements:

$$w = \frac{3}{2h} \int_{-h/2}^{h/2} w_0 \left[1 - \left(\frac{2z}{h}\right)^2\right] dz$$
$$\varphi_x = \frac{12}{h^2} \int_{-h/2}^{h/2} \frac{u_0 z}{h} \, dz \qquad (g)$$
$$\varphi_y = \frac{12}{h^2} \int_{-h/2}^{h/2} \frac{v_0 z}{h} \, dz$$

Using Eqs. (e) and observing Eq. (b), we are also able to express the stress components σ_x, σ_y, and τ_{xy} in terms of the actual displacements; we find[2]

$$\sigma_x = \frac{E}{1-\nu^2}\left(\frac{\partial u_0}{\partial x} + \nu \frac{\partial v_0}{\partial y}\right) - \frac{3q\nu}{4(1-\nu)}\left[\frac{2}{3} - \frac{2z}{h} + \frac{1}{3}\left(\frac{2z}{h}\right)^3\right]$$
$$\sigma_y = \frac{E}{1-\nu^2}\left(\frac{\partial v_0}{\partial y} + \nu \frac{\partial u_0}{\partial x}\right) - \frac{3q\nu}{4(1-\nu)}\left[\frac{2}{3} - \frac{2z}{h} + \frac{1}{3}\left(\frac{2z}{h}\right)^3\right] \qquad (h)$$
$$\tau_{xy} = \frac{E}{2(1+\nu)}\left(\frac{\partial u_0}{\partial y} + \frac{\partial v_0}{\partial x}\right)$$

---

[1] E. Reissner, in his treatment of the subject, makes use of Castigliano's principle of least work to introduce the conditions of compatibility in the analysis. The method here followed and leading to substantially the same results is due to A. E. Green, *Quart. Appl. Math.*, vol. 7, p. 223, 1949. See also M. Schäfer, *Z. angew. Math. Mech.*, vol. 32, p. 161, 1952.

[2] Terms with z^3 do not actually occur in the following expressions for σ_x and σ_y since they are canceled out by identical terms with opposite sign contained in $\partial u_0/\partial x$ and $\partial v_0/\partial y$.

Substituting this in Eqs. (*a*), multiplying the obtained equations by $12z\,dz/h^3$, integrating between $z = -h/2$ and $z = h/2$, and observing relations (*g*), we arrive at the expressions

$$M_x = D\left[\frac{\partial \varphi_x}{\partial x} + \nu \frac{\partial \varphi_y}{\partial y} + \frac{6\nu(1+\nu)}{5Eh}q\right]$$
$$M_y = D\left[\frac{\partial \varphi_y}{\partial y} + \nu \frac{\partial \varphi_x}{\partial x} + \frac{6\nu(1+\nu)}{5Eh}q\right] \quad (i)$$
$$M_{xy} = -\frac{D(1-\nu)}{2}\left(\frac{\partial \varphi_x}{\partial y} + \frac{\partial \varphi_y}{\partial x}\right)$$

in which D is defined, as before, by Eq. (3). In like manner, substituting expressions (*a*) for the stress components τ_{xz} and τ_{yz} in the last two equations (*e*), multiplying the result by $\frac{3}{2}[1 - (2z/h)^2]\,dz/h$, and integrating between the limits $z = \pm h/2$, we obtain

$$\varphi_x = -\frac{\partial w}{\partial x} + \frac{12}{5}\frac{1+\nu}{Eh}Q_x$$
$$\varphi_y = -\frac{\partial w}{\partial y} + \frac{12}{5}\frac{1+\nu}{Eh}Q_y \quad (j)$$

Now, eight unknown quantities, namely M_x, M_y, M_{xy}, Q_x, Q_y, w, φ_x, and φ_y, are connected by two equations (*j*), three equations (*i*), and, finally, by three equations of equilibrium (*c*) and (*d*).

In order to transform this set of equations into a form more convenient for analysis we eliminate the quantities φ_x and φ_y from Eqs. (*j*) and (*i*), and, taking into account Eq. (*c*), we obtain

$$M_x = -D\left(\frac{\partial^2 w}{\partial x^2} + \nu\frac{\partial^2 w}{\partial y^2}\right) + \frac{h^2}{5}\frac{\partial Q_x}{\partial x} - \frac{qh^2}{10}\frac{\nu}{1-\nu}$$
$$M_y = -D\left(\frac{\partial^2 w}{\partial y^2} + \nu\frac{\partial^2 w}{\partial x^2}\right) + \frac{h^2}{5}\frac{\partial Q_y}{\partial y} - \frac{qh^2}{10}\frac{\nu}{1-\nu} \quad (k)$$
$$M_{xy} = (1-\nu)D\frac{\partial^2 w}{\partial x\,\partial y} - \frac{h^2}{10}\left(\frac{\partial Q_x}{\partial y} + \frac{\partial Q_y}{\partial x}\right)$$

Substitution of these expressions in Eqs. (*d*) yields, if one observes Eq. (*c*), the result

$$Q_x - \frac{h^2}{10}\Delta Q_x = -D\frac{\partial(\Delta w)}{\partial x} - \frac{h^2}{10(1-\nu)}\frac{\partial q}{\partial x}$$
$$Q_y - \frac{h^2}{10}\Delta Q_y = -D\frac{\partial(\Delta w)}{\partial y} - \frac{h^2}{10(1-\nu)}\frac{\partial q}{\partial y} \quad (l)$$

in which, as before, the symbol Δ has the meaning (105). In the particular case of $h = 0$, that is, of an infinitely thin plate, the foregoing set of five equations, expressions (*k*) and (*l*), gives Eqs. (101) and (102) for the moments and Eqs. (108) for the shearing forces of the customary thin-plate theory.

To obtain the more complete differential equation for the deflection of the plate we only have to substitute expressions (l) in Eq. (c); thus we obtain

$$D\Delta\Delta w = q - \frac{h^2}{10}\frac{2-\nu}{1-\nu}\Delta q \tag{169}$$

We can satisfy this equation by taking w, that is, the "average deflection" at (x,y), in the form

$$w = w' + w'' \tag{m}$$

in which w' is a particular solution of the equation

$$D\Delta\Delta w' = q - \frac{h^2}{10}\frac{2-\nu}{1-\nu}\Delta q \tag{n}$$

and w'' is the general solution of the equation

$$\Delta\Delta w'' = 0 \tag{o}$$

Therefore, using Eq. (169), we are able, just as in the ordinary thin-plate theory, to satisfy four boundary conditions in all. We can obtain a supplementary differential equation, however, by introducing into consideration the shearing forces Q_x and Q_y. Equation of equilibrium (c) is satisfied, in fact, if we express these forces in a form suggested by the form of Eqs. (l), i.e.,

$$\begin{aligned} Q_x &= -D\frac{\partial(\Delta w)}{\partial x} + \frac{\partial \psi}{\partial y} \\ Q_y &= -D\frac{\partial(\Delta w)}{\partial y} - \frac{\partial \psi}{\partial x} \end{aligned} \tag{p}$$

or

$$\begin{aligned} Q_x &= Q'_x - D\frac{\partial(\Delta w'')}{\partial x} + \frac{\partial \psi}{\partial y} \\ Q_y &= Q'_y - D\frac{\partial(\Delta w'')}{\partial y} - \frac{\partial \psi}{\partial x} \end{aligned} \tag{q}$$

In these expressions ψ denotes some new stress function, whereas Q'_x and Q'_y must satisfy the relations

$$\begin{aligned} Q'_x - \frac{h^2}{10}\Delta Q'_x &= -D\frac{\partial(\Delta w')}{\partial x} - \frac{h^2}{10(1-\nu)}\frac{\partial q}{\partial x} \\ Q'_y - \frac{h^2}{10}\Delta Q'_y &= -D\frac{\partial(\Delta w')}{\partial y} - \frac{h^2}{10(1-\nu)}\frac{\partial q}{\partial y} \end{aligned} \tag{r}$$

as we can conclude from Eqs. (l) and (n). Differentiating the foregoing equations with respect to x and y, respectively, and adding the results we arrive at the condition of equilibrium

$$\frac{\partial Q'_x}{\partial x} + \frac{\partial Q'_y}{\partial y} + q = 0 \tag{s}$$

To establish a differential equation for the stress function ψ we substitute expressions (q) in Eqs. (l) with the result

$$\frac{\partial}{\partial y}\left(\psi - \frac{h^2}{10}\Delta\psi\right) = -\frac{\partial}{\partial x}\left(\psi - \frac{h^2}{10}\Delta\psi\right) = 0 \tag{t}$$

from which we conclude that the expressions in parentheses are constants. Making these constants equal to zero we have the relation

$$\Delta\psi - \frac{10}{h^2}\psi = 0 \tag{170}$$

which, still assuming that $h \neq 0$, yields a second fundamental equation of the generalized theory of bending, in addition to Eq. (169).

Having established two differential equations, one of which is of the fourth and the other of the second order, we now are able to satisfy three conditions, instead of only two, on the edge of the plate. Considering the general case of an element of the cylindrical boundary of the plate given by the directions of the normal n and the tangent t (Fig. 54) we can, for instance, fix the position of the element by the equations

$$w = \bar{w} \qquad \varphi_n = \bar{\varphi}_n \qquad \varphi_t = \bar{\varphi}_t \tag{u}$$

Herein $\bar{w}$ is the given average deflection and $\bar{\varphi}_n$ and $\bar{\varphi}_t$ are the given average rotations of the element with respect to the axes t and n respectively. In the particular case of a built-in edge the conditions are $w = 0$, $\varphi_n = 0$, and $\varphi_t = 0$. Instead of displacements some values $\bar{Q}_n$, $\bar{M}_n$, $\bar{M}_{nt}$ of the resultants may be prescribed on the boundary, and the corresponding edge conditions would be

$$Q_n = \bar{Q}_n \qquad M_n = \bar{M}_n \qquad M_{nt} = \bar{M}_{nt} \tag{v}$$

Hence the conditions along a free edge are expressed by equations $Q_n = 0$, $M_n = 0$, $M_{nt} = 0$, and for a simply supported edge the conditions are $w = 0$, $M_n = 0$, $M_{nt} = 0$. In the latter case we obtain no concentrated reactions at the corners of the plate, which act there according to the customary theory and are in obvious contradiction to the disregard of the shear deformation postulated by this theory.

As an illustration of the refined theory let us consider a plate in form of a semi-infinite rectangle bounded by two parallel edges $y = 0$, $y = a$ and the edge $x = 0$. We assume that there is no load acting on the plate, that the deflections w and the bending moments M_y vanish along the edges $y = 0$, $y = a$, and that the edge $x = 0$ is subjected to bending and twisting moments and to shearing forces given by

$$\bar{M}_x = M_0 \sin \frac{n\pi y}{a}$$

$$\bar{M}_{xy} = H_0 \cos \frac{n\pi y}{a} \tag{w}$$

$$\bar{Q}_x = Q_0 \sin \frac{n\pi y}{a}$$

where M_0, H_0, Q_0 are constants and n is an integer. Then, in view of $q = 0$, we have $w' = 0$ by Eq. (n) and $w = w''$ by Eq. (m). We can satisfy Eq. (o) and the condition of vanishing deflections at $x = \infty$ by taking

$$w = w'' = \sin \frac{n\pi y}{a} \left(A + \frac{n\pi x}{a} B \right) \frac{e^{-n\pi x/a}}{D}$$

A and B being any constants. Next, assuming for ψ a solution of the form

$$\psi = X \cos \frac{n\pi y}{a}$$

where X is a function of x alone, and substituting this in Eq. (170) we obtain

$$\psi = Ce^{-x\beta} \cos \frac{n\pi y}{a}$$

In this last expression

$$\beta = \sqrt{\frac{n^2\pi^2}{a^2} + \frac{10}{h^2}}$$

and C is a constant. From Eqs. (r) we have $Q_x' = Q_y' = 0$ and Eqs. (q) give

$$Q_x = -\left[2B \left(\frac{n\pi}{a} \right)^3 e^{-n\pi x/a} + C \frac{n\pi}{a} e^{-x\beta} \right] \sin \frac{n\pi y}{a}$$

$$Q_y = \left[2B \left(\frac{n\pi}{a} \right)^3 e^{-n\pi x/a} + C\beta e^{-x\beta} \right] \cos \frac{n\pi y}{a}$$

Finally, Eqs. (k) yield the following expressions for moments acting along the edge $x = 0$:

$$(M_x)_{x=0} = \left[-A(1 - \nu) + 2B \left(1 + \frac{n^2\pi^2 h^2}{5a^2} \right) + C \frac{\beta a h^2}{5n\pi} \right] \frac{n^2\pi^2}{a^2} \sin \frac{n\pi y}{a}$$

$$(M_{xy})_{x=0} = \left[-A(1 - \nu) + B \left(1 - \nu + \frac{2}{5} \frac{n^2\pi^2 h^2}{a^2} \right) + C \left(\frac{a^2}{n^2\pi^2} + \frac{h^2}{5} \right) \right] \frac{n^2\pi^2}{a^2} \cos \frac{n\pi y}{a}$$

Equating these expressions, together with the expression for the shearing force

$$(Q_x)_{x=0} = -\left[2B \left(\frac{n\pi}{a} \right)^3 + C \frac{n\pi}{a} \right] \sin \frac{n\pi y}{a}$$

to the expressions (w), respectively, we obtain a set of three equations sufficient to calculate the unknown constants A, B, and C. In this way, by using the refined plate theory, all three conditions at the edge $x = 0$ are satisfied.

Considering now the edges $y = 0$ we see that w vanishes along those edges, and M_y also vanishes there, as can be proved by substituting the expression for Q_y into the second of equations (k).

Another theory of plates that takes into account the transversal shear deformation has been advanced by A. Kromm.[1] This theory neglects the transverse contraction ϵ_z but, in return, does not restrict the mode of distribution of bending stresses across the thickness of the plate to a linear law. Applying this theory to the case of a uniformly loaded, simply supported square plate with $a/h = 20$, Kromm found the distribution of shear forces acting along the edge as shown in Fig. 81. For comparison the results of customary theory (Fig. 63) are also shown by the dashed line and the

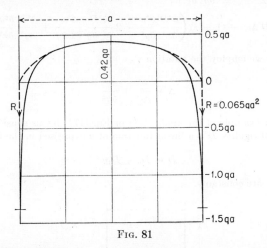

FIG. 81

forces R. We see that, as soon as the transversal shear deformation is taken into account, no concentrated reaction is obtained at the corner point of the plate. The corresponding negative forces are distributed instead over a small portion of the boundary adjacent to the corner, yielding at the corner itself a finite pressure acting downward. The moments M_{xy} on the four sides of the plate are zero in that solution.

Still another approach to the theory of shear deformation can be found in a paper of H. Hencky.[2]

40. Rectangular Plates of Variable Thickness.[3] In deriving the differential equation of equilibrium of plates of variable thickness, we assume that there is no abrupt variation in thickness so that the expressions for bending and twisting moments derived for plates of constant thickness apply with sufficient accuracy to this case also. Then

$$M_x = -D\left(\frac{\partial^2 w}{\partial x^2} + \nu\frac{\partial^2 w}{\partial y^2}\right) \qquad M_y = -D\left(\frac{\partial^2 w}{\partial y^2} + \nu\frac{\partial^2 w}{\partial y^2}\right)$$

$$M_{xy} = -M_{yx} = D(1 - \nu)\frac{\partial^2 w}{\partial x\,\partial y}$$

(a)

---

[1] A. Kromm, *Ingr.-Arch.*, vol. 21, p. 266, 1953; *Z. angew. Math. Mech.*, vol. 35, p. 231, 1955.

[2] *Ingr.-Arch.*, vol. 16, p. 72, 1947.

[3] This problem was discussed by R. Gran Olsson, *Ingr.-Arch.*, vol. 5, p. 363, 1934; see also E. Reissner, *J. Math. and Phys.*, vol. 16, p. 43, 1937.

Substituting these expressions in the differential equation of equilibrium of an element [Eq. (100), page 81],

$$\frac{\partial^2 M_x}{\partial x^2} - 2\frac{\partial^2 M_{xy}}{\partial x\, \partial y} + \frac{\partial^2 M_y}{\partial y^2} = -q \qquad (b)$$

and observing that the flexural rigidity D is no longer a constant but a function of the coordinates x and y, we obtain

$$D\Delta\Delta w + 2\frac{\partial D}{\partial x}\frac{\partial}{\partial x}\Delta w + 2\frac{\partial D}{\partial y}\frac{\partial}{\partial y}\Delta w$$
$$+ \Delta D\, \Delta w - (1-\nu)\left(\frac{\partial^2 D}{\partial x^2}\frac{\partial^2 w}{\partial y^2} - 2\frac{\partial^2 D}{\partial x\, \partial y}\frac{\partial^2 w}{\partial x\, \partial y} + \frac{\partial^2 D}{\partial y^2}\frac{\partial^2 w}{\partial x^2}\right) = q \qquad (171)$$

where, as before, we employ the notation

$$\Delta = \frac{\partial^2}{\partial x^2} + \frac{\partial^2}{\partial y^2}$$

As a particular example of the application of Eq. (171) let us consider the case in which the flexural rigidity D is a linear function of y expressed in the form

$$D = D_0 + D_1 y \qquad (c)$$

where D_0 and D_1 are constants.

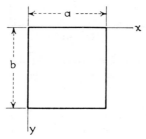

Fig. 82

In such a case Eq. (171) becomes

$$(D_0 + D_1 y)\Delta\Delta w + 2D_1\frac{\partial}{\partial y}\Delta w = q$$

or
$$\Delta[(D_0 + D_1 y)\, \Delta w] = q \qquad (172)$$

Let us consider the case in which the intensity of the load q is proportional to the flexural rigidity D. We shall assume the deflection of the plate (Fig. 82) in the form

$$w = w_1 + w_2$$

and let w_1 equal the deflection of a strip parallel to the x axis cut from the plate and loaded with a load of intensity

$$q = q_0\left(1 + \frac{D_1}{D_0} y\right) \qquad (d)$$

SIMPLY SUPPORTED RECTANGULAR PLATES

This deflection can be represented, as before, by the trigonometric series

$$w_1 = \frac{4q_0\left(1 + \frac{D_1}{D_0}y\right)a^4}{(D_0 + D_1 y)\pi^5} \sum_{m=1,3,5,\ldots}^{\infty} \frac{1}{m^5} \sin\frac{m\pi x}{a} = \frac{4q_0 a^4}{\pi^5 D_0} \sum_{m=1,3,5,\ldots}^{\infty} \frac{1}{m^5} \sin\frac{m\pi x}{a} \quad (e)$$

By substitution we can readily show that this expression for w_1 satisfies Eq. (172). It satisfies also the boundary conditions $w_1 = 0$ and $\partial^2 w_1/\partial x^2 = 0$ along the supported edges $x = 0$ and $x = a$.

The deflection w_2 must then satisfy the homogeneous equation

$$\Delta[(D_0 + D_1 y)\,\Delta w_2] = 0 \quad (f)$$

We take it in the form of a series

$$w_2 = \sum_{m=1,3,5,\ldots}^{\infty} Y_m \sin\frac{m\pi x}{a} \quad (g)$$

Substituting this series in Eq. (f), we find that the functions Y_m satisfy the following ordinary differential equation:

$$\left(\frac{\partial^2}{\partial y^2} - \frac{m^2\pi^2}{a^2}\right)\left[(D_0 + D_1 y)\left(Y_m'' - \frac{m^2\pi^2}{a^2}Y_m\right)\right] = 0 \quad (h)$$

Using the notation

$$f_m = (D_0 + D_1 y)\left(Y_m'' - \frac{m^2\pi^2}{a^2}Y_m\right) \qquad \frac{m\pi}{a} = \alpha_m \quad (i)$$

we find, from Eq. (h),

$$f_m = A_m e^{\alpha_m y} + B_m e^{-\alpha_m y}$$

Then, from Eq. (i), we obtain

$$Y_m'' - \alpha_m^2 Y_m = \frac{A_m e^{\alpha_m y} + B_m e^{-\alpha_m y}}{D_0 + D_1 y} \quad (j)$$

The general solution of this equation is

$$Y_m = C_m e^{\alpha_m y} + D_m e^{-\alpha_m y} + g_m \quad (k)$$

in which g_m is a particular integral of Eq. (j). To find this particular integral we use the Lagrange method of variation of constants; i.e., we assume that g_m has the form

$$g_m = E_m e^{\alpha_m y} + F_m e^{-\alpha_m y} \quad (l)$$

in which E_m and F_m are functions of y. These functions have to be determined from the following equations:[1]

$$E_m' e^{\alpha_m y} + F_m' e^{-\alpha_m y} = 0$$

$$E_m' e^{\alpha_m y} - F_m' e^{-\alpha_m y} = \frac{A_m e^{\alpha_m y} + B_m e^{-\alpha_m y}}{\alpha_m(D_0 + D_1 y)}$$

[1] E_m' and F_m' in these equations are the derivatives with respect to y of E_m and F_m.

from which

$$E'_m = \frac{A_m + B_m e^{-2\alpha_m y}}{2\alpha_m(D_0 + D_1 y)}$$

$$F'_m = -\frac{A_m e^{2\alpha_m y} + B_m}{2\alpha_m(D_0 + D_1 y)}$$

Integrating these equations, we find

$$E_m = \int \frac{A_m + B_m e^{-2\alpha_m y}}{2\alpha_m(D_0 + D_1 y)} dy = \frac{A_m}{2\alpha_m D_1} \log \frac{2\alpha_m}{D_1}(D_0 + D_1 y)$$

$$+ \frac{B_m}{2\alpha_m D_1} e^{\frac{2\alpha_m D_0}{D_1}} \int \frac{e^{\frac{-2\alpha_m(D_0+D_1 y)}{D_1}}}{2\alpha_m(D_0 + D_1 y)} d[2\alpha_m(D_0 + D_1 y)]$$

$$F_m = -\int \frac{A_m e^{2\alpha_m y} + B_m}{2\alpha_m(D_0 + D_1 y)} dy = -\frac{B_m}{2\alpha_m D_1} \log \frac{2\alpha_m}{D_1}(D_0 + D_1 y)$$

$$- \frac{A_m}{2\alpha_m D_1} e^{\frac{-2\alpha_m D_0}{D_1}} \int \frac{e^{\frac{2\alpha_m(D_0+D_1 y)}{D_1}}}{2\alpha_m(D_0 + D_1 y)} d[2\alpha_m(D_0 + D_1 y)]$$

Substituting these expressions in Eqs. (*l*) and (*k*) and using the notation[1]

$$E_i(u) = \int_{-\infty}^{u} \frac{e^u}{u} du \qquad E_i(-u) = \int_{\infty}^{u} \frac{e^{-u}}{u} du$$

we represent functions Y_m in the following form:

$$Y_m = A'_m \left\{ \log \frac{2\alpha_m}{D_1}(D_0 + D_1 y) - e^{\frac{-2\alpha_m}{D_1}(D_0+D_1 y)} E_i \left[\frac{2\alpha_m(D_0 + D_1 y)}{D_1} \right] \right\} e^{\alpha_m y}$$

$$- B'_m \left\{ e^{\frac{-2\alpha_m}{D_1}(D_0+D_1 y)} \log \frac{2\alpha_m}{D_1}(D_0 + D_1 y) - E_i \left[\frac{-2\alpha_m(D_0 + D_1 y)}{D_1} \right] \right\} e^{\alpha_m y}$$

$$+ C_m e^{\alpha_m y} + D_m e^{-\alpha_m y} \quad (m)$$

The four constants of integration A'_m, B'_m, C_m, D_m are obtained from the boundary conditions along the sides $y = 0$ and $y = b$. In the case of simply supported edges these are

$$(w)_{y=0} = 0 \qquad \left(\frac{\partial^2 w}{\partial y^2}\right)_{y=0} = 0$$

$$(w)_{y=b} = 0 \qquad \left(\frac{\partial^2 w}{\partial y^2}\right)_{y=b} = 0$$

The numerical results for a simply supported square plate obtained by taking only the first two terms of the series (*g*) are shown in Fig. 83.[2] The deflections and the moments M_x and M_y along the line $x = a/2$ for the plate of variable thickness are shown by full lines; the same quantities calculated for a plate of constant flexural rigidity $D = \frac{1}{2}(D_0 + D_1 b)$ are shown by dashed lines. It was assumed in the calculation that $D_1 b = 7 D_0$ and $\nu = 0.16$.

---

[1] The integral $E_i(u)$ is the so-called *exponential integral* and is a tabulated function; see, for instance, Jahnke-Emde, "Tables of Functions," 4th ed., pp. 1 and 6, Dover Publications, 1945; or "Tables of Sine, Cosine and Exponential Integrals," National Bureau of Standards, New York, 1940.

[2] These results are taken from R. Gran Olsson, *loc. cit.*

SIMPLY SUPPORTED RECTANGULAR PLATES

Finally, let us consider the case in which the thickness of the plate is a linear function of y alone and the intensity of the load is any function of y (Fig. 82). Denoting the thickness of the plate along the line $y = b/2$ by h_0 and the corresponding flexural rigidity by

$$D_0 = \frac{E h_0^3}{12(1 - \nu^2)} \quad (n)$$

we have at any point of the plate

$$D = D_0 \frac{h^3}{h_0^3} \quad \text{and} \quad h = \left[1 + \lambda \left(\frac{2y}{b} - 1\right)\right] h_0 \quad (o)$$

where λ is some constant. This yields $h = (1 - \lambda) h_0$ at $y = 0$ and $h = (1 + \lambda) h_0$ at $y = b$.

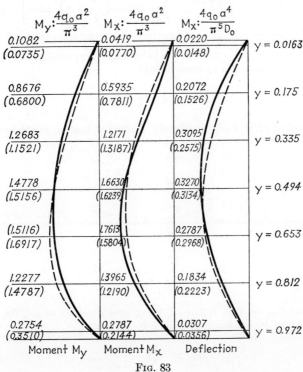

FIG. 83

The following method[1] introducing the quantity λ as a parameter proves to be most efficient in handling the present problem. Considering the deflection w as a function of the variables x, y, and λ, we can express $w(x,y,\lambda)$ in form of the power series

$$w = \sum_{m=0}^{\infty} w_m \lambda^m \quad (p)$$

in which m is an integer and the coefficients w_m are merely functions of x and y.

[1] See H. Favre and B. Gilg, *Z. angew. Math. u. Phys.*, vol. 3, p. 354, 1952.

Substituting expressions (o) and (p) in Eq. (171) and equating to zero the coefficients of successive powers of λ, we obtain a sequence of differential equations

$$\Delta\Delta w_0 = \frac{q}{D_0}$$

$$\Delta\Delta w_1 = -3\left[\frac{4}{b}\frac{\partial}{\partial y}\Delta w_0 + \left(\frac{2y}{b}-1\right)\Delta\Delta w_0\right]$$

$$\begin{aligned}\Delta\Delta w_2 = &-3\left[\frac{4}{b}\frac{\partial}{\partial y}\Delta w_1 + \left(\frac{2y}{b}-1\right)\Delta\Delta w_1\right] \\ &-3\left\{\frac{8}{b^2}\left[\Delta w_0 - (1-\nu)\frac{\partial^2 w_0}{\partial x^2}\right]\right. \\ &\left.+\frac{8}{b}\left(\frac{2y}{b}-1\right)\frac{\partial}{\partial y}\Delta w_0 + \left(\frac{2y}{b}-1\right)^2\Delta\Delta w_0\right\}\end{aligned} \quad (q)$$

. . . .

We assume the edges $x = 0$ and $x = a$ to be simply supported, and we shall restrict the problem to the case of a hydrostatic load

$$q = \frac{q_0 y}{b} \quad (r)$$

Using the method of M. Lévy we take the solution of Eqs. (q) in the form

$$w_0 = \sum_{n=1,3,\ldots}^{\infty} Y_{on} \sin\frac{n\pi x}{a} \quad (s)$$

$$w_1 = \sum_{n=1,3,\ldots}^{\infty} Y_{1n} \sin\frac{n\pi x}{a} \cdots \quad (t)$$

$$w_m = \sum_{n=1,3,\ldots}^{\infty} Y_{mn} \sin\frac{n\pi x}{a} \quad (u)$$

the coefficients Y_{mn} ($m = 0, 1, 2, \ldots$) being some functions of y. We can, finally, represent the load (r) in analogous manner by putting

$$q = \frac{4q_0 y}{\pi b}\sum_{n=1,3,\ldots}^{\infty}\frac{1}{n}\sin\frac{n\pi x}{a} \quad (v)$$

Substitution of expressions (s) and (v) in the first of the equations (q) enables us to determine the functions Y_{on}, the boundary conditions being $Y_{on} = 0$, $Y''_{on} = 0$ at $y = 0$ and $y = b$ if these edges are simply supported. The substitution of expressions (s) and (t) in the second of the equations (q) yields the function Y_{1n}. In like manner any function w_m is found by substitution of $w_0, w_1, \ldots, w_{m-1}$ in that differential equation of the system (q) which contains w_m at the left-hand side. The procedure remains substantially the same if the edges $y = 0, b$ are built-in or free instead of being simply supported.

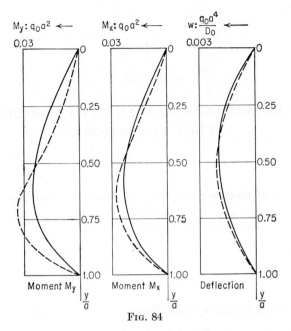

Fig. 84

Numerical results obtained by H. Favre and B. Gilg[1] for the deflections and the bending moments along the center line $x = a/2$ of a simply supported plate with $\lambda = 0.2$ and $\nu = 0.25$ under hydrostatic pressure (r) are shown in Fig. 84. Full lines give results obtained by taking three terms in the series (p), while the dashed lines hold for the result of the first approximations.

[1] *Ibid.*

CHAPTER 6

RECTANGULAR PLATES WITH VARIOUS EDGE CONDITIONS

41. Bending of Rectangular Plates by Moments Distributed along the Edges. Let us consider a rectangular plate supported along the edges and bent by moments distributed along the edges $y = \pm b/2$ (Fig. 85). The deflections w must satisfy the homogeneous differential equation

$$\frac{\partial^4 w}{\partial x^4} + 2 \frac{\partial^4 w}{\partial x^2 \, \partial y^2} + \frac{\partial^4 w}{\partial y^4} = 0 \tag{a}$$

and the following boundary conditions:

$$w = 0 \qquad \frac{\partial^2 w}{\partial x^2} = 0 \qquad \text{for } x = 0 \text{ and } x = a \tag{b}$$

$$w = 0 \qquad \text{for } y = \pm \frac{b}{2} \tag{c}$$

$$-D \left(\frac{\partial^2 w}{\partial y^2} \right)_{y=b/2} = f_1(x) \qquad -D \left(\frac{\partial^2 w}{\partial y^2} \right)_{y=-b/2} = f_2(x) \tag{d}$$

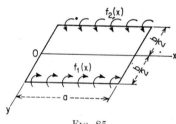

Fig. 85

in which f_1 and f_2 represent the bending moment distributions along the edges $y = \pm b/2$.

We take the solution of Eq. (a) in the form of the series

$$w = \sum_{m=1}^{\infty} Y_m \sin \frac{m\pi x}{a} \tag{e}$$

each term of which satisfies the boundary conditions (b). The functions Y_m we take, as before, in the form

$$Y_m = A_m \sinh \frac{m\pi y}{a} + B_m \cosh \frac{m\pi y}{a} + C_m \frac{m\pi y}{a} \sinh \frac{m\pi y}{a}$$

$$+ D_m \frac{m\pi y}{a} \cosh \frac{m\pi y}{a} \tag{f}$$

which satisfies Eq. (a).

To simplify the discussion let us begin with the two particular cases:
1. The symmetrical case in which $(M_y)_{y=b/2} = (M_y)_{y=-b/2}$
2. The antisymmetrical case in which $(M_y)_{y=b/2} = -(M_y)_{y=-b/2}$

The general case can be obtained by combining these two particular cases.

In the case of symmetry Y_m must be an even function of y, and it is necessary to put $A_m = D_m = 0$ in expression (f). Then we obtain, from Eq. (e),

$$w = \sum_{m=1}^{\infty} \left(B_m \cosh \frac{m\pi y}{a} + C_m \frac{m\pi y}{a} \sinh \frac{m\pi y}{a} \right) \sin \frac{m\pi x}{a} \qquad (g)$$

To satisfy the boundary condition (c) we must put

$$B_m \cosh \alpha_m + C_m \alpha_m \sinh \alpha_m = 0$$

where, as before,

$$\alpha_m = \frac{m\pi b}{2a}$$

Hence $\quad B_m = -C_m \alpha_m \tanh \alpha_m$

and the deflection in the symmetrical case is

$$w = \sum_{m=1}^{\infty} C_m \left(\frac{m\pi y}{a} \sinh \frac{m\pi y}{a} - \alpha_m \tanh \alpha_m \cosh \frac{m\pi y}{a} \right) \sin \frac{m\pi x}{a} \qquad (h)$$

We use the boundary conditions (d) to determine the constants C_m. Representing the distribution of bending moments along the edges $y = \pm b/2$ by a trigonometric series, we have in the case of symmetry

$$f_1(x) = f_2(x) = \sum_{m=1}^{\infty} E_m \sin \frac{m\pi x}{a} \qquad (i)$$

where the coefficients E_m can be calculated in the usual way for each particular case. For instance, in the case of a uniform distribution of the bending moments we have (see page 151)

$$(M_y)_{y=b/2} = \frac{4M_0}{\pi} \sum_{m=1,3,5,\ldots}^{\infty} \frac{1}{m} \sin \frac{m\pi x}{a} \qquad (j)$$

Substituting expressions (h) and (i) into conditions (d), we obtain

$$-2D \sum_{m=1}^{\infty} \frac{m^2 \pi^2}{a^2} C_m \cosh \alpha_m \sin \frac{m\pi x}{a} = \sum_{m=1}^{\infty} E_m \sin \frac{m\pi x}{a}$$

from which
$$C_m = -\frac{a^2 E_m}{2Dm^2\pi^2 \cosh \alpha_m}$$

and

$$w = \frac{a^2}{2\pi^2 D} \sum_{m=1}^{\infty} \frac{\sin \frac{m\pi x}{a}}{m^2 \cosh \alpha_m} E_m \left(\alpha_m \tanh \alpha_m \cosh \frac{m\pi y}{a} - \frac{m\pi y}{a} \sinh \frac{m\pi y}{a} \right) \quad (173)$$

In the particular case of uniformly distributed moments of intensity M_0 we obtain, by using expression (j),

$$w = \frac{2M_0 a^2}{\pi^3 D} \sum_{m=1,3,5,\ldots}^{\infty} \frac{1}{m^3 \cosh \alpha_m} \left(\alpha_m \tanh \alpha_m \cosh \frac{m\pi y}{a} - \frac{m\pi y}{a} \sinh \frac{m\pi y}{a} \right) \sin \frac{m\pi x}{a}$$

The deflection along the axis of symmetry ($y = 0$) is

$$(w)_{y=0} = \frac{2M_0 a^2}{\pi^3 D} \sum_{m=1,3,5,\ldots}^{\infty} \frac{1}{m^3} \frac{\alpha_m \tanh \alpha_m}{\cosh \alpha_m} \sin \frac{m\pi x}{a} \quad (k)$$

When a is very large in comparison with b, we can put $\tanh \alpha_m \approx \alpha_m$ and $\cosh \alpha_m \approx 1$. Then, by using series (j), we obtain

$$(w)_{y=0} = \frac{M_0 b^2}{2\pi D} \sum_{m=1,3,5,\ldots}^{\infty} \frac{1}{m} \sin \frac{m\pi x}{a} = \frac{1}{8} \frac{M_0 b^2}{D}$$

This is the deflection at the middle of a strip of length b bent by two equal and opposite couples applied at the ends.

When a is small in comparison with b, $\cosh \alpha_m$ is a large number, and the deflection of the plate along the x axis is very small.

For any given ratio between the lengths of the sides of the rectangle the deflection at the center of the plate, from expression (k), is

$$(w)_{y=0, x=a/2} = \frac{M_0 ab}{\pi^2 D} \sum_{m=1,3,5,\ldots}^{\infty} (-1)^{(m-1)/2} \frac{1}{m^2} \frac{\tanh \alpha_m}{\cosh \alpha_m}$$

Having expression (173) for deflections, we can obtain the slope of the deflection surface at the boundary by differentiation, and we can calculate the bending moments by forming the second derivatives of w.

Some values of the deflections and the bending moments computed in this way are given in Table 28. It is seen, for example, that the deflection of a strip of a width a is about $3\frac{1}{2}$ times that of a square plate of dimensions a. While the transverse section at the middle of a strip transmits the entire moment M_0 applied at the ends, the bending moment M_y at the center of the plate decreases rapidly as compared with M_0, with an increasing ratio b/a. This is due to a damping effect of the edges $x = 0$ and $x = a$ not exposed to couples.

TABLE 28. DEFLECTIONS AND BENDING MOMENTS AT THE CENTER OF RECTANGULAR PLATES SIMPLY SUPPORTED AND SUBJECTED TO COUPLES UNIFORMLY DISTRIBUTED ALONG THE EDGES $y = \pm b/2$ (FIG. 85)
$\nu = 0.3$

b/a	w	M_x	M_y
0	$0.1250 M_0 b^2/D$	$0.300 M_0$	$1.000 M_0$
0.50	$0.0964 M_0 b^2/D$	$0.387 M_0$	$0.770 M_0$
0.75	$0.0620 M_0 b^2/D$	$0.424 M_0$	$0.476 M_0$
1.00	$0.0368 M_0 a^2/D$	$0.394 M_0$	$0.256 M_0$
1.50	$0.0280 M_0 a^2/D$	$0.264 M_0$	$0.046 M_0$
2.00	$0.0174 M_0 a^2/D$	$0.153 M_0$	$-0.010 M_0$

Let us consider now the antisymmetrical case in which

$$f_1(x) = -f_2(x) = \sum_{m=1}^{\infty} E_m \sin \frac{m\pi x}{a}$$

In this case the deflection surface is an odd function of y, and we must put $B_m = C_m = 0$ in expression (f). Hence,

$$w = \sum_{m=1}^{\infty} \left(A_m \sinh \frac{m\pi y}{a} + D_m \frac{m\pi y}{a} \cosh \frac{m\pi y}{a} \right) \sin \frac{m\pi x}{a}$$

From the boundary conditions (c) it follows that

$$A_m \sinh \alpha_m + D_m \alpha_m \cosh \alpha_m = 0$$

whence
$$D_m = -\frac{1}{\alpha_m} \tanh \alpha_m A_m$$

and

$$w = \sum_{m=1}^{\infty} A_m \left(\sinh \frac{m\pi y}{a} - \frac{1}{\alpha_m} \tanh \alpha_m \frac{m\pi y}{a} \cosh \frac{m\pi y}{a} \right) \sin \frac{m\pi x}{a}$$

The constants A_m are obtained from conditions (d), from which it follows that

$$\frac{2\pi^2 D}{a^2} \sum_{m=1}^{\infty} A_m \frac{m^2}{\alpha_m} \sinh \alpha_m \tanh \alpha_m \sin \frac{m\pi x}{a} = \sum_{m=1}^{\infty} E_m \sin \frac{m\pi x}{a}$$

Hence
$$A_m = \frac{a^2}{2\pi^2 D} E_m \frac{\alpha_m}{m^2 \sinh \alpha_m \tanh \alpha_m}$$

and

$$w = \frac{a^2}{2\pi^2 D} \sum_{m=1}^{\infty} \frac{E_m}{m^2 \sinh \alpha_m} \left(\alpha_m \coth \alpha_m \sinh \frac{m\pi y}{a} - \frac{m\pi y}{a} \cosh \frac{m\pi y}{a} \right) \sin \frac{m\pi x}{a} \quad (174)$$

We can obtain the deflection surface for the general case represented by the boundary conditions (d) from solutions (173) and (174) for the symmetrical and the antisymmetrical cases. For this purpose we split the given moment distributions into a symmetrical moment distribution M'_y and an antisymmetrical distribution M''_y, as follows:

$$(M'_y)_{y=b/2} = (M'_y)_{y=-b/2} = \tfrac{1}{2}[f_1(x) + f_2(x)]$$
$$(M''_y)_{y=b/2} = -(M''_y)_{y=-b/2} = \tfrac{1}{2}[f_1(x) - f_2(x)]$$

These moments can be represented, as before, by the trigonometric series

$$(M'_y)_{y=b/2} = \sum_{m=1}^{\infty} E'_m \sin \frac{m\pi x}{a}$$
$$(M''_y)_{y=b/2} = \sum_{m=1}^{\infty} E''_m \sin \frac{m\pi x}{a} \quad (l)$$

and the total deflection is obtained by using expressions (173) and (174) and superposing the deflections produced by each of the two foregoing moment distributions (l). Hence

$$w = \frac{a^2}{2\pi^2 D} \sum_{m=1}^{\infty} \frac{\sin \frac{m\pi x}{a}}{m^2} \left[\frac{E'_m}{\cosh \alpha_m} \left(\alpha_m \tanh \alpha_m \cosh \frac{m\pi x}{a} - \frac{m\pi y}{a} \sinh \frac{m\pi y}{a} \right) + \frac{E''_m}{\sinh \alpha_m} \left(\alpha_m \coth \alpha_m \sinh \frac{m\pi y}{a} - \frac{m\pi y}{a} \cosh \frac{m\pi y}{a} \right) \right] \quad (175)$$

If the bending moments $M_y = \sum_{m=1}^{\infty} E_m \sin (m\pi x/l)$ are distributed only along the edge $y = b/2$, we have $f_2(x) = 0$, $E'_m = E''_m = \tfrac{1}{2}E_m$; and the deflection in this case becomes

$$w = \frac{a^2}{4\pi^2 D} \sum_{m=1}^{\infty} \frac{E_m \sin \dfrac{m\pi x}{a}}{m^2} \left[\frac{1}{\cosh \alpha_m} \left(\alpha_m \tanh \alpha_m \cosh \frac{m\pi y}{a} - \frac{m\pi y}{a} \sinh \frac{m\pi y}{a} \right) + \frac{1}{\sinh \alpha_m} \left(\alpha_m \coth \alpha_m \sinh \frac{m\pi y}{a} - \frac{m\pi y}{a} \cosh \frac{m\pi y}{a} \right) \right] \quad (176)$$

Solutions (173) to (176) of this article will be applied in the investigation of plates with various edge conditions.

Moments M_0 distributed along only one edge, say $y = b/2$, would produce, at the center of the plate, one-half the deflections and bending moments given in Table 28. In case of a simultaneous action of couples along the entire boundary of the plate, the deflections and moments can be obtained by suitable superposition of the results obtained above for a partial loading.[1]

42. Rectangular Plates with Two Opposite Edges Simply Supported and the Other Two Edges Clamped. Assume that the edges $x = 0$ and $x = a$ of the rectangular plate, shown in Fig. 86, are simply supported and that the other two edges are clamped. The deflection of the plate under any lateral load can be obtained by first solving the problem on the assumption that all edges are simply supported and then applying bending moments along the edges $y = \pm b/2$ of such a magnitude as to eliminate the rotations produced along these edges by the action of the lateral load. In this manner many problems can be solved by combining the solutions given in Chap. 5 with the solution of the preceding article.

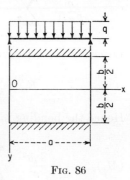

Fig. 86

Uniformly Loaded Plates.[2] Assuming that the edges of the plate are simply supported, the deflection is [see Eq. (139), page 116]

---

[1] Bending by edge couples was also discussed by H. Bay, *Ingr.-Arch.*, vol. 8, p. 4, 1937, and by U. Wegner, *Z. angew. Math. Mech.*, vol. 36, p. 340, 1956.

[2] Extensive numerical data regarding rectangular plates with uniform load and sides simply supported or clamped in any combination may be found in a paper by F. Czerny; see *Bautech.-Arch.*, vol. 11, p. 33, Berlin, 1955.

$$w = \frac{4qa^4}{\pi^5 D} \sum_{m=1,3,5,\ldots}^{\infty} \frac{1}{m^5} \sin \frac{m\pi x}{a} \left(1 - \frac{\alpha_m \tanh \alpha_m + 2}{2 \cosh \alpha_m} \cosh \frac{m\pi y}{a} \right.$$
$$\left. + \frac{1}{2 \cosh \alpha_m} \frac{m\pi y}{a} \sinh \frac{m\pi y}{a}\right) \quad (a)$$

and the slope of the deflection surface along the edge $y = b/2$ is

$$\left(\frac{\partial w}{\partial y}\right)_{y=b/2} = \frac{2qa^3}{\pi^4 D} \sum_{m=1,3,5,\ldots}^{\infty} \frac{1}{m^4} \sin \frac{m\pi x}{a} [\alpha_m - \tanh \alpha_m (1 + \alpha_m \tanh \alpha_m)] \quad (b)$$

To eliminate this slope and thus to satisfy the actual boundary conditions we distribute along the edges $y = \pm b/2$ the bending moments M_y given by the series

$$(M_y)_{y=\pm b/2} = \sum_{m=1}^{\infty} E_m \sin \frac{m\pi x}{a} \quad (c)$$

and we determine the coefficients E_m so as to make the slope produced by these moments equal and opposite to that given by expression (b). Using expression (173)[1] for the deflection produced by the moments, we find that the corresponding slope along the edge $y = b/2$ is

$$\frac{a}{2\pi D} \sum_{m=1,3,5,\ldots}^{\infty} \frac{\sin \frac{m\pi x}{a}}{m} E_m[\tanh \alpha_m (\alpha_m \tanh \alpha_m - 1) - \alpha_m] \quad (d)$$

Equating the negative of this quantity to expression (b), we find that

$$E_m = \frac{4qa^2}{\pi^3 m^3} \frac{\alpha_m - \tanh \alpha_m (1 + \alpha_m \tanh \alpha_m)}{\alpha_m - \tanh \alpha_m (\alpha_m \tanh \alpha_m - 1)} \quad (e)$$

Hence the bending moments along the built-in edges are

$$(M_y)_{y=\pm b/2} = \frac{4qa^2}{\pi^3} \sum_{m=1,3,5,\ldots}^{\infty} \frac{\sin \frac{m\pi x}{a}}{m^3} \frac{\alpha_m - \tanh \alpha_m (1 + \alpha_m \tanh \alpha_m)}{\alpha_m - \tanh \alpha_m (\alpha_m \tanh \alpha_m - 1)} \quad (f)$$

The maximum numerical value of this moment occurs at the middle of the sides, where $x = a/2$. Series (f) converges rapidly, and the maximum moment can be readily calculated in each particular case. For

---

[1] From the symmetry of the deflection surface produced by the uniform load it can be concluded that only odd numbers 1, 3, 5, . . . must be taken for m in expression (173).

example, the first three terms of series (f) give $-0.070qa^2$ as the maximum moment in a square plate. In the general case this moment can be represented by the formula γqa^2, where γ is a numerical factor the magnitude of which depends on the ratio a/b of the sides of the plate. Several values of this coefficient are given in Table 29.

Substituting the values (e) of the coefficients E_m in expression (173), we obtain the deflection surface produced by the moments M_y distributed

TABLE 29. CONSTANTS α, β_1, β_2, γ FOR A RECTANGULAR PLATE WITH TWO EDGES SIMPLY SUPPORTED AND TWO EDGES CLAMPED (FIG. 86)
$\nu = 0.3$
$b < a$

$\dfrac{a}{b}$	$x = \dfrac{a}{2}, y = 0$ $w_{\max} = \alpha \dfrac{qb^4}{D}$	$x = \dfrac{a}{2}, y = 0$ $M_x = \beta_1 qb^2$	$x = \dfrac{a}{2}, y = 0$ $M_y = \beta_2 qb^2$	$x = \dfrac{a}{2}, y = \dfrac{b}{2}$ $M_y = \gamma qb^2$
	α	β_1	β_2	γ
∞	0.00260	0.0125	0.0417	-0.0833
2	0.00260	0.0142	0.0420	-0.0842
1.5	0.00247	0.0179	0.0406	-0.0822
1.4	0.00240	0.0192	0.0399	-0.0810
1.3	0.00234	0.0203	0.0388	-0.0794
1.2	0.00223	0.0215	0.0375	-0.0771
1.1	0.00209	0.0230	0.0355	-0.0739

$b > a$

$\dfrac{b}{a}$	$w_{\max} = \alpha \dfrac{qa^4}{D}$	$x = \dfrac{a}{2}, y = 0$ $M_x = \beta_1 qa^2$	$x = \dfrac{a}{2}, y = 0$ $M_y = \beta_2 qa^2$	$x = \dfrac{a}{2}, y = \dfrac{b}{2}$ $M_y = \gamma qa^2$
	α	β_1	β_2	γ
1	0.00192	0.0244	0.0332	-0.0697
1.1	0.00251	0.0307	0.0371	-0.0787
1.2	0.00319	0.0376	0.0400	-0.0868
1.3	0.00388	0.0446	0.0426	-0.0938
1.4	0.00460	0.0514	0.0448	-0.0998
1.5	0.00531	0.0585	0.0460	-0.1049
1.6	0.00603	0.0650	0.0469	-0.1090
1.7	0.00668	0.0712	0.0475	-0.1122
1.8	0.00732	0.0768	0.0477	-0.1152
1.9	0.00790	0.0821	0.0476	-0.1174
2.0	0.00844	0.0869	0.0474	-0.1191
3.0	0.01168	0.1144	0.0419	-0.1246
∞	0.01302	0.1250	0.0375	-0.1250

along the edges

$$w_1 = -\frac{2qa^4}{\pi^5 D} \sum_{m=1,3,5,\ldots}^{\infty} \frac{\sin \frac{m\pi x}{a}}{m^5 \cosh \alpha_m}$$

$$\frac{\alpha_m - \tanh \alpha_m (1 + \alpha_m \tanh \alpha_m)}{\alpha_m - \tanh \alpha_m (\alpha_m \tanh \alpha_m - 1)} \left(\frac{m\pi y}{a} \sinh \frac{m\pi y}{a} \right.$$

$$\left. - \alpha_m \tanh \alpha_m \cosh \frac{m\pi y}{a} \right) \quad (g)$$

The deflection at the center is obtained by substituting $x = a/2$, $y = 0$ in expression (g). Then

$$(w_1)_{\max} = \frac{2qa^4}{\pi^5 D} \sum_{m=1,3,5,\ldots}^{\infty} \frac{(-1)^{(m-1)/2}}{m^5} \frac{\alpha_m \tanh \alpha_m}{\cosh \alpha_m}$$

$$\frac{\alpha_m - \tanh \alpha_m (1 + \alpha_m \tanh \alpha_m)}{\alpha_m - \tanh \alpha_m (\alpha_m \tanh \alpha_m - 1)}$$

This is a rapidly converging series, and the deflection can be obtained with a high degree of accuracy by taking only a few terms. In the case of a square plate, for example, the first term alone gives the deflection correct to three significant figures, and we obtain

$$w_1 = 0.00214 \frac{qa^4}{D}$$

Subtracting this deflection from the deflection produced at the center by the uniform load (Table 8, page 120), we obtain finally for the deflection of a uniformly loaded square plate with two simply supported and two clamped edges the value

$$w = 0.00192 \frac{qa^4}{D}$$

In the general case the deflection at the center can be represented by the formula

$$w = \alpha \frac{qa^4}{D}$$

Several values of the numerical factor α are given in Table 29.

Substituting expression (g) for deflections in the known formulas (101) for the bending moments, we obtain

$$M_x = -\frac{2qa^2}{\pi^3} \sum_{m=1,3,5,\ldots}^{\infty} \frac{\sin \frac{m\pi x}{a}}{m^3 \cosh \alpha_m} \frac{\alpha_m - \tanh \alpha_m (1 + \alpha_m \tanh \alpha_m)}{\alpha_m - \tanh \alpha_m (\alpha_m \tanh \alpha_m - 1)}$$

$$\left\{ (1 - \nu) \frac{m\pi y}{a} \sinh \frac{m\pi y}{a} - [2\nu + (1 - \nu)\alpha_m \tanh \alpha_m] \cosh \frac{m\pi y}{a} \right\} \quad (h)$$

$$M_y = \frac{2qa^2}{\pi^3} \sum_{m=1,3,5,\ldots}^{\infty} \frac{\sin \frac{m\pi x}{a}}{m^3 \cosh \alpha_m} \frac{\alpha_m - \tanh \alpha_m(1 + \alpha_m \tanh \alpha_m)}{\alpha_m - \tanh \alpha_m(\alpha_m \tanh \alpha_m - 1)}$$
$$\left\{ (1-\nu)\frac{m\pi y}{a} \sinh \frac{m\pi y}{a} + [2 - (1-\nu)\alpha_m \tanh \alpha_m] \cosh \frac{m\pi y}{a} \right\} \quad (i)$$

The values of these moments at the center of the plate are

$$M_x = \frac{2qa^2}{\pi^3} \sum_{m=1,3,5,\ldots}^{\infty} \frac{(-1)^{(m-1)/2}}{m^3 \cosh \alpha_m}$$
$$\frac{\alpha_m - \tanh \alpha_m(1 + \alpha_m \tanh \alpha_m)}{\alpha_m - \tanh \alpha_m(\alpha_m \tanh \alpha_m - 1)} [2\nu + (1-\nu)\alpha_m \tanh \alpha_m]$$

$$M_y = \frac{2qa^2}{\pi^3} \sum_{m=1,3,5,\ldots}^{\infty} \frac{(-1)^{(m-1)/2}}{m^3 \cosh \alpha_m}$$
$$\frac{\alpha_m - \tanh \alpha_m(1 + \alpha_m \tanh \alpha_m)}{\alpha_m - \tanh \alpha_m(\alpha_m \tanh \alpha_m - 1)} [2 - (1-\nu)\alpha_m \tanh \alpha_m]$$

These series converge rapidly so that sufficiently accurate values for the moments are found by taking only the first two terms in the series. Superposing these moments on the moments in a simply supported plate (Table 8), the final values of the moments at the center of the plate can be represented as follows:

$$M_x = \beta_1 qa^2 \qquad M_y = \beta_2 qa^2 \qquad (j)$$

where β_1 and β_2 are numerical factors the magnitude of which depends on the ratio b/a. Several values of these coefficients are given in Table 29.

Taking the case of a square plate, we find that at the center the moments are

$$M_x = 0.0244qa^2 \quad \text{and} \quad M_y = 0.0332qa^2$$

They are smaller than the moments $M_x = M_y = 0.0479qa^2$ at the center of the simply supported square plate. But the moments M_y at the middle of the built-in edges are, as we have seen, larger than the value $0.0479qa^2$. Hence, because of the constraint of the two edges, the magnitude of the maximum stress in the plate is increased. When the built-in sides of a rectangular plate are the longer sides ($b < a$), the bending moments at the middle of these sides and the deflections at the center of the plate rapidly approach the corresponding values for a strip with built-in ends as the ratio b/a decreases.

Plates under Hydrostatic Pressure (Fig. 87). The deflection surface of a simply supported rectangular plate submitted to the action of a hydrostatic pressure, as shown in Fig. 66 (Art. 31), is

$$w = \frac{q_0 a^4}{\pi^5 D} \sum_{m=1}^{\infty} \frac{(-1)^{m+1}}{m^5} \left(2 - \frac{2 + \alpha_m \tanh \alpha_m}{\cosh \alpha_m} \cosh \frac{m\pi y}{a} \right.$$
$$\left. + \frac{1}{\cosh \alpha_m} \frac{m\pi y}{a} \sinh \frac{m\pi y}{a} \right) \sin \frac{m\pi x}{a} \quad (k)$$

The slope of the deflection surface along the edge $y = b/2$ is

$$\left(\frac{\partial w}{\partial y} \right)_{y=b/2} = \frac{q_0 a^3}{\pi^4 D} \sum_{m=1}^{\infty} \frac{(-1)^{m+1}}{m^4}$$
$$[\alpha_m - \tanh \alpha_m (1 + \alpha_m \tanh \alpha_m)] \sin \frac{m\pi x}{a} \quad (l)$$

This slope is eliminated by distributing the moments M_y given by series (c) along the edges $y = \pm b/2$ and determining the coefficients E_m of that series so as to make the slope produced by the moments equal and opposite to that given by expression (l). In this way we obtain

$$E_m = \frac{2q_0 a^2 (-1)^{m+1}}{\pi^3 m^3} \frac{\alpha_m - \tanh \alpha_m (1 + \alpha_m \tanh \alpha_m)}{\alpha_m - \tanh \alpha_m (\alpha_m \tanh \alpha_m - 1)}$$

Substituting this in series (c), the expression for bending moments along the built-in edges is found to be

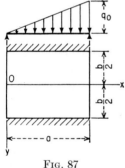

FIG. 87

$$(M_y)_{y=\pm b/2} = \frac{2q_0 a^2}{\pi^3} \sum_{m=1}^{\infty} \frac{(-1)^{m+1} \sin \frac{m\pi x}{a}}{m^3}$$
$$\frac{\alpha_m - \tanh \alpha_m (1 + \alpha_m \tanh \alpha_m)}{\alpha_m - \tanh \alpha_m (\alpha_m \tanh \alpha_m - 1)} \quad (m)$$

The terms in series (m) for which m is even vanish at the middle of the built-in sides where $x = a/2$, and the value of the series, as it should, becomes equal to one-half that for a uniformly loaded plate [see Eq. (f)]. The series converges rapidly, and the value of the bending moment at any point of the edge can be readily obtained. Several values of this moment together with those of the bending moments along the middle line $y = 0$ of the plate are given in Table 30.

Concentrated Force Acting on the Plate.[1] In this case again the deflection of the plate is obtained by superposing on the deflection of a simply supported plate (Art. 34) the deflection produced by moments distributed

---

[1] See S. Timoshenko, *Bauingenieur*, 1922, p. 51.

TABLE 30. BENDING MOMENTS IN RECTANGULAR PLATES WITH HYDROSTATIC LOAD, TWO EDGES SIMPLY SUPPORTED AND TWO EDGES CLAMPED (FIG. 87)
$\nu = 0.3$

b/a	$x = a/2, y = 0$		$x = 3a/4, y = 0$		$x = a/2, y = b/2$	$x = 3a/4, y = b/2$
	M_x	M_y	M_x	M_y	M_y	M_y
0.50	$0.007q_0b^2$	$0.021q_0b^2$	$0.018q_0b^2$	$0.029q_0b^2$	$-0.042q_0b^2$	$-0.062q_0b^2$
0.75	$0.011q_0b^2$	$0.020q_0b^2$	$0.018q_0b^2$	$0.021q_0b^2$	$-0.040q_0b^2$	$-0.045q_0b^2$
1.00	$0.013q_0a^2$	$0.017q_0a^2$	$0.017q_0a^2$	$0.015q_0a^2$	$-0.035q_0a^2$	$-0.035q_0a^2$
1.25	$0.021q_0a^2$	$0.021q_0a^2$	$0.024q_0a^2$	$0.019q_0a^2$	$-0.045q_0a^2$	$-0.043q_0a^2$
1.50	$0.030q_0a^2$	$0.023q_0a^2$	$0.031q_0a^2$	$0.020q_0a^2$	$-0.051q_0a^2$	$-0.048q_0a^2$
2	$0.043q_0a^2$	$0.024q_0a^2$	$0.042q_0a^2$	$0.020q_0a^2$	$-0.060q_0a^2$	$-0.053q_0a^2$
∞	$0.063q_0a^2$	$0.019q_0a^2$	$0.055q_0a^2$	$0.017q_0a^2$	$-0.063q_0a^2$	$-0.055q_0a^2$

along the clamped edges. Taking the case of a centrally loaded plate and assuming that the edges $y = \pm b/2$ are clamped, we obtain the following expression for the deflection under the load:

$$w_{\max} = \frac{Pb^2}{2\pi^3 D}\left[\frac{a^2}{b^2}\sum_{m=1,3,5,\ldots}^{\infty}\frac{1}{m^3}\left(\tanh\alpha_m - \frac{\alpha_m}{\cosh^2\alpha_m}\right)\right.$$
$$\left. - \frac{\pi^2}{4}\sum_{m=1,3,5,\ldots}^{\infty}\frac{1}{m}\frac{\tanh^2\alpha_m}{\sinh\alpha_m\cosh\alpha_m + \alpha_m}\right] \quad (n)$$

The first sum in the brackets corresponds to the deflection of a simply supported plate [see Eq. (147), page 143], and the second represents the deflection due to the action of the moments along the clamped edges. For the ratios $b/a = 2, 1, \frac{1}{2}$, and $\frac{1}{3}$ the values of the expression in the brackets in Eq. (n) are 0.238, 0.436, 0.448, and 0.449, respectively.

To obtain the maximum stress under the load we have to superpose on the stresses calculated for the simply supported plate the stresses produced by the following moments:

$$m_x = -P\sum_{m=1,3,5,\ldots}^{\infty}\frac{b}{4a}\frac{\tanh\alpha_m}{\sinh\alpha_m\cosh\alpha_m + \alpha_m}$$
$$[2\nu + (1-\nu)\alpha_m\tanh\alpha_m] \quad (o)$$
$$m_y = -P\sum_{m=1,3,5,\ldots}^{\infty}\frac{b}{4a}\frac{\tanh\alpha_m}{\sinh\alpha_m\cosh\alpha_m + \alpha_m}$$
$$[2 - (1-\nu)\alpha_m\tanh\alpha_m]$$

TABLE 31. CORRECTION BENDING MOMENTS AT $x = a/2$, $y = 0$, DUE TO CONSTRAINT AT $y = \pm b/2$ IN CASE OF A CENTRAL LOAD P (FIG. 71)
$\nu = 0.3$

b/a	$m_x = \beta_1 P$ β_1	$m_y = \beta_2 P$ β_2	b/a	$m_x = \beta_1 P$ β_1	$m_y = \beta_2 P$ β_2
0	−0.0484	−0.0742	1.0	−0.0505	−0.0308
0.5	−0.0504	−0.0708	1.2	−0.0420	−0.0166
0.6	−0.0524	−0.0656	1.4	−0.0319	−0.0075
0.7	−0.0540	−0.0580	1.6	−0.0227	−0.0026
0.8	−0.0544	−0.0489	1.8	−0.0155	−0.0002
0.9	−0.0532	−0.0396	2.0	−0.0101	+0.0007

Putting those correction moments equal to

$$m_x = \beta_1 P \qquad m_y = \beta_2 P \tag{p}$$

the numerical factors β_1 and β_2 for various values of the ratio b/a are given in Table 31. When the central load P is distributed over the area of a small circle or rectangle, we have only to add the moments (p) to bending moments obtained for the simply supported plate by means of the logarithmical expressions (157) and (167), respectively. The moment M_y at the middle of the clamped edges of a square plate is

$$M_y = -0.166P$$

The calculations show that this moment changes only slightly as the length of the clamped edges increases. It becomes equal to $-0.168P$ when $b/a = 0.5$ and drops to the value of $-0.155P$ when $b/a = 1.2$.*

It should be noted that the clamping moment with the numerically largest possible value of $-P/\pi = -0.3183P$ is produced by a load concentrated near the built-in edge of the plate rather than by a central load (see Art. 51). In the case of several movable loads the influence surface for the clamping moment may be used to obtain its maximum value with certainty (see Art. 76).

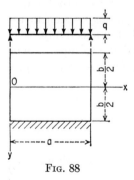

FIG. 88

43. Rectangular Plates with Three Edges Simply Supported and One Edge Built In. Let us consider a rectangular plate built in along the edge $y = b/2$ and simply supported along the other edges (Fig. 88). The deflection of the plate under any lateral load can be obtained by combining the solution for the plate with

* For further data regarding the plate with two opposite edges built in, see A. Pucher, *Ingr.-Arch.*, vol. 14, p. 246, 1943–1944.

all sides simply supported, with solution (176) for the case where bending moments are distributed along one side of the plate.

Uniformly Loaded Plates. The slope along the edge $y = b/2$ produced by a uniformly distributed load is

$$\left(\frac{\partial w}{\partial y}\right)_{y=b/2} = \frac{2qa^3}{\pi^4 D} \sum_{m=1,3,5,\ldots}^{\infty} \frac{1}{m^4} \sin \frac{m\pi x}{a} [\alpha_m - \tanh \alpha_m (1 + \alpha_m \tanh \alpha_m)] \quad (a)$$

The moments $M_y = \Sigma E_m \sin(m\pi x/a)$ distributed along the side $y = b/2$ produce the slope[1] [see Eq. (176)]

$$\left(\frac{\partial w_1}{\partial y}\right)_{y=b/2} = \frac{a}{4\pi D} \sum_{m=1,3,5,\ldots}^{\infty} \frac{1}{m} \sin \frac{m\pi x}{a} E_m(\alpha_m \tanh^2 \alpha_m$$
$$- \tanh \alpha_m + \alpha_m \coth^2 \alpha_m - \coth \alpha_m - 2\alpha_m) \quad (b)$$

From the condition of constraint these two slopes are equal in magnitude and of opposite signs. Hence

$$E_m = -\frac{8qa^2}{\pi^3 m^3} \frac{\alpha_m - \tanh \alpha_m (1 + \alpha_m \tanh \alpha_m)}{\alpha_m \tanh^2 \alpha_m - \tanh \alpha_m + \alpha_m \coth^2 \alpha_m - \coth \alpha_m - 2\alpha_m} \quad (c)$$

and the expression for the bending moments along the side $y = b/2$ is

$$(M_y)_{y=b/2} = \frac{8qa^2}{\pi^3} \sum_{m=1,3,5,\ldots}^{\infty} \frac{1}{m^3} \sin \frac{m\pi x}{a}$$
$$\frac{\alpha_m - \tanh \alpha_m (1 + \alpha_m \tanh \alpha_m)}{2\alpha_m - \tanh \alpha_m (\alpha_m \tanh \alpha_m - 1) - \coth \alpha_m (\alpha_m \coth \alpha_m - 1)} \quad (d)$$

Taking a square plate, as an example, the magnitude of the bending moment at the middle of the built-in edge from expression (d) is found to be

$$(M_y)_{y=b/2,x=a/2} = -0.084qa^2$$

This moment is numerically larger than the moment $-0.070qa^2$ which was found in the preceding article for a square plate with two edges built in. Several values of the moment at the middle of the built-in side for various values of the ratio a/b are given in Table 32.

Substituting the values (c) of the constants E_m into expression (176), we obtain the deflection surface produced by the moments of constraint, from which the deflection at the center of the plate is

$$(w_1)_{x=a/2,y=0} = \frac{a^2}{4\pi^2 D} \sum_{m=1,3,5,\ldots}^{\infty} \frac{(-1)^{(m-1)/2}}{m^2} \frac{E_m \alpha_m \tanh \alpha_m}{\cosh \alpha_m} \quad (e)$$

[1] Only odd numbers must be taken for m in this symmetrical case.

TABLE 32. DEFLECTIONS AND BENDING MOMENTS IN A RECTANGULAR PLATE WITH ONE EDGE BUILT IN AND THE THREE OTHERS SIMPLY SUPPORTED
(Fig. 88)
$\nu = 0.3$

b/a	$(w)_{x=a/2, y=0}$	$(M_y)_{x=a/2, y=b/2}$	$(M_x)_{x=a/2, y=0}$	$(M_y)_{x=a/2, y=0}$
∞	$0.0130qa^4/D$	$-0.125qa^2$	$0.125qa^2$	$0.037qa^2$
2	$0.0093qa^4/D$	$-0.122qa^2$	$0.094qa^2$	$0.047qa^2$
1.5	$0.0064qa^4/D$	$-0.112qa^2$	$0.069qa^2$	$0.048qa^2$
1.4	$0.0058qa^4/D$	$-0.109qa^2$	$0.063qa^2$	$0.047qa^2$
1.3	$0.0050qa^4/D$	$-0.104qa^2$	$0.056qa^2$	$0.045qa^2$
1.2	$0.0043qa^4/D$	$-0.098qa^2$	$0.049qa^2$	$0.044qa^2$
1.1	$0.0035qa^4/D$	$-0.092qa^2$	$0.041qa^2$	$0.042qa^2$
1.0	$0.0028qa^4/D$	$-0.084qa^2$	$0.034qa^2$	$0.039qa^2$
1/1.1	$0.0032qb^4/D$	$-0.092qb^2$	$0.033qb^2$	$0.043qb^2$
1/1.2	$0.0035qb^4/D$	$-0.098qb^2$	$0.032qb^2$	$0.047qb^2$
1/1.3	$0.0038qb^4/D$	$-0.103qb^2$	$0.031qb^2$	$0.050qb^2$
1/1.4	$0.0040qb^4/D$	$-0.108qb^2$	$0.030qb^2$	$0.052qb^2$
1/1.5	$0.0042qb^4/D$	$-0.111qb^2$	$0.028qb^2$	$0.054qb^2$
0.5	$0.0049qb^4/D$	$-0.122qb^2$	$0.023qb^2$	$0.060qb^2$
0	$0.0052qb^4/D$	$-0.125qb^2$	$0.019qb^2$	$0.062qb^2$

For a square plate the first two terms of this series give

$$(w_1)_{x=a/2, y=0} = 0.00127 \frac{qa^4}{D}$$

Subtracting this deflection from the deflection of the simply supported square plate (Table 8), we find that the deflection at the center of a uniformly loaded square plate with one edge built in is

$$(w)_{x=a/2, y=0} = 0.00279 \frac{qa^4}{D}$$

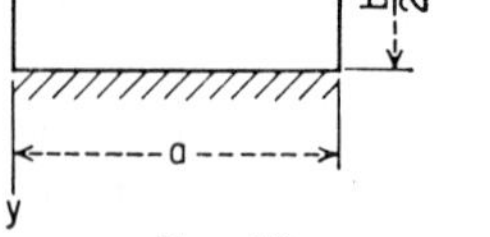

Values of deflection and bending moments for several other values of the ratio a/b obtained in a similar way are given in Table 32.

Plates under Hydrostatic Pressure. If the plate is under a hydrostatic pressure, as shown in Fig. 89, the slope along the edge $y = b/2$, in the case of simply supported edges, is (see page 190)

Fig. 89

$$\left(\frac{\partial w}{\partial y}\right)_{y=b/2} = \frac{q_0 a^3}{\pi^4 D} \sum_{m=1}^{\infty} \frac{(-1)^{m+1}}{m^4} (\alpha_m - \tanh \alpha_m - \alpha_m \tanh^2 \alpha_m) \sin \frac{m\pi x}{a} \quad (f)$$

The slope produced by bending moments distributed along the edge $y = b/2$ is

$$\left(\frac{\partial w_1}{\partial y}\right)_{y=b/2} = \frac{a}{4\pi D} \sum_{m=1}^{\infty} \frac{1}{m} \sin \frac{m\pi x}{a} E_m(\alpha_m \tanh^2 \alpha_m - \tanh \alpha_m + \alpha_m \coth^2 \alpha_m - \coth \alpha_m - 2\alpha_m) \quad (g)$$

From the condition of constraint along this edge, we find by equating expression (g) to expression (f) with negative sign

$$E_m = -\frac{4qa^2}{\pi^3} \frac{(-1)^{m+1}}{m^3} \frac{\alpha_m - \tanh \alpha_m(1 + \alpha_m \tanh \alpha_m)}{\alpha_m \tanh^2 \alpha_m - \tanh \alpha_m + \alpha_m \coth^2 \alpha_m - \coth \alpha_m - 2\alpha_m}$$

Hence the expression for the bending moment M_y along the edge $y = b/2$ is

$$(M_y)_{y=b/2} = \frac{4q_0 a^2}{\pi^3} \sum_{m=1}^{\infty} \frac{(-1)^{m+1}}{m^3} \sin \frac{m\pi x}{a} \frac{\alpha_m - \tanh \alpha_m(1 + \alpha_m \tanh \alpha_m)}{2\alpha_m - \tanh \alpha_m(\alpha_m \tanh \alpha_m - 1) - \coth \alpha_m(\alpha_m \coth \alpha_m - 1)} \quad (h)$$

This series converges rapidly, and we can readily calculate the value of the moment at any point of the built-in edge. Taking, for example, a square plate and putting $x = a/2$, we obtain for the moment at the middle of the built-in edge the value

$$(M_y)_{y=b/2, x=a/2} = -0.042 q_0 a^2$$

This is equal to one-half the value of the moment in Table 32 for a uniformly loaded square plate, as it should be. Values of the moment $(M_y)_{y=b/2}$ for several points of the built-in edge and for various values of the ratio b/a are given in Table 33. It is seen that as the ratio b/a decreases, the value of M_y along the built-in edge rapidly approaches the

TABLE 33. VALUES OF THE MOMENT M_y ALONG THE BUILT-IN EDGE $y = b/2$ OF RECTANGULAR PLATES UNDER HYDROSTATIC LOAD $g_0 x/a$ (FIG. 89)

b/a	$x = a/4$	$x = a/2$	$x = \frac{3}{4}a$
∞	$-0.039 q_0 a^2$	$-0.062 q_0 a^2$	$-0.055 q_0 a^2$
2	$-0.038 q_0 a^2$	$-0.061 q_0 a^2$	$-0.053 q_0 a^2$
$\frac{3}{2}$	$-0.034 q_0 a^2$	$-0.056 q_0 a^2$	$-0.050 q_0 a^2$
1	$-0.025 q_0 a^2$	$-0.042 q_0 a^2$	$-0.040 q_0 a^2$
$\frac{2}{3}$	$-0.030 q_0 b^2$	$-0.056 q_0 b^2$	$-0.060 q_0 b^2$
$\frac{1}{2}$	$-0.031 q_0 b^2$	$-0.061 q_0 b^2$	$-0.073 q_0 b^2$
0	$-0.031 q_0 b^2$	$-0.062 q_0 b^2$	$-0.094 q_0 b^2$

value $-q_0 b^2 x/8a$, which is the moment at the built-in end of a strip of length b uniformly loaded with a load of intensity $q_0 x/a$.

Now let us consider a plate subjected to a hydrostatic load just as before, this time, however, having the edge $x = a$ built in (Fig. 90).

In applying the method of M. Lévy to this case we take the deflection surface of the plate in the form

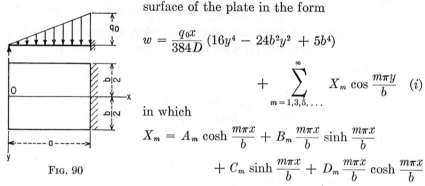

Fig. 90

$$w = \frac{q_0 x}{384D}(16y^4 - 24b^2 y^2 + 5b^4) + \sum_{m=1,3,5,\ldots}^{\infty} X_m \cos\frac{m\pi y}{b} \quad (i)$$

in which

$$X_m = A_m \cosh\frac{m\pi x}{b} + B_m \frac{m\pi x}{b} \sinh\frac{m\pi x}{b} + C_m \sinh\frac{m\pi x}{b} + D_m \frac{m\pi x}{b} \cosh\frac{m\pi x}{b}$$

Expression (i) satisfies the differential equation of the bent plate and the edge conditions at $y = \pm b/2$ as well. Expanding the expression in parentheses in Eq. (i) in the series

$$\frac{1{,}536 b^4}{\pi^5} \sum_{m=1,3,5,\ldots}^{\infty} (-1)^{(m-1)/2} \frac{1}{m^5} \cos\frac{m\pi y}{b}$$

we obtain the coefficients $A_m, B_m, \ldots$ from the conditions on both other edges; i.e.,

$$(w)_{x=0} = 0 \quad \left(\frac{\partial^2 w}{\partial x^2}\right)_{x=0} = 0 \quad (w)_{x=a} = 0 \quad \left(\frac{\partial w}{\partial x}\right)_{x=a} = 0 \quad (j)$$

Substitution of the coefficients in expression (i) makes the solution complete. Deflections and bending moments obtained from the latter equation are given in Table 34.

TABLE 34. DEFLECTIONS AND BENDING MOMENTS IN RECTANGULAR PLATES CLAMPED AT $x = a$ AND CARRYING HYDROSTATIC LOAD (FIG. 90)
$\nu = 0.3$

b/a	$(w)_{x=a/2, y=0}$	$(M_x)_{x=a/2, y=0}$	$(M_y)_{x=a/2, y=0}$	$(M_x)_{x=a, y=0}$
∞	$0.0024 q_0 a^4/D$	$0.029 q_0 a^2$	$0.009 q_0 a^2$	$-0.067 q_0 a^2$
2	$0.0023 q_0 a^4/D$	$0.029 q_0 a^2$	$0.011 q_0 a^2$	$-0.063 q_0 a^2$
1.5	$0.0019 q_0 a^4/D$	$0.026 q_0 a^2$	$0.013 q_0 a^2$	$-0.061 q_0 a^2$
1.0	$0.0013 q_0 a^4/D$	$0.019 q_0 a^2$	$0.016 q_0 a^2$	$-0.048 q_0 a^2$
$\frac{2}{3}$	$0.0030 q_0 b^4/D$	$0.028 q_0 b^2$	$0.034 q_0 b^2$	$-0.071 q_0 b^2$
0.5	$0.0045 q_0 b^4/D$	$0.024 q_0 b^2$	$0.046 q_0 b^2$	$-0.084 q_0 b^2$
0	$0.0065 q_0 b^4/D$	$0.019 q_0 b^2$	$0.062 q_0 b^2$	$-0.125 q_0 b^2$

44. Rectangular Plates with All Edges Built In.[1]

In discussing this problem, we use the same method as in the cases considered previously. We start with the solution of the problem for a simply supported rectangular plate and superpose on the deflection of such a plate the deflection of the plate by moments distributed along the edges (see Art. 41). These moments we adjust in such a manner as to satisfy the condition $\partial w/\partial n = 0$ at the boundary of the clamped plate. The method can be applied to any kind of lateral loading.

Uniformly Loaded Plates. To simplify our discussion we begin with the case of a uniformly distributed load. The deflections and the moments in this case will be symmetrical with respect to the coordinate axes shown in Fig. 91. The deflection of a simply supported plate, as given by Eq. (139) (page 116), is represented for the new coordinates in the following form:

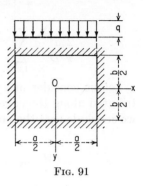

Fig. 91

$$w = \frac{4qa^4}{\pi^5 D} \sum_{m=1,3,5,\ldots}^{\infty} \frac{(-1)^{(m-1)/2}}{m^5} \cos \frac{m\pi x}{a} \left(1 - \frac{\alpha_m \tanh \alpha_m + 2}{2 \cosh \alpha_m} \cosh \frac{m\pi y}{a} + \frac{1}{2 \cosh \alpha_m} \frac{m\pi y}{a} \sinh \frac{m\pi y}{a} \right) \quad (a)$$

---

[1] For the mathematical literature on this subject see "Encyklopädie der mathematischen Wissenschaften," vol. 4, art. 25 (Tedone-Timpe), pp. 165 and 186. Other references on this subject are given in the paper by A. E. H. Love, *Proc. London Math. Soc.*, vol. 29, p. 189. The first numerical results for calculating stresses and deflections in clamped rectangular plates were obtained by B. M. Koyalovich in his doctor's dissertation, St. Petersburg, 1902. Further progress was made by I. G. Boobnov, who calculated the tables for deflections and moments in uniformly loaded rectangular plates with clamped edges; see his "Theory of Structures of Ships," vol. 2, p. 465, St. Petersburg, 1914, and "Collected Papers on the Theory of Plates," p. 144, Moscow, 1953. The same problem was discussed also by H. Hencky in his dissertation "Der Spannungszustand in rechteckigen Platten," Münich, 1913. Hencky's method was used by I. A. Wojtaszak, *J. Appl. Mechanics*, vol. 4, p. 173, 1937. The numerical results obtained by Wojtaszak in this way for a uniformly loaded plate coincide with the values given in Boobnov's table. Further solutions for the same plate and various cases of loading are due to H. Leitz, *Z. Math. Phys.*, vol. 64, p. 262, 1917; A. Nádai, *Z. angew. Math. Mech.*, vol. 2, p. 14, 1922; A. Weinstein and D. H. Rock, *Quart. Appl. Math.*, vol. 2, p. 262, 1944; P. Funk and E. Berger, "Federhofer-Girkmann-Festschrift," p. 199, Vienna, 1950; G. A. Grinberg, *Doklady Akad. Nauk. S.S.S.R.*, vol. 76, p. 661, 1951; K. Girkmann and E. Tungl, *Österr. Bauzeitschrift*, vol. 8, p. 47, 1953. An experimental investigation of the problem is due to B. C. Laws, *Phil. Mag.*, vol. 24, p. 1072, 1937. Our further discussion makes use of the method developed by S. Timoshenko, *Proc. Fifth Intern. Congr. Appl. Mech.*, Cambridge, Mass., 1938; the method is more general than most of those previously mentioned; it can be applied to any kind of loading, including the case of a concentrated load.

where $\alpha_m = m\pi b/2a$. The rotation at the edge $y = b/2$ of the plate is

$$\left(\frac{\partial w}{\partial y}\right)_{y=b/2} = \frac{2qa^3}{\pi^4 D} \sum_{m=1,3,5,\ldots}^{\infty} \frac{(-1)^{(m-1)/2}}{m^4} \cos \frac{m\pi x}{a}$$
$$[\alpha_m - \tanh \alpha_m (1 + \alpha_m \tanh \alpha_m)]$$

$$= \frac{2qa^3}{\pi^4 D} \sum_{m=1,3,5,\ldots}^{\infty} \frac{(-1)^{(m-1)/2}}{m^4} \cos \frac{m\pi x}{a} \left(\frac{\alpha_m}{\cosh^2 \alpha_m} - \tanh \alpha_m\right) \quad (b)$$

Let us consider now the deflection of the plate by the moments distributed along the edges $y = \pm b/2$. From considerations of symmetry we conclude that the moments can be represented by the following series:

$$(M_y)_{y=\pm b/2} = \sum_{m=1,3,5,\ldots}^{\infty} (-1)^{(m-1)/2} E_m \cos \frac{m\pi x}{a} \quad (c)$$

The corresponding deflection w_1 is obtained from expression (173) by substituting $x + a/2$ for x and taking $m = 1, 3, 5, \ldots$. Then

$$w_1 = -\frac{a^2}{2\pi^2 D} \sum_{m=1,3,5,\ldots}^{\infty} E_m \frac{(-1)^{(m-1)/2}}{m^2 \cosh \alpha_m} \cos \frac{m\pi x}{a} \left(\frac{m\pi y}{a} \sinh \frac{m\pi y}{a}\right.$$
$$\left. - \alpha_m \tanh \alpha_m \cosh \frac{m\pi y}{a}\right) \quad (d)$$

The rotation at the edge $y = b/2$, corresponding to this deflection, is

$$\left(\frac{\partial w_1}{\partial y}\right)_{y=b/2} = -\frac{a}{2\pi D} \sum_{m=1,3,5,\ldots}^{\infty} E_m \frac{(-1)^{(m-1)/2}}{m} \cos \frac{m\pi x}{a} \left(\tanh \alpha_m\right.$$
$$\left. + \frac{\alpha_m}{\cosh^2 \alpha_m}\right) \quad (e)$$

In our further discussion we shall need also the rotation at the edges parallel to the y axis. Forming the derivative of the expression (d) with respect to x and putting $x = a/2$, we obtain

$$\left(\frac{\partial w_1}{\partial x}\right)_{x=a/2} = \frac{a}{2\pi D} \sum_{m=1,3,5,\ldots}^{\infty} E_m \frac{1}{m \cosh \alpha_m} \left(\frac{m\pi y}{a} \sinh \frac{m\pi y}{a}\right.$$
$$\left. - \alpha_m \tanh \alpha_m \cosh \frac{m\pi y}{a}\right) = -\frac{1}{4D} \sum_{m=1,3,5,\ldots}^{\infty} \frac{E_m}{\cosh^2 \alpha_m}$$
$$\left(b \sinh \alpha_m \cosh \frac{m\pi y}{a} - 2y \cosh \alpha_m \sinh \frac{m\pi y}{a}\right) \quad (f)$$

The expression in parentheses is an even function of y which vanishes at the edges $y = \pm b/2$. Such a function can be represented by the series

$$\sum_{i=1,3,5,\ldots}^{\infty} A_i \cos \frac{i\pi y}{b} \qquad (g)$$

in which the coefficients A_i are calculated by using the formula

$$A_i = \frac{2}{b} \int_{-b/2}^{+b/2} \left(b \sinh \alpha_m \cosh \frac{m\pi y}{a} - 2y \cosh \alpha_m \sinh \frac{m\pi y}{a} \right) \cos \frac{i\pi y}{b} \, dy$$

from which it follows that

$$A_i = \frac{16ia(-1)^{(i-1)/2}}{m^3 \pi^2} \frac{b^2}{a^2} \frac{1}{\left(\dfrac{b^2}{a^2} + \dfrac{i^2}{m^2}\right)^2} \cosh^2 \alpha_m$$

Substituting this in expressions (g) and (f), we obtain

$$\left(\frac{\partial w_1}{\partial x}\right)_{x=a/2} = -\frac{4b^2}{\pi^2 Da} \sum_{m=1,3,5,\ldots}^{\infty} \frac{E_m}{m^3} \sum_{i=1,3,5,\ldots}^{\infty} \frac{i(-1)^{(i-1)/2}}{\left(\dfrac{b^2}{a^2} + \dfrac{i^2}{m^2}\right)^2} \cos \frac{i\pi y}{b} \qquad (h)$$

In a similar manner expressions can be obtained for the deflections w_2 and for the rotation at edges for the case where moments M_x are distributed along the edges $x = \pm a/2$. Assuming a symmetrical distribution and taking

$$(M_x)_{x=\pm a/2} = \sum_{m=1,3,5,\ldots}^{\infty} (-1)^{(m-1)/2} F_m \cos \frac{m\pi y}{b} \qquad (i)$$

we find for this case, by using expressions (e) and (h), that

$$\left(\frac{\partial w_2}{\partial x}\right)_{x=a/2} = -\frac{b}{2\pi D} \sum_{m=1,3,5,\ldots}^{\infty} F_m \frac{(-1)^{(m-1)/2}}{m} \cos \frac{m\pi y}{b} \left(\tanh \beta_m + \frac{\beta_m}{\cosh^2 \beta_m} \right) \qquad (j)$$

where $\beta_m = m\pi a/2b$, and that

$$\left(\frac{\partial w_2}{\partial y}\right)_{y=b/2} = -\frac{4a^2}{\pi^2 Db} \sum_{m=1,3,5,\ldots}^{\infty} \frac{F_m}{m^3} \sum_{i=1,3,5,\ldots}^{\infty} \frac{i(-1)^{(i-1)/2}}{\left(\dfrac{a^2}{b^2} + \dfrac{i^2}{m^2}\right)^2} \cos \frac{i\pi x}{a} \qquad (k)$$

When the moments (c) and (i) act simultaneously, the rotation at the edges of the plate is obtained by the method of superposition. Taking, for example, the edge $y = b/2$, we find

$$\left(\frac{\partial w_1}{\partial y} + \frac{\partial w_2}{\partial y}\right)_{y=b/2} = -\frac{a}{2\pi D} \sum_{m=1,3,5,\ldots}^{\infty} E_m \frac{(-1)^{(m-1)/2}}{m} \cos \frac{m\pi x}{a}$$
$$\left(\tanh \alpha_m + \frac{\alpha_m}{\cosh^2 \alpha_m}\right)$$
$$-\frac{4a^2}{\pi^2 Db} \sum_{m=1,3,5,\ldots}^{\infty} \frac{F_m}{m^3} \sum_{i=1,3,5,\ldots}^{\infty} \frac{i(-1)^{(i-1)/2}}{\left(\frac{a^2}{b^2} + \frac{i^2}{m^2}\right)^2} \cos \frac{i\pi x}{a} \quad (l)$$

Having expressions (b) and (l), we can now derive the equations for calculating the constants E_m and F_m in series (c) and (i) which represent the moments acting along the edges of a clamped plate. In the case of a clamped plate the edges do not rotate. Hence, for the edges $y = \pm b/2$, we obtain

$$\left(\frac{\partial w}{\partial y}\right)_{y=b/2} + \left(\frac{\partial w_1}{\partial y} + \frac{\partial w_2}{\partial y}\right)_{y=b/2} = 0 \quad (m)$$

In a similar manner, for the edges $x = \pm a/2$, we find

$$\left(\frac{\partial w}{\partial x}\right)_{x=a/2} + \left(\frac{\partial w_1}{\partial x} + \frac{\partial w_2}{\partial x}\right)_{x=a/2} = 0 \quad (n)$$

If we substitute expressions (b) and (l) in Eq. (m) and group[1] together the terms that contain the same $\cos(i\pi x/a)$ as a factor and then observe that Eq. (m) holds for any value of x, we can conclude that the coefficient by which $\cos(i\pi x/a)$ is multiplied must be equal to zero for each value of i. In this manner we obtain a system that consists of an infinite number of linear equations for calculating the coefficients E_i and F_i as follows:

$$\frac{4qa^2}{\pi^3}\frac{1}{i^4}\left(\frac{\alpha_i}{\cosh^2 \alpha_i} - \tanh \alpha_i\right)$$
$$-\frac{E_i}{i}\left(\tanh \alpha_i + \frac{\alpha_i}{\cosh^2 \alpha_i}\right) - \frac{8ia}{\pi b} \sum_{m=1,3,5,\ldots}^{\infty} \frac{F_m}{m^3} \frac{1}{\left(\frac{a^2}{b^2} + \frac{i^2}{m^2}\right)^2} = 0 \quad (o)$$

A similar system of equations is obtained also from Eq. (n). The constants $E_1, E_3, \ldots, F_1, F_3, \ldots$ can be determined in each particular case from these two systems of equations by the method of successive approximations.

To illustrate this method let us consider the case of a square plate. In such a case the distribution of the bending moments along all sides of the square is the same. Hence $E_i = F_i$, and the two systems of equa-

---

[1] It is assumed that the order of summation in expression (l) is interchangeable.

tions, mentioned above, are identical. The form of the equations is

$$\frac{E_i}{i}\left(\tanh \alpha_i + \frac{\alpha_i}{\cosh^2 \alpha_i}\right) + \frac{8i}{\pi}\sum_{m=1,3,5,\ldots}^{\infty}\frac{E_m}{m^3}\frac{1}{\left(1+\dfrac{i^2}{m^2}\right)^2}$$
$$= \frac{4qa^2}{\pi^3}\frac{1}{i^4}\left(\frac{\alpha_i}{\cosh^2 \alpha_i} - \tanh \alpha_i\right)$$

Substituting the numerical values of the coefficients in these equations and considering only the first four coefficients, we obtain the following system of four equations with four unknowns E_1, E_3, E_5, and E_7:

$$\begin{array}{l}
1.8033E_1 \;|\;+0.0764E_3 \quad +0.0188E_5 \quad +0.0071E_7 = 0.6677K\\
0.0764E_1 \;+0.4045E_3 \;|\;+0.0330E_5 \quad +0.0159E_7 = 0.01232K\\
0.0188E_1 \;+0.0330E_3 \quad +0.2255E_5\;|\;+0.0163E_7 = 0.00160K\\
0.0071E_1 \;+0.0159E_3 \quad +0.0163E_5 \quad +0.1558E_7 = 0.00042K
\end{array} \quad (p)$$

where $K = -4qa^2/\pi^3$. It may be seen that the terms along the diagonal have the largest coefficients. Hence we obtain the first approximations of the constants $E_1, \ldots, E_7$ by considering on the left-hand sides of Eqs. (p) only the terms to the left of the heavy line. In such a way we obtain from the first of the equations $E_1 = 0.3700K$. Substituting this in the second equation, we obtain $E_3 = -0.0395K$. Substituting the values of E_1 and E_3 in the third equation, we find $E_5 = -0.0180K$. From the last equation we then obtain $E_7 = -0.0083K$. Substituting these first approximations in the terms to the right of the heavy line in Eqs. (p), we can calculate the second approximations, which are $E_1 = 0.3722K$, $E_3 = -0.0380K$, $E_5 = -0.0178K$, $E_7 = -0.0085K$. Repeating the calculations again, we shall obtain the third approximation, and so on.

Substituting the calculated values of the coefficients $E_1, E_3, \ldots$ in series (c), we obtain the bending moments along the clamped edges of the plate. The maximum of the absolute value of these moments is at the middle of the sides of the square. With the four equations (p) taken, this value is

$$|M_y|_{y=b/2, x=0} = |E_1 - E_3 + E_5 - E_7| = 0.0517qa^2$$

The comparison of this result with Boobnov's table, calculated with a much larger number of equations similar to Eqs. (p), shows that the error in the maximum bending moment, by taking only four equations (p), is less than 1 per cent. It may be seen that we obtain for the moment a series with alternating signs, and the magnitude of the error depends on the magnitude of the last of the calculated coefficients $E_1, E_3, \ldots$.

Substituting the values of $E_1, E_3, \ldots$ in expression (d), we obtain the deflection of the plate produced by the moments distributed along

TABLE 35. DEFLECTIONS AND BENDING MOMENTS IN A UNIFORMLY LOADED RECTANGULAR PLATE WITH BUILT-IN EDGES (FIG. 91)
$\nu = 0.3$

b/a	$(w)_{x=0,y=0}$	$(M_x)_{x=a/2,y=0}$	$(M_y)_{x=0,y=b/2}$	$(M_x)_{x=0,y=0}$	$(M_y)_{x=0,y=0}$
1.0	$0.00126qa^4/D$	$-0.0513qa^2$	$-0.0513qa^2$	$0.0231qa^2$	$0.0231qa^2$
1.1	$0.00150qa^4/D$	$-0.0581qa^2$	$-0.0538qa^2$	$0.0264qa^2$	$0.0231qa^2$
1.2	$0.00172qa^4/D$	$-0.0639qa^2$	$-0.0554qa^2$	$0.0299qa^2$	$0.0228qa^2$
1.3	$0.00191qa^4/D$	$-0.0687qa^2$	$-0.0563qa^2$	$0.0327qa^2$	$0.0222qa^2$
1.4	$0.00207qa^4/D$	$-0.0726qa^2$	$-0.0568qa^2$	$0.0349qa^2$	$0.0212qa^2$
1.5	$0.00220qa^4/D$	$-0.0757qa^2$	$-0.0570qa^2$	$0.0368qa^2$	$0.0203qa^2$
1.6	$0.00230qa^4/D$	$-0.0780qa^2$	$-0.0571qa^2$	$0.0381qa^2$	$0.0193qa^2$
1.7	$0.00238qa^4/D$	$-0.0799qa^2$	$-0.0571qa^2$	$0.0392qa^2$	$0.0182qa^2$
1.8	$0.00245qa^4/D$	$-0.0812qa^2$	$-0.0571qa^2$	$0.0401qa^2$	$0.0174qa^2$
1.9	$0.00249qa^4/D$	$-0.0822qa^2$	$-0.0571qa^2$	$0.0407qa^2$	$0.0165qa^2$
2.0	$0.00254qa^4/D$	$-0.0829qa^2$	$-0.0571qa^2$	$0.0412qa^2$	$0.0158qa^2$
∞	$0.00260qa^4/D$	$-0.0833qa^2$	$-0.0571qa^2$	$0.0417qa^2$	$0.0125qa^2$

the edges $y = \pm b/2$. For the center of the plate ($x = y = 0$) this deflection is

$$(w_1)_{x=y=0} = \frac{a^2}{2\pi^2 D} \sum_{m=1,3,5,\ldots}^{\infty} E_m(-1)^{(m-1)/2} \frac{\alpha_m \tanh \alpha_m}{m^2 \cosh \alpha_m} = -0.00140 \frac{qa^4}{D}$$

Doubling this result, to take into account the action of the moments distributed along the sides $x = \pm a/2$, and adding to the deflection of the simply supported square plate (Table 8), we obtain for the deflection at the center of a uniformly loaded square plate with clamped edges

$$(w)_{\max} = (0.00406 - 0.00280) \frac{qa^4}{D} = 0.00126 \frac{qa^4}{D} \quad (q)$$

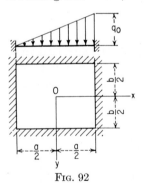

FIG. 92

Similar calculations can be made for any ratio of the sides of a rectangular plate. The results of these calculations are given in Table 35.[1]

Plates under Hydrostatic Pressure. Representing the intensity of the pressure distributed according to Fig. 92 in the form

---

[1] The table was calculated by T. H. Evans; see *J. Appl. Mechanics*, vol. 6, p. A-7, 1939.

$$q_0 = \frac{q_0}{2} + \frac{q_0}{2}\frac{x}{a}$$

we see that the effect of the term $q_0/2$ on the deflections of the plate is already given by the previous solution. Thus it remains to consider the pressure $q_0 x/2a$. The deflection surface of a simply supported plate carrying such a load is readily obtained by combining the expression (k) on page 190 with the expression (a) on page 186. Putting $q = -q_0/2$ in this latter expression and replacing x by $x + a/2$ in both expressions in accordance with new coordinates, we obtain the deflection surface

$$w = \frac{2q_0 a^4}{\pi^5 D} \sum_{m=2,4,6,\ldots}^{\infty} \frac{(-1)^{m/2+1}}{m^5} \left(2 - \frac{2 + \alpha_m \tanh \alpha_m}{\cosh \alpha_m} \cosh \frac{m\pi y}{a} \right.$$
$$\left. + \frac{1}{\cosh \alpha_m} \frac{m\pi y}{a} \sinh \frac{m\pi y}{a} \right) \sin \frac{m\pi x}{a} \quad (r)$$

symmetrical with respect to the x axis and antisymmetrical with respect to the y axis. Consequently, to eliminate the slope along the boundary of the plate we have to apply edge moments of the following form:

$$(M_x)_{x=\pm a/2} = \pm \sum_{m=1,3,5,\ldots}^{\infty} (-1)^{(m-1)/2} E_m \cos \frac{m\pi y}{b}$$
$$(M_y)_{y=\pm b/a} = \sum_{m=2,4,6,\ldots}^{\infty} (-1)^{m/2-1} F_m \sin \frac{m\pi x}{a} \quad (s)$$

Proceeding just as in the case of the uniformly distributed load, we calculate the coefficients E_m and F_m from a system of linear equations. The deflections due to the simultaneous action of the load $q_0 x/2a$ and the moments (s) must be added, finally, to the deflections of the clamped plate loaded uniformly with $q_0/2$. Numerical results obtained by such a procedure are given in Table 36.[1]

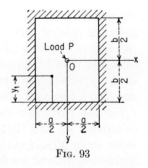

Fig. 93

Plates under Central Load. As a third example let us consider the bending of a rectangular plate with clamped edges under the action of a load P concentrated at the center (Fig. 93). Again we go back to the case of a simply supported plate. Substituting into expression (146) $a/2$ for ξ, and $x + a/2$

---

[1] See Dana Young, *J. Appl. Mechanics*, vol. 7, p. A-139, 1940. More extensive tables were computed, by means of the method of finite differences, by E. G. Odley, *J. Appl. Mechanics*, vol. 14, p. A-289, 1947.

TABLE 36. DEFLECTIONS AND BENDING MOMENTS IN RECTANGULAR PLATES WITH BUILT-IN EDGES AND HYDROSTATIC LOAD (FIG. 92)
$\nu = 0.3$

$\frac{b}{a}$	$x = 0, y = 0$			$x = a/2, y = 0$	$x = -a/2, y = 0$	$x = 0, y = \pm b/2$
	$w = \alpha \dfrac{q_0 a^4}{D}$	$M_x = \beta_1 q_0 a^2$	$M_y = \beta_2 q_0 a^2$	$M_x = \gamma_1 q_0 a^2$	$M_x = \gamma_2 q_0 a^2$	$M_y = \delta q_0 a^2$
	α	β_1	β_2	γ_1	γ_2	δ
0.5	0.000080	0.00198	0.00515	−0.0115	−0.0028	−0.0104
$\frac{2}{3}$	0.000217	0.00451	0.00817	−0.0187	−0.0066	−0.0168
1.0	0.00063	0.0115	0.0115	−0.0334	−0.0179	−0.0257
1.5	0.00110	0.0184	0.0102	−0.0462	−0.0295	−0.0285
∞	0.00130	0.0208	0.0063	−0.0500	−0.0333	

for x, we arrive at the deflection surface (valid for $y > 0$)

$$w = \frac{Pa^2}{2\pi^3 D} \sum_{m=1,3,5,\ldots}^{\infty} \frac{1}{m^3} \cos \frac{m\pi x}{a} \left[\left(\tanh \alpha_m - \frac{\alpha_m}{\cosh^2 \alpha_m} \right) \cosh \frac{m\pi y}{a} - \sinh \frac{m\pi y}{a} \right.$$
$$\left. - \frac{m\pi y}{a} \tanh \alpha_m \sinh \frac{m\pi y}{a} + \frac{m\pi y}{a} \cosh \frac{m\pi y}{a} \right]$$

The angle of rotation along the edge $y = b/2$ is

$$\left(\frac{\partial w}{\partial y} \right)_{y=b/2} = -\frac{Pa}{2\pi^2 D} \sum_{m=1,3,5,\ldots}^{\infty} \frac{1}{m^2} \cos \frac{m\pi x}{a} \frac{\alpha_m \tanh \alpha_m}{\cosh \alpha_m} \quad (t)$$

To calculate the bending moments along the clamped edges we proceed as in the case of uniform load and obtain the same two systems of Eqs. (m) and (n). The expressions for w_1 and w_2 are the same as in the former case, and it will be necessary to change only the first term of these equations by substituting expression (t) instead of $(\partial w/\partial y)_{y=b/2}$ in Eq. (m), and also a corresponding expression for $(\partial w/\partial x)_{x=a/2}$ in Eq. (n).

For the particular case of a square plate, limiting ourselves to four equations, we find that the left-hand side of the equations will be the same as in Eqs. (p). The right-hand sides will be obtained from the expression (t), and we find

$$1.8033 E_1 + 0.0764 E_3 + 0.0188 E_5 + 0.0071 E_7 = -0.1828 P$$
$$0.0764 E_1 + 0.4045 E_3 + 0.0330 E_5 + 0.0159 E_7 = +0.00299 P$$
$$0.0188 E_1 + 0.0330 E_3 + 0.2255 E_5 + 0.0163 E_7 = -0.000081 P$$
$$0.0071 E_1 + 0.0159 E_3 + 0.0163 E_5 + 0.1558 E_7 = +0.000005 P$$

Solving this system of equations by successive approximations, as before, we find

$$E_1 = -0.1025P \qquad E_3 = 0.0263P$$
$$E_5 = 0.0042P \qquad E_7 = 0.0015P$$

Substituting these values in expression (c), the bending moment for the middle of the side $y = b/2$ can be obtained. A more accurate calculation[1] gives

$$(M_y)_{y=b/2, x=0} = -0.1257P$$

Comparing this result with that for the uniformly loaded square plate, we conclude that the uniform load produces moments at the middle of the sides that are less than half of that which the same load produces if concentrated at the center.

Having the moments along the clamped edges, we can calculate the corresponding deflections by using Eq. (d). Superposing deflections produced by the moments on the deflections of a simply supported plate, we obtain the deflections of the plate with built-in edges. By the same method of superposition the other information regarding deflection of plates with built-in edges under a central concentrated load can be obtained.[2] Thus, if the load P is distributed uniformly over the area of a small circle or rectangle, the bending moments at the center of the loaded area $x = y = 0$ can be obtained by combining the results valid for simply supported plates [see Eqs. (157) and (167)] with some additional moments

$$m_x = \beta_1 P \qquad m_y = \beta_2 P$$

given in Table 37 along with data regarding the maximum deflection of the plate and the numerically largest clamping moment. This latter moment, however, can reach the value of $-P/\pi = -0.3183P$, as mentioned on page 192, in the case of a movable load.

45. Rectangular Plates with One Edge or Two Adjacent Edges Simply Supported and the Other Edges Built In. Let us begin with the case of a plate simply supported at the edge $y = 0$ and clamped along the other edges (Fig. 94). No matter how the load may be distributed over the

---

[1] In this calculation seven equations, instead of the four equations taken above, were used.

[2] Calculated by Dana Young, *J. Appl. Mechanics*, vol. 6, p. A-114, 1939. To obtain the moments with the four correct figures it was necessary to use in this calculation seven coefficients E and seven coefficients F in Eqs. (m) and (n). Further solutions of the problem were given by H. Marcus "Die Theorie elastischer Gewebe," 2d ed., p. 155, Berlin, 1932; J. Barta, *Z. angew. Math. Mech.*, vol. 17, p. 184, 1937; G. Pickett, *J. Appl. Mechanics*, vol. 6, p. A-168, 1939; C. J. Thorne and J. V. Atanasoff, *Iowa State Coll. J. Sci.*, vol. 14, p. 333, 1940. The case was investigated experimentally by R. G. Sturm and R. L. Moore, *J. Appl. Mechanics*, vol. 4, p. A-75, 1937.

TABLE 37. BENDING MOMENTS AT THE MIDDLE OF LONGER SIDES AND DEFLECTIONS AND ADDITIONAL MOMENTS AT THE CENTER OF RECTANGULAR PLATES LOADED AT THE CENTER (FIG. 93)

$\nu = 0.3$

			Correction moments	
b/a	$(w)_{x=y=0} = \alpha \dfrac{Pa^2}{D}$	$(M_y)_{x=0, y=b/2} = \gamma P$	$(m_x)_{x=y=0} = \beta_1 P$	$(m_y)_{x=y=0} = \beta_2 P$
	α	γ	β_1	β_2
1.0	0.00560	−0.1257	−0.0536	−0.0536
1.2	0.00647	−0.1490	−0.0579	−0.0526
1.4	0.00691	−0.1604	−0.0618	−0.0517
1.6	0.00712	−0.1651	−0.0653	−0.0510
1.8	0.00720	−0.1667	−0.0683	−0.0504
2.0	0.00722	−0.1674	−0.0710	−0.0500
∞	0.00725	−0.168	−0.0742	−0.0484

given plate $sstt$, we can consider this plate as one-half of a plate $rrtt$ having all edges clamped and carrying a load antisymmetrical with respect to the line ss. The deflections and the bending moments then are zero along that line. Thus the problem under consideration is reduced to the

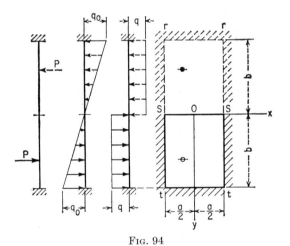

FIG. 94

problem already solved in Art. 44. Some numerical data concerning two cases of load distribution are given in Table 38.[1] A more extensive table

[1] The tabulated results are due to Dana Young, *J. Appl. Mechanics*, vol. 7, p. A-139, 1940, and to C. P. Siess and N. M. Newmark, *Univ. Illinois Bull.*, vol. 47, p. 98, 1950. Y. S. Uflyand used quite a different method in treating this problem; see *Doklady Akad. Nauk. S.S.S.R.*, vol. 72, p. 655, 1950.

TABLE 38. DEFLECTIONS AND BENDING MOMENTS IN RECTANGULAR PLATES
WITH ONE EDGE SIMPLY SUPPORTED AND THREE EDGES BUILT IN
(FIG. 94)

Load	b/a	$(w)_{x=0, y=b/2}$	$(M_x)_{x=a/2, y=b/2}$	$(M_y)_{x=0, y=b}$
Uniform pressure q	0.5	$0.00449qb^4/D$	$-0.0786qb^2$	$-0.1148qb^2$
	0.75	$0.00286qb^4/D$	$-0.0730qb^2$	$-0.0838qb^2$
	1.0	$0.00157qb^4/D$	$-0.0601qb^2$	$-0.0551qb^2$
	$\tfrac{4}{3}$	$0.00215qa^4/D$	$-0.0750qa^2$	$-0.0571qa^2$
	2	$0.00257qa^4/D$	$-0.0837qa^2$	$-0.0571qa^2$
Hydrostatic pressure q_0y/b	0.5	$0.00202q_0b^4/D$	$-0.0368q_0b^2$	$-0.0623q_0b^2$
	0.75	$0.00132q_0b^4/D$	$-0.0344q_0b^2$	$-0.0484q_0b^2$
	1.0	$0.00074q_0b^4/D$	$-0.0287q_0b^2$	$-0.0347q_0b^2$

of bending moments is given on page 244 in connection with a design method for floor slabs.

The rectangular plate *rsut* (Fig. 95) with two adjacent edges $x = 0$ and $y = 0$ simply supported and two other edges clamped can be regarded in like manner as an integral part of the plate bounded by $x = \pm a$, $y = \pm b$ with all edges built in.

Let us consider a load uniformly distributed over the area *rsut* of the given plate.[1] A checkerboard loading distributed over the area $2a$ by $2b$ as shown in Fig. 95 then yields the conditions of a simply supported edge along the lines $x = 0$ and $y = 0$. Thus the problem of bending a plate with two adjacent edges simply supported and two others clamped is again reduced to the problem, already solved in Art. 44, of a plate with all edges built in. Calculations show that the numerically largest moment is produced near the mid-point of the long side of the plate. The values of this clamping moment prove to be $-0.1180qb^2$ for $b/a = 0.5$ and $-0.0694qb^2$ for $b/a = 1.0$. The maximum bending moment near the center of a square plate has the value of $0.034qa^2$ (for $\nu = 0.3$) and the corresponding deflection is given by $0.0023qa^4/D$. Further numerical data regarding bending moments in this case are given on page 243.

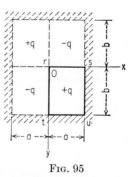

FIG. 95

---

[1] A modification of Timoshenko's method was applied in handling this case by Siess and Newmark, *loc. cit.* For use of the energy method see W. B. Stiles, *J. Appl. Mechanics*, vol. 14, p. A-55, 1947. See also M. K. Huang and H. D. Conway, *J. Appl. Mechanics*, vol. 19, p. 451, 1952.

46. Rectangular Plates with Two Opposite Edges Simply Supported, the Third Edge Free, and the Fourth Edge Built In or Simply Supported.[1]

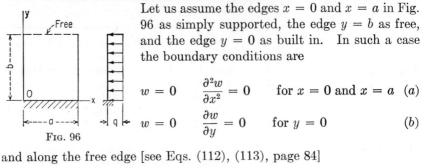

Fig. 96

Let us assume the edges $x = 0$ and $x = a$ in Fig. 96 as simply supported, the edge $y = b$ as free, and the edge $y = 0$ as built in. In such a case the boundary conditions are

$$w = 0 \qquad \frac{\partial^2 w}{\partial x^2} = 0 \qquad \text{for } x = 0 \text{ and } x = a \quad (a)$$

$$w = 0 \qquad \frac{\partial w}{\partial y} = 0 \qquad \text{for } y = 0 \quad (b)$$

and along the free edge [see Eqs. (112), (113), page 84]

$$\left(\frac{\partial^2 w}{\partial y^2} + \nu \frac{\partial^2 w}{\partial x^2}\right)_{y=b} = 0 \qquad \left[\frac{\partial^3 w}{\partial y^3} + (2-\nu)\frac{\partial^3 w}{\partial x^2 \, \partial y}\right]_{y=b} = 0 \quad (c)$$

In the particular case of a uniformly distributed load we proceed as in Art. 30 and assume that the total deflection consists of two parts, as follows:

$$w = w_1 + w_2$$

where w_1 represents the deflection of a uniformly loaded and simply supported strip of length a which can be expressed by the series

$$w_1 = \frac{4qa^4}{\pi^5 D} \sum_{m=1,3,5,\ldots}^{\infty} \frac{1}{m^5} \sin \frac{m\pi x}{a} \quad (d)$$

and w_2 is represented by the series

$$w_2 = \sum_{m=1,3,5,\ldots}^{\infty} Y_m \sin \frac{m\pi x}{a} \quad (e)$$

where

$$Y_m = \frac{qa^4}{D}\left(A_m \cosh \frac{m\pi y}{a} + B_m \frac{m\pi y}{a} \sinh \frac{m\pi y}{a} \right.$$
$$\left. + C_m \sinh \frac{m\pi y}{a} + D_m \frac{m\pi y}{a} \cosh \frac{m\pi y}{a}\right) \quad (f)$$

Series (d) and (e) satisfy the boundary conditions (a), and the four constants in expression (f) must be determined so as to satisfy the boundary

---

[1] This case was discussed by Boobnov; see the English translation of his work in *Trans. Inst. Naval Arch.*, vol. 44, p. 15, 1902, and his "Theory of Structure of Ships," vol. 2, p. 545, St. Petersburg, 1914. It was also discussed by K. Goriupp, *Ingr.-Arch.*, vol. 16, p. 77, 1947, and by V. Bogunović, "On the Bending of a Rectangular Plate with One Edge Free," Belgrade, 1953.

conditions (b) and (c). Using the conditions (b), we obtain

$$A_m = -\frac{4}{\pi^5 m^5} \qquad C_m = -D_m \qquad (g)$$

From the remaining two conditions (c) we find

$$B_m = \frac{4}{\pi^5 m^5}$$
$$\frac{(3+\nu)(1-\nu)\cosh^2 \beta_m + 2\nu \cosh \beta_m - \nu(1-\nu)\beta_m \sinh \beta_m - (1-\nu^2)}{(3+\nu)(1-\nu)\cosh^2 \beta_m + (1-\nu)^2 \beta_m^2 + (1+\nu)^2}$$

$$C_m = \frac{4}{\pi^5 m^5} \qquad (h)$$
$$\frac{(3+\nu)(1-\nu)\sinh \beta_m \cosh \beta_m + \nu(1+\nu)\sinh \beta_m - \nu(1-\nu)\beta_m \cosh \beta_m - (1-\nu)^2 \beta_m}{(3+\nu)(1-\nu)\cosh^2 \beta_m + (1-\nu)^2 \beta_m^2 + (1+\nu)^2}$$

where $\beta_m = m\pi b/a$.

Substituting the constants (g) and (h) in Eq. (f) and using series (e) and (d), we obtain the expression for the deflection surface. The maximum deflection occurs in this case at the middle of the unsupported edge. If the length b is very large in comparison with a, that is, if the free edge is far away from the built-in edge, the deflection of the free edge is the same as that of a uniformly loaded and simply supported strip of length a multiplied by the constant factor $(3-\nu)(1+\nu)/(3+\nu)$. Owing to the presence of this factor, the maximum deflection is larger than that of the strip by 6.4 per cent for $\nu = 0.3$. This fact can be readily explained if we observe that near the free edge the plate has an anticlastic deflection surface.

Taking another extreme case, when a is very large in comparison with b, the maximum deflection of the plate evidently is the same as for a uniformly loaded strip of length b built in at one end and free at the other. Several values of the maximum deflection calculated[1] for various values of the ratio b/a are given in Table 39. This table also gives the maximum values of bending moments which can be readily calculated from the expression for the deflection surface. The calculations show that $(M_x)_{\max}$ occurs at the middle of the unsupported edge. The numerical maximum of the moment M_y occurs at the middle of the built-in edge.

The case of the hydrostatic load distributed according to the law $q_0(1 - y/b)$ can be treated in the same manner as the foregoing case. Let the deflection be expressed by

$$w = \frac{4q_0(1-y/b)a^4}{\pi^5 D} \sum_{m=1,3,5,\ldots}^{\infty} \frac{1}{m^5} \sin \frac{m\pi x}{a} + \sum_{m=1,3,5,\ldots}^{\infty} Y_m \sin \frac{m\pi x}{a} \qquad (i)$$

[1] This table was calculated by Boobnov, *op. cit.*

TABLE 39. DEFLECTIONS AND BENDING MOMENTS FOR A UNIFORMLY LOADED PLATE WITH TWO OPPOSITE EDGES SIMPLY SUPPORTED, THE THIRD EDGE FREE, AND THE FOURTH BUILT IN (FIG. 96)
$\nu = 0.3$

b/a	w_{max}	$x = a/2, y = b$	$x = a/2, y = 0$
		M_x	M_y
0	$0.125qb^4/D$	0	$-0.500qb^2$
$\frac{1}{3}$	$0.094qb^4/D$	$0.0078qa^2$	$-0.428qb^2$
$\frac{1}{2}$	$0.0582qb^4/D$	$0.0293qa^2$	$-0.319qb^2$
$\frac{2}{3}$	$0.0335qb^4/D$	$0.0558qa^2$	$-0.227qb^2$
1	$0.0113qb^4/D$	$0.0972qa^2$	$-0.119qb^2$
$\frac{3}{2}$	$0.0141qa^4/D$	$0.123qa^2$	$-0.124qa^2$
2	$0.0150qa^4/D$	$0.131qa^2$	$-0.125qa^2$
3	$0.0152qa^4/D$	$0.133qa^2$	$-0.125qa^2$
∞	$0.0152qa^4/D$	$0.133qa^2$	$-0.125qa^2$

in which Y_m is of the form (f), only with the constant q_0 instead of q. Proceeding as before, we obtain the four constants $A_m, B_m, \ldots, D_m$ from the boundary conditions (a), (b), and (c).

If the plate is bent by a load distributed along the free edge, instead of by a load distributed over the surface, the second of the boundary conditions (c) must be modified by putting the intensity of the load distributed along the free edge instead of zero on the right-hand side of the equation. The particular case of a concentrated force applied at the free edge of a very long plate was investigated (Fig. 97).[1] It was found that the deflection along the free edge can be represented by the formula

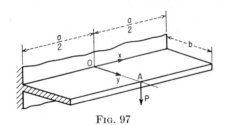

FIG. 97

$$(w)_{y=b} = \alpha \frac{Pb^2}{D}$$

The factor α rapidly diminishes as the distance from the point A of appli-

[1] See C. W. MacGregor, *Mech. Eng.*, vol. 57, p. 225, 1935; D. L. Holl, *J. Appl. Mechanics*, vol. 4, p. 8, 1937; T. J. Jaramillo, *J. Appl. Mechanics*, vol. 17, p. 67, 1950; and K. Girkmann, "Flächentragwerke," 4th ed., p. 233, Vienna, 1956. The case of a cantilever plate having three edges free and carrying a uniformly distributed load was discussed by W. A. Nash, *J. Appl. Mechanics*, vol. 19, p. 33, 1952. See also the investigation of such a plate by W. T. Koiter and J. B. Alblas with numerical results given in *Proc. Koninkl. Ned. Akad. Wetenschap. Amsterdam*, vol. 60, p. 173, 1957.

cation of the load increases. Several values of this factor are given in Table 40. The numerically largest values of the clamping moment produced by a load acting at the middle of the free edge of a plate of a finite length a are given in Table 41.[1]

TABLE 40

$x =$	0	$b/4$	$b/2$	b	$2b$
$\alpha =$	0.168	0.150	0.121	0.068	0.016

TABLE 41. BENDING MOMENTS $M = \beta P$, AT $x = 0$, $y = 0$, DUE TO A LOAD P ACTING AT $x = 0$, $y = b$ AND THE EDGES $x = \pm a/2$ BEING SIMPLY SUPPORTED (FIG. 97)
$\nu = 0.3$

$b/a =$	4	2	1.5	1	$\tfrac{2}{3}$	0.5	$\tfrac{1}{3}$	0.25	0
$\beta =$	−0.000039	−0.0117	−0.0455	−0.163	−0.366	−0.436	−0.498	−0.507	−0.509

The case of a uniformly loaded rectangular plate simply supported along three edges and free along the edge $y = b$ (Fig. 98) can be treated in the same manner as the preceding case in which the edge $y = 0$ was built in. It is necessary only to replace the second of the boundary conditions (b) by the condition

$$\left[\left(\frac{\partial^2 w}{\partial y^2}\right) + \nu \left(\frac{\partial^2 w}{\partial x^2}\right)\right]_{y=0} = 0$$

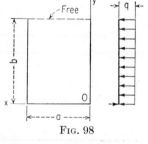

FIG. 98

Omitting the derivations, we give here only the final numerical results obtained for this case. The maximum deflection occurs at the middle of the free edge. At the same point the maximum bending moment M_x takes place. These values of deflections w_{max} and $(M_x)_{max}$ are given in the second and third column of Table 42.[2] The last two columns give the bending moments at the center of the plate.

Table 43, in a similar manner, contains the values of deflections and bending moments produced at the middle of the free edge and at the center of the plate by a hydrostatic load.

47. Rectangular Plates with Three Edges Built In and the Fourth Edge Free. Plates with such boundary conditions are of particular interest as an integral part of rectangular tanks or retaining walls. Con-

---

[1] This table was calculated by V. Bogunović, *loc. cit.* See also Art. 78.

[2] This table and Table 43 were calculated by B. G. Galerkin; see *Bull. Polytech. Inst.*, vol. 26, p. 124, St. Petersburg, 1915.

Table 42. Deflections and Bending Moments in Uniformly Loaded Rectangular Plates with Three Edges Simply Supported and the Fourth Edge Free (Fig. 98)

$\nu = 0.3$

b/a	$x = a/2, y = b$		$x = a/2, y = b/2$	
	$w_{\max}$	$(M_x)_{\max}$	M_x	M_y
$\frac{1}{2}$	$0.00710qa^4/D$	$0.060qa^2$	$0.039qa^2$	$0.022qa^2$
$\frac{2}{3}$	$0.00968qa^4/D$	$0.083qa^2$	$0.055qa^2$	$0.030qa^2$
1/1.4	$0.01023qa^4/D$	$0.088qa^2$	$0.059qa^2$	$0.032qa^2$
1/1.3	$0.01092qa^4/D$	$0.094qa^2$	$0.064qa^2$	$0.034qa^2$
1/1.2	$0.01158qa^4/D$	$0.100qa^2$	$0.069qa^2$	$0.036qa^2$
1/1.1	$0.01232qa^4/D$	$0.107qa^2$	$0.074qa^2$	$0.037qa^2$
1	$0.01286qa^4/D$	$0.112qa^2$	$0.080qa^2$	$0.039qa^2$
1.1	$0.01341qa^4/D$	$0.117qa^2$	$0.085qa^2$	$0.040qa^2$
1.2	$0.01384qa^4/D$	$0.121qa^2$	$0.090qa^2$	$0.041qa^2$
1.3	$0.01417qa^4/D$	$0.124qa^2$	$0.094qa^2$	$0.042qa^2$
1.4	$0.01442qa^4/D$	$0.126qa^2$	$0.098qa^2$	$0.042qa^2$
1.5	$0.01462qa^4/D$	$0.128qa^2$	$0.101qa^2$	$0.042qa^2$
2	$0.01507qa^4/D$	$0.132qa^2$	$0.113qa^2$	$0.041qa^2$
3	$0.01520qa^4/D$	$0.133qa^2$	$0.122qa^2$	$0.039qa^2$
∞	$0.01522qa^4/D$	$0.133qa^2$	$0.125qa^2$	$0.037qa^2$

Table 43. Deflections and Bending Moments in Hydrostatically Loaded Rectangular Plates with Three Edges Simply Supported and the Fourth Edge Free (Fig. 99)

$\nu = 0.3$

b/a	$x = a/2, y = b$		$x = a/2, y = b/2$		
	w	M_x	w	M_x	M_y
$\frac{1}{2}$	$0.00230q_0a^4/D$	$0.0197q_0a^2$	$0.00135q_0a^4/D$	$0.0145q_0a^2$	$0.0120q_0a^2$
$\frac{2}{3}$	$0.00304q_0a^4/D$	$0.0265q_0a^2$	$0.00207q_0a^4/D$	$0.0220q_0a^2$	$0.0156q_0a^2$
1	$0.00368q_0a^4/D$	$0.0325q_0a^2$	$0.00313q_0a^4/D$	$0.0331q_0a^2$	$0.0214q_0a^2$
1.5	$0.00347q_0a^4/D$	$0.0308q_0a^2$	$0.00445q_0a^4/D$	$0.0453q_0a^2$	$0.0231q_0a^2$
2.0	$0.00291q_0a^4/D$	$0.0258q_0a^2$	$0.00533q_0a^4/D$	$0.0529q_0a^2$	$0.0222q_0a^2$
∞	0	0	$0.00651q_0a^4/D$	$0.0625q_0a^2$	$0.0187q_0a^2$

sequently, the uniformly distributed and the hydrostatic load must be considered first of all in that case.

Let the boundary of the plate be clamped at $y = 0$ and $x = \pm a/2$ and free along $y = b$ (Fig. 100). Assuming first a uniformly distributed load of intensity q, the expression for deflections may be taken in the form

$$w = w_1 + w_2 + w_3 \tag{a}$$

The expressions for

$$w_1 = \frac{4qa^4}{\pi^5 D} \sum_{m=1,3,5,\ldots}^{\infty} \frac{(-1)^{(m-1)/2}}{m^5} \cos \frac{m\pi x}{a} \tag{b}$$

and

$$w_2 = \sum_{m=1,3,5,\ldots}^{\infty} Y_m (-1)^{(m-1)/2} \cos \frac{m\pi x}{a} \tag{c}$$

contained in Eq. (a) are identical with expressions (d) and (e) of the preceding article if one considers the new position of the origin.

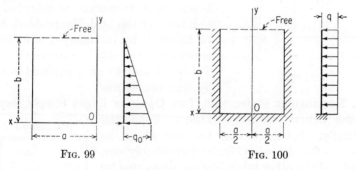

Fig. 99   Fig. 100

A suitable form for the additional deflections w_3 due to the additional constraint on the edges $x = \pm a/2$ is[1]

$$w_3 = \frac{qa^4}{D} \sum_{n=1,3,5,\ldots}^{\infty} \left(F_n \gamma_n \tanh \gamma_n \cosh \frac{n\pi x}{2b} - F_n \frac{n\pi x}{2b} \sinh \frac{n\pi x}{2b} \right) \sin \frac{n\pi y}{2b}$$

$$+ \frac{qa^4}{D} \sum_{m=1,3,5,\ldots}^{\infty} \left(G_m \sinh \frac{m\pi y}{a} + H_m \frac{m\pi y}{a} \cosh \frac{m\pi y}{a} \right.$$

$$\left. + I_m \frac{m\pi y}{a} \sinh \frac{m\pi y}{a} \right) \cos \frac{m\pi x}{a} \tag{d}$$

in which $F_n, \ldots, I_m$ are some constants and $\gamma_n = n\pi a/4b$.

[1] This method of solution essentially is due to Goriupp, *op. cit.*, p. 153 1948. See also W. J. Van der Eb, *Ingenieur*, vol. 26, p. 31, 1950.

As $w_3 = 0$ for $y = 0$ and $x = \pm a/2$, the boundary conditions still to be satisfied by deflections (d) are the following:

$$\left(\frac{\partial^2 w_3}{\partial y^2} + \nu \frac{\partial^2 w_3}{\partial x^2}\right)_{y=b} = 0 \qquad \left[\frac{\partial^3 w_3}{\partial y^3} + (2-\nu) \frac{\partial^3 w_3}{\partial x^2 \partial y}\right]_{y=b} = 0$$

$$\left(\frac{\partial w_3}{\partial y}\right)_{y=0} = 0 \qquad \left[\frac{\partial (w_1 + w_2 + w_3)}{\partial x}\right]_{x=\pm a/2} = 0 \qquad (e)$$

Now we expand all noncircular functions of x contained in expression (a) in a series of the form $\Sigma a_m \cos (m\pi x/a)$ and all similar functions of y in a series of the form $\Sigma b_n \sin (n\pi y/2b)$. A set of linear equations for F_n, G_n, ... , I_m is then readily obtained from conditions (e). Solving the equations we are able to express those unknown constants by the known values of A_m, ... , D_m (see page 209).

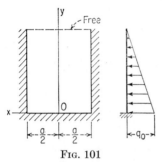

Fig. 101

In the case of a hydrostatic pressure acting in accordance with Fig. 101, we have to superpose solution (i) of the preceding article on the solution of form (d) and, besides that, to proceed as indicated above.

Whatever the load, the problem can also be handled[1] by the method of finite differences (see Art. 83). Numerical values of Tables 44 and 45 are computed essentially by that procedure.[1]

48. Rectangular Plates with Two Opposite Edges Simply Supported and the Other Two Edges Free or Supported Elastically. Let us consider the case where the edges $x = 0$ and $x = a$ (Fig. 102) are simply supported and the other two edges are supported by elastic beams. Assuming that the load is uniformly distributed and that the beams are identical, the deflection surface of the plate will be symmetrical with respect to the x axis, and we have to consider only the conditions along the side $y = b/2$. Assuming that the beams resist bending in vertical planes only and do not resist torsion, the boundary conditions along the edge $y = b/2$, by using Eq. (114), are

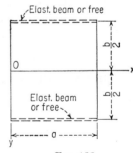

Fig. 102

$$\left(\frac{\partial^2 w}{\partial y^2} + \nu \frac{\partial^2 w}{\partial x^2}\right)_{y=b/2} = 0$$

$$D\left[\frac{\partial^3 w}{\partial y^3} + (2-\nu) \frac{\partial^3 w}{\partial x^2 \partial y}\right]_{y=b/2} = \left(EI \frac{\partial^4 w}{\partial x^4}\right)_{y=b/2} \qquad (a)$$

[1] See A. Smotrov, "Solution for Plates Loaded According to the Law of Trapeze," Moscow, 1936.

TABLE 44. DEFLECTIONS, BENDING MOMENTS, AND REACTIONS OF UNIFORMLY LOADED RECTANGULAR PLATES WITH THREE EDGES BUILT IN AND A FOURTH EDGE FREE (FIG. 100)
$\nu = \frac{1}{6}$

$\frac{b}{a}$	$x=0, y=b$		$x=0, y=b/2$				$x=a/2, y=b$		$x=a/2, y=b/2$		$x=0, y=0$	
	$w = \alpha_1 \frac{qa^4}{D}$	$M_x = \beta_1 qa^2$	$w = \alpha_2 \frac{qa^4}{D}$	$M_x = \beta_2 qa^2$	$M_y = \beta_2' qa^2$		$M_x = \beta_3 qa^2$	$V_x = \gamma_3 qa$	$M_x = \beta_4 qa^2$	$V_x = \gamma_4 qa$	$M_y = \beta_5 qa^2$	$V_y = \gamma_5 qa$
	α_1	β_1	α_2	β_2	β_2'		β_3	γ_3	β_4	γ_4	β_5	γ_5
0.6	0.00271	0.0336	0.00129	0.0168	0.0074		−0.0745	0.750	−0.0365	0.297	−0.0554	0.416
0.7	0.00292	0.0371	0.00159	0.0212	0.0097		−0.0782	0.717	−0.0439	0.346	−0.0545	0.413
0.8	0.00308	0.0401	0.00185	0.0252	0.0116		−0.0812	0.685	−0.0505	0.385	−0.0535	0.410
0.9	0.00323	0.0425	0.00209	0.0287	0.0129		−0.0836	0.656	−0.0563	0.414	−0.0523	0.406
1.0	0.00333	0.0444	0.00230	0.0317	0.0138		−0.0853	0.628	−0.0614	0.435	−0.0510	0.401
1.25	0.00345	0.0467	0.00269	0.0374	0.0142		−0.0867	0.570	−0.0708	0.475	−0.0470	0.388
1.5	0.00335	0.0454	0.00290	0.0402	0.0118		−0.0842	0.527	−0.0755	0.491	−0.0418	0.373

TABLE 45. DEFLECTIONS, BENDING MOMENTS, AND REACTIONS OF HYDROSTATICALLY LOADED RECTANGULAR PLATES WITH THREE EDGES BUILT IN AND A FOURTH EDGE FREE (FIG. 101)

$$\nu = \tfrac{1}{6}$$

$\dfrac{b}{a}$	$x = 0, y = b$		$x = 0, y = b/2$			$x = a/2, y = b$		$x = a/2, y = b/2$		$x = 0, y = 0$	
	$w = \alpha_1 \dfrac{q_0 a^4}{D}$	$M_x = \beta_1 q_0 a^2$	$w = \alpha_2 \dfrac{q_0 a^4}{D}$	$M_x = \beta_2 q_0 a^2$	$M_y = \beta'_2 q_0 a^2$	$M_x = \beta_3 q_0 a^2$	$V_x = \gamma_3 q_0 a$	$M_x = \beta_4 q_0 a^2$	$V_x = \gamma_4 q_0 a$	$M_y = \beta_5 q_0 a^2$	$V_y = \gamma_5 q_0 a$
	α_1	β_1	α_2	β_2	β'_2	β_3	γ_3	β_4	γ_4	β_5	γ_5
0.6	0.00069	0.0089	0.00044	0.0060	0.0062	−0.0179	0.093	−0.0131	0.136	−0.0242	0.248
0.7	0.00069	0.0093	0.00058	0.0080	0.0074	−0.0172	0.081	−0.0170	0.158	−0.0261	0.262
0.8	0.00068	0.0096	0.00072	0.0100	0.0083	−0.0164	0.069	−0.0206	0.177	−0.0278	0.275
0.9	0.00067	0.0096	0.00085	0.0118	0.0090	−0.0156	0.057	−0.0239	0.194	−0.0290	0.286
1.0	0.00065	0.0095	0.00097	0.0135	0.0094	−0.0146	0.045	−0.0269	0.209	−0.0299	0.295
1.25	0.00056	0.0085	0.00121	0.0169	0.0092	−0.0119	0.018	−0.0327	0.234	−0.0306	0.309
1.5	0.00042	0.0065	0.00138	0.0191	0.0075	−0.0087	−0.006	−0.0364	0.245	−0.0291	0.311

where EI denotes the flexural rigidity of the supporting beams. Proceeding as in Art. 46, we take the deflection surface in the form

$$w = w_1 + w_2 \qquad (b)$$

where
$$w_1 = \frac{4qa^4}{\pi^5 D} \sum_{m=1,3,5,\ldots}^{\infty} \frac{1}{m^5} \sin \frac{m\pi x}{a} \qquad (c)$$

and
$$w_2 = \sum_{m=1,3,5,\ldots}^{\infty} Y_m \sin \frac{m\pi x}{a} \qquad (d)$$

From symmetry it can be concluded that in expression (f) of Art. 46 we must put $C_m = D_m = 0$ and take

$$Y_m = \frac{qa^4}{D}\left(A_m \cosh \frac{m\pi y}{a} + B_m \frac{m\pi y}{a} \sinh \frac{m\pi y}{a}\right) \qquad (e)$$

The remaining two constants A_m and B_m are found from the boundary conditions (a), from which, using the notations

$$\frac{m\pi b}{2a} = \alpha_m \qquad \frac{EI}{aD} = \lambda$$

we obtain

$$A_m(1-\nu)\cosh \alpha_m + B_m[2\cosh \alpha_m + (1-\nu)\alpha_m \sinh \alpha_m] = \frac{4\nu}{m^5\pi^5}$$

$$-A_m[(1-\nu)\sinh \alpha_m + m\pi\lambda \cosh \alpha_m] + B_m[(1+\nu)\sinh \alpha_m$$
$$- (1-\nu)\alpha_m \cosh \alpha_m - m\pi\lambda\alpha_m \sinh \alpha_m] = \frac{4\lambda}{m^4\pi^4}$$

Solving these equations, we find

$$A_m = \frac{4}{m^5\pi^5} \frac{\nu(1+\nu)\sinh \alpha_m - \nu(1-\nu)\alpha_m \cosh \alpha_m - m\pi\lambda(2\cosh \alpha_m + \alpha_m \sinh \alpha_m)}{(3+\nu)(1-\nu)\sinh \alpha_m \cosh \alpha_m - (1-\nu)^2\alpha_m + 2m\pi\lambda \cosh^2 \alpha_m} \qquad (f)$$

$$B_m = \frac{4}{m^5\pi^5} \frac{\nu(1-\nu)\sinh \alpha_m + m\pi\lambda \cosh \alpha_m}{(3+\nu)(1-\nu)\sinh \alpha_m \cosh \alpha_m - (1-\nu)^2\alpha_m + 2m\pi\lambda \cosh^2 \alpha_m} \qquad (g)$$

The deflection surface of the plate is found by substituting these values of the constants in the expression

$$w = w_1 + w_2 = \frac{qa^4}{D} \sum_{m=1,3,5,\ldots}^{\infty} \left(\frac{4}{\pi^5 m^5} + A_m \cosh \frac{m\pi y}{a} + B_m \frac{m\pi y}{a} \sinh \frac{m\pi y}{a}\right) \sin \frac{m\pi x}{a} \qquad (h)$$

If the supporting beams are absolutely rigid, $\lambda = \infty$ in expressions (f) and (g), and A_m and B_m assume the same value as in Art. 30 for a plate all four sides of which are supported on rigid supports.

Substituting $\lambda = 0$ in expressions (f) and (g), we obtain the values of the constants in series (h) for the case where two sides of the plate are simply supported and the other two are free.

Except for the case of very small values of λ the maximum deflection and the maximum bending moments are at the center of the plate. Several values of these quantities calculated for a square plate and for various values of λ are given in Table 46.[1]

TABLE 46. DEFLECTIONS AND BENDING MOMENTS AT THE CENTER OF A UNIFORMLY LOADED SQUARE PLATE WITH TWO EDGES SIMPLY SUPPORTED AND THE OTHER TWO SUPPORTED BY ELASTIC BEAMS (FIG. 102)
$\nu = 0.3$

$\lambda = EI/aD$	$w_{\max}$	$(M_x)_{\max}$	$(M_y)_{\max}$
∞	$0.00406qa^4/D$	$0.0479qa^2$	$0.0479qa^2$
100	$0.00409qa^4/D$	$0.0481qa^2$	$0.0477qa^2$
30	$0.00416qa^4/D$	$0.0486qa^2$	$0.0473qa^2$
10	$0.00434qa^4/D$	$0.0500qa^2$	$0.0465qa^2$
6	$0.00454qa^4/D$	$0.0514qa^2$	$0.0455qa^2$
4	$0.00472qa^4/D$	$0.0528qa^2$	$0.0447qa^2$
2	$0.00529qa^4/D$	$0.0571qa^2$	$0.0419qa^2$
1	$0.00624qa^4/D$	$0.0643qa^2$	$0.0376qa^2$
0.5	$0.00756qa^4/D$	$0.0744qa^2$	$0.0315qa^2$
0	$0.01309qa^4/D$	$0.1225qa^2$	$0.0271qa^2$

The particular case $\lambda = 0$ of a plate with two opposite edges simply supported and the other two free deserves some consideration. As Table 47[2] shows, the deflections and the largest moments of such a plate loaded uniformly differ but little from the deflections and moments of a plate bent to a cylindrical surface.

49. Rectangular Plates Having Four Edges Supported Elastically or Resting on Corner Points with All Edges Free. Let us consider a plate subjected to a uniform pressure and supported along the boundary by four flexible beams. All beams are supposed to have rigid supports at the corners of the plate, and two beams parallel to each other may have the same flexural rigidity (Fig. 103).

[1] The table was calculated by K. A. Čališev, *Mem. Inst. Engrs. Ways Commun.*, St. Petersburg, 1914. More recently the problem was discussed by E. Müller, *Ingr.-Arch.*, vol. 2, p. 606, 1932. The tables for nonsymmetrical cases are calculated in this paper. Various cases of rectangular and continuous plates supported by flexible beams were discussed by V. P. Jensen, *Univ. Illinois Bull.*, 81, 1938.

[2] These results are due to D. L. Holl, *Iowa State Coll. Eng. Exp. Sta. Bull.* 129, 1936. For the case of a concentrated load see also R. Ohlig, *Ingr.-Arch.*, vol. 16, p. 51, 1947. Both authors also discuss the effect of clamping the supported edges.

TABLE 47. DEFLECTIONS AND BENDING MOMENTS IN UNIFORMLY LOADED RECTANGULAR PLATES WITH THE EDGES $x = 0$, $x = a$ SIMPLY SUPPORTED AND THE OTHER TWO FREE (FIG. 102)

$\nu = 0.3$

b/a	$x = a/2, y = 0$			$x = a/2, y = \pm b/2$	
	$w = \alpha \dfrac{qa^4}{D}$	$M_x = \beta_1 qa^2$	$M_y = \beta'_1 qa^2$	$w = \alpha_2 \dfrac{qa^4}{D}$	$M_x = \beta_2 qa^2$
	α_1	β_1	β'_1	α_2	β_2
0.5	0.01377	0.1235	0.0102	0.01443	0.1259
1.0	0.01309	0.1225	0.0271	0.01509	0.1318
2.0	0.01289	0.1235	0.0364	0.01521	0.1329
∞	0.01302	0.1250	0.0375	0.01522	0.1330

By writing the deflections in the form

$$w = \frac{q}{384D(\gamma + \delta)} [\gamma(16x^4 - 24a^2x^2 + 5a^4) + \delta(16y^4 - 24b^2y^2 + 5b^4)]$$
$$+ \sum A_n \cosh \frac{n\pi y}{a} \cos \frac{n\pi x}{a} + \sum B_n \cosh \frac{n\pi x}{b} \cos \frac{n\pi y}{b}$$
$$+ \sum C_n y \sinh \frac{n\pi y}{a} \cos \frac{n\pi x}{a} + \sum D_n x \sinh \frac{n\pi x}{b} \cos \frac{n\pi y}{b} \quad (a)$$

where δ/γ and $A_n, \ldots, D_n$ are some constants and $n = 1, 3, 5, \ldots$, we satisfy the differential equation $\Delta\Delta w = q/D$ of the plate and also the conditions of symmetry.[1]
Next, let us develop the algebraic and the hyperbolic functions contained in expression (a) in cosine series. Then, using for $x = a/2$ and $y = b/2$ the edge conditions similar to conditions (a) of the preceding article, we arrive at a set of equations for the constants $A_n, \ldots, D_n$ of expression (a).

Making, in particular, $\delta/\gamma = 0$ and $E_b I_b = \infty$, we would arrive at the solution of the problem already discussed in Art. 48.

Let us consider now the bending of a square plate $(a = b)$ supported by four identical beams. We have then, by symmetry, $\delta/\gamma = 1$, and $A_n = B_n$ and $C_n = D_n$. The unknown coefficients A_n are eliminated by equating to zero the edge moments. Taking, then, only four terms ($n = 1, 3, 5,$ and 7) in series (a), we arrive at four linear equations for $C_1, C_3, C_5,$ and C_7. The results of numerical calculations carried out in this way are given in Table 48.

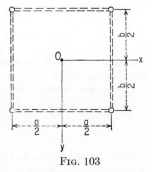

FIG. 103

[1] This method of solution is due to B. G. Galerkin; see his "Collected Papers," vol. 2, p. 15, Moscow, 1953. The boundary conditions under consideration are easily realizable and thus appropriate for the verification of the theory by tests. See N. Dimitrov, *Bauingenieur*, vol. 32, p. 359, 1957.

TABLE 48. DEFLECTIONS AND BENDING MOMENTS OF A SQUARE PLATE WITH FOUR SIDES SUPPORTED ELASTICALLY (Fig. 103)
$\nu = 0.25$

$\gamma = \dfrac{EI}{aD}$	$x = 0, y = 0$		$x = 0, y = a/2$
	$w = \alpha \dfrac{qa^4}{D}$	$M_x = M_y = \beta_1 q a^2$	$M_x = \beta_2 q a^2$
	α	β_1	β_2
∞	0.00406	0.0460	0
100	0.00412	0.0462	
50	0.00418	0.0463	
25	0.00429	0.0467	0.0002
10	0.00464	0.0477	0.0024
5	0.00519	0.0494	0.0065
4	0.00546	0.0502	0.0085
3	0.00588	0.0515	0.0117
2	0.00668	0.0539	0.0177
1	0.00873	0.0601	0.0332
0.5	0.01174	0.0691	0.0559
0	0.0257	0.1109	0.1527

In the particular case of $EI = 0$ we have a square plate carrying a uniformly distributed load and supported only at the corners. The value of ν has but little influence on the deflections and moments at the center of the plate; its effect on the edge moments is more considerable. Taking, for example, $\nu = 0.3$ the values given in the last line of Table 48 for $\nu = 0.25$ should be replaced by 0.249, 0.1090, and 0.1404 respectively.[1]

The problem of bending of a centrally loaded square plate fixed only at the corners has also been discussed.[2] If the load P is distributed uniformly over a small area of a rectangular or circular outline, an expression can be deducted[3] for moments taking place at the center of the loaded area. Taking, for example, a square loaded area u by u, those moments for $\nu = 0.3$ can be expressed in the form

$$M_x = M_y = \left(0.1034 \log \frac{a}{u} + 0.129\right) P \quad (b)$$

Having this solution and also the solution for the uniformly loaded square plate supported at the corners, the problem shown in Fig. 104a can be treated by the method of superposition. It is seen that if a square plate with free edges is supported by the

[1] See H. Marcus, "Die Theorie elasticher Gewebe," 2d ed., p. 173, Berlin, 1932; various cases of plates fixed at points were discussed by A. Nádai, *Z. angew. Math. Mech.*, vol. 2, p. 1, 1922, and also by C. J. Thorne, *J. Appl. Mechanics*, vol. 15, p. 73, 1948.

[2] See Marcus, *ibid.*

[3] See S. Woinowsky-Krieger, *Ingr.-Arch.*, vol. 23, p. 349, 1955.

uniformly distributed reactions, the bending moments at the center are obtained by subtracting from expression (b) the value $M_x = M_y = 0.1090qa^2$, given above for the uniformly loaded square plate supported at the corners and having $\nu = 0.3$. In this way we obtain

$$M_x = M_y = \left(0.1034 \log \frac{a}{u} + 0.020\right) P \qquad (c)$$

valid for $\nu = 0.3$. The distribution of bending moments along the middle line of the footing slab is shown in Fig. 104b for $u/a = 0.1$ and $u/a = 0.2$. A uniform distribution of the pressure may be assumed for a very rigid footing slab resting on soft subgrade. More general hypotheses regarding the law of distribution of that pressure will be postulated in Chap. 8.

50. Semi-infinite Rectangular Plates under Uniform Pressure. The deflection surface and the stress distribution near the short side of long rectangular plates are practically the same as those at the ends of semi-infinite plates, as shown in Fig. 105. It is mainly for this reason that the simple theory of these latter plates deserves consideration.

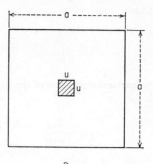
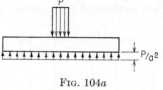

Fig. 104a

Let the load be uniformly dis-

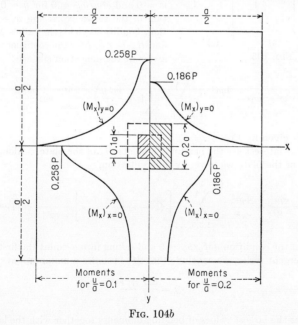

Fig. 104b

tributed over the area of the entire plate and let the edges $x = 0$, $x = a$ be simply supported.[1]

---

[1] The following solutions of the problem are due to A. Nádai; see his book "Elastische Platten," p. 72, Berlin, 1925.

The deflection surface of the plate may be expressed by

$$w = w_1 + w_2 \tag{a}$$

in which

$$w_1 = \frac{q}{24D}(x^4 - 2ax^3 + a^3x) = \frac{4qa^4}{\pi^5 D} \sum_{m=1,3,5,\ldots}^{\infty} \frac{1}{m^5} \sin\frac{m\pi x}{a} \tag{b}$$

is the particular solution of the equation $\Delta\Delta w = q/D$, q being the intensity of the load, and

$$w_2 = \frac{4qa^4}{\pi^5 D} \sum_{m=1,3,5,\ldots}^{\infty} \left(A_m + B_m \frac{m\pi y}{a} \right) e^{-m\pi y/a} \sin\frac{m\pi x}{a} \tag{c}$$

is a solution of the equation $\Delta\Delta w = 0$, yielding zero deflections at $y = \infty$. The coefficients A_m and B_m, which are still at our disposal, must be determined so as to satisfy the respective conditions along the edge $y = 0$ of the plate. The following three cases may be considered.

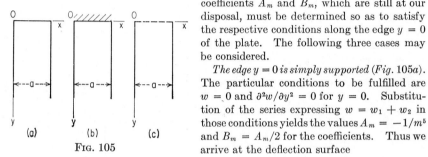

Fig. 105

The edge $y = 0$ is simply supported (Fig. 105a). The particular conditions to be fulfilled are $w = 0$ and $\partial^2 w/\partial y^2 = 0$ for $y = 0$. Substitution of the series expressing $w = w_1 + w_2$ in those conditions yields the values $A_m = -1/m^5$ and $B_m = A_m/2$ for the coefficients. Thus we arrive at the deflection surface

$$w = w_1 - \frac{4qa^4}{\pi^5 D} \sum_{m=1,3,5,\ldots}^{\infty} \left(1 + \frac{m\pi y}{2a}\right) \frac{e^{-m\pi y/a}}{m^5} \sin\frac{m\pi x}{a} \tag{d}$$

in which w_1 is given by Eq. (b).

Of particular interest are the bending moments M_y of the plate. Along the middle line $x = a/2$ of the plate we have, by differentiation,

$$M_y = \frac{\nu q a^2}{8} + \frac{4qa^2}{\pi^3} \sum_{m=1,3,5,\ldots}^{\infty} \left[(1-\nu)\frac{m\pi y}{2a} - \nu\right] \frac{e^{-m\pi y/a}}{m^3} (-1)^{(m-1)/2} \tag{e}$$

Making use of the condition $\partial M_y/\partial y = 0$ and taking into account the first term of the rapidly convergent series, we conclude that M_y becomes a maximum at

$$y = \frac{a}{\pi} \frac{1+\nu}{1-\nu}$$

Table 49 gives the largest values of bending moments together with the largest values of the edge reactions V_y and the forces R acting downward at the corners of the plate.

It should be noted that the value $0.0364qa^2$ exceeds by 45 per cent the value $0.0250qa^2$ of the largest moment M_y of an infinitely long plate, the value of Poisson's constant being the same in both cases.

TABLE 49. LARGEST BENDING MOMENTS AND REACTIONS OF A UNIFORMLY LOADED SEMI-INFINITE PLATE WITH ALL EDGES SIMPLY SUPPORTED
(FIG. 105a)

ν	$(M_x)_{\max}$	$(M_y)_{\max}$	$(V_y)_{\max}$	R
0.2	$0.1250qa^2, x = \frac{a}{2}, y = \infty$	$0.0364qa^2, x = \frac{a}{2}, y = 0.48a$	$0.520qa, x = \frac{a}{2}, y = 0$	$0.1085qa^2$
0.3	$0.1250qa^2, x = \frac{a}{2}, y = \infty$	$0.0445qa^2, x = \frac{a}{2}, y = 0.59a$	$0.502qa, x = \frac{a}{2}, y = 0$	$0.0949qa^2$

The edge $y = 0$ *is built in (Fig. 105b).* Following the general procedure described above, but using this time the edge conditions $w = 0$, $\partial w/\partial y = 0$ on $y = 0$, we obtain, instead of expression (d), the result

$$w = w_1 - \frac{4qa^4}{\pi^5 D} \sum_{m=1,3,5,\ldots}^{\infty} \left(1 + \frac{m\pi y}{a}\right) \frac{e^{-m\pi y/a}}{m^5} \sin \frac{m\pi x}{a} \quad (f)$$

in which w_1 again is given by Eq. (b). The corresponding bending moment

$$M_y = \frac{\nu q a^2}{8} + \frac{4qa^2}{\pi^3} \sum_{m=1,3,5,\ldots}^{\infty} \left[(1-\nu)\frac{m\pi y}{a} - 1 - \nu\right] \frac{e^{-m\pi y/a}}{m^3} \sin \frac{m\pi x}{a} \quad (g)$$

becomes a maximum at $x = a/2$ and $y = 2a/\pi(1-\nu)$. Assuming $\nu = 0.3$ we obtain $y = 0.91a$ and $(M_y)_{\max} = 0.0427qa^2$, whereas the assumption of $\nu = 0.2$ yields the values of $0.0387qa^2$ and $y = 0.80a$, respectively. It can be shown, also, that the variation of the clamping moments along the short side $y = 0$ of the plate obeys the simple law

$$(M_y)_{y=0} = -\frac{q}{2}(ax - x^2)$$

Observing that at large values of y the deflection surface of the plate can be assumed cylindrical, we have there

$$M_x = \frac{q}{2}(ax - x^2) \qquad M_y = \nu \frac{q}{2}(ax - x^2)$$

Thus, the distribution of the edge moments (g) is identical with the distribution of the moments M_x across the plate at $y = \infty$ but with opposite sign.

The edge $y = 0$ *is free (Fig. 105c).* If the conditions prescribed at $y = 0$ are

$$\nu \frac{\partial^2 w}{\partial x^2} + \frac{\partial^2 w}{\partial y^2} = 0 \qquad \frac{\partial^3 w}{\partial y^3} + (2-\nu)\frac{\partial^3 w}{\partial x^2 \partial y} = 0$$

then, making use of expressions (a), (b), and (c), we arrive at the deflection surface

$$w = w_1 + \frac{4\nu q a^4}{(3+\nu)\pi^5} \sum_{m=1,3,5,\ldots}^{\infty} \left(\frac{1+\nu}{1-\nu} - \frac{m\pi y}{a}\right) \frac{e^{-m\pi y/a}}{m^5} \sin \frac{m\pi x}{a} \quad (h)$$

The deflection and the bending moment M_x are largest at the middle of the free edge. It can be proved that

$$(w)_{y=0} = \frac{3-\nu}{(1-\nu)(3+\nu)} w_1$$

and

$$(M_x)_{y=0} = \frac{(3-\nu)(1+\nu)}{3+\nu} (M_x)_1$$

w_1 and $(M_x)_1$ being the deflections and the moments of an infinite simply supported plate. We have therefore

$$(M_x)_{\max} = \frac{(3-\nu)(1+\nu)}{3+\nu} \frac{qa^2}{8}$$

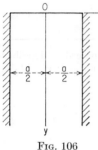

Fig. 106

As a last example, leading to a different form of solution, let us consider a uniformly loaded semi-infinite plate with the edge $y = 0$ simply supported and the edges $x = \pm a/2$ built in (Fig. 106). The solution can be obtained by substituting $b = \infty$ in a suitably chosen expression for the deflections of a finite rectangular plate simply supported on the edges $y = 0, b$ and clamped on the edges $x = \pm a/2$. The result of such a derivation, which is omitted here, is

$$w = \frac{4qa^4}{\pi D} \int_0^\infty \left[\frac{1}{2} - \frac{\left(\sinh\frac{\beta}{2} + \frac{\beta}{2}\cosh\frac{\beta}{2}\right)\cosh\frac{\beta x}{a} - \frac{\beta x}{a}\sinh\frac{\beta}{2}\sinh\frac{\beta x}{a}}{\sinh\beta + \beta} \right] \frac{\sin\frac{\beta y}{a}}{\beta^5} d\beta \quad (i)$$

Differentiating expression (i) and observing that

$$\int_0^\infty \frac{\sin\frac{\beta y}{a}}{\beta} d\beta = \frac{\pi}{2} \quad \text{for } y > 0$$

we obtain

$$\Delta\Delta w = \frac{2q}{\pi D} \int_0^\infty \frac{\sin\frac{\beta y}{a}}{\beta} d\beta = \frac{q}{D}$$

Thus the differential equation for bending of plates is satisfied. It can be shown that the required boundary conditions at $y = 0$ and $x = \pm a/2$ are also satisfied by solution (i).

The expressions for the bending moments of the plate again involve infinite integrals, which can be evaluated. Once more the moments M_y are of interest. Assuming, for example, $\nu = 0.2$, we arrive at a value of $(M_y)_{\max} = 0.0174 qa^2$, occurring at $y = 0.3a$, whereas the moment $M_y = \nu qa^2/24$ of an infinite plate does not exceed $0.00833 qa^2$ for the same value of ν.

It should be noted that the properties of the semi-infinite plates can be used as a basis for calculating the deflections and bending moments of finite rectangular plates with simply supported or built-in edges in any given combination.[1]

[1] For this approach to the theory of rectangular plates see W. Koepcke, *Ingr.-Arch.*, vol. 18, p. 106, 1950.

51. Semi-infinite Rectangular Plates under Concentrated Loads.

Assuming the edges $x = 0$ and $x = a$ of the plate to be simply supported, let us consider, regarding the third side $(y = 0)$, the following two cases: (1) the edge $y = 0$ is simply supported, and (2) the edge $y = 0$ is clamped.

The edge $y = 0$ is simply supported (Fig. 107). Assuming that the given load P is applied at point $x = \xi$, $y = \eta$ (Fig. 107), we first consider an infinite plate supported only at the edges $x = 0$ and $x = a$. In order to use the method of images (see page 156), we assume a second load $-P$ acting at the point $x = \xi$, $y = -\eta$ of the infinite plate. The line $y = 0$ becomes then a nodal line of the deflection surface of the plate. Thus the required bending of the semi-infinite plate is obtained by superposing the

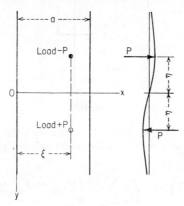

FIG. 107

deflections [see Eq. (148), page 145] produced in the infinite plate by both concentrated loads. In this way we arrive at the deflection surface

$$w_1 = \frac{Pa^2}{2\pi^3 D} \sum_{m=1}^{\infty} \frac{e^{-\frac{m\pi}{a}(\eta-y)}}{m^3} \left[1 + \frac{m\pi}{a}(\eta-y)\right] \sin\frac{m\pi\xi}{a} \sin\frac{m\pi x}{a}$$

$$- \frac{Pa^2}{2\pi^3 D} \sum_{m=1}^{\infty} \frac{e^{-\frac{m\pi}{a}(\eta+y)}}{m^3} \left[1 + \frac{m\pi}{a}(\eta+y)\right] \sin\frac{m\pi\xi}{a} \sin\frac{m\pi x}{a}$$

or, after some rearrangement,

$$w_1 = \frac{Pa^2}{\pi^3 D} \sum_{m=1}^{\infty} \frac{e^{-m\pi\eta/a}}{m^3} \left[\left(1 + \frac{m\pi\eta}{a}\right) \sinh\frac{m\pi y}{a} - \frac{m\pi y}{a} \cosh\frac{m\pi y}{a}\right] \sin\frac{m\pi\xi}{a} \sin\frac{m\pi x}{a}$$

(a)

an expression valid for $0 \leq y \leq \eta$ and yielding $w_1 = 0$, $\partial^2 w_1/\partial y^2 = 0$ at $y = 0$. The deflections in the range of $y > \eta$ may be obtained in a similar manner.

If we distribute the single load over a small area, the moments M_x at the center of that area and the corresponding deflections prove to be smaller than those of an infinite plate without the transverse edge at $y = 0$. But the moment M_y is again an

226   THEORY OF PLATES AND SHELLS

exception. Let us write this moment in the form $M_y = M_{y0} + m_y$, where M_{y0} is the moment of the infinite plate. The correction m_y, representing the effect of the load $-P$ in Fig. 107, is then readily found by means of the second of the equations (151) (see page 145). Assuming, for example, $\nu = 0.3$ we obtain $m_y = 0.0065P$ as the largest value of the correction, the corresponding position of the load being given by $x = a/2$, $y = 0.453a$.

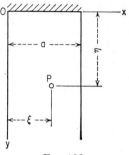

Fig. 108

The edge $y = 0$ is built in (Fig. 108). We begin with the calculation of the slope of the elastic surface (a) at $y = 0$, for which differentiation gives

$$\left(\frac{\partial w_1}{\partial y}\right)_{y=0} = \frac{P\eta}{\pi D} \sum_{m=1}^{\infty} \frac{e^{-m\pi\eta/a}}{m} \sin\frac{m\pi\xi}{a} \sin\frac{m\pi x}{a} \tag{b}$$

Next let us submit the simply supported semi-infinite plate to couples distributed along the edge $y = 0$ in accordance with the law

$$(M_y)_{y=0} = f(x) = \sum_{m=1}^{\infty} E_m \sin\frac{m\pi x}{a}$$

The corresponding deflections, vanishing at $y = \infty$, we take in the form

$$w_2 = \sum_{m=1}^{\infty} (A_m + B_m y) e^{-m\pi y/a} \sin\frac{m\pi x}{a} \tag{c}$$

The coefficients A_m and B_m in this expression are readily obtained from the conditions

$$(w_2)_{y=0} = 0 \qquad -D\left(\frac{\partial^2 w_2}{\partial y^2}\right)_{y=0} = f(x) \tag{d}$$

This yields $A_m = 0$, $B_m = E_m a/2m\pi D$, and, finally,

$$w_2 = \frac{ay}{2\pi D} \sum_{m=1}^{\infty} \frac{E_m e^{-m\pi y/a}}{m} \sin\frac{m\pi x}{a} \tag{e}$$

Since we have to eliminate the slope (b), the edge condition is

$$\left(\frac{\partial w_1}{\partial y}\right)_{y=0} + \left(\frac{\partial w_2}{\partial y}\right)_{y=0} = 0 \qquad (f)$$

Substitution of expressions (b) and (e) in Eq. (f) gives

$$E_m = -\frac{2P\eta}{a} e^{-m\pi\eta/a} \sin\frac{m\pi\xi}{a}$$

and expression (e) becomes accordingly

$$w_2 = -\frac{P y \eta}{\pi D} \sum_{m=1}^{\infty} \frac{e^{-\frac{m\pi}{a}(y+\eta)}}{m} \sin\frac{m\pi\xi}{a} \sin\frac{m\pi x}{a} \qquad (g)$$

The deflection surface of the semi-infinite plate clamped on $y = 0$ then is given by

$$w = w_1 + w_2 \qquad (h)$$

where w_1 denotes expression (a). As for the series (g), it can be represented in a closed form. We have only to express the sine functions contained in (g) in terms of the exponential functions

$$e^{\pm(m\pi\xi i/a)} \qquad \text{and} \qquad e^{\pm(m\pi x i/a)}$$

and to observe the expansion

$$\log(1 \pm e^z) = \pm e^z - \frac{e^{2z}}{2} \pm \frac{e^{3z}}{3} - \cdots$$

If we proceed in this manner, expression (g) finally appears in the simpler form

$$w_2 = \frac{P y \eta}{4\pi D} \log \frac{\cosh\frac{\pi}{a}(y+\eta) - \cos\frac{\pi}{a}(x-\xi)}{\cosh\frac{\pi}{a}(y+\eta) - \cos\frac{\pi}{a}(x+\xi)} \qquad (i)$$

The value of the clamping moments at $y = 0$ is readily obtained by differentiation of expression (i), and the result is

$$(M_y)_{y=0} = -\frac{P\eta}{2a} \sinh\frac{\pi\eta}{a} \left(\frac{1}{\cosh\frac{\pi\eta}{a} - \cos\frac{\pi}{a}(x-\xi)} - \frac{1}{\cosh\frac{\pi\eta}{a} - \cos\frac{\pi}{a}(x+\xi)} \right) \qquad (j)$$

When the concentrated load approaches the built-in edge $y = 0$, the value given by expression (j) tends to zero in general. If, however, $\xi = x$ and $\eta \to 0$ simultaneously, then Eq. (j) yields

$$(M_y)_{y=0} = -\lim_{\eta \to 0} \left(\frac{P\eta}{2a} \coth\frac{\pi\eta}{2a} \frac{1 - \cos\frac{2\pi x}{a}}{\cosh\frac{\pi\eta}{a} - \cos\frac{2\pi x}{a}} \right) = -\frac{P}{\pi} \qquad (k)$$

If, finally, $\eta = 0$, the moment M_y becomes zero.

In conclusion let us consider a single load P (Fig. 109) uniformly distributed over a straight-line segment of some length u. The moment caused by such a load at the mid-point of the built-in edge is readily found by means of expression (j). Substitut-

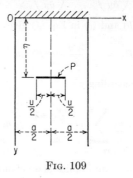

Fig. 109

ing $x = a/2$ and $P\, d\xi/u$ for P in this expression and integrating we obtain for the required moment

$$(M_y)_{x=a/2,\, y=0} = -\frac{2P\eta}{au}\sinh\frac{\pi\eta}{a}\int_{(a-u)/2}^{(a+u)/2}\frac{\sin\dfrac{\pi\xi}{a}\,d\xi}{\cosh\dfrac{2\pi\eta}{a}+\cos\dfrac{2\pi\xi}{a}}$$

$$= -\frac{2P\eta}{\pi u}\arctan\frac{\sin\dfrac{\pi u}{2a}}{\sinh\dfrac{\pi\eta}{a}}$$

Table 50 gives the position of the load producing the numerically largest clamping moment and the value of that moment for various values of the ratio u/a.

TABLE 50. LARGEST CLAMPING MOMENTS AT $x = a/2$ DUE TO A SINGLE LOAD DISTRIBUTED OVER A LENGTH u (FIG. 109)

u/a	0	0.1	0.2	0.4	0.6	0.8	1.0
η/a	0	0.147	0.203	0.272	0.312	0.321	0.343
M_y/P	-0.318	-0.296	-0.275	-0.237	-0.204	-0.172	-0.143

CHAPTER 7

CONTINUOUS RECTANGULAR PLATES

52. Simply Supported Continuous Plates. Floor slabs used in buildings, besides being supported by exterior walls, often have intermediate supports in the form of beams and partitions or in the form of columns. In the first case we have to deal with proper continuous plates; in the case of columns without intermediate beams we have to deal with *flat slabs*. The floor slab is usually subdivided by its supports into several

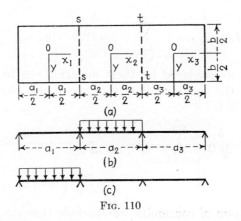

Fig. 110

panels. Only continuous plates with panels of rectangular shape will be considered in this chapter.

We begin with a case allowing a rigorous solution by methods already used in the foregoing chapter. A rectangular plate of width b and length $a_1 + a_2 + a_3$, supported along the edges and also along the intermediate lines ss and tt, as shown in Fig. 110, forms a simply supported continuous plate over three spans. We suppose that the intermediate supports neither yield to the pressure in the transverse direction nor offer any resistance to the rotation of the plate with respect to the axes ss and tt. With these assumptions, the bending of each span of the plate can be readily investigated by combining the known solutions for laterally loaded, simply supported rectangular plates with those for rectangular plates bent by moments distributed along the edges.

229

Let us begin with the symmetrical case in which

$$a_1 = a_2 = a_3 = a$$

and the middle span is uniformly loaded while the side spans are without load (Fig. 110b). Considering the middle span as a simply supported rectangular plate and using expression (b) of Art. 44 (see page 198), we conclude that the slope of the deflection surface along the edge $x_2 = a/2$ is

$$\left(\frac{\partial w}{\partial x_2}\right)_{x_2=a/2} = \frac{2qb^3}{\pi^4 D} \sum_{m=1,3,5,\ldots}^{\infty} \frac{(-1)^{(m-1)/2}}{m^4} \cos\frac{m\pi y}{b}\left(\frac{\beta_m}{\cosh^2 \beta_m} - \tanh \beta_m\right) \quad (a)$$

where $\beta_m = m\pi a/2b$. Owing to the continuity of the plate, bending moments M_x are distributed along the edges $x_2 = \pm a/2$. From symmetry it is seen that these moments can be represented by the following series:

$$(M_x)_{x_2=\pm a/2} = \sum_{m=1,3,5,\ldots}^{\infty} (-1)^{(m-1)/2} E_m \cos\frac{m\pi y}{b} \quad (b)$$

The deflections w_1 produced by these moments can be obtained from Eq. (173), and the corresponding slope along the edge $x_2 = a/2$ [see Eq. (e), page 198] is

$$\left(\frac{\partial w_1}{\partial x_2}\right)_{x_2=a/2} = -\frac{b}{2\pi D} \sum_{m=1,3,5,\ldots}^{\infty} E_m \frac{(-1)^{(m-1)/2}}{m}$$

$$\cos\frac{m\pi y}{b}\left(\tanh \beta_m + \frac{\beta_m}{\cosh^2 \beta_m}\right) \quad (c)$$

From the condition of continuity we conclude that the sum of expressions (a) and (c) representing the slope of the plate along the line $x_2 = a/2$ must be equal to the slope along the same line of the deflection surface of the plate in the adjacent span. Considering this latter span as a simply supported rectangular plate bent by the moments (b) distributed along the edge $x_3 = -a/2$, we find the corresponding deflection w_2 of the plate by using Eq. (176) (see page 185), from which follows

$$w_2 = \frac{b^2}{4\pi^2 D} \sum_{m=1,3,5,\ldots}^{\infty} E_m \cos\frac{m\pi y}{b} \frac{(-1)^{(m-1)/2}}{m^2}$$

$$\left[\frac{1}{\cosh \beta_m}\left(\beta_m \tanh \beta_m \cosh\frac{m\pi x_3}{b} - \frac{m\pi x_3}{b}\sinh\frac{m\pi x_3}{b}\right)\right.$$

$$\left. - \frac{1}{\sinh \beta_m}\left(\beta_m \coth \beta_m \sinh\frac{m\pi x_3}{b} - \frac{m\pi x_3}{b}\cosh\frac{m\pi x_3}{b}\right)\right] \quad (d)$$

The corresponding slope along the edge $x_3 = -a/2$ is

$$\left(\frac{\partial w_2}{\partial x_3}\right)_{x_3=-a/2} = \frac{b}{4\pi D} \sum_{m=1,3,5,\ldots}^{\infty} \frac{E_m}{m} (-1)^{(m-1)/2}$$
$$\cosh \frac{m\pi y}{b} \left(\tanh \beta_m + \coth \beta_m + \frac{\beta_m}{\cosh^2 \beta_m} - \frac{\beta_m}{\sinh^2 \beta_m}\right) \quad (e)$$

The equation for calculating the coefficients E_m is

$$\left(\frac{\partial w}{\partial x_2}\right)_{x_2=a/2} + \left(\frac{\partial w_1}{\partial x_2}\right)_{x_2=a/2} = \left(\frac{\partial w_2}{\partial x_3}\right)_{x_3=-a/2}$$

Since this equation holds for any value of y, we obtain for each value of m the following equation:

$$\frac{2qb^3}{\pi^4 D} \frac{1}{m^4}\left(\frac{\beta_m}{\cosh^2 \beta_m} - \tanh \beta_m\right) - \frac{b}{2\pi D} \frac{E_m}{m}\left(\tanh \beta_m + \frac{\beta_m}{\cosh^2 \beta_m}\right)$$
$$= \frac{b}{4\pi D} \frac{E_m}{m}\left(\tanh \beta_m + \coth \beta_m + \frac{\beta_m}{\cosh^2 \beta_m} - \frac{\beta_m}{\sinh^2 \beta_m}\right) \quad (f)$$

from which

$$E_m = \frac{8qb^2}{\pi^3 m^3} \frac{\beta_m - \tanh \beta_m \cosh^2 \beta_m}{3 \tanh \beta_m \cosh^2 \beta_m + \coth \beta_m \cosh^2 \beta_m + 3\beta_m - \beta_m \coth^2 \beta_m} \quad (g)$$

It is seen that E_m decreases rapidly as m increases and approaches the value $-2qb^2/\pi^3 m^3$. Having the coefficients E_m calculated from (g), we obtain the values of the bending moments M_x along the line tt from expression (b). The value of this moment at $y = 0$, that is, at the middle of the width of the plate, is

$$(M_x)_{x_2=\pm a/2, y=0} = \sum_{m=1,3,5,\ldots}^{\infty} E_m(-1)^{(m-1)/2}$$

Taking, as an example, $b = a$, we have $\beta_m = m\pi/2$, and the formula (g) gives

$$E_1 = -\frac{8qa^2}{\pi^3} 0.1555 \qquad E_3 = -\frac{8qa^2}{\pi^3} 0.0092 \qquad E_5 = -\frac{8qa^2}{\pi^3} 0.0020$$

$$(M_x)_{x_2=\pm a/2, y=0} = -0.0381 q a^2$$

The bending moments at the center of the middle span can be readily obtained by combining bending moments of a simply supported plate, bent by uniform load, with moments corresponding to the deflections w_1. Taking, for example, $a = b$ and $\nu = 0.2$, which is a convenient value for concrete, we get for the first of these moments the values of

$$(M_x)_0 = (M_y)_0 = 0.0479 \times \frac{1.2}{1.3} qa^2 = 0.0442 qa^2$$

(see Table 8, page 120) and for the second moments the values

$$(M_x)_1 = -0.0067 qa^2 \quad \text{and} \quad (M_y)_1 = -0.0125 qa^2$$

Therefore

$$(M_x)_{x_2=0, y=0} = 0.0375 qa^2 \quad (M_y)_{x_2=0, y=0} = 0.0317 qa^2$$

If a side span is uniformly loaded, as shown in Fig. 110c, the deflection surface is no longer symmetrical with respect to the vertical axis of symmetry of the plate, and the bending moment distributions along the lines ss and tt are not identical. Let

$$(M_x)_{x_1=a_1/2} = \sum_{m=1,3,5,\ldots}^{\infty} (-1)^{(m-1)/2} E_m \cos \frac{m\pi y}{b}$$

$$(M_x)_{x_2=a_2/2} = \sum_{m=1,3,5,\ldots}^{\infty} (-1)^{(m-1)/2} F_m \cos \frac{m\pi y}{b} \tag{h}$$

To calculate the coefficients E_m and F_m we derive two systems of equations from the conditions of continuity of the deflection surface of the plate along the lines ss and tt. Considering the loaded span and using expressions (a) and (e), we find that the slope of the deflection surface at the points of the support ss, for $a_1 = a_2 = a_3 = a$, is

$$\left(\frac{\partial w}{\partial x_1}\right)_{x_1=a/2} = \frac{2qb^3}{\pi^4 D} \sum_{m=1,3,5,\ldots}^{\infty} \frac{(-1)^{(m-1)/2}}{m^4} \cos \frac{m\pi y}{b} \left(\frac{\beta_m}{\cosh^2 \beta_m} - \tanh \beta_m\right)$$

$$- \frac{b}{4\pi D} \sum_{m=1,3,5,\ldots}^{\infty} E_m \frac{(-1)^{(m-1)/2}}{m} \cosh \frac{m\pi y}{b} \left(\tanh \beta_m\right.$$

$$\left. + \coth \beta_m + \frac{\beta_m}{\cosh^2 \beta_m} - \frac{\beta_m}{\sinh^2 \beta_m}\right) \tag{i}$$

Considering now the middle span as a rectangular plate bent by the moments M_x distributed along the lines ss and tt and given by the series (h), we find, by using Eq. (175) (see page 184),

$$\left(\frac{\partial w}{\partial x_2}\right)_{x_2=-a/2} = \frac{b}{4\pi D} \sum_{m=1,3,5,\ldots}^{\infty} \frac{(-1)^{(m-1)/2}}{m} \cos \frac{m\pi y}{b} \left[(E_m + F_m) \right.$$

$$\left. \left(\frac{\beta_m}{\cosh^2 \beta_m} + \tanh \beta_m\right) + (E_m - F_m) \left(\coth \beta_m - \frac{\beta_m}{\sinh^2 \beta_m}\right)\right] \tag{j}$$

From expressions (i) and (j) we obtain the following system of equations for calculating coefficients E_m and F_m:

$$A_m \frac{8qb^2}{\pi^3 m^3} + E_m(B_m + C_m) = -B_m(E_m + F_m) - C_m(E_m - F_m) \quad (k)$$

where the following notations are used:

$$A_m = \frac{\beta_m}{\cosh^2 \beta_m} - \tanh \beta_m \qquad B_m = -\left(\frac{\beta_m}{\cosh^2 \beta_m} + \tanh \beta_m\right) \quad (l)$$

$$C_m = \frac{\beta_m}{\sinh^2 \beta_m} - \coth \beta_m$$

The slope of the deflection surface of the middle span at the supporting line tt, by using expression (j), is

$$\left(\frac{\partial w}{\partial x_2}\right)_{x_2 = a/2} = -\frac{b}{4\pi D} \sum_{m=1,3,5,\ldots}^{\infty} \frac{(-1)^{(m-1)/2}}{m} \cos \frac{m\pi y}{b} \bigg[(E_m + F_m)$$
$$\left(\frac{\beta_m}{\cosh^2 \beta_m} + \tanh \beta_m\right) + (F_m - E_m)\left(\coth \beta_m - \frac{\beta_m}{\sinh^2 \beta_m}\right) \bigg]$$

This slope must be equal to the slope in the adjacent unloaded span which is obtained from expression (c) by substituting F_m for E_m. In this way we find the second system of equations which, using notations (l), can be written in the following form:

$$B_m(E_m + F_m) + C_m(F_m - E_m) = -(B_m + C_m)F_m \quad (m)$$

From this equation we obtain

$$F_m = E_m \frac{C_m - B_m}{2(B_m + C_m)} \quad (n)$$

Substituting in Eq. (k), we find

$$E_m = A_m \frac{8qa^2}{\pi^3 m^3} \frac{2(B_m + C_m)}{(C_m - B_m)^2 - 4(B_m + C_m)^2} \quad (o)$$

Substituting in each particular case for A_m, B_m, and C_m their numerical values, obtained from Eqs. (l), we find the coefficients E_m and F_m; and then, from expressions (h), we obtain the bending moments along the lines ss and tt. Take, as an example, $b = a$. Then $\beta_m = m\pi/2$, and we find from Eqs. (l)

$$A_1 = -0.6677 \qquad B_1 = -1.1667 \qquad C_1 = -0.7936$$
$$A_3 = -0.9983 \qquad B_3 = -1.0013 \qquad C_3 = -0.9987$$

For m larger than 3 we can take with sufficient accuracy

$$A_m = B_m = C_m = -1$$

Substituting these values in Eq. (o), we obtain

$$E_1 = -\frac{8qa^2}{\pi^3}0.1720 \qquad E_3 = -\frac{8qa^2}{\pi^3 3^3}0.2496 \qquad E_5 = -\frac{8qa^2}{\pi^3 5^3}0.2500$$

The moment at the middle of the support ss is

$$(M_x)_{x_1=a/2, y=0} = E_1 - E_3 + E_5 - \cdots = -0.0424qa^2$$

For the middle of the support tt we obtain

$$(M_x)_{x_2=a/2, y=0} = F_1 - F_3 + F_5 - \cdots = 0.0042qa^2$$

Having the bending moments along the lines of support, the deflections of the plate in each span can readily be obtained by superposing on the deflections produced by the lateral load the deflections due to the moments at the supports.

The bending moments in the panels of the continuous plate can be obtained in a similar manner. Calculating, for example, the moments at the center of the middle span and taking $\nu = 0.2$, we arrive at the values

$$(M_x)_{x_2=0, y=0} = -0.0039qa^2$$
$$(M_y)_{x_2=0, y=0} = -0.0051qa^2$$

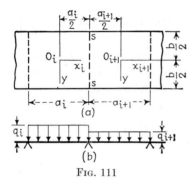

Fig. 111

The equations obtained for three spans can readily be generalized and expanded for the case of any number of spans. In this way an equation similar to the three-moment equation of continuous beams will be obtained.[1] Let us consider two adjacent spans i and $i+1$ of the length a_i and a_{i+1}, respectively (Fig. 111). The corresponding values of the functions (l) are denoted by A_m^i, B_m^i, C_m^i and A_m^{i+1}, B_m^{i+1}, C_m^{i+1}. The bending moments along the three consecutive lines of support can be represented by the series

$$M_x^{i-1} = \sum_{m=1,3,5,\ldots}^{\infty} (-1)^{(m-1)/2} E_m^{i-1} \cos \frac{m\pi y}{b}$$

$$M_x^i = \sum_{m=1,3,5,\ldots}^{\infty} (-1)^{(m-1)/2} E_m^i \cos \frac{m\pi y}{b}$$

$$M_x^{i+1} = \sum_{m=1,3,5,\ldots}^{\infty} (-1)^{(m-1)/2} E_m^{i+1} \cos \frac{m\pi y}{b}$$

[1] This problem in a somewhat different way was discussed by B. G. Galerkin; see his "Collected Papers," vol. 2, p. 410, Moscow, 1953.

Considering the span $i+1$ and using expressions (a) and (j), we find

$$\left(\frac{\partial w}{\partial x_{i+1}}\right)_{x_{i+1}=-(a_{i+1})/2} = -\frac{2q_{i+1}b^3}{\pi^4 D} \sum_{m=1,3,5,\ldots}^{\infty} \frac{(-1)^{(m-1)/2}}{m^4} \cos\frac{m\pi y}{b} A_m^{i+1}$$

$$-\frac{b}{4\pi D} \sum_{m=1,3,5,\ldots}^{\infty} \frac{(-1)^{(m-1)/2}}{m} \cos\frac{m\pi y}{b} [(E_m^i + E_m^{i+1})B_m^{i+1}$$

$$- (E_m^{i+1} + E_m^i)C_m^{i+1}] \quad (p)$$

In the same manner, considering the span i, we obtain

$$\left(\frac{\partial w}{\partial x_i}\right)_{x_i=a_i/2} = \frac{2q_i b^3}{\pi^4 D} \sum_{m=1,3,5,\ldots}^{\infty} \frac{(-1)^{(m-1)/2}}{m^4} \cos\frac{m\pi y}{b} A_m^i$$

$$+ \frac{b}{4\pi D} \sum_{m=1,3,5,\ldots}^{\infty} \frac{(-1)^{(m-1)/2}}{m} \cos\frac{m\pi y}{b} [(E_m^{i-1} + E_m^i)B_m^i$$

$$+ (E_m^i - E_m^{i-1})C_m^i] \quad (q)$$

From the condition of continuity we conclude that

$$\left(\frac{\partial w}{\partial x_{i+1}}\right)_{x_{i+1}=-(a_{i+1})/2} = \left(\frac{\partial w}{\partial x_i}\right)_{x_i=a_i/2}$$

Substituting expressions (p) and (q) in this equation and observing that it must be satisfied for any value of y, we obtain the following equation for calculating E_m^{i-1}, E_m^i, and E_m^{i+1}:

$$E_m^{i-1}(B_m^i - C_m^i) + E_m^i(B_m^i + C_m^i + B_m^{i+1} + C_m^{i+1})$$
$$+ E_m^{i+1}(B_m^{i+1} - C_m^{i+1}) = -\frac{8b^2}{\pi^3 m^3}(q_{i+1}A_m^{i+1} + q_i A_m^i) \quad (177)$$

Equations (k) and (m), which we obtained previously, are particular cases of this equation. We can write as many Eqs. (177) as there are intermediate supports, and there is no difficulty in calculating the moments at the intermediate supports if the ends of the plate are simply supported. The left-hand side of Eq. (177) holds not only for uniform load but also for any type of loading that is symmetrical in each span with respect to the x and y axes. The right-hand side of Eq. (177), however, has a different value for each type of loading, as in the three-moment equation for beams.

The problem of continuous plates carrying single loads can be treated in a similar manner. In the particular case of an infinite number of equal spans with a single load applied at any point of only one span, the deflection of the plate may be obtained by resolving an equation with finite differences for the unknown coefficient E_m^i as functions of the index i.[1]

If the intermediate supports are elastic, the magnitude of the coefficients E_m^i is governed by the five-term equations, similar to the five-moment equations of the theory of continuous beams.[2] The torsional rigidity of supporting beams, tending to reduce the rotations of the plate along the support, can also be taken into account in considering the bending of continuous plates.[3]

As the simplest example of a continuous plate carrying a concentrated load, let us consider an infinitely long plate simply supported along the sides $x = 0$, $x = a$, continuous over the support $y = 0$, and submitted to a concentrated load P at some point $x = \xi$, $y = \eta$ (Fig. 112a). The load and boundary conditions under consideration can be readily satisfied by superposition of cases shown in Fig. 112b and c. In the case of Fig. 112b each panel of the plate is simply supported along the line $y = 0$, and the elastic surface is given by the expression $\pm w_1/2$, in which the sign must be chosen according to whether y is greater or less than zero, w_1 denotes the deflections (a) of Art. 51, and $|y| \leq |\eta|$. In the case shown in Fig. 112c, each panel is clamped along the edge $y = 0$, and the corresponding deflections are $w/2$, w being given by expression (h) in Art. 51. We have therefore

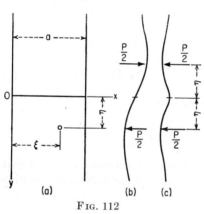

Fig. 112

$$w = w_1 + \frac{w_2}{2} \quad \text{for } \eta \geq y > 0$$

$$= \frac{w_2}{2} \quad \text{for } y < 0$$

and the moments along the edge $y = 0$ become equal to one-half of the clamping moments of a semi-infinite plate with one edge built in, these latter moments being given by expression (j) of Art. 51.

53. Approximate Design of Continuous Plates with Equal Spans.[4]

The layout of a floor slab usually involves continuity not only in one direction, as assumed in Art. 52, but rather in two perpendicular directions. A continuous slab of this kind is shown in Fig. 113. The spans and the thickness of the plate are equal for all rectangular panels. Each

---

[1] See S. Woinowsky-Krieger, *Ingr.-Arch.*, vol. 9, p. 396, 1938.

[2] Continuous plates on elastic beams were considered by V. P. Jensen, *Univ. Illinois Bull.* 81, 1938, and by N. M. Newmark, *Univ. Illinois Bull.* 84, 1938.

[3] See K. Girkmann, "Flächentragwerke," 4th ed., p. 274, Vienna, 1956.

[4] The method given below is substantially due to H. Marcus; see his book "Die vereinfachte Berechnung biegsamer Platten," Berlin, 1929. The coefficients of Tables 51 to 56 are, however, based on solutions considered in Chap. 6 and on the value of Poisson's ratio $\nu = 0.2$, whereas Marcus uses for the same purpose a simplified theory of rectangular plates and assumes $\nu = 0$.

panel may carry a dead load q_0 and, possibly, a live load p, both distributed uniformly over the area of the panel, the largest intensity of the load being $q = q_0 + p$.

Let us begin with the computation of bending moments at the intermediate supports of the floor plate. Calculations show that these moments depend principally on the loading of the two adjacent panels, and the effect of loading panels farther on is negligible. It is justifiable, therefore, to calculate the moments on supports by assuming the load q uniformly distributed over the entire floor slab (Fig. 114a).

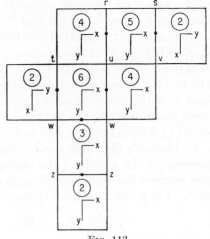

Fig. 113

Neglecting, at first, the rotations of the plate along the intermediate supports, each panel in Fig. 114a will have the same conditions as a rectangular plate clamped along the intermediate supports and simply supported at the external boundary of the floor slab.

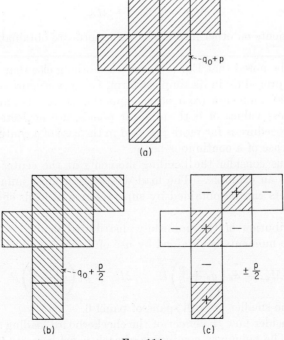

Fig. 114

The maximum bending moments for plates with such boundary conditions have been tabulated (see Tables 51 to 56). Six possible combinations of simply supported and built-in edges of a rectangular plate are shown at the head of these tables. The direction of the x and y axes in each panel of the slab (Fig. 113) must be chosen in accordance with Figs. 116 to 121; span a must be measured in the direction of the x axis and span b in the direction of the y axis of the respective panel. The six cases shown in Figs. 116 to 121 may be numbered 1 to 6, and the corresponding indices are attached to the coefficients of Tables 51 to 56.

To illustrate the application of the tables, let us calculate the bending moment at the middle of the support tw (Fig. 113). We calculate for this purpose the clamping moment of both panels adjacent to the support. For panel 2 we have to use the formula

$$\bar{M}_{2y} = \delta_2 q l^2 \qquad (a)$$

and Table 52, l being the smaller of spans a and b of the panel. In a similar manner we obtain the clamping moment of panel 6 from the expression

$$\bar{M}_{6x} = \gamma_6 q l^2 \qquad (b)$$

by making use of Table 56. The moment in question now is given with sufficient accuracy by

$$M_{tw} = \tfrac{1}{2}(\bar{M}_{2y} + \bar{M}_{6x}) \qquad (c)$$

and the moments on other intermediate supports are obtainable in a similar manner.

It should be noted that Eq. (c) expresses nothing else than a moment-distribution procedure in its simplest form, *i.e.*, a procedure in which the "carried-over" moments from other supports, as well as any difference in the stiffness values of both adjacent panels, are neglected. Such a simplified procedure is far more justified in the case of a continuous plate than in the case of a continuous beam.

Next, let us consider the bending moments at the center of panel 6 (Fig. 113) as an example. The load distribution most unfavorable for these moments can be obtained by superposition of loads shown in Fig. 114b and c.

The contribution of the uniformly distributed load $q_0 + p/2$ to the values of the moments is obtained by use of Table 56, which gives

$$M'_{6x} = \alpha_6 \left(q_0 + \frac{p}{2}\right) l^2 \qquad M'_{6y} = \beta_6 \left(q_0 + \frac{p}{2}\right) l^2 \qquad (d)$$

l denoting the smaller of both spans of panel 6.

Let us consider now the effect of the checkerboard loading as shown in Fig. 114c. The boundary conditions of each panel here are the same as those of a simply supported plate, and the moments at the center are readily computed by means of Table 51 for case 1. The load $+p/2$ acting in panel 6 yields

$$M''_{6x} = \frac{\alpha_1 p}{2} l^2 \qquad M''_{6y} = \frac{\beta_1 p}{2} l^2 \qquad (e)$$

and the largest moments at the center of panel 6 are

$$\begin{aligned} M_{6x} &= M'_{6x} + M''_{6x} \\ M_{6y} &= M'_{6y} + M''_{6y} \end{aligned} \qquad (f)$$

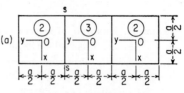

In order to calculate the largest negative moments at the same point we have only to alter the sign of the load in Fig. 114c. Still using results (d) and (e), we then have

$$\begin{aligned} M_{6x} &= M'_{6x} - M''_{6x} \\ M_{6y} &= M'_{6y} - M''_{6y} \end{aligned} \qquad (g)$$

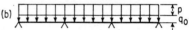

As a second example of the application of the approximate method, let us compute the bending moments of the continuous plate shown in Fig. 115, which was treated rigorously in Art. 52.

First we choose the direction of the x and y axes in accordance with Figs. 117 and 118. Assuming next a load $q = q_0 + p$ uniformly distributed over the entire surface of the plate (Fig. 115b) and using the coefficients given in Tables 52 and 53 for cases 2 and 3, with $b/a = 1$, we obtain at the center of the support ss the moment

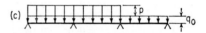

Fig. 115

$$M_{ss} = -\frac{0.0840 + 0.0697}{2}(q_0 + p)a^2 = -0.0769(q_0 + p)a^2 \qquad (h)$$

the procedure being the same as in the foregoing example [Eq. (c)]. Using the rigorous solution, the numerically largest moment at ss is produced by the load distribution shown in Fig. 115c. Superposing the bending moment obtained on page 231 upon those calculated on page 234, the exact minimum value of the moment M_{ss} proves to be

$$M_{ss} = -[0.0381(q_0 + p) + 0.0424(q_0 + p) - 0.0042q_0]a^2$$
or $$M_{ss} = -(0.0805q_0 + 0.0763p)a^2 \qquad (i)$$

Putting, for instance, $q_0 = q/3$, $p = 2q/3$, the result (i) yields $-0.0777qa^2$ as compared with the value $-0.0769qa^2$ obtained by the approximate method.

Finally, let us calculate the largest bending moment at the center of the middle panel, the most unfavorable load distribution being such as shown in Fig. 115d.

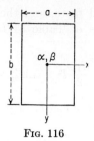

Fig. 116

TABLE 51. BENDING MOMENTS FOR UNIFORMLY LOADED PLATES IN CASE 1
$\nu = 0.2$, $l =$ the smaller of spans a and b

b/a	Center of plate		Factor
	$M_x = \alpha_1 q l^2$	$M_y = \beta_1 q l^2$	
	α_1	β_1	
0	0.0250*	0.1250	
0.5	0.0367	0.0999	
0.6	0.0406	0.0868	
0.7	0.0436	0.0742	qb^2
0.8	0.0446	0.0627	
0.9	0.0449	0.0526	
1.0	0.0442	0.0442	
1.1	0.0517	0.0449	
1.2	0.0592	0.0449	
1.3	0.0660	0.0444	
1.4	0.0723	0.0439	
1.5	0.0784	0.0426	
			qa^2
1.6	0.0836	0.0414	
1.7	0.0885	0.0402	
1.8	0.0927	0.0391	
1.9	0.0966	0.0378	
2.0	0.0999	0.0367	
∞	0.1250	0.0250†	

* $M_{\max} = 0.0364 q b^2$ at $0.48b$ from the short edge.
† $M_{\max} = 0.0364 q a^2$ at $0.48a$ from the short edge.

Combining the load in accordance with Fig. 115e and f and using the coefficients α and β of Tables 53 and 51, we arrive at the following expressions for these moments:

$$M_x = \left[0.0216\left(q_0 + \frac{p}{2}\right) + 0.0442\frac{p}{2}\right]a^2 = (0.0216 q_0 + 0.0329 p)a^2$$

$$M_y = \left[0.0316\left(q_0 + \frac{p}{2}\right) + 0.0442\frac{p}{2}\right]a^2 = (0.0316 q_0 + 0.0379 p)a^2$$

(j)

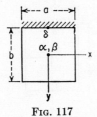

Fig. 117

Table 52. Bending Moments for Uniformly Loaded Plates in Case 2
$\nu = 0.2$; l = the smaller of spans a and b

b/a	Center of plate		Middle of fixed edge	Factor
	$M_x = \alpha_2 q l^2$	$M_y = \beta_2 q l^2$	$\bar{M}_y = \delta_2 q l^2$	
	α_2	β_2	δ_2	
0	0.0125	0.0625	−0.1250	
0.5	0.0177	0.0595	−0.1210	
0.6	0.0214	0.0562	−0.1156	
0.7	0.0249	0.0514	−0.1086	qb^2
0.8	0.0272	0.0465	−0.1009	
0.9	0.0294	0.0415	−0.0922	
1.0	0.0307	0.0367	−0.0840	
1.1	0.0378	0.0391	−0.0916	
1.2	0.0451	0.0404	−0.0983	
1.3	0.0525	0.0415	−0.1040	
1.4	0.0594	0.0418	−0.1084	
1.5	0.0661	0.0418	−0.1121	
1.6	0.0722	0.0414	−0.1148	qa^2
1.7	0.0780	0.0408	−0.1172	
1.8	0.0831	0.0399	−0.1189	
1.9	0.0879	0.0390	−0.1204	
2.0	0.0921	0.0382	−0.1216	
∞	0.1250	0.0250*	−0.1250	

* $M_{max} = 0.0387qa^2$ at $0.80a$ from the built-in edge.

It is of interest to verify the foregoing approximate values by use of the results obtained on pages 232 and 234. Distributing the load again as shown in Fig. 115d and interchanging the indices x and y in the results mentioned above, we have

$$M_x = 0.0317(q_0 + p)a^2 - (0.0051 + 0.0051)q_0 a^2$$
$$= (0.0215q_0 + 0.0317p)a^2$$
$$M_y = 0.0375(q_0 + p)a^2 - (0.0039 + 0.0039)q_0 a^2$$
$$= (0.0297q_0 + 0.0375p)a^2$$
(k)

Setting again $q_0 = q/3$ and $p = 2q/3$, we obtain for the moments the exact values of $0.0283qa^2$ and $0.0349qa^2$, respectively. Eqs. (j) yield for the same moments the approximate values of $0.0291qa^2$ and $0.0358qa^2$.

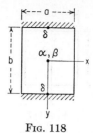

Fig. 118

Table 53. Bending Moments for Uniformly Loaded Plates in Case 3
$\nu = 0.2$, $l = $ the smaller of spans a and b

b/a	Center of plate		Middle of fixed edge	Factor
	$M_x = \alpha_3 q l^2$	$M_y = \beta_3 q l^2$	$\bar{M}_y = \delta_3 q l^2$	
	α_3	β_3	δ_3	
0	0.0083*	0.0417	−0.0833	
0.5	0.0100	0.0418	−0.0842	
0.6	0.0121	0.0410	−0.0834	
0.7	0.0152	0.0393	−0.0814	qb^2
0.8	0.0173	0.0371	−0.0783	
0.9	0.0196	0.0344	−0.0743	
1.0	0.0216	0.0316	−0.0697	
1.1	0.0276	0.0349	−0.0787	
1.2	0.0344	0.0372	−0.0868	
1.3	0.0414	0.0391	−0.0938	
1.4	0.0482	0.0405	−0.0998	
1.5	0.0554	0.0411	−0.1049	
1.6	0.0620	0.0413	−0.1090	qa^2
1.7	0.0683	0.0412	−0.1122	
1.8	0.0741	0.0408	−0.1152	
1.9	0.0795	0.0401	−0.1174	
2.0	0.0846	0.0394	−0.1191	
∞	0.1250	0.0250†	−0.1250	

* $M_{\max} = 0.0174 q b^2$ at $0.30b$ from the supported edge.

† $M_{\max} = 0.0387 q a^2$ at $0.80a$ from the built-in edge.

The largest error of the approximate method ensues from the fact that the largest positive moments do not always occur at the center of the panel. This is especially far from being true in the case of distinctly oblong rectangular panels. If b, for example, is much larger than a, the largest moment M_y occurs near the short side of the rectangular plate. Some values of these largest moments are given in footnotes to the tables, and they should be considered as the least possible values of the corresponding columns, regardless of the actual ratio b/a.

It should be noted, finally, that in the unsymmetrical case 4 neither M_x nor M_y

CONTINUOUS RECTANGULAR PLATES

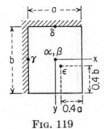

Fig. 119

Table 54. Bending Moments for Uniformly Loaded Plates in Case 4*
$\nu = 0.2$, l = the smaller of spans a and b

b/a	Center of plate		Middle of fixed edge		At $x = 0.1a$, $y = 0.1b$	Factor
	$M_x = \alpha_4 q l^2$	$M_y = \beta_4 q l^2$	$M_x = \gamma_4 q l^2$	$M_y = \delta_4 q l^2$	$M_{max} = \epsilon_4 q l^2$	
	α_4	β_4	γ_4	δ_4	ϵ_4	
0.5	0.0191	0.0574	−0.0787	−0.1180	0.0662	
0.6	0.0228	0.0522	−0.0781	−0.1093	0.0570	
0.7	0.0257	0.0460	−0.0767	−0.0991	0.0501	qb^2
0.8	0.0275	0.0396	−0.0746	−0.0882	0.0430	
0.9	0.0282	0.0336	−0.0715	−0.0775	0.0363	
1.0	0.0281	0.0281	−0.0678	−0.0678	0.0305	
1.1	0.0330	0.0283	−0.0766	−0.0709	0.0358	
1.2	0.0376	0.0279	−0.0845	−0.0736	0.0407	
1.3	0.0416	0.0270	−0.0915	−0.0754	0.0452	
1.4	0.0451	0.0260	−0.0975	−0.0765	0.0491	
1.5	0.0481	0.0248	−0.1028	−0.0772	0.0524	qa^2
1.6	0.0507	0.0236	−0.1068	−0.0778	0.0553	
1.7	0.0529	0.0224	−0.1104	−0.0782	0.0586	
1.8	0.0546	0.0213	−0.1134	−0.0785	0.0608	
1.9	0.0561	0.0202	−0.1159	−0.0786	0.0636	
2.0	0.0574	0.0191	−0.1180	−0.0787	0.0662	

* The authors are indebted to the National Research Council of Canada for a grant which greatly facilitated the computation of the table.

is the largest bending moment at the center of the plate. Table 54 shows, however, that the difference between M_{max} and the largest of the values of M_x and M_y does not exceed 10 per cent of the latter values and that the general procedure described on page 238 is justified in case 4 as well.

For the purpose of the design of isolated panels without continuity (Fig. 119), Table 54 contains the values of the largest moments M_{max} acting at $x = 0.1a$, $y = 0.1b$; for rectangular plates the direction of σ_{max} is practically that of the shorter span and for square plates that of the diagonal $x = -y$. For the sake of a greater security those values of M_{max} may also be used in calculating *continuous* panels of oblong shape.

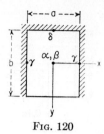

Fig. 120

Table 55. Bending Moments for Uniformly Loaded Plates in Case 5*
$\nu = 0.2$, l = smaller of spans a and b

b/a	Center of plate		Middle of fixed edge		Factor
	$M_x = \alpha_5 q l^2$	$M_y = \beta_5 q l^2$	$M_x \gamma_5 q l^2$	$M_y = \delta_5 q l^2$	
	α_5	β_5	γ_5	δ_5	
0.5	0.0206	0.0554	−0.0783	−0.114	
0.6	0.0245	0.0481	−0.0773	−0.102	
0.7	0.0268	0.0409	−0.0749	−0.0907	
0.8	0.0277	0.0335	−0.0708	−0.0778	qb^2
0.9	0.0274	0.0271	−0.0657	−0.0658	
1.0	0.0261	0.0213	−0.0600	−0.0547	
1.1	0.0294	0.0204	−0.0659	−0.0566	
1.2	0.0323	0.0192	−0.0705	−0.0573	
1.3	0.0346	0.0179	−0.0743	−0.0574	
1.4	0.0364	0.0166	−0.0770	−0.0576	
1.5	0.0378	0.0154	−0.0788	−0.0569	
1.6	0.0390	0.0143	−0.0803	−0.0568	qa^2
1.7	0.0398	0.0133	−0.0815	−0.0567	
1.8	0.0405	0.0125	−0.0825	−0.0567	
1.9	0.0410	0.0118	−0.0831	−0.0566	
2.0	0.0414	0.0110	−0.0833	−0.0566	
∞	0.0417	0.0083	−0.0833	−0.0566	

* The data of this table are due substantially to F. Czerny, *Bautech.-Arch.*, vol. 11, p. 33, W. Ernst & Sohn, Berlin, 1955.

The method given in this article is still applicable if the spans, the flexural rigidities, or the intensity of the load differs only slightly from panel to panel of the continuous plate. Otherwise more exact methods should be used.

It should be noted, however, that the application of the rigorous methods to the design of continuous floor slabs often leads to cumbersome calculations and that the accuracy thus obtained is illusory on account of many more or less indeterminable factors affecting the magnitude of the moments of the plate. Such factors are, for example, the flexibility and the torsional rigidity of the supporting beams, the restrain-

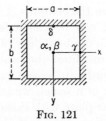

Fig. 121

Table 56. Bending Moments for Uniformly Loaded Plates in Case 6
$\nu = 0.2$, $l =$ the smaller of spans a and b

b/a	Center of plate		Middle of fixed edge		Factor
	$M_x = \alpha_6 q l^2$ α_6	$M_y = \beta_6 q l^2$ β_6	$\bar{M}_x = \gamma_6 q l^2$ γ_6	$\bar{M}_y = \delta_6 q l^2$ δ_6	
0	0.0083	0.0417	−0.0571	−0.0833	
0.5	0.0118	0.0408	−0.0571	−0.0829	
0.6	0.0150	0.0381	−0.0571	−0.0793	qb^2
0.7	0.0178	0.0344	−0.0569	−0.0736	
0.8	0.0198	0.0299	−0.0559	−0.0664	
0.9	0.0209	0.0252	−0.0540	−0.0588	
1.0	0.0213	0.0213	−0.0513	−0.0513	
1.1	0.0248	0.0210	−0.0581	−0.0538	
1.2	0.0284	0.0203	−0.0639	−0.0554	
1.3	0.0313	0.0193	−0.0687	−0.0563	
1.4	0.0337	0.0181	−0.0726	−0.0568	
1.5	0.0358	0.0169	−0.0757	−0.0570	
1.6	0.0372	0.0157	−0.0780	−0.0571	qa^2
1.7	0.0385	0.0146	−0.0799	−0.0571	
1.8	0.0395	0.0136	−0.0812	−0.0571	
1.9	0.0402	0.0126	−0.0822	−0.0571	
2.0	0.0408	0.0118	−0.0829	−0.0571	
∞	0.0417	0.0083	−0.0833	−0.0571	

ing effect of the surrounding walls, the anisotropy of the plate itself, and the inaccuracy in estimating the value of such constants as the Poisson ratio ν.

However, we can simplify the procedure of calculation by restricting the Fourier series, representing a bending moment in the plate, to its initial term or by replacing the actual values of moments or slopes along some support of the plate by their average values or, finally, by use of a moment distribution procedure.[1]

54. Bending of Plates Supported by Rows of Equidistant Columns—(Flat Slabs).
If the dimensions of the plate are large in comparison with

[1] For such methods see C. P. Siess and N. M. Newmark, *Univ. Illinois Bull.* 43, 1950, where a further bibliography on the subject is given. See also the paper of H. M. Westergaard, *Proc. Am. Concrete Inst.*, vol. 22, 1926, which contains valuable conclusions regarding the design of continuous floor slabs.

the distances a and b between the columns (Fig. 122) and the lateral load is uniformly distributed, it can be concluded that the bending in all panels, which are not close to the boundary of the plate, may be assumed to be identical, so that we can limit the problem to the bending of one panel only. Taking the coordinate axes parallel to the rows of columns and the origin at the center of a panel, we may consider this panel as a uniformly loaded rectangular plate with sides a and b. From symmetry we conclude that the deflection surface of the plate is as shown by the dashed lines in Fig. 122b. The maximum deflection is at the center of the plate, and the deflection at the corners is zero. To simplify the problem we assume that the cross-sectional dimensions of the columns are small and can be neglected in so far as deflection and moments at

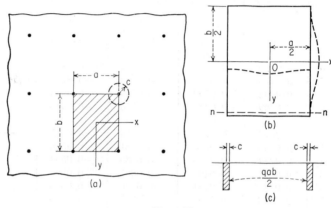

Fig. 122

the center of the plate are concerned.[1] We then have a uniformly loaded rectangular plate supported at the corners, and we conclude from symmetry that the slope of the deflection surface in the direction of the normal to the boundary and the shearing force are zero at all points along the edges of the plate except at the corners.[2]

Proceeding as in the case of a simply supported plate (Art. 30), we take the total deflection w in the form

$$w = w_1 + w_2 \qquad (a)$$

where
$$w_1 = \frac{qb^4}{384D}\left(1 - \frac{4y^2}{b^2}\right)^2 \qquad (b)$$

[1] In this simplified form the problem was discussed by several authors; see, for example, A. Nádai, Über die Biegung durchlaufender Platten, *Z. angew. Math. Mech.*, vol. 2, p. 1, 1922, and B. G. Galerkin, "Collected Papers," vol. 2, p. 29, Moscow, 1953.

[2] The equating to zero of the twisting moment M_{xy} along the boundary follows from the fact that the slope in the direction of the normal to the boundary is zero.

represents the deflection of a uniformly loaded strip clamped at the ends $y = \pm b/2$ and satisfies the differential equation (103) of the plate as well as the boundary conditions

$$\left(\frac{\partial w_1}{\partial x}\right)_{x=\pm a/2} = 0 \quad (Q_x)_{x=\pm a/2} = -D\frac{\partial}{\partial x}\left(\frac{\partial^2 w_1}{\partial x^2} + \frac{\partial^2 w_1}{\partial y^2}\right)_{x=\pm a/2} = 0 \quad (c)$$

The deflection w_2 is taken in the form of the series

$$w_2 = A_0 + \sum_{m=2,4,6,\ldots}^{\infty} Y_m \cos\frac{m\pi x}{a} \quad (d)$$

each term of which satisfies the conditions (c). The functions Y_m must be chosen so as to satisfy the homogeneous equation

$$\Delta\Delta w_2 = 0 \quad (e)$$

and so as to make w satisfy the boundary conditions at the edges $y = \pm b/2$. Equation (e) and the conditions of symmetry are satisfied by taking series (d) in the form

$$w_2 = A_0 + \sum_{m=2,4,6,\ldots}^{\infty} \left(A_m \cosh\frac{m\pi y}{a} + B_m \frac{m\pi y}{a}\sinh\frac{m\pi y}{a}\right)\cos\frac{m\pi x}{a} \quad (f)$$

where the constants A_0, A_m, and B_m are to be determined from the boundary conditions along the edge $y = b/2$. From the condition concerning the slope, viz., that

$$\left(\frac{\partial w}{\partial y}\right)_{y=b/2} = \left(\frac{\partial w_1}{\partial y} + \frac{\partial w_2}{\partial y}\right)_{y=b/2} = 0$$

we readily find that

$$B_m = -A_m \frac{\tanh \alpha_m}{\alpha_m + \tanh \alpha_m} \quad (g)$$

in which, as before,

$$\alpha_m = \frac{m\pi b}{2a} \quad (h)$$

Considering now the boundary condition concerning the shearing force, we see that on a normal section nn (Fig. 122b) of the plate infinitely close to the boundary $y = b/2$, the shearing force Q_y is equal to zero at all points except those which are close to the column, and at these points Q_y must be infinitely large in order to transmit the finite load $\tfrac{1}{2}qab$ to the column (Fig. 122c) along an infinitely small distance between $x = a/2 - c$ and $x = a/2 + c$. Representing Q_y by a trigonometric series which, from symmetry, has the form

$$Q_y = C_0 + \sum_{m=2,4,6,\ldots}^{\infty} C_m \cos \frac{m\pi x}{a} \qquad (i)$$

and observing that

$$Q_y = 0 \quad \text{for } 0 < x < \frac{a}{2} - c$$

and
$$\int_{a/2-c}^{a/2} Q_y \, dx = -\frac{qab}{4}$$

we find, by applying the usual method of calculation, that

$$C_0 = -\frac{qab}{2a} = -\frac{P}{2a}$$

and
$$C_m = \frac{4}{a} \int_0^{a/2} Q_y \cos \frac{m\pi x}{a} \, dx = -\frac{P}{a}(-1)^{m/2}$$

where $P = qab$ is the total load on one panel of the plate. Substituting these values of the coefficients C_0 and C_m in series (i), the required boundary condition takes the following form:

$$(Q_y)_{y=b/2} = -D\left(\frac{\partial^3 w}{\partial y^3} + \frac{\partial^3 w}{\partial x^2 \, \partial y}\right)_{y=b/2}$$

$$= -\frac{P}{a} \sum_{m=2,4,6,\ldots}^{\infty} (-1)^{m/2} \cos \frac{m\pi x}{a} - \frac{P}{2a}$$

Substituting expression (a) for w and observing that the second term in parentheses vanishes, on account of the boundary condition $\partial w/\partial y = 0$, we obtain

$$-D\left(\frac{\partial^3 w_2}{\partial y^3}\right)_{y=b/2} = -\frac{P}{a} \sum_{m=2,4,6,\ldots}^{\infty} (-1)^{m/2} \cos \frac{m\pi x}{a}$$

from which, by using expression (f), we find that

$$D\frac{m^3\pi^3}{a^3}[(A_m + 3B_m)\sinh \alpha_m + B_m \alpha_m \cosh \alpha_m] = \frac{P}{a}(-1)^{m/2} \qquad (j)$$

Solving Eqs. (g) and (j) for the constants A_m and B_m, we obtain

$$A_m = -\frac{Pa^2}{2m^3\pi^3 D}(-1)^{m/2}\frac{\alpha_m + \tanh \alpha_m}{\sinh \alpha_m \tanh \alpha_m}$$

$$B_m = \frac{Pa^2}{2m^3\pi^3 D}(-1)^{m/2}\frac{1}{\sinh \alpha_m} \qquad (k)$$

The deflection of the plate takes the form

$$w = \frac{qb^4}{384D}\left(1 - \frac{4y^2}{b^2}\right)^2 + A_0 + \frac{qa^3b}{2\pi^3 D}\sum_{m=2,4,6,\ldots}^{\infty}\frac{(-1)^{m/2}\cos\frac{m\pi x}{a}}{m^3 \sinh\alpha_m \tanh\alpha_m}$$
$$\left[\tanh\alpha_m\frac{m\pi y}{a}\sinh\frac{m\pi y}{a} - (\alpha_m + \tanh\alpha_m)\cosh\frac{m\pi y}{a}\right] \quad (l)$$

The constant A_0 can now be determined from the condition that the deflection vanishes at the corners of the plate. Hence

$$(w)_{x=a/2, y=b/2} = 0$$

and $$A_0 = -\frac{qa^3b}{2\pi^3 D}\sum_{m=2,4,6,\ldots}^{\infty}\frac{1}{m^3}\left(\alpha_m - \frac{\alpha_m + \tanh\alpha_m}{\tanh^2\alpha_m}\right) \quad (m)$$

The deflection at any point of the plate can be calculated by using expressions (l) and (m). The maximum deflection is evidently at the center of the plate, at which point we have

$$(w)_{x=0, y=0} = \frac{qb^4}{384D} - \frac{qa^3b}{2\pi^3 D}\sum_{m=2,4,6,\ldots}^{\infty}\frac{(-1)^{m/2}}{m^3}\frac{\alpha_m + \tanh\alpha_m}{\sinh\alpha_m \tanh\alpha_m}$$
$$- \frac{qa^3b}{2\pi^3 D}\sum_{m=2,4,6,\ldots}^{\infty}\frac{1}{m^3}\left(\alpha_m - \frac{\alpha_m + \tanh\alpha_m}{\tanh^2\alpha_m}\right) \quad (n)$$

Values of this deflection calculated for several values of the ratio b/a are given in Table 57. Values of the bending moments $(M_x)_{x=0, y=0}$ and $(M_y)_{x=0, y=0}$ calculated by using formulas (101) and expression (l) for deflection are also given. It is seen that for $b > a$ the maximum bend-

TABLE 57. DEFLECTIONS AND MOMENTS AT THE CENTER OF A PANEL (Fig. 122)
$\nu = 0.2$

b/a	$w = \alpha\dfrac{qb^4}{D}$	$M_x = \beta qb^2$	$M_y = \beta_1 qb^2$
	α	β	β_1
1	0.00581	0.0331	0.0331
1.1	0.00487	0.0261	0.0352
1.2	0.00428	0.0210	0.0363
1.3	0.00387	0.0175	0.0375
1.4	0.00358	0.0149	0.0384
1.5	0.00337	0.0131	0.0387
2.0	0.00292	0.0092	0.0411
∞	0.00260	0.0083	0.0417

ing moment at the center of the plate does not differ much from the moment at the middle of a uniformly loaded strip of length b clamped at the ends.

Concentrated reactions are acting at the points of support of the plate, and the moments calculated from expression (l) become infinitely large. We can, however, assume the reactive forces to be distributed uniformly over the area of a circle representing the cross section of the column. The bending moments arising at the center of the supporting area remain finite in such a case and can be calculated by a procedure similar to that used in the case of rectangular plates and described on page 147. With reference to Fig. 122, the result can be expressed by the formulas[1]

$$(M_x)_{x=a/2, y=b/2} = M_0 - \frac{qb^2}{4}\left[\frac{\nu}{3} + (1 - \nu)\sum_{n=1}^{\infty}\frac{1}{\sinh^2\frac{n\pi b}{a}}\right]$$

$$(M_y)_{x=a/2, y=b/2} = M_0 + \frac{qb^2}{4}\left[-\frac{1}{3} + (1 - \nu)\sum_{n=1}^{\infty}\frac{1}{\sinh^2\frac{n\pi b}{a}} + (1 - \nu)\frac{a}{\pi b}\right]$$

(o)

In these expressions

$$M_0 = -\frac{qab}{4\pi}\left[(1 + \nu)\log\frac{a}{2\pi c(1 - q^2)^2(1 - q^4)^2 \cdots} + 1\right]$$

$q = e^{-\pi b/a}$, and c denotes the radius of the circle, supposed to be small compared with spans a and b of the panel. Carrying out the required calculations, we can reduce Eqs. (o) to the form

$$(M_x)_{x=a/2, y=b/2} = -\frac{qab}{4\pi}\left[(1 + \nu)\log\frac{a}{c} - (\alpha + \beta\nu)\right]$$

$$(M_y)_{x=a/2, y=b/2} = -\frac{qab}{4\pi}\left[(1 + \nu)\log\frac{a}{c} - (\beta + \alpha\nu)\right]$$

(p)

in which α and β are coefficients given for several values of the ratio b/a in Table 58.

TABLE 58. VALUES OF COEFFICIENTS α AND β IN EQS. (p) FOR MOMENTS ON SUPPORT

b/a	1	1.1	1.2	1.3	1.4	1.5	2.0
α	0.811	0.822	0.829	0.833	0.835	0.836	0.838
β	0.811	0.698	0.588	0.481	0.374	0.268	−0.256

The bending moments corresponding to the centers of columns of rectangular cross section also can be calculated by assuming that the reactions are uniformly distributed over the rectangles, shown shaded in Fig.

[1] Given by A. Nádai in his book "Elastische Platten," p. 154, Berlin, 1925.

TABLE 59. BENDING MOMENTS AND LARGEST SHEAR FORCE OF A SQUARE PANEL OF A UNIFORMLY LOADED PLATE (Fig. 123)
$\nu = 0.2$

$u/a = k$	$(M)_{x=y=a/2} = \beta qa^2$	$(M)_{x=y=0} = \beta_1 qa^2$	$(M_x)_{x=a/2,y=0} = \beta_2 qa^2$	$(M_y)_{x=a/2,y=0} = \beta_3 qa^2$	$Q_{max} = \gamma qa$
	β	β_1	β_2	β_3	γ
0	$-\infty$	0.0331	-0.0185	0.0512	∞
0.1	-0.196	0.0329	-0.0182	0.0508	2.73
0.2	-0.131	0.0321	-0.0178	0.0489	
0.3	-0.0933	0.0308	-0.0170	0.0458	0.842
0.4	-0.0678	0.0289	-0.0158	0.0415	
0.5	-0.0487	0.0265	-0.0140	0.0361	0.419

123, that represent the cross sections of the columns.[1] In the case of square panels and square columns we have $u/a = v/b = k$, and the moments at the centers of the columns and at the centers of the panels are given by the following formulas:

$$(M_x)_{x=y=a/2} = (M_y)_{x=y=a/2} = -\frac{(1+\nu)qa^2}{4}\left[\frac{(1-k)(2-k)}{12}\right.$$
$$\left. + \frac{1}{\pi^3 k^2}\sum_{m=1}^{\infty}\frac{2}{m^3 \sinh m\pi}\sinh\frac{m\pi k}{2}\cosh\frac{m\pi(2-k)}{2}\sin m\pi k\right] \quad (q)$$

$$(M_x)_{x=y=0} = (M_y)_{x=y=0}$$
$$= \frac{(1+\nu)qa^2}{4}\left[\frac{1-k^2}{12} + \frac{1}{\pi^3 k^2}\sum_{m=1}^{\infty}(-1)^{m+1}\frac{\sinh m\pi k \sin m\pi k}{m^3 \sinh m\pi}\right] \quad (r)$$

The values of these moments, together with values of moments at half a distance between columns, obtained from the same solution and calculated for various values of k and for $\nu = 0.2$, are given in Table 59.

It is seen that the moments at the columns are much larger than the moments at the panel center and that their magnitude depends very much on the cross-sectional dimensions of the columns. The moments at the panel center remain practically constant for ratios up to $k = 0.2$. Hence the previous solution, obtained on

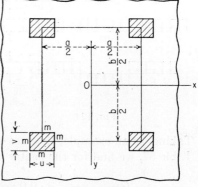

FIG. 123

[1] This case was investigated by S. Woinowsky-Krieger; see *Z. angew. Math. Mech.*, vol. 14, p. 13, 1934. See also the papers by V. Lewe, *Bauingenieur*, vol. 1, p. 631, 1920, and by K. Frey, *Bauingenieur*, vol. 7, p. 21, 1926.

the assumption that the reactions are concentrated at the panel corners, is sufficiently accurate for the central portion of the panel.

An approximate calculation of moments given by Eq. (q) in the form of a series can also be made by means of expressions (p). Using for this purpose Eq. (c), Art. 37, we substitute

$$c = \frac{u}{\sqrt{2}} e^{\pi/4-1} = 0.57u$$

i.e., the radius of a circle equivalent to the given square area u by u, in Eqs. (p). In the particular case of square panels numerical results obtained in this manner are but slightly different from those given in the second column of Table 59.

The shearing forces have their maximum value at the middle of the sides of the columns, at points m in Fig. 123. This value, for the case of square panels, depends on the value of the ratio k and can be represented by the formula $Q = \gamma q a^2$. Several numerical values of the factor γ are given in Table 59. It is interesting to note that there is a difference of only about 10 per cent between these values and the average values obtained by dividing the total column load $qa^2(1 - k^2)$ by the perimeter $4ka$ of the cross section of the column.

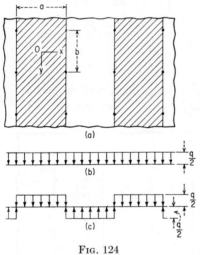

Fig. 124

Uniform loading of the entire plate gives the most unfavorable condition at the columns. To get the maximum bending moment at the center of a panel, the load must be distributed as shown by the shaded areas in Fig. 124a. The solution for this case is readily obtained by combining the uniform load distribution of intensity $q/2$ shown in Fig. 124b with the load $q/2$ alternating in sign in consecutive spans shown in Fig. 124c. The deflection surface for the latter case is evidently the same as that for a uniformly loaded strip of length a simply supported at the ends.

Taking, as an example, the case of square panels and using the values in Table 57, we find for the center of a panel (Fig. 124a):

$$(w)_{x=y=0} = \frac{1}{2} q \cdot 0.00581 \frac{a^4}{D} + \frac{5}{384} \frac{q}{2} \frac{a^4}{D} = 0.00942 \frac{qa^4}{D}$$

$$(M_x)_{x=y=0} = \frac{1}{2} q \cdot 0.0331 a^2 + \frac{1}{16} qa^2 = 0.0791 qa^2$$

$$(M_y)_{x=y=0} = \frac{1}{2} q \cdot 0.0331 a^2 + \frac{0.2}{16} qa^2 = 0.0291 qa^2$$

From Table 59 we conclude, furthermore, that

$$(M_x)_{x=0,y=b/2} = \tfrac{1}{2}q \cdot 0.0512a^2 + \tfrac{1}{16}qa^2 = 0.0881qa^2$$

The foregoing results are obtained in assuming that the plate is free to rotate at the points of support. Usually the columns are in rigid connection with the plate, and, in the case of the load distribution shown in Fig. 124, they produce not only vertical reactions but also couples with a restraining effect of those couples on the bending of the panels. A frame analysis extended on the flat slab and the columns as a joint structure therefore becomes necessary in order to obtain more accurate values of bending moments under alternate load.[1]

The case in which one panel is uniformly loaded while the four adjacent panels are not loaded is obtained by superposing on a uniform load $q/2$ the load $q/2$, the sign of which alternates as shown in Fig. 125. In this latter case each panel is in the same condition as a simply supported plate, and all necessary information regarding bending can be taken from Table 8. Taking the case of a square panel, we find for the center of a panel that

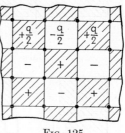

Fig. 125

$$(w)_{x=y=0} = \frac{1}{2}q \cdot 0.00581 \frac{a^4}{D} + \frac{1}{2}q \cdot 0.00406 \frac{a^4}{D} = 0.00494 \frac{qa^4}{D}$$

$$(M_x)_{x=y=0} = (M_y)_{x=y=0} = \frac{1}{2}q \cdot 0.0331a^2 + \frac{1}{2}q \cdot 0.0479 \cdot \frac{1.2}{1.3}a^2 = 0.0387qa^2$$

The case of an infinitely large slab subjected to equal concentrated loads centrally applied in all panels can be handled substantially in the same manner as in the preceding case, *i.e.*, by using the double periodicity in the deflections of the plate.[2]

The problem of bending of a uniformly loaded flat slab with skew panels has also been discussed.[3]

55. Flat Slab Having Nine Panels and Slab with Two Edges Free. So far, an infinite extension of the slab has always been assumed. Now let us consider a plate simply supported by exterior walls, forming the square boundary of the plate, together with four intermediate columns (Fig. 126). From symmetry we conclude that a uniformly distributed

---

[1] The procedure to be used is discussed in several publications; see, for instance, H. Marcus, "Die Theorie elastischer Gewebe," p. 310, Berlin, 1932.

[2] This problem was discussed by V. Lewe in his book "Pilzdecken und andere trägerlose Eisenbetonplatten," Berlin, 1926, and also by P. Pozzati, *Riv. math. Univ. Parma*, vol. 2, p. 123, 1951.

[3] See V. I. Blokh, *Doklady Akad. Nauk S.S.S.R.*, n. s., vol. 73, p. 45, 1950.

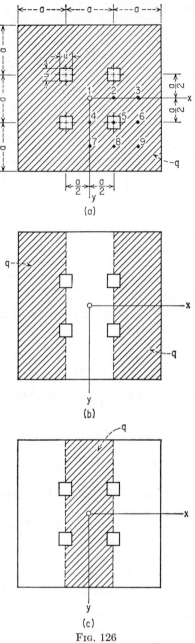

Fig. 126

load of intensity q produces equal column reactions R, which we may consider as redundant in the given statically indeterminate structure. Removing all columns, we obtain a simply supported square plate carrying merely the given load q. The deflections w_0 produced by this load at the center of the columns can easily be calculated by means of the theory given in Chap. 5. Next, removing the load q and distributing a load $R = 1$ (acting downward) uniformly over each area u by u, we obtain some new deflections w_1 at the same points $x = \pm a/2$, $y = \pm a/2$ as before. From the condition that in the actual case these points do not deflect, we conclude that $w_0 - Rw_1 = 0$, which yields $R = w_0/w_1$. Now it remains only to combine the effect of the uniform load q with the effect of four known reactions on the bending moments of the square plate of the size $3a$ by $3a$.

In the case of a partial loading, such as shown in Fig. 126b and c, we have to superpose one-half of the moments previously obtained on the moments of a simply supported plate with the area a by $3a$, carrying a uniformly distributed load $\pm q/2$. Calculations of this kind carried out by Marcus[1] led to the values of bending moments given in Table 60. The reaction of a column is $R = 1.196qa^2$ in this case. The bending of an infinite plate which is supported not only along both its parallel sides

---

[1] "Die Theorie elastischer Gewebe"; see also Lewe, op. cit. The case of a square plate with one intermediate support was discussed by N. J. Nielsen, "Bestemmelse af Spændinger I Plader," p. 217, Copenhagen, 1920.

TABLE 60. COEFFICIENTS β FOR CALCULATION OF BENDING MOMENTS $M = \beta q a^2$ OF A SIMPLY SUPPORTED SQUARE PLATE WITH FOUR INTERMEDIATE COLUMNS (Fig. 126)
$u/a = 0.25$, $\nu = 0.2$

Point	$\dfrac{x}{a}$	$\dfrac{y}{a}$	Load a		Load b		Load c	
			M_x	M_y	M_x	M_y	M_x	M_y
1	0	0	0.021	0.021	−0.048	−0.004	0.069	0.025
2	0.5	0	−0.040	0.038	−0.020	0.019	−0.020	0.019
3	1.0	0	0.069	0.025	0.093	0.027	−0.024	−0.002
4	0	0.5	0.038	−0.040	−0.036	−0.036	0.074	−0.004
5	0.5	0.5	−0.140	−0.140	−0.070	−0.070	−0.070	−0.070
6	1.0	0.5	0.074	−0.004	0.092	0.014	−0.018	−0.018
7	0	1.0	0.025	0.069	−0.028	0.017	0.052	0.052
8	0.5	1.0	−0.004	0.074	−0.002	0.037	−0.002	0.037
9	1.0	1.0	0.053	0.053	0.066	0.044	−0.013	0.009

but also by one or several rows of equidistant columns[1] can be discussed in a similar manner.

The case of bending of a long rectangular plate supported only by the two parallel rows of equidistant columns (Fig. 127) can also be solved without any difficulty for several types of loading. We begin with the case in which the plate is bent by the moments M_y represented by the series

$$(M_y)_{y=\pm b/2} = M_0 + \sum_{m=2,4,6,\ldots}^{\infty} E_m \cos \frac{m\pi x}{a} \qquad (a)$$

Since there is no lateral load, the deflection surface of the plate can be taken in the form of the series

$$w = A_0 + A_1\left(y^2 - \frac{b^2}{4}\right) + \sum_{m=2,4,6,\ldots}^{\infty} \left(A_m \cosh \frac{m\pi y}{a} + B_m \frac{m\pi y}{a} \sinh \frac{m\pi y}{a}\right) \cos \frac{m\pi x}{a} \qquad (b)$$

the coefficients of which are to be determined from the following boundary conditions:

$$-D\left(\frac{\partial^2 w}{\partial y^2} + \nu \frac{\partial^2 w}{\partial x^2}\right)_{y=\pm b/2} = M_0 + \sum_{m=2,4,6,\ldots}^{\infty} E_m \cos \frac{m\pi x}{a}$$

$$D\left[\frac{\partial^3 w}{\partial y^3} + (2-\nu)\frac{\partial^3 w}{\partial y\, \partial x^2}\right]_{y=\pm b/2} = 0 \qquad (c)$$

[1] This problem has been considered by K. Grein, "Pilzdecken," Berlin, 1948.

and from the condition that the deflection vanishes at the columns. Substituting series (b) in Eqs. (c), we find that

$$A_1 = -\frac{M_0}{2D}$$

$$A_m = -\frac{a^2 E_m}{\pi^2 m^2 D} \frac{(1+\nu)\sinh\alpha_m - (1-\nu)\alpha_m \cosh\alpha_m}{(3+\nu)(1-\nu)\sinh\alpha_m \cosh\alpha_m - \alpha_m(1-\nu)^2} \quad (d)$$

$$B_m = -\frac{a^2 E_m}{\pi^2 m^2 D} \frac{\sinh\alpha_m}{(3+\nu)\sinh\alpha_m \cosh\alpha_m - \alpha_m(1-\nu)}$$

Combining this solution with solution (l), Art. 54, we can investigate the bending of the plate shown in Fig. 127a under the action of a uniformly

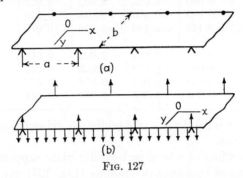

Fig. 127

distributed load. For this purpose we calculate the bending moments M_y from expression (l) by using formula (101) and obtain

$$(M_y)_{y=\pm b/2} = -\frac{qb^2}{12}$$

$$-\frac{qab}{2\pi} \sum_{m=2,4,6,\ldots}^{\infty} \frac{(-1)^{m/2}}{m} \left[\frac{1+\nu}{\tanh\alpha_m} - \frac{\alpha_m(1-\nu)}{\sinh^2\alpha_m}\right] \cos\frac{m\pi x}{a} \quad (e)$$

Equating this moment to the moment (a) taken with the negative sign, we obtain the values of M_0 and E_m which are to be substituted in Eqs. (d) for the constants A_1, A_m, and B_m in expression (b). Adding expression (b) with these values of the constants to expression (l), Art. 54, we obtain the desired solution for the uniformly loaded plate shown in Fig. 127a.

Combining this solution with that for a uniformly loaded and simply supported strip of length b which is given by the equation

$$w = -\frac{q}{24D}\left(\frac{b^2}{4} - y^2\right)\left(\frac{5}{4}b^2 - y^2\right)$$

we obtain the solution for the case in which the plate is bent by the load uniformly distributed along the edges of the plate as shown in Fig. 127b.

56. Effect of a Rigid Connection with Column on Moments of the Flat Slab. In discussing the bending of a flat slab it has always been assumed that the column reactions are concentrated at some points or distributed uniformly over some areas corresponding to the cross section of the columns or their capitals. As a rule, however, concrete slabs are rigidly connected with the columns, as shown in Fig. 128.

In discussing moments at such rigid joints, let us begin with the case of a circular column and let c be the radius of its cross section. The calculation of bending

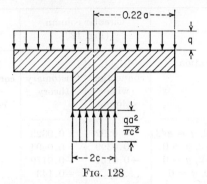

Fig. 128

moments using expression (l) in Art. 54 shows[1] that, in the case of a square panel ($a = b$) and small values of c/a, the bending moments in the radial direction practically vanish along a circle of radius $e = 0.22a$ (Fig. 122a). Thus the portion of the plate around the column and inside such a circle is in the state of an annular plate simply supported along the circle $r = 0.22a$ and clamped along the circle $r = c$, with a transverse displacement of one circle with respect to the other. Hence the maximum

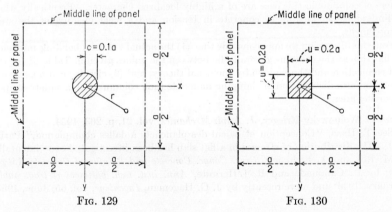

Fig. 129   Fig. 130

bending stress around the column can be obtained by using formulas (75), previously derived for circular plates (see page 61), and combining cases 3 and 8 in Fig. 36.

A more elaborate discussion of the same problem is due to F. Tölke.[2] Numerical results obtained by F. Tölke for a square panel and $c/a = 0.1$ (Fig. 129) are given in Table 61, together with values of bending moments calculated for the same case on the

---

[1] Such calculations were made by A. Nádai; see his book "Elastische Platten," p. 156, Berlin, 1925.

[2] F. Tölke, *Ingr.-Arch.*, vol. 5, p. 187, 1934.

basis of the customary theory. It is seen that a rigid connection between slab and column tends to increase numerically the moments on support and to reduce the positive moments of the slab.

TABLE 61. COEFFICIENTS β FOR CALCULATION OF BENDING MOMENTS $M = \beta q a^2$ OF A UNIFORMLY LOADED SQUARE PANEL OF A FLAT SLAB
$\nu = 0.2$

Bending moment	Location	Circular column (Fig. 129)		Square column (Fig. 130)	
		Rigid connection with column	Customary theory	Rigid connection with column	Customary theory
$M_x = M_y$	$x = a/2, y = a/2$	0.0292	0.0323	0.0264	0.0321
M_x	$x = a/2, y = 0$	0.0399	0.0494	0.0348	0.0487
M_y	$x = a/2, y = 0$	−0.0161	−0.0179	−0.0146	−0.0178
$M_x = M_y$	$x = 0, y = 0$	....	−0.143	....	−0.131
M_x	$x = u/2, y = 0$	....	....	−0.0626	−0.0803
M_r	$x = u/2, y = u/2$	....	....	− ∞	−0.0480
M_r	$r = c$	−0.1682	−0.0629		

The same table also gives moments for a flat slab rigidly connected with a column of a square cross section[1] (Fig. 130). The infinitely large stresses occurring at the corners of columns in this case are of a highly localized character. Practically, they are limited by a cracking of concrete in tension and a local yielding of the steel reinforcement.

From this discussion we may conclude that (1) the actual values of bending moments of a flat slab at the columns generally lie between the values given in Table 61 for the rigid connection and those given by the usual theory, and (2) circular columns secure a more uniform distribution of clamping moments than columns with a square-shaped supporting area.[2]

[1] See S. Woinowsky-Krieger, *J. Appl. Mechanics*, vol. 21, p. 263, 1954.

[2] See T. Haas, "Conception et calcul des planchets à dalles champignon," Paris, 1950. The distribution of stresses in a flat slab has been investigated experimentally by M. Roš and A. Eichinger, *Proc. Congr. Concrete and Reinforced Concrete*, Liége, 1930; by R. Caminade and R. L'Hermite, *Ann. inst. tech. bâtiment et trav. publ.*, February, 1936; and more recently by J. G. Hageman, *Ingenieur*, vol. 65, June, 1953.

CHAPTER 8

PLATES ON ELASTIC FOUNDATION

57. Bending Symmetrical with Respect to a Center. A laterally loaded plate may rest on an elastic foundation, as in the case of a concrete road, an airport runway, or a mat. We begin the discussion of such problems with the simplest assumption that the intensity of the reaction of the subgrade is proportional to the deflections w of the plate. This intensity is then given by the expression kw. The constant k, expressed in pounds per square inch per inch of deflection, is called the *modulus of the foundation*. The numerical value of the modulus depends largely on the properties of the subgrade; in the case of a pavement slab or a mat of greater extension this value may be estimated by means of the diagram in Table 62.[1]

TABLE 62. VALUES OF THE MODULUS OF SUBGRADE

Modulus "k" in lb/sq in./in.							
100	150	200	250	500	800		
General soil rating as subgrade, subbase or base							
Very poor subgrade	Poor subgrade	Fair to good subgrade	Excellent subgrade	Good subbase	Good base	Best base	

```
G - Gravel              P - Poorly graded
S - Sand                L - Low to med. compressibility
M -"Mo",very fine sand,silt  H - High compressibility
C - Clay
F - Fines, material
    less than 0.1mm
O - Organic
W - Well graded
```

Soil types (positioned along the k-scale): GW, GC, GP, GF, SW, SC, SP, SF, ML, CL, OL, CH, OH, MH

Let us begin with the case of a circular plate in which the load is distributed symmetrically with respect to the center. In using Eq. (58),

[1] Based on Casagrande's soil classification. The table should not be regarded as a substitute for plate bearing tests. For further information see *Trans. Am. Soc. Civ. Engrs.*, vol. 113, p. 901, 1948. See also K. Terzaghi, *Geotechnique*, vol. 5, p. 297, 1955 (Harvard Soil Mechanics Series, no. 51).

we add the load $-kw$, due to the reaction of the subgrade, to the given lateral load q. Thus we arrive at the following differential equation for the bent plate:

$$\left(\frac{d^2}{dr^2} + \frac{1}{r}\frac{d}{dr}\right)\left(\frac{d^2w}{dr^2} + \frac{1}{r}\frac{dw}{dr}\right) = \frac{q - kw}{D} \tag{178}$$

In the particular case of a plate loaded at the center with a load P,* q is equal to zero over the entire surface of the plate except at the center. By introducing the notation

$$\frac{k}{D} = \frac{1}{l^4} \tag{a}$$

Eq. (178) becomes

$$l^4\left(\frac{d^2}{dr^2} + \frac{1}{r}\frac{d}{dr}\right)\left(\frac{d^2w}{dr^2} + \frac{1}{r}\frac{dw}{dr}\right) + w = 0 \tag{b}$$

Since k is measured in pounds per cubic inch and D in pound-inches, the quantity l has the dimension of length. To simplify our further discussion it is advantageous to introduce dimensionless quantities by using the following notations:

$$\frac{w}{l} = z \qquad \frac{r}{l} = x \tag{c}$$

Then Eq. (b) becomes

$$\left(\frac{d^2}{dx^2} + \frac{1}{x}\frac{d}{dx}\right)\left(\frac{d^2z}{dx^2} + \frac{1}{x}\frac{dz}{dx}\right) + z = 0 \tag{d}$$

Using the symbol Δ for

$$\frac{d^2}{dx^2} + \frac{1}{x}\frac{d}{dx}$$

we then write

$$\Delta\Delta z + z = 0 \tag{e}$$

This is a linear differential equation of the fourth order, the general solution of which can be represented in the following form:

$$z = A_1 X_1(x) + A_2 X_2(x) + A_3 X_3(x) + A_4 X_4(x) \tag{f}$$

where $A_1, \ldots, A_4$ are constants of integration and the functions $X_1, \ldots, X_4$ are four independent solutions of Eq. (e).

We shall now try to find a solution of Eq. (e) in the form of a power

* This problem was discussed by H. Hertz, *Wiedemann's Ann. Phys. u. Chem.*, vol. 22, p. 449, 1884; see also his "Gesammelte Werke," vol. 1, p. 288, 1895, and A. Föppl, "Vorlesungen über technische Mechanik," vol. 5, p. 103, 1922. It is worth noting that Hertz's investigation deals with the problem of a floating plate rather than with that of a plate on an elastic foundation. Thus, in this case the assumption regarding the constancy of k is fulfilled, k being the unit weight of the liquid.

series. Let $a_n x^n$ be a term of this series. Then, by differentiation, we find

$$\Delta(a_n x^n) = n(n-1)a_n x^{n-2} + n a_n x^{n-2} = n^2 a_n x^{n-2}$$

and $\quad \Delta\Delta(a_n x^n) = n^2(n-2)^2 a_n x^{n-4}$

To satisfy Eq. (e) it is necessary that each term $a_n x^n$ in the series have a corresponding term $a_{n-4} x^{n-4}$ such that

$$n^2(n-2)^2 a_n x^{n-4} + a_{n-4} x^{n-4} = 0 \tag{g}$$

Following this condition, all terms cancel when the series is substituted in Eq. (e); hence the series, if it is a convergent one, represents a particular solution of the equation. From Eq. (g) it follows that

$$a_n = -\frac{a_{n-4}}{n^2(n-2)^2} \tag{h}$$

Observing also that

$$\Delta\Delta(a_0) = 0 \quad \text{and} \quad \Delta\Delta(a_2 x^2) = 0 \tag{i}$$

we can conclude that there are two series satisfying Eq. (e), viz.,

$$X_1(x) = 1 - \frac{x^4}{2^2 \cdot 4^2} + \frac{x^8}{2^2 \cdot 4^2 \cdot 6^2 \cdot 8^2}$$

$$- \frac{x^{12}}{2^2 \cdot 4^2 \cdot 6^2 \cdot 8^2 \cdot 10^2 \cdot 12^2} + \cdots$$

and $\tag{j}$

$$X_2(x) = x^2 - \frac{x^6}{4^2 \cdot 6^2} + \frac{x^{10}}{4^2 \cdot 6^2 \cdot 8^2 \cdot 10^2}$$

$$- \frac{x^{14}}{4^2 \cdot 6^2 \cdot 8^2 \cdot 10^2 \cdot 12^2 \cdot 14^2} + \cdots$$

It may be seen from the notations (c) that for small values of the distance r, that is, for points that are close to the point of application of the load P, the quantity x is small, and series (j) are rapidly convergent. It may be seen also that the consecutive derivatives of series (j) remain finite at the point of application of the load ($x = 0$). This indicates that these series alone are not sufficient to represent the stress conditions at the point of application of the load where, as we know from previously discussed cases, the bending moments become infinitely large.

For this reason the particular solution X_3 of Eq. (e) will be taken in the following form:

$$X_3 = X_1 \log x + F_3(x) \tag{k}$$

in which $F_3(x)$ is a function of x which can again be represented by a power series. By differentiation we find

$$\Delta\Delta X_3 = \frac{4}{x} \frac{d^3 X_1}{dx^3} + \log x \, \Delta\Delta X_1 + \Delta\Delta F_3(x)$$

and substituting X_3 for z in Eq. (e), we obtain

$$\frac{4}{x}\frac{d^3X_1}{dx^3} + \log x(\Delta\Delta X_1 + X_1) + \Delta\Delta F_3(x) + F_3(x) = 0$$

Since X_1 satisfies Eq. (e) and is represented by the first of the series (j), we obtain the following equation for determining $F_3(x)$:

$$\Delta\Delta F_3(x) + F_3(x) = -\frac{4}{x}\frac{d^3X_1}{dx^3} = -4\left(-\frac{2\cdot 3\cdot 4}{2^2\cdot 4^2}\right.$$
$$\left. + \frac{6\cdot 7\cdot 8\cdot x^4}{2^2\cdot 4^2\cdot 6^2\cdot 8^2} - \frac{10\cdot 11\cdot 12\cdot x^8}{2^2\cdot 4^2\cdot 6^2\cdot 8^2\cdot 10^2\cdot 12^2} + \cdots\right) \quad (l)$$

Taking $F_3(x)$ in the form of the series

$$F_3(x) = b_4 x^4 + b_8 x^8 + b_{12} x^{12} + \cdots \quad (m)$$

and substituting this series in Eq. (l), we determine the coefficients b_4, b_8, b_{12}, . . . so that the resulting equation will be satisfied. Observing that

$$\Delta\Delta(b_4 x^4) = 4^2 \cdot 2^2 \cdot b_4$$

we find, by equating to zero the sum of the terms that do not contain x, that

$$4^2 \cdot 2^2 \cdot b_4 = 4 \cdot \frac{2\cdot 3\cdot 4}{2^2\cdot 4^2}$$

or

$$b_4 = \frac{2\cdot 3\cdot 4^2}{2^4\cdot 4^4} = \frac{3}{128}$$

Equating to zero the sum of the terms containing x^4, we find

$$b_8 = -\frac{25}{1{,}769{,}472}$$

In general, we find

$$b_n = (-1)^{n/4-1}\frac{1}{n^2(n-2)^2}\left[b_{n-4} + \frac{n(n-1)(n-2)}{2^2\cdot 4^2\cdot 6^2\cdots n^2}\right]$$

Thus the third particular solution of Eq. (e) is

$$X_3 = X_1 \log x + \frac{3}{128} x^4 - \frac{25}{1{,}769{,}472} x^8 + \cdots \quad (n)$$

The fourth particular integral X_4 of Eq. (e) is obtained in a similar manner by taking

$$X_4 = X_2 \log x + F_4(x) = X_2 \log x + 4 \cdot \frac{4\cdot 5\cdot 6}{4^4\cdot 6^4} x^6$$
$$- \frac{1}{10^2\cdot 8^2}\left(4\cdot \frac{4\cdot 5\cdot 6}{4^4\cdot 6^4} + \frac{10\cdot 9\cdot 8}{4^2\cdot 6^2\cdots 10^2}\right) x^{10} + \cdots \quad (o)$$

By substituting the particular solutions (j), (n), and (o) in expression (f) we obtain the general solution of Eq. (e) in the following form:

$$z = A_1 \left(1 - \frac{x^4}{2^2 \cdot 4^2} + \frac{x^8}{2^2 \cdot 4^2 \cdot 6^2 \cdot 8^2} - \cdots \right)$$
$$+ A_2 \left(x^2 - \frac{x^6}{4^2 \cdot 6^2} + \frac{x^{10}}{4^2 \cdot 6^2 \cdot 8^2 \cdot 10^2} - \cdots \right)$$
$$+ A_3 \left[\left(1 - \frac{x^4}{2^2 \cdot 4^2} + \frac{x^8}{2^2 \cdot 4^2 \cdot 6^2 \cdot 8^2} - \cdots \right) \log x + \frac{3}{128} x^4 \right.$$
$$\left. - \frac{25}{1{,}769{,}472} x^8 + \cdots \right] + A_4 \left[\left(x^2 - \frac{x^6}{4^2 \cdot 6^2} + \frac{x^{10}}{4^2 \cdot 6^2 \cdot 8^2 \cdot 10^2} \right.\right.$$
$$\left.\left. - \cdots \right) \log x + \frac{5}{3{,}456} x^6 - \frac{1{,}054 \cdot 10^{-4}}{442{,}368} x^{10} + \cdots \right] \quad (p)$$

It remains now to determine in each particular case the constants of integration $A_1, \ldots, A_4$ so as to satisfy the boundary conditions.

Let us consider the case in which the edge of a circular plate of radius a is entirely free. Making use of expression (52) for the radial moments and expression (55) for the radial shear force Q_r, we write the boundary conditions as

$$\left(\frac{d^2w}{dr^2} + \nu \frac{1}{r} \frac{dw}{dr}\right)_{r=a} = 0$$
$$\frac{d}{dr}\left(\frac{d^2w}{dr^2} + \frac{1}{r} \frac{dw}{dr}\right)_{r=a} = 0 \quad (q)$$

In addition to these two conditions we have two more conditions that hold at the center of the plate; viz., the deflection at the center of the plate must be finite, and the sum of the shearing forces distributed over the lateral surface of an infinitesimal circular cylinder cut out of the plate at its center must balance the concentrated force P. From the first of these two conditions it follows that the constant A_3 in the general solution (p) vanishes. The second condition gives

$$\left(\int_0^{2\pi} Q_r r \, d\theta\right)_{r=\epsilon} + P = 0 \quad (r)$$

or, by using notation (a),

$$-kl^4 \frac{d}{dr}\left(\frac{d^2w}{dr^2} + \frac{1}{r}\frac{dw}{dr}\right)_{r=\epsilon} 2\pi\epsilon + P = 0 \quad (s)$$

where ϵ is the radius of the infinitesimal cylinder. Substituting lz for w in this equation and using for z expression (p), we find that for an infinitely small value of x equal to ϵ/l the equation reduces to

$$-kl^4 \frac{4A_4}{l\epsilon} 2\pi\epsilon + P = 0$$

from which

$$A_4 = \frac{P}{8\pi k l^3} \qquad (t)$$

Having the values of the constants A_3 and A_4, the remaining two constants A_1 and A_2 can be found from Eqs. (q). For given dimensions of the plate and given moduli of the plate and of the foundation these equations furnish two linear equations in A_1 and A_2.

Let us take, as an example, a plate of radius $a = 5$ in. and of such rigidity that

$$l = \sqrt[4]{\frac{D}{k}} = 5 \text{ in.}$$

We apply at the center a load P such that

$$A_4 = \frac{P}{8\pi k l^3} = 102 \cdot 10^{-5}$$

Using this value of A_4 and substituting lz for w, we find, by using expression (p) and taking $x = a/l = 1$, that Eqs. (q) give

$$0.500 A_1 + 0.250 A_2 = 4.062 A_4 = 4.062 \cdot 102 \cdot 10^{-5}$$
$$0.687 A_1 - 8.483 A_2 = 11.09 A_4 = 11.09 \cdot 102 \cdot 10^{-5}$$

These equations give

$$A_1 = 86 \cdot 10^{-4} \qquad A_2 = -64 \cdot 10^{-5}$$

Substituting these values in expression (p) and retaining only the terms that contain x to a power not larger than the fourth, we obtain the following expression for the deflection:

$$w = lz = 5 \left[86 \cdot 10^{-4} \left(1 - \frac{x^4}{2^2 \cdot 4^2} \right) - 64 \cdot 10^{-5} x^2 + 102 \cdot 10^{-5} x^2 \log x \right]$$

The deflection at the center $(x = 0)$ is then

$$w_{\max} = 43 \cdot 10^{-3} \text{ in.}$$

and the deflection at the boundary $(x = 1)$ is

$$w_{\min} = 39.1 \cdot 10^{-3} \text{ in.}$$

The difference of these deflections is comparatively small, and the pressure distribution over the foundation differs only slightly from a uniform distribution.

If we take the radius of the plate two times larger ($a = 10$ in.) and retain the previous values for the rigidities D and k, x becomes equal to 2 at the boundary, and Eqs. (q) reduce to

$$0.826 A_1 + 1.980 A_2 = 1.208 A_4$$
$$2.665 A_1 - 5.745 A_2 = 16.37 A_4$$

These equations give

$$A_1 = 3.93 A_4 = 400 \cdot 10^{-5} \qquad A_2 = -1.03 A_4 = -105 \cdot 10^{-5} \qquad (u)$$

The deflection is obtained from expression (p) as

$$w = lz = 5\left\{400\cdot 10^{-5}\left(1 - \frac{x^4}{2^2\cdot 4^2}\right) - 105\cdot 10^{-5}\left(x^2 - \frac{x^6}{576}\right)\right.$$
$$\left. + 102\cdot 10^{-5}\left[\log x\left(x^2 - \frac{x^6}{4^2\cdot 6^2}\right) + \frac{5}{3,456}x^6\right]\right\}$$

The deflections at the center and at the boundary of the plate are, respectively,

$$w_{\max} = 2.10^{-2} \text{ in.} \quad \text{and} \quad w_{\min} = 0.88\cdot 10^{-2} \text{ in.}$$

It is thus seen that, if the radius of the plate is twice as large as the quantity l, the distribution of pressure over the foundation is already far from uniform. The application of the strain energy method to the problem of bending of a plate on elastic subgrade will be shown in Art. 80.

58. Application of Bessel Functions to the Problem of the Circular Plate. The general solution (f) of Eq. (e) in the preceding article can also be represented in terms of Bessel functions. To this end we introduce into Eq. (e) a new variable $\xi = x\sqrt{i}$; thus we arrive at the equation

$$\Delta'\Delta'z - z = 0 \qquad (a)$$

in which the symbol Δ' stands for

$$\frac{d^2}{d\xi^2} + \frac{1}{\xi}\frac{d}{d\xi}$$

Now Eq. (a) is equivalent to equation

$$\Delta'(\Delta'z + z) - (\Delta'z + z) = 0 \qquad (b)$$

and also to

$$\Delta'(\Delta'z - z) + (\Delta'z - z) = 0 \qquad (c)$$

Hence Eq. (a) is satisfied by the solutions of the Bessel differential equation

$$\Delta'z + z = \frac{d^2z}{d\xi^2} + \frac{1}{\xi}\frac{dz}{d\xi} + z = 0 \qquad (d)$$

as well as by the solutions of the equation

$$\Delta'z - z = \frac{d^2z}{d\xi^2} + \frac{1}{\xi}\frac{dz}{d\xi} - z = 0 \qquad (e)$$

which is transformable into Eq. (d) by substituting ξi for ξ. Thus the combined solution of Eqs. (d) and (e) can be written as

$$z = B_1 I_0(x\sqrt{i}) + B_2 I_0(xi\sqrt{i}) + B_3 K_0(x\sqrt{i}) + B_4 K_0(xi\sqrt{i}) \qquad (f)$$

I_0 and K_0 being Bessel functions of the first and second kind, respectively, and of imaginary argument, whereas $B_1, B_2, \ldots$ are arbitrary constants. The argument x being real, all functions contained in Eq. (f) appear in a complex form. To single out the real part of the solution, it is convenient to introduce four other functions, first used by Lord Kelvin and defined by the relations[1]

$$I_0(x\sqrt{\pm i}) = \text{ber } x \pm \text{bei } x$$
$$K_0(x\sqrt{\pm i}) = \text{ker } x \pm \text{kei } x \qquad (g)$$

---

[1] See, for instance, G. N. Watson, "Theory of Bessel Functions," p. 81, Cambridge, 1948.

Setting, furthermore,
$$B_1 + B_2 = C_1 l \quad B_1 - B_2 = -C_2 i l$$
$$B_3 + B_4 = C_4 l \quad B_3 - B_4 = -C_3 i l$$

where the new constants C_1, C_2, ... are real, we obtain the following expression for the deflections of the plate:

$$w = C_1 \operatorname{ber} x + C_2 \operatorname{bei} x + C_3 \operatorname{kei} x + C_4 \operatorname{ker} x \tag{h}$$

All functions herein contained are tabulated functions,[1] real for real values of the argument.

For small values of the argument we have

$$\begin{aligned}
\operatorname{ber} x &= 1 - x^4/64 + \cdots \\
\operatorname{bei} x &= x^2/4 - x^6/2{,}304 + \cdots \\
\operatorname{ker} x &= -\log x + \log 2 - \gamma + \pi x^2/16 + \cdots \\
\operatorname{kei} x &= -(x^2/4) \log x - \pi/4 + (1 + \log 2 - \gamma) x^2/4 + \cdots
\end{aligned} \tag{i}$$

in which $\gamma = 0.5772157 \cdots$ is Euler's constant and $\log 2 - \gamma = 0.11593 \cdots$. For large values of the argument the following asymptotic expressions hold:

$$\begin{aligned}
\operatorname{ber} x &\sim \frac{e^\sigma}{\sqrt{2\pi x}} \cos\left(\sigma - \frac{\pi}{8}\right) \\
\operatorname{bei} x &\sim \frac{e^\sigma}{\sqrt{2\pi x}} \sin\left(\sigma - \frac{\pi}{8}\right) \\
\operatorname{ker} x &\sim \frac{e^{-\sigma}}{\sqrt{2x/\pi}} \cos\left(\sigma + \frac{\pi}{8}\right) \\
\operatorname{kei} x &\sim -\frac{e^{-\sigma}}{\sqrt{2x/\pi}} \sin\left(\sigma + \frac{\pi}{8}\right)
\end{aligned} \tag{j}$$

in which $\sigma = x/\sqrt{2}$.

The general solution (h) can be used for the analysis of any symmetrical bending of a circular plate, with or without a hole, resting on an elastic foundation. The four constants C, corresponding in the most general case to four boundary conditions, must be determined in each particular case.[2]

[1] See "Tables of Bessel Functions $J_0(z)$ and $J_1(z)$ for Complex Arguments," Columbia University Press, New York, 1943, and "Tables of Bessel Functions $Y_0(z)$ and $Y_1(z)$ for Complex Arguments," Columbia University Press, New York, 1950. We have

$$\operatorname{ber} x = \operatorname{Re}\left[J_0(xe^{i\pi/4})\right] \quad \operatorname{bei} x = -\operatorname{Im}\left[J_0(xe^{i\pi/4})\right]$$

$$\operatorname{ker} x = -\frac{\pi}{2} \operatorname{Re}\left[Y_0(xe^{i\pi/4})\right] - \frac{\pi}{2} \operatorname{Im}\left[J_0(xe^{i\pi/4})\right]$$

$$\operatorname{kei} x = \frac{\pi}{2} \operatorname{Im}\left[Y_0(xe^{i\pi/4})\right] - \frac{\pi}{2} \operatorname{Re}\left[J_0(xe^{i\pi/4})\right]$$

[2] Many particular solutions of this problem are given by F. Schleicher in his book "Kreisplatten auf elastischer Unterlage," Berlin, 1926, which also contains tables of functions $Z_1(x) = \operatorname{ber} x$, $Z_2(x) = -\operatorname{bei} x$, $Z_3(x) = -(2/\pi) \operatorname{kei} x$, and $Z_4(x) = -(2/\pi) \operatorname{ker} x$ as well as the first derivatives of those functions. An abbreviated table of the functions Z and their first derivatives is given in Art. 118, where they are denoted by the symbol ψ.

We shall confine ourselves to the case of an infinitely extended plate carrying a single load P at the point $x = 0$. Now, from the four functions forming solution (h), the first two functions increase indefinitely with increasing argument in accordance with Eqs. (j); and the function ker x becomes infinitely large at the origin, as we can conclude from Eqs. (i). Accordingly, setting $C_1 = C_2 = C_4 = 0$, solution (h) is reduced to

$$w = C_3 \text{ kei } x \tag{k}$$

In order to determine the constant C_3, we calculate, by means of Eqs. (i), the shearing force [see Eqs. (193)]

$$Q_r = -\frac{D}{l^3}\frac{d}{dx}\left(\frac{d^2w}{dx^2} + \frac{1}{x}\frac{dw}{dx}\right) = \frac{C_3 D}{l^3}\left(\frac{1}{x} - \frac{\pi x}{8} + \cdots\right)$$

As x decreases, the value of Q_r tends to $C_3 D/l^3 x = C_3 D/l^2 r$. On the other hand, upon distributing the load P uniformly over the circumference with radius r, we have $Q_r = -P/2\pi r$. Equating both expressions obtained for Q_r, we have

$$C_3 = -\frac{Pl^2}{2\pi D} \tag{l}$$

Substitution of C_3 into Eq. (k) yields, finally, the complete solution of Hertz's problem in the form

$$w = -\frac{Pl^2}{2\pi D} \text{ kei } x \tag{179}$$

and the corresponding reaction of the subgrade is given by $p = kw = \dfrac{wD}{l^4}$. The variation of these quantities along a meridional section through the deflection surface of the plate is shown in Fig. 131, together with similar curves based on a theory which will be discussed in Art. 61.

At the origin we have kei $x = -\pi/4$ and the deflection under the load becomes

$$w_{\max} = \frac{Pl^2}{8D} \tag{180}$$

For the reaction of the subgrade at the same point we obtain

$$p_{\max} = \frac{P}{8l^2} \tag{181}$$

If we take an infinitely large plate with the conditions of rigidity and loading assumed on page 264, the deflection under the load becomes

$$w_{\max} = \frac{Pl^2}{8D} = \frac{P}{8kl^2} = \pi l A_4 = (3.14)(5)(102 \cdot 10^{-5}) = 0.016 \text{ in.}$$

as compared with the value of 0.02 in. obtained for a finite circular plate with the radius $a = 2l$.

The distribution of the bending moments due to the concentrated load is shown in

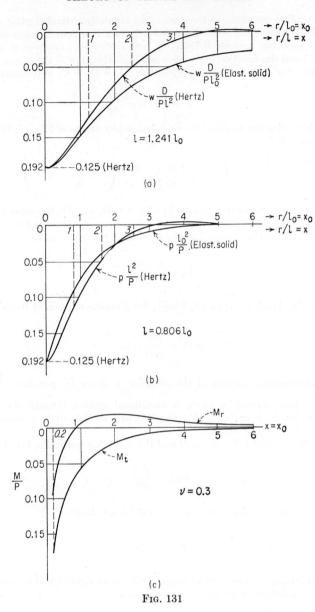

Fig. 131

Fig. 131c. It is seen that the radial moments become negative at some distance from the load, their numerically largest value being about $-0.02P$. The positive moments are infinitely large at the origin, but at a small[1] distance from the point of application of the load they can be easily calculated by taking the function kei x in the form (i). Upon applying formulas (52) and (53) to expression (179), we arrive at the results

[1] As compared with the characteristic length $l = \sqrt[4]{D/k}$.

$$M_r = \frac{P}{4\pi}\left[(1+\nu)\left(\log\frac{2l}{r} - \gamma\right) - \frac{1}{2}(1-\nu)\right]$$
$$M_t = \frac{P}{4\pi}\left[(1+\nu)\left(\log\frac{2l}{r} - \gamma\right) + \frac{1}{2}(1-\nu)\right] \quad (182)$$

A comparison of the foregoing expressions with Eqs. (90) and (91) shows that the stress condition in a plate in the vicinity of the load in Hertz's case is identical with that of a simply supported circular plate with a radius $a = 2le^{-\gamma} = 1.123l$, except for a moment $M'_r = M'_t = -\frac{P}{8\pi}(1-\nu)$, which is superimposed on the moments of the circular plate.

Let us consider now the case in which the load P is distributed over the area of a circle with a radius c, small in comparison with l. The bending moments at the center of a circular plate carrying such a load are

$$M_r = M_t = \frac{P}{4\pi}\left[(1+\nu)\log\frac{a}{c} + 1\right] \quad (m)$$

This results from Eq. (83), if we neglect there the term c^2/a^2 against unity. By substituting $a = 2le^{-\gamma}$ into Eq. (m) and adding the moment $-P/8\pi(1-\nu)$, we obtain at the center of the loaded circle of the infinitely large plate the moments

$$M_{max} = \frac{(1+\nu)P}{4\pi}\left(\log\frac{2l}{c} - \gamma + \frac{1}{2}\right) \quad (n)$$

or

$$M_{max} = \frac{(1+\nu)P}{4\pi}\left(\log\frac{l}{c} + 0.616\right) \quad (183)$$

Stresses resulting from Eq. (183) must be corrected by means of the thick-plate theory in the case of a highly concentrated load. Such a corrected stress formula is given on page 275.

In the case of a load uniformly distributed over the area of a small rectangle, we may proceed as described in Art. 37. The equivalent of a square area, in particular, is a circle with the radius $c = 0.57u$, u being the length of the side of the square (see page 162). Substituting this into Eq. (183) we obtain

$$M_{max} = \frac{1+\nu}{4\pi}P\left(\log\frac{l}{u} + 1.177\right) \quad (o)$$

The effect of any group of concentrated loads on the deflections of the infinitely large plate can be calculated by summing up the deflections produced by each load separately.

59. Rectangular and Continuous Plates on Elastic Foundation.

An example of a plate resting on elastic subgrade and supported at the same time along a rectangular boundary is shown in Fig. 132, which represents a beam of a rectangular tubular cross section pressed into an elastic foundation by the loads P. The bottom plate of the beam, loaded by the elastic reactions of the foundation, is supported by the vertical sides of the tube and by the transverse diaphragms indicated in the figure by dashed lines. It is assumed again that the intensity of the reaction p at any point of the bottom plate is proportional to the deflection w at that point, so that $p = kw$, k being the modulus of the foundation.

In accordance with this assumption, the differential equation for the deflection, written in rectangular coordinates, becomes

$$\frac{\partial^4 w}{\partial x^4} + 2 \frac{\partial^4 w}{\partial x^2 \, \partial y^2} + \frac{\partial^4 w}{\partial y^4} = \frac{q}{D} - \frac{kw}{D} \quad (a)$$

where q, as before, is the intensity of the lateral load.

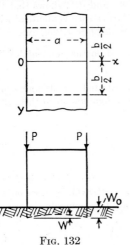

Fig. 132

Let us begin with the case shown in Fig. 132. If w_0 denotes the deflection of the edges of the bottom plate, and w the deflection of this plate with respect to the plane of its boundary, the intensity of the reaction of the foundation at any point is $k(w_0 - w)$, and Eq. (a) becomes

$$\Delta \Delta w = \frac{k}{D}(w_0 - w) \quad (b)$$

Taking the coordinate axes as shown in the figure and assuming that the edges of the plate parallel to the y axis are simply supported and the other two edges are clamped, the boundary conditions are

$$(w)_{x=0, x=a} = 0 \qquad \left(\frac{\partial^2 w}{\partial x^2}\right)_{x=0, x=a} = 0 \quad (c)$$

$$(w)_{y=\pm b/2} = 0 \qquad \left(\frac{\partial w}{\partial y}\right)_{y=\pm b/2} = 0 \quad (d)$$

The deflection w can be taken in the form of a series:

$$w = \frac{4kw_0}{D\pi} \sum_{m=1,3,5,\ldots}^{\infty} \frac{\sin \frac{m\pi x}{a}}{m\left(\frac{m^4 \pi^4}{a^4} + \frac{k}{D}\right)} + \sum_{m=1,3,5,\ldots}^{\infty} Y_m \sin \frac{m\pi x}{a} \quad (e)$$

The first series on the right-hand side is a particular solution of Eq. (b) representing the deflection of a simply supported strip resting on an elastic foundation. The second series is the solution of the homogeneous equation

$$\Delta \Delta w + \frac{k}{D} w = 0 \quad (f)$$

Hence the functions Y_m have to satisfy the ordinary differential equation

$$Y_m^{IV} - 2 \frac{m^2 \pi^2}{a^2} Y_m'' + \left(\frac{m^4 \pi^4}{a^4} + \frac{k}{D}\right) Y_m = 0 \quad (g)$$

Using notations

$$\frac{m\pi}{a} = \mu_m \qquad \frac{k}{D} = \lambda^4 \qquad (h)$$

$$2\beta_m^2 = \sqrt{\mu_m^4 + \lambda^4} + \mu_m^2 \qquad 2\gamma_m^2 = \sqrt{\mu_m^4 + \lambda^4} - \mu_m^2 \qquad (i)$$

and taking the solution of Eq. (g) in the form e^{ry}, we obtain for r the following four roots:

$$\beta + i\gamma \qquad -\beta + i\gamma \qquad \beta - i\gamma \qquad -\beta - i\gamma$$

The corresponding four independent particular solutions of Eq. (g) are

$$e^{\beta_m y} \cos \gamma_m y \qquad e^{-\beta_m y} \cos \gamma_m y \qquad e^{\beta_m y} \sin \gamma_m y \qquad e^{-\beta_m y} \sin \gamma_m y \qquad (j)$$

which can be taken also in the following form:

$$\begin{array}{ll} \cosh \beta_m y \cos \gamma_m y & \sinh \beta_m y \cos \gamma_m y \\ \cosh \beta_m y \sin \gamma_m y & \sinh \beta_m y \sin \gamma_m y \end{array} \qquad (k)$$

From symmetry it can be concluded that Y_m in our case is an even function of y. Hence, by using integrals (k), we obtain

$$Y_m = A_m \cosh \beta_m y \cos \gamma_m y + B_m \sinh \beta_m y \sin \gamma_m y$$

and the deflection of the plate is

$$w = \sum_{m=1,3,5,\ldots}^{\infty} \sin \frac{m\pi x}{a} \left[\frac{4kw_0}{D\pi} \frac{1}{m\left(\frac{m^4\pi^4}{a^4} + \frac{k}{D}\right)} \right.$$
$$\left. + A_m \cosh \beta_m y \cos \gamma_m y + B_m \sinh \beta_m y \sin \gamma_m y \right] \qquad (l)$$

This expression satisfies the boundary conditions (c). To satisfy conditions (d) we must choose the constants A_m and B_m so as to satisfy the equations

$$\frac{4kw_0}{D\pi} \frac{1}{m\left(\frac{m^4\pi^4}{a^4} + \frac{k}{D}\right)} + A_m \cosh \frac{\beta_m b}{2} \cos \frac{\gamma_m b}{2}$$
$$+ B_m \sinh \frac{\beta_m b}{2} \sin \frac{\gamma_m b}{2} = 0 \qquad (m)$$

$$(A_m \beta_m + B_m \gamma_m) \sinh \frac{\beta_m b}{2} \cos \frac{\gamma_m b}{2}$$
$$- (A_m \gamma_m - B_m \beta_m) \cosh \frac{\beta_m b}{2} \sin \frac{\gamma_m b}{2} = 0$$

Substituting these values of A_m and B_m in expression (l), we obtain the required deflection of the plate.

The problem of the plate with all four edges simply supported can be solved by using Eq. (a). Taking the coordinate axes as shown in Fig. 59

(page 105) and using the Navier solution, the deflection of the plate is

$$w = \sum_{m=1}^{\infty}\sum_{n=1}^{\infty} A_{mn} \sin\frac{m\pi x}{a} \sin\frac{n\pi y}{b} \qquad (n)$$

In similar manner let the series

$$q = \sum_{m=1}^{\infty}\sum_{n=1}^{\infty} a_{mn} \sin\frac{m\pi x}{a} \sin\frac{n\pi y}{b} \qquad (o)$$

represent the distribution of the given load, and the series

$$p = kw = \sum\sum k A_{mn} \sin\frac{m\pi x}{a} \sin\frac{n\pi y}{b} \qquad (p)$$

represent the reaction of the subgrade. Substituting the series (n) in the left-hand side and the series (o) and (p) in the right-hand side of Eq. (a), we obtain

$$A_{mn} = \frac{a_{mn}}{\pi^4 D \left(\frac{m^2}{a^2} + \frac{n^2}{b^2}\right)^2 + k} \qquad (q)$$

As an example, let us consider the bending of the plate by a force P concentrated at some point (ξ,η). In such a case

$$a_{mn} = \frac{4P}{ab} \sin\frac{m\pi\xi}{a} \sin\frac{n\pi\eta}{b} \qquad (r)$$

by Eq. (b) on page 111. By substitution of expressions (q) and (r) into Eq. (n) we finally obtain

$$w = \frac{4P}{ab} \sum_{m=1}^{\infty}\sum_{n=1}^{\infty} \frac{\sin\frac{m\pi\xi}{a}\sin\frac{n\pi\eta}{b}}{\pi^4 D\left(\frac{m^2}{a^2}+\frac{n^2}{b^2}\right)^2 + k} \sin\frac{m\pi x}{a} \sin\frac{n\pi y}{b} \qquad (s)$$

Having the deflection of the plate produced by a concentrated force, the deflection produced by any kind of lateral loading is obtained by the method of superposition. Take, as an example, the case of a uniformly distributed load of the intensity q. Substituting $q\,d\xi\,d\eta$ for P in expression (s) and integrating between the limits 0 and a and between 0 and b, we obtain

$$w = \frac{16q}{\pi^2} \sum_{m=1,3,5,\ldots}^{\infty} \sum_{n=1,3,5,\ldots}^{\infty} \frac{\sin\frac{m\pi x}{a}\sin\frac{n\pi y}{b}}{mn\left[\pi^4 D\left(\frac{m^2}{a^2}+\frac{n^2}{b^2}\right)^2 + k\right]} \qquad (t)$$

When k is equal to zero, this deflection reduces to that given in Navier solution (131) for the deflection of a uniformly loaded plate.[1]

Let us consider now the case represented in Fig. 133. A large plate which rests on an elastic foundation is loaded at equidistant points along the x axis by forces P.* We shall take the coordinate axes as shown in

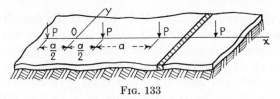

FIG. 133

the figure and use Eq. (f), since there is no distributed lateral load. Let us consider a solution of this equation in the form of the series

$$w = w_0 + \sum_{m=2,4,6,\ldots}^{\infty} Y_m \cos \frac{m\pi x}{a} \qquad (u)$$

in which the first term

$$w_0 = \frac{P\lambda}{2\sqrt{2}\,ak} e^{-\lambda y/\sqrt{2}} \left(\cos \frac{\lambda y}{\sqrt{2}} + \sin \frac{\lambda y}{\sqrt{2}} \right)$$

represents the deflection of an infinitely long strip of unit width parallel to the y axis loaded at $y = 0$ by a load P/a [see Eq. (283), page 471]. The other terms of the series satisfy the requirement of symmetry that the tangent to the deflection surface in the x direction shall have a zero slope at the loaded points and at the points midway between the loads. We take for functions Y_m those of the particular integrals (j) which vanish for infinite values of y. Hence,

$$Y_m = A_m e^{-\beta_m y} \cos \gamma_m y + B_m e^{-\beta_m y} \sin \gamma_m y$$

To satisfy the symmetry condition $(\partial w/\partial y)_{y=0} = 0$ we must take in this expression

$$B_m = \frac{\beta_m A_m}{\gamma_m}$$

---

[1] The case of a rectangular plate with prescribed deflections and moments on two opposite edges and various boundary conditions on two others was discussed by H. J. Fletcher and C. J. Thorne, *J. Appl. Mechanics*, vol. 19, p. 361, 1952. Many graphs are given in that paper.

* This problem has been discussed by H. M. Westergaard; see *Ingeniøren*, vol. 32, p. 513, 1923. Practical applications of the solution of this problem in concrete road design are discussed by H. M. Westergaard in the journal *Public Roads*, vol. 7, p. 25, 1926; vol. 10, p. 65, 1929; and vol. 14, p. 185, 1933.

Hence, by introducing the new constants $A'_m = A_m/\gamma_m$, we represent the deflections (u) in the following form:

$$w = w_0 + \sum_{m=2,4,6,\ldots}^{\infty} A'_m \cos \frac{m\pi x}{a} e^{-\beta_m y} (\gamma_m \cos \gamma_m y + \beta_m \sin \gamma_m y) \quad (v)$$

In order to express the constants A'_m in terms of the magnitude of loads P, we consider the shearing force Q_y acting along the normal section of the plate through the x axis. From symmetry we conclude that this force vanishes at all points except the points of application of the loads P, at which points the shearing forces must give resultants equal to $-P/2$. It was shown in the discussion of a similar distribution of shearing forces in Art. 54 (see page 248) that the shear forces can be represented by the series

$$Q_y = -\frac{P}{2a} - \frac{P}{a} \sum_{m=2,4,6,\ldots}^{\infty} (-1)^{m/2} \cos \frac{m\pi x}{a}$$

The shearing force, as calculated from expression (v), is

$$Q_y = -D \frac{\partial}{\partial y}\left(\frac{\partial^2 w}{\partial x^2} + \frac{\partial^2 w}{\partial y^2}\right)_{y=0}$$

$$= -\frac{P}{2a} - 2D \sum_{m=2,4,6,\ldots}^{\infty} A'_m \beta_m \gamma_m (\beta_m^2 + \gamma_m^2) \cos \frac{m\pi x}{a}$$

Comparing these two expressions for the shearing force, we find

$$A'_m = \frac{P(-1)^{m/2}}{2aD\beta_m\gamma_m(\beta_m^2 + \gamma_m^2)}$$

or, by using notations (i),

$$A'_m = \frac{P(-1)^{m/2}}{aD\lambda \sqrt{\lambda_m^4 + \mu_m^4}}$$

Substituting this in expression (v), we finally obtain

$$w = w_0 + \frac{P\lambda^2}{ak} \sum_{m=2,4,6,\ldots}^{\infty} \frac{(-1)^{m/2}}{\sqrt{\lambda^4 + \mu_m^4}} \cos \frac{m\pi x}{a} e^{-\beta_m y}(\gamma_m \cos \gamma_m y$$
$$+ \beta_m \sin \gamma_m y) \quad (w)$$

The maximum deflection is evidently under the loads P and is obtained by substituting $x = a/2$, $y = 0$ in expression (w), which gives

$$w_{\max} = \frac{P\lambda}{2\sqrt{2}\,ak} + \frac{P\lambda^2}{ak} \sum_{m=2,4,6,\ldots}^{\infty} \frac{\gamma_m}{\sqrt{\lambda^4 + \mu_m^4}} \quad (184)$$

PLATES ON ELASTIC FOUNDATION 275

The deflection in the particular case of one isolated load P acting on an infinitely large plate can also be obtained by setting $a = \infty$ in formula (184). In such a case the first term in the formula vanishes, and by using notations (i) we obtain

$$w_{\max} = \frac{P\lambda^2}{2\sqrt{2}\,\pi k} \sum_{m=2,4,6,\ldots} \frac{2\pi}{a} \sqrt{\frac{\sqrt{\lambda^4 + \mu_m^4} - \mu_m^2}{\lambda^4 + \mu_m^4}}$$

$$= \frac{P\lambda^2}{2\sqrt{2}\,\pi k} \int_0^\infty \sqrt{\frac{\sqrt{\lambda^4 + \mu^4} - \mu^2}{\lambda^4 + \mu^4}}\, d\mu$$

Using the substitution

$$\frac{\mu^2}{\lambda^2} = \frac{1}{2u\sqrt{u^2 + 1}}$$

we find

$$w_{\max} = \frac{P\lambda^2}{2\sqrt{2}\,\pi k} \int_0^\infty \frac{1}{\sqrt{2}} \frac{du}{1 + u^2} \approx \frac{P\lambda^2}{8k} \quad (185)$$

in accordance with the result (180). With this magnitude of the deflection, the maximum pressure on the elastic foundation is

$$(p)_{\max} = kw_{\max} = \frac{P\lambda^2}{8} = \frac{P}{8}\sqrt{\frac{k}{D}} \quad (186)$$

The maximum tensile stress is at the bottom of the plate under the point of application of the load. The theory developed above gives an infinite value for the bending moment at this point, and recourse should be had to the theory of thick plates (see Art. 26). In the above-mentioned investigation by Westergaard the following formula for calculating maximum tensile stress at the bottom of the plate is established by using the thick-plate theory:

$$(\sigma_r)_{\max} = 0.275(1 + \nu)\frac{P}{h^2}\log_{10}\frac{Eh^3}{kb^4} \quad (x)$$

Here h denotes the thickness of the plate, and

$$b = \sqrt{1.6c^2 + h^2} - 0.675h \quad \text{when } c < 1.724h$$
$$= c \quad \text{when } c > 1.724h$$

where c is the radius of the circular area over which the load P is assumed to be uniformly distributed. For $c = 0$ the case of the concentrated force is obtained.

In the case of a square loaded area u by u, we have to replace c by $0.57u$ (see page 162).

The case of equidistant loads P applied along the edge of a semi-infinite plate, as shown in Fig. 134, can also be treated in a similar way. The final formula for the maximum tensile stress at the bottom of the plate under the load when the distance a is large is

$$(\sigma_x)_{\max} = 0.529(1 + 0.54\nu)\frac{P}{h^2}\left[\log_{10}\left(\frac{Eh^3}{kb^4}\right) - 0.71\right] \qquad (y)$$

where b is calculated as in the previous case, and c is the radius of the semicircular area over which the load P is assumed to be uniformly distributed. Formulas (x) and (y) have proved very useful in the design of concrete roads, in which case the circle of radius c represents the area of contact of the wheel tire with the road surface.[1]

60. Plate Carrying Rows of Equidistant Columns.
As a last example, let us consider an infinite plate or mat resting on elastic subgrade and carrying equidistant and equal loads P, each load being distributed uniformly over the area u by v of a rectangle, as shown in Fig. 135. The

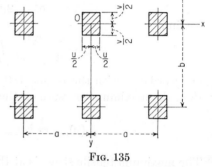

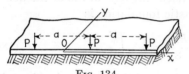

Fig. 134

Fig. 135

bending of such a "reversed flat slab" may be treated by means of the previously discussed Westergaard's solution, using simple series.[2] Much simpler, however, and, except for the case of a highly concentrated load, also adequate is the solution in double series, making use of Navier's method.

Conditions of symmetry compel us to represent the lateral load due to the columns in form of a cosine series:

$$q = \sum_{m=0}^{\infty}\sum_{n=0}^{\infty} a_{mn} \cos\frac{2m\pi x}{a} \cos\frac{2n\pi y}{b} \qquad (a)$$

The intensity of the given load is equal to P/uv within the shadowed rectangles in Fig. 135 and is zero elsewhere. Thus, proceeding in the usual manner, i.e., multiplying Eq. (a) by $\cos\dfrac{2m\pi x}{a}\cos\dfrac{2n\pi y}{b}\,dx\,dy$ and integrating between the limits $-a/2$, $+a/2$

---

[1] The problem of stress distribution near the load applied at a corner of a large plate has not yet been solved with the same reliability as the problems discussed above. Several empiric and semiempiric stress formulas regarding that case may be found in "Concrete Pavement Design," p. 79, Portland Cement Association, Chicago, 1951. Noteworthy experimental results concerning this problem were obtained by M. Dantu, Ann. ponts et chaussées, vol. 122, p. 337, 1952. See also L. D. Black, Trans. Eng. Inst. Canada, vol. 2, p. 129, 1958, and D. E. Nevel, ibid., p. 132.

[2] See W. Müller, Ingr.-Arch., vol. 20, p. 278, 1952, and Österr. Ingr.-Arch., vol. 6, p. 404, 1952.

for x; $-b/2$, $+b/2$ for y, we have

$$a_{mn} = \frac{4P\epsilon_{mn}}{\pi^2 mnuv} \sin \frac{m\pi u}{a} \sin \frac{n\pi v}{b} \qquad (b)$$

where $\epsilon_{mn} = 1$ for $m \neq 0, n \neq 0$
$\epsilon_{mn} = \frac{1}{2}$ for $m = 0, n \neq 0$ or $m \neq 0, n = 0$
$\epsilon_{mn} = \frac{1}{4}$ for $m = n = 0$

In the particular case of $m = 0$ or $n = 0$ the coefficient itself is readily obtained as a limit value of the expression (b).

Now, in accordance with Eq. (a) we take for deflections the series

$$w = \sum_{m=0}^{\infty} \sum_{n=0}^{\infty} A_{mn} \cos \frac{2m\pi x}{a} \cos \frac{2n\pi y}{b} \qquad (c)$$

and the relation between the coefficients a_{mn} and A_{mn} is easily established by the same reasoning as before (see page 272). Thus, using the notation

$$\alpha_m = \frac{2m\pi}{a} \qquad \beta_n = \frac{2n\pi}{b} \qquad \gamma_{mn}^2 = \alpha_m^2 + \beta_n^2 \qquad (d)$$

we obtain

$$A_{mn} = \frac{a_{mn}}{D\gamma_{mn}^4 + k} \qquad (e)$$

Substituting this in the series (c) and observing Eq. (b) we have the final result[1]

$$w = \frac{4P}{\pi^2 uv} \sum_{m=0}^{\infty} \sum_{n=0}^{\infty} \frac{\epsilon_{mn} \sin \dfrac{m\pi u}{a} \sin \dfrac{n\pi v}{b} \cos \alpha_m x \cos \beta_n y}{mn(D\gamma_{mn}^4 + k)} \qquad (f)$$

The bending moments of the plate are now obtained by the usual differentiation, and the distribution of the pressure between the plate and the subgrade is found by multiplication of expression (f) by the modulus k.

The particular case $k = 0$ corresponds to a uniformly distributed reaction of the subgrade, i.e., to the case of a "reversed flat slab" uniformly loaded with $q = P/ab$. It is seen from Eq. (f) that the introduction of the modulus tends to reduce the deflections and also the bending moments of the plate.

The case of a rectangular plate of finite dimensions resting on an elastic foundation and submitted to the action of a concentrated load has been discussed by H. Happel.[2] The Ritz method (see page 344) has been used to determine the deflections of this plate, and it was shown in the particular example of a centrally loaded square plate that the series representing the deflection converges rapidly and that the deflection can be calculated with sufficient accuracy by taking only the first few terms of the series.[3]

---

[1] Due to V. Lewe, *Bauingenieur*, vol. 3, p. 453, 1923.

[2] *Math. Z.*, vol. 6, p. 203, 1920. See also F. Halbritter, *Bautechnik*, vol. 26, p. 181, 1949.

[3] The problem of a square plate on an elastic foundation has also been investigated experimentally; see the paper by J. Vint and W. N. Elgood, *Phil. Mag.*, ser. 7, vol. 19, p. 1, 1935; and that by G. Murphy, *Iowa State Coll. Eng. Expt. Sta. Bull.* 135, 1937.

61. Bending of Plates Resting on a Semi-infinite Elastic Solid. So far, the settling of the subgrade at some point of its surface has been assumed as proportional to the pressure between the plate and the subgrade at the same point, and consequently as independent of the pressure elsewhere. This is correct in the case of a floating plate, considered by Hertz (see page 260), but in the case of a coherent subgrade such a hypothesis approximates but crudely the actual behavior of the subgrade; a better approximation can sometimes be obtained on the basis of the following assumptions:

1. The foundation has the properties of a semi-infinite elastic body.
2. The plate rests on the subgrade without friction.
3. A perfect contact between the plate and foundation also exists in the case of a negative mutual pressure.

This last supposition appears arbitrary; however, a negative pressure between plate and subgrade actually is compensated, more or less, by the weight of the plate.

The elastic properties of the elastic foundation may be characterized, if isotropy is assumed, by a Young modulus E_0 and a Poisson ratio ν_0. The approximate numerical values[1] of these constants, depending on the nature of the subgrade and based on results of dynamical tests, are given in Table 63, together with the value of the constant

$$k_0 = \frac{E_0}{2(1 - \nu_0^2)} \tag{a}$$

used in the following.

TABLE 63. VALUES OF ELASTIC CONSTANTS DEPENDING ON NATURE OF FOUNDATION

Subgrade	E_0, psi	ν_0	k_0, psi
Clay..................	11,000	0.17	5,700
Loess and clay...........	13,000	0.42	7,900
Medium sand...........	14,000–18,500	0.33–0.23	7,900–9,800
Sand and gravel.........	40,000	0.31	22,000
Liassic plastic clay.......	38,000	0.44	23,500
Lime (air-slaked)........	165,000–190,000	0.32–0.38	92,000–110,000
Sandstone...............	1,600,000	0.26	860,000

We restrict the further consideration to the case of an infinitely large plate in a state of axial symmetry. Using polar coordinates r, θ, we can write the plate equation as

$$D\Delta\Delta w(r) = q(r) - p(r) \tag{b}$$

where $q(r)$ denotes the given surface loading and $p(r)$ the reaction of the subgrade.

Let $K_0(r,\rho,\varphi)$ be the deflection at the point $(r,0)$ of the subgrade surface due to a normal unit load applied on this surface (ρ,φ). The form of the "influence function" K_0 depends merely upon the nature of the foundation. Making use of some properties of the Bessel functions, it can be shown[2] that Eq. (b) is satisfied by the expression

$$w(r) = \int_0^\infty \frac{Q(\alpha)K(\alpha)J_0(\alpha r)\alpha\, d\alpha}{1 + D\alpha^4 K(\alpha)} \tag{c}$$

[1] Due to E. Schultze and H. Muhs, "Bodenuntersuchungen für Ingenieurbauten," Berlin, 1950. See also *Veröffentl. Degebo*, Heft 4, p. 37, 1936.

[2] The solution of the problem in this general form is due to D. L. Holl, *Proc. Fifth Intern. Congr. Appl. Mech.*, Cambridge, Mass., 1938.

In Eq. (c) J_0 denotes the Bessel function of zero order; the term depending on the nature of the subgrade is

$$K(\alpha) = \int_0^\infty 2\pi s K_0(s) J_0(\alpha s)\, ds \tag{d}$$

in which the form of K_0 is defined by

$$K_0(s) = K_0[(r^2 + \rho^2 - 2r\rho \cos \varphi)^{\frac{1}{2}}]$$

s being the distance between points $(r,0)$ and (ρ,φ). Finally

$$Q(\alpha) = \int_0^\infty q(\rho) J_0(\alpha\rho) \rho\, d\rho \tag{e}$$

is the term depending on the intensity $q(\rho)$ of the symmetrical loading at $r = \rho$.

In the particular case of a load P uniformly distributed along the periphery of a circle with a radius c, we have

$$Q(\alpha) = \frac{P}{2\pi} J_0(\alpha c) \tag{f}$$

In the case of the load P distributed uniformly over the area of the same circle, Eq. (e) yields

$$Q(\alpha) = \frac{P}{\pi c \alpha} J_1(\alpha c) \tag{g}$$

where the Bessel function is of the order one. Finally, where a load is concentrated at the origin ($\rho = 0$), we obtain from Eq. (f)

$$Q(\alpha) = \frac{P}{2\pi} \tag{h}$$

As for the distribution of the reactive pressure, the respective function $p(r)$ is obtained from Eq. (b), the term

$$q(r) = \int_0^\infty Q(\alpha) J_0(\alpha r) \alpha\, d\alpha \tag{i}$$

being previously expressed through its Fourier-Bessel transform (e). Thus, we obtain

$$p(r) = \int_0^\infty \frac{Q(\alpha) J_0(\alpha r) \alpha\, d\alpha}{1 + D\alpha^4 K(\alpha)} \tag{j}$$

Now let us consider two particular cases with respect to the physical nature of the subgrade. For a *floating plate* (Art. 57) the influence function $K_0(s)$ is zero everywhere except at $s = 0$, where the unit force is applied. With regard to Eq. (d) the quantity $K_0(\alpha)$ then must be a constant. In order to get from Eq. (c) the expression $w(r) = p(r)/k$, this in accordance with the definition of the modulus, we have to assume $K_0(\alpha) = 1/k$. Using the previous notation $l^4 = D/k$ (page 260), we obtain from Eq. (c) the expression

$$w(r) = \frac{1}{k} \int_0^\infty \frac{Q(\alpha) J_0(\alpha r) \alpha\, d\alpha}{1 + \alpha^4 l^4} \tag{k}$$

which actually satisfies the differential equation (178) of the floating plate.

In the case of an isotropic semi-infinite medium we have, by a result due to Boussinesq,[1] $K_0(s) = (1 - \nu_0^2)/\pi E_0 s$ and, by Eq. (d), $K(\alpha) = 2(1 - \nu_0^2)/E_0 \alpha$, or

$$K(\alpha) = \frac{1}{(k_0 \alpha)}$$

where k_0 is the elastic constant defined by Eq. (a). Writing for brevity,

$$\frac{k_0}{D} = \frac{E_0}{2D(1 - \nu_0^2)} = \frac{1}{l_0^3} \tag{l}$$

we finally obtain the solution (c) in the more special form[2]

$$w(r) = \frac{1}{k_0} \int_0^\infty \frac{Q(\alpha) J_0(\alpha r) \, d\alpha}{1 + \alpha^3 l_0^3} \tag{m}$$

In the particular case of a load concentrated at the origin, expression (m) in connection with (h) yields

$$w = \frac{P l_0^2}{2\pi D} \int_0^\infty \frac{J_0\left(\frac{\lambda r}{l_0}\right) d\lambda}{1 + \lambda^3} \tag{187}$$

where λ is written for αl_0. Therefore, the deflection under the load is

$$w_{\max} = \frac{P l_0^2}{2\pi D} \int_0^\infty \frac{d\lambda}{1 + \lambda^3} = \frac{P l_0^2 \sqrt{3}}{9D} = 0.192 \frac{P l_0^2}{D} \tag{188}$$

against the result $0.125 P l^2/D$ of Hertz. The distribution of the pressure is readily obtained from the general expression (j). We have at any point

$$p = \frac{P}{2\pi l_0^2} \int_0^\infty \frac{J_0\left(\frac{\lambda r}{l_0}\right) \lambda \, d\lambda}{1 + \lambda^3} \tag{189}$$

and especially under the load

$$p_{\max} = \frac{P}{2\pi l_0^2} \int_0^\infty \frac{\lambda \, d\lambda}{1 + \lambda^3} = \frac{P \sqrt{3}}{9 l_0^2} = 0.192 \frac{P}{l_0^2} \tag{190}$$

in comparison with the value of $0.125 P/l^2$ obtained by Hertz. If we assume equal values of $w_{\max}$ in both cases, formula (190) yields a value for $p_{\max}$ which is 2.37 times as large as the value from Hertz's formula (181). In such a case the relation $l = 1.241 l_0$ must hold, and curves of the respective deflections as calculated from Eqs. (179) and (187) are shown in Fig. 131a. Figure 131b shows in like manner the variation of the pressure; this time, in order to obtain equal values for $p_{\max}$ in both cases, it must be assumed that $l = 0.806 l_0$.

It can be shown, finally, that the magnitude of bending moments in the vicinity of

[1] See, for example, S. Timoshenko and J. N. Goodier, "Theory of Elasticity," 2d ed., p. 365, New York, 1951.

[2] For this result see also S. Woinowsky-Krieger, *Ingr.-Arch.*, vol. 3, p. 250, 1932, and vol. 17, p. 142, 1949; K. Marguerre, *Z. angew. Math. Mech.*, vol. 17, p. 229, 1937; A. H. A. Hogg, *Phil. Mag.*, vol. 25, p. 576, 1938.

the concentrated load is the same for foundations of both kinds if expressed in terms of the dimensionless argument $x = r/l$ and $x = r/l_0$, respectively. We conclude from this fact that expressions for bending moments, such as given by Eq. (183), can also be used for a plate resting on an isotropic elastic medium if we replace l by l_0. Proceeding in this manner with the stress formula (x) of Westergaard (page 275), we arrive at the formula

$$\sigma_{max} = 0.366(1 + \nu)\frac{P}{h^2}\left[\log_{10}\left(\frac{Eh^3}{k_0 b^3}\right) - 0.266\right] \quad (n)$$

in which k_0 is given by Eq. (a), and b denotes the same quantity as on page 275.

The problem of the bending of a finite circular plate leads to an infinite set of linear equations for the coefficients of the series, which has to represent the deflections of such a plate.[1]

The use of the method of finite differences should also be considered in handling the problem of finite circular plates.[2]

The bending of an infinite plate supported by an elastic layer, which rests in its turn on a perfectly rigid base,[3] and the problem of a semi-infinite pavement slab[4] have also been discussed.

Stresses due to a highly concentrated surface load should be corrected in accordance with the general theory of thick plates. However, a special theory of thick plates supported elastically has also been established.[5]

[1] See H. Borowicka, *Ingr.-Arch.*, vol. 10, p. 113, 1939; A. G. Ishkova, *Doklady Akad. Nauk S.S.S.R.*, vol. 56, p. 129, 1947; G. Pickett and F. J. McCormick, *Proc. First U.S. Natl. Congr. Appl. Mech.*, p. 331, Chicago, 1951. The effect of raising the outer portion of the plate submitted to a central load was discussed by H. Jung, *Ingr.-Arch.*, vol. 20, p. 8, 1952. For bending of rectangular plates see M. I. Gorbounov-Posadov, *Priklad. Mat. Mekhan.*, vol. 4, p. 68, 1940.

[2] A. Habel, *Bauingenieur*, vol. 18, p. 188, 1937; for application to rectangular plates see G. Pickett, W. C. Janes, M. E. Raville, and F. J. McCormick, *Kansas State Coll. Eng. Expt. Sta. Bull.* 65, 1951.

[3] A. H. A. Hogg, *Phil. Mag.*, vol. 35, p. 265, 1944.

[4] G. Pickett and S. Badaruddin, *Proc. Ninth Intern. Congr. Appl. Mech.*, vol. 6, p. 396, Brussels, 1957.

[5] The first discussion of the statical and dynamical behavior of such plates is due to K. Marguerre, *Ingr.-Arch.*, vol. 4, p. 332, 1933; see also I. Szabó, *Ingr.-Arch.*, vol. 19, pp. 128, 342, 1951; *Z. angew. Math. Mech.*, vol. 32, p. 145, 1952. For application of E. Reissner's theory see P. M. Naghdi and J. C. Rowley, *Proc. First Midwest Conf. Solid Mech.* (*Univ. Illinois*), 1953, p. 119, and D. Frederick, *J. Appl. Mechanics*, vol. 23, p. 195, 1956.

CHAPTER 9

PLATES OF VARIOUS SHAPES

62. Equations of Bending of Plates in Polar Coordinates. In the discussion of symmetrical bending of circular plates polar coordinates were used (Chap. 3). The same coordinates can also be used to advantage in the general case of bending of circular plates.

If the r and θ coordinates are taken, as shown in Fig. 136a, the relation between the polar and cartesian coordinates is given by the equations

$$r^2 = x^2 + y^2 \qquad \theta = \arctan \frac{y}{x} \qquad (a)$$

from which it follows that

$$\frac{\partial r}{\partial x} = \frac{x}{r} = \cos\theta \qquad \frac{\partial r}{\partial y} = \frac{y}{r} = \sin\theta$$

$$\frac{\partial \theta}{\partial x} = -\frac{y}{r^2} = -\frac{\sin\theta}{r} \qquad \frac{\partial \theta}{\partial y} = \frac{x}{r^2} = \frac{\cos\theta}{r} \qquad (b)$$

Using these expressions, we obtain the slope of the deflection surface of a plate in the x direction as

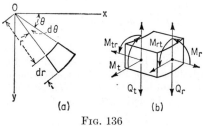

Fig. 136

$$\frac{\partial w}{\partial x} = \frac{\partial w}{\partial r}\frac{\partial r}{\partial x} + \frac{\partial w}{\partial \theta}\frac{\partial \theta}{\partial x}$$

$$= \frac{\partial w}{\partial r}\cos\theta - \frac{1}{r}\frac{\partial w}{\partial \theta}\sin\theta \qquad (c)$$

A similar expression can be written for the slope in the y direction. To obtain the expression for curvature in polar coordinates the second derivatives are required. Repeating twice the operation indicated in expression (c), we find

$$\frac{\partial^2 w}{\partial x^2} = \left(\frac{\partial}{\partial r}\cos\theta - \frac{1}{r}\sin\theta\frac{\partial}{\partial \theta}\right)\left(\frac{\partial w}{\partial r}\cos\theta - \frac{1}{r}\frac{\partial w}{\partial \theta}\sin\theta\right)$$

$$= \frac{\partial^2 w}{\partial r^2}\cos^2\theta - 2\frac{\partial^2 w}{\partial \theta\, \partial r}\frac{\sin\theta\cos\theta}{r} + \frac{\partial w}{\partial r}\frac{\sin^2\theta}{r}$$

$$+ 2\frac{\partial w}{\partial \theta}\frac{\sin\theta\cos\theta}{r^2} + \frac{\partial^2 w}{\partial \theta^2}\frac{\sin^2\theta}{r^2} \qquad (d)$$

In a similar manner we obtain

$$\frac{\partial^2 w}{\partial y^2} = \frac{\partial^2 w}{\partial r^2}\sin^2\theta + 2\frac{\partial^2 w}{\partial \theta\, \partial r}\frac{\sin\theta\cos\theta}{r} + \frac{\partial w}{\partial r}\frac{\cos^2\theta}{r}$$
$$- 2\frac{\partial w}{\partial \theta}\frac{\sin\theta\cos\theta}{r^2} + \frac{\partial^2 w}{\partial \theta^2}\frac{\cos^2\theta}{r^2} \quad (e)$$

$$\frac{\partial^2 w}{\partial x\, \partial y} = \frac{\partial^2 w}{\partial r^2}\sin\theta\cos\theta + \frac{\partial^2 w}{\partial r\, \partial \theta}\frac{\cos 2\theta}{r} - \frac{\partial w}{\partial \theta}\frac{\cos 2\theta}{r^2}$$
$$- \frac{\partial w}{\partial r}\frac{\sin\theta\cos\theta}{r} - \frac{\partial^2 w}{\partial \theta^2}\frac{\sin\theta\cos\theta}{r^2} \quad (f)$$

With this transformation of coordinates we obtain

$$\Delta w = \frac{\partial^2 w}{\partial x^2} + \frac{\partial^2 w}{\partial y^2} = \frac{\partial^2 w}{\partial r^2} + \frac{1}{r}\frac{\partial w}{\partial r} + \frac{1}{r^2}\frac{\partial^2 w}{\partial \theta^2} \quad (g)$$

Repeating this operation twice, the differential equation (103) for the deflection surface of a laterally loaded plate transforms in polar coordinates to the following form:

$$\Delta\Delta w = \left(\frac{\partial^2}{\partial r^2} + \frac{1}{r}\frac{\partial}{\partial r} + \frac{1}{r^2}\frac{\partial^2}{\partial \theta^2}\right)\left(\frac{\partial^2 w}{\partial r^2} + \frac{1}{r}\frac{\partial w}{\partial r} + \frac{1}{r^2}\frac{\partial^2 w}{\partial \theta^2}\right) = \frac{q}{D} \quad (191)$$

When the load is symmetrically distributed with respect to the center of the plate, the deflection w is independent of θ, and Eq. (191) coincides with Eq. (58) (see page 54), which was obtained in the case of symmetrically loaded circular plates.

Let us consider an element cut out of the plate by two adjacent axial planes forming an angle $d\theta$ and by two cylindrical surfaces of radii r and $r + dr$, respectively (Fig. 136b). We denote the bending and twisting moments acting on the element per unit length by M_r, M_t, and M_{rt} and take their positive directions as shown in the figure. To express these moments by the deflection w of the plate we assume that the x axis coincides with the radius r. The moments M_r, M_t, and M_{rt} then have the same values as the moments M_x, M_y, and M_{xy} at the same point, and by substituting $\theta = 0$ in expressions (d), (e), and (f), we obtain

$$M_r = -D\left(\frac{\partial^2 w}{\partial x^2} + \nu\frac{\partial^2 w}{\partial y^2}\right)_{\theta=0} = -D\left[\frac{\partial^2 w}{\partial r^2} + \nu\left(\frac{1}{r}\frac{\partial w}{\partial r} + \frac{1}{r^2}\frac{\partial^2 w}{\partial \theta^2}\right)\right]$$
$$M_t = -D\left(\frac{\partial^2 w}{\partial y^2} + \nu\frac{\partial^2 w}{\partial x^2}\right)_{\theta=0} = -D\left(\frac{1}{r}\frac{\partial w}{\partial r} + \frac{1}{r^2}\frac{\partial^2 w}{\partial \theta^2} + \nu\frac{\partial^2 w}{\partial r^2}\right) \quad (192)$$
$$M_{rt} = (1-\nu)D\left(\frac{\partial^2 w}{\partial x\, \partial y}\right)_{\theta=0} = (1-\nu)D\left(\frac{1}{r}\frac{\partial^2 w}{\partial r\, \partial \theta} - \frac{1}{r^2}\frac{\partial w}{\partial \theta}\right)$$

In a similar manner, from formulas (108), we obtain the expressions for the shearing forces[1]

$$Q_r = -D\frac{\partial}{\partial r}(\Delta w) \quad \text{and} \quad Q_t = -D\frac{\partial(\Delta w)}{r\,\partial\theta} \tag{193}$$

where Δw is given by expression (g).

In the case of a clamped edge the boundary conditions of a circular plate of radius a are

$$(w)_{r=a} = 0 \qquad \left(\frac{\partial w}{\partial r}\right)_{r=a} = 0 \tag{h}$$

In the case of a simply supported edge

$$(w)_{r=a} = 0 \qquad (M_r)_{r=a} = 0 \tag{i}$$

In the case of a free edge (see page 87)

$$(M_r)_{r=a} = 0 \qquad V = \left(Q_r - \frac{\partial M_{rt}}{r\,\partial\theta}\right)_{r=a} = 0 \tag{j}$$

The general solution of Eq. (191) can be taken, as before, in the form of a sum

$$w = w_0 + w_1 \tag{k}$$

in which w_0 is a particular solution of Eq. (191) and w_1 is the solution of the homogeneous equation

$$\left(\frac{\partial^2}{\partial r^2} + \frac{1}{r}\frac{\partial}{\partial r} + \frac{1}{r^2}\frac{\partial^2}{\partial\theta^2}\right)\left(\frac{\partial^2 w_1}{\partial r^2} + \frac{1}{r}\frac{\partial w_1}{\partial r} + \frac{1}{r^2}\frac{\partial^2 w_1}{\partial\theta^2}\right) = 0 \tag{194}$$

This latter solution we take in the form of the following series:[2]

$$w_1 = R_0 + \sum_{m=1}^{\infty} R_m \cos m\theta + \sum_{m=1}^{\infty} R'_m \sin m\theta \tag{195}$$

in which $R_0, R_1, \ldots, R'_1, R'_2, \ldots$ are functions of the radial distance r only. Substituting this series in Eq. (194), we obtain for each of these functions an ordinary differential equation of the following kind:

$$\left(\frac{d^2}{dr^2} + \frac{1}{r}\frac{d}{dr} - \frac{m^2}{r^2}\right)\left(\frac{d^2 R_m}{dr^2} + \frac{1}{r}\frac{dR_m}{dr} - \frac{m^2 R_m}{r^2}\right) = 0$$

The general solution of this equation for $m > 1$ is

$$R_m = A_m r^m + B_m r^{-m} + C_m r^{m+2} + D_m r^{-m+2} \tag{l}$$

---

[1] The direction of Q_r in Fig. 136b is opposite to that used in Fig. 28. This explains the minus sign in Eq. (193).

[2] This solution was given by A. Clebsch in his "Theorie der Elasticität fester Körper," 1862.

For $m = 0$ and $m = 1$ the solutions are

$$R_0 = A_0 + B_0 r^2 + C_0 \log r + D_0 r^2 \log r$$
and
$$R_1 = A_1 r + B_1 r^3 + C_1 r^{-1} + D_1 r \log r \qquad (m)$$

Similar expressions can be written for the functions R'_m. Substituting these expressions for the functions R_m and R'_m in series (195), we obtain the general solution of Eq. (194). The constants $A_m, B_m, \ldots, D_m$ in each particular case must be determined so as to satisfy the boundary conditions. The solution R_0, which is independent of the angle θ, represents symmetrical bending of circular plates. Several particular cases of this kind have already been discussed in Chap. 3.

63. Circular Plates under a Linearly Varying Load. If a circular plate is acted upon by a load distributed as shown in Fig. 137, this load can always be divided into two parts: (1) a uniformly distributed load of intensity $\frac{1}{2}(p_2 + p_1)$ and (2) a linearly varying load having zero intensity along the diameter CD of the plate and the intensities $-p$ and $+p$ at the ends A and B of the diameter AB. The case of uniform load has already been discussed in Chap. 3. We have to consider here only the nonuniform load represented in the figure by the two shaded triangles.[1]

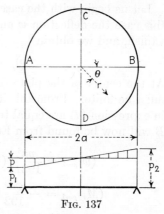

FIG. 137

The intensity of the load q at any point with coordinates r and θ is

$$q = \frac{pr \cos \theta}{a} \qquad (a)$$

The particular solution of Eq. (191) can thus be taken in the following form:

$$w_0 = A \frac{pr^5 \cos \theta}{a}$$

This, after substitution in Eq. (191), gives

$$A = \frac{1}{192D}$$

Hence
$$w_0 = \frac{pr^5 \cos \theta}{192aD} \qquad (b)$$

As the solution of the homogeneous equation (194) we take only the term of series (195) that contains the function R_1 and assume

$$w_1 = (A_1 r + B_1 r^3 + C_1 r^{-1} + D_1 r \log r) \cos \theta \qquad (c)$$

[1] This problem has been discussed by W. Flügge, *Bauingenieur*, vol. 10, p. 221, 1929.

Since it is advantageous to work with dimensionless quantities, we introduce, in place of r, the ratio

$$\rho = \frac{r}{a}$$

With this new notation the deflection of the plate becomes

$$w = w_0 + w_1 = \frac{pa^4}{192D}(\rho^5 + A\rho + B\rho^3 + C\rho^{-1} + D\rho \log \rho)\cos\theta \quad (d)$$

where ρ varies from zero to unity. The constants $A, B, \ldots$ in this expression must now be determined from the boundary conditions.

Let us begin with the case of a simply supported plate (Fig. 137). In this case the deflection w and the bending moment M_r at the boundary vanish, and we obtain

$$(w)_{\rho=1} = 0 \qquad (M_r)_{\rho=1} = 0 \quad (e)$$

At the center of the plate ($\rho = 0$) the deflection w and the moment M_r must be finite. From this it follows at once that the constants C and D in expression (d) are equal to zero. The remaining two constants A and B will now be found from Eqs. (e), which give

$$(w)_{\rho=1} = \frac{pa^4}{192D}(1 + A + B)\cos\theta = 0$$

$$(M_r)_{\rho=1} = -\frac{pa^2}{192}[4(5 + \nu) + 2(3 + \nu)B]\cos\theta = 0$$

Since these equations must be fulfilled for any value of θ, the factors before $\cos\theta$ must vanish. This gives

$$1 + A + B = 0$$
$$4(5 + \nu) + 2(3 + \nu)B = 0$$

and we obtain

$$B = -\frac{2(5 + \nu)}{3 + \nu} \qquad A = \frac{7 + \nu}{3 + \nu}$$

Substituting these values in expression (d), we obtain the deflection w of the plate in the following form:

$$w = \frac{pa^4\rho(1 - \rho^2)}{192(3 + \nu)D}[7 + \nu - (3 + \nu)\rho^2]\cos\theta \quad (f)$$

For calculating the bending moments and the shearing forces we substitute expression (f) in Eqs. (192) and (193), from which

$$M_r = \frac{pa^2}{48}(5 + \nu)\rho(1 - \rho^2)\cos\theta$$

$$M_t = \frac{pa^2}{48(3 + \nu)}\rho[(5 + \nu)(1 + 3\nu) - (1 + 5\nu)(3 + \nu)\rho^2]\cos\theta$$

(g)

$$Q_r = \frac{pa}{24(3+\nu)}[2(5+\nu) - 9(3+\nu)\rho^2]\cos\theta$$
$$Q_t = -\frac{pa}{24(3+\nu)}\rho[2(5+\nu) - 3(3+\nu)\rho^2]\sin\theta \qquad (h)$$

It is seen that $(M_r)_{\max}$ occurs at $\rho = 1/\sqrt{3}$ and is equal to
$$(M_r)_{\max} = \frac{pa^2(5+\nu)}{72\sqrt{3}}$$

The maximum value of M_t occurs at
$$\rho = \sqrt{(5+\nu)(1+3\nu)}/\sqrt{3(1+5\nu)(3+\nu)}$$
and is equal to
$$(M_t)_{\max} = \frac{pa^2}{72}\frac{(5+\nu)(1+3\nu)}{3+\nu}$$

The value of the intensity of the vertical reaction at the boundary is[1]
$$-V = -Q_r + \frac{\partial M_{rt}}{r\,\partial\theta} = \frac{pa}{4}\cos\theta$$

The moment of this reaction with respect to the diameter CD of the plate (Fig. 137) is
$$4\int_0^{\pi/2} \frac{pa}{4}\cos\theta\, a^2 \cos\theta\, d\theta = \frac{\pi a^3 p}{4}$$

This moment balances the moment of the load distributed over the plate with respect to the same diameter.

As a second example, let us consider the case of a circular plate with a free boundary. Such a condition is encountered in the case of a circular foundation slab supporting a chimney. As the result of wind pressure, a moment M will be transmitted to the slab (Fig. 138). Assuming that the reactions corresponding to this moment are distributed following a linear law, as shown in the figure, we obtain the same kind of loading as in the previous case; and the general solution can be taken in the same form (d) as before. The boundary conditions at the outer boundary of the plate, which is free from forces, are

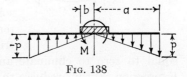

Fig. 138

$$(M_r)_{\rho=1} = 0 \qquad (V)_{\rho=1} = \left(Q_r - \frac{\partial M_{rt}}{r\,\partial\theta}\right)_{\rho=1} = 0 \qquad (i)$$

The inner portion of the plate of radius b is considered absolutely rigid. It is also assumed that the edge of the plate is clamped along the circle

---

[1] The reaction in the upward direction is taken as positive.

of radius b. Hence for $\rho = b/a = \beta$ the following boundary condition must be satisfied:

$$\left(\frac{\partial w}{\partial \rho}\right)_{\rho=\beta} = \left(\frac{w}{\rho}\right)_{\rho=\beta} \qquad (j)$$

Substituting expression (d) in Eqs. (i) and (j), we obtain the following equations for the determination of the constants:

$$4(5 + \nu) + 2(3 + \nu)B + 2(1 - \nu)C + (1 + \nu)D = 0$$
$$4(17 + \nu) + 2(3 + \nu)B + 2(1 - \nu)C - (3 - \nu)D = 0$$
$$4\beta^4 + 2\beta^2 B - 2\beta^{-2}C + D = 0$$

From these equations

$$B = -2\frac{4(2 + \nu) + (1 - \nu)\beta^2(3 + \beta^4)}{(3 + \nu) + (1 - \nu)\beta^4}$$

$$C = -2\frac{4(2 + \nu)\beta^4 - (3 + \nu)\beta^2(3 + \beta^4)}{(3 + \nu) + (1 - \nu)\beta^4} \qquad D = 12$$

Substituting these values in expression (d) and using Eqs. (192) and (193), we can obtain the values of the moments and of the shearing forces. The constant A does not appear in these equations. The corresponding term in expression (d) represents the rotation of the plate as a rigid body with respect to the diameter perpendicular to the plane of Fig. 138. Provided the modulus of the foundation is known, the angle of rotation can be calculated from the condition of equilibrium of the given moment M and the reactions of the foundation.

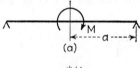

(a)

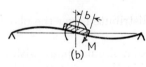

(b)

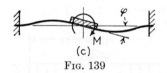

(c)

Fig. 139

Using expression (d), the case of a simply supported circular plate loaded by a moment M at the center (Fig. 139a) can be readily solved. In this case we have to omit the term containing ρ^5, which represents the distributed load. The constant C must be taken equal to zero to eliminate an infinitely large deflection at the center. Expression (d) thus reduces to

$$w = (A\rho + B\rho^3 + D\rho \log \rho) \cos \theta \qquad (k)$$

The three constants A, B, and D will now be determined from the following boundary conditions:

$$(w)_{\rho=1} = 0 \qquad (M_r)_{\rho=1} = 0$$
$$-a \int_{-\pi}^{+\pi} (M_{rt})_{\rho=1} \sin \theta \, d\theta + a^2 \int_{-\pi}^{+\pi} (Q_r)_{\rho=1} \cos \theta \, d\theta + M = 0 \qquad (l)$$

The first two of these equations represent the conditions at a simply supported edge; the last states the condition of equilibrium of the forces and moments acting at the boundary of the plate and the external moment M. From Eqs. (l) we obtain

$$A = -\frac{1+\nu}{3+\nu}\frac{Ma}{8\pi D} \qquad B = \frac{1+\nu}{3+\nu}\frac{Ma}{8\pi D} \qquad D = -\frac{Ma}{4\pi D}$$

Hence

$$w = -\frac{Ma}{8\pi D(3+\nu)}\rho[(1+\nu)(1-\rho^2) + 2(3+\nu)\log\rho]\cos\theta \qquad (m)$$

Because of the logarithmic term in the brackets, the slope of the deflection surface calculated from expression (m) becomes infinitely large. To eliminate this difficulty the central portion of radius b of the plate may be considered as absolutely rigid.[1] Assuming the plate to be clamped along this inner boundary, which rotates under the action of the moment M (Fig. 139b), we find

$$w = \frac{Ma}{8\pi D[(3+\nu) + (1-\nu)\beta^4]}\{-[(1+\nu) + (1-\nu)\beta^4]\rho^3$$
$$+ (1+\nu)(1-\beta^2)^2\rho + 2[(3+\nu) + (1-\nu)\beta^4]\rho\log\rho$$
$$- \beta^2[(1+\nu)\beta^2 - (3+\nu)]\rho^{-1}\}\cos\theta \qquad (n)$$

where $\beta = b/a$. When β is equal to zero, Eq. (n) reduces to Eq. (m), previously obtained. By substituting expression (n) in Eq. (192) the bending moments M_r and M_t can be calculated.

The case in which the outer boundary of the plate is clamped (Fig. 139c) can be discussed in a similar manner. This case is of practical interest in the design of elastic couplings of shafts.[2] The maximum radial stresses at the inner and at the outer boundaries and the angle of rotation φ of the central rigid portion for this case are

$$(\sigma_r)_{r=b} = \alpha\frac{h}{a}E\varphi \qquad (\sigma_r)_{r=a} = \alpha_1\frac{h}{a}E\varphi \qquad \varphi = \frac{M}{\alpha_2 E h^3}$$

where the constants α, α_1, and α_2 have the values given in Table 64.

TABLE 64

$\beta = b/a$	α	α_1	α_2
0.5	14.17	7.10	12.40
0.6	19.54	12.85	28.48
0.7	36.25	25.65	77.90
0.8	82.26	66.50	314.00

[1] Experiments with such plates were made by R. J. Roark, *Univ. Wisconsin Bull.* 74, 1932.

[2] H. Reissner, *Ingr.-Arch.*, vol. 1, p. 72, 1929.

64. Circular Plates under a Concentrated Load.

The case of a load applied at the center of the plate has already been discussed in Art. 19. Here we shall assume that the load P is applied at point A at distance b from the center O of the plate (Fig. 140).[1] Dividing the plate into two parts by the cylindrical section of radius b as shown in the figure by the dashed line, we can apply solution (195) for each of these portions of the plate. If the angle θ is measured from the radius OA, only the terms containing $\cos m\theta$ should be retained. Hence for the outer part of the plate we obtain

$$w = R_0 + \sum_{m=1}^{\infty} R_m \cos m\theta \qquad (a)$$

where
$$\begin{aligned} R_0 &= A_0 + B_0 r^2 + C_0 \log r + D_0 r^2 \log r \\ R_1 &= A_1 r + B_1 r^3 + C_1 r^{-1} + D_1 r \log r \\ &\cdots\cdots\cdots\cdots\cdots\cdots\cdots\cdots\cdots\cdots\cdots \\ R_m &= A_m r^m + B_m r^{-m} + C_m r^{m+2} + D_m r^{-m+2} \end{aligned} \qquad (b)$$

Similar expressions can also be written for the functions R_0', R_1', R_m' corresponding to the inner portion of the plate. Using the symbols A_m', B_m', . . . instead of A_m, B_m, . . . for the constants of the latter portion of the plate, from the condition that the deflection, the slope, and the moments must be finite at the center of the plate, we obtain

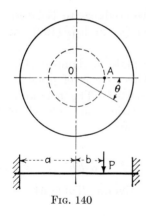

Fig. 140

$$\begin{aligned} C_0' &= D_0' = 0 \\ C_1' &= D_1' = 0 \\ &\cdots\cdots\cdots\cdots \\ B_m' &= D_m' = 0 \end{aligned}$$

Hence for each term of series (a) we have to determine four constants for the outer portion of the plate and two for the inner portion.

The six equations necessary for this determination can be obtained from the boundary conditions at the edge of the plate and from the continuity conditions along the circle of radius b. If the outer edge of the plate is assumed to be clamped, the corresponding boundary con-

---

[1] This problem was solved by Clebsch, *op. cit.* See also A. Föppl, *Sitzber. bayer. Akad. Wiss., Jahrg.*, 1912, p. 155. The discussion of the same problem by using bipolar coordinates was given by E. Melan, *Eisenbau*, 1920, p. 190, and by W. Flügge, "Die strenge Berechnung von Kreisplatten unter Einzellasten," Berlin, 1928. See also the paper by H. Schmidt, *Ingr.-Arch.*, vol. 1, p. 147, 1930, and W. Müller, *Ingr.-Arch.*, vol. 13, p. 355, 1943.

ditions are

$$(w)_{r=a} = 0 \qquad \left(\frac{\partial w}{\partial r}\right)_{r=a} = 0 \qquad (c)$$

Denoting the deflection of the inner portion of the plate by w_1 and observing that there are no external moments applied along the circle of radius b, we write the continuity conditions along that circle as

$$w = w_1 \qquad \frac{\partial w}{\partial r} = \frac{\partial w_1}{\partial r} \qquad \frac{\partial^2 w}{\partial r^2} = \frac{\partial^2 w_1}{\partial r^2} \qquad \text{for } r = b \qquad (d)$$

The last equation is obtained from a consideration of the shearing force Q_r along the dividing circle. This force is continuous at all points of the circle except point A, where it has a discontinuity due to concentrated force P. Using for this force the representation in form of the series[1]

$$\frac{P}{\pi b}\left(\frac{1}{2} + \sum_{m=1}^{\infty} \cos m\theta\right) \qquad (e)$$

and for the shearing force the first of the expressions (193), we obtain

$$D\frac{\partial}{\partial r}(\Delta w)_{r=b} - D\frac{\partial}{\partial r}(\Delta w_1)_{r=b} = \frac{P}{\pi b}\left(\frac{1}{2} + \sum_{m=1}^{\infty} \cos m\theta\right) \qquad (f)$$

From the six equations (c), (d), and (f), the six constants can be calculated, and the functions R_m and R'_m can be represented in the following form:

$$R_0 = \frac{P}{8\pi D}\left[(r^2 + b^2)\log\frac{r}{a} + \frac{(a^2 + b^2)(a^2 - r^2)}{2a^2}\right]$$

$$R'_0 = \frac{P}{8\pi D}\left[(r^2 + b^2)\log\frac{b}{a} + \frac{(a^2 + r^2)(a^2 - b^2)}{2a^2}\right]$$

$$R_1 = -\frac{Pb^3}{16\pi D}\left[\frac{1}{r} + \frac{2(a^2 - b^2)r}{a^2b^2} - \frac{(2a^2 - b^2)r^3}{a^4b^2} - \frac{4r}{b^2}\log\frac{a}{r}\right]$$

$$R'_1 = -\frac{Pb^3}{16\pi D}\left[\frac{2(a^2 - b^2)r}{a^2b^2} + \frac{(a^2 - b^2)^2r^3}{a^4b^4} - \frac{4r}{b^2}\log\frac{a}{b}\right]$$

$$R_m = \frac{Pb^m}{8m(m-1)\pi D}\left\{\frac{r^m}{a^{2m}}\left[(m-1)b^2 - ma^2 + (m-1)r^2\right.\right.$$
$$\left.\left. - \frac{m(m-1)}{m+1}\frac{b^2r^2}{a^2}\right] + \frac{1}{r^m}\left(r^2 - \frac{m-1}{m+1}b^2\right)\right\}$$

$$R'_m = \frac{Pb^m}{8m(m-1)\pi D}\left\{\frac{r^m}{a^{2m}}\left[(m-1)b^2 - ma^2 + \frac{a^{2m}}{b^{2m-2}}\right]\right.$$
$$\left. + (m-1)\frac{r^{m+2}}{a^{2m}}\left[1 - \frac{m}{m+1}\frac{b^2}{a^2} - \frac{1}{m+1}\left(\frac{a}{b}\right)^{2m}\right]\right\}$$

---

[1] This series is analogous to the series that was used in the case of continuous plates (see p. 248).

Using these functions, we obtain the deflection under the load as

$$(w)_{r=b,\theta=0} = \frac{P}{16\pi D}\frac{(a^2-b^2)^2}{a^2} \qquad (196)$$

For $b = 0$ this formula coincides with formula (92) for a centrally loaded plate. The case of the plate with simply supported edge can be treated in a similar manner.

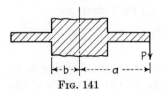

Fig. 141

The problem in which a circular ring plate is clamped along the inner edge ($r = b$) and loaded by a concentrated force P at the outer boundary (Fig. 141) can also be solved by using series (a). In this case the boundary conditions for the clamped inner boundary are

$$(w)_{r=b} = 0 \qquad \left(\frac{\partial w}{\partial r}\right)_{r=b} = 0 \qquad (g)$$

For the outer boundary, which is loaded only in one point, the conditions are

$$(M_r)_{r=a} = 0$$

$$(V)_{r=a} = \frac{P}{\pi a}\left(\frac{1}{2} + \sum_{m=1}^{\infty}\cos m\theta\right) \qquad (h)$$

Calculations made for a particular case $b/a = \tfrac{2}{3}$ show[1] that the largest bending moment M_r at the inner boundary is

$$(M_r)_{r=b,\theta=0} = -4.45\frac{P}{2\pi}$$

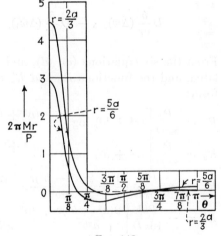

Fig. 142

The variation of the moment along the inner edge and also along a circle of radius $r = 5a/6$ is shown in Fig. 142. It can be seen that this moment diminishes rapidly as the angle θ, measured from the point of application of the load, increases.

The general solution of the form (a) may be used to advantage in handling circular plates with a system of single loads distributed symmetrically with respect to the center of the plate,[2] and also in the case of

---

[1] H. Reissner, *loc. cit.*

[2] By combining such reactive loads with a given uniform loading, we may solve the problem of a flat slab bounded by a circle; see K. Hajnal-Konyi, "Berechnung von kreisförmig begrenzten Pilzdecken," Berlin, 1929.

annular plates. For circular plates having no hole and carrying but one eccentric load, simpler solutions can be obtained by the method of complex variables,[1] or, when the plate is clamped, by the method of inversion.[2] In this latter case the deflection surface of the plate is obtained in the form

$$w = \frac{Pa^2}{16\pi D} \left[(1 - x^2)(1 - \xi^2) + (x^2 + \xi^2 - 2x\xi \cos \theta) \log \frac{x^2 + \xi^2 - 2x\xi \cos \theta}{1 + x^2\xi^2 - 2x\xi \cos \theta} \right] \quad (197)$$

where $x = r/a$ and $\xi = b/a$ (Fig. 140). Expression (197) holds throughout the whole plate and yields for $x = \xi$, $\theta = 0$, that is, under the load, the value (196), previously obtained by the series method.

65. Circular Plates Supported at Several Points along the Boundary.

Considering the case of a load symmetrically distributed with respect to the center of the plate, we take the general expression for the deflection surface in the following form:[3]

$$w = w_0 + w_1 \quad (a)$$

in which w_0 is the deflection of a plate simply supported along the entire boundary, and w_1 satisfies the homogeneous differential equation

$$\Delta\Delta w_1 = 0 \quad (b)$$

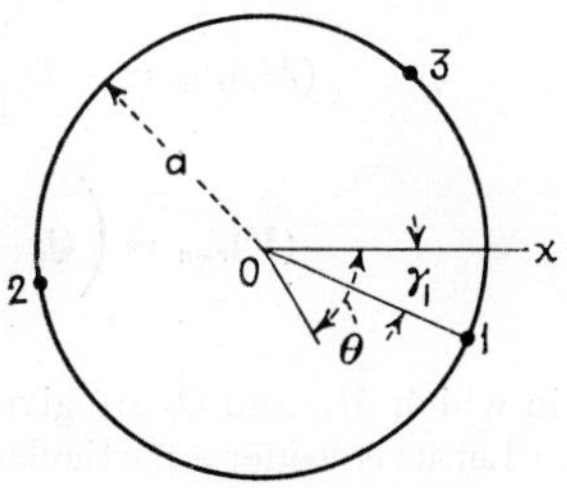

Denoting the concentrated reactions at the points of support 1, 2, 3, . . . by $N_1, N_2, \ldots, N_i$ and using series (h) of the previous article for representation of concentrated forces, we have for each reaction N_i the expression

FIG. 143

$$\frac{N_i}{\pi a}\left(\frac{1}{2} + \sum_{m=1}^{\infty} \cos m\theta_i\right) \quad (c)$$

where
$$\theta_i = \theta - \gamma_i$$

γ_i being the angle defining the position of the support i (Fig. 143). The intensity of the reactive forces at any point of the boundary is then given by the expression

---

[1] The simply supported plate was treated in that manner by E. Reissner, *Math. Ann.*, vol. 111, p. 777, 1935; for the application of Muschelišvili's method see A. I. Lourye, *Bull. Polytech. Inst., Leningrad*, vol. 31, p. 305, 1928, and *Priklad. Mat. Mekhan.*, vol. 4, p. 93, 1940. See also K. Nasitta, *Ingr.-Arch.*, vol. 23, p. 85, 1955, and R. J. Roark, *Wisconsin Univ. Eng. Expt. Sta. Bull.* 74, 1932.

[2] J. H. Michell, *Proc. London Math. Soc.*, vol. 34, p. 223, 1902.

[3] Several problems of this kind were discussed by A. Nádai, *Z. Physik*, vol. 23, p. 366, 1922. Plates supported at several points were also discussed by W. A. Bassali, *Proc. Cambridge Phil. Soc.*, vol. 53, p. 728, 1957, and circular plates with mixed boundary conditions by G. M. L. Gladwell, *Quart. J. Mech. Appl. Math.*, vol. 11, p. 159, 1958.

$$\sum_1^i \frac{N_i}{\pi a}\left(\frac{1}{2} + \sum_{m=1}^{\infty} \cos m\theta_i\right) \qquad (d)$$

in which the summation is extended over all the concentrated reactions (c).

The general solution of the homogeneous equation (b) is given by expression (195) (page 284). Assuming that the plate is solid and omitting the terms that give infinite deflections and moments at the center, we obtain from expression (195)

$$w_1 = A_0 + B_0 r^2 + \sum_{m=1}^{\infty} (A_m r^m + C_m r^{m+2}) \cos m\theta$$

$$+ \sum_{m=1}^{\infty} (A'_m r^m + C'_m r^{m+2}) \sin m\theta \qquad (e)$$

For determining the constants we have the following conditions at the boundary:

$$(M_r)_{r=a} = -D\left[\frac{\partial^2 w}{\partial r^2} + \nu\left(\frac{1}{r}\frac{\partial w}{\partial r} + \frac{1}{r^2}\frac{\partial^2 w}{\partial \theta^2}\right)\right]_{r=a} = 0$$

$$(V)_{r=a} = \left(Q_r - \frac{\partial M_{rt}}{r\,\partial\theta}\right)_{r=a} = -\sum_1^i \frac{N_i}{\pi a}\left(\frac{1}{2} + \sum_{m=1}^{\infty} \cos m\theta_i\right) \qquad (f)$$

in which M_{rt} and Q_r are given by Eqs. (192) and (193).

Let us consider a particular case in which the plate is supported at two points which are the ends of a diameter. We shall measure θ from this diameter. Then $\gamma_1 = 0$, $\gamma_2 = \pi$, and we obtain

$$w = w_0 + \frac{Pa^2}{2\pi(3+\nu)D}\left\{2\log 2 - 1 + \frac{1+\nu}{1-\nu}\left(2\log 2 - \frac{\pi^2}{12}\right)\right.$$

$$\left. - \sum_{m=2,4,6,\ldots}^{\infty}\left[\frac{1}{m(m-1)} + \frac{2(1+\nu)}{(1-\nu)(m-1)m^2} - \frac{\rho^2}{m(m+1)}\right]\rho^m \cos m\theta\right\} \qquad (g)$$

in which w_0 is the deflection of the simply supported and symmetrically loaded plate, P is the total load on the plate, and $\rho = r/a$. When the load is applied at the center, we obtain from expression (g), by assuming $\nu = 0.25$,

$$(w)_{\rho=0} = 0.116\frac{Pa^2}{D}$$

$$(w)_{\rho=1,\theta=\pi/2} = 0.118\frac{Pa^2}{D}$$

For a uniformly loaded plate we obtain

$$(w)_{\rho=0} = 0.269\frac{qa^4}{D}$$

$$(w)_{\rho=1,\theta=\pi/2} = 0.371\frac{qa^4}{D}$$

By combining two solutions of the type (g), the case shown in Fig. 144 can also be obtained.

When a circular plate is supported at three points 120° apart, the deflection produced at the center of the plate, when the load is applied at the center, is

$$(w)_{\rho=0} = 0.0670 \frac{Pa^2}{D}$$

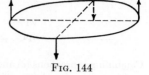

Fig. 144

When the load is uniformly distributed, the deflection at the center is

$$(w)_{\rho=0} = 0.0362 \frac{Pa^2}{D}$$

where $P = \pi a^2 q$.

The case of a circular plate supported at three points was investigated by experiments with glass plates. These experiments showed a very satisfactory agreement with the theory.[1]

66. Plates in the Form of a Sector. The general solution developed for circular plates (Art. 62) can also be adapted for a plate in the form of a sector, the straight edges of which are simply supported.[2] Take, as an example, a plate in the form of a semicircle simply supported along the diameter AB and uniformly loaded (Fig. 145). The deflection of this plate is evidently the same as that of the circular plate indicated by the dashed line and loaded as shown in Fig. 145b. The distributed load is represented in such a case by the series

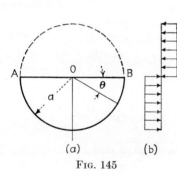

(a)

Fig. 145

$$q = \sum_{m=1,3,5,\ldots}^{\infty} \frac{4q}{m\pi} \sin m\theta \qquad (a)$$

and the differential equation of the deflection surface is

$$\Delta\Delta w = \frac{1}{D} \sum_{m=1,3,5,\ldots}^{\infty} \frac{4q}{m\pi} \sin m\theta \qquad (b)$$

The particular solution of this equation that satisfies the boundary conditions along the diameter AB is

$$w_0 = \sum_{m=1,3,5,\ldots}^{\infty} \frac{4qr^4}{\pi m(16-m^2)(4-m^2)D} \sin m\theta \qquad (c)$$

The solution of the homogeneous differential equation (194) that satisfies the condi-

---

[1] These experiments were made by Nádai, *ibid.*

[2] Problems of this kind were discussed by Nádai, *Z. Ver. deut. Ing.*, vol. 59, p. 169, 1915. See also B. G. Galerkin, "Collected Papers," vol. 2, p. 320, Moscow, 1953, which gives numerical tables for such cases.

tions along the diameter AB is

$$w_1 = \sum_{m=1,3,5,\ldots}^{\infty} (A_m r^m + B_m r^{m+2}) \sin m\theta \qquad (d)$$

Combining expressions (c) and (d), we obtain the complete expression for the deflection w of a semicircular plate. The constants A_m and B_m are determined in each particular case from the conditions along the circular boundary of the plate.

In the case of a simply supported plate we have

$$(w)_{r=a} = 0$$
$$\left[\frac{\partial^2 w}{\partial r^2} + \nu\left(\frac{1}{r}\frac{\partial w}{\partial r} + \frac{1}{r^2}\frac{\partial^2 w}{\partial \theta^2}\right)\right]_{r=a} = 0 \qquad (e)$$

Substituting the sum of series (c) and (d) for w in these equations, we obtain the following equations for calculating A_m and B_m:

$$A_m a^m + B_m a^{m+2} = -\frac{4qa^2}{m\pi(16-m^2)(4-m^2)D}$$

$$A_m a^m[m(m-1) - \nu m(m-1)] + B_m a^{m+2}(m+1)[m+2+\nu(2-m)]$$
$$= -\frac{4qa^2[12+\nu(4-m^2)]}{m\pi(16-m^2)(4-m^2)D}$$

From these equations,

$$A_m = \frac{qa^4(m+5+\nu)}{a^m m\pi(16-m^2)(2+m)[m+\frac{1}{2}(1+\nu)]D}$$

$$B_m = -\frac{qa^4(m+3+\nu)}{a^{m+2}m\pi(4+m)(4-m^2)[m+\frac{1}{2}(1+\nu)]D}$$

With these values of the constants the expression for the deflection of the plate becomes

$$w = \frac{qa^4}{D}\sum_{m=1,3,5,\ldots}^{\infty}\left\{\frac{4r^4}{a^4}\frac{1}{m\pi(16-m^2)(4-m^2)}\right.$$
$$+ \frac{r^m}{a^m}\frac{m+5+\nu}{m\pi(16-m^2)(2+m)[m+\frac{1}{2}(1+\nu)]}$$
$$\left. - \frac{r^{m+2}}{a^{m+2}}\frac{m+3+\nu}{m\pi(4+m)(4-m^2)[m+\frac{1}{2}(1+\nu)]}\right\}\sin m\theta$$

With this expression for the deflection, the bending moments are readily obtained from Eqs. (192).

In a similar manner we can obtain the solution for any sector with an angle π/k, k being a given integer. The final expressions for the deflections and bending moments at a given point can be represented in each particular case by the following formulas:

$$w = \alpha\frac{qa^4}{D} \qquad M_r = \beta qa^2 \qquad M_t = \beta_1 qa^2 \qquad (f)$$

in which α, β, and β_1 are numerical factors. Several values of these factors for points taken on the axis of symmetry of a sector are given in Table 65.

TABLE 65. VALUES OF THE FACTORS α, β, AND β_1 FOR VARIOUS ANGLES π/k OF A SECTOR SIMPLY SUPPORTED AT THE BOUNDARY
$\nu = 0.3$

π/k	$r/a = \frac{1}{4}$			$r/a = \frac{1}{2}$			$r/a = \frac{3}{4}$			$r/a = 1$		
	α	β	β_1	α	β	β_1	α	β	β_1	α	β	β_1
$\pi/4$	0.00006	−0.0015	0.0093	0.00033	0.0069	0.0183	0.00049	0.0161	0.0169	0	0	0.0025
$\pi/3$	0.00019	−0.0025	0.0177	0.00080	0.0149	0.0255	0.00092	0.0243	0.0213	0	0	0.0044
$\pi/2$	0.00092	0.0036	0.0319	0.00225	0.0353	0.0352	0.00203	0.0381	0.0286	0	0	0.0088
π	0.00589	0.0692	0.0357	0.00811	0.0868	0.0515	0.00560	0.0617	0.0468	0	0	0.0221

The case in which a plate in the form of a sector is clamped along the circular boundary and simply supported along the straight edges can be treated by the same method of solution as that used in the preceding case. The values of the coefficients α and β for the points taken along the axis of symmetry of the sector are given in Table 66.

TABLE 66. VALUES OF THE COEFFICIENTS α AND β FOR VARIOUS ANGLES π/k OF A SECTOR CLAMPED ALONG THE CIRCULAR BOUNDARY AND SIMPLY SUPPORTED ALONG THE STRAIGHT EDGES
$\nu = 0.3$

π/k	$r/a = \frac{1}{4}$		$r/a = \frac{1}{2}$		$r/a = \frac{3}{4}$		$r/a = 1$	
	α	β	α	β	α	β	α	β
$\pi/4$	0.00005	−0.0008	0.00026	0.0087	0.00028	0.0107	0	−0.0250
$\pi/3$	0.00017	−0.0006	0.00057	0.0143	0.00047	0.0123	0	−0.0340
$\pi/2$	0.00063	0.0068	0.00132	0.0272	0.00082	0.0113	0	−0.0488
π	0.00293	0.0473	0.00337	0.0446	0.00153	0.0016	0	−0.0756

It can be seen that in this case the maximum bending stress occurs at the mid-point of the circular edge of the sector.

If the circular edge of a uniformly loaded plate having the form of a sector is entirely free, the maximum deflection occurs at the mid-point of the unsupported circular edge. For the case when $\pi/k = \pi/2$ we obtain

$$w_{max} = 0.0633 \frac{qa^4}{D}$$

The bending moment at the same point is

$$M_t = 0.1331 qa^2$$

In the general case of a plate having the form of a circular sector with radial edges clamped or free, approximate methods must be applied.[1] However, the particular problem of a wedge-shaped plate carrying a lateral load can be solved rigorously (see Art. 78). Another problem which allows an exact solution is that of bending of a plate clamped along two circular arcs.[2] Bipolar coordinates must be introduced in that case and data regarding the clamped semicircular plate in particular are given in Table 67.

TABLE 67. VALUES OF THE FACTORS α, β, AND β_1 [EQS. (f)] FOR A SEMICIRCULAR PLATE CLAMPED ALONG THE BOUNDARY (Fig. 145a)
$\nu = 0.3$

Load distribution	$r/a = 0$ β	$r/a = 0.483$ β_{max}	$r/a = 0.486$ α_{max}	$r/a = 0.525$ $\beta_{1\,max}$	$r/a = 1$ β
Uniform load q	−0.0731	0.0355	0.00202	0.0194	−0.0584
Hydrostatic load qy/a	−0.0276	. . . . . .	. . . . . .	. . . . . .	−0.0355

Bipolar coordinates can also be used to advantage in case of a plate clamped between an outer and an inner (eccentric) circle and carrying a single load.[3]

67. Circular Plates of Nonuniform Thickness. Circular plates of nonuniform thickness are sometimes encountered in the design of machine parts, such as diaphragms of steam turbines and pistons of reciprocating engines. The thickness of such plates is usually a function of the radial distance, and the acting load is symmetrical with respect to the center of the plate. We shall limit our further discussion to this symmetrical case.

Proceeding as explained in Art. 15 and using the notations of that article, from the condition of equilibrium of an element as shown in Fig. 28 (page 52) we derive the following equation:

$$M_r + \frac{dM_r}{dr} r - M_t + Qr = 0 \qquad (a)$$

---

[1] See G. F. Carrier and F. S. Shaw, *Proc. Symposia Appl. Math.*, vol. 3, p. 125, 1950; H. D. Conway and M. K. Huang, *J. Appl. Mechanics*, vol. 19, p. 5, 1952; H. R. Hassé, *Quart. Mech. Appl. Math.*, vol. 3, p. 271, 1950. The case of a concentrated load has been discussed by T. Sekiya and A. Saito, *Proc. Fourth Japan. Congr. Appl. Mech.*, 1954, p. 195. For plates bounded by two radii and two arcs and clamped see G. F. Carrier, *J. Appl. Mechanics*, vol. 11, p. A-134, 1944. The same problem with various edge conditions was discussed by L. I. Deverall and C. J. Thorne, *J. Appl. Mechanics*, vol. 18, p. 359, 1951. The bending of a uniformly loaded semicircular plate simply supported around the curved edge and free along the diameter (a "diaphragm" of a steam turbine) has been discussed in detail by D. F. Muster and M. A. Sadowsky, *J. Appl. Mechanics*, vol. 23, p. 329, 1956. A similar case, however, with a curved edge clamped, has been handled by H. Müggenburg, *Ingr.-Arch.*, vol. 24, p. 308, 1956.

[2] Green's function for these boundary conditions has been obtained by A. C. Dixon, *Proc. London Math. Soc.*, vol. 19, p. 373, 1920. For an interesting limiting case see W. R. Dean, *Proc. Cambridge Phil. Soc.*, vol. 49, p. 319, 1953. In handling distributed loads the use of the rather cumbersome Green function may be avoided; see S. Woinowsky-Krieger, *J. Appl. Mechanics*, vol. 22, p. 129, 1955, and *Ingr.-Arch.*, vol. 24, p. 48, 1956.

[3] This problem was discussed by N. V. Kudriavtzev, *Doklady Akad. Nauk S.S.S.R.*, vol. 53, p. 203, 1946.

in which, as before,

$$M_r = D\left(\frac{d\varphi}{dr} + \frac{\nu}{r}\varphi\right)$$
$$M_t = D\left(\frac{\varphi}{r} + \nu\frac{d\varphi}{dr}\right) \quad (b)$$

where
$$\varphi = -\frac{dw}{dr} \quad (c)$$

and Q is the shearing force per unit length of a circular section of radius r. In the case of a solid plate, Q is given by the equation

$$Q = \frac{1}{2\pi r}\int_0^r q\, 2\pi r\, dr \quad (d)$$

in which q is the intensity of the lateral load.

Substituting expressions (b), (c), and (d) in Eq. (a) and observing that the flexural rigidity D is no longer constant but varies with the radial distance r, we obtain the following equation:

$$D\frac{d}{dr}\left(\frac{d\varphi}{dr} + \frac{\varphi}{r}\right) + \frac{dD}{dr}\left(\frac{d\varphi}{dr} + \nu\frac{\varphi}{r}\right) = -\frac{1}{r}\int_0^r q\,r\,dr \quad (e)$$

Thus the problem of bending of circular symmetrically loaded plates reduces to the solution of a differential equation (e) of the second order with variable coefficients. To represent the equation in dimensionless form, we introduce the following notations:

a = outer radius of plate
h = thickness of plate at any point
h_0 = thickness of plate at center

then
$$\frac{r}{a} = x \qquad \frac{h}{h_0} = y \quad (f)$$

We also assume that the load is uniformly distributed. Using the notation

$$p = \frac{6(1-\nu^2)a^3q}{Eh_0^3} \quad (g)$$

Eq. (e) then becomes

$$\frac{d^2\varphi}{dx^2} + \left(\frac{1}{x} + \frac{d\log y^3}{dx}\right)\frac{d\varphi}{dx} - \left(\frac{1}{x^2} - \frac{\nu}{x}\frac{d\log y^3}{dx}\right)\varphi = -\frac{px}{y^3} \quad (198)$$

In many cases the variation of the plate thickness can be represented with sufficient accuracy by the equation[1]

$$y = e^{-\beta x^2/6} \quad (h)$$

in which β is a constant that must be chosen in each particular case so as to approximate as closely as possible the actual proportions of the plate. The variation of thickness

[1] The first investigation of bending of circular plates of nonuniform thickness was made by H. Holzer, *Z. ges. Turbinenwesen*, vol. 15, p. 21, 1918. The results given in this article are taken from O. Pichler's doctor's dissertation, "Die Biegung kreissymmetrischer Platten von veränderlicher Dicke," Berlin, 1928. See also the paper by R. Gran Olsson, *Ingr.-Arch.*, vol. 8, p. 81, 1937.

along a diameter of a plate corresponding to various values of the constant β is shown in Fig. 146. Substituting expression (h) in Eq. (198), we find

$$\frac{d^2\varphi}{dx^2} + \left(\frac{1}{x} - \beta x\right)\frac{d\varphi}{dx} - \left(\frac{1}{x^2} + \nu\beta\right)\varphi = -pxe^{\beta x^2/2} \qquad (i)$$

It can be readily verified that

$$\varphi_0 = -\frac{p}{(3-\nu)\beta} xe^{\beta x^2/2} \qquad (j)$$

is a particular solution of Eq. (i). One of the two solutions of the homogeneous equation corresponding to Eq. (i) can be taken in the form of a power series:

$$\varphi_1 = a_1\left[x + \sum_{n=1}^{\infty} \frac{\beta^n(1+\nu)(3+\nu) \cdots (2n-1+\nu)}{2 \cdot 4 \cdot 4 \cdot 6 \cdot 6 \cdots 2n \cdot 2n(2n+2)} x^{2n+1}\right] \qquad (k)$$

in which a_1 is an arbitrary constant. The second solution of the same equation becomes infinitely large at the center of the plate, i.e., for $x = 0$, and therefore should

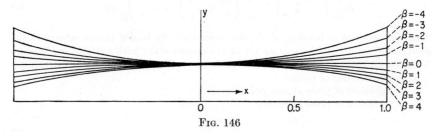

Fig. 146

not be considered in the case of a plate without a hole at the center. If solutions (j) and (k) are combined, the general solution of Eq. (i) for a solid plate can be put in the following form:

$$\varphi = p\left[C\varphi_1 - \frac{x}{(3-\nu)\beta}e^{\beta x^2/2}\right] \qquad (l)$$

The constant C in each particular case must be determined from the condition at the boundary of the plate. Since series (k) is uniformly convergent, it can be differentiated, and the expressions for the bending moments can be obtained by substitution in Eqs. (b). The deflections can be obtained from Eq. (c).

In the case of a plate *clamped at the edge*, the boundary conditions are

$$(w)_{x=1} = 0 \qquad (\varphi)_{x=1} = 0 \qquad (m)$$

and the constant C in solution (l) is

$$C = \frac{e^{\beta/2}}{(3-\nu)\beta(\varphi_1)_{x=1}} \qquad (n)$$

To get the numerical value of C for a given value of β, which defines the shape of the diametrical section of the plate (see Fig. 146), the sum of series (k) must be calculated for $x = 1$. The results of such calculations are given in the above-mentioned paper by Pichler. This paper also gives the numerical values for the derivative and for the integral of series (k) by the use of which the moments and the deflections of a plate can be calculated.

The deflection of the plate at the center can be represented by the formula

$$w_{\max} = \alpha a p = \alpha \frac{6(1-\nu^2)a^4 q}{E h_0^3} \qquad (o)$$

in which α is a numerical factor depending on the value of the constant β. Several values of this factor, calculated for $\nu = 0.3$, are given in the first line of Table 68.

TABLE 68. NUMERICAL FACTORS α AND α' FOR CALCULATING DEFLECTIONS AT THE CENTER OF CIRCULAR PLATES OF VARIABLE THICKNESS
$\nu = 0.3$

β	4	3	2	1	0	-1	-2	-3	-4
α	0.0801	0.0639	0.0505	0.0398	0.0313	0.0246	0.0192	0.0152	0.01195
α'	0.2233	0.1944	0.1692	0.1471	0.1273	0.1098	0.0937	0.0791	0.06605

The maximum bending stresses at various radial distances can be represented by the formulas

$$(\sigma_r)_{\max} = \pm \gamma \frac{3qa^2}{h_0^2} \qquad (\sigma_t)_{\max} = \pm \gamma_1 \frac{3qa^2}{h_0^2} \qquad (p)$$

The values of the numerical factors γ and γ_1 for various proportions of the plate and for various values of $x = r/a$ are given by the curves in Figs. 147 and 148, respectively.

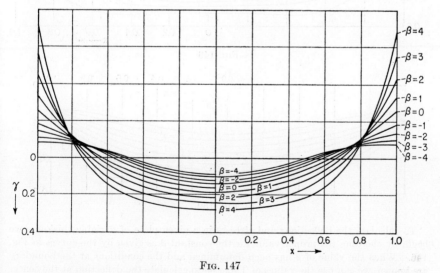

Fig. 147

For $\beta = 0$ these curves give the same values of stresses as were previously obtained for plates of uniform thickness (see Fig. 29, page 56).

In the case of a plate *simply supported along the edge*, the boundary conditions are

$$(w)_{x=1} = 0 \qquad (M_r)_{x=1} = 0 \qquad (q)$$

Investigation shows that the deflections and maximum stresses can be represented again by equations analogous to Eqs. (o) and (p). The notations α', γ', and γ_1' will be used for constants in this case, instead of α, γ, and γ_1 as used for clamped plates. The values of α' are given in the last line of Table 68, and the values of γ' and γ_1' are represented graphically in Figs. 149 and 150, respectively.

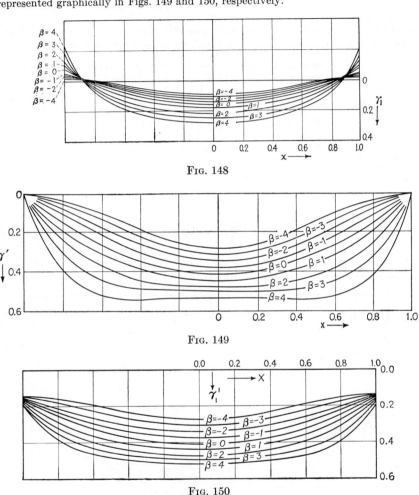

Fig. 148

Fig. 149

Fig. 150

To calculate the deflections and stresses in a given plate of variable thickness we begin by choosing the proper value for the constant β as given by the curves in Fig. 146. When the value of β has been determined and the conditions at the boundary are known, we can use the values of Table 68 to calculate the deflection at the center and the curves in Figs. 147, 148 or 149, 150 to calculate the maximum stress. If the shape of the diametrical section of the given plate cannot be represented with satisfactory accuracy by one of the curves in Fig. 146, an approximate method of solving the problem can always be used. This method consists in dividing the plate by concentric circles into several rings and using for each ring formulas developed for a ring plate of constant thickness. The procedure of calculation is then similar to that proposed by R. Grammel for calculating stresses in rotating disks.[1]

68. Annular Plates with Linearly Varying Thickness. Let us consider a circular plate with a concentric hole and a thickness varying as shown in Fig. 151. The plate carries a uniformly distributed surface load q and a line load $p = P/2\pi b$ uniformly distributed along the edge of the hole.[2] Letting $D_0 = Eh_0^3/12(1 - \nu^2)$ be the flexural rigidity of the plate at $r = b$, we have at any distance r from the center

$$D = \frac{D_0 r^3}{b^3} \qquad (a)$$

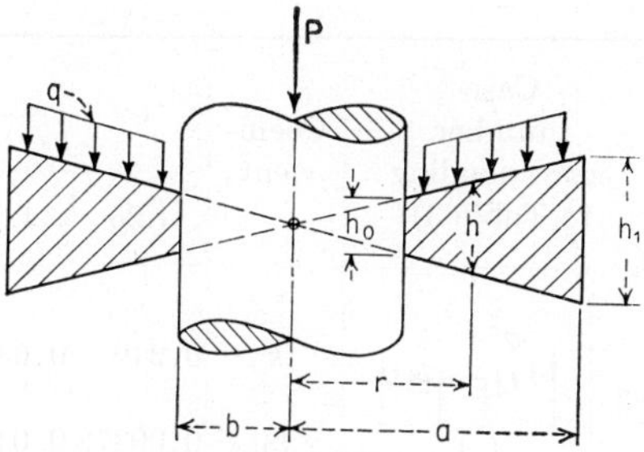

Fig. 151

Substituting this in Eq. (e) of Art. 67 and taking into account the additional shear force $P/2\pi r$ due to the edge load, we arrive at the differential equation

$$r^2 \frac{d^2\varphi}{dr^2} + 4r \frac{d\varphi}{dr} + (3\nu - 1)\varphi = -\frac{qb^3}{2D_0}\left(1 - \frac{b^2}{r^2}\right) - \frac{Pb^3}{2\pi D_0 r^2} \qquad (b)$$

The solution of the homogeneous equation corresponding to Eq. (b) is readily obtained by setting $\varphi = r^\alpha$. Combining this solution with a particular solution of Eq. (b), we get

$$\varphi = A r^{\alpha_1} + B r^{\alpha_2} + \frac{qb^3}{2D_0(1 - 3\nu)} - \frac{qb^5}{6(1 - \nu)D_0 r^2} + \frac{Pb^3}{6\pi(1 - \nu)D_0 r^2} \qquad (c)$$

in which

$$\alpha_1 = -1.5 + \sqrt{3.25 - 3\nu} \qquad \alpha_2 = -1.5 - \sqrt{3.25 - 3\nu} \qquad (d)$$

In the special case $\nu = \frac{1}{3}$, expression (c) has to be replaced by

$$\varphi = A + \frac{B}{r^3} - \frac{qb^3}{6D_0}\log\frac{r}{b} - \frac{qb^5}{4D_0 r^2} + \frac{Pb^3}{4\pi D_0 r^2} \qquad (e)$$

The arbitrary constants A and B must be determined from the respective conditions on the boundary of the plate. Writing, for brevity, φ_b for $(\varphi)_{r=b}$, and M_b for $(M_r)_{r=b}$, and introducing likewise φ_a, M_a, the last column of Table 69 contains the boundary conditions and the special values of q and P assumed in six different cases. The same table gives the values of coefficients k and k_1 calculated by means of the solution (e) and defined by the following expressions for the numerically largest stress and the largest deflection of the plate:

[1] R. Grammel, *Dinglers Polytech. J.*, vol. 338, p. 217, 1923. The analogy between the problem of a rotating disk and the problem of lateral bending of a circular plate of variable thickness was indicated by L. Föppl, *Z. angew. Math. Mech.*, vol. 2, p. 92, 1922. Nonsymmetrical bending of circular plates of nonuniform thickness is discussed by R. Gran Olsson, *Ingr.-Arch.*, vol. 10, p. 14, 1939.

[2] This case has been discussed by H. D. Conway, *J. Appl. Mechanics*, **vol. 15, p. 1,** 1948. Numerical results given in Table 69 are taken from that paper.

TABLE 69. VALUES OF COEFFICIENTS IN EQS. (f) FOR VARIOUS VALUES OF THE RATIO a/b (Fig. 151)

$\nu = \frac{1}{3}$

Case (number corresponding to Table 3)	Coefficient	a/b						Boundary conditions
		1.25	1.5	2	3	4	5	
3	k	0.249	0.638	3.96	13.64	26.0	40.6	$P = Q^*$ $\varphi_b = 0$ $M_a = 0$
	k_1	0.00372	0.0453	0.401	2.12	4.25	6.28	
4	k	0.149	0.991	2.23	5.57	7.78	9.16	$P = 0$ $\varphi_b = 0$ $M_a = 0$
	k_1	0.00551	0.0564	0.412	1.673	2.79	3.57	
5	k	0.1275	0.515	2.05	7.97	17.35	30.0	$P = Q^*$ $\varphi_b = 0$ $\varphi_a = 0$
	k_1	0.00105	0.0115	0.0934	0.537	1.261	2.16	
6	k	0.159	0.396	1.091	3.31	6.55	10.78	$q = 0$ $\varphi_b = 0$ $\varphi_a = 0$
	k_1	0.00174	0.0112	0.0606	0.261	0.546	0.876	
8	k	0.353	0.933	2.63	6.88	11.47	16.51	$q = 0$ $\varphi_b = 0$ $M_a = 0$
	k_1	0.00816	0.0583	0.345	1.358	2.39	3.27	
—	k	0.0785	0.208	0.52	1.27	1.94	2.52	$P = 0$ $\varphi_b = 0$ $\varphi_a = 0$
	k_1	0.00092	0.008	0.0495	0.193	0.346	0.482	

* Where $Q = \pi q(a^2 - b^2)$.

$$(\sigma_r)_{\max} = k\frac{qa^2}{h_1^2} \quad \text{or} \quad (\sigma_r)_{\max} = k\frac{P}{h_1^2}$$
$$w_{\max} = k_1\frac{qa^4}{Eh_1^3} \quad \text{or} \quad w_{\max} = k_1\frac{Pa^2}{Eh_1^3} \tag{f}$$

Numerical results valid for similar plates with constant thickness have been given in Table 3.

69. Circular Plates with Linearly Varying Thickness. In discussing the bending of the circular plate shown in Fig. 152,[1] we have to consider two portions of the plate separately.

1. The annular area $b < r < a$. Provided $\nu \neq \frac{1}{3}$, the slope $\varphi = dw/dr$ again is given by the expression (c) of Art. 68 without, however, its next to last term.

2. The inner area $r < b$. Here we have $dD/dr = 0$, and Eq. (e) of Art. 67 is reduced to

$$r^2\frac{d^2\varphi_i}{dr^2} + r\frac{d\varphi_i}{dr} - \varphi_i = -\frac{qr^3}{2D_0} - \frac{Pr}{2\pi D_0} \tag{a}$$

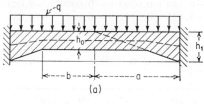

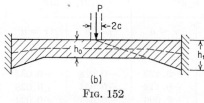

Fig. 152

where the subscript i refers to the inner portion of the plate. The general solution of Eq. (a) is

$$\varphi_i = A_i r + \frac{B_i}{r} - \frac{qr^3}{16D_0}$$
$$- \frac{Pr}{8\pi D_0}(2\log r + 1) \tag{b}$$

The constants A, B in Eq. (c) of Art. 68, and A_i, B_i in Eq. (b) above can be obtained from the boundary condition

$$(\varphi)_{r=a} = 0$$

and the conditions of continuity

$$(\varphi_i)_{r=0} = 0 \quad (\varphi - \varphi_i)_{r=b} = 0 \quad \left(\frac{d\varphi}{dr} - \frac{d\varphi_i}{dr}\right)_{r=b} = 0$$

Tables 70 and 71 give the deflection $w_{\max}$ and values of bending moments of the plate in two cases of loading. To calculate the bending moment at the center in the case of a central load P, we may assume a uniform distribution of that load over a small circular area of a radius c. The moment $M_r = M_t$ at $r = 0$ then can be expressed in the form

$$M_{\max} = M_0 - \frac{P}{4\pi}\left(1 - \frac{c^2}{2a^2}\right) + \gamma_1 P \tag{c}$$

In this formula M_0 is given by Eq. (83), which holds for a supported plate of constant thickness; the second term represents the effect of the edge moment; and the third term, due to the nonuniformity of the thickness of the plate, is given by Table 71.

[1] Clamped and simply supported plates of such a shape were discussed by H. Favre, *Bull. Tech. Suisse romande*, vol. 75, 1949. Numerical results given below are due substantially to H. Favre and E. Chabloz, *Bull. Tech. Suisse romande*, vol. 78, 1952.

TABLE 70. DEFLECTIONS AND BENDING MOMENTS OF CLAMPED CIRCULAR PLATES LOADED UNIFORMLY (Fig. 152a)
$\nu = 0.25$

$\dfrac{b}{a}$	$w_{\max} = \alpha \dfrac{qa^4}{Eh_0^3}$	$M_r = \beta qa^2$			$M_t = \beta_1 qa^2$		
		$r = 0$	$r = b$	$r = a$	$r = 0$	$r = b$	$r = a$
	α	β	β	β	β_1	β_1	β_1
0.2	0.008	0.0122	0.0040	−0.161	0.0122	0.0078	−0.040
0.4	0.042	0.0332	0.0007	−0.156	0.0332	0.0157	−0.039
0.6	0.094	0.0543	−0.0188	−0.149	0.0543	0.0149	−0.037
0.8	0.148	0.0709	−0.0591	−0.140	0.0709	0.0009	−0.035
1.0	0.176	0.0781	−0.125	−0.125	0.0781	−0.031	−0.031

TABLE 71. DEFLECTIONS AND BENDING MOMENTS OF CLAMPED CIRCULAR PLATES UNDER A CENTRAL LOAD (Fig. 152b)
$\nu = 0.25$

$\dfrac{b}{a}$	$w_{\max} = \alpha \dfrac{Pa^2}{Eh_0^3}$	$M_r = M_t$ $r = 0$	$M_r = \beta P$		$M_t = \beta_1 P$	
			$r = b$	$r = a$	$r = b$	$r = a$
	α	γ_1^*	β	β	β_1	β_1
0.2	0.031	−0.114	−0.034	−0.129	−0.028	−0.032
0.4	0.093	−0.051	−0.040	−0.112	−0.034	−0.028
0.6	0.155	−0.021	−0.050	−0.096	−0.044	−0.024
0.8	0.203	−0.005	−0.063	−0.084	−0.057	−0.021
1.0	0.224	0	−0.080	−0.080	−0.020	−0.020

* In Eq. (c).

In the case of a highly concentrated load requiring the use of the thick-plate theory, the stress at the center of the bottom surface of the plate is given by the expression

$$\sigma_{\max} = \sigma_0 + \frac{6P\gamma_1}{h_0^2} \qquad (d)$$

in which σ_0 may be calculated by means of expression (97).

Assuming next a variation of the flexural rigidity of the plate in accordance with the law

$$D = D_0 \left(1 - \frac{r}{a_0}\right)^m \qquad (e)$$

where a_0 denotes a length at least equal to the radius of the plate, we arrive in general at a slope φ expressible in terms of the hypergeometric function.[1] The particular assumption $m = 1/\nu$ leads, however, to a solution in a closed form. Taking, in addition, $\nu = \tfrac{1}{3}$ we arrive again at a plate with linearly variable thickness.[2]

[1] R. Gran Olsson, *Ingr.-Arch.*, vol. 8, p. 270, 1937.
[2] See especially H. D. Conway, *J. Appl. Mechanics*, vol. 18, p. 140, 1951, and vol. 20, p. 564, 1953.

Symmetrical deformation of plates such as shown in Fig. 153 also can be investigated by means of a parameter method akin to that described in Art. 39. Some numerical results[1] obtained in that way are given in Tables 72 and 73.

For bending moments and tensile stresses under central load P (Fig. 153b) expressions

$$M_{\max} = M_0 + \gamma_2 P \qquad (f)$$

and

$$\sigma_{\max} = \sigma_0 + \frac{6P\gamma_2}{h_0^2} \qquad (g)$$

analogous to Eqs. (c) and (d) may be used. M_0 again is given by expression (83), σ_0 denotes the value calculated by means of expression (96), and γ_2 is given in Table 73.

Of practical interest is also a combination of loadings shown in Fig. 153a and b. Taking $q = -P/\pi a^2$, we have the state of equilibrium of a circular footing carrying a central load P and submitted at the same time to a uniformly distributed soil reaction (Fig. 153c). Some data regarding this case, in particular the values of the factor γ_2, to be used in formulas (f) and (g), are given in Table 74.[2]

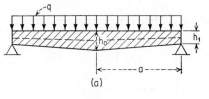
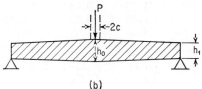
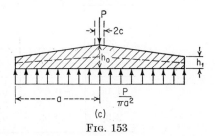

FIG. 153

TABLE 72. DEFLECTIONS AND BENDING MOMENTS OF SIMPLY SUPPORTED PLATES UNDER UNIFORM LOAD (Fig. 153a)
$\nu = 0.25$

$\dfrac{h_0}{h_1}$	$w_{\max} = \alpha \dfrac{qa^4}{Eh_0^3}$	$M_r = \beta q a^2$			$M_t = \beta_1 q a^2$		
		$r = 0$	$r = a/2$	$r = a$	$r = 0$	$r = a/2$	$r = a$
	α	β	β		β_1	β_1	β_1
1.00	0.738	0.203	0.152		0.203	0.176	0.094
1.50	1.26	0.257	0.176		0.257	0.173	0.054
2.33	2.04	0.304	0.195		0.304	0.167	0.029

[1] Due, as well as the method itself, to H. Favre and E. Chabloz, *Z. angew. Math. u. Phys.*, vol. 1, p. 317, 1950, and *Bull. Tech. Suisse romande*, vol. 78, 1952.

[2] For further results concerning circular plates with varying thickness see W. Gittleman, *Aircraft Eng.*, vol. 22, p. 224, 1950, and J. Paschoud, *Schweiz. Arch.*, vol. 17, p. 305, 1951. A graphical method of design has been given by P. F. Chenea and P. M. Naghdi, *J. Appl. Mechanics*, vol. 19, p. 561, 1952.

TABLE 73. DEFLECTIONS AND BENDING MOMENTS OF SIMPLY SUPPORTED
CIRCULAR PLATES UNDER CENTRAL LOAD (Fig. 153b)
$\nu = 0.25$

$\dfrac{h_0}{h_1}$	$w_{max} = \alpha \dfrac{Pa^2}{Eh_0^3}$	$M_r = M_t$	$M_r = \beta P$	$M_t = \beta_1 P$	
		$r = 0$	$r = a/2$	$r = a/2$	$r = a$
	α	γ_2	β	β_1	β_1
1.00	0.582	0	0.069	0.129	0.060
1.50	0.93	0.029	0.088	0.123	0.033
2.33	1.39	0.059	0.102	0.116	0.016

TABLE 74. BENDING MOMENTS OF A CIRCULAR FOOTING PLATE WITH CENTRAL
LOAD AND UNIFORMLY DISTRIBUTED SOIL PRESSURE (Fig. 153c)
$\nu = 0.25$

$\dfrac{h_0}{h_1}$	$M_r = M_t$	$M_r = \beta P$	$M_t = \beta_1 P$	
	$r = 0$	$r = a/2$	$r = a/2$	$r = a$
	γ_2	β	β_1	β_1
1.00	-0.065	0.021	0.073	0.030
1.50	-0.053	0.032	0.068	0.016
2.33	-0.038	0.040	0.063	0.007

70. Nonlinear Problems in Bending of Circular Plates.

From the theory of bending of bars it is known that, if the conditions at the supports of a bar or the loading conditions are changing with the deflection of the bar, this deflection will no longer be proportional to the load, and the principle of superposition cannot be applied.[1] Similar problems are also encountered in the case of bending of plates.[2] A simple example of this kind is shown in Fig. 154. A circular plate of radius a is pressed by a uniform load q against an absolutely rigid horizontal foundation. If moments of an intensity M_a are applied along the edge of the plate, a ring-shaped portion of the plate may be bent as shown in the figure,

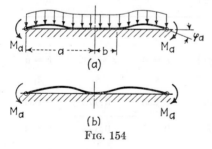

FIG. 154

---

[1] An example of such problems is discussed in S. Timoshenko, "Strength of Materials," part II, 3d ed., p. 69, 1956.

[2] See K. Girkmann, *Stahlbau*, vol. 18, 1931. Several examples of such problems are discussed also in a paper by R. Hofmann, *Z. angew. Math. Mech.*, vol. 18, p. 226, 1938.

whereas a middle portion of radius b may remain flat. Such conditions prevail, for example, in the bending of the bottom plate of a circular cylindrical container filled with liquid. The moments M_a represent in this case the action of the cylindrical wall of the container, which undergoes a local bending at the bottom. Applying to the ring-shaped portion of the bottom plate the known solution for a uniformly loaded circular plate [see expressions (m) in Art. 62], we obtain the deflection

$$w = C_1 + C_2 \log r + C_3 r^2 + C_4 r^2 \log r + \frac{qr^4}{64D} \qquad (a)$$

For determining the constants of integration $C_1, \ldots, C_4$ we have the following boundary conditions at the outer edge:

$$(w)_{r=a} = 0 \qquad (M_r)_{r=a} = -M_a \qquad (b)$$

Along the circle of radius b the deflection and the slope are zero. The bending moment M_r also must be zero along this circle, since the inner portion of the plate remains flat. Hence the conditions at the circle of radius b are

$$(w)_{r=b} = 0 \qquad \left(\frac{dw}{dr}\right)_{r=b} = 0 \qquad (M_r)_{r=b} = 0 \qquad (c)$$

By applying conditions (b) and (c) to expression (a) we obtain the five following equations:

$$C_1 + C_2 \log a + C_3 a^2 + C_4 a^2 \log a = -\frac{qa^4}{64D}$$

$$C_1 + C_2 \log b + C_3 b^2 + C_4 b^2 \log b = -\frac{qb^4}{64D}$$

$$C_2 \frac{\nu-1}{a^2} + C_3 2(\nu+1)$$
$$+ C_4(3 + 2\log a + 2\nu \log a + \nu) = -\frac{qa^2}{16D}(3+\nu) + \frac{M_a}{D} \qquad (d)$$

$$C_2 \frac{\nu-1}{b^2} + C_3 2(\nu+1)$$
$$+ C_4(3 + 2\log b + 2\nu \log b + \nu) = -\frac{qb^2}{16D}(3+\nu)$$

$$C_2 \frac{1}{b} + C_3 2b + C_4 b(2\log b + 1) = -\frac{qb^3}{16D}$$

By eliminating the constants $C_1, \ldots, C_4$ from these equations we obtain an equation connecting M_a and the ratio b/a, from which the radius b of the flat portion of the plate can be calculated for each given value of M_a. With this value of b the constants of integration can be evaluated, and the expression for the deflection of the plate can be obtained from Eq. (a). Representing the moment M_a and the angle of rotation φ_a of the edge of the plate by the equations

$$M_a = \alpha \frac{qa^2}{32} \quad \text{and} \quad \varphi_a = \beta \frac{qa^3}{32D} \tag{e}$$

and repeating the above-mentioned calculations for several values of the moment M_a, we can represent the relation between the constant factors α and β graphically, as shown in Fig. 155, for the particular case[1] $\nu = 0$.

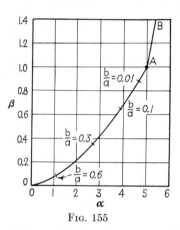

Fig. 155

It is seen from this figure that β does not vary in proportion to α and that the resistance to rotation of the edge of the plate decreases as the ratio b/a decreases. This condition holds up to the value $\alpha = 5$, at which value $\beta = 1$, $b/a = 0$, and the plate touches the foundation only at the center, as shown in Fig. 154b. For larger values of α, that is, for moments larger than $M_a = 5qa^2/32$, the plate does not touch the foundation, and the relation between α and β is represented by the straight line AB. The value $M_a = 5qa^2/32$ is that value at which the deflection at the center of the plate produced by the moments M_a is numerically equal to the deflection of a uniformly loaded plate simply supported along the edge [see Eq. (68)].

Another example of the same kind is shown in Fig. 156. A uniformly loaded circular plate is simply supported along the edge and rests at the center upon an absolutely rigid foundation. Again the ring-shaped portion of the plate with outer radius a and inner radius b can be treated as

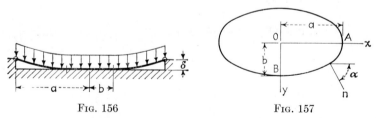

Fig. 156 Fig. 157

a uniformly loaded plate, and solution (a) can be used. The ratio b/a depends on the deflection δ and the intensity of the load q.

71. Elliptical Plates. *Uniformly Loaded Elliptical Plate with a Clamped Edge.* Taking the coordinates as shown in Fig. 157, the equation of the boundary of the plate is

$$\frac{x^2}{a^2} + \frac{y^2}{b^2} - 1 = 0 \tag{a}$$

---

[1] This case is discussed in the paper by Hofmann, *op. cit.*

The differential equation

$$\Delta\Delta w = \frac{q}{D} \qquad (b)$$

and the boundary conditions for the clamped edge, i.e.,

$$w = 0 \quad \text{and} \quad \frac{\partial w}{\partial n} = 0 \qquad (c)$$

are satisfied by taking for the deflection w the expression[1]

$$w = w_0\left(1 - \frac{x^2}{a^2} - \frac{y^2}{b^2}\right)^2 \qquad (d)$$

It is noted that this expression and its first derivatives with respect to x and y vanish at the boundary by virtue of Eq. (a). Substituting expression (d) in Eq. (b), we see that the equation is also satisfied provided

$$w_0 = \frac{q}{D\left(\dfrac{24}{a^4} + \dfrac{24}{b^4} + \dfrac{16}{a^2b^2}\right)} \qquad (199)$$

Thus, since expression (d) satisfies Eq. (b) and the boundary conditions, it represents the rigorous solution for a uniformly loaded elliptical plate with a clamped edge. Substituting $x = y = 0$ in expression (d), we find that w_0, as given by Eq. (199), is the deflection of the plate at the center. If $a = b$, we obtain for the deflection the value previously derived for a clamped circular plate [Eq. (62), page 55]. If $a = \infty$, the deflection w_0 becomes equal to the deflection of a uniformly loaded strip with clamped ends and having the span $2b$.

The bending and twisting moments are obtained by substituting expression (d) in Eqs. (101) and (102). In this way we find

$$M_x = -D\left(\frac{\partial^2 w}{\partial x^2} + \nu\frac{\partial^2 w}{\partial y^2}\right) = -4w_0 D\left[\frac{3x^2}{a^4} + \frac{y^2}{a^2b^2} - \frac{1}{a^2}\right.$$
$$\left. + \nu\left(\frac{x^2}{a^2b^2} + \frac{3y^2}{b^4} - \frac{1}{b^2}\right)\right] \qquad (e)$$

For the center of the plate and for the ends of the horizontal axis we obtain, respectively,

$$(M_x)_{x=0,y=0} = 4w_0 D\left(\frac{1}{a^2} + \frac{\nu}{b^2}\right) \quad \text{and} \quad (M_x)_{x=a,y=0} = -\frac{8w_0 D}{a^2} \qquad (f)$$

[1] This solution and the solution for a uniformly varying load q are obtained by G. H. Bryan; see A. E. H. Love's book, "Theory of Elasticity," 4th ed., p. 484. The case of an elliptical plate of variable thickness is discussed by R. Gran Olsson, *Ingr.-Arch.*, vol. 9, p. 108, 1938.

Similarly, for the moments M_y at the center and at the ends of the vertical axis we find, respectively,

$$(M_y)_{x=0,y=0} = 4w_0 D \left(\frac{1}{b^2} + \frac{\nu}{a^2}\right) \quad \text{and} \quad (M_y)_{x=0,y=b} = -\frac{8w_0 D}{b^2} \quad (g)$$

It is seen that the maximum bending stress is obtained at the ends of the shorter principal axis of the ellipse. Having the moments M_x, M_y, and M_{xy}, the values of the bending moment M_n and the twisting moment M_{nt} at any point on the boundary are obtained from Eqs. (c) (Art. 22, page 87) by substituting in these equations

$$\cos \alpha = \frac{dy}{ds} = \frac{b^2 x}{\sqrt{a^4 y^2 + b^4 x^2}} \quad \sin \alpha = -\frac{dx}{ds} = \frac{a^2 y}{\sqrt{a^4 y^2 + b^4 x^2}} \quad (h)$$

The shearing forces Q_x and Q_y at any point are obtained by substituting expression (d) in Eqs. (106) and (107). At the boundary the shearing force Q_n is obtained from Eq. (d) (Art. 22, page 87), and the reaction V_n from Eq. (g) of the same article. In this manner we find that the intensity of the reaction is a maximum at the ends of the minor axis of the ellipse and that its absolute value is

$$(V_n)_{\max} = \frac{a^2 b(3a^2 + b^2) q}{3a^4 + 3b^4 + 2a^2 b^2} \quad \text{for } a > b \quad (i)$$

The smallest absolute value of V_n is at the ends of the major axis of the ellipse where

$$(V_n)_{\min} = \frac{ab^2(a^2 + 3b^2) q}{3a^4 + 3b^4 + 2a^2 b^2} \quad (j)$$

For a circle, $a = b$, and we find $(V_n)_{\max} = (V_n)_{\min} = qa/2$.

Elliptical Plate with a Clamped Edge and Bent by a Linearly Varying Pressure. Assuming that $q = q_0 x$, we find that Eq. (b) and the boundary conditions (c) are satisfied by taking

$$w = \frac{q_0 x}{24 D} \frac{\left(1 - \dfrac{x^2}{a^2} - \dfrac{y^2}{b^2}\right)^2}{\dfrac{5}{a^4} + \dfrac{1}{b^4} + \dfrac{2}{a^2 b^2}} \quad (200)$$

From this expression the bending moments and the reactions at the boundary can be calculated as in the previous case.

Uniformly Loaded Elliptical Plate with Simply Supported Edge. The solution for this case is more complicated than in the case of clamped edges;[1] therefore we give here only some final numerical results. Assuming that $a/b > 1$, we represent the deflection and the bending moments at the center by the formulas

$$(w)_{x=y=0} = \alpha \frac{qb^4}{Eh^3} \quad M_x = \beta q b^2 \quad M_y = \beta_1 q b^2 \quad (k)$$

---

[1] See B. G. Galerkin, *Z. angew. Math. Mech.*, vol. 3, p. 113, 1923.

The values of the constant factors α, β, and β_1 for various values of the ratio a/b and for $\nu = 0.3$ are given in Table 75.

TABLE 75. FACTORS α, β, β_1 IN FORMULAS (k) FOR UNIFORMLY LOADED AND SIMPLY SUPPORTED ELLIPTICAL PLATES
$\nu = 0.3$

a/b	1	1.1	1.2	1.3	1.4	1.5	2	3	4	5	∞
α	0.70	0.83	0.96	1.07	1.17	1.26	1.58	1.88	2.02	2.10	2.28
β	0.206	0.215	0.219	0.223	0.223	0.222	0.210	0.188	0.184	0.170	0.150
β_1	0.206	0.235	0.261	0.282	0.303	0.321	0.379	0.433	0.465	0.480	0.500

Comparison of these numerical values with those previously obtained for rectangular plates (Table 8, page 120) shows that, for equal values of the ratio of the sides of rectangular plates and the ratio a/b of the semiaxes of elliptical plates, the values of the deflections and the moments at the center in the two kinds of plate do not differ appreciably. The case of a plate having the form of half an ellipse bounded by the transverse axis has also been discussed.[1]

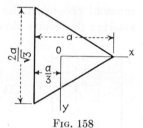

FIG. 158

72. Triangular Plates. *Equilateral Triangular Plate Simply Supported at the Edges.* The bending of such a triangular plate by moments M_n uniformly distributed along the boundary has already been discussed (see page 94). It was shown that in such a case the deflection surface of the plate is the same as that of a uniformly stretched and uniformly loaded membrane and is represented by the equation

$$w = \frac{M_n}{4aD}\left[x^3 - 3y^2x - a(x^2 + y^2) + \frac{4}{27}a^3\right] \quad (a)$$

in which a denotes the height of the triangle, and the coordinate axes are taken as shown in Fig. 158.

In the case of a uniformly loaded plate the deflection surface is[2]

$$w = \frac{q}{64aD}\left[x^3 - 3y^2x - a(x^2 + y^2) + \frac{4}{27}a^3\right]\left(\frac{4}{9}a^2 - x^2 - y^2\right) \quad (201)$$

---

[1] B. G. Galerkin, *Messenger Math.*, vol. 52, p. 99, 1923. For bending of clamped elliptical plates by concentrated forces see H. Happel, *Math. Z.*, vol. 6, p. 203, 1920, and C. L. Perry, *Proc. Symposia Appl. Math.*, vol. 3, p. 131, 1950. See also H. M. Sengupta, *Bull. Calcutta Math. Soc.*, vol. 41, p. 163, 1949, and vol. 43, p. 123, 1950; this latter paper also contains a correction to the former one. By means of curvilinear coordinates, solutions for plates clamped along some other contour lines and submitted to a uniform load have been obtained by B. Sen, *Phil. Mag.*, vol. 33, p. 294, 1942.

[2] The problem of bending of a plate having the form of an equilateral triangle was solved by S. Woinowsky-Krieger, *Ingr.-Arch.*, vol. 4, p. 254, 1933.

By differentiation we find

$$\Delta w = -\frac{q}{4aD}\left[x^3 - 3y^2 x - a(x^2 + y^2) + \frac{4}{27}a^3\right] \quad (b)$$

It may be seen from (201) and (b) that the deflection and the bending moment at the boundary vanish, since the expression in the brackets is zero at the boundary. Further differentiation gives

$$\Delta\Delta w = \frac{q}{D} \quad (c)$$

Hence the differential equation of the deflection surface is also satisfied, and expression (201) represents the solution of the problem. Having the expression for deflections, the expressions for the bending moments and the shearing forces can be readily obtained. The maximum bending moment occurs on the lines bisecting the angles of the triangle. Considering the points along the x axis and taking $\nu = 0.3$, we find

$$\begin{aligned}(M_x)_{\max} &= 0.0248qa^2 \quad \text{at } x = -0.062a \\ (M_y)_{\max} &= 0.0259qa^2 \quad \text{at } x = 0.129a\end{aligned} \quad (202)$$

At the center of the plate

$$M_x = M_y = (1 + \nu)\frac{qa^2}{54} \quad (203)$$

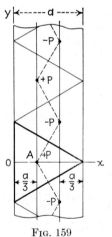

Fig. 159

The case of a concentrated force acting on the plate can be solved by using the *method of images* (see page 156). Let us take a case in which the point of application of the load is at the center A of the plate (Fig. 159). Considering the plate, shown in the figure by the heavy lines, as a portion of an infinitely long rectangular plate of width a, we apply the fictitious loads P with alternating signs as shown in the figure. The nodal lines of the deflection surface, produced by such loading, evidently divide the infinitely long plate into equilateral triangles each of which is in exactly the same condition as the given plate. Thus our problem is reduced to that of bending of an infinitely long rectangular plate loaded by the two rows of equidistant loads $+P$ and $-P$. Knowing the solution for one concentrated force (see Art. 36) and using the method of superposition, the deflection at point A and the stresses near that point can be readily calculated, since the effect of the fictitious forces on bending decreases rapidly as their distance from point A increases. In this manner we find the deflection at A:

$$w_0 = 0.00575\frac{Pa^2}{D} \quad (204)$$

The bending moments at a small distance c from A are given by the expressions

$$M_x = \frac{(1+\nu)P}{4\pi}\left(\log\frac{a\sqrt{3}}{\pi c} - 0.379\right) - \frac{(1-\nu)P}{8\pi}$$
$$M_y = \frac{(1+\nu)P}{4\pi}\left(\log\frac{a\sqrt{3}}{\pi c} - 0.379\right) + \frac{(1-\nu)P}{8\pi}$$
(205)

Since for a simply supported and centrally loaded circular plate of radius a_0 the radial and the tangential moments at a distance c from the center are, respectively (see page 68),

and
$$M_r = \frac{(1+\nu)P}{4\pi}\log\frac{a_0}{c}$$
$$M_t = \frac{(1+\nu)P}{4\pi}\log\frac{a_0}{c} + \frac{(1-\nu)P}{4\pi}$$
(d)

it can be concluded that the first terms on the right-hand side of Eqs. (205) are identical with the logarithmical terms for a circular plate with a radius

$$a_0 = \frac{a\sqrt{3}}{\pi}e^{-0.379} \quad (e)$$

Hence the local stresses near the point of application of the load can be calculated by using the thick-plate theory developed for circular plates (see Art. 19).

Equilateral Triangular Plates with Two or Three Edges Clamped. Triangular plates are used sometimes as bottom slabs of bunkers and silos. In such a case each triangular plate is rigidly clamped along both its inclined edges and clamped elastically along its third, horizontal edge (Fig. 160). Only the uniform and the hydrostatic distribution of the load is of practical interest. The largest bending moment of the panel and the clamping moments at the middle of a built-in edge may be represented as

$$M = \beta q a^2 \quad \text{or} \quad M = \beta_1 q_0 a^2 \quad (f)$$

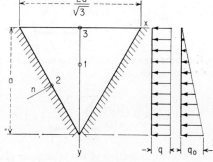

Fig. 160

according to the type of loading (Fig. 160). The values of coefficients β and β_1, obtained by the method of finite differences,[1] are given in Table 76.

It should be noted, finally, that a plate in form of a triangle with angles $\pi/2$, $\pi/3$, and $\pi/6$ and having all edges simply supported can be considered as one-half of the equilateral plate (Fig. 158), this latter being loaded antisymmetrically above the axis

[1] See A. Smotrov, "Solutions for Plates Loaded According to the Law of Trapeze," Moscow, 1936.

TABLE 76. VALUES OF THE FACTORS β, β_1 IN EQS. (f) FOR EQUILATERAL TRIANGULAR PLATES (Fig. 160)
$\nu = 0.20$

Load distribution		Edge $y = 0$ simply supported				Edge $y = 0$ clamped			
		M_{x1}	M_{y1}	M_{n2}	M_{y3}	M_{x1}	M_{y1}	M_{n2}	M_{y3}
Uniform.........	β	0.0126	0.0147	−0.0285	0	0.0113	0.0110	−0.0238	−0.0238
Hydrostatic.....	β_1	0.0053	0.0035	−0.0100	0	0.0051	0.0034	−0.0091	−0.0060

x. The problem of bending of such a plate can be solved in several ways—for example, by the method of images.[1]

Plate in the Form of an Isosceles Right Triangle with Simply Supported Edges. Such a plate may be considered as one-half of a square plate, as indicated in Fig. 161 by dashed lines, and the methods previously developed for rectangular plates can be applied.[2] If a load P is applied at a point A with coordinates ξ, η (Fig. 161), we assume a fictitious load $-P$ applied at A', which is the image of the point A with respect to the line BC. These two loads evidently produce a deflection of the square plate such that the diagonal BC becomes a nodal line. Thus the portion OBC of the square plate is in exactly the same condition as a simply supported triangular plate OBC. Considering the load $+P$ and using the Navier solution for a square plate (page 111), we obtain the deflection

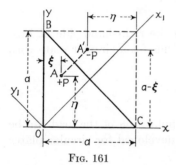

Fig. 161

$$w_1 = \frac{4Pa^2}{\pi^4 D} \sum_{m=1}^{\infty} \sum_{n=1}^{\infty} \frac{\sin\dfrac{m\pi\xi}{a} \sin\dfrac{n\pi\eta}{a}}{(m^2+n^2)^2} \sin\frac{m\pi x}{a} \sin\frac{n\pi y}{a} \qquad (g)$$

In the same manner, considering the load $-P$ and taking $a - \eta$ instead of ξ and $a - \xi$ instead of η, we obtain

$$w_2 = -\frac{4Pa^2}{\pi^4 D} \sum_{m=1}^{\infty} \sum_{n=1}^{\infty} (-1)^{m+n} \frac{\sin\dfrac{m\pi\eta}{a} \sin\dfrac{n\pi\xi}{a}}{(m^2+n^2)^2} \sin\frac{m\pi x}{a} \sin\frac{n\pi y}{a} \qquad (h)$$

The complete deflection of the triangular plate is obtained by summing up expressions (g) and (h):

$$w = w_1 + w_2 \qquad (i)$$

[1] For the solution of this problem in double series, see R. Girtler, *Sitzber. Akad. Wiss. Wien*, vol. 145, p. 61, 1936. The bending of equilateral triangular plates with variable thickness has been discussed by H. Göttlicher, *Ingr.-Arch.*, vol. 9, p. 12, 1938.

[2] This method of solution was given by A. Nádai, "Elastische Platten," p. 178, 1925. Another way of handling the same problem was given by B. G. Galerkin, *Bull. acad. sci. Russ.*, p. 223, 1919, and *Bull. Polytech. Inst.*, vol. 28, p. 1, St. Petersburg, 1919.

PLATES OF VARIOUS SHAPES

To obtain the deflection of the triangular plate produced by a uniformly distributed load of intensity q, we substitute $q\,d\xi\,d\eta$ for P and integrate expression (i) over the area of the triangle OBC. In this manner we obtain

$$w = \frac{16qa^4}{\pi^6 D}\left[\sum_{m=1,3,5,\ldots}^{\infty}\sum_{n=2,4,6,\ldots}^{\infty}\frac{n\sin\dfrac{m\pi x}{a}\sin\dfrac{n\pi y}{a}}{m(n^2-m^2)(m^2+n^2)^2} \right.$$
$$\left.+ \sum_{m=2,4,6,\ldots}^{\infty}\sum_{n=1,3,5,\ldots}^{\infty}\frac{m\sin\dfrac{m\pi x}{a}\sin\dfrac{n\pi y}{a}}{n(m^2-n^2)(m^2+n^2)^2}\right] \quad (j)$$

This is a rapidly converging series and can be used to calculate the deflection and the bending moments at any point of the plate. Taking the axis of symmetry of the triangle in Fig. 161 as the x_1 axis and representing the deflections and the moments M_{x_1} and M_{y_1} along this axis by the formulas

$$w = \alpha\frac{qa^4}{Eh^3} \qquad M_{x_1} = \beta qa^2 \qquad M_{y_1} = \beta_1 qa^2 \quad (k)$$

the values of the numerical factors α, β, and β_1 are as given in Figs. 162 and 163. By comparing these results with those given in Table 8 for a uniformly loaded square

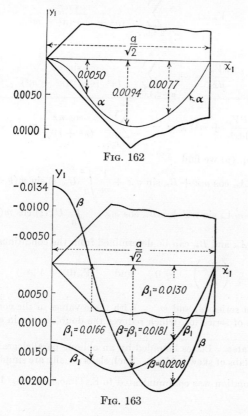

Fig. 162

Fig. 163

plate, it can be concluded that for the same value of a the maximum bending moment for a triangular plate is somewhat less than half the maximum bending moment for a square plate.

To simplify the calculation of the deflections and moments, the double series (j) can be transformed into simple series.[1] For this purpose we use the known series

$$U_m(x) = \sum_{n=2,4,6,\ldots}^{\infty} \frac{\cos nx}{(n^2 + m^2)^2} = -\frac{2}{m^4} + \frac{\pi}{2m^3} \frac{\cosh m\left(\frac{\pi}{2} - x\right)}{\sinh \frac{\pi m}{2}}$$

$$+ \frac{\pi^2}{4m^2} \frac{\cosh mx}{\sinh^2 \frac{\pi m}{2}} + \frac{\pi x}{2m^2} \frac{\sinh m\left(\frac{\pi}{2} - x\right)}{\sinh \frac{m\pi}{2}} \quad (l)$$

which can be represented in the following form

$$U_m(x) = (\alpha_m + \beta_m x)\cosh mx + (\gamma_m + \delta_m x)\sinh mx - \frac{2}{m^4} \quad (m)$$

Considering now the series

$$V_m(x) = \sum_{n=2,4,6,\ldots}^{\infty} \frac{\cos nx}{(n^2 + m^2)^2(n^2 - m^2)} \quad (n)$$

we obtain

$$\frac{dV_m}{dx} = -\sum_{n=2,4,6,\ldots}^{\infty} \frac{n \sin nx}{(n^2 + m^2)^2(n^2 - m^2)} \quad (o)$$

and

$$\frac{d^2V_m}{dx^2} + m^2 V_m = -\sum_{n=2,4,6,\ldots}^{\infty} \frac{\cos nx}{(n^2 + m^2)^2} = -U_m \quad (p)$$

By integrating Eq. (p) we find

$$V_m = A_m \cos mx + B_m \sin mx + \frac{1}{m} \int_0^x U_m(\xi) \sin m(\xi - x)\, d\xi \quad (q)$$

and

$$\frac{dV_m}{dx} = -mA_m \sin mx + mB_m \cos mx - \int_0^x U_m(\xi) \cos m(\xi - x)\, d\xi \quad (r)$$

The constants A_m and B_m can be determined from the conditions

$$\left(\frac{dV_m}{dx}\right)_{x=0} = 0 \quad \text{and} \quad V_m(0) = V_m(\pi) \quad (s)$$

which follow from series (o) and (n). With these values of the constants expression (r) gives the sum of series (o), which reduces the double series in expression (j) to a simple series.

73. Skewed Plates. Plates bounded by an oblique parallelogram have been used recently as floor slabs of skew bridges. Such slabs usually are simply supported along

---

[1] This transformation was communicated to S. Timoshenko by J. V. Uspensky.

the abutments, whereas both other sides remain free or are supported elastically by "curbs" or beams.

In the most general case the use of an oblique system of coordinates chosen in accordance with the given angle of skew should be recommended; in certain particular cases rectangular coordinates may also be used to advantage in dealing with skew plates, and the method of finite differences appears, in general, to be the most promising. The following numerical data for uniformly loaded skewed plates were obtained in that way.[1] At the center of a skew plate with all edges simply supported (Fig. 164a), let

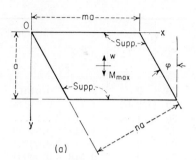

$$w = \alpha \frac{qa^4}{D} \qquad M_{\max} = \beta q a^2 \qquad (a)$$

The bending moment $M_{\max}$ acts very nearly in the direction of the short span of the plate.

If the edges $y = 0$ and $y = a$ are free and the other two edges are simply supported (Fig. 164b), the central portion of the plate carries the load in the direction normal to the abutments. Letting w_0 and $(M_0)_{\max}$ be, respectively, the deflection and bending moment at the center of the plate, and $(w_1)_{\max}$ and $(M_1)_{\max}$ the corresponding quantities at the free edge, we may express these quantities in the form

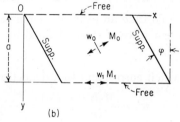

Fig. 164

$$w_0 = \alpha_0 \frac{qa^4}{D} \qquad (M_0)_{\max} = \beta_0 q a^2$$

$$(w_1)_{\max} = \alpha_1 \frac{qa^4}{D} \qquad (M_1)_{\max} = \beta_1 q a^2$$

(b)

The numerical values of the coefficients are given in Table 77.

74. Stress Distribution around Holes. In order to investigate the stress distribution around a hole, it is simplest to consider a very large plate; results obtained in this way prove to be applicable without appreciable inaccuracy to plates of any shape, provided the width of the hole remains small as compared with the over-all dimensions of the plate.

---

[1] The most data are due to V. P. Jensen, *Univ. Illinois Bull.* 332, 1941, and V. P. Jensen and J. W. Allen, *Univ. Illinois Bull.* 369, 1947. See also C. P. Siess, *Proc. ASCE*, vol. 74, p. 323, 1948. Analytical methods have been applied by H. Favre, *Schweiz. Bauztg.*, vol. 60, p. 35, 1942; P. Lardy, *Schweiz. Bauztg.*, vol. 67, p. 207, 1949; and also by J. Krettner, *Ingr.-Arch.*, vol. 22, p. 47, 1954, where further bibliography is given. For use of energy methods see also A. M. Guzmán and C. J. Luisoni, *Publs. Univ. Nacl. Buenos Aires*, p. 452, 1953. Pure bending of skewed plates has been discussed by E. Reissner, *Quart. Appl. Math.*, vol. 10, p. 395, 1953. Models of skewed plates were tested by L. Schmerber, G. Brandes, and H. Schambeck, *Bauingenieur*, vol. 33, p. 174, 1958. For use of finite differences see also Art. 83.

Table 77. Values of Coefficients in Eqs. (a) and (b) for Deflections and Bending Moments of Uniformly Loaded Skewed Plates
$\nu = 0.2$

$\varphi°$	m	n	Plate in Fig. 164a		Plate in Fig. 164b			
			α	β	α_0	α_1	β_0	β_1
0	2	2	0.01013	0.0999	0.214	0.224	0.495	0.508
30	2.02	1.75	0.01046	0.0968				
30	1.92	1.67	......	......	0.1183	0.1302	0.368	0.367
45	2	1.414	0.00938	0.0898	0.0708	0.0869	0.291	0.296
60	2	1	0.00796	0.0772	0.0186	0.0396	0.166	0.152
75	2	0.518	0.00094	0.0335				

To take an example, let us consider an infinitely large plate in a uniform state of stress defined by the bending moments

$$M'_x = M_0 \qquad M'_y = 0 \tag{a}$$

which correspond to a deflection surface

$$w' = \frac{M_0(x^2 - \nu y^2)}{2D(1 - \nu^2)} = -\frac{M_0 r^2}{4D(1 - \nu^2)}[1 - \nu + (1 + \nu)\cos 2\theta] \tag{b}$$

To obtain the disturbance produced in such a state of pure bending by a circular hole with a radius a (Fig. 165), we assume the material to be removed inside the periphery of the circle. Then we have to replace the action of the initial stresses along the periphery of the hole by the action of the external couples and forces:

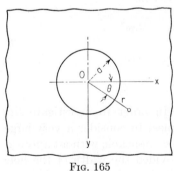

Fig. 165

$$(M'_r)_{r=a} = \frac{M_0}{2}(1 + \cos 2\theta)$$
$$(V'_r)_{r=a} = \frac{M_0}{a}\cos 2\theta \tag{c}$$

which are readily obtained by differentiation of expression (b) in accordance with Eqs. (192).

On the initial state of stress we superimpose now an additional state of stress such that (1) the combined couples and forces vanish at $r = a$ and (2) the superimposed stresses taken alone vanish at infinity ($r = \infty$).

We can fulfill both conditions by choosing the additional deflection in the form

$$w'' = -\frac{M_0 a^2}{2D}\left[A \log r + \left(B + C\frac{a^2}{r^2}\right)\cos 2\theta\right] \tag{d}$$

This expression also satisfies the homogeneous differential equation (194) and yields the following stress resultants on the periphery of the hole:

$$(M_r'')_{r=a} = -\frac{M_0}{2}\{(1-\nu)A + [4\nu B - 6(1-\nu)C]\cos 2\theta\}$$
$$(V_r'')_{r=a} = \frac{M_0}{a}[(6-2\nu)A + 6(1-\nu)C]\cos 2\theta \qquad (e)$$

Since expressions (c) and (e) for M_r contain a constant term as well as a term proportional to $\cos 2\theta$, while both expressions for V_r contain only one term, three equations are needed to satisfy the required conditions $M_r' + M_r'' = 0$ and $V_r' + V_r'' = 0$ on the periphery of the hole. Resolving these equations with respect to the unknown coefficients A, B, and C, we obtain the final deflections $w = w' + w''$ and the following stress resultants along the periphery of the plate:

$$M_t = M_0\left[1 - \frac{2(1+\nu)}{3+\nu}\cos 2\theta\right]$$
$$Q_t = \frac{4M_0}{(3+\nu)a}\sin 2\theta \qquad (f)$$

For $\theta = \pi/2$ and $\theta = \pi/4$, respectively, we obtain

$$(M_t)_{\max} = \frac{5+3\nu}{3+\nu}M_0$$
$$(Q_t)_{\max} = \frac{4}{(3+\nu)a}M_0 \qquad (g)$$

It is usual to represent the largest value of a stress component due to a local disturbance in the form

$$\sigma_{\max} = k\sigma \qquad (h)$$

where σ denotes the average value of the respective component in the same section and k is the so-called factor of stress concentration. Having in mind the largest bending stress along the periphery of the hole, we can also write $k = (M_t)_{\max}/M_0$, M_0 being the initial value of the stress couples at $\theta = \pi/2$, where this largest stress occurs. Thus in the event of pure bending we have

$$k = \frac{5+3\nu}{3+\nu} \qquad (i)$$

equal to about 1.80 for steel ($\nu = \frac{1}{3}$).

Factors of stress concentration could be obtained in a similar manner for various modes of a uniform state of stress and also for holes of other than circular shape.[1] All such results, however, prove to be of relatively little value for the following reason.

---

[1] See J. N. Goodier, *Phil. Mag.*, vol. 22, p. 69, 1936, and G. N. Savin, "Stress Concentration around Holes," Moscow, 1951.

While the bending stresses (to take only the previously discussed case) do not exceed the value of $\sigma_{max} = 6M_0k/h^2$, the largest value of the corresponding shearing stresses is given by

$$\tau_{max} = \frac{3}{2h} Q_{max} = \frac{6M_0}{(3+\nu)ah} = \frac{\sigma_{max}}{(3+\nu)k} \frac{h}{a} \qquad (j)$$

Thus, by decreasing the ratio a/h we can increase the ratio τ_{max}/σ_{max} at will. In this way we soon arrive at transverse shearing stresses of such a magnitude that their effect on the plate deformation ceases to be negligible in comparison with the effect of the couples. Consequently, to

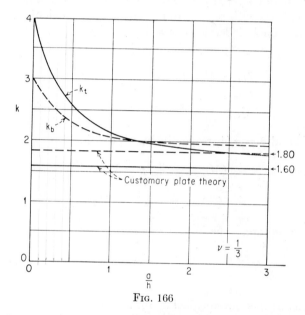

Fig. 166

assure reliable results regarding the stress distribution around holes, we have to resort to special theories which take the shear deformation into account.

Stress-concentration factors obtained[1] by means of E. Reissner's theory (see Art. 39) are plotted in Fig. 166 versus the value of a/h. The curve k_b holds in the case of pure bending considered above; the curve k_t gives the stress concentration in the event of a uniform twist, produced by couples $M_x = M_0$, $M_y = -M_0$ in the initial state of stress. The values

---

[1] E. Reissner, *J. Appl. Mechanics*, vol. 12, p. A-75, 1945. The case is discussed most rigorously by J. B. Alblas, "Theorie van de driedimensionale Spanningstoestand in een doorborde plaat," Amsterdam, 1957. For bending of a square plate with a circular hole, see M. El-Hashimy, "Ausgewählte Plattenprobleme," Zürich, 1956, where customary theory is applied.

$k_b = 1.80$ and $k_t = 1.60$ given for these cases by the customary theory appear, if plotted, as straight lines which approach both respective curves asymptotically as the ratio a/h increases indefinitely. It is seen from the graph that even for holes three times as wide as the plate is thick the error resulting from the application of the usual theory exceeds 10 per cent of the true value of k_b. It is also noteworthy that for vanishing hole diameter the limit value $k_b = 3$ of the stress-concentration factor in pure bending becomes equal to the value of the same factor in plane stress when uniform tension in one direction is assumed.

If the hole (Fig. 165) is filled up with an elastic material other than that of the plate, we have to deal with an "elastic inclusion." The unfilled hole and the rigid inclusion

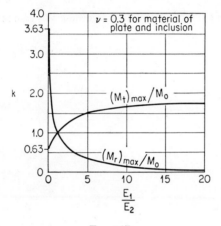

Fig. 167

can be regarded as the limiting cases of the elastic inclusion, Young's modulus of the filling being zero in the former and infinitely large in the latter case. In the following, the effect of a rigid inclusion is briefly considered.

Just as in the case of a hole, we have to combine an initial state of stress with a supplementary one; however, the conditions now to be fulfilled on the periphery of the circle $r = a$ are (in the symmetrical case)

$$(w)_{r=a} = 0 \qquad \left(\frac{\partial w}{\partial r}\right)_{r=a} = 0 \qquad (k)$$

where w is the combined deflection of the plate. From the expressions (192) for the stress resultants, we readily conclude that on the periphery of the inclusion the relation $M_t = \nu M_r$ must hold, whereas the moments M_{rt} become zero.

In the particular case of pure bending, assumed on page 43, we obtain a distribution of radial moments around the rigid inclusion given by[1]

[1] M. Goland, *J. Appl. Mechanics*, vol. 10, p. A-69, 1943; Fig. 167 is taken from this paper. See also Yi-Yuan Yu, *Proc. Second U.S. Natl. Congr. Appl. Mech.*, Ann Arbor, Mich., 1954, p. 381.

$$M_r = M_0 \left(\frac{1}{1+\nu} + \frac{2\cos 2\theta}{1-\nu} \right) \tag{l}$$

The respective stress concentration factor is equal to $k = (3 + \nu)/(1 - \nu^2)$, that is, to 3.63 for steel. The effect of the transverse shear deformation, however, is not implied in this result, which consequently holds only for large values of a/h.

It is seen that in the vicinity of a rigid inclusion the radial couples M_r far exceed the tangential couples M_t; this is in strong contrast to the stress state around a hole, where the couples M_t dominate the couples M_r. Both moments are balanced best in their magnitude in the case of an elastic inclusion, as shown in Fig. 167. Here E_1 denotes Young's modulus of the plate and E_2 that of the filling.

An inclusion with elastic filling may be replaced, without substantially changing its effect on the plate, by an annular elastic inclusion. Reinforcing a hole with a ring of properly chosen stiffness can therefore considerably reduce the stress concentration in the material of the plate around the hole.[1]

---

[1] For stress analysis and numerical data regarding this case see Savin, *op. cit.*

CHAPTER 10

SPECIAL AND APPROXIMATE METHODS IN THEORY OF PLATES

75. Singularities in Bending of Plates. The state of stress in a plate is said to have a singularity at a point[1] (x_0,y_0) if any of the stress components at that point becomes infinitely large. From expressions (101), (102), and (108) for moments and shearing forces we see that a singularity does not occur as long as the deflection $w(x,y)$ and its derivatives up to the order four are continuous functions of x and y.

Singularities usually occur at points of application of concentrated forces and couples. In certain cases a singularity due to reactive forces can occur at a corner of a plate, irrespective of the distribution of the surface loading.

In the following discussion, let us take the origin of the coordinates at the point of the plate where the singularity occurs. The expressions for the deflection given below yield (after appropriate differentiations) stresses which are large in comparison with the stresses resulting from loading applied elsewhere or from edge forces, provided x and y are small.

Single Force at an Interior Point of a Plate. If the distance of the point under consideration from the boundary and from other concentrated loads is sufficiently large, we have approximately a state of axial symmetry around the single load P. Consequently, the radial shearing force at distance r from the load P is

$$Q_r = -\frac{P}{2\pi r}$$

Observing the expression (193) for Q_r we can readily verify that the respective deflection is given by

$$w_0 = \frac{P}{8\pi D} r^2 \log \frac{r}{a} \qquad (206)$$

in which a is an arbitrary length. The corresponding term $r^2 \log a$ yields negligible stresses when the ratio r/a remains small.

Single Couple at an Interior Point of a Plate. Let us apply a single

---

[1] More exactly, at a point (x_0,y_0,z).

force $-M_1/\Delta x$ at the origin and a single force $+M_1/\Delta x$ at the point $(-\Delta x,0)$, assuming that M_1 is a known couple. From the previous result [Eq. (206)] the deflection due to the combined action of both forces is

$$w = \frac{M_1}{8\pi D} \frac{(x+\Delta x)^2 + y^2}{\Delta x} \log \frac{[(x+\Delta x)^2 + y^2]^{\frac{1}{2}}}{a}$$
$$- \frac{M_1}{8\pi D} \frac{x^2 + y^2}{\Delta x} \log \frac{(x^2+y^2)^{\frac{1}{2}}}{a} \quad (a)$$

As Δx approaches zero, we obtain the case of a couple M_1 concentrated at the origin (Fig. 168a) and the deflection is

$$w_1 = \lim\,[w]_{\Delta x \to 0} = \frac{M_1}{P} \frac{\partial w_0}{\partial x}$$

where w_0 is the deflection given by expression (206). Performing the differentiation we obtain

$$w_1 = \frac{M_1 x}{8\pi D}\left(\log \frac{x^2+y^2}{a^2} + 1\right) \quad (b)$$

If we omit the second term $M_1 x/8\pi D$, which gives no stresses, and use polar coordinates, this expression becomes

$$w_1 = \frac{M_1}{4\pi D} r \log \frac{r}{a} \cos \theta \quad (207)$$

In the case of the couple M_2 shown in Fig. 168b we have only to replace θ by $\theta + \pi/2$ in the previous formula to obtain the corresponding deflection.

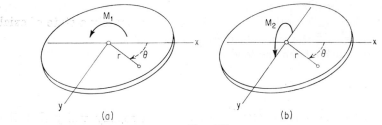

Fig. 168

Double Couple at an Interior Point of a Plate. Next we consider the combined action of two equal and opposite couples acting in two parallel planes Δx apart, as shown in Fig. 169. Putting $M_1\,\Delta x = H_1$ and fixing the value of H_1 we proceed in essentially the same manner as before and arrive at the deflection

$$w_2 = \frac{H_1}{M_1} \frac{\partial w_1}{\partial x} = \frac{H_1}{P} \frac{\partial^2 w_0}{\partial x^2} \quad (c)$$

due to a singularity of a higher order than that corresponding to a couple.[1] Substitution of expression (206), where rectangular coordinates may be used temporarily, yields the deflection

$$w_2 = \frac{H_1}{8\pi D}\left(2\log\frac{r}{a} + 2 + \cos 2\theta\right) \quad (208)$$

Expressions containing a singularity are also obtainable in the case of a couple acting at the corner of a wedge-shaped plate with both edges free, as well as in the case of a semi-infinite plate submitted to the action of a transverse force or a couple at some point along the free edge.[2]

Single Load Acting in the Vicinity of a Built-in

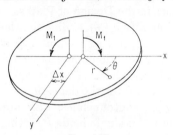

Fig. 169

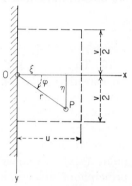

Fig. 170

Edge (Fig. 170). The deflection of a semi-infinite cantilever plate carrying a single load P at some point (ξ, η) is given by the expression

$$w = \frac{P}{16\pi D}\left[4x\xi - r^2 \log\frac{(x+\xi)^2 + (y-\eta)^2}{r_1^2}\right] \quad (d)$$

where $r_1^2 = (x - \xi)^2 + (y - \eta)^2$. We confine ourselves to the consideration of the clamping moment at the origin. Due differentiation of expression (d) yields

$$M_x = -\frac{P}{\pi}\cos^2\varphi \quad (209)$$

at $x = y = 0$, provided ξ and η do not vanish simultaneously. It is seen that in general the clamping moment M_x depends only on the ratio η/ξ.

---

[1] To make the nature of such a loading clear, let us assume a simply supported beam of a span L and a rigidity EI with a rectangular moment diagram Δx by M, symmetrical to the center of beam and due to two couples M applied at a distance Δx from each other. Proceeding as before, *i.e.*, making $\Delta x \to 0$, however fixing the value of $H = M\,\Delta x$, we would arrive at a diagram of magnitude H concentrated at the middle of beam. Introducing a fictitious central load H/EI and using Mohr's method, we would also obtain a triangular deflection diagram of the beam with a maximum ordinate $HL/4EI$. A similar deflection diagram would result from a load applied at the center of a perfectly flexible string.

[2] See A. Nádai, "Elastische Platten," p. 203, Berlin, 1925.

If, however, $\xi = \eta = 0$ the moment M_x vanishes, and thus the function $M_x(\xi,\eta)$ proves to be discontinuous at the origin.

Of similar character is the action of a single load near any edge rigidly or elastically clamped, no matter how the plate may be supported elsewhere. This leads also to the characteristic shape of influence surfaces plotted for moments on the boundary of plates clamped or continuous along that boundary (see Figs. 171 and 173).

For the shearing, or reactive, force acting at $x = y = 0$ in Fig. 170 we obtain in similar manner

$$Q_x = \frac{2P}{\pi r} \cos^3 \varphi \tag{210}$$

where $r^2 = \xi^2 + \eta^2$.

76. The Use of Influence Surfaces in the Design of Plates. In Art. 29 we considered an influence function $K(x,y,\xi,\eta)$ giving the deflection at some point (x,y) when a unit load is applied at a point (ξ,η) of a simply supported rectangular plate. Similar functions may be constructed for any other boundary conditions and for plates of any shape. We may also represent the influence surface $K(\xi,\eta)$ for the deflection at some fixed point (x,y) graphically by means of contour lines. By applying the principle of superposition to a group of n single loads P_i acting at points (ξ_i,η_i) we find the total deflection at (x,y) as

$$w = \sum_{i=1}^{n} P_i K(x,y,\xi_i,\eta_i) \tag{a}$$

In a similar manner, a load of intensity $p(\xi,\eta)$ distributed over an area A of the surface of the plate gives the deflection

$$w = \iint_A p(\xi,\eta) K(x,y,\xi,\eta) \, d\xi \, d\eta \tag{b}$$

By Maxwell's reciprocal law we also have the symmetry relation

$$K(x,y,\xi,\eta) = K(\xi,\eta,x,y) \tag{c}$$

i.e., the influence surface for the deflection at some point (x,y) may be obtained as the deflection surface $w(\xi,\eta)$ due to a unit load acting at (x,y). The surface $w(\xi,\eta)$ is given therefore by the differential equation $\Delta\Delta w(\xi,\eta) = 0$, and the solution of this equation not only must fulfill the boundary conditions but also must contain a singularity of the kind represented in Eq. (206) at $\xi = x$, $\eta = y$.

Of special practical interest are the influence surfaces for stress resultants[1] given by a combination of partial derivatives of $w(x,y)$ with respect to x and y. To take an

---

[1] Such surfaces have been used first by H. M. Westergaard, *Public Roads*, vol. 11, 1930. See also F. M. Baron, *J. Appl. Mechanics*, vol. 8, p. A-3, 1941.

example, let us consider the influence surfaces for the quantity

$$-D\frac{\partial^2 w}{\partial x^2} = -D\frac{\partial^2}{\partial x^2}K(x,y,\xi,\eta) \qquad (d)$$

By result (c) of Art. 75 this latter expression yields the ordinates of a deflection surface in coordinates ξ, η containing at $\xi = x$, $\eta = y$ a singularity due to a "couple of second order" $H = 1$ which acts at that point in accordance with Fig. 169.

The procedure of the construction and the use of influence surfaces may be illustrated by the following examples.[1]

Influence Surface for the Edge Moment of a Clamped Circular Plate[2] (Fig. 171). By representing the deflection (197), page 293, in the form $w = PK(x,0,\xi,\theta)$, we can consider K as the influence function for the deflection at some point $(x,0)$, the momentary position of the unit load being (ξ,θ). In calculating the edge couple M_r at $x = r/a = 1$, $y = 0$ we observe that all terms of the respective expressions (192), except for the following one, vanish along the clamped edge $x = 1$. The only remaining term yields

$$M_r = -\frac{D}{a^2}\left(\frac{\partial^2 K}{\partial x^2}\right)_{x=1} = -\frac{1}{4\pi}\frac{(1-\xi^2)^2}{\xi^2 - 2\xi\cos\theta + 1} \qquad (e)$$

For brevity let us put $\xi^2 - 2\xi\cos\theta + 1 = \eta^2$ and, furthermore, introduce the angle φ (Fig. 171a). Then we have $\xi^2 = 1 - 2\eta\cos\varphi + \eta^2$ and

$$M_r = -\frac{1}{4\pi}(2\cos\varphi - \eta)^2$$

which, for negligible values of η, coincides with the expression (209). The influence surface for the moment M_r is represented by the contour map in Fig. 171b, with the ordinates multiplied by 4π.

Influence Surface for the Bending Moment M_x at the Center of a Simply Supported Square Plate.[3] It is convenient to use the influence surfaces for the quantities $M_{x0} = -D\,\partial^2 w/\partial x^2$ and $M_{y0} = -D\,\partial^2 w/\partial y^2$ with the purpose of obtaining the final result by means of Eqs. (101).

The influence surface for M_{x0} may be constructed on the base of Fig. 76. The influence of the single load $P = 1$ acting at point 0 is given by the first of the equations (151) and by Eq. (152). This latter expression also contains the required singularity of the type given by Eq. (206), located at the point 0. The effect of other loads may be calculated by means of the first of the equations (149), the series being rapidly convergent. The influence surface is shown in Fig. 172 with ordinates multiplied by 8π.

Let us calculate the bending moment M_x for two single loads P_1 and $P_2 \leq P_1$ at a fixed distance of $0.25a$ from each other, each load being distributed uniformly over

---

[1] For details of the so-called *singularity method* see A. Pucher, *Ingr.-Arch.*, vol. 12, p. 76, 1941.

[2] Several influence surfaces for the clamped circular plate are given by M. El-Hashimy, "Ansgewählte Plattenprobleme," Zürich, 1956.

[3] The most extensive set of influence surfaces for rectangular plates with various edge conditions is due to A. Pucher, "Einflussfelder elastischer Platten," 2d ed., Vienna, 1958. See also his paper in "Federhofer-Girkmann-Festschrift," p. 303, Vienna, 1950. For influence surfaces of continuous plates, see G. Hoeland, *Ingr.-Arch.*, vol. 24, p. 124, 1956.

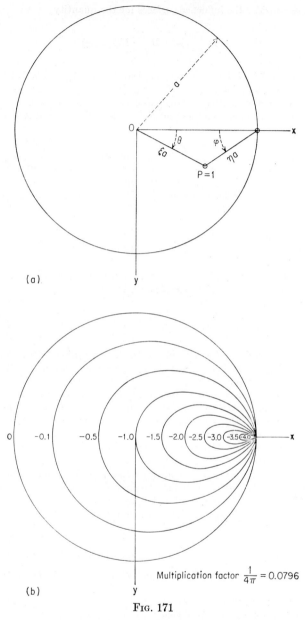

Fig. 171

an area $0.1a \cdot 0.1a$. Outside those areas the plate may carry a uniformly distributed live load of an intensity $q < P_2/0.01a^2$.

The influence surface (Fig. 172) holds for M_{x0}, and the distribution of the loading which yields the largest value of M_{x0} is given in this figure by full lines. Because of the singularity, the ordinates of the surface are infinitely large at the center of the plate;

SPECIAL AND APPROXIMATE METHODS IN THEORY OF PLATES   331

therefore it is simplest to calculate the effect of the load P_1 separately, by means of Eqs. (163) and (165), in connection with Tables 26 and 27.[1] For this case we have $\nu = 0$, $v/u = k = 1$, $\varphi = 1.5708$, $\psi = 0$, $\lambda = 2.669$, and $\mu = 0$, which yields $N = 0$ and a value of M calculated hereafter. As for the effect of the load P_2, it can be assumed as proportional to the ordinate 2.30 of the surface at the center of the loaded

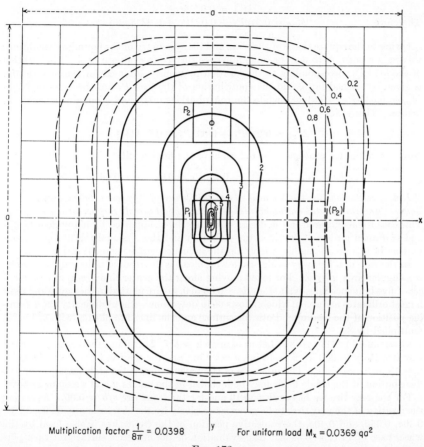

Multiplication factor $\frac{1}{8\pi} = 0.0398$ For uniform load $M_x = 0.0369\, qa^2$

Fig. 172

area. Introducing only the excesses of both single loads over the respective loads due to q, we have to sum up the following contributions to the value of M_{x0}:

1. Load P_1: from Eqs. (163), (165), with $\xi = a/2$, $d = 0.1\sqrt{2}\,a$,

$$M'_{x0} = \frac{M}{2} = \frac{P_1 - 0.01qa^2}{8\pi}\left(2\log\frac{4}{0.1\pi\sqrt{2}} + 2.669 - 1.571\right)$$
$$= 0.219(P_1 - 0.01qa^2)$$

[1] The effect of the central load may also be calculated by means of influence lines similar to those used in the next example or by means of Table 20.

2. Load P_2:

$$M''_{x0} = \frac{1}{8\pi} 2.30(P_2 - 0.01qa^2) = 0.092(P_2 - 0.01qa^2)$$

3. Uniform load q: from data on Fig. 172,

$$M'''_{x0} = 0.0369qa^2$$

Therefore $\quad M_{x0} = 0.219P_1 + 0.092P_2 + 0.0338qa^2$

Owing to the square shape of the plate and the symmetry of the boundary conditions we are in a position to use the same influence surface to evaluate M_{y0}. The location of the load P_2 corresponding to the location previously assumed for the surface M_{x0} is given by dashed lines, and the contribution of the load P_2 now becomes equal to $M''_{y0} = 0.035(P_2 - 0.01qa^2)$, while the contributions of P_1 and q remain the same as before. This yields

$$M_{y0} = 0.219P_1 + 0.035P_2 + 0.0344qa^2$$

Now assuming, for example, $\nu = 0.2$ we have the final result

$$M_x = M_{x0} + 0.2M_{y0} = 0.263P_1 + 0.099P_2 + 0.0407qa^2$$

Influence Surface for the Moment M_x at the Center of Support between Two Interior Square Panels of a Plate Continuous in the Direction x and Simply Supported at $y = \pm b/2$. This case is encountered in the design of bridge slabs supported by many floor beams and two main girders. Provided the deflection and the torsional rigidity of all supporting beams are negligible, we obtain the influence surface shown[1] in Fig. 173.

In the case of a highway bridge each wheel load is distributed uniformly over some rectangular area u by v. For loads moving along the center line $y = 0$ of the slab a set of five influence lines (valid for $v/b = 0.05$ to 0.40) are plotted in the figure and their largest ordinates are given, which allows us to determine without difficulty the governing position of the loading. Both the surface and the lines are plotted with ordinates multiplied by 8π.

EXAMPLE OF EVALUATION. Let us assume $a = b = 24$ ft 0 in.; furthermore, for the rear tire $P_r = 16,000$ lb, $u = 18$ in., $v = 30$ in., and for the front tire $P_f = 4,000$ lb, $u = 18$ in., $v = 15$ in. The influence of the pavement and the slab thickness on the distribution of the single loads may be included in the values u and v assumed above.

For the rear tire we have $v/b \approx 0.10$ and for the front tire $v/b \approx 0.05$. Assuming the position of the rear tires to be given successively by the abscissas $\xi = 0.20a$, $0.25a$, $0.30a$, $0.35a$, and $0.40a$, the respective position of the front tires is also fixed by the wheel base of 14 ft $= 0.583a$. The evaluation of the influence surface for each particular location of the loading gives a succession of values of the moment plotted in Fig. 173 versus the respective values of ξ by a dashed line. The curve proves to have a maximum at about $\xi = 0.30a$. The procedure of evaluation may be shown for this latter position only.

The influence lines marked 0.10 and 0.05, respectively, yield the contribution of both central loads (at $y = 0$) equal to

$$-(16{,}000 \cdot 3.24 + 4{,}000 \cdot 3.32) = -65{,}100 \text{ lb}$$

and the influence surface gives the contribution of the remaining six loads as

$$-16{,}000(1.66 + 2.25 + 0.44) - 4{,}000(1.59 + 2.25 + 0.41) = -86{,}600 \text{ lb}$$

[1] For methods of its construction see references given in Art. 52.

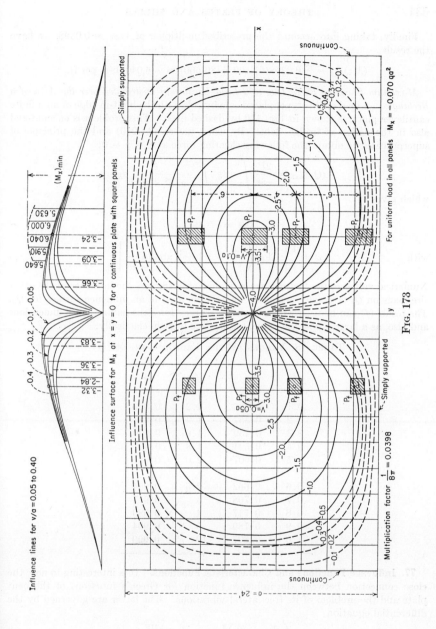

Fig. 173

Finally, taking into account the prescribed multiplier of $1/8\pi = 0.0398$, we have the result

$$(M_x)_{\min} = -0.0398(65{,}100 + 86{,}600) = -6{,}040 \text{ lb-ft per ft}$$

Maximum Shearing Force Due to a Load Uniformly Distributed over the Area of a Rectangle. A load of this type, placed side by side with the built-in edge of an infinite cantilever plate, is shown in Fig. 170 by dashed lines. This problem is encountered also in the design of bridge slabs. By using the result (210) and the principle of superposition we obtain the following shearing force at $x = y = 0$:

$$(Q_x)_{\max} = \frac{2P}{\pi u v} \int_0^u d\xi \int_{-v/2}^{v/2} \frac{\xi^3}{(\xi^2 + \eta^2)^2} \, d\eta$$

which gives

$$(Q_x)_{\max} = \alpha \frac{P}{v} \tag{f}$$

with

$$\alpha = \frac{1}{\pi} \left[\frac{v}{u} \log\left(\frac{4u^2}{v^2} + 1\right) + 2 \arctan \frac{v}{2u} \right] \tag{g}$$

Numerical values of the factor α are given in Table 78. As the influence of the other tire loads on Q_x is usually negligible we have no need of an influence surface for Q_x. The result (f) can be used with sufficient accuracy for slabs having finite dimensions and also, as a largest possible value, for an edge built in elastically.

TABLE 78. VALUES OF THE FACTOR α IN EQ. (f)

v/u	α	v/u	α
0.1	0.223	1.2	0.852
0.2	0.357	1.4	0.884
0.3	0.459	1.6	0.909
0.4	0.541	1.8	0.927
0.5	0.607	2.0	0.941
0.6	0.662	2.5	0.964
0.7	0.708	3	0.977
0.8	0.747	4	0.989
0.9	0.780	5	0.994
1.0	0.807	10	0.999

77. Influence Functions and Characteristic Functions. It is interesting to note the close connection between the influence function (or Green's function) of the bent plate and the problem of its free lateral vibrations. The latter are governed by the differential equation

$$\left(\frac{\partial^2}{\partial x^2} + \frac{\partial^2}{\partial y^2}\right)^2 W = -\frac{\mu}{D}\frac{\partial^2 W}{\partial t^2} \tag{a}$$

where $W(x,y,t)$ is the deflection, μ the mass of the plate per unit area, and t the time. With the assumption $W = w(x,y) \cos pt$ we obtain for the function w the differential equation

$$D\Delta\Delta w - \lambda w = 0 \tag{b}$$

in which $\lambda = p^2\mu$. For some specific boundary conditions, solutions of Eq. (b) exist only for a definite set of values $\lambda_1, \lambda_2, \ldots, \lambda_k, \ldots$ of the parameter λ, the so-called *characteristic numbers* (or *eigenvalues*) of the problem. The respective solutions form a set of characteristic functions $w_1(x,y), w_2(x,y), \ldots, w_k(x,y), \ldots$. These functions are mutually orthogonal; i.e.,

$$\iint_A w_i(x,y)w_k(x,y)\,dx\,dy = 0 \qquad (c)$$

for $i \neq k$, the integral being extended over the surface of the plate. As the functions $w_k(x,y)$ are defined except for a constant factor, we can "normalize" them by choosing this factor such as to satisfy the condition

$$\iint_A w_k^2(x,y)\,dx\,dy = a^2b^2 \qquad (d)$$

The form chosen for the right-hand side of (d) is appropriate in the case of a rectangular plate with the sides a and b, but whatever the contour of the plate may be, the dimension of a length must be secured for w_k. The set of numbers λ_k and the corresponding set of normalized functions $w_k(x,y)$ being established, it can be shown[1] that the expansion

$$K(x,y,\xi,\eta) = \frac{1}{a^2b^2}\sum_{k=1}^{\infty}\frac{w_k(x,y)w_k(\xi,\eta)}{\lambda_k} \qquad (e)$$

holds for the influence function of the plate with boundary conditions satisfied by the characteristic functions.

By applying Eqs. (a) and (b) of the previous article to the result (e) we conclude that, no matter what the distribution of the loading may be, the deflection of the plate can always be represented by a linear combination of its characteristic functions.

As an example, let us take the rectangular plate with simply supported edges (Fig. 59). Eigenfunctions which satisfy Eq. (b) along with the boundary conditions $w = \Delta w = 0$ and the condition (d) are

$$w_k = 2\sqrt{ab}\,\sin\frac{m\pi x}{a}\sin\frac{n\pi y}{b} \qquad (f)$$

m and n being two arbitrary integers. The respective eigenvalue, from Eq. (b), is

$$\lambda_k = \pi^4 D\left(\frac{m^2}{a^2} + \frac{n^2}{b^2}\right)^2 \qquad (g)$$

Substitution of this in the expansion (e) immediately leads to the result (134). For rectangular plates with only two opposite edges supported, the conditions on the other edges being arbitrary, influence functions may be obtained in a similar manner. However, in such a case a preliminary computation of the values of λ_k from the respective transcendental frequency equation becomes necessary. A further example of an influence function obtainable in the form of an expansion is the case of a circular plate,

---

[1] See, for instance, R. Courant and D. Hilbert, "Methods of Mathematical Physics," vol. 1, p. 370, New York, 1953.

for which the modes of vibration, expressible in terms of Bessel functions, are well known.

78. The Use of Infinite Integrals and Transforms. Another method of treating the problems of bending of plates is the use of various transforms.[1] A few such transforms will be discussed in this article.

Fourier Integrals. In the case of infinite or semi-infinite strips with arbitrary conditions on the two parallel edges the method of M. Lévy, described on page 113, can be used, but in doing so the Fourier series necessarily must be replaced by the respective infinite integrals. In addition to the example considered in Art. 50, the problem of an infinite cantilever plate (Fig. 174) carrying a single load P may be solved in this way.[2]

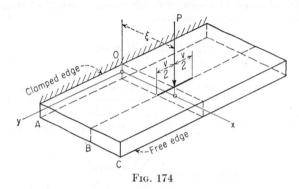

Fig. 174

Let w_1 be the deflection of the portion AB and w_2 the deflection of the portion BC of the plate of width $AC = a$. Then we have to satisfy the boundary conditions

$$w_1 = 0 \qquad \frac{\partial w_1}{\partial x} = 0 \qquad \text{on } x = 0$$

$$\frac{\partial^2 w_2}{\partial x^2} + \nu \frac{\partial^2 w_2}{dy^2} = 0 \qquad \frac{\partial^3 w_2}{\partial x^3} + (2 - \nu) \frac{\partial^3 w_2}{dx\,dy^2} = 0 \qquad \text{on } x = a \qquad (a)$$

together with the conditions of continuity

$$w_1 = w_2 \qquad \frac{\partial w_1}{\partial x} = \frac{\partial w_2}{\partial x} \qquad \Delta w_1 = \Delta w_2 \qquad \text{on } x = \xi \qquad (b)$$

The single force P may be distributed uniformly over a length v. Now, any even function of y can be represented by the Fourier integral

$$f(y) = \frac{2}{\pi} \int_0^\infty \cos \alpha y \, d\alpha \int_0^\infty f(\eta) \cos \alpha \eta \, d\eta \qquad (c)$$

Since the intensity of the loading is given by $f(\eta) = P/v$ for $-v/2 < \eta < v/2$ and by

---

[1] For their theory and application see I. N. Sneddon, "Fourier Transforms," New York, 1951.

[2] The solution and numerical results hereafter given are due to T. J. Jaramillo, *J. Appl. Mechanics*, vol. 17, p. 67, 1950. Making use of the Fourier transform, H. Jung treated several problems of this kind; see *Math. Nachr.*, vol. 6, p. 343, 1952.

zero elsewhere, we have

$$f(y) = \frac{2P}{\pi v} \int_0^\infty \frac{\sin \frac{\alpha v}{2} \cos \alpha y}{\alpha} \, d\alpha \qquad (d)$$

On the other hand, the function $f(y)$ is equal to the difference of the shearing forces Q_x at both sides of the section $x = \xi$. Thus, by Eqs. (108), we have

$$D \frac{\partial}{\partial x} (\Delta w_1 - \Delta w_2) = f(y) \qquad (e)$$

on $x = \xi$. In accordance with Eq. (d) we represent the deflections w_1 and w_2 by the integrals

$$w_i = \int_0^\infty X_i(x,\alpha) \cos \alpha y \, d\alpha \qquad i = 1, 2 \qquad (f)$$

in which the function

$$X_i(x,\alpha) = (A_i + B_i x) \cosh \alpha x + (C_i + D_i x) \sinh \alpha x$$

is of the same form as the function Y_m on page 114.

It remains now to substitute expressions (f) into Eqs. (a), (b), and (e) in order to determine the coefficients $A_1, B_1, \ldots, D_2$, independent of y but depending on α.

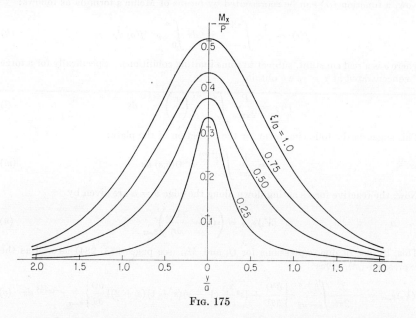

Fig. 175

The distribution of bending moments along the built-in edge, as computed from the foregoing solution for various positions of the single load and for $v = 0$, $\nu = 0.3$, is shown in Fig. 175.

Mellin Transform. The application of this transform is suitable in the case of a wedge-shaped plate with any homogeneous conditions along the edges $\theta = 0$ and

$\theta = \alpha$ (Fig. 176). To take an example let us consider the edge $\theta = 0$ as clamped and the edge $\theta = \alpha$, except for a single load P at $r = r_0$, as free.[1]

We use polar coordinates (see Art. 62) and begin by taking the general solution of the differential equation $\Delta\Delta w = 0$ in the form

$$W(s) = r^{-s}\Theta(\theta,s) \tag{g}$$

where s is a parameter and

$$\Theta(\theta,s) = A(s)\cos s\theta + B(s)\sin s\theta + C(s)\cos(s+2)\theta + D(s)\sin(s+2)\theta \tag{h}$$

The deflection and the slope along the clamped edge vanish if

$$[W(s)]_{\theta=0} = 0 \qquad \frac{1}{r}\left[\frac{\partial W(s)}{\partial \theta}\right]_{\theta=0} = 0 \tag{i}$$

The bending moment M_t on the free edge vanishes on the condition that

$$\left[\nu\frac{\partial^2 W(s)}{\partial r^2} + \frac{1}{r}\frac{\partial W(s)}{\partial r} + \frac{1}{r^2}\frac{\partial^2 W(s)}{\partial \theta^2}\right]_{\theta=\alpha} = 0 \tag{j}$$

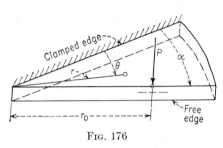

Fig. 176

Now, a function $f(r)$ can be represented by means of Mellin's formula as follows:

$$f(r) = \frac{1}{2\pi i}\int_{\sigma-\infty i}^{\sigma+\infty i} r^{-s}\,ds \int_0^\infty \rho^{s-1}f(\rho)\,d\rho \tag{k}$$

where σ is a real constant, subject to some limiting conditions. Specifically for a force P concentrated at $r = r_0$ we obtain

$$f(r) = \frac{P}{2\pi i r_0}\int_{\sigma-\infty i}^{\sigma+\infty i}\left(\frac{r}{r_0}\right)^{-(s+3)}ds \tag{l}$$

This suggests the following form for the deflection of the plate:

$$w = \frac{1}{2\pi i}\int_{\sigma-\infty i}^{\sigma+\infty i} r^{-s}\Theta(\theta,s)\,ds \tag{m}$$

Now, the reactive forces acting down along the edge $\theta = \alpha$ are given by

$$(V_t)_{\theta=\alpha} = \left(Q_t - \frac{\partial M_{rt}}{\partial r}\right)_{\theta=\alpha} \tag{n}$$

This, by use of due expressions for Q_t and M_{rt} (see pages 283, 284) as well as the expression (m), gives

$$(V_t)_{\theta=\alpha} = -\frac{D}{2\pi i}\int_{\sigma-\infty i}^{\sigma+\infty i}\left\{\frac{\partial^3\Theta}{\partial\theta^3} + [s^2 + (1-\nu)(s+1)(s+2)]\frac{\partial\Theta}{\partial\theta}\right\}_{\theta=\alpha} r^{-(s+3)}\,ds \tag{o}$$

---

[1] The problem was discussed by S. Woinowsky-Krieger, *Ingr.-Arch.*, vol. 20, p. 391, 1952. Some corrections are due to W. T. Koiter, *Ingr.-Arch.*, vol. 21, p. 381, 1953. For a plate with two clamped edges see Y. S. Uflyand, *Doklady Akad. Nauk S.S.S.R.*, vol. 84, p. 463, 1952. See also W. T. Koiter and J. B. Alblas, *Proc. Koninkl. Ned. Akad. Wetenschap.*, ser. B, vol. 57, no. 2, p. 259, 1954.

SPECIAL AND APPROXIMATE METHODS IN THEORY OF PLATES

We finally equate expressions (l) and (o) and thus obtain, in addition to Eqs. (i) and (j), a fourth condition to determine the quantities $A(s)$, $B(s)$, $C(s)$, and $D(s)$. Substitution of these coefficients in the expressions (h) and (m) and introduction of a new variable $u = -(s + 1)i$, where $i = \sqrt{-1}$, yields the following expression for the deflection of the plate:

$$w = \frac{2Prr_0}{\pi D} \int_0^\infty \frac{G \cos\left(u \log \frac{r}{r_0}\right) + H \sin\left(u \log \frac{r}{r_0}\right)}{Nu(1 + u^2)} du \qquad (p)$$

in which G and H are some functions of α, θ, and u, and N is a function of α and u.

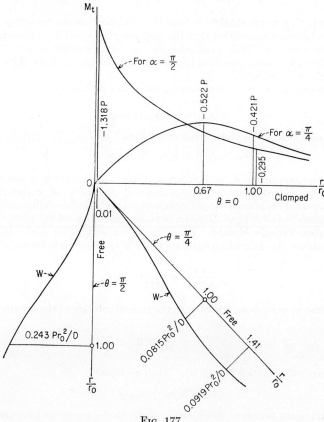

Fig. 177

The variation of the deflections along the free edge and the distribution of the moments M_t along the edge $\theta = 0$ in the particular case of $\alpha = \pi/4$ and $\alpha = \pi/2$ is shown in Fig. 177.

Hankel Transform. Let a circular plate with a radius a be bent to a surface of revolution by a symmetrically distributed load $q(r)$. We multiply the differential equation $\Delta\Delta w = q/D$ of such a plate by $rJ_0(\lambda r) \, dr$ and integrate by parts between

$r = 0$ and $r = \infty$. Provided $w = 0$ for $r > a$, the result is

$$\lambda^4 \int_0^\infty w(r) r J_0(\lambda r)\, dr = g(\lambda) \tag{q}$$

where

$$g(\lambda) = (C_1 + \lambda^2 C_2) J_0(\lambda a) + (\lambda C_3 + \lambda^3 C_4) J_1(\lambda a) + \frac{1}{D} \int_0^a q(\rho) \rho J_0(\lambda \rho)\, d\rho \tag{r}$$

J_0 and J_1 are Bessel functions of the order zero and one, and C_i are constants. Application of the Hankel inversion theorem to Eq. (q) gives

$$w = \int_0^\infty g(\lambda) \frac{1}{\lambda^3} J_0(\lambda r)\, d\lambda \tag{s}$$

The constants C_i now are obtainable from the conditions on the boundary $r = a$ of the plate and from the condition that the function $g(\lambda)/\lambda^4$ must be bounded. The expression (r) must be slightly modified in the case of an annular plate.[1] Examples of the application of solutions of the type (s) to the problem of elastically supported plates are given in Art. 61.

Sine Transform. In the case of rectangular plates we have used solutions of the form

$$w(x,y) = \Sigma Y(y,\alpha) \sin \alpha x$$

and in the case of sectorial plates those of the form

$$w(r,\theta) = \Sigma R(r,\beta) \sin \beta \theta$$

The finite sine transforms of the function w, taken with respect to x and θ, respectively, and introduced together with transformed derivatives of w and the transformed differential equation of the plate, then prove useful in calculating the constants of the functions Y and R from the given boundary conditions of the plate.[2]

79. Complex Variable Method. By taking $z = x + iy$ and $\bar{z} = x - iy$ for independent variables the differential equation (104) of the bent plate becomes

$$\frac{\partial^4 w}{\partial z^2\, \partial \bar{z}^2} = \frac{1}{16D} q(z,\bar{z}) \tag{a}$$

Let us assume $w = w_0 + w_1$, where w_1 is the general solution of the equation

$$\frac{\partial^4 w}{\partial z^2\, \partial \bar{z}^2} = 0$$

and w_0 a particular solution of Eq. (a). Then we have[3]

$$w_1 = \Re[\bar{z}\varphi(z) + \chi(z)] \tag{b}$$

where φ and χ are functions which are analytic in the region under consideration. Usually the derivative $\psi = \partial \chi / \partial z$ is introduced along with χ.

[1] For the foundation of the method and an extensive list of transforms needed in its application see H. Jung, *Z. angew. Math. Mech.*, vol. 32, p. 46, 1952.

[2] The application of the method is due to L. I. Deverall and C. J. Thorne, *J. Appl. Mechanics*, vol. 18, pp. 152, 359, 1951.

[3] $\Re$ denotes the real part of the solution. This form of the solution of the bipotential equation is due to E. Goursat, *Bull. Soc. Math. France*, vol. 26, p. 236, 1898.

In the case of a single load P acting at $z_0 = x_0 + iy_0$ the solution w_0 may be chosen in the form

$$w_0 = \frac{P}{16\pi D} (z - z_0)(\bar{z} - \bar{z}_0) \log [(z - z_0)(\bar{z} - \bar{z}_0)] \tag{c}$$

which is substantially equivalent to expression (206). For a uniform load

$$w_0 = \frac{qz^2\bar{z}^2}{64D}$$

would be a suitable solution.

If the outer or the inner boundary of the plate is a circle we always can replace it by a unit circle $z = e^{i\theta}$, or briefly $z = \sigma$. The boundary conditions on $z = \sigma$ must be expressed in complex form also. The functions φ and ψ may be taken in the form of a power series, with additional terms, if necessary, depending on the value of stress resultants taken along the inner edge of the plate. Multiplication of the boundary conditions by the factor $[2\pi i(\sigma - z)]^{-1} d\sigma$ and integration along $z = \sigma$ then yields the required functions φ and ψ.*

For boundaries other than a circle a mapping function $z = \omega(\zeta) = \omega(\rho e^{i\varphi})$ may be used so as to map the given boundary line onto the unit circle $\zeta = e^{i\varphi} = \sigma$. The determination of the functions $\varphi_1(\zeta) = \varphi(z)$ and $\psi_1(\zeta) = \psi(z)$ from the boundary conditions on $\zeta = \sigma$ then is reduced to the problem already considered. The Muschelišhvili method outlined above is especially efficient in cases concerning stress distribution around holes;[1] the function $\omega(\zeta)$ then has to map the infinite region of the plate into the interior of the unit circle.

The complex variable method also allows us to express Green's functions of a circular plate with various boundary conditions in closed form.[2] In other cases, such as that of a clamped square plate, we must rely on an approximate determination of the Green functions.[3]

When expressible by a double trigonometric series, the deformation of the plate can also be represented in a simpler form by making use of the doubly periodic properties of the elliptic functions. For the quantity Δw, satisfying the potential equation $\Delta(\Delta w) = 0$, such a representation becomes particularly convenient because of the close connection between the Green function for the expression Δw and the mapping function of the region of the given plate into the unit circle.[4] Once Δw is determined

---

*For evaluation of integrals of the Cauchy type implied in this procedure see N. I. Muschelišhvili, "Some Basic Problems of the Mathematical Theory of Elasticity," Groningen, 1953.

[1] An extensive application of the method to the problem of stress concentration is due to G. N. Savin; see his "Stress Concentration around Holes," Moscow, 1951. See also Yi-Yuan Yu, *J. Appl. Mechanics*, vol. 21, p. 129, 1954, and *Proc. Ninth Intern. Congr. Appl. Mech.*, vol. 6, p. 378, Brussels, 1957; also L. I. Deverall, *J. Appl. Mechanics*, vol. 24, p. 295, 1957. A somewhat different method, applicable as well to certain problems of the thick-plate theory, was used by A. C. Stevenson, *Phil. Mag.*, vol. 33, p. 639, 1942.

[2] E. Reissner, *Math. Ann.*, vol. 111, p. 777, 1935; A. Lourye, *Priklad. Mat. Mekhan.*, vol. 4, p. 93, 1940.

[3] F. Schultz-Grunow, *Z. angew. Math. Mech.*, vol. 33, p. 227, 1953.

[4] Courant and Hilbert, *op. cit.*, vol. 1, p. 377. Elliptic functions have been used in particular by A. Nádai, *Z. angew. Math. Mech.*, vol. 2, p. 1, 1922 (flat slabs); by F. Tölke, *Ingr.-Arch.*, vol. 5, p. 187, 1934 (rectangular plates); and also by B. D. Aggarwala, *Z. angew. Math. Mech.*, vol. 34, p. 226, 1954 (polygonal plates and, in particular, triangular plates).

the shearing forces of the plate are readily given by the derivatives of that function by virtue of Eqs. (108).

80. Application of the Strain Energy Method in Calculating Deflections. Let us consider again the problem of the simply supported rectangular plate. From the discussion in Art. 28 it is seen that the deflection of such a plate (Fig. 59) can always be represented in the form of a double trigonometric series:[1]

$$w = \sum_{m=1}^{\infty} \sum_{n=1}^{\infty} a_{mn} \sin \frac{m\pi x}{a} \sin \frac{n\pi y}{b} \qquad (a)$$

The coefficients a_{mn} may be considered as the coordinates defining the shape of the deflection surface, and for their determination the principle of virtual displacements may be used. In the application of this principle we need the expression for strain energy (see page 88):

$$V = \frac{1}{2} D \int_0^a \int_0^b \left\{ \left(\frac{\partial^2 w}{\partial x^2} + \frac{\partial^2 w}{\partial y^2} \right)^2 - 2(1-\nu) \left[\frac{\partial^2 w}{\partial x^2} \frac{\partial^2 w}{\partial y^2} - \left(\frac{\partial^2 w}{\partial x \, dy} \right)^2 \right] \right\} dx \, dy \qquad (b)$$

Substituting series (a) for w, the first term under the integral sign in (b) becomes

$$\frac{1}{2} D \int_0^a \int_0^b \left[\sum_{m=1}^{\infty} \sum_{n=1}^{\infty} a_{mn} \left(\frac{m^2 \pi^2}{a^2} + \frac{n^2 \pi^2}{b^2} \right) \sin \frac{m\pi x}{a} \sin \frac{n\pi y}{b} \right]^2 dx \, dy \qquad (c)$$

Observing that

$$\int_0^a \sin \frac{m\pi x}{a} \sin \frac{m'\pi x}{a} dx = \int_0^b \sin \frac{n\pi y}{b} \sin \frac{n'\pi y}{b} dy = 0$$

if $m \neq m'$ and $n \neq n'$, we conclude that in calculating the integral (c) we have to consider only the squares of terms of the infinite series in the parentheses. Using the formula

$$\int_0^a \int_0^b \sin^2 \frac{m\pi x}{a} \sin^2 \frac{n\pi y}{b} dx \, dy = \frac{ab}{4}$$

the calculation of the integral (c) gives

$$\frac{\pi^4 ab}{8} D \sum_{m=1}^{\infty} \sum_{n=1}^{\infty} a_{mn}^2 \left(\frac{m^2}{a^2} + \frac{n^2}{b^2} \right)^2$$

---

[1] The terms of this series are characteristic functions of the plate under consideration (see Art. 77).

From the fact that

$$\int_0^a \int_0^b \sin^2 \frac{m\pi x}{a} \sin^2 \frac{n\pi y}{b} \, dx \, dy = \int_0^a \int_0^b \cos^2 \frac{m\pi x}{a} \cos^2 \frac{n\pi y}{b} \, dx \, dy = \frac{ab}{4}$$

it can be concluded that the second term under the integral sign in expression (b) is zero after integration. Hence the total strain energy in this case is given by expression (c) and is

$$V = \frac{\pi^4 ab}{8} D \sum_{m=1}^{\infty} \sum_{n=1}^{\infty} a_{mn}^2 \left(\frac{m^2}{a^2} + \frac{n^2}{b^2} \right)^2 \quad (d)$$

Let us consider the deflection of the plate (Fig. 59) by a concentrated force P perpendicular to the plate and applied at a point $x = \xi$, $y = \eta$. To get a virtual displacement satisfying boundary conditions we give to any coefficient $a_{m'n'}$ of series (a) an infinitely small variation $\delta a_{m'n'}$. As a result of this the deflection (a) undergoes a variation

$$\delta w = \delta a_{m'n'} \sin \frac{m'\pi x}{a} \sin \frac{n'\pi y}{b}$$

and the concentrated load P produces a virtual work

$$P \, \delta a_{m'n'} \sin \frac{m'\pi \xi}{a} \sin \frac{n'\pi \eta}{b}$$

From the principle of virtual displacements it follows that this work must be equal to the change in potential energy (d) due to the variation $\delta a_{m'n'}$. Hence

$$P \, \delta a_{m'n'} \sin \frac{m'\pi \xi}{a} \sin \frac{n'\pi \eta}{b} = \frac{\partial V}{\partial a_{m'n'}} \delta a_{m'n'}$$

Substituting expression (d) for V, we obtain

$$P \, \delta a_{m'n'} \sin \frac{m'\pi \xi}{a} \sin \frac{n'\pi \eta}{b} = \frac{\pi^4 ab}{4} D a_{m'n'} \left(\frac{m'^2}{a^2} + \frac{n'^2}{b^2} \right)^2 \delta a_{m'n'} \quad (e)$$

from which

$$a_{m'n'} = \frac{4P \sin \frac{m'\pi \xi}{a} \sin \frac{n'\pi \eta}{b}}{\pi^4 ab D \left(\frac{m'^2}{a^2} + \frac{n'^2}{b^2} \right)^2} \quad (f)$$

Substituting this into expression (a), we obtain once more the result (133).

Instead of using the principle of virtual displacements in calculating coefficients a_{mn} in expression (a) for the deflection, we can obtain the same result from the consideration of the total energy of the system. If a system is in a position of stable equilibrium, its total energy is a minimum. Applying this statement to the investigation of bending of plates, we observe that the total energy in such cases consists of two parts: the strain energy of bending, given by expression (b), and the potential energy of the load distributed over the plate. Defining the position of the element $q\,dx\,dy$ of the load by its vertical distance w from the horizontal plane xy, the corresponding potential energy may be taken equal to $-wq\,dx\,dy$, and the potential energy of the total load is

$$- \iint wq\,dx\,dy \tag{g}$$

The total energy of the system then is

$$I = \iint \left(\frac{D}{2} \left\{ \left(\frac{\partial^2 w}{\partial x^2} + \frac{\partial^2 w}{\partial y^2} \right)^2 - 2(1-\nu) \left[\frac{\partial^2 w}{\partial x^2} \frac{\partial^2 w}{\partial y^2} - \left(\frac{\partial^2 w}{\partial x\,\partial y} \right)^2 \right] \right\} - wq \right) dx\,dy \tag{h}$$

The problem of bending of a plate reduces in each particular case to that of finding a function w of x and y that satisfies the given boundary conditions and makes the integral (h) a minimum. If we proceed with this problem by the use of the calculus of variations, we obtain for w the partial differential equation (104), which was derived before from the consideration of the equilibrium of an element of the plate. The integral (h), however, can be used advantageously in an approximate investigation of bending of plates. For that purpose we replace the problem of variational calculus with that of finding the minimum of a certain function by assuming that the deflection w can be represented in the form of a series

$$w = a_1 \varphi_1(x,y) + a_2 \varphi_2(x,y) + a_3 \varphi_3(x,y) + \cdots + a_n \varphi_n(x,y) \tag{211}$$

in which the functions $\varphi_1, \varphi_2, \ldots, \varphi_n$ are chosen so as to be suitable[1] for representation of the deflection surface w and at the same time to satisfy the boundary conditions. Substituting expression (211) in the integral (h), we obtain, after integration, a function of second degree with coefficients $a_1, a_2, \ldots$ These coefficients must now be chosen so as to make the integral (h) a minimum, from which it follows that

$$\frac{\partial I}{\partial a_1} = 0 \qquad \frac{\partial I}{\partial a_2} = 0 \qquad \cdots \qquad \frac{\partial I}{\partial a_n} = 0 \tag{i}$$

This is a system of n linear equations in $a_1, a_2, \ldots, a_n$, and these quantities can readily be calculated in each particular case. If the functions φ are of such a kind that series (211) can represent any arbi-

---

[1] From experience we usually know approximately the shape of the deflection surface, and we should be guided by this information in choosing suitable functions φ.

trary function within the boundary of the plate,[1] this method of calculating deflections w brings us to a closer and closer approximation as the number n of the terms of the series increases, and by taking n infinitely large we obtain an exact solution of the problem.

Applying the method to the case of a simply supported rectangular plate, we take the deflection in the form of the trigonometric series (a). Then, by using expression (d) for the strain energy, the integral (h) is represented in the following form:

$$I = \frac{\pi^4 abD}{8} \sum_{m=1}^{\infty} \sum_{n=1}^{\infty} a_{mn}^2 \left(\frac{m^2}{a^2} + \frac{n^2}{b^2}\right)^2$$

$$- \int_0^a \int_0^b q \sum_{m=1}^{\infty} \sum_{n=1}^{\infty} a_{mn} \sin \frac{m\pi x}{a} \sin \frac{n\pi y}{b} \, dx \, dy \quad (j)$$

and Eqs. (i) have the form

$$\frac{\pi^4 abD}{4} a_{mn} \left(\frac{m^2}{a^2} + \frac{n^2}{b^2}\right)^2 - \int_0^a \int_0^b q \sin \frac{m\pi x}{a} \sin \frac{n\pi y}{b} \, dx \, dy = 0 \quad (k)$$

In the case of a load P applied at a point with the coordinates ξ, η, the intensity q of the load is zero at all points except the point ξ, η, where we have to put $q \, dx \, dy = P$. Then Eq. (k) coincides with Eq. (e), previously derived by the use of the principle of virtual displacements. For practical purposes it should be noted that the integral

$$\iint \left[\frac{\partial^2 w}{\partial x^2} \frac{\partial^2 w}{\partial y^2} - \left(\frac{\partial^2 w}{\partial x \, \partial y}\right)^2\right] dx \, dy \quad (l)$$

contained in expressions (b) and (h) vanishes for a plate rigidly clamped on the boundary. The same simplification holds for a polygonal plate if one of the boundary conditions is either $w = 0$ or $\partial w/\partial n = 0$, where n = direction normal to the edge.[2]

If polar coordinates instead of rectangular coordinates are used and axial symmetry of loading and deformation is assumed, Eq. (h) has to be replaced by

$$I = \iint \left\{\frac{D}{2}\left[\left(\frac{\partial^2 w}{\partial r^2} + \frac{1}{r}\frac{\partial w}{\partial r}\right)^2 - \frac{2(1-\nu)}{r}\frac{\partial w}{\partial r}\frac{\partial^2 w}{\partial r^2}\right] - wq\right\} r \, dr \, d\theta \quad (m)$$

---

[1] We have seen that a double trigonometrical series (a) possesses this property with respect to deflections w of a simply supported rectangular plate. Hence it can be used for obtaining an exact solution of the problem. The method of solving the bending problems of plates by the use of the integral (h) was developed by W. Ritz; see *J. reine angew. Math.*, vol. 135, 1908; and *Ann. Physik*, ser. 4, vol. 28, p. 737, 1909.

[2] See, for instance, E. R. Berger, *Österr. Ingr.-Arch.*, vol. 7, p. 41, 1953.

The contribution of the term containing the factor $1 - \nu$ again is zero for a plate clamped along the boundary.

The strain energy method can also be used for calculating the deflection of a circular plate resting on an elastic foundation. For example, to obtain a rough approximation for the case of a circular plate, we take for the deflection the expression

$$w = A + Br^2 \qquad (n)$$

in which A and B are two constants to be determined from the condition that the to al energy of the system in stable equilibrium is minimum.

The strain energy of the plate of radius a as given by Eq. (m) is

$$V_1 = 4B^2 D \pi a^2 (1 + \nu)$$

The strain energy of the deformed elastic foundation is

$$V_2 = \int_0^{2\pi} \int_0^a \frac{kw^2}{2} r\, dr\, d\theta = \pi k \left(\frac{1}{2} A^2 a^2 + \frac{1}{2} ABa^4 + \frac{1}{6} B^2 a^6 \right)$$

The total energy of the system for the case of a load P applied at the center is

$$V = 4B^2 D \pi a^2 (1 + \nu) + \pi k (\tfrac{1}{2} A^2 a^2 + \tfrac{1}{2} ABa^4 + \tfrac{1}{6} B^2 a^6) - PA$$

Taking the derivatives of this expression with respect to A and B and equating them to zero, we obtain

$$A + Ba^2 \left[\frac{2}{3} + \frac{16D(1+\nu)}{ka^4} \right] = 0$$

$$A + \frac{1}{2} Ba^2 = \frac{P}{\pi k a^2}$$

In accordance with the numerical example on page 264 we take

$$l = a \qquad \frac{D}{ka^4} = 1 \qquad \frac{P}{8\pi ka^3} = 102 \cdot 10^{-5}$$

and obtain

$$w_{\max} = A = 41.8 \cdot 10^{-3} \text{ in.}$$

This result is about 3 per cent less than the result $43 \cdot 10^{-3}$ obtained from the differential equation of a plate resting on elastic foundation. For greater accuracy more terms should be taken in expression (n).

If the stress distribution around the single load, not merely the deflection, were desired, a term of the form

$$\frac{P}{8\pi D} r^2 \log r$$

should be included in expression (n) in accordance with the type of singularity here required [see Eq. (206)].

When using polar coordinates in the most general case the integral (h) assumes the form

$$I = \iint \left\{ \frac{D}{2} \left[\left(\frac{\partial^2 w}{\partial r^2} + \frac{1}{r} \frac{\partial w}{\partial r} + \frac{1}{r^2} \frac{\partial^2 w}{\partial \theta^2} \right)^2 - 2(1-\nu) \frac{\partial^2 w}{\partial r^2} \left(\frac{1}{r} \frac{\partial w}{\partial r} + \frac{1}{r^2} \frac{\partial^2 w}{\partial \theta^2} \right) \right. \right.$$
$$\left. \left. + 2(1-\nu) \left(\frac{1}{r} \frac{\partial^2 w}{\partial r\, \partial \theta} - \frac{1}{r^2} \frac{\partial w}{\partial \theta} \right)^2 \right] - wq \right\} r\, dr\, d\theta \qquad (o)$$

81. Alternative Procedure in Applying the Strain Energy Method.

The calculation of the coefficients $a_1, a_2, \ldots, a_n$ in expression (211), which had to satisfy the boundary conditions but not the differential equation of the problem, may also be carried out without actually determining the potential energy of the system.

Let us assume a virtual deflection δw of the plate; then, we can calculate the respective work of the loading q either directly, by means of the integral

$$(\delta V)_1 = \iint q \, \delta w \, dx \, dy \qquad (a)$$

or indirectly, using the expression

$$(\delta V)_2 = \iint D \Delta \Delta w \, \delta w \, dx \, dy \qquad (b)$$

If w were the exact solution of the differential equation $D \Delta \Delta w = q$ of the plate, then the expressions (a) and (b) would be identical. For an approximate solution, which Eq. (211) represents, this is certainly not the case. We can succeed, however, in equalizing the expressions for the work for a particular set of virtual deflections, namely for $\delta w_1 = \varphi_1 \, \delta a_1$, $\delta w_2 = \varphi_2 \, \delta a_2, \ldots, \delta w_n = \varphi_n \, \delta a_n$. Substituting these expressions consecutively in the equation $(\delta V)_1 = (\delta V)_2$ or, what is the same, in the equation

$$\iint q \, \delta w \, dx \, dy = \iint D \Delta \Delta w \, \delta w \, dx \, dy \qquad (c)$$

we obtain the following system of equations:[1]

$$\iint \left(\Delta \Delta w - \frac{q}{D} \right) \varphi_1 \, dx \, dy = 0$$
$$\iint \left(\Delta \Delta w - \frac{q}{D} \right) \varphi_2 \, dx \, dy = 0$$
$$\cdots\cdots\cdots\cdots\cdots\cdots\cdots\cdots\cdots\cdots$$
$$\iint \left(\Delta \Delta w - \frac{q}{D} \right) \varphi_n \, dx \, dy = 0$$

$\qquad(d)$

It remains only to substitute the expression (211) in Eqs. (d) and to resolve them with respect to the unknown coefficients $a_1, a_2, \ldots, a_n$. This leads to the final expression for the deflection (211).

To illustrate the application of the method let us consider a uniformly loaded rectangular plate with all edges built in (Fig. 91). Writing for brevity $2x/a = u$, $2y/b = v$, we shall use the expressions

$$U_1 = u^4 - 2u^2 + 1 \qquad V_1 = v^4 - 2v^2 + 1$$
$$U_2 = u^6 - 2u^4 + u^2 \qquad V_2 = v^6 - 2v^4 + v^2$$

$\qquad(e)$

[1] The principle leading to these so-called Galerkin equations was indicated by W. Ritz; see "Gesammelte Werke," p. 228, 1911.

The set of functions

$$\varphi_1 = U_1V_1 \qquad \varphi_2 = U_1V_2 \qquad \varphi_3 = U_2V_1 \qquad \varphi_4 = U_2V_2 \qquad (f)$$

then fulfills the required conditions

$$w = \frac{\partial w}{\partial u} = 0 \qquad \text{on } u = \pm 1$$

and

$$w = \frac{\partial w}{\partial v} = 0 \qquad \text{on } v = \pm 1$$

Let us carry out the computation for the particular case of the square plate. As x and y now are interchangeable, we have $a_2 = a_3$ and, consequently,

$$\varphi_2 = \varphi_3 = U_1V_2 + U_2V_1$$

Putting $qa^4/16D = N$ we take expression (211) in the form

$$w = a_1U_1V_1 + a_2(U_1V_2 + U_2V_1) + a_4U_2V_2 \qquad (g)$$

Substituting this consecutively in Eqs. (d) with the factors φ_1, φ_2, and φ_4 and observing notation (e) we have then to evaluate the integrals between the limits $u = \pm 1, v = \pm 1$. Thus we arrive at the following system of equations:

$$\begin{aligned} 6.687345a_1 + 1.215879a_2 + 0.0675488a_4 &= 0.1422221N \\ 1.215879a_1 + 2.743525a_2 + 0.218235a_4 &= 0.0406349N \\ 0.0675488a_1 + 0.218235a_2 + 0.00590462a_4 &= 0.00290249N \end{aligned} \qquad (h)$$

For the first approximation we have

$$a_1 = \frac{0.1422221}{6.687345} N = 0.02127N$$

Resolving the whole system (h) we have

$$a_1 = 0.02023N \qquad a_2 = 0.00535N \qquad a_4 = 0.00625N$$

for the third approximation.

Numerical results obtained by means of the expression (g) for the deflection at the center, the moments $M_x = M_y$ at the center, and the moment M_x at $x = a/2, y = 0$, respectively, are the following:

First approx. $0.001329qa^4/D$, $0.0276qa^2$, $-0.0425qa^2$
Third approx. $0.001264qa^4/D$, $0.0228qa^2$, $-0.0512qa^2$

For comparison, Table 35 gives the values

$$0.00126qa^4/D, \; 0.0231qa^2, \; -0.0513qa^2$$

The moments at the center are calculated for $\nu = 0.3$.

It is seen that, whereas the first approximation is not yet satisfactory, the third approximation appears quite sufficient even for the bending moments concerned.

82. Various Approximate Methods. A Combined Method.[1] The procedure described in the foregoing article may be restricted as well to one variable, say y, thus obtaining for the other variable, x, an ordinary differential equation. Let us consider again the bending of a clamped square plate under uniform load (Fig. 91).

---

[1] Due to L. V. Kantorovich, *Izvest. Akad. Nauk S.S.S.R.*, no. 5, 1933.

SPECIAL AND APPROXIMATE METHODS IN THEORY OF PLATES 349

In confining ourselves to the first approximation we take, this time,

$$w = \varphi(x)\psi(y) = \varphi(x)(a^4 - 8a^2y^2 + 16y^4) \tag{a}$$

the boundary conditions $w = dw/dy = 0$ on $y = \pm a/2$ thus being fulfilled by the function $\psi(y)$. Now we try to satisfy the condition (c) of Art. 81 by choosing the variation in the form

$$\delta w = \psi(y)\,\delta\varphi(x) \tag{b}$$

This, after substitution in Eq. (c) of Art. 81, yields

$$\int\left[\iint\left(\Delta\Delta w - \frac{q}{D}\right)\psi(y)\,dy\right]dx\,\delta\varphi(x) = 0 \tag{c}$$

which is fulfilled if

$$\int_{-a/2}^{a/2}\left(\Delta\Delta w - \frac{q}{D}\right)\psi(y)\,dy = 0 \tag{d}$$

Next, we substitute expression (a) in this latter equation and obtain the following differential equation for the unknown function $\varphi(x)$:

$$\frac{a^4}{504}\frac{d^4\varphi}{dx^4} - \frac{a^2}{21}\frac{d^2\varphi}{dx^2} + \varphi = \frac{q}{384D} \tag{e}$$

An obvious particular solution of this equation is $\varphi = q/384D$. For the homogeneous equation resulting from Eq. (e), when $q = 0$, we have to assume $\varphi = e^{\lambda x/a}$. This yields $\lambda = \pm\alpha \pm \beta i$, with $\alpha = 4.1503$ and $\beta = 2.2858$. In view of the symmetry of the deflection surface about the y axis, solutions of Eq. (e) must be even functions in x; accordingly we have

$$\varphi = \frac{q}{384D}\left(1 + C_1\cosh\frac{\alpha x}{a}\cos\frac{\beta x}{a} + C_2\sinh\frac{\alpha x}{a}\sin\frac{\beta x}{a}\right) \tag{f}$$

To calculate the constants C_1 and C_2 we use the boundary conditions $\varphi = \partial\varphi/\partial x = 0$ on $x = \pm a/2$. Thus we obtain $C_1 = -0.50227$, $C_2 = -0.04396$, which establishes definitively the form of the function (f) and the solution (a).

We derive from this latter the following numerical results for the center of the plate: $w = 0.001296qa^4/D$ and (for $\nu = 0.3$) $M_x = 0.0241qa^2$ and $M_y = 0.0261qa^2$.

Owing to the partial use of the differential equation the results of the first approximation prove to be more exact than those of Art. 81, where a pure strain energy method was applied. To improve the accuracy still further, we have to assume

$$w = \varphi_1(x)\psi_1(x) + \varphi_2(x)\psi_2(\psi) + \cdots \tag{g}$$

where all the functions $\psi(y)$ have to fulfill the boundary conditions on $y = \pm a/2$. The use of Eq. (c) in conjunction with the variations $\delta w_1 = \psi_1\,\delta\varphi_1$, $\delta w_2 = \psi_2\,\delta\varphi_2$, ... would lead this time to a system of linear differential equations with constant coefficients for the functions $\varphi_1(x)$, $\varphi_2(x)$, ... The handling of such a system, though simple in principle, may become troublesome for higher approximations; the second approximation, however, should be adequate for the most practical purposes.

The Method of Reversion. Solution (211), fulfilling only the boundary conditions of the problem, may also be used in the following manner. Instead of calculating the deflections from a given load distribution by means of the differential equation (103) we use the same equation to calculate the loading

$$\bar{q} = D\Delta\Delta w \tag{h}$$

resulting from the tentative expression (211) for the deflection. According to our hypothesis, expression (211) does not represent the rigorous solution of the problem and, therefore, the loading (h) will never be identical with the given loading q. We can, however, choose the parameters $a_1, a_2, \ldots$ in Eq. (211) so as to equalize the functions q and $\bar{q}$ on the average over some portions of the area of the plate.

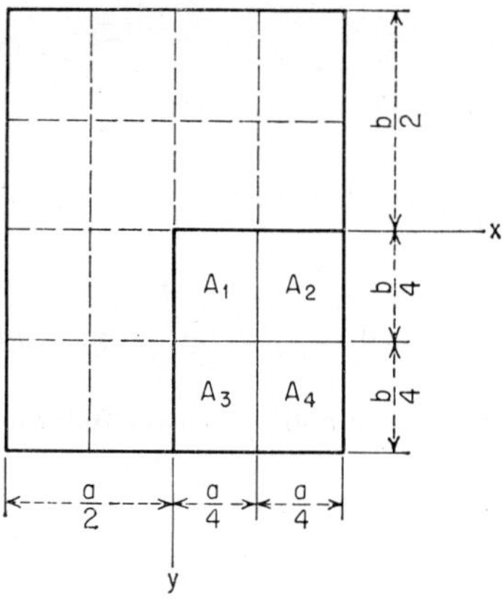

Consider, for example, a rectangular plate (Fig. 178) with boundary conditions and a distribution of loading symmetrical about both axes x and y. Having subdivided the plate into 16 equal rectangles, we need, because of the symmetry, to consider only four partial areas, such as A_1, A_2, A_3, and A_4. Expression (211) can be restricted accordingly to four terms, i.e., to

$$w = a_1\varphi_1 + a_2\varphi_2 + a_3\varphi_3 + a_4\varphi_4 \qquad (i)$$

Now let q and q' undergo in each of the partial areas the condition

$$\iint_{A_n} (q - \bar{q})\, dx\, dy = 0 \qquad n = 1, 2, 3, 4 \qquad (j)$$

Fig. 178

This gives four linear equations for the four parameters a_n and the resolution of these equations establishes the expression (i) in its final form.[1]

Methods Approximating the Boundary Conditions. If we succeed in finding a solution which fulfills the differential equation (103) together with one of the boundary conditions, the second prescribed condition may be satisfied by determination of a set of suitably chosen parameters. In solving the problem stated in Art. 44 coefficients of the two trigonometric series representing the variation of the edge moments of the plate were introduced as such parameters. Expansion of the slope $\partial w/\partial N$ in Fourier series[2] along the boundary was used in order to let this slope vanish in accordance with the requirements of the problem. In using the latter condition the parameters could be calculated. Some minimum principle—for example, the method of least squares—may be used as well in order to satisfy approximately the conditions on the boundary. The application of such a principle needs more detailed consideration when two boundary conditions must be simultaneously fulfilled.[3]

In using a solution which satisfies only the differential equation of the problem it sometimes proves simplest to fulfill the boundary conditions merely at a number of points suitably chosen along the boundary. The symmetry of the deformation of the plate, if such a symmetry exists, should be taken into account in locating those points. In order to satisfy all boundary conditions at m points we must introduce $2m$ unknown parameters.

In the most general case[4] we may use an expression for the deflection which satisfies neither the differential equation of the bent plate nor the boundary conditions of the

---

[1] An illustrative example for the application of the method may be found in C. B. Biezeno and R. Grammel, "Technische Dynamik," 2d ed., vol. 1, p. 147, Berlin, 1953.

[2] A more general system of functions orthogonalized along an edge was used by A. Nádai to fulfill a boundary condition; see "Elastische Platten," p. 180, Berlin, 1925.

[3] An important contribution to this question is due to E. Berger, *op. cit.*, p. 39.

[4] The method was discussed by C. J. Thorne and J. V. Atanasoff, *Iowa State Coll. J. Sci.*, vol. 14, p. 333, 1940.

problem. A number of points, say n, will be chosen then on and inside the boundary of the plate in which the differential equation must be satisfied exactly. Therefore a total of $2m + n$ parameters will be needed to obtain the solution of the problem.

Weinstein's Method.[1] In the specific case of a plate built in along the boundary we may seek at first a solution of the differential equation $\Delta\Delta w_1 = q/D$ such that the solution is valid for the given loading q and for the boundary conditions $w_1 = 0$, $\Delta w_1 = 0$, instead of the actual conditions. It has been shown in Art. 24 that this latter procedure is equivalent to solving in succession two problems, each dealing with the equilibrium of a loaded membrane.

The solution of the actual problem may be taken in the form

$$w = w_1 + \sum_{k=1}^{m} a_k \varphi_k \qquad (k)$$

where a_k are some coefficients and φ_k functions of x, y, vanishing at the boundary and obeying the differential equation $\Delta\Delta\varphi_k = 0$. The required condition $\partial w/\partial N$ at the boundary (where N is the normal to the boundary) can be modified by means of Green's theorem, which leads to the following system of m linear equations for the parameters a_k:

$$\iint \frac{q\varphi_1}{D} dx\, dy + \sum_{k=1}^{m} a_k \iint \Delta\varphi_1\, \Delta\varphi_k\, dx\, dy = 0$$

$$\iint \frac{q\varphi_2}{D} dx\, dy + \sum_{k=1}^{m} a_k \iint \Delta\varphi_2\, \Delta\varphi_k\, dx\, dy = 0 \qquad (l)$$

$$\cdots\cdots\cdots\cdots\cdots\cdots\cdots\cdots\cdots\cdots\cdots$$

where all integrals are taken over the entire area of the plate. The method may be used to advantage when the boundary conditions $w = 0$, $\Delta w = 0$ suggest a much simpler solution of the problem than the actual conditions $w = 0$, $\partial w/\partial N = 0$.

83. Application of Finite Differences Equations to the Bending of Simply Supported Plates.

In our previous discussion (see Art. 24) it was shown that the differential equation for the bending of plates can be replaced by two equations each of which has the form of the equation for the deflection of a uniformly stretched membrane. It was mentioned also that this latter equation can be solved with sufficient accuracy by replacing it by a finite differences equation. To illustrate this method of solution let us begin with the case of a uniformly loaded long rectangular plate. At a considerable distance from the short sides of the plate the deflection surface in this case may be considered cylindrical. Then, by taking the x axis parallel to the short sides of the plate, the differential equations (120) become

$$\frac{\partial^2 M}{\partial x^2} = -q$$

$$\frac{\partial^2 w}{\partial x^2} = -\frac{M}{D} \qquad (a)$$

---

[1] A. Weinstein and D. H. Rock, *Quart. Appl. Math.*, vol. 2, p. 262, 1944.

Both these equations have the same form as the equation for the deflection of a stretched and laterally loaded flexible string.

Let AB (Fig. 179) represent the deflection curve of a string stretched by forces S and uniformly loaded with a vertical load of intensity q. In deriving the equation of this curve we consider the equilibrium of an infinitesimal element mn. The tensile forces at points m and n have the

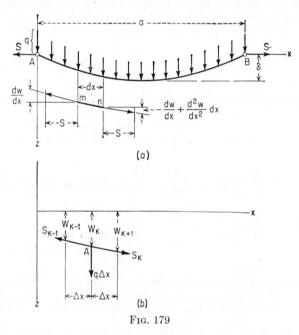

Fig. 179

directions of tangents to the deflection curve at these points; and, by projecting these forces and also the load $q\,dx$ on the z axis, we obtain

$$-S\frac{dw}{dx} + S\left(\frac{dw}{dx} + \frac{\partial^2 w}{\partial x^2}\,dx\right) + q\,dx = 0 \qquad (b)$$

from which

$$\frac{\partial^2 w}{\partial x^2} = -\frac{q}{S} \qquad (c)$$

This equation has the same form as Eqs. (a) derived for an infinitely long plate. The deflection curve is now obtained by integrating Eq. (c), which gives the parabolic curve

$$w = \frac{4\delta x\,(a - x)}{a^2} \qquad (d)$$

satisfying the conditions $w = 0$ at the ends and having a deflection δ at the middle.

The same problem can be solved graphically by replacing the uniform load by a system of equidistant concentrated forces $q\,\Delta x$, Δx being the distance between two adjacent forces, and constructing the funicular polygon for these forces. If A (Fig. 179) is one of the apexes of this funicular polygon and S_{k-1} and S_k are the tensile forces in the two adjacent sides of the polygon, the horizontal projections of these forces are equal to S and the sum of their vertical projections is in equilibrium with the load $q\,\Delta x$, which gives

$$-S\frac{w_k - w_{k-1}}{\Delta x} + S\frac{w_{k+1} - w_k}{\Delta x} + q\,\Delta x = 0 \qquad (e)$$

In this equation w_{k-1}, w_k, and w_{k+1} are the ordinates corresponding to the three consecutive apexes of the funicular polygon, and $(w_k - w_{k-1})/\Delta x$ and $(w_{k+1} - w_k)/\Delta x$ are the slopes of the two adjacent sides of the polygon. Equation (e) can be used in calculating the consecutive ordinates $w_1, w_2, \ldots, w_{k-1}, w_k, w_{k+1}, \ldots, w_n$ of the funicular polygon. For this purpose let us construct Table (f).

0	w_0			
		Δw_0		
Δx	w_1			
...	...	...	...	
...	...	...	...	
$(k-1)\,\Delta x$	w_{k-1}			
		Δw_{k-1}		
$k\,\Delta x$	w_k		$\Delta^2 w_k$	
		Δw_k		
$(k+1)\,\Delta x$	w_{k+1}			
...	...	...	...	

(f)

The abscissas of the consecutive division points of the span are entered in the first column of the table. In the second column are the consecutive ordinates of the apexes of the polygon. Forming the differences of the consecutive ordinates, such as $w_1 - w_0, \ldots, w_k - w_{k-1}, w_{k+1} - w_k, \ldots$, we obtain the so-called *first differences* denoted by $\Delta w_0, \ldots, \Delta w_{k-1}, \Delta w_k, \ldots$, which we enter in the third column of the table. The *second differences* are obtained by forming the differences between the consecutive numbers of the third column. For example, for the point k with the abscissa $k\,\Delta x$ the second difference is

$$\Delta^2 w_k = \Delta w_k - \Delta w_{k-1}$$
$$= w_{k+1} - w_k - (w_k - w_{k-1}) = w_{k+1} - 2w_k + w_{k-1} \qquad (g)$$

With this notation Eq. (*e*) can be written in the following form:

$$\frac{\Delta^2 w}{\Delta x^2} = -\frac{q}{S} \qquad (h)$$

This is a finite differences equation which corresponds to the differential equation (*c*) and approaches it closer and closer as the number of division points of the span increases.

In a similar manner the differential equations (*a*) can be replaced by the following finite differences equations:

$$\begin{aligned} \frac{\Delta^2 M}{\Delta x^2} &= -q \\ \frac{\Delta^2 w}{\Delta x^2} &= -\frac{M}{D} \end{aligned} \qquad (i)$$

To illustrate the application of these equations in calculating the deflections of the plate let us divide the span, say, into eight equal parts, *i.e.*, let $\Delta x = \tfrac{1}{8} a$. Then Eqs. (*i*) become

$$\Delta^2 M = -\frac{qa^2}{64}$$

$$\Delta^2 w = -\frac{Ma^2}{64D}$$

Forming the second differences for the consecutive division points w_1, w_2, w_3, and w_4 in accordance with Eq. (*g*) and observing that in our case $w_0 = 0$ and $M_0 = 0$ and from symmetry $w_3 = w_5$ and $M_3 = M_5$, we obtain the two following groups of linear equations:

$$\begin{aligned} M_2 - 2M_1 &= -\frac{qa^2}{64} & w_2 - 2w_1 &= -\frac{M_1 a^2}{64D} \\ M_3 - 2M_2 + M_1 &= -\frac{qa^2}{64} & w_3 - 2w_2 + w_1 &= -\frac{M_2 a^2}{64D} \\ M_4 - 2M_3 + M_2 &= -\frac{qa^2}{64} & w_4 - 2w_3 + w_2 &= -\frac{M_3 a^2}{64D} \\ M_3 - 2M_4 + M_3 &= -\frac{qa^2}{64} & w_3 - 2w_4 + w_3 &= -\frac{M_4 a^2}{64D} \end{aligned} \qquad (j)$$

Solving the first group, we obtain the following values for M:

$$M_1 = \frac{7}{2}\frac{qa^2}{64} \qquad M_2 = 6\frac{qa^2}{64} \qquad M_3 = \frac{15}{2}\frac{qa^2}{64} \qquad M_4 = 8\frac{qa^2}{64} \qquad (k)$$

These values coincide exactly with the values of the bending moments for a uniformly loaded strip, calculated from the known equation

$$M = \frac{qa}{2} x - \frac{qx^2}{2}$$

Substituting the values (k) for the moments in the second group of Eqs. (j), we obtain

$$w_2 - 2w_1 = -\tfrac{7}{2}N$$
$$w_3 - 2w_2 + w_1 = -6N$$
$$w_4 - 2w_3 + w_2 = -\tfrac{15}{2}N$$
$$w_3 - 2w_4 + w_3 = -8N$$

where
$$N = \frac{qa^4}{64^2 D}$$

Solving these equations, we obtain the following deflections at the division points:

$$w_1 = 21N \qquad w_2 = 38.5N \qquad w_3 = 50N \qquad w_4 = 54N \qquad (l)$$

The exact values of these deflections as obtained from the known equation

$$w = \frac{qx}{24D}(a^3 - 2ax^2 + x^3)$$

for the deflection of a uniformly loaded strip of length a, for purposes of comparison, are

$$w_1 = 20.7N \qquad w_2 = 38N \qquad w_3 = 49.4N \qquad w_4 = 53.3N$$

It is seen that by dividing the span into eight parts, the error in the magnitude of the maximum deflection as obtained from the finite differences equations (i) is about 1.25 per cent. By increasing the number of division points the accuracy of our calculations can be increased; but this will require more work, since the number of equations in the system (j) increases as we increase the number of divisions.

Let us consider next a rectangular plate of finite length. In this case the deflections are functions of both x and y, and Eqs. (a) must be replaced by the general equations (120). In replacing these equations by the finite differences equations we have to consider the differences corresponding to the changes of both the coordinates x and y.

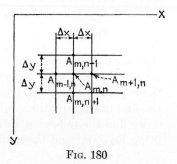

Fig. 180

We shall use the following notations for the first differences at a point A_{mn} with coordinates $m\,\Delta x$ and $n\,\Delta y$. The notation used in designating adjacent points is shown in Fig. 180.

$$\Delta_x w_{m-1,n} = w_{mn} - w_{m-1,n} \qquad \Delta_x w_{mn} = w_{m+1,n} - w_{mn}$$
$$\Delta_y w_{m,n-1} = w_{mn} - w_{m,n-1} \qquad \Delta_y w_{mn} = w_{m,n+1} - w_{mn}$$

Having the first differences, we can form the three kinds of second differences as follows:

$$\Delta_{xx}w_{mn} = \Delta_x w_{mn} - \Delta_x w_{m-1,n} = w_{m+1,n} - w_{mn} - (w_{mn} - w_{m-1,n})$$
$$= w_{m+1,n} - 2w_{mn} + w_{m-1,n}$$
$$\Delta_{yy}w_{mn} = \Delta_y w_{mn} - \Delta_y w_{m,n-1} = w_{m,n+1} - w_{mn} - (w_{mn} - w_{m,n-1}) \quad (m)$$
$$= w_{m,n+1} - 2w_{mn} + w_{m,n-1}$$
$$\Delta_{xy}w_{mn} = \Delta_y w_{mn} - \Delta_y w_{m-1,n} = w_{m,n+1} - w_{mn} - (w_{m-1,n+1} - w_{m-1,n})$$
$$= w_{m,n+1} - w_{mn} - w_{m-1,n+1} + w_{m-1,n}$$

With these notations the differential equations (120) will be replaced by the following differences equations:

$$\frac{\Delta_{xx}M}{\Delta x^2} + \frac{\Delta_{yy}M}{\Delta y^2} = -q$$
$$\frac{\Delta_{xx}w}{\Delta x^2} + \frac{\Delta_{yy}w}{\Delta y^2} = -\frac{M}{D} \quad (n)$$

In the case of a simply supported rectangular plate, M and w are equal to zero at the boundary, and we can solve Eqs. (n) in succession without any difficulty.

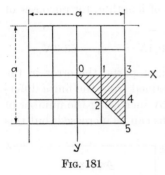

Fig. 181

To illustrate the process of calculating moments and deflections let us take the very simple case of a uniformly loaded square plate (Fig. 181). A rough approximation for M and w will be obtained by dividing the plate into 16 small squares, as shown in the figure, and by taking $\Delta x = \Delta y = a/4$ in Eqs. (n). It is evident from symmetry that the calculations need be extended over an area of one-eighth of the plate only, as shown in the figure by the shaded triangle. In this area we have to make the calculations only for the three points 0, 1, 2, for which M and w are different from zero. At the remaining points 3, 4, 5, these quantities are zero from the boundary conditions. Beginning with the first of the equations (n) and considering the center of the plate, point 0, we find the following values of the second differences for this point by using Eqs. (m) and the conditions of symmetry:

$$\Delta_{xx}M_0 = 2M_1 - 2M_0$$
$$\Delta_{yy}M_0 = 2M_1 - 2M_0$$

in which M_1 and M_0 are the values of M at points 1 and 0, respectively. Similarly for point 1 we obtain

$$\Delta_{xx}M_1 = M_3 - 2M_1 + M_0 = -2M_1 + M_0$$
$$\Delta_{yy}M_1 = 2M_2 - 2M_1$$

The second differences at point 2 can be calculated in the same way. Substituting these expressions for the second differences in the first of the equations (n), we obtain for points 0, 1, and 2 the following three equations:

$$4M_1 - 4M_0 = -\frac{qa^2}{16}$$

$$2M_2 - 4M_1 + M_0 = -\frac{qa^2}{16}$$

$$-4M_2 + 2M_1 = -\frac{qa^2}{16}$$

from which we find

$$M_0 = \frac{9}{2}\frac{qa^2}{64} \qquad M_1 = \frac{7}{2}\frac{qa^2}{64} \qquad M_2 = \frac{11}{4}\frac{qa^2}{64}$$

Substituting these values of moments in the second of the equations (n), we obtain the following three equations for calculating deflections w_0, w_1, and w_2:

$$4w_1 - 4w_0 = -\tfrac{9}{2}N$$
$$2w_2 - 4w_1 + w_0 = -\tfrac{7}{2}N$$
$$-4w_2 + 2w_1 = -\tfrac{11}{4}N$$

where $\qquad N = \dfrac{qa^4}{16 \cdot 64 D}$

From these equations we find the following values of the deflections:

$$w_0 = \tfrac{66}{16}N \qquad w_1 = \tfrac{48}{16}N \qquad w_2 = \tfrac{35}{16}N$$

For the deflection at the center we obtain

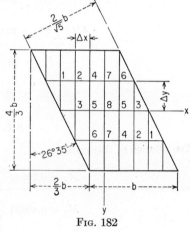

Fig. 182

$$w_0 = \frac{66}{16}N = \frac{66qa^4}{16 \cdot 16 \cdot 64 D} = 0.00403\,\frac{qa^4}{D}$$

Comparing this with the value $0.00406qa^4/D$ given in Table 8, it can be concluded that the error of the calculated maximum deflection is less than 1 per cent. For the bending moment at the center of the plate we find

$$M_x = M_y = \frac{M_0(1+\nu)}{2} = \frac{1.3}{2}\frac{9}{2}\frac{qa^2}{64} = 0.0457qa^2$$

which is less than the exact value $0.0479qa^2$ by about $4\tfrac{1}{2}$ per cent. It can be seen that in this case a small number of subdivisions of the plate gives an accuracy sufficient for practical applications. By taking twice the number of subdivisions, i.e., by making $\Delta x = \Delta y = \tfrac{1}{8}a$, the value of the bending moment will differ from the exact value by less than 1 per cent.

As a second problem let us consider the bending of a simply supported skew plate carrying a uniform load of intensity q (Fig. 182). The subdivisions in this case are

$\Delta x = b/6$ and $\Delta y = b/3$. Therefore the first of the equations (n) can be written as

$$4\Delta_{xx}M + \Delta_{yy}M = -\frac{qb^2}{9} \qquad (o)$$

Applying this equation to points 1 to 8 successively and using expressions (m) for the differences, we obtain the following system of linear equations:

$$-10M_1 + 4M_2 = -\frac{qb^2}{9}$$

$$4M_1 - 10M_2 + M_3 + 4M_4 = -\frac{qb^2}{9}$$

$$M_2 - 10M_3 + 4M_5 = -\frac{qb^2}{9}$$

$$4M_2 - 10M_4 + M_5 + 4M_7 = -\frac{qb^2}{9}$$

$$4M_3 + M_4 - 10M_5 + M_6 + 4M_8 = -\frac{qb^2}{9} \qquad (p)$$

$$M_5 - 10M_6 + 4M_7 = -\frac{qb^2}{9}$$

$$4M_4 + 4M_6 - 10M_7 + M_8 = -\frac{qb^2}{9}$$

$$8M_5 + 2M_7 - 10M_8 = -\frac{qb^2}{9}$$

The solution of this system is

$$M_1 = 0.29942 \frac{qb^2}{9} \qquad M_5 = 0.66191 \frac{qb^2}{9}$$

$$M_2 = 0.49854 \frac{qb^2}{9} \qquad M_6 = 0.39387 \frac{qb^2}{9}$$

$$M_3 = 0.41462 \frac{qb^2}{9} \qquad M_7 = 0.56920 \frac{qb^2}{9} \qquad (q)$$

$$M_4 = 0.59329 \frac{qb^2}{9} \qquad M_8 = 0.74337 \frac{qb^2}{9}$$

The second of the equations (n) now becomes

$$4\Delta_{xx}w + \Delta_{yy}w = -\frac{Mb^2}{9D} \qquad (r)$$

Taking into account the result (q) this gives a second group of equations:

$$\begin{aligned}
-10w_1 + 4w_2 &= -0.29942N \\
4w_1 - 10w_2 + w_3 + 4w_4 &= -0.49854N \\
w_2 - 10w_3 + 4w_5 &= -0.41462N \\
4w_2 - 10w_4 + w_5 + 4w_7 &= -0.59329N \\
4w_3 + w_4 - 10w_5 + w_6 + 4w_8 &= -0.66191N \\
w_5 - 10w_6 + 4w_7 &= -0.39387N \\
4w_4 + 4w_6 - 10w_7 + w_8 &= -0.56920N \\
8w_5 + 2w_7 - 10w_8 &= -0.74337N
\end{aligned} \qquad (s)$$

in which
$$N = \frac{qb^4}{81D}$$
This yields the deflections

$$\begin{aligned}
w_1 &= 0.13176N & w_5 &= 0.38549N \\
w_2 &= 0.25455N & w_6 &= 0.20293N \\
w_3 &= 0.22111N & w_7 &= 0.31249N \\
w_4 &= 0.32469N & w_8 &= 0.44523N
\end{aligned} \quad (t)$$

It should be noted that the integration of the differential equation of the bent plate by analytic methods would encounter considerable difficulties in this case.

To calculate the moments at the middle point 8 of the plate we have to use expressions (101) and (102), in which the derivatives first must be replaced by the respective differences. Thus, making use[1] of expressions (m) and using the values (t) for the deflections, and also taking $\nu = 0.2$, we obtain

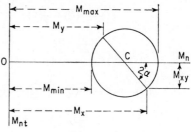

FIG. 183

$$(M_x)_8 = -D\left(\frac{w_5 - 2w_8 + w_5}{\Delta x^2} + \nu\,\frac{w_7 - 2w_8 + w_7}{\Delta y^2}\right) = 0.0590qb^2$$

$$(M_y)_8 = -D\left(\frac{w_7 - 2w_8 + w_7}{\Delta y^2} + \nu\,\frac{w_5 - 2w_8 + w_5}{\Delta x^2}\right) = 0.0401qb^2$$

$$(M_{xy})_8 = (1-\nu)D\,\frac{w_4 - w_6 - w_6 + w_4}{4\Delta x\,\Delta y} = 0.0108qb^2$$

Mohr's circle (Fig. 183) now gives[2] the following principal moments at point 8:

$$M_{\max} = \frac{M_x + M_y}{2} + \sqrt{\left(\frac{M_x - M_y}{2}\right)^2 + M_{xy}^2} = 0.0639qb^2$$

$$M_{\min} = \frac{M_x + M_y}{2} - \sqrt{\left(\frac{M_x - M_y}{2}\right)^2 + M_{xy}^2} = 0.0352qb^2$$

The direction of stresses due to these moments with respect to the coordinate axes x and y, respectively, is given by

$$\alpha = \frac{1}{2}\arctan\frac{2M_{xy}}{M_x - M_y} = 24°25'$$

From Fig. 182 we conclude that the stresses due to $M_{\max}$ at the center are acting almost exactly in the direction of the short span of the plate.

The plan of the plate in Fig. 182 was such that we could use a rectangular network

[1] See also the diagrams in Fig. 184 for the particular case $\Delta x = \Delta y$.

[2] Note the difference of notations in Figs. 183 and 22. The principal moments in Fig. 183 are denoted by $M_{\max}$ and $M_{\min}$. Note also that if in both diagrams the point on the circle moves in the clockwise direction, the normal to corresponding section will move in the same direction.

360 THEORY OF PLATES AND SHELLS

with constant subdivisions Δx and Δy. In a more general case a triangular network[1] must be used for the analysis of a skew slab.

The method of finite differences can also be applied to plates with edges built in or free and, finally, to plates with mixed boundary conditions.[2] Since in the general

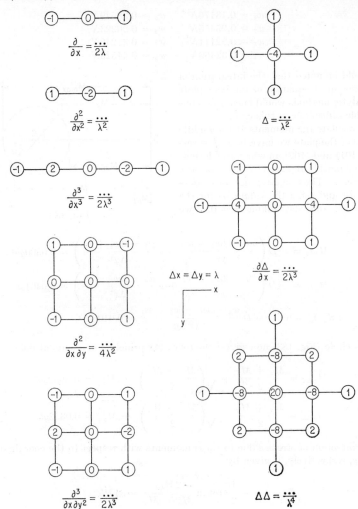

Fig. 184

case the value of M is not fixed on the boundary, and accordingly the use of M becomes less advantageous, the deflections w may be calculated directly by means of a sequence

[1] Extensive use of such networks is made by V. P. Jensen in *Univ. Illinois Bull.* 332, 1941, and the previous numerical example is taken therefrom.

[2] Many numerical examples of this kind may be found in the book by H. Marcus, "Die Theorie elastischer Gewebe," 2d ed., Berlin, 1932; see also N. J. Nielsen, "Bestemmelse af Spændinger i Plader," Copenhagen, 1920.

SPECIAL AND APPROXIMATE METHODS IN THEORY OF PLATES 361

of difference equations equivalent to the differential equation $\Delta\Delta w = q/D$ of the bent plate. For convenience the finite difference equivalent of the operator $\Delta\Delta(\cdots)$ is represented in Fig. 184 together with the other useful operators. The diagram is based on the assumption $\Delta x = \Delta y = \lambda$. Each number has to be multiplied by the symbol w_k denoting the deflection at the respective point k and the sum of such products then divided by an expression given in the caption.

In order to formulate the boundary conditions for an edge with vanishing deflections let us establish the equation for an interior point 7, next to the edge (Fig. 185). Applying the operator $\Delta\Delta(\cdots)$ we have

$$[w_1 + w_5 + w_9 + w_{13} + 2(w_2 + w_4 + w_{10} + w_{12})$$
$$- 8(w_3 + w_6 + w_8 + w_{11}) + 20w_7]\frac{1}{\lambda^4} = \frac{q_7}{D} \quad (u)$$

in which $w_2 = w_3 = w_4 = 0$. Next we have to eliminate the deflection w_1 at a fictive point 1, obtained by continuation of the network beyond the boundary of the plate.

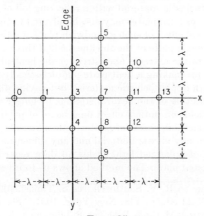

Fig. 185

This is readily done by means of the relation $w_1 = -w_7$ when the plate is simply supported at point 3 and by means of $w_1 = w_7$ when the plate is built in. Thus, there remain only the deflections of the interior points in Eq. (u) and the total number of such unknown deflections will not exceed the number of the equations of the type (u) at our disposal.

In the case of a free edge the number of such difference equations will be increased by the number of such points 2, 3, 4, . . . on the boundary at which the deflections do not vanish. The respective operators $\Delta\Delta w$ now must be extended over the exterior point at the distance λ and also 2λ from the free edge. Corresponding to each pair of such unknown deflections w_0, w_1, there will be two boundary conditions

$$\frac{\partial^2 w}{\partial x^2} + \nu \frac{\partial^2 w}{\partial y^2} = 0 \qquad \frac{\partial^3 w}{\partial x^3} + (2-\nu)\frac{\partial^3 w}{\partial x\,\partial y^2} = 0$$

expressed by means of the differences and written for point 3, opposite to both exterior points 0 and 1. Hence the total number of equations will still be the same as the number of unknown deflections.

When the values of M in the interior of the plate are no longer independent of the deflections w, the difference equations for the deflections become more involved than was the case in the two previous examples. In solving such equations the method of relaxation can sometimes be used to great advantage.[1]

84. Experimental Methods. For irregularly shaped plates or plates with irregularly varying thickness or weakened by many holes, experimental methods of investigation become more efficient than purely analytical methods. Conventional devices, such as electrical strain gauges and extensometers of all kinds, can be used for determination of strain in a bent plate.[2] The following brief review is restricted to methods which are appropriate to special conditions connected with the bending of thin elastic plates.

Use of Photoelasticity.[3] This method, usually applied to problems of plane stress, must be necessarily altered if employed in the case of bending of plates. In fact, the normal stresses in a thin bent plate are equal in magnitude but opposite in sign for two fibers symmetrical with respect to the middle plane of the plate. Accordingly, the optical effect produced in the zone of tension on a beam of polarized light passing through the plate is nullified by an opposite effect due to the zone of compression.

The influence of the second zone can be eliminated by cementing together two identical plates of photoelastic material with a reflecting foil of metal between them. The inner surface of one or both plates may also be silvered to the same end.[4] Calculations show that the optical effect of such a sandwich plate of a thickness h is about the same as the effect of a single plate of the thickness $h/2$ if this latter plate is submitted to a plane stress equal to the extreme fiber stress of the bent plate.

Another alternative[5] for making a bent plate photoelastically effective is to cement together two plates, both of photoelastic material, but having different elastic properties. The law of distribution of the flexural stress is no longer linear in such a plate. Hence, being bent, it yields an optical effect on a beam of polarized light.

According to a third method, sheets of photoelastic material are bonded on a reflective surface of a plate of any elastic material and any dimensions.[6] The behavior of such sheets in a beam of polarized light yields all data regarding the strain in the extreme fibers of the tested plate. The method allows us to investigate the strain in a

---

[1] For this method, due to R. V. Southwell, see S. Timoshenko and J. N. Goodier, "Theory of Elasticity," 2d ed., p. 468, New York, 1951. See also F. S. Shaw, "An Introduction to Relaxation Methods," Dover Publications, New York, 1953, where further bibliography is given. Another method of successive approximation in using the finite differences equation was developed by H. Liebman, Die angenäherte Ermittlung harmonischer Funktionen und konformer Abbildungen, *Sitzber. München. Akad.*, p. 385, 1918. The convergency of this method was discussed by F. Wolf, *Z. angew. Math. Mech.*, vol. 6, p. 118, 1926, and by R. Courant, *Z. angew. Math. Mech.*, vol. 6, p. 322, 1926. For an improved method see also R. Zurmühl, *Z. angew. Math. Mech.*, vol. 37, p. 1, 1957.

[2] An electromechanical method in measuring curvatures of a bent slab was used by W. Andrä, F. Leonhardt, and R. Krieger, *Bauingenieur*, vol. 33, p. 407, 1958.

[3] See for instance Timoshenko and Goodier, *op. cit.*, p. 131.

[4] See J. N. Goodier and G. H. Lee, *J. Appl. Mechanics*, vol. 8, p. A-27, 1941, and M. Dantu, *Ann. ponts et chaussées*, p. 281, 1952.

[5] See H. Favre, *Schweiz. Bauztg.*, 1950. For application of the method to a cantilever plate of variable thickness see H. Schwieger and G. Haberland, *Z. angew. Math. Mech.*, vol. 36, p. 287, 1956.

[6] This *photostress* method is due in principle to A. Mesnager (1930), but its practical application has been realized only recently; see, for example, F. Zandman and M. R. Wood, *Prod. Eng.*, September, 1956. For application of the so-called *freeze* procedure to plates, see D. C. Drucker, *J. Appl. Mechanics.*, vol. 9, p. A-161, 1942.

slab which is part of an actual structure and subjected to the actual loading, rather than being restricted to a model of the slab.

Use of Reflected Light.[1] The effect of a reflective surface of a strained plate on the direction of two adjacent light beams can be used to calculate the surface curvatures $\partial^2 w/\partial x^2$, $\partial^2 w/\partial y^2$, and $\partial^2 w/\partial x\, \partial y$, and, consequently, also the values of the flexural and torsional moments of the plate. For the same purpose the distortion of a luminous rectangular mesh projected on the initially plane surface of the plate may be used. Especially valuable are results obtained in this way for plates on elastic foundation, whose mechanical properties never can be expressed in a perfect manner analytically.

The Interference Method. Similar to the classic method used for determination of Poisson's ratio on beams, the interference method has also been applied to measure the deflections of a bent plate.[2]

Analogy between Plane Stress and Plate Bending.[3] There is an analogy between the plate deflection, governed by the differential equation $\Delta\Delta w = 0$ on the particular case of edge forces acting alone, and Airy's stress function φ satisfying the equation $\Delta\Delta\varphi = 0$. Whereas the function w yields the curvatures of the deformed plate, Airy's function yields the components $\sigma_x = \partial^2\varphi/\partial y^2$, $\sigma_y = \partial^2\varphi/\partial x^2$, and $\tau_{xy} = -\partial^2\varphi/\partial x\, \partial y$ of the plane stress in an elastic solid. Provided the contour, say $f(x,y) = 0$, is the same in both cases, we can put

$$\frac{\partial^2 w}{\partial x^2} = K\sigma_y \qquad \frac{\partial^2 w}{\partial y^2} = K\sigma_x \qquad \frac{\partial^2 w}{\partial x\, \partial y} = -K\tau_{xy}$$

where K is an arbitrary constant, such that the curvatures remain small.

Measured deflections w can be used for computation of the components of the plane stress and vice versa if certain conditions of analogy are satisfied both on the boundary of the plate and on that of the elastic solid.[4]

[1] For theory of the method and its application to various problems of bending of plates see M. Dantu, *Ann. ponts et chaussées*, 1940 and 1952. See also G. Bowen, *Eng. News-Record*, vol. 143, p. 70, 1949.

[2] See R. Landwehr and G. Grabert, *Ingr.-Arch.*, vol. 18, p. 1, 1950.

[3] Established by K. Wieghardt, *Mitt. Forschungsarb. Ingenieurwesens*, vol. 49, 1908. For a further extension of the analogy see H. Schaefer, *Abhandl. Braunschweig. wiss. Ges.*, vol. 8, p. 142, 1956.

[4] A simple formulation of those conditions is due to M. Dantu, *Ann. ponts et chaussées*, p. 386, 1952. For experimental methods based on analogy with electrical phenomena see R. H. MacNeal, *J. Appl. Mechanics*, vol. 18, p. 59, 1951, and K. Wotruba, *Czechoslov. J. Phys.*, vol. 2, p. 56, 1953. Further information on various experimental methods may be found in L. Föppl and E. Mönch, "Praktische Spannungsoptik," 2d ed., Berlin, 1959.

CHAPTER 11

BENDING OF ANISOTROPIC PLATES

85. Differential Equation of the Bent Plate. In our previous discussions we have assumed that the elastic properties of the material of the plate are the same in all directions. There are, however, cases in which an anisotropic material must be assumed if we wish to bring the theory of plates into agreement with experiments.[1] Let us assume that the material of the plate has three planes of symmetry with respect to its elastic properties.[2] Taking these planes as the coordinate planes, the relations between the stress and strain components for the case of plane stress in the xy plane can be represented by the following equations:

$$\sigma_x = E'_x \epsilon_x + E'' \epsilon_y$$
$$\sigma_y = E'_y \epsilon_y + E'' \epsilon_x \qquad (a)$$
$$\tau_{xy} = G \gamma_{xy}$$

It is seen that in the case of plane stress, four constants, E'_x, E'_y, E'', and G, are needed to characterize the elastic properties of a material.

Considering the bending of a plate made of such a material, we assume, as before, that linear elements perpendicular to the middle plane (xy plane) of the plate before bending remain straight and normal to the deflection surface of the plate after bending.[3] Hence we can use our previous expressions for the components of strain:

$$\epsilon_x = -z \frac{\partial^2 w}{\partial x^2} \qquad \epsilon_y = -z \frac{\partial^2 w}{\partial y^2} \qquad \gamma_{xy} = -2z \frac{\partial^2 w}{\partial x \, \partial y} \qquad (b)$$

---

[1] The case of a plate of anisotropic material was discussed by J. Boussinesq, *J. math.*, ser. 3, vol. 5, 1879. See also Saint Venant's translation of "Théorie de l'élasticité des corps solides," by A. Clebsch, note 73, p. 693.

[2] Such plates sometimes are called "orthotropic." The bending of plates with more general elastic properties has been considered by S. G. Lechnitzky in his book "Anisotropic Plates," 2d ed., Moscow, 1957.

[3] The effect of transverse shear in the case of anisotropy has been considered by K. Girkmann and R. Beer, *Österr. Ingr.-Arch.*, vol. 12, p. 101, 1958.

The corresponding stress components, from Eqs. (a), are

$$\sigma_x = -z\left(E'_x \frac{\partial^2 w}{\partial x^2} + E'' \frac{\partial^2 w}{\partial y^2}\right)$$

$$\sigma_y = -z\left(E'_y \frac{\partial^2 w}{\partial y^2} + E'' \frac{\partial^2 w}{\partial x^2}\right) \qquad (c)$$

$$\tau_{xy} = -2Gz \frac{\partial^2 w}{\partial x\, \partial y}$$

With these expressions for stress components the bending and twisting moments are

$$M_x = \int_{-h/2}^{h/2} \sigma_x z\, dz = -\left(D_x \frac{\partial^2 w}{\partial x^2} + D_1 \frac{\partial^2 w}{\partial y^2}\right)$$

$$M_y = \int_{-h/2}^{h/2} \sigma_y z\, dz = -\left(D_y \frac{\partial^2 w}{\partial y^2} + D_1 \frac{\partial^2 w}{\partial x^2}\right) \qquad (212)$$

$$M_{xy} = -\int_{-h/2}^{h/2} \tau_{xy} z\, dz = 2D_{xy} \frac{\partial^2 w}{\partial x\, \partial y}$$

in which

$$D_x = \frac{E'_x h^3}{12} \qquad D_y = \frac{E'_y h^3}{12} \qquad D_1 = \frac{E'' h^3}{12} \qquad D_{xy} = \frac{G h^3}{12} \qquad (d)$$

Substituting expressions (212) in the differential equation of equilibrium (100), we obtain the following equation for anisotropic plates:

$$D_x \frac{\partial^4 w}{\partial x^4} + 2(D_1 + 2D_{xy}) \frac{\partial^4 w}{\partial x^2\, \partial y^2} + D_y \frac{\partial^4 w}{\partial y^4} = q$$

Introducing the notation

$$H = D_1 + 2D_{xy} \qquad (e)$$

we obtain

$$D_x \frac{\partial^4 w}{\partial x^4} + 2H \frac{\partial^4 w}{\partial x^2\, \partial y^2} + D_y \frac{\partial^4 w}{\partial y^4} = q \qquad (213)$$

The corresponding expressions for the shearing forces are readily obtained from the conditions of equilibrium of an element of the plate (Fig. 48) and the previous expressions for the moments. Thus, we have

$$Q_x = -\frac{\partial}{\partial x}\left(D_x \frac{\partial^2 w}{\partial x^2} + H \frac{\partial^2 w}{\partial y^2}\right)$$

$$Q_y = -\frac{\partial}{\partial y}\left(D_y \frac{\partial^2 w}{\partial y^2} + H \frac{\partial^2 w}{\partial x^2}\right) \qquad (214)$$

In the particular case of isotropy we have

$$E'_x = E'_y = \frac{E}{1-\nu^2} \qquad E'' = \frac{\nu E}{1-\nu^2} \qquad G = \frac{E}{2(1+\nu)}$$

Hence

$$D_x = D_y = \frac{Eh^3}{12(1-\nu^2)}$$

$$H = D_1 + 2D_{xy} = \frac{h^3}{12}\left(\frac{\nu E}{1-\nu^2} + \frac{E}{1+\nu}\right) = \frac{Eh^3}{12(1-\nu^2)} \quad (f)$$

and Eq. (213) reduces to our previous Eq. (103).

Equation (213) can be used in the investigation of the bending of plates of nonisotropic and even nonhomogeneous material, such as reinforced concrete slabs,[1] which has different flexural rigidities in two mutually perpendicular directions.

86. Determination of Rigidities in Various Specific Cases. The expressions (d) given for the rigidities in the preceding article are subject to slight modifications according to the nature of the material employed. In particular, all values of torsional rigidity D_{xy} based on purely theoretical considerations should be regarded as a first approximation, and a direct test as shown in Fig. 25c must be recommended in order to obtain more reliable values of the modulus G. Usual values of the rigidities in some cases of practical interest are given below.

Reinforced Concrete Slabs. Let E_s be Young's modulus of steel, E_c that of the concrete, ν_c Poisson's ratio for concrete, and $n = E_s/E_c$. In terms of the elastic constants introduced in Art. 85 we have approximately $\nu_c = E''/\sqrt{E'_x E'_y}$. For a slab with two-way reinforcement in the directions x and y we can assume

$$\begin{aligned} D_x &= \frac{E_c}{1-\nu_c^2}[I_{cx} + (n-1)I_{sx}] \\ D_y &= \frac{E_c}{1-\nu_c^2}[I_{cy} + (n-1)I_{sy}] \\ D_1 &= \nu_c \sqrt{D_x D_y} \\ D_{xy} &= \frac{1-\nu_c}{2}\sqrt{D_x D_y} \end{aligned} \quad (a)$$

In these equations, I_{cx} is the moment of inertia of the slab material, I_{sx} that of the reinforcement taken about the neutral axis in the section $x = $ constant, and I_{cy} and I_{sy} are the respective values for the section $y = $ constant.

With the expression given for D_{xy} (also recommended by Huber) we obtain

$$H = \sqrt{D_x D_y} \quad (b)$$

and the differential equation

$$D_x \frac{\partial^4 w}{\partial x^4} + 2\sqrt{D_x D_y}\frac{\partial^4 w}{\partial x^2 \partial y^2} + D_y \frac{\partial^4 w}{\partial y^4} = q \quad (c)$$

---

[1] The application of the theory of anisotropic plates to reinforced concrete slabs is due to M. T. Huber, who published a series of papers on this subject; see *Z. Österr. Ing. u. Architektur Ver.*, 1914, p. 557. The principal results are collected in his books: "Teorya Plyt," Lvov, 1922, and "Probleme der Statik technisch wichtiger orthotroper Platten," Warsaw, 1929. Abstracts of his papers are given in *Compt. rend.*, vol. 170, pp. 511 and 1305, 1920; and vol. 180, p. 1243, 1925.

which can readily be reduced to the form (103) by introducing $y_1 = y \sqrt[4]{D_x/D_y}$ as a new variable.

It is obvious that the values (a) are not independent of the state of the concrete. For instance, any difference of the reinforcement in the directions x and y will affect the ratio D_x/D_y much more after cracking of the concrete than before.

Plywood. For a plate glued together of three or five plies, the x axis supposed to be parallel to the face grain, we may use the constants given in Table 79.

TABLE 79. ELASTIC CONSTANTS FOR PLYWOOD
Unit = 10^6 psi

Material	E'_x	E'_y	E''	G
Maple,* 5-ply...............	1.87	0.60	0.073	0.159
Afara,* 3-ply................	1.96	0.165	0.043	0.110
Gaboon* (Okoumé), 3-ply.....	1.28	0.11	0.014	0.085
Birch,† 3- and 5-ply..........	2.00	0.167	0.077	0.17
Birch† with bakelite membranes..	1.70	0.85	0.061	0.10

* By R. F. S. Hearmon and E. H. Adams, *Brit. J. Appl. Phys.*, vol. 3, p. 155, 1952.
† By S. G. Lechnitzky, "Anisotropic Plates," p. 40, Moscow, 1947.

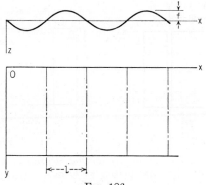

FIG. 186

Corrugated Sheet. Let E and ν be the elastic constants of the material of the sheet, h its thickness,

$$z = f \sin \frac{\pi x}{l}$$

the form of the corrugation, and s the length of the arc of one-half a wave (Fig. 186). Then we have[1]

$$D_x = \frac{l}{s} \frac{Eh^3}{12(1 - \nu^2)}$$
$$D_y = EI$$
$$D_1 \sim 0$$
$$H = 2D_{xy} = \frac{s}{l} \frac{Eh^3}{12(1 + \nu)}$$

[1] See E. Seydel, *Ber. deut. Versuchsanstalt Luftfahrt*, 1931.

in which, approximately,

$$s = l\left(1 + \frac{\pi^2 f^2}{4l^2}\right)$$

$$I = \frac{f^2 h}{2}\left[1 - \frac{0.81}{1 + 2.5\left(\dfrac{f}{2l}\right)^2}\right]$$

Plate Reinforced by Equidistant Stiffeners in One Direction. For a plate reinforced symmetrically with respect to its middle plane, as shown in Fig. 187, we may take[1]

$$D_x = H = \frac{Eh^3}{12(1 - \nu^2)}$$

$$D_y = \frac{Eh^3}{12(1 - \nu^2)} + \frac{E'I}{a_1}$$

in which E and ν are the elastic constants of the material of the plating, E' the Young modulus, and I the moment of inertia of a stiffener, taken with respect to the middle axis of the cross section of the plate.

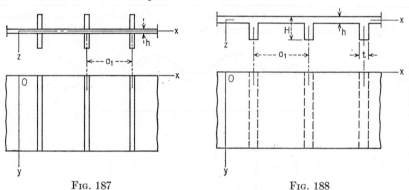

Fig. 187   Fig. 188

Plate Cross-stiffened by Two Sets of Equidistant Stiffeners. Provided the reinforcement is still symmetrical about the plating we have

$$D_x = \frac{Eh^3}{12(1 - \nu^2)} + \frac{E'I_1}{b_1}$$

$$D_y = \frac{Eh^3}{12(1 - \nu^2)} + \frac{E'I_2}{a_1}$$

$$H = \frac{Eh^3}{12(1 - \nu^2)}$$

I_1 being the moment of inertia of one stiffener and b_1 the spacing of the stiffeners in direction x, and I_2 and a_1 being the respective values for the stiffening in direction y.

Slab Reinforced by a Set of Equidistant Ribs. In the case shown in Fig. 188 the theory established in Art. 85 can give only a rough idea of the actual state of stress and

---

[1] Recommended by Lechnitzky, *op. cit.* For more exact values see N. J. Huffington, *J. Appl. Mechanics*, vol. 23, p. 15, 1956. An experimental determination of the rigidities of stiffened and grooved plates was carried out by W. H. Hoppmann, N. J. Huffington, and L. S. Magness, *J. Appl. Mechanics*, vol. 23, p. 343, 1956.

strain of the slab. Let E be the modulus of the material (for instance, concrete), I the moment of inertia of a T section of width a_1, and $\alpha = h/H$. Then we may assume

$$D_x = \frac{Ea_1 h^3}{12(a_1 - t + \alpha^3 t)}$$
$$D_y = \frac{EI}{a_1}$$
$$D_1 = 0$$

The effect of the transverse contraction is neglected in the foregoing formulas. The torsional rigidity, finally, may be calculated by means of the expression

$$D_{xy} = D'_{xy} + \frac{C}{2a_1}$$

in which D'_{xy} is the torsional rigidity of the slab without the ribs and C the torsional rigidity of one rib.[1]

87. Application of the Theory to the Calculation of Gridworks. Equation (213) can also be applied to the gridwork system shown in Fig. 189.

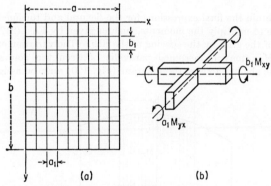

FIG. 189

This consists of two systems of parallel beams spaced equal distances apart in the x and y directions and rigidly connected at their points of intersection. The beams are supported at the ends, and the load is applied normal to the xy plane. If the distances a_1 and b_1 between the beams are small in comparison with the dimensions a and b of the grid, and if the flexural rigidity of each of the beams parallel to the x axis is equal to B_1 and that of each of the beams parallel to y axis is equal to B_2, we can substitute in Eq. (213)

$$D_x = \frac{B_1}{b_1} \qquad D_y = \frac{B_2}{a_1} \qquad (a)$$

[1] For a more exact theory concerning slabs with ribs in one or two directions and leading to a differential equation of the eighth order for the deflection see K. Trenks, *Bauingenieur*, vol. 29, p. 372, 1954; see also A. Pflüger, *Ingr.-Arch.*, vol. 16, p. 111, 1947.

The quantity D_1 in this case is zero, and the quantity D_{xy} can be expressed in terms of the torsional rigidities C_1 and C_2 of the beams parallel to the x and y axes, respectively. For this purpose we consider the twist of an element as shown in Fig. 189b and obtain the following relations between the twisting moments and the twist $\partial^2 w / \partial x\, \partial y$:

$$M_{xy} = \frac{C_1}{b_1} \frac{\partial^2 w}{\partial x\, \partial y} \qquad M_{yx} = -\frac{C_2}{a_1} \frac{\partial^2 w}{\partial x\, \partial y} \tag{b}$$

Substituting these expressions in the equation of equilibrium (e) on page 81, we find that in the case of the system represented in Fig. 189a the differential equation of the deflection surface is

$$\frac{B_1}{b_1} \frac{\partial^4 w}{\partial x^4} + \left(\frac{C_1}{b_1} + \frac{C_2}{a_1}\right) \frac{\partial^4 w}{\partial x^2\, \partial y^2} + \frac{B_2}{a_1} \frac{\partial^4 w}{\partial y^4} = q \tag{215}$$

which is of the same form as Eq. (213).

In order to obtain the final expressions for the flexural and torsional moments of a rib we still have to multiply the moments, such as given by Eqs. (212) and valid for the unit width of the grid, by the spacing of the ribs. The variation of the moments, say M_x and M_{xy}, may be assumed parabolic between the points $(m - 1)$ and $(m + 1)$ and the shaded area of the diagram (Fig. 190) may be assigned to the rib

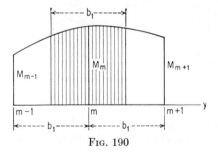

Fig. 190

(m) running in the direction x. Then, observing the expressions (212), we obtain the following approximate formulas for both moments of the rib (m):

$$M_x = -\frac{B_1}{24}\left[\left(\frac{\partial^2 w}{\partial x^2}\right)_{m-1} + 22\left(\frac{\partial^2 w}{\partial x^2}\right)_m + \left(\frac{\partial^2 w}{\partial x^2}\right)_{m+1}\right]$$
$$M_{xy} = \frac{C_1}{24}\left[\left(\frac{\partial^2 w}{\partial x\, \partial y}\right)_{m-1} + 22\left(\frac{\partial^2 w}{\partial x\, \partial y}\right)_m + \left(\frac{\partial^2 w}{\partial x\, \partial y}\right)_{m+1}\right] \tag{c}$$

For ribs of the direction y we have to interchange x and y in the foregoing expressions and replace B_1 by B_2 and C_1 by C_2; $(m - 1)$, (m), and $(m + 1)$ then denote three successive joints on a rib having the direction x.

Two parameters largely defining the elastic properties of a grid and often used in calculation are

$$\lambda = \sqrt[4]{\frac{B_2 b_1}{B_1 a_1}} \qquad \mu = \frac{1}{2} \frac{\dfrac{C_1}{b_1} + \dfrac{C_2}{a_1}}{\sqrt{\dfrac{B_1 B_2}{a_1 b_1}}} \qquad (d)$$

The parameter λ multiplied by the side ratio a/b (Fig. 189) yields the relative carrying capacity of a rectangular plate in the directions y and x, whereas the parameter μ characterizes the torsional rigidity of a grid as compared with its flexural rigidity.

Equation (215) has been extensively used in investigating the distribution of an arbitrarily located single load between the main girders of a bridge stiffened in the transverse direction by continuous floor beams.[1]

88. Bending of Rectangular Plates. When the plate is simply supported on all sides Eq. (213) can be solved by the methods used in the case of an isotropic plate. Let us apply the Navier method (see Art. 28) and assume that the plate is uniformly loaded. Taking the coordinate axes as shown in Fig. 59 and representing the load in the form of a double trigonometric series, the differential equation (213) becomes

$$D_x \frac{\partial^4 w}{\partial x^4} + 2H \frac{\partial^4 w}{\partial x^2 \, \partial y^2} + D_y \frac{\partial^4 w}{\partial y^4}$$

$$= \frac{16 q_0}{\pi^2} \sum_{m=1,3,5,\ldots}^{\infty} \sum_{n=1,3,5,\ldots}^{\infty} \frac{1}{mn} \sin \frac{m\pi x}{a} \sin \frac{n\pi y}{b} \qquad (a)$$

A solution of this equation that satisfies the boundary conditions can be taken in the form of the double trigonometrical series

$$w = \sum_{m=1,3,5,\ldots}^{\infty} \sum_{n=1,3,5,\ldots}^{\infty} a_{mn} \sin \frac{m\pi x}{a} \sin \frac{n\pi y}{b} \qquad (b)$$

Substituting this series in Eq. (a), we find the following expression for the coefficients a_{mn}:

$$a_{mn} = \frac{16 q_0}{\pi^6} \frac{1}{mn \left(\dfrac{m^4}{a^4} D_x + \dfrac{2 m^2 n^2}{a^2 b^2} H + \dfrac{n^4}{b^4} D_y \right)}$$

---

[1] Factors giving the distribution of a single load have been calculated for $\mu = 0$ by Y. Guyon, *Ann. ponts et chaussées*, vol. 116, p. 553, 1946, and for $\mu \neq 0$ by C. Massonnet, *Publs. Intern. Assoc. Bridge and Structural Engrs.*, vol. 10, p. 147, 1950. For verification of calculated results by test see K. Sattler, *Bauingenieur*, vol. 30, p. 77, 1955, and also M. Naruoka and H. Yonezawa, *Publs. Intern. Assoc. Bridge and Structural Engrs.*, vol. 16, 1956. For skewed grids see S. Woinowsky-Krieger, *Ingr.-Arch.*, vol. 25, p. 350, 1957.

Hence the solution of Eq. (a) is

$$w = \frac{16q_0}{\pi^6} \sum_{m=1,3,5,\ldots}^{\infty} \sum_{n=1,3,5,\ldots}^{\infty} \frac{\sin \frac{m\pi x}{a} \sin \frac{n\pi y}{b}}{mn \left(\frac{m^4}{a^4} D_x + \frac{2m^2 n^2}{a^2 b^2} H + \frac{n^4}{b^4} D_y\right)} \quad (c)$$

In the case of an isotropic material $D_x = D_y = H = D$, and this solution coincides with that given on page 110.

Furthermore, let us consider the particular case of $H = \sqrt{D_x D_y}$ already mentioned on page 366. Comparing expression (c) with the corresponding expression (131) for the isotropic plate, we conclude that the deflection at the center of such an orthotropic plate with rigidities D_x, D_y, and the sides a, b is the same as that of an isotropic plate having a rigidity D and the sides $a_0 = a \sqrt[4]{D/D_x}$ and $b_0 = b \sqrt[4]{D/D_y}$. In like manner the curvatures of the orthotropic plate may be expressed by those of a certain isotropic plate. The deflection and the bending moments at the center of the orthotropic plate obtained in this way can be expressed by the formulas

$$w = \alpha \frac{q_0 b^4}{D_y}$$

$$M_x = \left(\beta_1 + \beta_2 \frac{E''}{E'_x} \sqrt{\frac{D_x}{D_y}}\right) \frac{q_0 a^2}{\epsilon} \quad (d)$$

$$M_y = \left(\beta_2 + \beta_1 \frac{E''}{E'_y} \sqrt{\frac{D_y}{D_x}}\right) q_0 b^2$$

where α, β_1, and β_2 are numerical coefficients[1] given in Table 80 and

$$\epsilon = \frac{a}{b} \sqrt[4]{\frac{D_y}{D_x}} \quad (e)$$

As a second example let us consider an infinitely long plate (Fig. 74) and assume that the load is distributed along the x axis following the sinusoidal relation

$$q = q_0 \sin \frac{m\pi x}{a} \quad (f)$$

In this case Eq. (213) for the unloaded portions of the plate becomes

$$D_x \frac{\partial^4 w}{\partial x^4} + 2H \frac{\partial^4 w}{\partial x^2 \, \partial y^2} + D_y \frac{\partial^4 w}{\partial y^4} = 0 \quad (g)$$

---

[1] Calculated by M. T. Huber, "Probleme der Statik technisch wichtiger orthotroper Platten," p. 74, Warsaw, 1929. For numerical data regarding uniformly loaded rectangular plates with various edge conditions and various torsion coefficients, see H. A. Schade, *Trans. Soc. Naval Architects Marine Engrs.*, vol. 49, pp. 154, 180, 1941.

TABLE 80. Constants α, β_1, and β_2 for a Simply Supported Rectangular Orthotropic Plate with $H = \sqrt{D_x D_y}$, Eqs. (d), (e) (Fig. 59)

ϵ	α	β_1	β_2	ϵ	α	β_1	β_2
1	0.00407	0.0368	0.0368	1.8	0.00932	0.0214	0.0884
1.1	0.00488	0.0359	0.0447	1.9	0.00974	0.0191	0.0929
1.2	0.00565	0.0344	0.0524	2.0	0.01013	0.0174	0.0964
1.3	0.00639	0.0324	0.0597	2.5	0.01150	0.0099	0.1100
1.4	0.00709	0.0303	0.0665	3	0.01223	0.0055	0.1172
1.5	0.00772	0.0280	0.0728	4	0.01282	0.0015	0.1230
1.6	0.00831	0.0257	0.0785	5	0.01297	0.0004	0.1245
1.7	0.00884	0.0235	0.0837	∞	0.01302	0	0.1250

A solution of this equation, satisfying the boundary conditions at the sides parallel to the y axis, can be taken in the following form:

$$w = Y_m \sin \frac{m\pi x}{a} \tag{h}$$

where Y_m is a function of y only. Substituting this in Eq. (g), we obtain the following equation for determining the function Y_m:

$$D_y Y_m^{IV} - 2H \frac{m^2 \pi^2}{a^2} Y_m^{II} + D_x \frac{m^4 \pi^4}{a^4} Y_m = 0 \tag{i}$$

The roots of the corresponding characteristic equation are

$$r_{1,2,3,4} = \pm \frac{m\pi}{a} \sqrt{\frac{H}{D_y} \pm \sqrt{\frac{H^2}{D_y^2} - \frac{D_x}{D_y}}} \tag{j}$$

Using, in accordance with Eq. (d), Art. 87, the notation

$$\lambda = \sqrt[4]{\frac{D_y}{D_x}} \qquad \mu = \frac{H}{\sqrt{D_x D_y}} \tag{k}$$

we have to consider the following three cases:

Case 1, $\mu > 1$:
$$H^2 > D_x D_y$$

Case 2, $\mu = 1$:
$$H^2 = D_x D_y \tag{l}$$

Case 3, $\mu < 1$:
$$H^2 < D_x D_y$$

In the first case all the roots of Eq. (j) are real. Considering the part of the plate with positive y and observing that the deflection w and its derivatives must vanish at large distances from the load, we can retain only the negative roots. Using the notation

$$\alpha = \frac{a\lambda}{\pi} \sqrt{\mu + \sqrt{\mu^2 - 1}}$$
$$\beta = \frac{a\lambda}{\pi} \sqrt{\mu - \sqrt{\mu^2 - 1}}$$
(m)

the integral of Eq. (i) becomes

$$Y_m = A_m e^{-my/\alpha} + B_m e^{-my/\beta}$$

and expression (h) can be represented in the form

$$w = (A_m e^{-my/\alpha} + B_m e^{-my/\beta}) \sin \frac{m\pi x}{a}$$

From symmetry we conclude that along the x axis

$$\left(\frac{\partial w}{\partial y}\right)_{y=0} = 0$$

and we find

$$B_m = -\frac{\beta}{\alpha} A_m$$

and

$$w = A_m \left(e^{-my/\alpha} - \frac{\beta}{\alpha} e^{-my/\beta}\right) \sin \frac{m\pi x}{a}$$
(n)

The coefficient A_m is obtained from the condition relating to the shearing force Q_y along the x axis, which gives

$$-\frac{\partial}{\partial y}\left(D_y \frac{\partial^2 w}{\partial y^2} + H \frac{\partial^2 w}{\partial x^2}\right) = -\frac{q_0}{2} \sin \frac{m\pi x}{a}$$

Substituting for w its expression (n), we obtain

$$A_m = \frac{q_0 \alpha^3 \beta^2}{2m^3 D_y (\alpha^2 - \beta^2)} = \frac{\alpha q_0 a^4}{2\pi^4 m^3 D_x (\alpha^2 - \beta^2)}$$

and the final expression (n) for the deflection becomes

$$w = \frac{q_0 a^4}{2\pi^4 m^3 D_x (\alpha^2 - \beta^2)} (\alpha e^{-my/\alpha} - \beta e^{-my/\beta}) \sin \frac{m\pi x}{a}$$
(o)

In the second of the three cases (l) the characteristic equation has two double roots, and the function Y_m has the same form as in the case of an isotropic plate (Art. 36). In the third of the cases (l) we use the notation

$$\alpha' = \frac{a\lambda}{\pi} \sqrt{\frac{2}{1-\mu}}$$
$$\beta' = \frac{a\lambda}{\pi} \sqrt{\frac{2}{1+\mu}}$$
(p)

and thus obtain the solution

$$w = \frac{q_0 a^2}{4\pi^2 m^3 \sqrt{D_x D_y}} \left(\alpha' \sin \frac{my}{\alpha'} + \beta' \cos \frac{my}{\alpha'} \right) e^{-mx/\beta'} \sin \frac{m\pi x}{a} \quad (q)$$

We can also shift from case 1 to case 2 by using the complex relations

$$\frac{1}{\alpha} = \frac{1}{\beta'} - i\frac{1}{\alpha'}$$
$$\frac{1}{\beta} = \frac{1}{\beta'} + i\frac{1}{\alpha'} \quad (r)$$

Having the deflection surface for the sinusoidal load (f), the deflection for any other kind of load along the x axis can be obtained by expanding the load in the series

$$q = \sum_{m=1}^{\infty} a_m \sin \frac{m\pi x}{a}$$

and using the solution obtained for the load (f) for each term of this series. The following expressions hold when, for instance, a load P is concentrated at a point $x = \xi$, $y = 0$ of the infinite strip (Fig. 72):

Case 1, $\mu > 1$:

$$w = \frac{Pa^3}{\pi^4 D_x} \frac{1}{\alpha^2 - \beta^2} \sum_{m=1}^{\infty} \frac{1}{m^3} (\alpha e^{-my/\alpha} - \beta e^{-my/\beta}) \sin \frac{m\pi \xi}{a} \sin \frac{m\pi x}{a} \quad (s)$$

Case 2, $\mu = 1$:

$$w = \frac{Pa^2}{2\pi^3 D_x \lambda} \sum_{m=1}^{\infty} \frac{1}{m^3} \left(1 + \frac{m\pi y}{a\lambda}\right) e^{-my/\beta'} \sin \frac{m\pi \xi}{a} \sin \frac{m\pi x}{a} \quad (t)$$

Case 3, $\mu < 1$:

$$w = \frac{Pa}{2\pi^2 \sqrt{D_x D_y}} \sum_{m=1}^{\infty} \frac{1}{m^3} \left(\alpha' \sin \frac{my}{\alpha'} + \beta' \cos \frac{my}{\alpha'} \right) e^{-my/\beta'} \sin \frac{m\pi \xi}{a} \sin \frac{m\pi x}{a}$$
$$(u)$$

Expressions in closed form[1] can be obtained for bending moments due to a single load in a manner similar to that used for the isotropic plate in Art. 35.

Having this solution, the deflection of the plate by a load distributed

[1] See W. Nowacki, *Acta Tech. Acad. Sci. Hung.*, vol. 8, p. 109, 1954; S. Woinowsky-Krieger, *Ingr.-Arch.*, vol. 25, p. 90, 1957. Numerical results regarding influence surfaces of orthotropic rectangular plates may be found in H. Olsen and F. Reinitzhuber, "Die zweiseitig gelagerte Platte," Berlin, 1950, and in H. Homberg and J. Weinmeister, "Einflussflächen für Kreuzwerke," 2d ed., Berlin, 1956.

over a circular area can be obtained by integration, as was shown in the case of an isotropic plate (see Art. 35). By applying the method of images the solutions obtained for an infinitely long plate can be used in the investigation of the bending of plates of finite dimensions.[1]

89. Bending of Circular and Elliptic Plates. A simple solution of Eq. (213) can be obtained in the case of an elliptic plate clamped[2] on the boundary and carrying a uniform load of intensity q. Provided the principal directions x and y of the orthotropic material are parallel to the principal axes of the ellipse (Fig. 157) the expression

$$w = w_0 \left(1 - \frac{x^2}{a^2} - \frac{y^2}{b^2}\right)^2 \qquad (a)$$

in which

$$w_0 = \frac{q}{\dfrac{24D_x}{a^4} + \dfrac{16H}{a^2 b^2} + \dfrac{24D_y}{b^4}} \qquad (b)$$

satisfies Eq. (213) and the required conditions on the boundary. The bending moments of the plate are readily obtained by means of expressions (212). In the particular case of a clamped circular plate ($a = b$) we have the following results:

$$w = \frac{q(a^2 - r^2)^2}{64D'}$$

$$M_x = \frac{q}{16D'}[(D_x + D_1)(a^2 - r^2) - 2(D_x x^2 + D_1 y^2)]$$

$$M_y = \frac{q}{16D'}[(D_y + D_1)(a^2 - r^2) - 2(D_y y^2 + D_1 x^2)] \qquad (c)$$

$$M_{xy} = \frac{q}{4D'} D_{xy} xy$$

$$Q_x = -\frac{qx}{8D'}(3D_x + H)$$

$$Q_y = -\frac{qy}{8D'}(3D_y + H)$$

in which

$$r = \sqrt{x^2 + y^2} \quad \text{and} \quad D' = \tfrac{1}{8}(3D_x + 2H + 3D_y)$$

Since the twist is zero along the edge, the reactions of the support are given by a linear combination of the boundary values of the shearing forces Q_x and Q_y (see page 87).

A straightforward solution can also be obtained in the case of pure bending or pure twist of an orthotropic plate. Let such a plate be subjected to uniform couples $M_x = M_1$, $M_y = M_2$, and $M_{xy} = M_3$. By taking the deflection in the form

$$w = Ax^2 + Bxy + Cy^2 \qquad (d)$$

---

[1] Several examples of this kind are worked out in the books by M. T. Huber: "Teorya Plyt," Lvov, 1922, and "Probleme der Statik technisch wichtiger orthotroper Platten," Warsaw, 1929.

[2] For bending of a simply supported elliptical plate, see Y. Ōhasi, *Z. angew. Math. u. Phys.*, vol. 3, p. 212, 1952.

we obviously satisfy the differential equation (213). The constants A, B, and C then are given by the linear equations

$$D_x A + D_1 C = -\tfrac{1}{2} M_1$$
$$D_1 A + D_y C = -\tfrac{1}{2} M_2 \qquad (e)$$
$$D_{xy} B = \tfrac{1}{2} M_3$$

which ensue from the expressions (212).

The bending of a circular plate with cylindrical aeolotropy has been discussed too.[1] If, in addition to the elastic symmetry, the given load distribution is also symmetrical about the center of the plate, then the ordinary differential equation of the bent plate contains only two flexural rigidities, the radial and the tangential. Formal solutions of this equation for any boundary conditions are simple to obtain; the choice of the elastic constants of the material, however, requires special consideration since certain assumptions regarding these constants lead to infinite bending moments at the center of the plate even in the case of a continuously distributed loading.

Most of the special methods used in solving the problems of bending of an isotropic plate (Chap. 10) can be applied with some modifications to the case of an anisotropic plate as well.

If we take the complex variable method,[2] for example, the form of the solution proves to be different from that considered in Art. 79. As can be shown, it depends upon the roots ρ_1, ρ_2, $-\rho_1$, and $-\rho_2$ of the characteristic equation

$$D_y \rho^4 + 2H \rho^2 + D_x = 0$$

which are either imaginary or complex. These roots being determined, the solution of the homogeneous equation $D_x \, \partial^4 w_1/\partial x^4 + 2H \, \partial^4 w_1/\partial x^2 \, \partial y^2 + D_y \, \partial^4 w_1/\partial y^4 = 0$ can be represented either in the form

$$w_1 = \Re[\varphi_1(z_1) + \varphi_2(z_2)]$$

if $\rho_1 \neq \rho_2$, or else in the form

$$w_1 = \Re[\varphi_1(z_1) + \bar{z}_1 \varphi_2(z_1)]$$

if $\rho_1 = \rho_2$. In these expressions φ_1 and φ_2 are arbitrary analytic functions of the complex variables $z_1 = x + \rho_1 y$ and $z_2 = x + \rho_2 y$.

In using the Ritz method, expression (b) of Art. 80 for the strain energy has to be replaced by the expression

$$V = \frac{1}{2} \int_0^a \int_0^b \left[D_x \left(\frac{\partial^2 w}{\partial x^2} \right)^2 + 2D_1 \frac{\partial^2 w}{\partial x^2} \frac{\partial^2 w}{\partial y^2} + D_y \left(\frac{\partial^2 w}{\partial y^2} \right)^2 + 4 D_{xy} \left(\frac{\partial^2 w}{\partial x \, \partial y} \right)^2 \right] dx \, dy$$

while the rest of the procedure remains the same as in the case of the isotropic plate.

[1] G. F. Carrier, *J. Appl. Mechanics*, vol. 11, p. A-129, 1944, and Lechnitzky, *op. cit.*
[2] See S. G. Lechnitzky, *Priklad. Mat. Mekhan.*, vol. 2, p. 181, 1938, and V. Morcovin, *Quart. Appl. Math.*, vol. 1, p. 116, 1943. For application of the method to the problem of stress concentration, see also G. N. Savin, "Stress Concentration around Holes," Moscow, 1951, and S. G. Lechnitzky, *Inzhenernyi Sbornik*, vol. 17, p. 3, 1953. Stress concentration in isotropic and anisotropic plates was also discussed by S. Holgate, *Proc. Roy. Soc. London*, vol. 185A, pp. 35, 50, 1946.

CHAPTER 12

BENDING OF PLATES UNDER THE COMBINED ACTION OF LATERAL LOADS AND FORCES IN THE MIDDLE PLANE OF THE PLATE

90. Differential Equation of the Deflection Surface. In our previous discussion it has always been assumed that the plate is bent by lateral loads only. If in addition to lateral loads there are forces acting in the middle plane of the plate, these latter forces may have a considerable effect on the bending of the plate and must be considered in deriving the corresponding differential equation of the deflection surface. Proceeding as in the case of lateral loading (see Art. 21, page 79), we consider the equilibrium of a small element cut from the plate by two pairs of planes parallel to the xz and yz coordinate planes (Fig. 191). In addition to the forces discussed in Art. 21 we now have forces acting in the middle plane of the plate. We denote the magnitude of these forces per unit length by N_x, N_y, and $N_{xy} = N_{yx}$, as shown in the figure. Projecting these forces on the x and y axes and assuming that there are no body forces or tangential forces acting in those directions at the faces of the plate, we obtain the following equations of equilibrium:

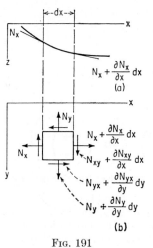

Fig. 191

$$\frac{\partial N_x}{\partial x} + \frac{\partial N_{xy}}{\partial y} = 0$$
$$\frac{\partial N_{xy}}{\partial x} + \frac{\partial N_y}{\partial y} = 0$$
(216)

These equations are entirely independent of the three equations of equilibrium considered in Art. 21 and can be treated separately, as will be shown in Art. 92.

In considering the projection of the forces shown in Fig. 191 on the z axis, we must take into account the bending of the plate and the resulting small angles between the forces N_x and N_y that act on the opposite sides of the element. As a result of this bending the projection

378

of the normal forces N_x on the z axis gives

$$-N_x\,dy\,\frac{\partial w}{\partial x} + \left(N_x + \frac{\partial N_x}{\partial x}\,dx\right)\left(\frac{\partial w}{\partial x} + \frac{\partial^2 w}{\partial x^2}\,dx\right)dy$$

After simplification, if the small quantities of higher than the second order are neglected, this projection becomes

$$N_x\,\frac{\partial^2 w}{\partial x^2}\,dx\,dy + \frac{\partial N_x}{\partial x}\,\frac{\partial w}{\partial x}\,dx\,dy \qquad (a)$$

In the same way the projection of the normal forces N_y on the z axis gives

$$N_y\,\frac{\partial^2 w}{\partial y^2}\,dx\,dy + \frac{\partial N_y}{\partial y}\,\frac{\partial w}{\partial y}\,dx\,dy \qquad (b)$$

Regarding the projection of the shearing forces N_{xy} on the z axis, we observe that the slope of the deflection surface in the y direction on the two opposite sides of the element is $\partial w/\partial y$ and $\partial w/\partial y + (\partial^2 w/\partial x\,\partial y)\,dx$. Hence the projection of the shearing forces on the z axis is equal to

$$N_{xy}\,\frac{\partial^2 w}{\partial x\,\partial y}\,dx\,dy + \frac{\partial N_{xy}}{\partial x}\,\frac{\partial w}{\partial y}\,dx\,dy$$

An analogous expression can be obtained for the projection of the shearing forces $N_{yx} = N_{xy}$ on the z axis. The final expression for the projection of all the shearing forces on the z axis then can be written as

$$2N_{xy}\,\frac{\partial^2 w}{\partial x\,\partial y}\,dx\,dy + \frac{\partial N_{xy}}{\partial x}\,\frac{\partial w}{\partial y}\,dx\,dy + \frac{\partial N_{xy}}{\partial y}\,\frac{\partial w}{\partial x}\,dx\,dy \qquad (c)$$

Adding expressions (a), (b), and (c) to the load $q\,dx\,dy$ acting on the element and using Eqs. (216), we obtain, instead of Eq. (100) (page 81), the following equation of equilibrium:

$$\frac{\partial^2 M_x}{\partial x^2} - 2\,\frac{\partial^2 M_{xy}}{\partial x\,\partial y} + \frac{\partial^2 M_y}{\partial y^2} = -\left(q + N_x\,\frac{\partial^2 w}{\partial x^2} + N_y\,\frac{\partial^2 w}{\partial y^2} + 2N_{xy}\,\frac{\partial^2 w}{\partial x\,\partial y}\right)$$

Substituting expressions (101) and (102) for M_x, M_y, and M_{xy}, we obtain

$$\frac{\partial^4 w}{\partial x^4} + 2\,\frac{\partial^4 w}{\partial x^2\,\partial y^2} + \frac{\partial^4 w}{\partial y^4}$$
$$= \frac{1}{D}\left(q + N_x\,\frac{\partial^2 w}{\partial x^2} + N_y\,\frac{\partial^2 w}{\partial y^2} + 2N_{xy}\,\frac{\partial^2 w}{\partial x\,\partial y}\right) \qquad (217)$$

This equation should be used instead of Eq. (103) in determining the deflection of a plate if in addition to lateral loads there are forces in the middle plane of the plate.

380 THEORY OF PLATES AND SHELLS

If there are body forces[1] acting in the middle plane of the plate or tangential forces distributed over the surfaces of the plate, the differential equations of equilibrium of the element shown in Fig. 191 become

$$\frac{\partial N_x}{\partial x} + \frac{\partial N_{xy}}{\partial y} + X = 0$$
$$\frac{\partial N_{xy}}{\partial x} + \frac{\partial N_y}{\partial y} + Y = 0 \tag{218}$$

Here X and Y denote the two components of the body forces or of the tangential forces per unit area of the middle plane of the plate.

Using Eqs. (218), instead of Eqs. (216), we obtain the following differential equation[2] for the deflection surface:

$$\frac{\partial^4 w}{\partial x^4} + 2\frac{\partial^4 w}{\partial x^2 \partial y^2} + \frac{\partial^4 w}{\partial y^4} = \frac{1}{D}\left(q + N_x \frac{\partial^2 w}{\partial x^2} + N_y \frac{\partial^2 w}{\partial y^2} + 2N_{xy} \frac{\partial^2 w}{\partial x \partial y} - X \frac{\partial w}{\partial x} - Y \frac{\partial w}{\partial y}\right) \tag{219}$$

Equation (217) or Eq. (219) together with the conditions at the boundary (see Art. 22, page 83) defines the deflection of a plate loaded laterally and submitted to the action of forces in the middle plane of the plate.

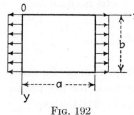

Fig. 192

91. Rectangular Plate with Simply Supported Edges under the Combined Action of Uniform Lateral Load and Uniform Tension. Assume that the plate is under uniform tension in the x direction, as shown in Fig. 192. The uniform lateral load q can be represented by the trigonometric series (see page 109).

$$q = \frac{16q}{\pi^2} \sum_{m=1,3,5,\ldots}^{\infty} \sum_{n=1,3,5,\ldots}^{\infty} \frac{1}{mn} \sin \frac{m\pi x}{a} \sin \frac{n\pi y}{b} \tag{a}$$

Equation (217) thus becomes

$$\frac{\partial^4 w}{\partial x^4} + 2\frac{\partial^4 w}{\partial x^2 \partial y^2} + \frac{\partial^4 w}{\partial y^4} - \frac{N_x}{D}\frac{\partial^2 w}{\partial x^2}$$
$$= \frac{16q}{D\pi^2} \sum_{m=1,3,5,\ldots}^{\infty} \sum_{n=1,3,5,\ldots}^{\infty} \frac{1}{mn} \sin \frac{m\pi x}{a} \sin \frac{n\pi y}{b} \tag{b}$$

This equation and the boundary conditions at the simply supported edges

---

[1] An example of a body force acting in the middle plane of the plate is the gravity force in the case of a vertical position of a plate.

[2] This differential equation has been derived by Saint Venant (see final note 73) in his translation of Clebsch, "Théorie de l'élasticité des corps solides," p. 704, 1883.

will be satisfied if we take the deflection w in the form of the series

$$w = \sum\sum a_{mn} \sin\frac{m\pi x}{a} \sin\frac{n\pi y}{b} \qquad (c)$$

Substituting this series in Eq. (b), we find the following values for the coefficients a_{mn}:

$$a_{mn} = \frac{16q}{D\pi^6 mn \left[\left(\dfrac{m^2}{a^2}+\dfrac{n^2}{b^2}\right)^2 + \dfrac{N_x m^2}{\pi^2 D a^2}\right]} \qquad (d)$$

in which m and n are odd numbers 1, 3, 5, . . . , and $a_{mn} = 0$ if m or n or both are even numbers. Hence the deflection surface of the plate is

$$w = \frac{16q}{\pi^6 D} \sum_{m=1,3,5,\ldots}^{\infty} \sum_{n=1,3,5,\ldots}^{\infty} \frac{1}{mn\left[\left(\dfrac{m^2}{a^2}+\dfrac{n^2}{b^2}\right)^2 + \dfrac{N_x m^2}{\pi^2 D a^2}\right]} \sin\frac{m\pi x}{a} \sin\frac{n\pi y}{b} \qquad (e)$$

Comparing this result with solution (131) (page 110), we conclude from the presence of the term $N_x m^2/\pi^2 D a^2$ in the brackets of the denominator that the deflection of the plate is somewhat diminished by the action of the tensile forces N_x. This is as would be expected.

By using M. Lévy's method (see Art. 30) a solution in simple series may be obtained which is equivalent to expression (e) but more convenient for numerical calculation. The maximum values of deflection and bending moments obtained in this way[1] for $\nu = 0.3$ can be represented in the form

$$w_{\max} = \alpha\frac{qb^4}{Eh^3} \qquad (M_x)_{\max} = \beta qb^2 \qquad (M_y)_{\max} = \beta_1 qb^2 \qquad (f)$$

The constants α, β, and β_1 depend upon the ratio a/b and a parameter

$$\gamma = \frac{N_x b^2}{4\pi^2 D}$$

and are plotted in Figs. 193, 194, and 195.

If, instead of tension, we have compression, the force N_x becomes

---

[1] H. D. Conway, *J. Appl. Mechanics*, vol. 16, p. 301, 1949, where graphs in the case of compression are also given; the case $N_x = N_y$ has been discussed by R. F. Morse and H. D. Conway, *J. Appl. Mechanics*, vol. 18, p. 209, 1951, and the case of a plate clamped all around by C. C. Chang and H. D. Conway, *J. Appl. Mechanics*, vol. 19, p. 179, 1952. For combined bending and compression, see also J. Lockwood Taylor, *The Shipbuilder and Marine Engine Builder*, no. 494, p. 15, 1950.

negative, and the deflections (e) become larger than those of the plate bent by lateral load only. It may be seen also in this case that at certain values of the compressive force N_x the denominator of one of the terms in series (e) may vanish. This indicates that at such values of N_x the plate may buckle laterally without any lateral loading.

92. Application of the Energy Method. The energy method, which was previously used in discussing bending of plates by lateral loading (see Art. 80, page 342), can be applied also to the cases in which the

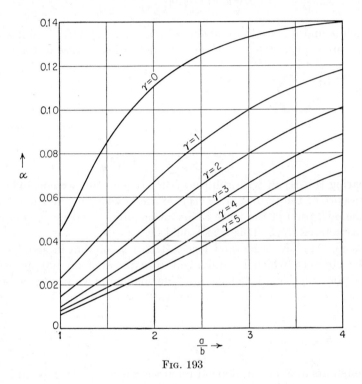

Fig. 193

lateral load is combined with forces acting in the middle plane of the plate. To establish the expression for the strain energy corresponding to the latter forces let us assume that these forces are applied first to the unbent plate. In this way we obtain a two-dimensional problem which can be treated by the methods of the theory of elasticity.[1] Assuming that this problem is solved and that the forces N_x, N_y, and N_{xy} are known at each point of the plate, the components of strain of the middle plane of the plate are obtained from the known formulas representing Hooke's

---

[1] See, for example, S. Timoshenko and J. N. Goodier, "Theory of Elasticity," 2d ed., p. 11, 1951.

law, viz.,

$$\epsilon_x = \frac{1}{hE}(N_x - \nu N_y) \qquad \epsilon_y = \frac{1}{hE}(N_y - \nu N_x)$$

$$\gamma_{xy} = \frac{N_{xy}}{hG}$$

(a)

The strain energy, due to stretching of the middle plane of the plate, is then

$$V_1 = \tfrac{1}{2}\iint (N_x \epsilon_x + N_y \epsilon_y + N_{xy}\gamma_{xy})\, dx\, dy$$

$$= \frac{1}{2hE} \iint [N_x^2 + N_y^2 - 2\nu N_x N_y + 2(1 + \nu)N_{xy}^2]\, dx\, dy \quad (220)$$

where the integration is extended over the entire plate.

Let us now apply the lateral load. This load will bend the plate and produce additional strain of the middle plane. In our previous discussion of bending of plates, this latter strain was always neglected. Here,

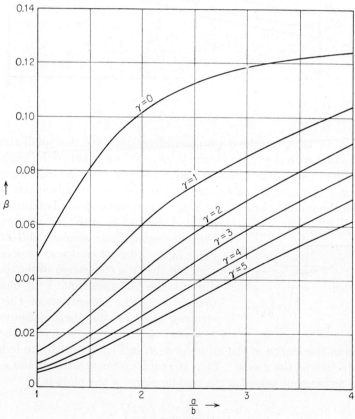

Fig. 194

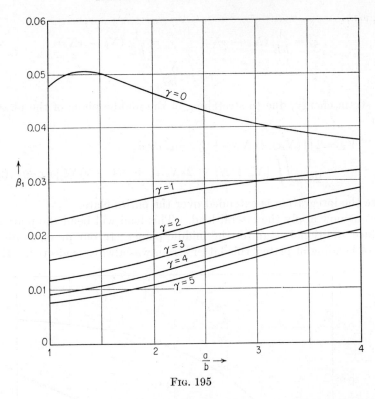

Fig. 195

however, we have to take it into consideration, since this small strain in combination with the finite forces N_x, N_y, N_{xy} may add to the expression for strain energy some terms of the same order as the strain energy of bending. The x, y, and z components of the small displacement that a point in the middle plane of the plate experiences during bending will be denoted by u, v, and w, respectively. Considering a linear element AB of that plane in the x direction, it may be seen from Fig. 196 that the elongation of the element due to the displacement u is equal to $(\partial u/\partial x)\,dx$. The elongation of the same element due to the displacement w is $\frac{1}{2}(\partial w/\partial x)^2\,dx$, as may be seen from the comparison of the length of the element A_1B_1 in Fig. 196 with the length of its projection on the x axis. Thus the total unit elongation in the x direction of an element taken in the middle plane of the plate is

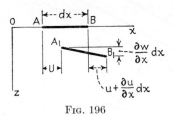

Fig. 196

$$\epsilon'_x = \frac{\partial u}{\partial x} + \frac{1}{2}\left(\frac{\partial w}{\partial x}\right)^2 \tag{221}$$

Similarly the strain in the y direction is

$$\epsilon'_y = \frac{\partial v}{\partial y} + \frac{1}{2}\left(\frac{\partial w}{\partial y}\right)^2 \qquad (222)$$

Considering now the shearing strain in the middle plane due to bending, we conclude as before (see Fig. 23) that the shearing strain due to the displacements u and v is $\partial u/\partial y + \partial v/\partial x$. To determine the shearing strain due to the displacement w we take two infinitely small linear elements OA and OB in the x and y directions, as shown in Fig. 197. Because of displacements in the z direction these elements come to the positions O_1A_1 and O_1B_1. The difference between the angle $\pi/2$ and the angle $A_1O_1B_1$ is the shearing strain corresponding to the displacement w. To determine this difference we consider the right angle $B_2O_1A_1$, in which B_2O_1 is parallel to BO. Rotating the plane $B_2O_1A_1$ about the axis O_1A_1 by the angle $\partial w/\partial y$, we bring the plane $B_2O_1A_1$ into coincidence with the plane $B_1O_1A_1$* and the point B_2 to position C. The displacement B_2C is equal to $(\partial w/\partial y)\, dy$ and is inclined to the vertical B_2B_1 by the angle $\partial w/\partial x$. Hence B_1C is equal to $(\partial w/\partial x)(\partial w/\partial y)\, dy$, and the angle CO_1B_1, which represents the shearing strain corresponding to the displacement w, is $(\partial w/\partial x)(\partial w/\partial y)$. Adding this shearing strain to the strain produced by the displacements u and v, we obtain

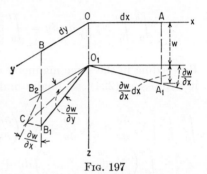

Fig. 197

$$\gamma'_{xy} = \frac{\partial u}{\partial y} + \frac{\partial v}{\partial x} + \frac{\partial w}{\partial x}\frac{\partial w}{\partial y} \qquad (223)$$

Formulas (221), (222), and (223) represent the components of the additional strain in the middle plane of the plate due to small deflections. Considering them as very small in comparison with the components ϵ_x, ϵ_y, and γ_{xy} used in the derivation of expression (220), we can assume that the forces N_x, N_y, N_{xy} remain unchanged during bending. With this assumption the additional strain energy of the plate, due to the strain produced in the middle plane by bending, is

$$V_2 = \iint (N_x \epsilon'_x + N_y \epsilon'_y + N_{xy} \gamma'_{xy})\, dx\, dy$$

Substituting expressions (221), (222), and (223) for ϵ'_x, ϵ'_y, and γ'_{xy}, we

* The angles $\partial w/\partial y$ and $\partial w/\partial x$ correspond to small deflections of the plate and are regarded as small quantities.

finally obtain

$$V_2 = \iint \left[N_x \frac{\partial u}{\partial x} + N_y \frac{\partial v}{\partial y} + N_{xy}\left(\frac{\partial u}{\partial y} + \frac{\partial v}{\partial x}\right) \right] dx\, dy$$
$$+ \frac{1}{2} \iint \left[N_x \left(\frac{\partial w}{\partial x}\right)^2 + N_y \left(\frac{\partial w}{\partial y}\right)^2 + 2N_{xy} \frac{\partial w}{\partial x} \frac{\partial w}{\partial y} \right] dx\, dy \quad (224)$$

It can be shown, by integration by parts, that the first integral on the right-hand side of expression (224) is equal to the work done during bending by the forces acting in the middle plane of the plate. Taking, for example, a rectangular plate with the coordinate axes directed, as shown in Fig. 192, we obtain for the first term of the integral

$$\int_0^b \int_0^a N_x \frac{\partial u}{\partial x} dx\, dy = \int_0^b \left| N_x u \right|_0^a dy - \int_0^b \int_0^a u \frac{\partial N_x}{\partial x} dx\, dy$$

Proceeding in the same manner with the other terms of the first integral in expression (224), we finally find

$$\int_0^b \int_0^a \left[N_x \frac{\partial u}{\partial x} + N_y \frac{\partial v}{\partial y} + N_{xy}\left(\frac{\partial u}{\partial y} + \frac{\partial v}{\partial x}\right) \right] dx\, dy$$
$$= \int_0^b \left(\left|N_x u\right|_0^a + \left|N_{xy} v\right|_0^a \right) dy + \int_0^a \left(\left|N_y v\right|_0^b + \left|N_{xy} u\right|_0^b \right) dx$$
$$- \int_0^b \int_0^a u\left(\frac{\partial N_x}{\partial x} + \frac{\partial N_{xy}}{\partial y}\right) dx\, dy - \int_0^b \int_0^a v\left(\frac{\partial N_{xy}}{\partial x} + \frac{\partial N_y}{\partial y}\right) dx\, dy$$

The first integral on the right-hand side of this expression is evidently equal to the work done during bending by the forces applied at the edges $x = 0$ and $x = a$ of the plate. Similarly, the second integral is equal to the work done by the forces applied at the edges $y = 0$ and $y = b$. The last two integrals, by virtue of Eqs. (218), are equal to the work done during bending by the body forces acting in the middle plane. These integrals each vanish in the absence of such corresponding forces.

Adding expressions (220) and (224) to the energy of bending [see Eq. (117), page 88], we obtain the total strain energy of a bent plate under the combined action of lateral loads and forces acting in the middle plane of the plate. This strain energy is equal to the work T_v done by the lateral load during bending of the plate plus the work T_h done by the forces acting in the middle plane of the plate. Observing that this latter work is equal to the strain energy V_1 plus the strain energy represented by the first integral of expression (224), we conclude that the work produced by the lateral forces is

$$T_v = \frac{1}{2} \iint \left[N_x \left(\frac{\partial w}{\partial x}\right)^2 + N_y \left(\frac{\partial w}{\partial y}\right)^2 + 2N_{xy} \frac{\partial w}{\partial x} \frac{\partial w}{\partial y} \right] dx\, dy$$
$$+ \frac{D}{2} \iint \left\{ \left(\frac{\partial^2 w}{\partial x^2} + \frac{\partial^2 w}{\partial y^2}\right)^2 - 2(1-\nu) \left[\frac{\partial^2 w}{\partial x^2} \frac{\partial^2 w}{\partial y^2} - \left(\frac{\partial^2 w}{\partial x\, \partial y}\right)^2 \right] \right\} dx\, dy \quad (225)$$

Applying the principle of virtual displacement, we now give a variation δw to the deflection w and obtain, from Eq. (225),

$$\delta T_v = \frac{1}{2} \delta \iint \left[N_x \left(\frac{\partial w}{\partial x}\right)^2 + N_y \left(\frac{\partial w}{\partial y}\right)^2 + 2N_{xy} \frac{\partial w}{\partial x} \frac{\partial w}{\partial y} \right] dx\, dy$$
$$+ \frac{D}{2} \delta \iint \left\{ \left(\frac{\partial^2 w}{\partial x^2} + \frac{\partial^2 w}{\partial y^2}\right)^2 - 2(1-\nu) \left[\frac{\partial^2 w}{\partial x^2} \frac{\partial^2 w}{\partial y^2} - \left(\frac{\partial^2 w}{\partial x\, \partial y}\right)^2 \right] \right\} dx\, dy \quad (226)$$

The left-hand side in this equation represents the work done during the virtual displacement by the lateral load, and the right-hand side is the corresponding change in the strain energy of the plate. The application of this equation will be illustrated by several examples in the next article.

93. Simply Supported Rectangular Plates under the Combined Action of Lateral Loads and of Forces in the Middle Plane of the Plate. Let us begin with the case of a rectangular plate uniformly stretched in the x direction (Fig. 192) and carrying a concentrated load P at a point with coordinates ξ and η. The general expression for the deflection that satisfies the boundary conditions is

$$w = \sum_{m=1,2,3,\ldots} \sum_{n=1,2,3,\ldots} a_{mn} \sin \frac{m\pi x}{a} \sin \frac{n\pi y}{b} \quad (a)$$

To obtain the coefficients a_{mn} in this series we use the general equation (226). Since $N_y = N_{xy} = 0$ in our case, the first integral on the right-hand side of Eq. (225), after substitution of series (a) for w, is

$$\frac{1}{2} \int_0^a \int_0^b N_x \left(\frac{\partial w}{\partial x}\right)^2 dx\, dy = \frac{ab}{8} N_x \sum_{m=1}^{\infty} \sum_{n=1}^{\infty} a_{mn}^2 \frac{m^2 \pi^2}{a^2} \quad (b)$$

The strain energy of bending representing the second integral in Eq. (225) is [see Eq. (d), page 343]

$$V = \frac{\pi^4 ab}{8} D \sum_{m=1}^{\infty} \sum_{n=1}^{\infty} a_{mn}^2 \left(\frac{m^2}{a^2} + \frac{n^2}{b^2}\right)^2 \quad (c)$$

To obtain a virtual deflection δw we give to a coefficient $a_{m_1 n_1}$ an increase $\delta a_{m_1 n_1}$. The corresponding deflection of the plate is

$$\delta w = \delta a_{m_1 n_1} \sin \frac{m_1 \pi x}{a} \sin \frac{n_1 \pi y}{b}$$

The work done during this virtual displacement by the lateral load P is

$$P \delta a_{m_1 n_1} \sin \frac{m_1 \pi \xi}{a} \sin \frac{n_1 \pi \eta}{b} \qquad (d)$$

The corresponding change in the strain energy consists of the two terms which are

$$\frac{1}{2} \delta \int_0^a \int_0^b N_x \left(\frac{\partial w}{\partial x} \right)^2 dx\, dy$$

$$= \frac{ab}{8} N_x \frac{\partial}{\partial a_{m_1 n_1}} \left(\sum_{m=1}^{\infty} \sum_{n=1}^{\infty} a_{mn}^2 \frac{m^2 \pi^2}{a^2} \right) \delta a_{m_1 n_1} = \frac{ab}{4} N_x a_{m_1 n_1} \frac{m_1^2 \pi^2}{a^2} \delta a_{m_1 n_1} \qquad (e)$$

and $\qquad \delta V = \dfrac{\partial V}{\partial a_{m_1 n_1}} \delta a_{m_1 n_1} = \dfrac{\pi^4 ab}{4} D a_{m_1 n_1} \left(\dfrac{m_1^2}{a^2} + \dfrac{n_1^2}{b^2} \right)^2 \delta a_{m_1 n_1}$

Substituting expressions (d) and (e) in Eq. (226), we obtain

$$P \delta a_{m_1 n_1} \sin \frac{m_1 \pi \xi}{a} \sin \frac{n_1 \pi \eta}{b} = \frac{ab}{4} N_x a_{m_1 n_1} \frac{m_1^2 \pi^2}{a^2} \delta a_{m_1 n_1}$$

$$+ \frac{\pi^4 ab}{4} D a_{m_1 n_1} \left(\frac{m_1^2}{a^2} + \frac{n_1^2}{b^2} \right)^2 \delta a_{m_1 n_1}$$

from which

$$a_{m_1 n_1} = \frac{4P \sin \dfrac{m_1 \pi \xi}{a} \sin \dfrac{n_1 \pi \eta}{b}}{ab \pi^4 D \left[\left(\dfrac{m_1^2}{a^2} + \dfrac{n_1^2}{b^2} \right)^2 + \dfrac{m_1^2 N_x}{\pi^2 a^2 D} \right]} \qquad (f)$$

Substituting these values of the coefficients $a_{m_1 n_1}$ in expression (a), we find the deflection of the plate to be

$$w = \frac{4P}{ab\pi^4 D} \sum_{m=1}^{\infty} \sum_{n=1}^{\infty} \frac{\sin \dfrac{m\pi \xi}{a} \sin \dfrac{n\pi \eta}{b}}{\left(\dfrac{m^2}{a^2} + \dfrac{n^2}{b^2} \right)^2 + \dfrac{m^2 N_x}{\pi^2 a^2 D}} \sin \frac{m\pi x}{a} \sin \frac{n\pi y}{b} \qquad (g)$$

If, instead of the tensile forces N_x, there are compressive forces of the same magnitude, the deflection of the plate is obtained by substituting $-N_x$ in place of N_x in expression (g). This substitution gives

$$w = \frac{4P}{ab\pi^4 D} \sum_{m=1}^{\infty} \sum_{n=1}^{\infty} \frac{\sin \dfrac{m\pi \xi}{a} \sin \dfrac{n\pi \eta}{b}}{\left(\dfrac{m^2}{a^2} + \dfrac{n^2}{b^2} \right)^2 - \dfrac{m^2 N_x}{\pi^2 a^2 D}} \sin \frac{m\pi x}{a} \sin \frac{n\pi y}{b} \qquad (h)$$

The smallest value of N_x at which the denominator of one of the terms in expression (h) becomes equal to zero is the *critical value* of the compressive force N_x. It is evident that this critical value is obtained by taking $n = 1$. Hence

$$(N_x)_{cr} = \frac{\pi^2 a^2 D}{m^2}\left(\frac{m^2}{a^2} + \frac{1}{b^2}\right)^2 = \frac{\pi^2 D}{b^2}\left(\frac{mb}{a} + \frac{a}{mb}\right)^2 \quad (227)$$

where m must be chosen so as to make expression (227) a minimum. Plotting the factor

$$k = \left(\frac{mb}{a} + \frac{a}{mb}\right)^2$$

against the ratio a/b, for various integral values of m, we obtain a system of curves shown in Fig. 198. The portions of the curves that must be

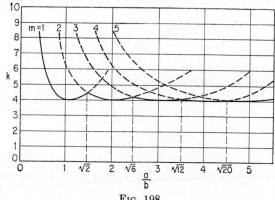

Fig. 198

used in determining k are indicated by heavy lines. It is seen that the factor k is equal to 4 for a square plate as well as for any plate that can be subdivided into an integral number of squares with the side b. It can also be seen that for long plates k remains practically constant at a value of 4.* Since the value of m in Eq. (227) may be other than 1 for oblong plates, such plates, being submitted to a lateral load combined with compression, do not generally deflect[1] in the form of a half wave in the direction of the longer side of the plate. If, for instance, $a/b = 2, 4, \ldots$ the respective elastic surface becomes markedly unsymmetrical with respect to the middle line $x = a/2$ (Fig. 192), especially so for values of N_x close to the critical value $(N_x)_{cr}$.

By using the deflection (g) produced by one concentrated load, the

---

* A more detailed discussion of this problem is given in S. Timoshenko, "Theory of Elastic Stability," p. 327, 1936.

[1] Several examples of such a deformation have been considered by K. Girkmann, *Stahlbau*, vol. 15, p. 57, 1942.

deflection produced by any lateral load can be obtained by superposition. Assuming, for example, that the plate is uniformly loaded by a load of intensity q, we substitute $q\,d\xi\,d\eta$ for P in expression (g) and integrate the expression over the entire area of the plate. In this way we obtain the same expression for the deflection of the plate under uniform load as has already been derived in another manner (see page 381).

If the plate laterally loaded by the force P is compressed in the middle plane by uniformly distributed forces N_x and N_y, proceeding as before we obtain

$$w = \frac{4P}{ab\pi^4 D} \sum_{m=1}^{\infty}\sum_{n=1}^{\infty} \frac{\sin\frac{m\pi\xi}{a}\sin\frac{n\pi\eta}{b}}{\left(\frac{m^2}{a^2}+\frac{n^2}{b^2}\right)^2 - \frac{m^2 N_x}{\pi^2 a^2 D} - \frac{n^2 N_y}{\pi^2 b^2 D}} \sin\frac{m\pi x}{a}\sin\frac{n\pi y}{b} \quad (i)$$

The critical value of the forces N_x and N_y is obtained from the condition[1]

$$\frac{m^2 (N_x)_{cr}}{\pi^2 a^2 D} + \frac{n^2 (N_y)_{cr}}{\pi^2 b^2 D} = \left(\frac{m^2}{a^2}+\frac{n^2}{b^2}\right)^2 \quad (j)$$

where m and n are chosen so as to make N_x and N_y a minimum for any given value of the ratio N_x/N_y. In the case of a square plate submitted to the action of a uniform pressure p in the middle plane we have $a = b$ and $N_x = N_y = p$. Equation (j) then gives

$$p_{cr} = \frac{\pi^2 D}{a^2}(m^2 + n^2)_{\min} \quad (k)$$

The critical value of p is obtained by taking $m = n = 1$, which gives

$$p_{cr} = \frac{2\pi^2 D}{a^2} \quad (228)$$

In the case of a plate in the form of an isosceles right triangle with simply supported edges (Fig. 161) the deflection surface of the buckled plate which satisfies all the boundary conditions is[2]

$$w = a\left(\sin\frac{\pi x}{a}\sin\frac{2\pi y}{a} + \sin\frac{2\pi x}{a}\sin\frac{\pi y}{a}\right)$$

Thus the critical value of the compressive stress is obtained by substituting $m = 1$, $n = 2$ or $m = 2$, $n = 1$ into expression (k). This gives

$$p_{cr} = \frac{5\pi^2 D}{a^2} \quad (229)$$

---

[1] A complete discussion of this problem is given in Timoshenko, "Elastic Stability," p. 333.

[2] This is the form of natural vibration of a square plate having a diagonal as a nodal line.

94. Circular Plates under Combined Action of Lateral Load and Tension or Compression. Consider a circular plate (Fig. 199) submitted to the simultaneous action of a symmetrical lateral load and a uniform compression $N_r = N_t = N$ in the middle plane of the plate. Owing to the slope φ of the deformed plate (Fig. 27) the radial compression N gives a transverse component $N\, d\varphi/dr$ which we have to add to the shearing force Q (Fig. 28) due to the lateral load. Hence the differential equation (54) becomes

$$\frac{d^2\varphi}{dr^2} + \frac{1}{r}\frac{d\varphi}{dr} + \left(\frac{k^2}{a^2} - \frac{1}{r^2}\right)\varphi = -\frac{Q}{D} \qquad (a)$$

in which

$$k^2 = \frac{Na^2}{D} \qquad (b)$$

In the case of a circular plate without a hole[1] the solution of Eq. (a) is of the form

$$\varphi = C_1 J_1\left(\frac{kr}{a}\right) + \varphi_0 \qquad (c)$$

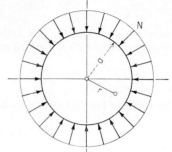

FIG. 199

where J_1 is the Bessel function of the order one, φ_0 a particular solution of Eq. (a) depending on Q, and C_1 a constant defined by the boundary conditions of the plate.

Let us take as an example a rigidly clamped[2] plate carrying a uniform load of intensity q. Then, as a particular solution, we use

$$\varphi_0 = -\frac{qra^2}{2k^2 D} = -\frac{qr}{2N}$$

and therefore

$$\varphi = \frac{dw}{dr} = C_1 J_1\left(\frac{kr}{a}\right) - \frac{qr}{2N} \qquad (d)$$

It follows, by integration, that

$$w = \frac{C_1 a}{k} J_0\left(\frac{kr}{a}\right) - \frac{qr^2}{4N} + C_2 \qquad (e)$$

where J_0 is the Bessel function of the order zero and C_2 a second constant. Having calculated C_1 from the condition $\varphi = 0$ on $r = a$, and C_2 from the condition $w = 0$ on $r = a$, we obtain the final solution[3]

$$w = \frac{qa^4\left[J_0\left(\frac{kr}{a}\right) - J_0(k)\right]}{2k^3 J_1(k) D} - \frac{qa^2(a^2 - r^2)}{4k^2 D} \qquad (f)$$

The deflections (f) become infinite for $J_1(k) = 0$. Denoting the zeros of the function J_1 in order of their magnitude by $j_1, j_2, \ldots$ we see that the condition $k = j_1$

[1] In the case of a concentric hole a term proportional to a Bessel function of second kind must be added to expression (c). The inner boundary must be submitted then to the same compression N, or else the problem becomes more complex because of the inconstancy of stresses N_r and N_t.
[2] The case of an elastic restraint without transverse load has been discussed by H. Reismann, *J. Appl. Mechanics*, vol. 19, p. 167, 1952.
[3] This result may be found in A. Nádai, "Elastische Platten," p. 255, Berlin, 1925.

defines the lowest critical value

$$N_{cr} = \frac{Dj_1^2}{a^2} \tag{g}$$

of the compressive stress N. Now, for the function $J_1(k)$ we have the expression

$$J_1(k) = \frac{k}{2}\left(1 - \frac{k^2}{j_1^2}\right)\left(1 - \frac{k^2}{j_2^2}\right)\cdots \tag{h}$$

in which $j_1 = 3.83171$, $j_2 = 7.01559$, ... As $k < j_1$ we can neglect the terms k^2/j^2 beginning with the second parentheses. Observing, furthermore, that

$$\frac{k^2}{j_1^2} = \frac{N}{N_{cr}}$$

by virtue of Eqs. (b) and (g) we have, approximately,

$$J_1(k) = \frac{k}{2}(1 - \alpha) \tag{i}$$

where

$$\alpha = \frac{N}{N_{cr}} \tag{j}$$

Making use of the expression (i), it can be shown that, approximately,[1]

$$w = \frac{w_q}{1 - \alpha}$$

where w_q is the deflection due to the load q alone. Cases with other boundary conditions and other laws of distribution of the lateral load may be handled in like manner. In the general case of a symmetrical lateral load combined with compression we can put, approximately, for the center of the plate $(r = 0)$

$$w_0 = \frac{(w_q)_0}{1 - \alpha}$$

$$\left(\frac{1}{r}\frac{dw}{dr}\right)_0 = \left(\frac{d^2w}{dr^2}\right)_0 = \frac{1 + c_0\alpha}{1 - \alpha}\left(\frac{1}{r}\frac{dw_q}{dr}\right)_0 \tag{k}$$

and on the boundary $(r = a)$

$$\left(\frac{1}{r}\frac{dw}{dr}\right)_a = \frac{1 + c'\alpha}{1 - \alpha}\left(\frac{1}{r}\frac{dw_q}{dr}\right)_a$$

$$\left(\frac{d^2w}{dr^2}\right)_a = \frac{1 + c''\alpha}{1 - \alpha}\left(\frac{d^2w_q}{dr^2}\right)_a \tag{l}$$

where w_q relates to a plate carrying the given lateral load alone and $\alpha = N/N_{cr}$ has the following meaning:

For a simply supported plate: $\quad \alpha = \dfrac{Na^2}{4.20D}$

For a clamped plate: $\quad \alpha = \dfrac{Na^2}{14.68D}$ \hfill (m)

[1] See O. Pettersson, *Acta Polytech.*, Stockholm, no. 138, 1954. The following results are taken from this paper, in which, more generally, an elastic restraint at the edge is assumed.

TABLE 81. VALUES OF CONSTANTS IN APPROXIMATE EXPRESSIONS (k) AND (l)
$\nu = 0.3$

Case	Load distribution	Boundary conditions	Constants
1	Uniform edge couples	Simply supported	$c_0 = 0.305$ $c' = -0.270$ $c'' = -1.219$
2	Uniform load	Simply supported	$c_0 = 0.0480$ $c' = c'' = -0.0327$
3		Clamped	$c_0 = 0.308$ $c'' = -0.473$
4	Central uniform load over area of radius ϵa	Simply supported	$c_0 = -1 + \dfrac{2.153}{1 - 1.3 \ln \epsilon}$ $c' = c'' = 0.205$
5		Clamped	$c_0 = -1 - \dfrac{1.308}{\ln \epsilon}$ $c'' = 0.0539$

the former value being valid for $\nu = 0.3$. The values of the constants c_0, c', and c'' are given in Table 81.

If the circular plate is subjected to a lateral load combined with a uniform tension N, instead of compression, then we have, approximately,

$$(w)_{r=0} = \frac{(w_q)_{r=0}}{1 + \alpha}$$

where α is the absolute value of N/N_{cr}. As for the curvatures, a factor

$$\frac{1}{1 + (1 + c)\alpha}$$

instead of the factor $(1 + c\alpha)/(1 - \alpha)$ must be used in expressions (k) and (l), the constant c having the meaning of c_0, c', and c'', respectively.

95. Bending of Plates with a Small Initial Curvature.[1] Assume that a plate has some initial warp of the middle surface so that at any point there is an initial deflection w_0 which is small in comparison with the thickness of the plate. If such a plate is submitted to the action of transverse loading, additional deflection w_1 will be produced, and the total deflection at any point of the middle surface of the plate will be $w_0 + w_1$. In calculating the deflection w_1 we use Eq. (103) derived for flat plates. This procedure is justifiable if the initial deflection w_0 is

[1] See S. Timoshenko's paper in *Mem. Inst. Ways Commun.*, vol. 89, St. Petersburg, 1915 (Russian).

small, since we may consider the initial deflection as produced by a fictitious load and apply the principle of superposition.[1] If in addition to lateral loads there are forces acting in the middle plane of the plate, the effect of these forces on bending depends not only on w_1 but also on w_0. To take this into account, in applying Eq. (217) we use the total deflection $w = w_0 + w_1$ on the right-hand side of the equation. It will be remembered that the left-hand side of the same equation was obtained from expressions for the bending moments in the plate. Since these moments depend not on the total curvature but only on the change in curvature of the plate, the deflection w_1 should be used instead of w in applying that side of the equation to this problem. Hence, for the case of an initially curved plate, Eq. (217) becomes

$$\frac{\partial^4 w_1}{\partial x^4} + 2\frac{\partial^4 w_1}{\partial x^2 \partial y^2} + \frac{\partial^4 w_1}{\partial y^4} = \frac{1}{D}\left[q + N_x \frac{\partial^2(w_0 + w_1)}{\partial x^2} + N_y \frac{\partial^2(w_0 + w_1)}{\partial y^2} + 2N_{xy}\frac{\partial^2(w_0 + w_1)}{\partial x \partial y}\right] \quad (230)$$

It is seen that the effect of an initial curvature on the deflection is equivalent to the effect of a fictitious lateral load of an intensity

$$N_x \frac{\partial^2 w_0}{\partial x^2} + N_y \frac{\partial^2 w_0}{\partial y^2} + 2N_{xy}\frac{\partial^2 w_0}{\partial x \partial y}$$

Thus a plate will experience bending under the action of forces in the xy plane alone provided there is an initial curvature.

Take as an example the case of a rectangular plate (Fig. 192), and assume that the initial deflection of the plate is defined by the equation

$$w_0 = a_{11} \sin \frac{\pi x}{a} \sin \frac{\pi y}{b} \quad (a)$$

If uniformly distributed compressive forces N_x are acting on the edges of this plate, Eq. (230) becomes

$$\frac{\partial^4 w_1}{\partial x^4} + 2\frac{\partial^4 w_1}{\partial x^2 \partial y^2} + \frac{\partial^4 w_1}{\partial y^4} = \frac{1}{D}\left(N_x \frac{a_{11}\pi^2}{a^2}\sin\frac{\pi x}{a}\sin\frac{\pi y}{b} - N_x \frac{\partial^2 w_1}{\partial x^2}\right) \quad (b)$$

Let us take the solution of this equation in the form

$$w_1 = A \sin \frac{\pi x}{a} \sin \frac{\pi y}{b} \quad (c)$$

Substituting this value of w_1 into Eq. (b), we obtain

$$A = \frac{a_{11}N_x}{\dfrac{\pi^2 D}{a^2}\left(1 + \dfrac{a^2}{b^2}\right)^2 - N_x}$$

[1] In the case of large deflections the magnitude of the deflection is no longer proportional to the load, and the principle of superposition is not applicable.

With this value of A expression (c) gives the deflection of the plate produced by the compressive forces N_x. Adding this deflection to the initial deflection (a), we obtain for the total deflection of the plate the following expression:

$$w = w_0 + w_1 = \frac{a_{11}}{1-\alpha} \sin\frac{\pi x}{a} \sin\frac{\pi y}{b} \qquad (d)$$

in which

$$\alpha = \frac{N_x}{\dfrac{\pi^2 D}{a^2}\left(1+\dfrac{a^2}{b^2}\right)^2} \qquad (e)$$

The maximum deflection will be at the center and will be

$$w_{\max} = \frac{a_{11}}{1-\alpha} \qquad (f)$$

This formula is analogous to that used for a bar with initial curvature.[1]

In a more general case we can take the initial deflection surface of the rectangular plate in the form of the following series:

$$w_0 = \sum_{m=1}^{\infty}\sum_{n=1}^{\infty} a_{mn} \sin\frac{m\pi x}{a} \sin\frac{n\pi y}{b} \qquad (g)$$

Substituting this series in Eq. (230), we find that the additional deflection at any point of the plate is

$$w_1 = \sum_{m=1}^{\infty}\sum_{n=1}^{\infty} b_{mn} \sin\frac{m\pi x}{a} \sin\frac{n\pi y}{b} \qquad (h)$$

in which

$$b_{mn} = \frac{a_{mn} N_x}{\dfrac{\pi^2 D}{a^2}\left(m + \dfrac{n^2}{m}\dfrac{a^2}{b^2}\right)^2 - N_x} \qquad (i)$$

It is seen that all the coefficients b_{mn} increase with an increase of N_x. Thus when N_x approaches the critical value, the term in series (h) that corresponds to the laterally buckled shape of the plate [see Eq. (227)] becomes the predominating one. We have here a complete analogy with the case of bending of initially curved bars under compression.

The problem can be handled in the same manner if, instead of compression, we have tension in the middle plane of the plate. In such a case it is necessary only to change the sign of N_x in the previous equations. Without any difficulty we can also obtain the deflection in the case when there are not only forces N_x but also forces N_y and N_{xy} uniformly distributed along the edges of the plate.

[1] See S. Timoshenko, "Strength of Materials," part II, 3d ed., p. 56, 1956.

CHAPTER 13

LARGE DEFLECTIONS OF PLATES

96. Bending of Circular Plates by Moments Uniformly Distributed along the Edge. In the previous discussion of pure bending of circular plates it was shown (see page 47) that the strain of the middle plane of the plate can be neglected in cases in which the deflections are small as compared with the thickness of the plate. In cases in which the deflections are no longer small in comparison with the thickness of the plate but are still small as compared with the other dimensions, the analysis of the problem must be extended to include the strain of the middle plane of the plate.[1]

We shall assume that a circular plate is bent by moments M_0 uniformly distributed along the edge of the plate (Fig. 200a). Since the deflection surface in such a case is symmetrical with respect to the center O, the displacement of a point in the middle plane of the plate can be resolved into two components: a component u in the radial direction and a component w perpendicular to the plane of the plate. Proceeding as previously indicated in Fig. 196 (page 384), we conclude that the strain in the radial direction is[2]

$$\epsilon_r = \frac{du}{dr} + \frac{1}{2}\left(\frac{dw}{dr}\right)^2 \qquad (a)$$

The strain in the tangential direction is evidently

$$\epsilon_t = \frac{u}{r} \qquad (b)$$

Denoting the corresponding tensile forces per unit length by N_r and

---

[1] This problem has been discussed by S. Timoshenko; see *Mem. Inst. Ways Commun.*, vol. 89, St. Petersburg, 1915.

[2] In the case of very large deflections we have

$$\epsilon_r = \frac{du}{dr} + \frac{1}{2}\left[\left(\frac{du}{dr}\right)^2 + \left(\frac{dw}{dr}\right)^2\right]$$

which modifies the following differential equations. See E. Reissner, *Proc. Symposia Appl. Math.*, vol. 1, p. 213, 1949.

N_t and applying Hooke's law, we obtain

$$N_r = \frac{Eh}{1-\nu^2}(\epsilon_r + \nu\epsilon_t) = \frac{Eh}{1-\nu^2}\left[\frac{du}{dr} + \frac{1}{2}\left(\frac{dw}{dr}\right)^2 + \nu\frac{u}{r}\right]$$
$$N_t = \frac{Eh}{1-\nu^2}(\epsilon_t + \nu\epsilon_r) = \frac{Eh}{1-\nu^2}\left[\frac{u}{r} + \nu\frac{du}{dr} + \frac{\nu}{2}\left(\frac{dw}{dr}\right)^2\right] \quad (c)$$

These forces must be taken into consideration in deriving equations of equilibrium for an element of the plate such as that shown in Fig. 200b

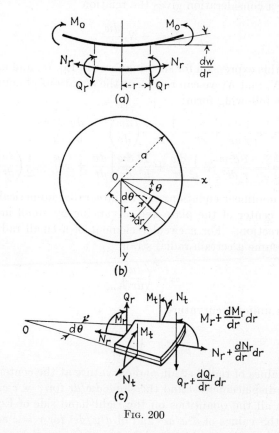

Fig. 200

and c. Taking the sum of the projections in the radial direction of all the forces acting on the element, we obtain

$$r\frac{dN_r}{dr}\,dr\,d\theta + N_r\,dr\,d\theta - N_t\,dr\,d\theta = 0$$

from which

$$N_r - N_t + r\frac{dN_r}{dr} = 0 \quad (d)$$

The second equation of equilibrium of the element is obtained by taking moments of all the forces with respect to an axis perpendicular to the radius in the same manner as in the derivation of Eq. (55) (page 53). In this way we obtain[1]

$$Q_r = -D\left(\frac{d^3w}{dr^3} + \frac{1}{r}\frac{d^2w}{dr^2} - \frac{1}{r^2}\frac{dw}{dr}\right) \tag{e}$$

The magnitude of the shearing force Q_r is obtained by considering the equilibrium of the inner circular portion of the plate of radius r (Fig. 200a). Such a consideration gives the relation

$$Q_r = -N_r \frac{dw}{dr} \tag{f}$$

Substituting this expression for shearing force in Eq. (e) and using expressions (c) for N_r and N_t we can represent the equations of equilibrium (d) and (e) in the following form:

$$\begin{aligned}\frac{d^2u}{dr^2} &= -\frac{1}{r}\frac{du}{dr} + \frac{u}{r^2} - \frac{1-\nu}{2r}\left(\frac{dw}{dr}\right)^2 - \frac{dw}{dr}\frac{d^2w}{dr^2} \\ \frac{d^3w}{dr^3} &= -\frac{1}{r}\frac{d^2w}{dr^2} + \frac{1}{r^2}\frac{dw}{dr} + \frac{12}{h^2}\frac{dw}{dr}\left[\frac{du}{dr} + \nu\frac{u}{r} + \frac{1}{2}\left(\frac{dw}{dr}\right)^2\right]\end{aligned} \tag{231}$$

These two nonlinear equations can be integrated numerically by starting from the center of the plate and advancing by small increments in the radial direction. For a circular element of a small radius c at the center, we assume a certain radial strain

$$\epsilon_0 = \left(\frac{du}{dr}\right)_{r=0}$$

and a certain uniform curvature

$$\frac{1}{\rho_0} = -\left(\frac{d^2w}{dr^2}\right)_{r=0}$$

With these values of radial strain and curvature at the center, the values of the radial displacement u and the slope dw/dr for $r = c$ can be calculated. Thus all the quantities on the right-hand side of Eqs. (231) are known, and the values of d^2u/dr^2 and of d^3w/dr^3 for $r = c$ can be calculated. As soon as these values are known, another radial step of length c can be made, and all the quantities entering in the right-hand side of Eqs. (231) can be calculated for $r = 2c$* and so on. The numerical

---

[1] The direction for Q_r is opposite to that used in Fig. 28. This explains the minus sign in Eq. (e).

* If the intervals into which the radius is divided are sufficiently small, a simple procedure, such as that used in S. Timoshenko's "Vibration Problems in Engineering," 3d ed., p. 143, can be applied. The numerical results represented in Fig. 201 are

values of u and w and their derivatives at the end of any interval being known, the values of the forces N_r and N_t can then be calculated from Eqs. (c) and the bending moments M_r and M_t from Eqs. (52) and (53) (see page 52). By such repeated calculations we proceed up to the radial distance $r = a$ at which the radial force N_r vanishes. In this way we obtain a circular plate of radius a bent by moments M_0 uniformly distributed along the edge. By changing the numerical values of ϵ_0 and

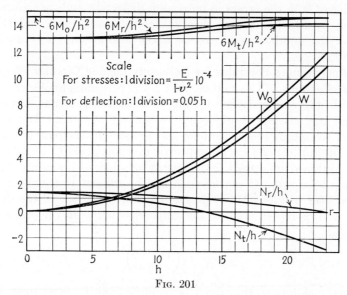

Fig. 201

$1/\rho_0$ at the center we obtain plates with various values of the outer radius and various values of the moment along the edge.

Figure 201 shows graphically the results obtained for a plate with

$$a \approx 23h \quad \text{and} \quad (M_r)_{r=a} = M_0 = 2.93 \cdot 10^{-3} \frac{D}{h}$$

It will be noted that the maximum deflection of the plate is $0.55h$, which is about 9 per cent less than the deflection w_0 given by the elementary theory which neglects the strain in the middle plane of the plate. The forces N_r and N_t are both positive in the central portion of the plate. In the outer portion of the plate the forces N_t become negative; i.e.,

---

obtained in this manner. A higher accuracy can be obtained by using the methods of Adams or Störmer. For an account of the Adams method see Francis Bashforth's book on forms of fluid drops, Cambridge University Press, 1883. Störmer's method is discussed in detail in A. N. Krilov's book "Approximate Calculations," published by the Russian Academy of Sciences, Moscow, 1935. See also L. Collatz, "Numerische Behandlung von Differentialgleichungen," Berlin, 1951.

compression exists in the tangential direction. The maximum tangential compressive stress at the edge amounts to about 18 per cent of the maximum bending stress $6M_0/h^2$. The bending stresses produced by the moments M_r and M_t are somewhat smaller than the stress $6M_0/h^2$ given by the elementary theory and become smallest at the center, at which point the error of the elementary theory amounts to about 12 per cent. From this numerical example it may be concluded that for deflections of the order of $0.5h$ the errors in maximum deflection and maximum stress as given by the elementary theory become considerable and that the strain of the middle plane must be taken into account to obtain more accurate results.

97. Approximate Formulas for Uniformly Loaded Circular Plates with Large Deflections. The method used in the preceding article can also be applied in the case of lateral loading of a plate. It is not, however, of practical use, since a considerable amount of numerical calculation is required to obtain the deflections and stresses in each particular case. A more useful formula for an approximate calculation of the deflections can be obtained by applying the energy method.[1] Let a circular plate of radius a be clamped at the edge and be subject to a uniformly distributed load of intensity q. Assuming that the shape of the deflected surface can be represented by the same equation as in the case of small deflections, we take

$$w = w_0 \left(1 - \frac{r^2}{a^2}\right)^2 \quad (a)$$

The corresponding strain energy of bending from Eq. (m) (page 345) is

$$V = \frac{D}{2} \int_0^{2\pi} \int_0^a \left[\left(\frac{\partial^2 w}{\partial r^2}\right)^2 + \frac{1}{r^2}\left(\frac{\partial w}{\partial r}\right)^2 + \frac{2\nu}{r}\frac{\partial w}{\partial r}\frac{\partial^2 w}{\partial r^2}\right] r\, dr\, d\theta = \frac{32\pi}{3} \frac{w_0^2}{a^2} D \quad (b)$$

For the radial displacements we take the expression

$$u = r(a - r)(C_1 + C_2 r + C_3 r^2 + \cdots) \quad (c)$$

each term of which satisfies the boundary conditions that u must vanish at the center and at the edge of the plate. From expressions (a) and (c) for the displacements, we calculate the strain components ϵ_r and ϵ_t of the middle plane as shown in the preceding article and obtain the strain energy due to stretching of the middle plane by using the expression

$$V_1 = 2\pi \int_0^a \left(\frac{N_r \epsilon_r}{2} + \frac{N_t \epsilon_t}{2}\right) r\, dr = \frac{\pi E h}{1 - \nu^2} \int_0^a (\epsilon_r^2 + \epsilon_t^2 + 2\nu \epsilon_r \epsilon_t) r\, dr \quad (d)$$

---

[1] See Timoshenko, "Vibration Problems," p. 452. For approximate formulas see also Table 82.

Taking only the first two terms in series (c), we obtain

$$V_1 = \frac{\pi E h a^2}{1-\nu^2}\left(0.250 C_1^2 a^2 + 0.1167 C_2^2 a^4 + 0.300 C_1 C_2 a^3 \right.$$
$$\left. - 0.00846 C_1 a \frac{8w_0^2}{a^2} + 0.00682 C_2 a^2 \frac{8w_0^2}{a^2} + 0.00477 \frac{64 w_0^4}{a^4}\right) \quad (e)$$

The constants C_1 and C_2 are now determined from the condition that the total energy of the plate for a position of equilibrium is a minimum. Hence

$$\frac{\partial V_1}{\partial C_1} = 0 \quad \text{and} \quad \frac{\partial V_1}{\partial C_2} = 0 \quad (f)$$

Substituting expression (e) for V_1, we obtain two linear equations for C_1 and C_2. From these we find that

$$C_1 = 1.185 \frac{w_0^2}{a^3} \quad \text{and} \quad C_2 = -1.75 \frac{w_0^2}{a^4}$$

Then, from Eq. (e) we obtain[1]

$$V_1 = 2.59 \pi D \frac{w_0^4}{a^2 h^2} \quad (g)$$

Adding this energy, which results from stretching of the middle plane, to the energy of bending (b), we obtain the total strain energy

$$V + V_1 = \frac{32}{3}\pi D \frac{w_0^2}{a^2}\left(1 + 0.244 \frac{w_0^2}{h^2}\right) \quad (h)$$

The second term in the parentheses represents the correction due to strain in the middle surface of the plate. It is readily seen that this correction is small and can be neglected if the deflection w_0 at the center of the plate is small in comparison with the thickness h of the plate.

The strain energy being known from expression (h), the deflection of the plate is obtained by applying the principle of virtual displacements. From this principle it follows that

$$\frac{d(V+V_1)}{dw_0}\delta w_0 = 2\pi \int_0^a q\,\delta w\, r\, dr = 2\pi q\, \delta w_0 \int_0^a \left(1 - \frac{r^2}{a^2}\right)^2 r\, dr$$

Substituting expression (h) in this equation, we obtain a cubic equation for w_0. This equation can be put in the form

$$w_0 = \frac{qa^4}{64D} \frac{1}{1 + 0.488 \frac{w_0^2}{h^2}} \quad (232)$$

The last factor on the right-hand side represents the effect of the stretching of the middle surface on the deflection. Because of this effect the deflection w_0 is no longer proportional to the intensity q of the load, and

---
[1] It is assumed that $\nu = 0.3$ in this calculation.

the rigidity of the plate increases with the deflection. For example, taking $w_0 = \frac{1}{2}h$, we obtain, from Eq. (232),

$$w_0 = 0.89 \frac{qa^4}{64D}$$

This indicates that the deflection in this case is 11 per cent less than that obtained by neglecting the stretching of the middle surface.

Up to now we have assumed the radial displacements to be zero on the periphery of the plate. Another alternative is to assume the edge as free to move in the radial direction. The expression (232) then has to be replaced by

$$w_0 = \frac{qa^4}{64D} \frac{1}{1 + 0.146 \frac{w_0^2}{h^2}} \tag{233}$$

a result[1] which shows that under the latter assumption the effect of the stretching of the plate is considerably less marked than under the former one. Taking, for instance, $w_0 = \frac{1}{2}h$ we arrive at $w_0 = 0.965(qa^4/64D)$, with an effect of stretching of only $3\frac{1}{2}$ per cent in place of 11 per cent obtained above.

Furthermore we can conclude from Eqs. (b) and (c) of Art. 96 that, if $N_r = 0$ on the edge, then the edge value of N_t becomes $N_t = Eh\epsilon_t = Ehu/r$, that is, negative. We can expect, therefore, that for a certain critical value of the lateral load the edge zone of the plate will become unstable.[2]

Another method for the approximate solution of the problem has been developed by A. Nádai.[3] He begins with equations of equilibrium similar to Eqs. (231). To derive them we have only to change Eq. (f), of the preceding article, to fit the case of lateral load of intensity q. After such a change the expression for the shearing force evidently becomes

$$Q_r = -N_r \frac{dw}{dr} - \frac{1}{r} \int_0^r qr \, dr \tag{i}$$

Using this expression in the same manner in which expression (f) was used in the preceding article, we obtain the following system of equations in place of Eqs. (231):

$$\frac{d^2u}{dr^2} + \frac{1}{r}\frac{du}{dr} - \frac{u}{r^2} = -\frac{1-\nu}{2r}\left(\frac{dw}{dr}\right)^2 - \frac{dw}{dr}\frac{d^2w}{dr^2}$$

$$\frac{d^3w}{dr^3} + \frac{1}{r}\frac{d^2w}{dr^2} - \frac{1}{r^2}\frac{dw}{dr} = \frac{12}{h^2}\frac{dw}{dr}\left[\frac{du}{dr} + \nu\frac{u}{r} + \frac{1}{2}\left(\frac{dw}{dr}\right)^2\right] + \frac{1}{Dr}\int_0^r qr \, dr \tag{234}$$

---

[1] Obtained by a method which will be described in Art. 100.
[2] The instability occurring in such a case has been investigated by D. Y. Panov and V. I. Feodossiev, *Priklad. Mat. Mekhan.*, vol. 12, p. 389, 1948.
[3] See his book "Elastische Platten," p. 288, 1925.

To obtain an approximate solution of the problem a suitable expression for the deflection w should be taken as a first approximation. Substituting it in the right-hand side of the first of the equations (234), we obtain a linear equation for u which can be integrated to give a first approximation for u. Substituting the first approximations for u and w in the right-hand side of the second of the equations (234), we obtain a linear differential equation for w which can be integrated to give a second approximation for w. This second approximation can then be used to obtain further approximations for u and w by repeating the same sequence of calculations.

In discussing bending of a uniformly loaded circular plate with a clamped edge, Nádai begins with the derivative dw/dr and takes as first approximation the expression

$$\frac{dw}{dr} = C\left[\frac{r}{a} - \left(\frac{r}{a}\right)^n\right] \tag{j}$$

which vanishes for $r = 0$ and $r = a$ in compliance with the condition at the built-in edge. The first of the equations (234) then gives the first approximation for u. Substituting these first approximations for u and dw/dr in the second of the equations (234) and solving it for q, we determine the constants C and n in expression (j) so as to make q as nearly a constant as possible. In this manner the following equation[1] for calculating the deflection at the center is obtained when $\nu = 0.25$:

$$\frac{w_0}{h} + 0.583\left(\frac{w_0}{h}\right)^3 = 0.176\frac{q}{E}\left(\frac{a}{h}\right)^4 \tag{235}$$

In the case of very thin plates the deflection w_0 may become very large in comparison with h. In such cases the resistance of the plate to bending can be neglected, and it can be treated as a flexible membrane. The general equations for such a membrane are obtained from Eqs. (234) by putting zero in place of the left-hand side of the second of the equations. An approximate solution of the resulting equations is obtained by neglecting the first term on the left-hand side of Eq. (235) as being small in comparison with the second term. Hence

$$0.583\left(\frac{w_0}{h}\right)^3 \approx 0.176\frac{q}{E}\left(\frac{a}{h}\right)^4 \quad \text{and} \quad w_0 = 0.665a\sqrt[3]{\frac{qa}{Eh}}$$

[1] Another method for the approximate solution of Eqs. (234) was developed by K. Federhofer, *Eisenbau*, vol. 9, p. 152, 1918; see also *Forschungsarb. VDI*, vol. 7, p. 148, 1936. His equation for w_0 differs from Eq. (235) only by the numerical value of the coefficient on the left-hand side; viz., 0.523 must be used instead of 0.583 for $\nu = 0.25$.

A more complete investigation of the same problem[1] gives

$$w_0 = 0.662a \sqrt[3]{\frac{qa}{Eh}} \qquad (236)$$

This formula, which is in very satisfactory agreement with experiments,[2] shows that the deflections are not proportional to the intensity of the load but vary as the cube root of that intensity. For the tensile stresses at the center of the membrane and at the boundary the same solution gives, respectively,

$$(\sigma_r)_{r=0} = 0.423 \sqrt[3]{\frac{Eq^2a^2}{h^2}} \qquad \text{and} \qquad (\sigma_r)_{r=a} = 0.328 \sqrt[3]{\frac{Eq^2a^2}{h^2}}$$

To obtain deflections that are proportional to the pressure, as is often required in various measuring instruments, recourse should be had to corrugated membranes[3] such as that shown in Fig. 202. As a result of the corrugations the deformation consists primarily in bending and thus increases in proportion to the pressure.[4] If the corrugation (Fig. 202) follows a sinusoidal law and the number of waves along a diameter is sufficiently large ($n > 5$) then, with the notation of Fig. 186, the following expression[5] for $w_0 = (w)_{\max}$ may be used:

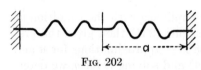

Fig. 202

$$8\left(\frac{w_0}{h}\right)\left[\frac{2}{3(1-\nu^2)} + \left(\frac{f}{h}\right)^2\right] + \frac{6}{7}\left(\frac{w_0}{h}\right)^3 = \frac{q}{E}\left(\frac{a}{h}\right)^4$$

98. Exact Solution for a Uniformly Loaded Circular Plate with a Clamped Edge.[6]

To obtain a more satisfactory solution of the problem of large deflections of a uniformly loaded circular plate with a clamped edge, it is necessary to solve Eqs. (234). To do this we first write the equations in a somewhat different form. As may be seen from its deri-

---

[1] The solution of this problem was given by H. Hencky, *Z. Math. Physik*, vol. 63, p. 311, 1915. For some peculiar effects arising at the edge zone of very thin plates see K. O. Friedrichs, *Proc. Symposia Appl. Math.*, vol. 1, p. 188, 1949.

[2] See Bruno Eck, *Z. angew. Math. Mech.*, vol. 7, p. 498, 1927. For tests on circular plates with clamped edges, see also A. McPherson, W. Ramberg, and S. Levy, *NACA Rept.* 744, 1942.

[3] The theory of deflection of such membranes is discussed by K. Stange, *Ingr.-Arch.*, vol. 2, p. 47, 1931.

[4] For a bibliography on diaphragms used in measuring instruments see M. D. Hersey's paper in *NACA Rept.* 165, 1923.

[5] A. S. Volmir, "Flexible Plates and Shells," p. 214, Moscow, 1956. This book also contains a comprehensive bibliography on large deflections of plates and shells.

[6] This solution is due to S. Way, *Trans. ASME*, vol. 56, p. 627, 1934.

vation in Art. 96, the first of these equations is equivalent to the equation

$$N_r - N_t + r\frac{dN_r}{dr} = 0 \tag{237}$$

Also, as is seen from Eq. (e) of Art. 96 and Eq. (i) of Art. 97, the second of the same equations can be put in the following form:

$$D\left(\frac{d^3w}{dr^3} + \frac{1}{r}\frac{d^2w}{dr^2} - \frac{1}{r^2}\frac{dw}{dr}\right) = N_r\frac{dw}{dr} + \frac{qr}{2} \tag{238}$$

From the general expressions for the radial and tangential strain (page 396) we obtain

$$\epsilon_r = \epsilon_t + r\frac{d\epsilon_t}{dr} + \frac{1}{2}\left(\frac{dw}{dr}\right)^2$$

Substituting

$$\epsilon_r = \frac{1}{hE}(N_r - \nu N_t) \quad \text{and} \quad \epsilon_t = \frac{1}{hE}(N_t - \nu N_r)$$

in this equation and using Eq. (237), we obtain

$$r\frac{d}{dr}(N_r + N_t) + \frac{hE}{2}\left(\frac{dw}{dr}\right)^2 = 0 \tag{239}$$

The three Eqs. (237), (238), and (239) containing the three unknown functions N_r, N_t, and w will now be used in solving the problem. We begin by transforming these equations to a dimensionless form by introducing the following notations:

$$p = \frac{q}{E} \qquad \xi = \frac{r}{h} \qquad S_r = \frac{N_r}{hE} \qquad S_t = \frac{N_t}{hE} \tag{240}$$

With this notation, Eqs. (237), (238), and (239) become, respectively,

$$\frac{d}{d\xi}(\xi S_r) - S_t = 0 \tag{241}$$

$$\frac{1}{12(1-\nu^2)}\frac{d}{d\xi}\left[\frac{1}{\xi}\frac{d}{d\xi}\left(\xi\frac{dw}{dr}\right)\right] = \frac{p\xi}{2} + S_r\frac{dw}{dr} \tag{242}$$

$$\xi\frac{d}{d\xi}(S_r + S_t) + \frac{1}{2}\left(\frac{dw}{dr}\right)^2 = 0 \tag{243}$$

The boundary conditions in this case require that the radial displacement u and the slope dw/dr vanish at the boundary. Using Eq. (b) of Art. 96 for the displacements u and applying Hooke's law, these conditions become

$$(u)_{r=a} = r(S_t - \nu S_r)_{r=a} = 0 \tag{244}$$

$$\left(\frac{dw}{dr}\right)_{r=a} = 0 \tag{a}$$

Assuming that S_r is a symmetrical function and dw/dr an antisymmetrical function of ξ, we represent these functions by the following power series:

$$S_r = B_0 + B_2\xi^2 + B_4\xi^4 + \cdots \qquad (b)$$

$$\frac{dw}{dr} = \sqrt{8}\,(C_1\xi + C_3\xi^3 + C_5\xi^5 + \cdots) \qquad (c)$$

in which B_0, B_2, ... and C_1, C_3, ... are constants to be determined later. Substituting the first of these series in Eq. (241), we find

$$S_t = B_0 + 3B_2\xi^2 + 5B_4\xi^4 + \cdots \qquad (d)$$

By integrating and differentiating Eq. (c), we obtain, respectively,

$$\frac{w}{h} = \sqrt{8}\left(C_1\frac{\xi^2}{2} + C_3\frac{\xi^4}{4} + C_5\frac{\xi^6}{6} + \cdots\right) \qquad (e)$$

$$\frac{d}{d\xi}\left(\frac{dw}{dr}\right) = \sqrt{8}\,(C_1 + 3C_3\xi^2 + 5C_5\xi^4 + \cdots) \qquad (f)$$

It is seen that all the quantities in which we are interested can be found if we know the constants B_0, B_2, ..., C_1, C_3, ... Substituting series (b), (c), and (d) in Eqs. (242) and (243) and observing that these equations must be satisfied for any value of ξ, we find the following relations between the constants B and C:

$$B_k = -\frac{4}{k(k+2)} \sum_{m=1,3,5,\ldots}^{k-1} C_m C_{k-m} \qquad k = 2, 4, 6, \ldots$$

$$C_k = \frac{12(1-\nu^2)}{k^2-1} \sum_{m=0,2,4,\ldots}^{k-3} B_m C_{k-2-m} \qquad k = 5, 7, 9, \ldots \qquad (g)$$

$$C_3 = \frac{3}{2}(1-\nu^2)\left(\frac{p}{2\sqrt{8}} + B_0 C_1\right)$$

It can be seen that when the two constants B_0 and C_1 are assigned, all the other constants are determined by relations (g). The quantities S_r, S_t, and dw/dr are then determined by series (b), (d), and (c) for all points in the plate. As may be seen from series (b) and (f), fixing B_0 and C_1 is equivalent to selecting the values of S_r and the curvature at the center of the plate.[1]

To obtain the following curves for calculating deflections and stresses in particular cases, the procedure used was: For given values of ν and

---

[1] The selection of these same quantities has already been encountered in the case of bending of circular plates by moments uniformly distributed along the edge (see page 398).

$p = q/E$ and for selected values of B_0 and C_1, a considerable number of numerical cases were calculated,[1] and the radii of the plates were determined so as to satisfy the boundary condition (a). For all these plates the values of S_r and S_t at the boundary were calculated, and the values of the radial displacements $(u)_{r=a}$ at the boundary were determined. Since all calculations were made with arbitrarily assumed values of B_0 and C_1, the boundary condition (244) was not satisfied. However, by interpolation it was possible to obtain all the necessary data for plates for which both conditions (244) and (a) are satisfied. The results of these calculations are represented graphically in Fig. 203. If the deflection of the

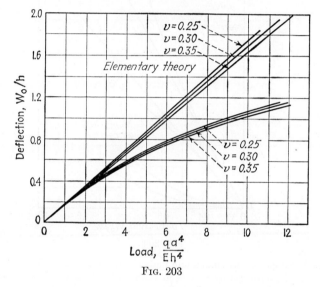

Fig. 203

plate is found from this figure, the corresponding stress can be obtained by using the curves of Fig. 204. In this figure, curves are given for the membrane stresses

$$\sigma_r = \frac{N_r}{h}$$

and for the bending stresses

$$\sigma'_r = \frac{6M_r}{h^2}$$

as calculated for the center and for the edge of the plate.[2] By adding together σ_r and σ'_r, the total maximum stress at the center and at the edge of the plate can be obtained. For purposes of comparison Figs. 203 and 204 also include straight lines showing the results obtained from

---

[1] Nineteen particular cases have been calculated by Way, *op. cit.*
[2] The stresses are given in dimensionless form.

the elementary theory in which the strain of the middle plane is neglected. It will be noted that the errors of the elementary theory increase as the load and deflections increase.

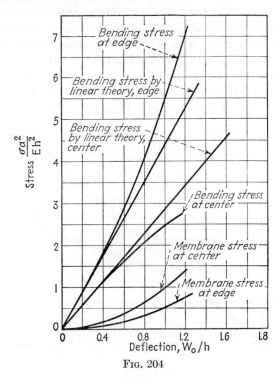

Fig. 204

99. A Simply Supported Circular Plate under Uniform Load. An exact solution of the problem[1] can be obtained by a series method similar to that used in the preceding article.

Because of the axial symmetry we have again $dw/dr = 0$ and $N_r = N_t$ at $r = 0$. Since the radial couples must vanish on the edge, a further condition is

$$\left[\frac{d}{dr}\left(\frac{dw}{dr}\right) + \frac{\nu}{r}\frac{dw}{dr}\right]_{r=a} = 0 \qquad (a)$$

With regard to the stress and strain in the middle plane of the plate two boundary conditions may be considered:

1. Assuming the edge is immovable we have, by Eq. (244), $S_t - \nu S_r = 0$, which, by Eq. (237), is equivalent to

$$\left[S_r(1-\nu) + r\frac{dS_r}{dr}\right]_{r=a} = 0 \qquad (b)$$

[1] K. Federhofer and H. Egger, *Sitzber. Akad. Wiss. Wien*, IIa, vol. 155, p. 15, 1946; see also M. Stippes and A. H. Hausrath, *J. Appl. Mechanics*, vol. 19, p. 287, 1952. The perturbation method used in this latter paper appears applicable in the case of a concentrated load as well.

2. Supposing the edge as free to move in the radial direction we simply have

$$(S_r)_{r=a} = 0 \qquad (c)$$

The functions S_r and dw/dr may be represented again in form of the series

$$S_r = \frac{h^2}{12(1-\nu^2)a^2\rho}(B_1\rho + B_3\rho^3 + B_5\rho^5 + \cdots) \qquad (d)$$

$$\frac{dw}{dr} = -\frac{h}{2a\sqrt{3}}(C_1\rho + C_3\rho^3 + C_5\rho^5 + \cdots) \qquad (e)$$

where $\rho = r/a$. Using these series and also Eqs. (241), (242), (243), from which the quantity S_r can readily be eliminated, we arrive at the following relations between the constants B and C:

$$B_k = -\frac{1-\nu^2}{2(k^2-1)}\sum_{m=1,3,5,\ldots}^{k-2} C_m C_{k-m-1} \qquad k = 3, 5, \ldots \qquad (f)$$

$$C_k = \frac{1}{k^2-1}\sum_{m=1,3,5,\ldots}^{k-2} C_m B_{k-m-1} \qquad k = 5, 7, \ldots \qquad (g)$$

$$8C_3 - B_1 C_1 + 12\sqrt{3}(1-\nu^2)\frac{pa^4}{h^4} = 0 \qquad (h)$$

where $p = q/E$, q being the intensity of the load.

Again, all constants can easily be expressed in terms of both constants B_1 and C_1, for which two additional relations, ensuing from the boundary conditions, hold:

In case 1 we have

$$\sum_{k=1,3,5,\ldots} B_k(k-\nu) = 0 \qquad \sum_{k=1,3,5,\ldots} C_k(k+\nu) = 0 \qquad (i)$$

and in case 2

$$\sum_{k=1,3,5,\ldots} B_k = 0 \qquad \sum_{k=1,3,5,\ldots} C_k(k+\nu) = 0 \qquad (j)$$

To start the resolution of the foregoing system of equations, suitable values of B_1 and C_1 may be taken on the basis of an approximate solution. Such a solution, satisfying condition (a), can be, for instance, of the form

$$\frac{dw}{dr} = C(\beta\rho^n - \rho) \qquad (k)$$

where C is a constant and $\beta = \dfrac{1+\nu}{n+\nu}$ ($n = 3, 5, \ldots$). Substituting this in Eqs. (241) and (243), in which ξ must be replaced by $\rho a/h$, and eliminating S_t we obtain

$$S_r = c_1 + \frac{c_2}{\rho^2} - \frac{C^2}{2}\left(\beta^2\frac{\rho^{2n}}{n_1} - 2\beta\frac{\rho^{n+1}}{n_2} + \frac{\rho^2}{8}\right) \qquad (l)$$

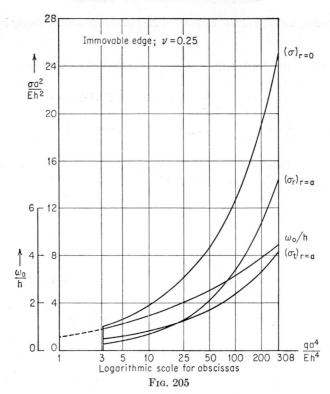

Fig. 205

Table 82. Data for Calculation of Approximate Values of Deflections w_0 and Stresses in Uniformly Loaded Plates

$\nu = 0.3$

Boundary conditions		A	B	Center		Edge			
				$\alpha_r = \alpha_t$	$\beta_r = \beta_t$	α_r	α_t	β_r	β_t
Plate clamped	Edge immovable	0.471	0.171	0.976	2.86	0.476	0.143	−4.40	−1.32
	Edge free to move	0.146	0.171	0.500	2.86	0	−0.333	−4.40	−1.32
Plate simply supported	Edge immovable	1.852	0.696	0.905	1.778	0.610	0.183	0	0.755
	Edge free to move	0.262	0.696	0.295	1.778	0	−0.427	0	0.755

Herein c_1 and c_2 are constants of integration and

$$n_1 = 4n(n+1) \qquad n_2 = (n+1)(n+3)$$

Let us, for example, assume the boundary conditions of case 2. Then we obtain

$$c_1 = \frac{C^2}{2}\left(\frac{\beta^2}{n_1} - \frac{2\beta}{n_2} + \frac{1}{8}\right) \qquad c_2 = 0 \qquad (m)$$

The constant C, finally, can be determined by some strain energy method—for example, that described in Art. 100. Using there Eqs. (m) or (o) we have only to replace

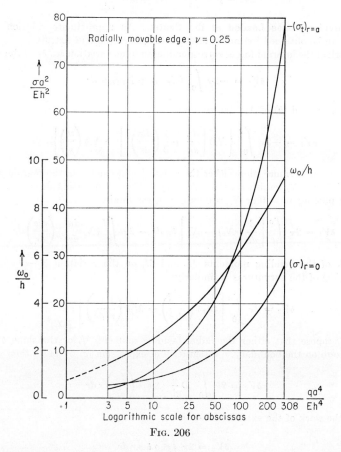

Fig. 206

$d\varphi/dr = rhES_r$ and dw/dr by approximate expressions in accordance with Eqs. (k) and (l) given above.

The largest values of deflections and of total stresses obtained by Federhofer and Egger from the exact solution are given in Fig. 205 for case 1 and in Fig. 206 for case 2. The calculation has been carried out for $\nu = 0.25$.

Table 82 may be useful for *approximate* calculations of the deflection w_0 at the center, given by an equation of the form

$$\frac{w_0}{h} + A\left(\frac{w_0}{h}\right)^3 = B\frac{q}{E}\left(\frac{a}{h}\right)^4 \quad (n)$$

also of the stresses in the middle plane, given by

$$\sigma_r = \alpha_r E \frac{w_0^2}{a^2} \qquad \sigma_t = \alpha_t E \frac{w_0^2}{a^2} \quad (o)$$

and of the extreme fiber bending stresses[1]

$$\sigma'_r = \beta_r E \frac{w_0 h}{a^2} \qquad \sigma'_t = \beta_t E \frac{w_0 h}{a^2} \quad (p)$$

100. Circular Plates Loaded at the Center. An approximate solution of this problem can be obtained by means of the method described in Art. 81.

The work of the internal forces corresponding to some variation $\delta\epsilon_r$, $\delta\epsilon_t$ of the strain is

$$\delta V_1 = -2\pi \int_0^a (N_r \delta\epsilon_r + N_t \delta\epsilon_t) r \, dr$$

Using Eqs. (a) and (b) of Art. 96 we have

$$\delta V_1 = -2\pi \int_0^a \left\{ N_r \delta\left[\frac{du}{dr} + \frac{1}{2}\left(\frac{dw}{dr}\right)^2\right] + N_t \delta\left(\frac{u}{r}\right) \right\} r \, dr \quad (a)$$

We assume, furthermore, that either the radial displacements in the middle plane or the radial forces N_r vanish on the boundary. Then, integrating expression (a) by parts and putting $\delta u = 0$ or $N_r = 0$ on $r = a$, we obtain

$$\delta V_1 = 2\pi \int_0^a \left[\frac{d}{dr}(rN_r) - N_t\right]\delta u \, dr - 2\pi \int_0^a rN_r \frac{dw}{dr}\delta\left(\frac{dw}{dr}\right) dr \quad (b)$$

The work of the bending moments M_r and M_t on the variation $\delta(-d^2w/dr^2)$ and $\delta(-\frac{1}{2}\,dw/dr)$ of the curvatures is similarly

$$\delta V_2 = 2\pi \int_0^a \left[M_r \delta\left(\frac{d^2w}{dr^2}\right) + M_t \delta\left(\frac{1}{r}\frac{dw}{dr}\right)\right] r \, dr \quad (c)$$

Now we suppose that either the radial bending moment M_r or the slope $\delta(dw/dr)$ becomes zero on the boundary. Integration of expression (c) by parts then yields

$$\delta V_2 = 2\pi \int_0^a D \frac{d}{dr}(\Delta w) \delta\left(\frac{dw}{dr}\right) r \, dr \quad (d)$$

Finally, the work of the external forces is

$$\delta V_3 = 2\pi \int_0^a q \, \delta w \, r \, dr$$

or, by putting

$$\psi = \frac{1}{r} \int_0^r qr \, dr \qquad q = \frac{1}{r}\frac{d}{dr}(r\psi) \quad (e)$$

[1] The sign is negative if the bottom of the plate is in compression.

we have

$$\delta V_2 = 2\pi \int_0^a \frac{d}{dr}(r\psi)\,\delta w\, r\, dr$$

Provided $\delta w = 0$ on the boundary we finally obtain

$$\delta V_3 = -2\pi \int_0^a r\psi \delta\left(\frac{dw}{dr}\right) r\, dr \qquad (f)$$

The condition $\delta(V_1 + V_2 + V_3) = 0$ now yields the equation

$$\int_0^a \left[D\frac{d}{dr}(\Delta w) - \psi - N_r \frac{dw}{dr}\right]\frac{d}{dr}(\delta w)\, r\, dr + \int_0^a \left[\frac{d}{dr}(rN_r) - N_t\right]\delta u\, dr = 0 \qquad (g)$$

We could proceed next by assuming both variations δw and δu as arbitrary. Thus we would arrive at the second of the differential equations (234), N_r being given by expression (c) of Art. 96, and at Eq. (d) of the same article. If we suppose only this latter equation of equilibrium to be satisfied, then we have still to fulfill the condition

$$\int_0^a \left[D\frac{d}{dr}(\Delta w) - \psi - \frac{1}{r}\frac{df}{dr}\frac{dw}{dr}\right]\frac{d}{dr}(\delta w)\, r\, dr = 0 \qquad (h)$$

in which f is a stress function defining

$$N_r = \frac{1}{r}\frac{df}{dr} \qquad N_t = \frac{d^2f}{dr^2} \qquad (i)$$

and governed by the differential equation

$$\frac{d}{dr}(\Delta f) = -\frac{Eh}{2r}\left(\frac{dw}{dr}\right)^2 \qquad (j)$$

which follows from Eq. (239). Integrating expression (h) by parts once more we obtain

$$\int_0^a \left[D\Delta\Delta w - q - \frac{1}{r}\frac{d}{dr}\left(\frac{df}{dr}\frac{dw}{dr}\right)\right]\delta w\, r\, dr = 0 \qquad (k)$$

With intent to use the method described in Art. 81 we take the deflection in the form

$$w = a_1\varphi_1(r) + a_2\varphi_2(r) + \cdots + a_n\varphi_n(r) \qquad (l)$$

Just as in the case of the expression (211) each function $\varphi_i(r)$ has to satisfy two boundary conditions prescribed for the deflection. Substituting expression (l) either in Eq. (h) or in Eq. (k) and applying the same reasoning as in Art. 81, we arrive at a sequence of equations of the form

$$\int_0^a X\frac{d\varphi_i}{dr} r\, dr \qquad i = 1, 2, \ldots, n \qquad (m)$$

in which

$$X = D\frac{d}{dr}(\Delta w) - \psi - \frac{1}{r}\frac{df}{dr}\frac{dw}{dr} \qquad (n)$$

or at a set of equations

$$\int_0^a Y\varphi_i r\, dr \qquad i = 1, 2, \ldots, n \tag{o}$$

where
$$Y = D\Delta\Delta w - q - \frac{1}{r}\frac{d}{dr}\left(\frac{df}{dr}\frac{dw}{dr}\right) \tag{p}$$

Now let us consider a clamped circular plate with a load P concentrated at $r = 0$. We reduce expression (l) to its first term by taking the deflection in the form

$$w = w_0\left(1 - \frac{r^2}{a^2} + 2\frac{r^2}{a^2}\log\frac{r}{a}\right) \tag{q}$$

which holds rigorously for a plate with small deflections. From Eq. (j) we obtain, by integration,

$$\frac{df}{dr} = -\frac{Ehw_0^2 r^3}{a^4}\left(\log^2\frac{r}{a} - \frac{3}{2}\log\frac{r}{a} + \frac{7}{8}\right) + C_1 r + \frac{C_2}{r} \tag{r}$$

Let there be a free radial displacement at the boundary. The constants of integration C_1 and C_2 then are determined by two conditions. The first, namely,

$$(N_r)_{r=a} = 0$$

can be rewritten as

$$\left(\frac{1}{r}\frac{df}{dr}\right)_{r=a} = 0 \tag{s}$$

and the second is

$$\left(\frac{df}{dr}\right)_{r=0} = 0 \tag{t}$$

This latter condition must be added in order to limit, at $r = 0$, the value of the stress N_r given by Eq. (i). Thus we obtain

$$C_1 = \frac{7}{8}\frac{Ehw_0^2}{a^2} \qquad C_2 = 0$$

The load function is equal to

$$\psi = \frac{P}{2\pi r}$$

in our case, and expressions (q) and (r) yield

$$X = D\frac{8w_0}{a^2 r} - \frac{P}{2\pi r} + \frac{4Ew_0^3 h}{a^3}\left(\frac{r^3}{a^3}\log^3\frac{r}{a} - \frac{3}{2}\frac{r^3}{a^3}\log^2\frac{r}{a} + \frac{7}{8}\frac{r^3}{a^3}\log\frac{r}{a} - \frac{7}{8}\frac{r}{a}\log\frac{r}{a}\right) \tag{u}$$

while φ_1 is given by the expression in the parentheses in Eq. (q). Substituting this in Eq. (m) we arrive at the relation

$$16\, Dw_0 + \frac{191}{648}Ehw_0^3 = \frac{Pa^2}{\pi} \tag{v}$$

The general expressions for the extreme fiber bending stresses corresponding to the deflection (q) and obtainable by means of Eqs. (101) are

$$\sigma'_r = \frac{2Ehw_0}{(1-\nu^2)a^2}\left[(1+\nu)\log\frac{a}{r} - 1\right]$$

$$\sigma'_t = \frac{2Ehw_0}{(1-\nu^2)a^2}\left[(1+\nu)\log\frac{a}{r} - \nu\right]$$

(w)

These expressions yield infinite values of stresses as r tends to zero. However, assuming the load P to be distributed uniformly over a circular area with a small radius $r = c$, we can use a simple relation existing in plates with small deflections between the stresses $\sigma''_r = \sigma''_t$ at the center of such an area and the stresses $\sigma'_r = \sigma'_t$ caused at $r = c$ by the same load P acting at the point $r = 0$. According to Nádai's result,[1] expressed in terms of stresses,

$$\sigma''_r = \sigma''_t = \sigma'_r + \frac{3}{2}\frac{P}{\pi h^2}$$

Applying this relation to the plate with large deflections we obtain, at the center of the loaded area with a radius c, approximately

$$\sigma''_r = \sigma''_t = \frac{2Ehw_0}{(1-\nu^2)a^2}\left[(1+\nu)\log\frac{a}{c} - 1\right] + \frac{3}{2}\frac{P}{\pi h^2} \quad (x)$$

The foregoing results hold for a circular plate with a clamped and movable edge. By introducing other boundary conditions we obtain for w_0 an equation

$$\frac{w_0}{h} + A\left(\frac{w_0}{h}\right)^3 = B\frac{Pa^2}{Eh^4} \quad (y)$$

which is a generalization of Eq. (v). The constants A and B are given in Table 83. The same table contains several coefficients[2] needed for calculation of stresses

$$\sigma_r = \alpha_r E \frac{w_0^2}{a^2} \qquad \sigma_t = \alpha_t E \frac{w_0^2}{a^2} \quad (z)$$

acting in the middle plane of the plate and the extreme fiber bending stresses

$$\sigma'_r = \beta_r E \frac{w_0 h}{a^2} \qquad \sigma'_t = \beta_t E \frac{w_0 h}{a^2} \quad (z')$$

The former are calculated using expressions (i), the latter by means of expressions (101) for the moments, the sign being negative if the compression is at the bottom.[3]

101. General Equations for Large Deflections of Plates.

In discussing the general case of large deflections of plates we use Eq. (219), which was

---

[1] A. Nádai, "Elastische Platten," p. 63, Berlin, 1925.
[2] All data contained in Table 82 are taken from A. S. Volmir, *op. cit.*
[3] For bending of the ring-shaped plates with large deflections see K. Federhofer, *Österr. Ingr.-Arch.*, vol. 1, p. 21, 1946; E. Reissner, *Quart. Appl. Math.*, vol. 10, p. 167, 1952, and vol. 11, p. 473, 1953. Large deflections of elliptical plates have been discussed by N. A. Weil and N. M. Newmark, *J. Appl. Mechanics*, vol. 23, p. 21, 1956.

TABLE 83. DATA FOR CALCULATION OF APPROXIMATE VALUES OF DEFLECTIONS w_0 AND STRESSES IN CENTRALLY LOADED PLATES
$\nu = 0.3$

Boundary conditions		A	B	Center	Edge			
				$\alpha_r = \alpha_t$	α_r	α_t	β_r	β_t
Plate clamped	Edge immovable	0.443	0.217	1.232	0.357	0.107	−2.198	−0.659
	Edge free to move	0.200	0.217	0.875	0	−0.250	−2.198	−0.659
Plate simply supported	Edge immovable	1.430	0.552	0.895	0.488	0.147	0	0.606
	Edge free to move	0.272	0.552	0.407	0	−0.341	0	0.606

derived by considering the equilibrium of an element of the plate in the direction perpendicular to the plate. The forces N_x, N_y, and N_{xy} now depend not only on the external forces applied in the xy plane but also on the strain of the middle plane of the plate due to bending. Assuming that there are no body forces in the xy plane and that the load is perpendicular to the plate, the equations of equilibrium of an element in the xy plane are

$$\frac{\partial N_x}{\partial x} + \frac{\partial N_{xy}}{\partial y} = 0$$
$$\frac{\partial N_{xy}}{\partial x} + \frac{\partial N_y}{\partial y} = 0 \qquad (a)$$

The third equation necessary to determine the three quantities N_x, N_y, and N_{xy} is obtained from a consideration of the strain in the middle surface of the plate during bending. The corresponding strain components [see Eqs. (221), (222), and (223)] are

$$\epsilon_x = \frac{\partial u}{\partial x} + \frac{1}{2}\left(\frac{\partial w}{\partial x}\right)^2$$
$$\epsilon_y = \frac{\partial v}{\partial y} + \frac{1}{2}\left(\frac{\partial w}{\partial y}\right)^2 \qquad (b)$$
$$\gamma_{xy} = \frac{\partial u}{\partial y} + \frac{\partial v}{\partial x} + \frac{\partial w}{\partial x}\frac{\partial w}{\partial y}$$

By taking the second derivatives of these expressions and combining the resulting expressions, it can be shown that

$$\frac{\partial^2 \epsilon_x}{\partial y^2} + \frac{\partial^2 \epsilon_y}{\partial x^2} - \frac{\partial^2 \gamma_{xy}}{\partial x\, \partial y} = \left(\frac{\partial^2 w}{\partial x\, \partial y}\right)^2 - \frac{\partial^2 w}{\partial x^2} \frac{\partial^2 w}{\partial y^2} \qquad (c)$$

By replacing the strain components by the equivalent expressions

$$\epsilon_x = \frac{1}{hE}(N_x - \nu N_y)$$

$$\epsilon_y = \frac{1}{hE}(N_y - \nu N_x) \qquad (d)$$

$$\gamma_{xy} = \frac{1}{hG} N_{xy}$$

the third equation in terms of N_x, N_y, and N_{xy} is obtained.

The solution of these three equations is greatly simplified by the introduction of a *stress function*.[1] It may be seen that Eqs. (a) are identically satisfied by taking

$$N_x = h\frac{\partial^2 F}{\partial y^2} \qquad N_y = h\frac{\partial^2 F}{\partial x^2} \qquad N_{xy} = -h\frac{\partial^2 F}{\partial x\, \partial y} \qquad (e)$$

where F is a function of x and y. If these expressions for the forces are substituted in Eqs. (d), the strain components become

$$\epsilon_x = \frac{1}{E}\left(\frac{\partial^2 F}{\partial y^2} - \nu \frac{\partial^2 F}{\partial x^2}\right)$$

$$\epsilon_y = \frac{1}{E}\left(\frac{\partial^2 F}{\partial x^2} - \nu \frac{\partial^2 F}{\partial y^2}\right) \qquad (f)$$

$$\gamma_{xy} = -\frac{2(1+\nu)}{E}\frac{\partial^2 F}{\partial x\, \partial y}$$

Substituting these expressions in Eq. (c), we obtain

$$\frac{\partial^4 F}{\partial x^4} + 2\frac{\partial^4 F}{\partial x^2\, \partial y^2} + \frac{\partial^4 F}{\partial y^4} = E\left[\left(\frac{\partial^2 w}{\partial x\, \partial y}\right)^2 - \frac{\partial^2 w}{\partial x^2}\frac{\partial^2 w}{\partial y^2}\right] \qquad (245)$$

The second equation necessary to determine F and w is obtained by substituting expressions (e) in Eq. (217), which gives

$$\frac{\partial^4 w}{\partial x^4} + 2\frac{\partial^4 w}{\partial x^2\, \partial y^2} + \frac{\partial^4 w}{\partial y^4} = \frac{h}{D}\left(\frac{q}{h} + \frac{\partial^2 F}{\partial y^2}\frac{\partial^2 w}{\partial x^2}\right.$$

$$\left. + \frac{\partial^2 F}{\partial x^2}\frac{\partial^2 w}{\partial y^2} - 2\frac{\partial^2 F}{\partial x\, \partial y}\frac{\partial^2 w}{\partial x\, \partial y}\right) \qquad (246)$$

---

[1] See S. Timoshenko and J. N. Goodier, "Theory of Elasticity," 2d ed., p. 26, 1951.

Equations (245) and (246), together with the boundary conditions, determine the two functions F and w.* Having the stress function F, we can determine the stresses in the middle surface of a plate by applying Eqs. (e). From the function w, which defines the deflection surface of the plate, the bending and the shearing stresses can be obtained by using the same formulas as in the case of plates with small deflection [see Eqs. (101) and (102)]. Thus the investigation of large deflections of plates reduces to the solution of the two nonlinear differential equations (245) and (246). The solution of these equations in the general case is unknown. Some approximate solutions of the problem are known, however, and will be discussed in the next article.

In the particular case of bending of a plate to a cylindrical surface[1] whose axis is parallel to the y axis, Eqs. (245) and (246) are simplified by observing that in this case w is a function of x only and that $\partial^2 F/\partial x^2$ and $\partial^2 F/\partial y^2$ are constants. Equation (245) is then satisfied identically, and Eq. (246) reduces to

$$\frac{\partial^4 w}{\partial x^4} = \frac{q}{D} + \frac{N_x}{D} \frac{\partial^2 w}{\partial x^2}$$

Problems of this kind have already been discussed fully in Chap. 1.

If polar coordinates, more convenient in the case of circular plates, are used, the system of equations (245) and (246) assumes the form

$$\Delta\Delta F = -\frac{E}{2} L(w,w)$$

$$\Delta\Delta w = \frac{h}{D} L(w,F) + \frac{q}{D}$$

in which

$$L(w,F) = \frac{\partial^2 w}{\partial r^2} \left(\frac{1}{r} \frac{\partial F}{\partial r} + \frac{1}{r^2} \frac{\partial^2 F}{\partial \theta^2} \right) + \left(\frac{1}{r} \frac{\partial w}{\partial r} + \frac{1}{r^2} \frac{\partial^2 w}{\partial \theta^2} \right) \frac{\partial^2 F}{\partial r^2} - 2 \frac{\partial}{\partial r} \left(\frac{1}{r} \frac{\partial F}{\partial \theta} \right) \frac{\partial}{\partial r} \left(\frac{1}{r} \frac{\partial w}{\partial \theta} \right)$$

and $L(w,w)$ is obtained from the foregoing expression if w is substituted for F.

In the case of very thin plates, which may have deflections many times larger than their thickness, the resistance of the plate to bending can be

---

* These two equations were derived by Th. von Kármán; see "Encyklopädie der Mathematischen Wissenschaften," vol. IV$_4$, p. 349, 1910. A general method of nonlinear elasticity has been applied to bending of plates by E. Koppe, *Z. angew. Math. Mech.*, vol. 36, p. 455, 1956.

[1] For a more general theory of plates (in particular of cantilever plates) bent, without extension, to a developable surface, see E. H. Mansfield, *Quart. J. Mech. Appl. Math.*, vol. 8, p. 338, 1955, and D. G. Ashwell, *Quart. J. Mech. Appl. Math.*, vol. 10, p. 169, 1957. A boundary-layer phenomenon arising along the free edges of such plates was considered by Y. C. Fung and W. H. Witrick, *Quart. J. Mech. Appl. Math.*, vol. 8, p. 191, 1955.

neglected; *i.e.*, the flexural rigidity D can be taken equal to zero, and the problem reduced to that of finding the deflection of a flexible membrane. Equations (245) and (246) then become[1]

$$\frac{\partial^4 F}{\partial x^4} + 2 \frac{\partial^4 F}{\partial x^2 \, \partial y^2} + \frac{\partial^4 F}{\partial y^4} = E \left[\left(\frac{\partial^2 w}{\partial x \, \partial y} \right)^2 - \frac{\partial^2 w}{\partial x^2} \frac{\partial^2 w}{\partial y^2} \right]$$
$$\frac{q}{h} + \frac{\partial^2 F}{\partial y^2} \frac{\partial^2 w}{\partial x^2} + \frac{\partial^2 F}{\partial x^2} \frac{\partial^2 w}{\partial y^2} - 2 \frac{\partial^2 F}{\partial x \, \partial y} \frac{\partial^2 w}{\partial x \, \partial y} = 0 \quad (247)$$

A numerical solution of this system of equations by the use of finite differences has been discussed by H. Hencky.[2]

The energy method affords another means of obtaining an approximate solution for the deflection of a membrane. The strain energy of a membrane, which is due solely to stretching of its middle surface, is given by the expression

$$V = \tfrac{1}{2} \iint (N_x \epsilon_x + N_y \epsilon_y + N_{xy} \gamma_{xy}) \, dx \, dy$$
$$= \frac{Eh}{2(1 - \nu^2)} \iint [\epsilon_x^2 + \epsilon_y^2 + 2\nu \epsilon_x \epsilon_y + \tfrac{1}{2}(1 - \nu) \gamma_{xy}^2] \, dx \, dy \quad (248)$$

Substituting expressions (221), (222), and (223) for the strain components ϵ_x, ϵ_y, γ_{xy}, we obtain

$$V = \frac{Eh}{2(1 - \nu^2)} \iint \left\{ \left(\frac{\partial u}{\partial x} \right)^2 + \frac{\partial u}{\partial x} \left(\frac{\partial w}{\partial x} \right)^2 + \left(\frac{\partial v}{\partial y} \right)^2 + \frac{\partial v}{\partial y} \left(\frac{\partial w}{\partial y} \right)^2 \right.$$
$$+ \frac{1}{4} \left[\left(\frac{\partial w}{\partial x} \right)^2 + \left(\frac{\partial w}{\partial y} \right)^2 \right]^2 + 2\nu \left[\frac{\partial u}{\partial x} \frac{\partial v}{\partial y} + \frac{1}{2} \frac{\partial v}{\partial y} \left(\frac{\partial w}{\partial x} \right)^2 + \frac{1}{2} \frac{\partial u}{\partial x} \left(\frac{\partial w}{\partial y} \right)^2 \right]$$
$$+ \frac{1 - \nu}{2} \left[\left(\frac{\partial u}{\partial y} \right)^2 + 2 \frac{\partial u}{\partial y} \frac{\partial v}{\partial x} + \left(\frac{\partial v}{\partial x} \right)^2 + 2 \frac{\partial u}{\partial y} \frac{\partial w}{\partial x} \frac{\partial w}{\partial y} \right.$$
$$\left. \left. + 2 \frac{\partial v}{\partial x} \frac{\partial w}{\partial x} \frac{\partial w}{\partial y} \right] \right\} dx \, dy \quad (249)$$

In applying the energy method we must assume in each particular case suitable expressions for the displacements u, v, and w. These expressions must, of course, satisfy the boundary conditions and will contain several arbitrary parameters the magnitudes of which have to be determined by the use of the principle of virtual displacements. To illustrate the method, let us consider a uniformly loaded square membrane[3] with sides of length $2a$ (Fig. 207). The displacements u, v, and w in this case must vanish at the boundary. Moreover, from symmetry, it can be concluded

---

[1] These equations were obtained by A. Föppl, "Vorlesungen über Technische Mechanik," vol. 5, p. 132, 1907.

[2] H. Hencky, *Z. angew. Math. Mech.*, vol. 1, pp. 81 and 423, 1921; see also R. Kaiser, *Z. angew. Math. Mech.*, vol. 16, p. 73, 1936.

[3] Calculations for this case are given in the book "Drang und Zwang" by August and Ludwig Föppl, vol. 1, p. 226, 1924; see also Hencky, *ibid.*

that w is an even function of x and y, whereas u and v are odd functions of x and of y, respectively. All these requirements are satisfied by taking the following expressions for the displacements:

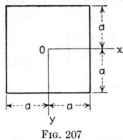

Fig. 207

$$w = w_0 \cos \frac{\pi x}{2a} \cos \frac{\pi y}{2a}$$
$$u = c \sin \frac{\pi x}{a} \cos \frac{\pi y}{2a} \quad (g)$$
$$v = c \sin \frac{\pi y}{a} \cos \frac{\pi x}{2a}$$

which contain two parameters w_0 and c. Substituting these expressions in Eq. (249), we obtain, for $\nu = 0.25$,

$$V = \frac{Eh}{7.5} \left[\frac{5\pi^4}{64} \frac{w_0^4}{a^2} - \frac{17\pi^2}{6} \frac{cw_0^2}{a} + c^2 \left(\frac{35\pi^2}{4} + \frac{80}{9} \right) \right] \quad (h)$$

The principle of virtual displacements gives the two following equations:[1]

$$\frac{\partial V}{\partial c} = 0 \quad (i)$$

$$\frac{\partial V}{\partial w_0} \delta w_0 = \int_{-a}^{+a} \int_{-a}^{+a} q \, \delta w_0 \cos \frac{\pi x}{2a} \cos \frac{\pi y}{2a} \, dx \, dy \quad (j)$$

Substituting expression (h) for V, we obtain from Eq. (i)

$$c = 0.147 \frac{w_0^2}{a}$$

and from Eq. (j)

$$w_0 = 0.802 a \sqrt[3]{\frac{qa}{Eh}} \quad (250)$$

This deflection at the center is somewhat larger than the value (236) previously obtained for a uniformly loaded circular membrane. The tensile strain at the center of the membrane as obtained from expressions (g) is

$$\epsilon_x = \epsilon_y = \frac{\pi c}{a} = 0.462 \frac{w_0^2}{a^2}$$

and the corresponding tensile stress is

$$\sigma = \frac{E}{1-\nu} 0.462 \frac{w_0^2}{a^2} = 0.616 \frac{Ew_0^2}{a^2} = 0.396 \sqrt[3]{\frac{q^2 E a^2}{h^2}} \quad (251)$$

Some application of these results to the investigation of large deflections of thin plates will be shown in the next article.

[1] The right-hand side of Eq. (i) is zero, since the variation of the parameter c produces only horizontal displacements and the vertical load does not produce work.

102. Large Deflections of Uniformly Loaded Rectangular Plates. We begin with the case of a plate with clamped edges. To obtain an approximate solution of the problem the energy method will be used.[1] The total strain energy V of the plate is obtained by adding to the energy of bending [expression (117), page 88] the energy due to strain of the middle surface [expression (249), page 419]. The principle of virtual displacements then gives the equation

$$\delta V - \delta \iint qw \, dx \, dy = 0 \qquad (a)$$

which holds for any variation of the displacements u, v, and w. By deriving the variation of V we can obtain from Eq. (a) the system of Eqs. (245) and (246), the exact solution of which is unknown. To find an approximate solution of our problem we assume for u, v, and w three functions satisfying the boundary conditions imposed by the clamped edges and containing several parameters which will be determined by using Eq. (a). For a rectangular plate with sides $2a$ and $2b$ and coordinate axes, as shown in Fig. 207, we shall take the displacements in the following form:

$$\begin{aligned} u &= (a^2 - x^2)(b^2 - y^2)x(b_{00} + b_{02}y^2 + b_{20}x^2 + b_{22}x^2y^2) \\ v &= (a^2 - x^2)(b^2 - y^2)y(c_{00} + c_{02}y^2 + c_{20}x^2 + c_{22}x^2y^2) \\ w &= (a^2 - x^2)^2(b^2 - y^2)^2(a_{00} + a_{02}y^2 + a_{20}x^2) \end{aligned} \qquad (b)$$

The first two of these expressions, which represent the displacements u and v in the middle plane of the plate, are odd functions in x and y, respectively, and vanish at the boundary. The expression for w, which is an even function in x and y, vanishes at the boundary, as do also its first derivatives. Thus all the boundary conditions imposed by the clamped edges are satisfied.

Expressions (b) contain 11 parameters $b_{00}, \ldots, a_{20}$, which will now be determined from Eq. (a), which must be satisfied for any variation of each of these parameters. In such a way we obtain 11 equations, 3 of the form

$$\frac{\partial}{\partial a_{mn}} \left(V - \iint qw \, dx \, dy \right) = 0 \qquad (c)$$

and 8 equations of the form[2]

$$\frac{\partial V}{\partial b_{mn}} = 0 \qquad \text{or} \qquad \frac{\partial V}{\partial c_{mn}} = 0 \qquad (d)$$

These equations are not linear in the parameters a_{mn}, b_{mn}, and c_{mn} as was true in the case of small deflections (see page 344). The three equations of the form (c) will contain terms of the third degree in the parameters a_{mn}. Equations of the form (d) will be linear in the parameters b_{mn} and c_{mn} and quadratic in the parameters a_{mn}. A solution is obtained by solving Eqs. (d) for the b_{mn}'s and c_{mn}'s in terms of the a_{mn}'s and then substituting these expressions in Eqs. (c). In this way we obtain three equa-

---

[1] Such a solution has been given by S. Way; see *Proc. Fifth Intern. Congr. Appl. Mech.*, Cambridge, Mass., 1938. For application of a method of successive approximation and experimental verification of results see Chien Wei-Zang and Yeh Kai-Yuan, *Proc. Ninth Intern. Congr. Appl. Mech.*, Brussels, vol. 6, p. 403, 1957. Large deflections of slightly curved rectangular plates under edge compression were considered by Syed Yusuff, *J. Appl. Mechanics*, vol. 19, p. 446, 1952.

[2] The zeros on the right-hand sides of these equations result from the fact that the lateral load does not do work when u or v varies.

tions of the third degree involving the parameters a_{mn} alone. These equations can then be solved numerically in each particular case by successive approximations.

Numerical values of all the parameters have been computed for various intensities of the load q and for three different shapes of the plate $b/a = 1$, $b/a = \frac{2}{3}$, and $b/a = \frac{1}{2}$ by assuming $\nu = 0.3$.

It can be seen from the expression for w that, if we know the constant a_{00}, we can at once obtain the deflection of the plate at the center. These deflections are graphically represented in Fig. 208, in which $w_{\max}/h$ is plotted against qb^4/Dh. For comparison the figure also includes the straight lines which represent the deflections calculated by using the theory of small deflections. Also included is the curve for $b/a = 0$, which represents deflections of an infinitely long plate calculated as explained in Art. 3 (see page 13). It can be seen that the deflections of finite plates with $b/a < \frac{2}{3}$ are very close to those obtained for an infinitely long plate.

Knowing the displacements as given by expressions (b), we can calculate the strain of the middle plane and the corresponding membrane stresses from Eqs. (b) of the

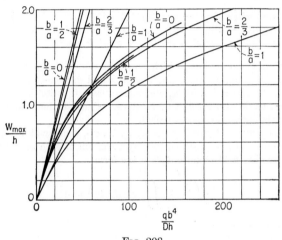

Fig. 208

preceding article. The bending stresses can then be found from Eqs. (101) and (102) for the bending and twisting moments. By adding the membrane and the bending stresses, we obtain the total stress. The maximum values of this stress are at the middle of the long sides of plates. They are given in graphical form in Fig. 209. For comparison, the figure also includes straight lines representing the stresses obtained by the theory of small deflections and a curve $b/a = 0$ representing the stresses for an infinitely long plate. It would seem reasonable to expect the total stress to be greater for $b/a = 0$ than for $b/a = \frac{1}{2}$ for any value of load. We see that the curve for $b/a = 0$ falls below the curves for $b/a = \frac{1}{2}$ and $b/a = \frac{2}{3}$. This is probably a result of approximations in the energy solution which arise out of the use of a finite number of constants. It indicates that the calculated stresses are in error on the safe side, i.e., that they are too large. The error for $b/a = \frac{1}{2}$ appears to be about 10 per cent.

The energy method can also be applied in the case of large deflections of simply supported rectangular plates. However, as may be seen from the foregoing discussion of the case of clamped edges, the application of this method requires a considerable amount of computation. To get an approximate solution for a simply supported rectangular plate, a simple method consisting of a combination of the known solutions given by the theory of small deflections and the membrane theory can be used.[1] This method will now be illustrated by a simple example of a square plate. We assume that the load q can be resolved into two parts q_1 and q_2 in such a manner that part q_1 is balanced by the bending and shearing stresses calculated by

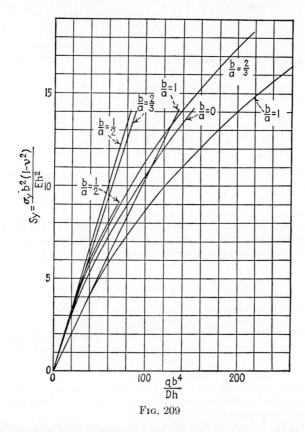

FIG. 209

the theory of small deflections, part q_2 being balanced by the membrane stresses. The deflection at the center as calculated for a square plate with sides $2a$ by the theory of small deflections is[2]

$$w_0 = 0.730 \frac{q_1 a^4}{Eh^3}$$

From this we determine

$$q_1 = \frac{w_0 Eh^3}{0.730 a^4} \qquad (e)$$

---

[1] This method is recommended by Föppl; see "Drang und Zwang," p. 345.
[2] The factor 0.730 is obtained by multiplying the number 0.00406, given in Table 8, by 16 and by $12(1 - \nu^2) = 11.25$.

Considering the plate as a membrane and using formula (250), we obtain

$$w_0 = 0.802a \sqrt[3]{\frac{q_2 a}{Eh}}$$

from which

$$q_2 = \frac{w_0^3 Eh}{0.516 a^4} \qquad (f)$$

The deflection w_0 is now obtained from the equation

$$q = q_1 + q_2 = \frac{w_0 Eh^3}{0.730 a^4} + \frac{w_0^3 Eh}{0.516 a^4}$$

which gives

$$q = \frac{w_0 Eh^3}{a^4} \left(1.37 + 1.94 \frac{w_0^2}{h^2}\right) \qquad (252)$$

After the deflection w_0 has been calculated from this equation, the loads q_1 and q_2 are found from Eqs. (e) and (f), and the corresponding stresses are calculated by using for q_1 the small deflection theory (see Art. 30) and for q_2, Eq. (251). The total stress is then the sum of the stresses due to the loads q_1 and q_2.

Another approximate method of practical interest is based on consideration of the expression (248) for the strain energy due to the stretching of the middle surface of the plate.[1] This expression can be put in the form

$$V = \frac{Eh}{2(1-\nu^2)} \iint [e^2 - 2(1-\nu)e_2]\, dx\, dy \qquad (g)$$

in which

$$e = \epsilon_x + \epsilon_y \qquad e_2 = \epsilon_x \epsilon_y - \tfrac{1}{4}\gamma_{xy}^2$$

A similar expression can be written in polar coordinates, e_2 being, in case of axial symmetry, equal to $\epsilon_r \epsilon_t$. The energy of bending must be added, of course, to the energy (g) in order to obtain the total strain energy of the plate. Yet an examination of exact solutions, such as described in Art. 98, leads to the conclusion that terms of the differential equations due to the presence of the term e_2 in expression (g) do not much influence the final result.

Starting from the hypothesis that the term containing e_2 actually can be neglected in comparison with e^2, we arrive at the differential equation of the bent plate

$$\Delta\Delta w - \alpha^2\, \Delta w = \frac{q}{D} \qquad (h)$$

in which the quantity

$$\alpha^2 = \frac{12}{h^2}\left[\frac{\partial u}{\partial x} + \frac{\partial v}{\partial y} + \frac{1}{2}\left(\frac{\partial w}{\partial x}\right)^2 + \frac{1}{2}\left(\frac{\partial w}{\partial y}\right)^2\right] \qquad (i)$$

proves to be a constant. From Eqs. (b) of Art. 101 it follows that the dilatation $e = \epsilon_x + \epsilon_y$ then also remains constant throughout the middle surface of the bent plate. The problem in question, simplified in this way, thus becomes akin to problems discussed in Chap. 12.

---

[1] H. M. Berger, J. Appl. Mechanics, vol. 22, p. 465, 1955.

For a circular plate under symmetrical loading, Eq. (*i*) must be replaced by

$$\alpha^2 = \frac{12}{h^2}\left[\frac{du}{dr} + \frac{u}{r} + \frac{1}{2}\left(\frac{dw}{dr}\right)^2\right] \qquad (j)$$

In this latter case the constants of integration of Eq. (*h*) along with the constant α allow us to fulfill all conditions prescribed on the boundary of the plate. However, for a more accurate calculation of the membrane stresses N_r, N_t from the deflections, the first of the equations (231) should be used in place of the relation (*j*).

The calculation of the membrane stresses in rectangular plates proves to be relatively more cumbersome. As a whole, however, the procedure still remains much simpler than the handling of the exact equations (245) and (246), and the numerical results, in cases discussed till now, prove to have an accuracy satisfactory for technical purposes. Nevertheless some reservation appears opportune in application of this method as long as the hypothesis providing its basis lacks a straight mechanical interpretation.

103. Large Deflections of Rectangular Plates with Simply Supported Edges. An exact solution[1] of this problem, treated in the previous article approximately, can be established by starting from the simultaneous equations (245) and (246).

The deflection of the plate (Fig. 59) may be taken in the Navier form

$$w = \sum_{m=1}^{\infty}\sum_{n=1}^{\infty} w_{mn} \sin\frac{m\pi x}{a} \sin\frac{n\pi y}{b} \qquad (a)$$

the boundary conditions with regard to the deflections and the bending moments thus being satisfied by any, yet unknown, values of the coefficients w_{mn}. The given lateral pressure may be expanded in a double Fourier series

$$q = \sum_{m=1}^{\infty}\sum_{n=1}^{\infty} q_{mn} \sin\frac{m\pi x}{a} \sin\frac{n\pi y}{b} \qquad (b)$$

A suitable expression for the Airy stress function, then, is

$$F = \frac{P_x y^2}{2bh} + \frac{P_y x^2}{2ah} + \sum_{m=0}^{\infty}\sum_{n=0}^{\infty} f_{mn}\cos\frac{m\pi x}{a}\cos\frac{n\pi y}{b} \qquad (c)$$

where P_x and P_y denote the total tension load applied on the sides $x = 0, a$ and $y = 0, b$, respectively. Substituting the expressions (*a*) and (*c*) into Eq. (245), we arrive at the following relation between the coefficients of both series:

$$f_{mn} = \frac{E}{4(m^2 b/a + n^2 a/b)^2}\sum b_{rspq} w_{rs} w_{pq} \qquad (d)$$

---

[1] Due to S. Levy, *NACA Tech. Note* 846, 1942, and *Proc. Symposia Appl. Math.*, vol. 1, p. 197, 1949. For application of the same method to plates with clamped edges see this latter paper and *NACA Tech. Notes* 847 and 852, 1942; for application to slightly curved plates under edge compression see J. M. Coan, *J. Appl. Mechanics*, vol. 18, p. 143, 1951. M. Stippes has applied the Ritz method to the case where the membrane forces vanish on the boundary and two opposite edges are supported; see *Proc. First Natl. Congr. Appl. Mech.*, Chicago, 1952, p. 339.

The sum includes all products for which $r \pm p = m$ and $s \pm q = n$. The coefficients b_{rspq} are given by the expression

$$b_{rspq} = 2rspq \pm (r^2q^2 + s^2p^2) \tag{e}$$

where the sign is positive for $r + p = m$ and $s - q = n$ or for $r - p = m$ and $s + q = n$, and is negative otherwise. Taking, for example, a square plate ($a = b$), we obtain

$$f_{2,4} = \frac{E}{1{,}600} (-4w_{1,1}w_{1,3} + 36w_{1,1}w_{3,3} + 36w_{1,1}w_{1,5} + 64w_{1,2}w_{1,6} \cdots)$$

It still remains to establish a relation between the deflections, the stress function, and the lateral loading. Inserting expressions (a), (b), and (c) into Eq. (246), we arrive at the equation

$$q_{mn} = Dw_{mn}\pi^4 \left(\frac{m^2}{a^2} + \frac{n^2}{b^2}\right)^2 + P_x w_{mn} \frac{m^2\pi^2}{a^2 b} + P_y w_{mn} \frac{n^2\pi^2}{ab^2} + \frac{h\pi^4}{4a^2b^2} \sum C_{rspq} f_{rs} w_{pq} \tag{f}$$

The summation includes, this time, all products for which $r \pm p = m$ and $s \pm q = n$, and the coefficients are given by

$$C_{rspq} = \pm (rq \pm sp)^2 \quad \text{if } r \neq 0 \text{ and } s \neq 0 \tag{g}$$

and are twice this value otherwise. The first sign is positive if either $r - p = m$ or $s - q = n$ (but not simultaneously), and is negative in all other cases. The second sign is positive if $r + p = m$ and $s - q = n$ or $r - p = m$ and $s + q = n$, and is negative otherwise. For example,

$$q_{1,3} = Dw_{1,3}\pi^4 \left(\frac{1}{a^2} + \frac{9}{b^2}\right)^2 + P_x w_{1,3} \frac{\pi^2}{a^2 b} + P_y w_{1,3} \frac{9\pi^2}{ab^2}$$
$$+ \frac{h\pi^4}{4a^2b^2} (-8f_{0,2}w_{1,1} - 8f_{0,2}w_{1,5} + 100f_{2,4}w_{3,1} - 64f_{2,2}w_{3,1} + \cdots)$$

In accordance with conditions occurring in airplane structures the plate is considered rigidly framed, all edges thus remaining straight[1] after deformation. Then the elongation of the plate, say in the direction x, is independent of y. By Eqs. (b) and (f) of Art. 101 its value is equal to

$$\delta_x = \int_0^a \frac{\partial u}{\partial x} dx = \int_0^a \left[\frac{1}{E}\left(\frac{\partial^2 F}{\partial y^2} - \nu \frac{\partial^2 F}{\partial x^2}\right) - \frac{1}{2}\left(\frac{\partial w}{\partial x}\right)^2\right] dx \tag{h}$$

Using the series (a) and (c), this yields

$$\delta_x = \frac{P_x a}{bhE} - \frac{\nu P_y}{hE} - \frac{\pi^2}{8a} \sum_{m=1}^{\infty} \sum_{n=1}^{\infty} m^2 w_{mn}^2 \tag{i}$$

i.e., an expression which in fact does not include y. Similarly, one obtains

$$\delta_y = \frac{P_y b}{ahE} - \frac{\nu P_x}{hE} - \frac{\pi^2}{8b} \sum_{m=1}^{\infty} \sum_{n=1}^{\infty} n^2 w_{mn}^2 \tag{j}$$

[1] A solution due to Kaiser, *loc. cit.*, is free from this restriction.

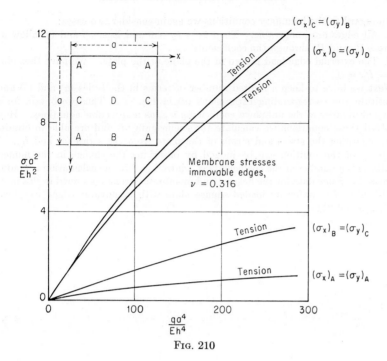

Fig. 210

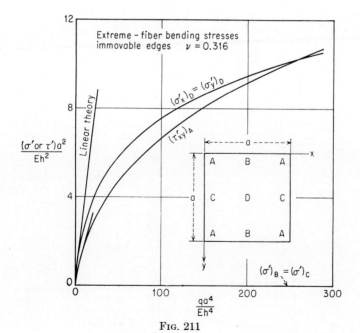

Fig. 211

With regard to the boundary conditions we again consider two cases:

1. All edges are immovable. Then $\delta_x = \delta_y = 0$ and Eqs. (*i*) and (*j*) allow us to express P_x and P_y through the coefficients w_{mn}.

2. The external edge load is zero in the plane of the plate. We have then simply $P_x = P_y = 0$.

Next we have to keep a limited number of terms in the series (*a*) and (*b*) and to substitute the corresponding expressions (*d*) in Eq. (*f*). Thus we obtain for any assumed number of the unknown coefficients w_{mn} as many cubic equations. Having resolved these equations we calculate the coefficients (*d*) and are able to obtain all data regarding the stress and strain of the plate from the series (*a*) and (*c*). The accuracy of the solution can be judged by observing the change in the numerical results as the number of the coefficients w_{mn} introduced in the calculation is gradually increased. Some data for the flexural and membrane stresses obtained in this manner in the case of a uniformly loaded square plate with immovable edges are given in Figs. 210 and 211.

CHAPTER 14

DEFORMATION OF SHELLS WITHOUT BENDING

104. Definitions and Notation. In the following discussion of the deformations and stresses in shells the system of notation is the same as that used in the discussion of plates. We denote the *thickness* of the shell by h, this quantity always being considered small in comparison with the other dimensions of the shell and with its radii of curvature. The surface that bisects the thickness of the plate is called the *middle surface*. By specifying the form of the middle surface and the thickness of the shell at each point, a shell is entirely defined geometrically.

To analyze the internal forces we cut from the shell an infinitely small element formed by two pairs of adjacent planes which are normal to the middle surface of the shell and which contain its principal curvatures (Fig. 212a). We take the coordinate axes x and y tangent at O to the lines of principal curvature and the axis z normal to the middle surface, as shown in the figure. The principal radii of curvature which lie in the xz and yz planes are denoted by r_x and r_y, respectively. The stresses acting on the plane faces of the element are resolved in the directions of the coordinate axes, and the stress components are denoted by our previous symbols σ_x, σ_y, $\tau_{xy} = \tau_{yx}$, τ_{xz}. With this notation[1] the resultant forces per unit length of the normal sections shown in Fig. 212b are

$$N_x = \int_{-h/2}^{+h/2} \sigma_x \left(1 - \frac{z}{r_y}\right) dz \qquad N_y = \int_{-h/2}^{+h/2} \sigma_y \left(1 - \frac{z}{r_x}\right) dz \qquad (a)$$

$$N_{xy} = \int_{-h/2}^{+h/2} \tau_{xy} \left(1 - \frac{z}{r_y}\right) dz \qquad N_{yx} = \int_{-h/2}^{+h/2} \tau_{yx} \left(1 - \frac{z}{r_x}\right) dz \qquad (b)$$

$$Q_x = \int_{-h/2}^{+h/2} \tau_{xz} \left(1 - \frac{z}{r_y}\right) dz \qquad Q_y = \int_{-h/2}^{+h/2} \tau_{yz} \left(1 - \frac{z}{r_x}\right) dz \qquad (c)$$

The small quantities z/r_x and z/r_y appear in expressions (a), (b), (c), because the lateral sides of the element shown in Fig. 212a have a trapezoidal form due to the curvature of the shell. As a result of this, the shearing forces N_{xy} and N_{yx} are generally not equal to each other, although

---

[1] In the cases of surfaces of revolution in which the position of the element is defined by the angles θ and φ (see Fig. 213) the subscripts θ and φ are used instead of x and y in notation for stresses, resultant forces, and resultant moments.

it still holds that $\tau_{xy} = \tau_{yx}$. In our further discussion we shall always assume that the thickness h is very small in comparison with the radii r_x, r_y and omit the terms z/r_x and z/r_y in expressions (a), (b), (c). Then $N_{xy} = N_{yx}$, and the resultant shearing forces are given by the same expressions as in the case of plates (see Art. 21).

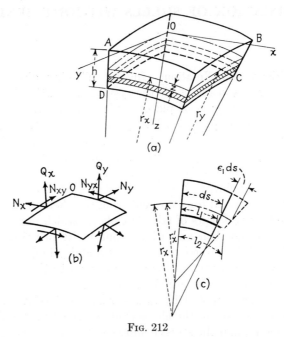

Fig. 212

The bending and twisting moments per unit length of the normal sections are given by the expressions

$$M_x = \int_{-h/2}^{+h/2} \sigma_x z \left(1 - \frac{z}{r_y}\right) dz \qquad M_y = \int_{-h/2}^{+h/2} \sigma_y z \left(1 - \frac{z}{r_x}\right) dz \qquad (d)$$

$$M_{xy} = -\int_{-h/2}^{+h/2} \tau_{xy} z \left(1 - \frac{z}{r_y}\right) dz \qquad M_{yx} = \int_{-h/2}^{+h/2} \tau_{yx} z \left(1 - \frac{z}{r_x}\right) dz \qquad (e)$$

in which the rule used in determining the directions of the moments is the same as in the case of plates. In our further discussion we again neglect the small quantities z/r_x and z/r_y, due to the curvature of the shell, and use for the moments the same expressions as in the discussion of plates.

In considering bending of the shell, we assume that linear elements, such as AD and BC (Fig. 212a), which are normal to the middle surface of the shell, remain straight and become normal to the deformed middle surface of the shell. Let us begin with a simple case in which, during bending, the lateral faces of the element $ABCD$ rotate only with respect to their lines of intersection with the middle surface. If r'_x and r'_y are the values of the radii of curvature after deformation, the unit elongations of a thin lamina at a distance z from the middle surface (Fig. 212a) are

$$\epsilon_x = -\frac{z}{1 - \dfrac{z}{r_x}}\left(\frac{1}{r'_x} - \frac{1}{r_x}\right) \qquad \epsilon_y = -\frac{z}{1 - \dfrac{z}{r_y}}\left(\frac{1}{r'_y} - \frac{1}{r_y}\right) \qquad (f)$$

If, in addition to rotation, the lateral sides of the element are displaced parallel to themselves, owing to stretching of the middle surface, and if the corresponding unit elongations of the middle surface in the x and y directions are denoted by ϵ_1 and ϵ_2, respectively, the elongation ϵ_x of the lamina considered above, as seen from Fig. 212c, is

$$\epsilon_x = \frac{l_2 - l_1}{l_1}$$

Substituting

$$l_1 = ds\left(1 - \frac{z}{r_x}\right) \qquad l_2 = ds(1 + \epsilon_1)\left(1 - \frac{z}{r'_x}\right)$$

we obtain

$$\epsilon_x = \frac{\epsilon_1}{1 - \dfrac{z}{r_x}} - \frac{z}{1 - \dfrac{z}{r_x}}\left[\frac{1}{(1 - \epsilon_1)r'_x} - \frac{1}{r_x}\right] \qquad (g)$$

A similar expression can be obtained for the elongation ϵ_y. In our further discussion the thickness h of the shell will be always assumed small in comparison with the radii of curvature. In such a case the quantities z/r_x and z/r_y can be neglected in comparison with unity. We shall neglect also the effect of the elongations ϵ_1 and ϵ_2 on the curvature.[1] Then, instead of such expressions as (g), we obtain

$$\epsilon_x = \epsilon_1 - z\left(\frac{1}{r'_x} - \frac{1}{r_x}\right) = \epsilon_1 - \chi_x z$$

$$\epsilon_y = \epsilon_2 - z\left(\frac{1}{r'_y} - \frac{1}{r_y}\right) = \epsilon_2 - \chi_y z$$

where χ_x and χ_y denote the changes of curvature. Using these expressions for the components of strain of a lamina and assuming that there are no normal stresses between laminae ($\sigma_z = 0$), the following expres-

---

[1] Similar simplifications are usually made in the theory of bending of thin curved bars. It can be shown in this case that the procedure is justifiable if the depth of the cross section h is small in comparison with the radius r, say $h/r < 0.1$; see S. Timoshenko, "Strength of Materials," part I, 3d ed., p. 370, 1955.

sions for the components of stress are obtained:

$$\sigma_x = \frac{E}{1-\nu^2}[\epsilon_1 + \nu\epsilon_2 - z(\chi_x + \nu\chi_y)]$$

$$\sigma_y = \frac{E}{1-\nu^2}[\epsilon_2 + \nu\epsilon_1 - z(\chi_y + \nu\chi_x)]$$

Substituting these expressions in Eqs. (a) and (d) and neglecting the small quantities z/r_x and z/r_y in comparison with unity, we obtain

$$N_x = \frac{Eh}{1-\nu^2}(\epsilon_1 + \nu\epsilon_2) \qquad N_y = \frac{Eh}{1-\nu^2}(\epsilon_2 + \nu\epsilon_1) \qquad (253)$$
$$M_x = -D(\chi_x + \nu\chi_y) \qquad M_y = -D(\chi_y + \nu\chi_x)$$

where D has the same meaning as in the case of plates [see Eq. (3)] and denotes the flexural rigidity of the shell.

A more general case of deformation of the element in Fig. 212 is obtained if we assume that, in addition to normal stresses, shearing stresses also are acting on the lateral sides of the element. Denoting by γ the shearing strain in the middle surface of the shell and by $\chi_{xy}\,dx$ the rotation of the edge BC relative to Oz about the x axis (Fig. 212a) and proceeding as in the case of plates [see Eq. (42)], we find

$$\tau_{xy} = (\gamma - 2z\chi_{xy})G$$

Substituting this in Eqs. (b) and (e) and using our previous simplifications, we obtain

$$N_{xy} = N_{yx} = \frac{\gamma hE}{2(1+\nu)} \qquad (254)$$
$$M_{xy} = -M_{yx} = D(1-\nu)\chi_{xy}$$

Thus assuming that during bending of a shell the linear elements normal to the middle surface remain straight and become normal to the deformed middle surface, we can express the resultant forces per unit length N_x, N_y, and N_{xy} and the moments M_x, M_y, and M_{xy} in terms of six quantities: the three components of strain ϵ_1, ϵ_2, and γ of the middle surface of the shell and the three quantities χ_x, χ_y, and χ_{xy} representing the changes of curvature and the twist of the middle surface.

In many problems of deformation of shells the bending stresses can be neglected, and only the stresses due to strain in the middle surface of the shell need be considered. Take, as an example, a thin spherical container submitted to the action of a uniformly distributed internal pressure normal to the surface of the shell. Under this action the middle surface of the shell undergoes a uniform strain; and since the thickness of the shell is small, the tensile stresses can be assumed as uniformly distributed across the thickness. A similar example is afforded by a thin circular cylindrical container in which a gas or a liquid is compressed by means of pistons which move freely along the axis of the cylinder. Under the action of a uniform internal pressure the hoop stresses that are produced in the cylindrical shell are uniformly distributed over the thickness of the shell. If the ends of the cylinder are built in along the edges, the shell is no longer free to expand laterally, and some bending must occur near the built-in edges when internal pressure is applied. A more complete investigation shows, however (see Art. 114), that this bending is of a local character and that the portion of the shell at some distance from the ends continues to remain cylindrical and undergoes only strain in the middle surface without appreciable bending.

If the conditions of a shell are such that bending can be neglected, the problem of stress analysis is greatly simplified, since the resultant moments (d) and (e) and the resultant shearing forces (c) vanish. Thus the only unknowns are the three quantities N_x, N_y, and $N_{xy} = N_{yx}$, which can be determined from the conditions of equilibrium of an element, such as shown in Fig. 212. Hence the problem becomes statically determinate if all the forces acting on the shell are known. The forces N_x, N_y, and N_{xy} obtained in this manner are sometimes called *membrane forces*, and the theory of shells based on the omission of bending stresses is called *membrane theory*. The application of this theory to various particular cases will be discussed in the remainder of this chapter.

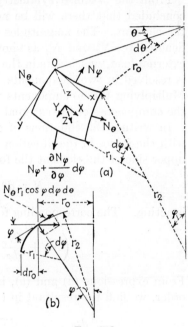

Fig. 213

105. Shells in the Form of a Surface of Revolution and Loaded Symmetrically with Respect to Their Axis. Shells that have the form of surfaces of revolution find extensive application in various kinds of containers, tanks, and domes. A surface of revolution is obtained by rotation of a plane curve about an axis lying in the plane of the curve. This curve is called the *meridian*, and its plane is a *meridian plane*. An element of a shell is cut out by two adjacent meridians and two parallel circles, as shown in Fig. 213a. The position of a meridian is defined by an angle θ, measured from some datum meridian plane; and the position of a parallel circle is defined by the angle φ, made by the normal to the surface and the axis of rotation. The meridian plane and the plane perpendicular to the meridian are the planes of principal curvature at a point of a surface of revolution, and the corresponding radii of curvature are denoted by r_1 and r_2, respectively. The radius of the parallel circle is denoted by r_0 so that the length of the sides of the element meeting at O, as shown in the figure, are $r_1 \, d\varphi$ and $r_0 \, d\theta = r_2 \sin \varphi \, d\theta$. The surface area of the element is then $r_1 r_2 \sin \varphi \, d\varphi \, d\theta$.

From the assumed symmetry of loading and deformation it can be concluded that there will be no shearing forces acting on the sides of the element. The magnitudes of the normal forces per unit length are denoted by N_φ and N_θ as shown in the figure. The intensity of the external load, which acts in the meridian plane, in the case of symmetry is resolved in two components Y and Z parallel to the coordinate axes. Multiplying these components with the area $r_1 r_2 \sin \varphi \, d\varphi \, d\theta$, we obtain the components of the external load acting on the element.

In writing the equations of equilibrium of the element, let us begin with the forces in the direction of the tangent to the meridian. On the upper side of the element the force

$$N_\varphi r_0 \, d\theta = N_\varphi r_2 \sin \varphi \, d\theta \qquad (a)$$

is acting. The corresponding force on the lower side of the element is

$$\left(N_\varphi + \frac{dN_\varphi}{d\varphi} d\varphi\right)\left(r_0 + \frac{dr_0}{d\varphi} d\varphi\right) d\theta \qquad (b)$$

From expressions (a) and (b), by neglecting a small quantity of second order, we find the resultant in the y direction to be equal to

$$N_\varphi \frac{dr_0}{d\varphi} d\varphi \, d\theta + \frac{dN_\varphi}{d\varphi} r_0 \, d\varphi \, d\theta = \frac{d}{d\varphi}(N_\varphi r_0) \, d\varphi \, d\theta \qquad (c)$$

The component of the external force in the same direction is

$$Y r_1 r_0 \, d\varphi \, d\theta \qquad (d)$$

The forces acting on the lateral sides of the element are equal to $N_\theta r_1 \, d\varphi$ and have a resultant in the direction of the radius of the parallel circle equal to $N_\theta r_1 \, d\varphi \, d\theta$. The component of this force in the y direction (Fig. 213b) is

$$-N_\theta r_1 \cos \varphi \, d\varphi \, d\theta \qquad (e)$$

Summing up the forces (c), (d), and (e), the equation of equilibrium in the direction of the tangent to the meridian becomes

$$\frac{d}{d\varphi}(N_\varphi r_0) - N_\theta r_1 \cos \varphi + Y r_1 r_0 = 0 \qquad (f)$$

DEFORMATION OF SHELLS WITHOUT BENDING

The second equation of equilibrium is obtained by summing up the projections of the forces in the z direction. The forces acting on the upper and lower sides of the element have a resultant in the z direction equal to

$$N_\varphi r_0 \, d\theta \, d\varphi \quad (g)$$

The forces acting on the lateral sides of the element and having the resultant $N_\theta r_1 \, d\varphi \, d\theta$ in the radial direction of the parallel circle give a component in the z direction of the magnitude

$$N_\theta r_1 \sin \varphi \, d\varphi \, d\theta \quad (h)$$

The external load acting on the element has in the same direction a component

$$Z r_1 r_0 \, d\theta \, d\varphi \quad (i)$$

Summing up the forces (g), (h), and (i), we obtain the second equation of equilibrium

$$N_\varphi r_0 + N_\theta r_1 \sin \varphi + Z r_1 r_0 = 0 \quad (j)$$

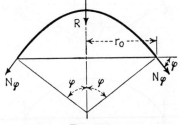

Fig. 214

From the two Eqs. (f) and (j) the forces N_θ and N_φ can be calculated in each particular case if the radii r_0 and r_1 and the components Y and Z of the intensity of the external load are given.

Instead of the equilibrium of an element, the equilibrium of the portion of the shell above the parallel circle defined by the angle φ may be considered (Fig. 214). If the resultant of the total load on that portion of the shell is denoted by R, the equation of equilibrium is

$$2\pi r_0 N_\varphi \sin \varphi + R = 0 \quad (255)$$

This equation can be used instead of the differential equation (f), from which it can be obtained by integration. If Eq. (j) is divided by $r_1 r_0$, it can be written in the form

$$\frac{N_\varphi}{r_1} + \frac{N_\theta}{r_2} = -Z \quad (256)$$

It is seen that when N_φ is obtained from Eq. (255), the force N_θ can be calculated from Eq. (256). Hence the problem of membrane stresses can be readily solved in each particular case. Some applications of these equations will be discussed in the next article.

106. Particular Cases of Shells in the Form of Surfaces of Revolution.[1]

Spherical Dome. Assume that a spherical shell (Fig. 215a) is submitted to the action of its own weight, the magnitude of which per unit area is constant and equal to q. Denoting the radius of the sphere by a, we have $r_0 = a \sin \varphi$ and

$$R = 2\pi \int_0^\varphi a^2 q \sin \varphi \, d\varphi = 2\pi a^2 q(1 - \cos \varphi)$$

Equations (255) and (256) then give

$$N_\varphi = -\frac{aq(1 - \cos \varphi)}{\sin^2 \varphi} = -\frac{aq}{1 + \cos \varphi}$$
$$N_\theta = aq \left(\frac{1}{1 + \cos \varphi} - \cos \varphi \right) \qquad (257)$$

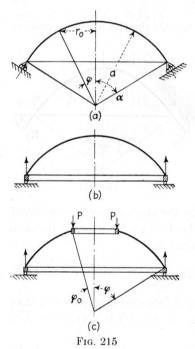

FIG. 215

It is seen that the forces N_φ are always negative. There is thus a compression along the meridians that increases as the angle φ increases. For $\varphi = 0$ we have $N_\varphi = -aq/2$, and for $\varphi = \pi/2$, $N_\varphi = -aq$. The forces N_θ are also negative for small angles φ. When

$$\frac{1}{1 + \cos \varphi} - \cos \varphi = 0$$

i.e., for $\varphi = 51°50'$, N_θ becomes equal to zero and, with further increase of φ, becomes positive. This indicates that for φ greater than $51°50'$ there are tensile stresses in the direction perpendicular to the meridians.

The stresses as calculated from (257) will represent the actual stresses in the shell with great accuracy[2] if the supports are of such a type that the reactions are tangent to meridians (Fig. 215a). Usually the arrangement is such that only vertical reactions are imposed on the dome by the supports, whereas the horizontal components of the forces N_φ are taken by a

---
[1] Examples of this kind can be found in the book by A. Pflüger, "Elementare Schalenstatik," Berlin, 1957; see also P. Forchheimer, "Die Berechnung ebener und gekrümmter Behälterböden," 3d ed., Berlin, 1931, and J. W. Geckeler's article in "Handbuch der Physik," vol. 6, Berlin, 1928.

[2] Small bending stresses due to strain of the middle surface will be discussed in Chap. 16.

supporting ring (Fig. 215b) which undergoes a uniform circumferential extension. Since this extension is usually different from the strain along the parallel circle of the shell, as calculated from expressions (257), some bending of the shell will occur near the supporting ring. An investigation of this bending[1] shows that in the case of a thin shell it is of a very localized character and that at a certain distance from the supporting ring Eqs. (257) continue to represent the stress conditions in the shell with satisfactory accuracy.

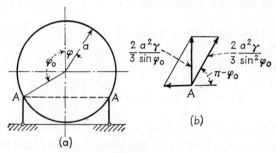

Fig. 216

Very often the upper portion of a spherical dome is removed, as shown in Fig. 215c, and an upper reinforcing ring is used to support the upper structure. If $2\varphi_0$ is the angle corresponding to the opening and P is the vertical load per unit length of the upper reinforcing ring, the resultant R corresponding to an angle φ is

$$R = 2\pi \int_{\varphi_0}^{\varphi} a^2 q \sin \varphi \, d\varphi + 2\pi Pa \sin \varphi_0$$

From Eqs. (255) and (256) we then find

$$N_\varphi = -aq \frac{\cos \varphi_0 - \cos \varphi}{\sin^2 \varphi} - P \frac{\sin \varphi_0}{\sin^2 \varphi}$$
$$N_\theta = aq \left(\frac{\cos \varphi_0 - \cos \varphi}{\sin^2 \varphi} - \cos \varphi \right) + P \frac{\sin \varphi_0}{\sin^2 \varphi} \quad (258)$$

As another example of a spherical shell let us consider a spherical tank supported along a parallel circle AA (Fig. 216) and filled with liquid of a specific weight γ. The inner pressure for any angle φ is given by the

[1] See Art. 131. It should be noted, however, that in the case of a negative or zero curvature of the shell ($r_1 r_2 \leq 0$) bending stresses due to the edge effect are not necessarily restricted to the edge zone of the shell. See, for instance, W. Flügge, "Statik und Dynamik der Schalen," p. 65, 2d ed., Berlin, 1957. The limitations of the membrane theory of shells are discussed in detail by A. L. Goldenveiser, "Theory of Elastic Thin Shells," p. 423, Moscow, 1953. The compatibility of a membrane state of stress under a given load with given boundary conditions was also discussed by E. Behlendorff, Z. angew. Math. Mech., vol. 36, p. 399, 1956.

expression[1]

$$p = -Z = \gamma a(1 - \cos \varphi)$$

The resultant R of this pressure for the portion of the shell defined by an angle φ is

$$R = -2\pi a^2 \int_0^\varphi \gamma a(1 - \cos \varphi) \sin \varphi \cos \varphi \, d\varphi$$
$$= -2\pi a^3 \gamma [\tfrac{1}{6} - \tfrac{1}{2} \cos^2 \varphi (1 - \tfrac{2}{3} \cos \varphi)]$$

Substituting in Eq. (255), we obtain

$$N_\varphi = \frac{\gamma a^2}{6 \sin^2 \varphi}[1 - \cos^2 \varphi(3 - 2 \cos \varphi)] = \frac{\gamma a^2}{6}\left(1 - \frac{2 \cos^2 \varphi}{1 + \cos \varphi}\right) \quad (259)$$

and from Eq. (256) we find that

$$N_\theta = \frac{\gamma a^2}{6}\left(5 - 6 \cos \varphi + \frac{2 \cos^2 \varphi}{1 + \cos \varphi}\right) \quad (260)$$

Equations (259) and (260) hold for $\varphi < \varphi_0$. In calculating the resultant R for larger values of φ, that is, for the lower portion of the tank, we must take into account not only the internal pressure but also the sum of the vertical reactions along the ring AA. This sum is evidently equal to the total weight of the liquid $4\pi a^3 \gamma/3$. Hence

$$R = -\tfrac{4}{3}\pi a^3 \gamma - 2\pi a^3 \gamma[\tfrac{1}{6} - \tfrac{1}{2} \cos^2 \varphi(1 - \tfrac{2}{3} \cos \varphi)]$$

Substituting in Eq. (255), we obtain

$$N_\varphi = \frac{\gamma a^2}{6}\left(5 + \frac{2 \cos^2 \varphi}{1 - \cos \varphi}\right) \quad (261)$$

and from Eq. (256),

$$N_\theta = \frac{\gamma a^2}{6}\left(1 - 6 \cos \varphi - \frac{2 \cos^2 \varphi}{1 - \cos \varphi}\right) \quad (262)$$

Comparing expressions (259) and (261), we see that along the supporting ring AA the forces N_φ change abruptly by an amount equal to $2\gamma a^2/(3 \sin^2 \varphi_0)$. The same quantity is also obtained if we consider the vertical reaction per unit length of the ring AA and resolve it into two components (Fig. 216b): one in the direction of the tangent to the meridian and the other in the horizontal direction. The first of these components is equal to the abrupt change in the magnitude of N_φ mentioned above; the horizontal component represents the reaction on the supporting ring which produces in it a uniform compression. This compression can be eliminated if we use members in the direction of tangents to the meridians instead of vertical supporting members, as shown in Fig. 216a. As may

---

[1] A uniform pressure producing a uniform tension in the spherical shell can be superposed without any complication on this pressure.

be seen from expressions (260) and (262), the forces N_θ also experience an abrupt change at the circle AA. This indicates that there is an abrupt change in the circumferential expansion on the two sides of the parallel circle AA. Thus the membrane theory does not satisfy the condition of continuity at the circle AA, and we may expect some local bending to take place near the supporting ring.

Conical Shell. In this case certain membrane stresses can be produced by a force applied at the top of the cone. If a force P is applied in the direction of the axis of the cone, the stress distribution is symmetrical, and from Fig. 217 we obtain

$$N_\varphi = -\frac{P}{2\pi r_0 \cos \alpha} \qquad (a)$$

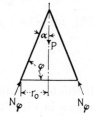

Fig. 217

Equation (256) then gives $N_\theta = 0$. The case of a force applied at the top in the direction of a generatrix will be discussed in Art. 110 and the loading of the shell by its weight in Art. 133.

If lateral forces are symmetrically distributed over the conical surface, the membrane stresses can be calculated by using Eqs. (255) and (256). Since the curvature of the meridian in the case of a cone is zero, $r_1 = \infty$; we can write these equations in the following form:

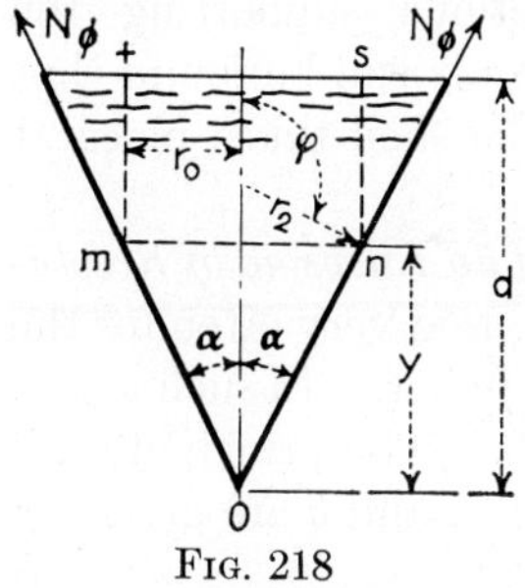

$$N_\varphi = -\frac{R}{2\pi r_0 \sin \varphi}$$
$$N_\theta = -Zr_2 = -\frac{Zr_0}{\sin \varphi} \qquad (b)$$

Fig. 218

Each of the resultant forces N_φ and N_θ can be calculated independently provided the load distribution is known. As an example, we take the case of the conical tank filled with a liquid of specific weight γ as shown in Fig. 218. Measuring the distances y from the bottom of the tank and denoting by d the total depth of the liquid in the tank, the pressure at any parallel circle mn is

$$p = -Z = \gamma(d - y)$$

Also, for such a tank $\varphi = (\pi/2) + \alpha$ and $r_0 = y \tan \alpha$. Substituting in the second of the equations (b), we obtain

$$N_\theta = \frac{\gamma(d-y)y \tan \alpha}{\cos \alpha} \qquad (c)$$

This force is evidently a maximum when $y = d/2$, and we find

$$(N_\theta)_{max} = \frac{\gamma d^2 \tan \alpha}{4 \cos \alpha}$$

In calculating the force N_φ we observe that the load R in the first of the equations (b) is numerically equal to the weight of the liquid in the conical part mno together with the weight of the liquid in the cylindrical part $mnst$. Hence

$$R = -\pi\gamma y^2(d - y + \tfrac{1}{3}y)\tan^2\alpha$$

and we obtain

$$N_\varphi = \frac{\gamma y(d - \tfrac{2}{3}y)\tan\alpha}{2\cos\alpha} \qquad (d)$$

This force becomes a maximum when $y = \tfrac{3}{4}d$, at which point

$$(N_\varphi)_{\max} = \frac{3}{16}\frac{d^2\gamma\tan\alpha}{\cos\alpha}$$

If the forces supporting the tank are in the direction of generatrices, as shown in Fig. 218, expressions (c) and (d) represent the stress conditions in the shell with great accuracy. Usually there will be a reinforcing ring along the upper edge of the tank. This ring takes the horizontal components of the forces N_φ; the vertical components of the same forces constitute the reactions supporting the tank. In such a case it will be found that a local bending of the shell takes place at the reinforcing ring.

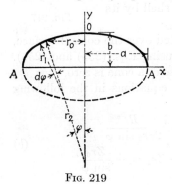

Fig. 219

Shell in the Form of an Ellipsoid of Revolution. Such a shell is used very often for the ends of a cylindrical boiler. In such a case a half of the ellipsoid is used, as shown in Fig. 219. The principal radii of curvature in the case of an ellipse with semiaxes a and b are given by the formulas

$$r_1 = \frac{a^2 b^2}{(a^2\sin^2\varphi + b^2\cos^2\varphi)^{\frac{3}{2}}} \qquad r_2 = \frac{a^2}{(a^2\sin^2\varphi + b^2\cos^2\varphi)^{\frac{1}{2}}} \qquad (e)$$

or, by using the orthogonal coordinates x and y shown in the figure,

$$r_1 = r_2^3\frac{b^2}{a^4} \qquad r_2 = \frac{(a^4 y^2 + b^4 x^2)^{\frac{1}{2}}}{b^2} \qquad (f)$$

If the principal curvatures are determined from Eqs. (e) or (f), the forces N_φ and N_θ are readily found from Eqs. (255) and (256). Let p be the uniform steam pressure in the boiler. Then for a parallel circle of a radius r_0 we have $R = -\pi p r_0^2$, and Eq. (255) gives

$$N_\varphi = \frac{pr_0}{2\sin\varphi} = \frac{pr_2}{2} \qquad (263)$$

Substituting in Eq. (256), we find

$$N_\theta = r_2 p - \frac{r_2}{r_1} N_\varphi = p \left(r_2 - \frac{r_2^2}{2r_1} \right) \quad (264)$$

At the top of the shell (point O) we have $r_1 = r_2 = a^2/b$, and Eqs. (263) and (264) give

$$N_\varphi = N_\theta = \frac{pa^2}{2b} \quad (g)$$

At the equator AA we have $r_1 = b^2/a$ and $r_2 = a$; hence

$$N_\varphi = \frac{pa}{2} \qquad N_\theta = pa \left(1 - \frac{a^2}{2b^2} \right) \quad (h)$$

It is seen that the forces N_φ are always positive, whereas the forces N_θ become negative at the equator if

$$a^2 > 2b^2 \quad (i)$$

In the particular case of a sphere, $a = b$; and we find in all points $N_\varphi = N_\theta = pa/2$.

Shell in Form of a Torus. If a torus is obtained by rotation of a circle of radius a about a vertical axis (Fig. 220), the forces N_φ are obtained by

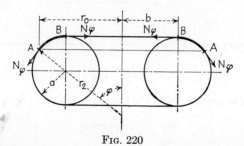

Fig. 220

considering the equilibrium of the ring-shaped portion of the shell represented in the figure by the heavy line AB. Since the forces N_φ along the parallel circle BB are horizontal, we need consider only the forces N_φ along the circle AA and the external forces acting on the ring when discussing equilibrium in the vertical direction. Assuming that the shell is submitted to the action of uniform internal pressure p, we obtain the equation of equilibrium

$$2\pi r_0 N_\varphi \sin \varphi = \pi p (r_0^2 - b^2)$$

from which

$$N_\varphi = \frac{p(r_0^2 - b^2)}{2r_0 \sin \varphi} = \frac{pa(r_0 + b)}{2r_0} \quad (265)$$

Substituting this expression in Eq. (256), we find[1]

$$N_\theta = \frac{pr_2(r_0 - b)}{2r_0} = \frac{pa}{2} \qquad (266)$$

A torus of an elliptical cross section may be treated in a similar manner.

107. Shells of Constant Strength. As a first example of a shell of constant strength, let us consider a dome of nonuniform thickness supporting its own weight. The weight of the shell per unit area of the middle surface is γh, and the two components of this weight along the coordinate axes are

$$Y = \gamma h \sin \varphi \qquad Z = \gamma h \cos \varphi \qquad (a)$$

In the case of a shell of constant strength the form of the meridians is determined in such a way that the compressive stress is constant and equal to σ in all the directions in the middle surface, *i.e.*, so that

$$N_\varphi = N_\theta = -\sigma h$$

Substituting in Eq. (256), we find

$$\sigma h \left(\frac{1}{r_1} + \frac{1}{r_2} \right) = \gamma h \cos \varphi \qquad (b)$$

or, by substituting $r_2 = r_0 \sin \varphi$ and solving for r_1,

$$r_1 = \frac{r_0}{\dfrac{\gamma}{\sigma} r_0 \cos \varphi - \sin \varphi} \qquad (c)$$

From Fig. 213b, we have

$$r_1 \, d\varphi = \frac{dr_0}{\cos \varphi}$$

Thus Eq. (c) can be represented in the form

$$\frac{dr_0}{d\varphi} = \frac{r_0 \cos \varphi}{\dfrac{\gamma}{\sigma} r_0 \cos \varphi - \sin \varphi} \qquad (d)$$

At the top of the dome where $\varphi = 0$, the right-hand side of the equation becomes indefinite. To remove this difficulty we use Eq. (b). Because of the conditions of symmetry at the top, $r_1 = r_2$, and we conclude that

$$r_1 = r_2 = \frac{2\sigma}{\gamma} \qquad \text{and} \qquad dr_0 = r_1 \, d\varphi = \frac{2\sigma}{\gamma} d\varphi$$

[1] Nevertheless, a consideration of the deformation of the shell shows that bending stresses inevitably must arise near the crown $r_0 = b$ of the shell, and this in spite of the lack of any singularity either in the shape of the shell surface or in the distribution of the loading. See W. R. Dean, *Phil. Mag.*, ser. 7, vol. 28, p. 452, 1939, and also Flügge, *op. cit.*, p. 81.

Hence, for the top of the dome we have

$$\frac{dr_0}{d\varphi} = \frac{2\sigma}{\gamma} \qquad (e)$$

Using Eqs. (e) and (d), we can obtain the shape of the meridian by numerical integration, starting from the top of the dome and calculating for each increment $\Delta\varphi$ of the angle φ the corresponding increment Δr_0 of the radius r_0. To find the variation of the thickness of the shell, Eq. (f), Art. 105, must be used. Substituting $N_\varphi = N_\theta = -\sigma h$ in this equation and observing that σ is constant, we obtain

$$-\frac{d}{d\varphi}(hr_0) + hr_1 \cos\varphi + \frac{\gamma}{\sigma} r_1 r_0 h \sin\varphi = 0 \qquad (f)$$

Substituting expression (c) for r_1, the following equation is obtained:

$$\frac{d}{d\varphi}(hr_0) = hr_0 \frac{\cos\varphi + \dfrac{\gamma}{\sigma} r_0 \sin\varphi}{\dfrac{\gamma}{\sigma} r_0 \cos\varphi - \sin\varphi} \qquad (g)$$

For $\varphi = 0$, we obtain from Eq. (f)

$$\frac{d}{d\varphi}(hr_0) \approx hr_1 = h\frac{dr_0}{d\varphi}$$

It is seen that for the first increment $\Delta\varphi$ of the angle φ any constant value for h can be taken. Then for the other points of the meridian the thickness is found by the numerical integration of Eq. (g). In Fig. 221 the result of such a calculation is represented.[1] It is seen that the condition

$$N_\theta = N_\varphi = -\sigma h$$

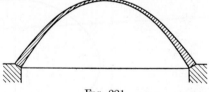

Fig. 221

brings us not only to a definite form of the middle surface of the dome but also to a definite law of variation of the thickness of the dome along the meridian.

In the case of a tank of equal strength that contains a liquid with a pressure γd at the upper point A (Fig. 222) we must find a shape of the meridian such that an internal pressure equal to γz will give rise at all points of the shell to forces[2]

$$N_\varphi = N_\theta = \text{const}$$

A similar problem is encountered in finding the shape of a drop of liquid resting on a horizontal plane. Because of the capillary forces a thin surface film of uniform tension is formed which envelops the liquid and prevents it from spreading over the supporting surface. Both problems are mathematically identical.

[1] This example has been calculated by Flügge, *op. cit.*, p. 38.
[2] A mathematical discussion of this problem is given in the book by C. Runge and H. König, "Vorlesungen über numerisches Rechnen," p. 320, Berlin, 1924.

In such cases, Eq. (256) gives

$$N_\varphi \left(\frac{1}{r_1} + \frac{1}{r_2} \right) = \gamma z \tag{h}$$

Taking the orthogonal coordinates as shown in the figure, we have

$$r_2 = \frac{x}{\sin \varphi} \qquad r_1 \, d\varphi = ds = \frac{dx}{\cos \varphi}$$

Hence

$$\frac{1}{r_2} = \frac{\sin \varphi}{x} \qquad \frac{1}{r_1} = \frac{\cos \varphi \, d\varphi}{dx} = \frac{d \sin \varphi}{dx}$$

and Eq. (h) gives

$$\frac{d \sin \varphi}{dx} + \frac{\sin \varphi}{x} = \frac{\gamma z}{N_\varphi} \tag{i}$$

Observing that

$$\tan \varphi = \frac{dz}{dx} \quad \text{and} \quad \sin \varphi = \frac{\tan \varphi}{\sqrt{1 + \tan^2 \varphi}} \tag{j}$$

it is possible to eliminate $\sin \varphi$ from Eq. (i) and obtain in this way a differential equation for z as a function of x. The equation obtained in this manner is very

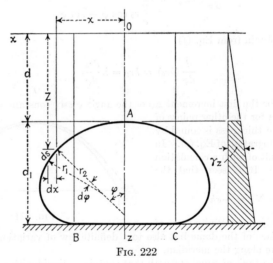

Fig. 222

complicated, and a simpler means of solving the problem is to introduce a new variable $u = \sin \varphi$. Making this substitution in Eqs. (i) and (j), we obtain

$$\frac{du}{dx} + \frac{u}{x} = \frac{\gamma z}{N_\varphi} \tag{k}$$

$$\frac{dz}{dx} = \frac{u}{\sqrt{1 - u^2}} \tag{l}$$

These equations can be integrated numerically starting from the upper point A of the tank. At this point, from symmetry, $r_1 = r_2$, and we find from Eq. (h) that

$$r_1 = \frac{2N_\varphi}{\gamma d}$$

By introducing the notation

$$\frac{N_\varphi}{\gamma} = a^2$$

we write

$$r_1 = \frac{2a^2}{d} \qquad (m)$$

With this radius we make the first element of the meridian curve $r_1 \Delta\varphi = \Delta x$, corresponding to the small angle $\Delta\varphi$. At the end of this arc we have, as for a small arc of a circle,

$$z \approx d + r_1 \frac{(\Delta\varphi)^2}{2} = d + \frac{(\Delta x)^2}{2r_1} = d\left[1 + \frac{(\Delta x)^2}{4a^2}\right]$$

$$u \approx \frac{\Delta x}{r_1} = d\frac{\Delta x}{2a^2} \qquad (n)$$

When the values u and z have been found from Eqs. (n), the values of du/dx and dz/dx for the same point are found from Eqs. (k) and (l). With these values of the derivatives we can calculate the values of z and u at the end of the next interval, and so on. Such calculations can be continued without difficulty up to an angle φ equal, say, to 50°, at which the value of u becomes approximately 0.75. From this point on and up to $\varphi = 140°$ the increments of z are much longer than the corresponding increments of x, and it is advantageous to take z as the independent variable instead of x. For $\varphi > 140°$, x must again be taken as the independent variable, and the calculation is continued up to point B, where the meridian curve has the horizontal tangent BC. Over the circular area BC the tank has a horizontal surface of contact with the foundation, and the pressure $\gamma(d + d_1)$ is balanced by the reaction of the foundation.

A tank designed in this manner[1] is a tank of constant strength only if the pressure at A is such as assumed in the calculations. For any other value of this pressure the forces N_θ and N_φ will no longer be constant but will vary along the meridian. Their magnitude can then be calculated by using the general equations (255) and (256). It will also be found that the equilibrium of the tank requires that vertical shearing forces act along the parallel circle BC. This indicates that close to this circle a local bending of the wall of the tank must take place.

108. Displacements in Symmetrically Loaded Shells Having the Form of a Surface of Revolution.

In the case of symmetrical deformation of a shell, a small displacement of a point can be resolved into two components: v in the direction of the tangent to the meridian and w in the direction of the normal to the middle surface. Considering an element AB of the meridian (Fig. 223), we see that the increase of the length of the element due to tangential displacements v and $v + (dv/d\varphi) d\varphi$ of its ends is equal to $(dv/d\varphi) d\varphi$. Because of the radial displacements w of the points A and B the length of the element decreases by an amount $w \, d\varphi$. The change in the length of the element due to the difference in the radial displacements of the points A and B can be neglected as a small quantity

---

[1] Tanks of this kind were constructed by the Chicago Bridge and Iron Works; see C. L. Day, *Eng. News-Record*, vol. 103, p. 416, 1929.

of higher order. Thus the total change in length of the element AB due to deformation is

$$\frac{dv}{d\varphi} d\varphi - w\, d\varphi$$

Dividing this by the initial length $r_1\, d\varphi$ of the element, we find the strain of the shell in the meridional direction to be

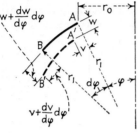

Fig. 223

$$\epsilon_\varphi = \frac{1}{r_1} \frac{dv}{d\varphi} - \frac{w}{r_1} \qquad (a)$$

Considering an element of a parallel circle it may be seen (Fig. 223) that owing to displacements v and w the radius r_0 of the circle increases by the amount

$$v \cos \varphi - w \sin \varphi$$

The circumference of the parallel circle increases in the same proportion as its radius; hence

$$\epsilon_\theta = \frac{1}{r_0} (v \cos \varphi - w \sin \varphi)$$

or, substituting $r_0 = r_2 \sin \varphi$,

$$\epsilon_\theta = \frac{v}{r_2} \cot \varphi - \frac{w}{r_2} \qquad (b)$$

Eliminating w from Eqs. (a) and (b), we obtain for v the differential equation

$$\frac{dv}{d\varphi} - v \cot \varphi = r_1 \epsilon_\varphi - r_2 \epsilon_\theta \qquad (c)$$

The strain components ϵ_φ and ϵ_θ can be expressed in terms of the forces N_φ and N_θ by applying Hooke's law. This gives

$$\epsilon_\varphi = \frac{1}{Eh}(N_\varphi - \nu N_\theta)$$
$$\epsilon_\theta = \frac{1}{Eh}(N_\theta - \nu N_\varphi) \qquad (d)$$

Substituting in Eq. (c), we obtain

$$\frac{dv}{d\varphi} - v \cot \varphi = \frac{1}{Eh}[N_\varphi(r_1 + \nu r_2) - N_\theta(r_2 + \nu r_1)] \qquad (267)$$

In each particular case the forces N_φ and N_θ can be found from the loading conditions, and the displacement v will then be obtained by integration of the differential equation (267). Denoting the right-hand side of this equation by $f(\varphi)$, we write

$$\frac{dv}{d\varphi} - v \cot \varphi = f(\varphi)$$

The general solution of this equation is

$$v = \sin \varphi \left[\int \frac{f(\varphi)}{\sin \varphi} d\varphi + C \right] \quad (e)$$

in which C is a constant of integration to be determined from the condition at the support.

Take, as an example, a spherical shell of constant thickness loaded by its own weight (Fig. 215a). In such a case $r_1 = r_2 = a$, N_φ and N_θ are given by expressions (257), and Eq. (267) becomes

$$\frac{dv}{d\varphi} - v \cot \varphi = \frac{a^2 q(1 + \nu)}{Eh} \left(\cos \varphi - \frac{2}{1 + \cos \varphi} \right)$$

The general solution (e) is then

$$v = \frac{a^2 q(1 + \nu)}{Eh} \left[\sin \varphi \log (1 + \cos \varphi) - \frac{\sin \varphi}{1 + \cos \varphi} \right] + C \sin \varphi \quad (f)$$

The constant C will now be determined from the condition that for $\varphi = \alpha$ the displacement v is zero (Fig. 215a). From this condition

$$C = \frac{a^2 q(1 + \nu)}{Eh} \left[\frac{1}{1 + \cos \alpha} - \log (1 + \cos \alpha) \right] \quad (g)$$

The displacement v is obtained by substitution in expression (f). The displacement w is readily found from Eq. (b). At the support, where $v = 0$, the displacement w can be calculated directly from Eq. (b), without using solution (f), by substituting for ϵ_θ its value from the second of the equations (d).

109. Shells in the Form of a Surface of Revolution under Unsymmetrical Loading. Considering again an element cut from a shell by two adjacent meridians and two parallel circles (Fig. 224), in the general case not only normal forces N_φ and N_θ but also shearing forces $N_{\varphi\theta} = N_{\theta\varphi}$ will act on the sides of the element. Taking the sum of the projections in the y direction of all forces acting on the element, we must add to the forces considered in Art. 105 the force

$$\frac{\partial N_{\theta\varphi}}{\partial \theta} r_1 \, d\theta \, d\varphi \quad (a)$$

representing the difference in the shearing forces acting on the lateral sides of the element. Hence, instead of Eq. (f), Art. 105, we obtain the equation

$$\frac{\partial}{\partial \varphi}(N_\varphi r_0) + \frac{\partial N_{\theta\varphi}}{\partial \theta} r_1 - N_\theta r_1 \cos\varphi + Y r_1 r_0 = 0 \qquad (268)$$

Considering the forces in the x direction, we must include the difference of the shearing forces acting on the top and bottom of the element as

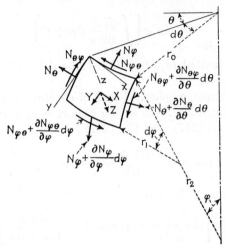

Fig. 224

given by the expression

$$N_{\varphi\theta} \frac{dr_0}{d\varphi} d\varphi\, d\theta + \frac{\partial N_{\varphi\theta}}{\partial \varphi} r_0\, d\varphi\, d\theta = \frac{\partial}{\partial \varphi}(r_0 N_{\varphi\theta})\, d\varphi\, d\theta \qquad (b)$$

the force

$$\frac{\partial N_\theta}{\partial \theta} r_1\, d\theta\, d\varphi \qquad (c)$$

due to variation of the force N_θ and the force

$$N_{\theta\varphi} r_1 \cos\varphi\, d\theta\, d\varphi \qquad (d)$$

due to the small angle $\cos\varphi\, d\theta$ between the shearing forces $N_{\theta\varphi}$ acting on the lateral sides of the element. The component in x direction of the external load acting on the element is

$$X r_0 r_1\, d\theta\, d\varphi \qquad (e)$$

Summing up all these forces, we obtain the equation

$$\frac{\partial}{\partial \varphi}(r_0 N_{\varphi\theta}) + \frac{\partial N_\theta}{\partial \theta} r_1 + N_{\theta\varphi} r_1 \cos\varphi + X r_0 r_1 = 0 \qquad (269)$$

The third equation of equilibrium is obtained by projecting the forces on the z axis. Since the projection of shearing forces on this axis vanishes, the third equation conforms with Eq. (256), which was derived for symmetrical loading.

The problem of determining membrane stresses under unsymmetrical loading reduces to the solution of Eqs. (268), (269), and (256) for given values of the components X, Y, and Z of the intensity of the external load. The application of these equations to the case of shells subjected to wind pressure will be discussed in the next article.

110. Stresses Produced by Wind Pressure.[1] As a particular example of the application of the general equations of equilibrium derived in the previous article, let us consider the action of wind pressure on a shell. Assuming that the direction of the wind is in the meridian plane $\theta = 0$ and that the pressure is normal to the surface, we take

$$X = Y = 0 \qquad Z = p \sin \varphi \cos \theta \qquad (a)$$

The equations of equilibrium then become

$$\frac{\partial}{\partial \varphi}(r_0 N_\varphi) + \frac{\partial N_{\theta\varphi}}{\partial \theta} r_1 - N_\theta r_1 \cos \varphi = 0$$

$$\frac{\partial}{\partial \varphi}(r_0 N_{\varphi\theta}) + \frac{\partial N_\theta}{\partial \theta} r_1 + N_{\theta\varphi} r_1 \cos \varphi = 0 \qquad (b)$$

$$N_\varphi r_0 + N_\theta r_1 \sin \varphi = -p r_0 r_1 \sin \varphi \cos \theta$$

By using the last of these equations we eliminate the force N_θ and obtain the following two differential equations[2] of the first order for determining N_φ and $N_{\theta\varphi} = N_{\varphi\theta}$:

$$\frac{\partial N_\varphi}{\partial \varphi} + \left(\frac{1}{r_0}\frac{dr_0}{d\varphi} + \cot \varphi\right) N_\varphi + \frac{r_1}{r_0}\frac{\partial N_{\theta\varphi}}{\partial \theta} = -p r_1 \cos \varphi \cos \theta$$

$$\frac{\partial N_{\theta\varphi}}{\partial \varphi} + \left(\frac{1}{r_0}\frac{dr_0}{d\varphi} + \frac{r_1}{r_2}\cot \varphi\right) N_{\theta\varphi} - \frac{1}{\sin \varphi}\frac{\partial N_\varphi}{\partial \theta} = -p r_1 \sin \theta \qquad (c)$$

Let us consider the particular problem of a spherical shell, in which case $r_1 = r_2 = a$. We take the solution of Eqs. (c) in the form

$$N_\varphi = S_\varphi \cos \theta \qquad N_{\theta\varphi} = S_{\theta\varphi} \sin \theta \qquad (d)$$

---

[1] The first investigation of this kind was made by H. Reissner, "Müller-Breslau-Festschrift," p. 181, Leipzig, 1912; see also F. Dischinger in F. von Emperger's "Handbuch für Eisenbetonbau," 4th ed., vol. 6, Berlin, 1928; E. Wiedemann, *Schweiz. Bauztg.*, vol. 108, p. 249, 1936; and K. Girkmann, *Stahlbau*, vol. 6, 1933. Further development of the theory of unsymmetrical deformation is due to C. Truesdell, *Trans. Am. Math. Soc.*, vol. 58, p. 96, 1945, and *Bull. Am. Math. Soc.*, vol. 54, p. 994, 1948; E. Reissner, *J. Math. and Phys.*, vol. 26, p. 290, 1948; and W. Zerna, *Ingr.-Arch.*, vol. 17, p. 223, 1949.

[2] The application of the stress function in investigating wind stresses was used by A. Pucher, *Publs. Intern. Assoc. Bridge Structural Engrs.*, vol. 5, p. 275, 1938; see also Art. 113.

in which S_φ and $S_{\theta\varphi}$ are functions of φ only. Substituting in Eqs. (c), we obtain the following ordinary differential equations for the determination of these functions:

$$\frac{dS_\varphi}{d\varphi} + 2\cot\varphi\, S_\varphi + \frac{1}{\sin\varphi} S_{\theta\varphi} = -pa\cos\varphi$$
$$\frac{dS_{\theta\varphi}}{d\varphi} + 2\cot\varphi\, S_{\theta\varphi} + \frac{1}{\sin\varphi} S_\varphi = -pa \qquad (e)$$

By adding and subtracting these equations and introducing the notation

$$U_1 = S_\varphi + S_{\theta\varphi} \qquad U_2 = S_\varphi - S_{\theta\varphi} \qquad (f)$$

the following two ordinary differential equations, each containing only one unknown, are obtained:

$$\frac{dU_1}{d\varphi} + \left(2\cot\varphi + \frac{1}{\sin\varphi}\right) U_1 = -pa(1 + \cos\varphi)$$
$$\frac{dU_2}{d\varphi} + \left(2\cot\varphi - \frac{1}{\sin\varphi}\right) U_2 = pa(1 - \cos\varphi) \qquad (g)$$

Applying the general rule for integrating differential equations of the first order, we obtain

$$U_1 = \frac{1+\cos\varphi}{\sin^3\varphi}\left[C_1 + pa\left(\cos\varphi - \frac{1}{3}\cos^3\varphi\right)\right]$$
$$U_2 = \frac{1-\cos\varphi}{\sin^3\varphi}\left[C_2 - pa\left(\cos\varphi - \frac{1}{3}\cos^3\varphi\right)\right] \qquad (h)$$

where C_1 and C_2 are constants of integration. Substituting in Eqs. (f) and using Eqs. (d), we finally obtain

$$N_\varphi = \frac{\cos\theta}{\sin^3\varphi}\left[\frac{C_1+C_2}{2} + \frac{C_1-C_2}{2}\cos\varphi + pa\left(\cos^2\varphi - \frac{1}{3}\cos^4\varphi\right)\right]$$
$$N_{\theta\varphi} = \frac{\sin\theta}{\sin^3\varphi}\left[\frac{C_1-C_2}{2} + \frac{C_1+C_2}{2}\cos\varphi + pa\left(\cos\varphi - \frac{1}{3}\cos^3\varphi\right)\right] \qquad (i)$$

To determine the constants of integration C_1 and C_2 let us consider a shell in the form of a hemisphere and put $\varphi = \pi/2$ in expressions (i). Then the forces along the equator of the shell are

$$N_\varphi = \frac{C_1+C_2}{2}\cos\theta \qquad N_{\theta\varphi} = \frac{C_1-C_2}{2}\sin\theta \qquad (j)$$

Since the pressure at each point of the sphere is in a radial direction, the moment of the wind forces with respect to the diameter of the sphere perpendicular to the plane $\theta = 0$ is zero. Using this fact and applying the first of the equations (j), we obtain

$$\int_0^{2\pi} N_\varphi a^2 \cos\theta\, d\theta = a^2\frac{C_1+C_2}{2}\int_0^{2\pi}\cos^2\theta\, d\theta = 0$$

which gives
$$C_1 = -C_2 \qquad (k)$$

The second necessary equation is obtained by taking the sum of the components of all forces acting on the half sphere in the direction of the horizontal diameter in the plane $\theta = 0$. This gives

$$\int_0^{2\pi} N_{\theta\varphi} a \sin \theta \, d\theta = -\int_0^{\pi/2} \int_0^{2\pi} p \sin \varphi \cos \theta \, a^2 \sin \varphi \sin \varphi \cos \theta \, d\varphi \, d\theta$$

or
$$a\pi \frac{C_1 - C_2}{2} = -pa^2 \frac{2}{3} \pi \qquad (l)$$

From (k) and (l) we obtain
$$C_1 = -\tfrac{2}{3}ap \qquad C_2 = \tfrac{2}{3}ap$$

Substituting these values for the constants in expressions (i) and using the third of the equations (b), we obtain

$$N_\varphi = -\frac{pa}{3} \frac{\cos \theta \cos \varphi}{\sin^3 \varphi} (2 - 3 \cos \varphi + \cos^3 \varphi)$$

$$N_\theta = \frac{pa}{3} \frac{\cos \theta}{\sin^3 \varphi} (2 \cos \varphi - 3 \sin^2 \varphi - 2 \cos^4 \varphi) \qquad (m)$$

$$N_{\theta\varphi} = -\frac{pa}{3} \frac{\sin \theta}{\sin^3 \varphi} (2 - 3 \cos \varphi + \cos^3 \varphi)$$

By using these expressions the wind stresses at any point of the shell can be readily calculated. If the shell is in the form of a hemisphere, there will be no normal forces acting along the edge of the shell, since $(N_\varphi)_{\varphi=\pi/2} = 0$. The shearing forces $N_{\theta\varphi}$ along the edge are different from zero and are equal and opposite to the horizontal resultant of the wind pressure. The maximum numerical value of these forces is found at the ends of the diameter perpendicular to the plane $\theta = 0$, at which point they are equal to $\pm 2pa/3$.

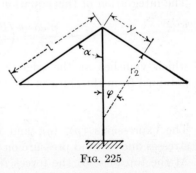

Fig. 225

As a second application of Eqs. (c) let us consider the case of a shell having the shape of a circular cone and supported by a column at the vertex (Fig. 225). In this case the radius r_1 is infinitely large. For an element dy of a meridian we can write $dy = r_1 \, d\varphi$. Hence

$$\frac{d}{d\varphi} = r_1 \frac{d}{dy}$$

In addition we have
$$r_0 = y \sin \alpha \qquad \frac{dr_0}{dy} = \sin \alpha \qquad r_2 = y \tan \alpha$$

Substituting in Eqs. (c), we obtain for a conical shell submitted to a wind pressure $Z = p \sin \varphi \cos \theta$ the equations

$$\frac{\partial N_\varphi}{\partial y} + \frac{N_\varphi}{y} + \frac{1}{y \sin \alpha} \frac{\partial N_{\theta\varphi}}{\partial \theta} = -p \sin \alpha \cos \theta$$
$$\frac{\partial N_{\theta\varphi}}{\partial y} + \frac{2N_{\theta\varphi}}{y} = -p \sin \theta \qquad (n)$$

The second equation can be readily integrated to obtain

$$N_{\theta\varphi} = -\frac{1}{y^2}\left(\frac{py^3}{3} + C\right) \sin \theta \qquad (o)$$

The edge of the shell $y = l$ is free from forces; hence the constant of integration in expression (o) is

$$C = -\frac{pl^3}{3}$$

and we finally obtain

$$N_{\theta\varphi} = \frac{p}{3} \frac{l^3 - y^3}{y^2} \sin \theta \qquad (p)$$

Substituting in the first of the equations (n), we find

$$\frac{\partial N_\varphi}{\partial y} + \frac{N_\varphi}{y} = -\left(\frac{p}{3} \frac{l^3 - y^3}{y^3 \sin \alpha} + p \sin \alpha\right) \cos \theta$$

The integration of this equation gives

$$N_\varphi = \frac{p \cos \theta}{\sin \alpha}\left(\frac{l^3 - y^3}{3y^2} - \frac{l^2 - y^2}{2y} \cos^2 \alpha\right) \qquad (q)$$

which vanishes at the edge $y = l$, as it should. The forces N_θ are obtained from the third of the equations (b), which gives

$$N_\theta = -py \sin \alpha \cos \theta \qquad (r)$$

The expressions (p), (q), and (r) give the complete solution for the stresses due to wind pressure on the conical shell represented in Fig. 225. At the top ($y = 0$) the forces N_φ and $N_{\theta\varphi}$ become infinitely large. To remove this difficulty we must assume a parallel circle corresponding to a certain finite value of y along which the conical shell is fastened to the column. The forces N_φ, $N_{\theta\varphi}$ distributed along this circle balance the wind pressure acting on the cone. It can be seen that, if the radius of the circle is not sufficient, these forces may become very large.

In the case of a transverse load Q applied at the top of the cone (Fig. 226a) we can satisfy Eqs. (n), in which the right-hand side becomes zero, by putting

$$N_\varphi = \frac{Q}{\pi y} \frac{\cos \theta}{\sin^2 \alpha} \qquad N_{\varphi\theta} = 0 \qquad (s)$$

It is readily verified by integration that the shearing force which results from the stresses N_φ for any section normal to the axis of the cone is equal to Q and that the moment of those stresses with respect to the axis $\theta = \pi/2$ of this section equalizes the moment $Qy \cos \alpha$ of the load. As for the stress components N_θ, they vanish throughout the shell, as ensues from the third of the equations (b), where we have to assume $r_1 = \infty$ and $p = 0$.

Should a load S act in the direction of the generatrix of the cone (Fig. 226b), we must combine the effect of both its components $P = S \cos \alpha$ (Fig. 217) and $Q = S \sin \alpha$ upon the forces N_φ.

The result is

$$N_\varphi = \frac{S}{2\pi r_0} (2 \cos \theta - 1)$$

which yields the extreme values of $S/2\pi r_0$ at $\theta = 0$ and $-3S/2\pi r_0$ at $\theta = \pi$, respectively.

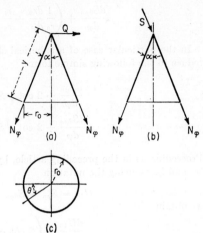

Fig. 226

111. Spherical Shell Supported at Isolated Points.[1] We begin with the general case of a shell having the form of a surface of revolution and consider the case when the forces are acting only along the edge of the shell so that $X = Y = Z = 0$. The general equations (b) of the preceding article then become

$$\frac{\partial}{\partial \varphi}(r_0 N_\varphi) + \frac{\partial N_{\theta\varphi}}{\partial \theta} r_1 - N_\theta r_1 \cos \varphi = 0$$

$$\frac{\partial}{\partial \varphi}(r_0 N_{\theta\varphi}) + \frac{\partial N_\theta}{\partial \theta} r_1 + N_{\theta\varphi} r_1 \cos \varphi = 0 \qquad (a)$$

$$N_\varphi r_0 + N_\theta r_1 \sin \varphi = 0$$

Let us take the solution of these equations in the form

$$\begin{aligned} N_\varphi &= S_{\varphi n} \cos n\theta \\ N_\theta &= S_{\theta n} \cos n\theta \\ N_{\theta\varphi} &= S_{\theta\varphi n} \sin n\theta \end{aligned} \qquad (b)$$

where $S_{\varphi n}$, $S_{\theta n}$, and $S_{\theta\varphi n}$ are functions of φ only and n is an integer. Substituting expressions (b) in Eqs. (a), we obtain

$$\frac{d}{d\varphi}(r_0 S_{\varphi n}) + n r_1 S_{\theta\varphi n} - r_1 S_{\theta n} \cos \varphi = 0$$

$$\frac{d}{d\varphi}(r_0 S_{\theta\varphi n}) - n r_1 S_{\theta n} + r_1 S_{\theta\varphi n} \cos \varphi = 0 \qquad (c)$$

$$S_{\varphi n} + \frac{r_1}{r_2} S_{\theta n} = 0$$

---

[1] Flügge, *op. cit.* For the application of the stress function in solving such problems, see the paper by Pucher, *op. cit.*

Using the third of these equations, we can eliminate the function $S_{\theta n}$ and thus obtain

$$\frac{dS_{\varphi n}}{d\varphi} + \left(\frac{1}{r_0}\frac{dr_0}{d\varphi} + \cot \varphi\right) S_{\varphi n} + n\frac{r_1}{r_2}\frac{S_{\theta \varphi n}}{\sin \varphi} = 0$$

$$\frac{dS_{\theta \varphi n}}{d\varphi} + \left(\frac{1}{r_0}\frac{dr_0}{d\varphi} + \frac{r_1}{r_2}\cot \varphi\right) S_{\theta \varphi n} + \frac{nS_{\varphi n}}{\sin \varphi} = 0 \tag{d}$$

In the particular case of a spherical shell $r_1 = r_2 = a$, $r_0 = a \sin \varphi$; and Eqs. (d) reduce to the following simple form:

$$\frac{dS_{\varphi n}}{d\varphi} + 2\cot \varphi\, S_{\varphi n} + \frac{n}{\sin \varphi} S_{\theta \varphi n} = 0$$

$$\frac{dS_{\theta \varphi n}}{d\varphi} + 2\cot \varphi\, S_{\theta \varphi n} + \frac{n}{\sin \varphi} S_{\varphi n} = 0 \tag{e}$$

Proceeding as in the preceding article, by taking the sum and the difference of Eqs. (e) and introducing the notation

$$U_{1n} = S_{\varphi n} + S_{\theta \varphi n} \qquad U_{2n} = S_{\varphi n} - S_{\theta \varphi n} \tag{f}$$

we obtain

$$\frac{dU_{1n}}{d\varphi} + \left(2\cot \varphi + \frac{n}{\sin \varphi}\right) U_{1n} = 0$$

$$\frac{dU_{2n}}{d\varphi} + \left(2\cot \varphi - \frac{n}{\sin \varphi}\right) U_{2n} = 0 \tag{g}$$

The solution of these equations is

$$U_{1n} = C_{1n} \frac{\left(\cot \frac{\varphi}{2}\right)^n}{\sin^2 \varphi} \qquad U_{2n} = C_{2n} \frac{\left(\tan \frac{\varphi}{2}\right)^n}{\sin^2 \varphi} \tag{h}$$

From Eqs. (f) we then obtain

$$S_{\varphi n} = \frac{U_{1n} + U_{2n}}{2} = \frac{1}{2\sin^2 \varphi}\left[C_{1n}\left(\cot \frac{\varphi}{2}\right)^n + C_{2n}\left(\tan \frac{\varphi}{2}\right)^n\right]$$

$$S_{\theta \varphi n} = \frac{U_{1n} - U_{2n}}{2} = \frac{1}{2\sin^2 \varphi}\left[C_{1n}\left(\cot \frac{\varphi}{2}\right)^n - C_{2n}\left(\tan \frac{\varphi}{2}\right)^n\right] \tag{i}$$

If we have a shell without an opening at the top, expressions (i) must be finite for $\varphi = 0$. This requires that the constant of integration $C_{1n} = 0$. Substituting this in Eqs. (i) and using Eqs. (b), we find

$$N_\varphi = -N_\theta = \frac{C_{2n}}{2\sin^2 \varphi}\left(\tan \frac{\varphi}{2}\right)^n \cos n\theta$$

$$N_{\theta \varphi} = -\frac{C_{2n}}{2\sin^2 \varphi}\left(\tan \frac{\varphi}{2}\right)^n \sin n\theta \tag{j}$$

Substituting for φ the angle φ_0 corresponding to the edge of the spherical shell, we shall obtain the normal and the shearing forces which must be distributed along the edge of the shell to produce in this shell the forces (j). Taking, as an example, the case when $\varphi_0 = \pi/2$, that is, when the shell is a hemisphere, we obtain, from expressions (j),

$$(N_\varphi)_{\varphi=\pi/2} = \frac{C_{2n}}{2} \cos n\theta$$

$$(N_{\theta\varphi})_{\varphi=\pi/2} = -\frac{C_{2n}}{2} \sin n\theta \qquad (k)$$

Knowing the stresses produced in a spherical shell by normal and shearing forces applied to the edge and proportional to $\cos n\theta$ and $\sin n\theta$, respectively, we can treat the problem of any distribution of normal forces along the edge by representing this

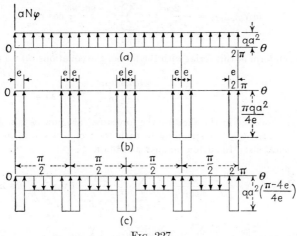

FIG. 227

distribution by a trigonometric series in which each term of the series is a solution similar to solution (j).[1] Take, as an example, the case of a hemispherical dome of radius a, carrying only its own weight of q psf and supported by four symmetrically located columns. If the dome is resting on a continuous foundation, the forces N_φ are uniformly distributed along the edge as shown in Fig. 227a, in which the intensity of force aN_φ per unit angle is plotted against the angle θ. In the case of four equidistant columns the distribution of reactions will be as shown in Fig. 227b, in which $2e$ denotes the angle corresponding to the circumferential distance supported by each column. Subtracting the force distribution of Fig. 227a from the force distribution of Fig. 227b, we obtain the distribution of Fig. 227c, representing a system of forces in equilibrium. This distribution can be represented in the form of a series

$$(aN_\varphi)_{\varphi=\pi/2} = \sum_{n=4,8,12,\ldots}^{\infty} A_n \cos n\theta \qquad (l)$$

[1] In using a series $N_\varphi = \frac{1}{2} \sum_{n=1,2,3,\ldots}^{\infty} C_{2n} \cos n\theta$ for normal forces we obtain a distribution of these forces symmetrical with respect to the diameter $\theta = 0$. In the general case the series will contain not only cosine terms but also sine terms. The solutions for sine terms can be obtained in exactly the same manner as used in our discussion of the cosine terms. It is only necessary to exchange the places of $\cos n\theta$ and $\sin n\theta$ in Eqs. (b).

in which only the terms $n = 4, 8, 12, \ldots$ must be considered, since the diagram 227c repeats itself after each interval of $\pi/2$ and has four complete periods in the angle 2π. Applying the usual method for calculating the coefficients of series (l), we find

$$A_n = -\frac{2qa^2}{ne} \sin (ne)$$

Hence the distribution shown by diagram 227c is represented by the series

$$(aN_\varphi)_{\varphi=\pi/2} = -\frac{2qa^2}{e} \sum_{n=4,8,12,\ldots}^{\infty} \frac{\sin ne}{n} \cos n\theta \qquad (m)$$

Comparing each term of this series with the first of the equations (k) we conclude that

$$C_{2n} = -\frac{4qa}{e} \frac{\sin ne}{n}$$

The stresses produced in the shell by the forces (m) are now obtained by taking a solution of the form (j) corresponding to each term of series (m) and then superposing these solutions. In such a manner we obtain

$$N_\varphi = -N_\theta = -\frac{2qa}{e \sin^2 \varphi} \sum_{n=4,8,12,\ldots}^{\infty} \frac{\sin ne}{n} \left(\tan \frac{\varphi}{2}\right)^n \cos n\theta$$

$$N_{\theta\varphi} = \frac{2qa}{e \sin^2 \varphi} \sum_{n=4,8,12,\ldots}^{\infty} \frac{\sin ne}{n} \left(\tan \frac{\varphi}{2}\right)^n \sin n\theta \qquad (n)$$

Superposing this solution on solution (257), which was previously obtained for a dome supported by forces uniformly distributed along the edge (Fig. 215a), we obtain formulas for calculating the stresses in a dome resting on four columns. It must be noted, however, that, whereas the above-mentioned superposition gives the necessary distribution of the reactive forces N_φ as shown in Fig. 227b, it also introduces shearing forces $N_{\theta\varphi}$ which do not vanish at the edge of the dome. Thus our solution does not satisfy all the conditions of the problem. In fact, so long as we limit ourselves to membrane theory, we shall not have enough constants to satisfy all the conditions and to obtain the complete solution of the problem. In actual constructions a reinforcing ring is usually put along the edge of the shell to carry the shearing forces $N_{\theta\varphi}$. In such cases the solution obtained by the combination of solutions (257) and (n) will be a sufficiently accurate representation of the internal forces produced in a spherical dome resting on four columns. For a more satisfactory solution of this problem the bending theory of shells must be used.[1]

The method discussed in this article can also be used in the case of a nonspherical dome. In such cases it is necessary to have recourse to Eqs. (d), which can be solved with sufficient accuracy by using numerical integration.[2]

[1] An example of such a solution is given in A. Aas-Jakobsen's paper, *Ingr.-Arch.*, vol. 8, p. 275, 1937.

[2] An example of such integration is given by Flügge, *op. cit.* On p. 51 of this book the calculation of membrane forces in an apsidal shell, due to the weight of the

112. Membrane Theory of Cylindrical Shells.

In discussing a cylindrical shell (Fig. 228a) we assume that the generator of the shell is horizontal and parallel to the x axis. An element is cut from the shell by two adjacent generators and two cross sections perpendicular to the x axis, and its position is defined by the coordinate x and the angle φ. The forces acting on the sides of the element are shown in Fig. 228b. In

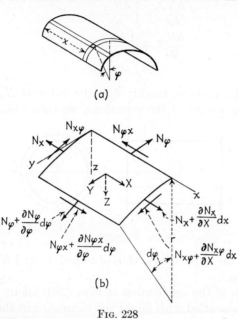

Fig. 228

addition a load will be distributed over the surface of the element, the components of the intensity of this load being denoted, as before, by X, Y, and Z. Considering the equilibrium of the element and summing up the forces in the x direction, we obtain

$$\frac{\partial N_x}{\partial x} r\, d\varphi\, dx + \frac{\partial N_{\varphi x}}{\partial \varphi} d\varphi\, dx + Xr\, d\varphi\, dx = 0 \qquad (a)$$

Similarly, the forces in the direction of the tangent to the normal cross section, *i.e.*, in the y direction, give as a corresponding equation of equilibrium

$$\frac{\partial N_{x\varphi}}{\partial x} r\, d\varphi\, dx + \frac{\partial N_\varphi}{\partial \varphi} d\varphi\, dx + Yr\, d\varphi\, dx = 0 \qquad (b)$$

The forces acting in the direction of the normal to the shell, *i.e.*, in the

---

shell, is also discussed. For application of the complex variable method to the stress analysis in spherical shells, see F. Martin, *Ingr.-Arch.*, vol. 17, p. 167, 1949; see also V. Z. Vlasov, *Priklad. Mat. Mekhan.*, vol. 11, p. 397, 1947.

z direction, give the equation

$$N_\varphi \, d\varphi \, dx + Zr \, d\varphi \, dx = 0 \tag{c}$$

After simplification, the three equations of equilibrium can be represented in the following form:

$$\begin{aligned}\frac{\partial N_x}{\partial x} + \frac{1}{r}\frac{\partial N_{x\varphi}}{\partial \varphi} &= -X \\ \frac{\partial N_{x\varphi}}{\partial x} + \frac{1}{r}\frac{\partial N_\varphi}{\partial \varphi} &= -Y \\ N_\varphi &= -Zr\end{aligned} \tag{270}$$

In each particular case we readily find the value of N_φ. Substituting this value in the second of the equations, we then obtain $N_{x\varphi}$ by inte-

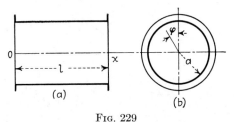

Fig. 229

gration. Using the value of $N_{x\varphi}$ thus obtained we find N_x by integrating the first equation.

As an example of the application of Eqs. (270) let us consider a horizontal circular tube filled with liquid and supported at the ends.[1] Measuring the angle φ as shown in Fig. 229b and denoting by p_0 the pressure at the axis of the tube, the pressure at any point is $p_0 - \gamma a \cos \varphi$. We thus obtain

$$X = Y = 0 \qquad Z = -p_0 + \gamma a \cos \varphi \tag{d}$$

Substituting in Eqs. (270), we find

$$N_\varphi = p_0 a - \gamma a^2 \cos \varphi \tag{e}$$

$$N_{x\varphi} = -\int \gamma a \sin \varphi \, dx + C_1(\varphi) = -\gamma a x \sin \varphi + C_1(\varphi) \tag{f}$$

$$\begin{aligned}N_x &= \int \gamma \cos \varphi \, x \, dx - \frac{1}{a}\int \frac{dC_1(\varphi)}{d\varphi} dx + C_2(\varphi) \\ &= \gamma \frac{x^2}{2}\cos\varphi - \frac{x}{a}\frac{dC_1(\varphi)}{d\varphi} + C_2(\varphi)\end{aligned} \tag{g}$$

The functions $C_1(\varphi)$ and $C_2(\varphi)$ must now be determined from the conditions at the edges.

Let us first assume that there are no forces N_x at the ends of the tube.

[1] This problem was discussed by D. Thoma, *Z. ges. Turbinenwesen*, vol. 17, p. 49, 1920.

Then
$$(N_x)_{x=0} = 0 \quad (N_x)_{x=l} = 0$$

We shall satisfy these conditions by taking
$$C_2(\varphi) = 0 \quad C_1(\varphi) = \frac{a\gamma l}{2} \sin \varphi + C$$

It is seen from expression (f) that the constant C represents forces $N_{x\varphi}$ uniformly distributed around the edge of the tube, as is the case when the tube is subjected to torsion. If there is no torque applied, we must take $C = 0$. Then the solution of Eqs. (270) in our particular case is

$$N_\varphi = p_0 a - \gamma a^2 \cos \varphi$$
$$N_{x\varphi} = \gamma a \left(\frac{l}{2} - x\right) \sin \varphi \quad (271)$$
$$N_x = -\frac{\gamma}{2} x(l - x) \cos \varphi$$

It is seen that $N_{x\varphi}$ and N_x are proportional, respectively, to the shearing force and to the bending moment of a uniformly loaded beam of span l and can be obtained by applying beam formulas to the tube carrying a uniformly distributed load of the magnitude[1] $\pi a^2 \gamma$ per unit length of the tube.

By a proper selection of the function $C_2(\varphi)$ we can also obtain a solution of the problem for a cylindrical shell with built-in edges. In such a case the length of the generator remains unchanged, and we have the condition
$$\int_0^l (N_x - \nu N_\varphi)\, dx = 0$$

Substituting
$$N_x = -\frac{\gamma}{2} x(l - x) \cos \varphi + C_2(\varphi) \quad N_\varphi = p_0 a - \gamma a^2 \cos \varphi$$

we obtain
$$C_2(\varphi) = \nu p_0 a + \left(\frac{l^2}{12} - \nu a^2\right) \gamma \cos \varphi$$

and $\quad N_x = -\dfrac{\gamma x}{2}(l - x) \cos \varphi + \nu p_0 a + \left(\dfrac{l^2}{12} - \nu a^2\right) \gamma \cos \varphi \quad (272)$

Owing to the action of the forces N_φ and N_x there will be a certain amount of strain in the circumferential direction at the end of the tube in contradiction to our assumption of built-in edges. This indicates that at the ends of the tube there will be some local bending, which is disregarded in the membrane theory. A more complete solution of the problem can be obtained only by considering membrane stresses together with bending stresses, as will be discussed in the next chapter.

[1] The weight of the tube is neglected in this discussion.

Sections of cylindrical shells, such as shown in Fig. 230, are sometimes used as coverings of various kinds of structures. These shells are usually supported only at the ends while the edges AB and CD are free. In calculating the membrane stresses for such shells Eqs. (270) can again be used. Take, for example, a shell of a semicircular cross section supporting its own weight, which is assumed to be uniformly distributed over the surface of the shell. In such a case we have

$$X = 0 \quad Y = p \sin \varphi \quad Z = p \cos \varphi$$

The third of the equations (270) gives

$$N_\varphi = -pa \cos \varphi \qquad (h)$$

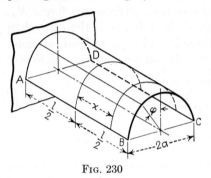

Fig. 230

which vanishes along the edges AB and CD, as it should. It is seen that this condition will also be satisfied if some other curve is taken instead of a semicircle, provided that $\varphi = \pm \pi/2$ at the edges. Substituting expression (h) in the second of the equations (270), we find

$$N_{x\varphi} = -2px \sin \varphi + C_1(\varphi) \qquad (i)$$

By putting the origin of the coordinates at the middle of the span and assuming the same end conditions at both ends, $x = \pm l/2$ of the tube, it can be concluded from symmetry that $C_1(\varphi) = 0$. Hence

$$N_{x\varphi} = -2px \sin \varphi \qquad (j)$$

It is seen that this solution does not vanish along the edges AB and CD as it should for free edges. In structural applications, however, the edges are usually reinforced by longitudinal members strong enough to resist the tension produced by shearing force (j). Substituting expression (j) in the first of the equations (270), we obtain

$$N_x = \frac{px^2}{a} \cos \varphi + C_2(\varphi) \qquad (k)$$

If the ends of the shell are supported in such a manner that the reactions act in the planes of the end cross sections, the forces N_x must vanish at the ends. Hence $C_2(\varphi) = -(pl^2 \cos \varphi)/4a$, and we obtain

$$N_x = -\frac{p \cos \varphi}{4a}(l^2 - 4x^2) \qquad (l)$$

Expressions (h), (j), and (l) represent the solution of Eqs. (270) for our particular case (Fig. 230) satisfying the conditions at the ends and also one of the conditions along the edges AB and CD. The second condition, which concerns the shearing forces $N_{x\varphi}$, cannot be satisfied by using the membrane stresses alone. In practical applications it is assumed that the forces $N_{x\varphi}$ will be taken by the longitudinal members that reinforce the edges. It can be expected that this assumption will be satisfactory in those cases in which the length of the shell is not large, say $l \leqq 2a$, and that the membrane theory will give an approximate picture of the stress distribution in such cases. For longer shells a satisfactory solution can be obtained only by considering bending as well as membrane stresses. This problem will be discussed in the next chapter (see Arts. 124 and 126).

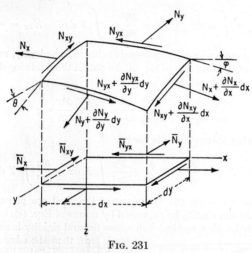

Fig. 231

113. The Use of a Stress Function in Calculating Membrane Forces of Shells.

In the general case of a shell given by the equation $z = f(x,y)$ of its middle surface the use of a stress function[1] defining all three stress components may be convenient. Let us consider an element of a shell submitted to a loading the magnitude of which per unit area in xy plane is given by its components X, Y, Z (Fig. 231). The static equilibrium of the element then can be expressed by the equations

$$\frac{\partial \bar{N}_x}{\partial x} + \frac{\partial \bar{N}_{yx}}{\partial y} + X = 0$$

$$\frac{\partial \bar{N}_{xy}}{\partial x} + \frac{\partial \bar{N}_y}{\partial y} + Y = 0 \qquad (a)$$

$$\frac{\partial}{\partial x}\left(\bar{N}_x \frac{\partial z}{\partial x} + \bar{N}_{xy}\frac{\partial z}{\partial y}\right) + \frac{\partial}{\partial y}\left(\bar{N}_y \frac{\partial z}{\partial y} + \bar{N}_{xy}\frac{\partial z}{\partial x}\right) + Z = 0 \qquad (b)$$

---

[1] The introduction of the function considered here is due to A. Pucher, *op. cit.*, and *Beton u. Eisen*, vol. 33, p. 298, 1934; see also *Proc. Fifth Intern. Congr. Appl. Mech.*, Cambridge, Mass., 1938. Cylindrical coordinates, more suitable for shells in the form of a surface of revolution, are also used by Pucher. For a general theory of deformation following Pucher's approach, see W. Flügge and F. Geyling, *Proc. Ninth Intern. Congr. Appl. Mech.*, vol. 6, p. 250, Brussels, 1957.

in which the following notation is used:

$$N_x = \bar{N}_x \frac{\cos \theta}{\cos \varphi} \qquad N_y = \bar{N}_y \frac{\cos \varphi}{\cos \theta} \qquad N_{xy} = \bar{N}_{xy} \qquad (c)$$

where $\tan \varphi = \partial z/\partial x$ and $\tan \theta = \partial z/\partial y$. Carrying out the differentiation as indicated in Eq. (b) and taking into account Eqs. (a), we obtain

$$\bar{N}_x \frac{\partial^2 z}{\partial x^2} + 2\bar{N}_{xy} \frac{\partial^2 z}{\partial x\, \partial y} + \bar{N}_y \frac{\partial^2 z}{\partial y^2} = -Z + X \frac{\partial z}{\partial x} + Y \frac{\partial z}{\partial y} \qquad (d)$$

We can satisfy both equations (a) by introducing a stress function $F(x,y)$ such that

$$\bar{N}_x = \frac{\partial^2 F}{\partial y^2} - \int X\, dx \qquad \bar{N}_y = \frac{\partial^2 F}{\partial x^2} - \int Y\, dy \qquad \bar{N}_{xy} = -\frac{\partial^2 F}{\partial x\, \partial y} \qquad (e)$$

the lower and the upper limits of the integrals being x_0, x and y_0, y, respectively, with x_0 and y_0 fixed. Substituting this in Eq. (d) we obtain the following differential equation governing the stress function F:

$$\frac{\partial^2 F}{\partial x^2} \frac{\partial^2 z}{\partial y^2} - 2 \frac{\partial^2 F}{\partial x\, \partial y} \frac{\partial^2 z}{\partial x\, \partial y} + \frac{\partial^2 F}{\partial y^2} \frac{\partial^2 z}{\partial x^2} = q \qquad (f)$$

in which the following abbreviation is used:

$$q = -Z + X \frac{\partial z}{\partial x} + Y \frac{\partial z}{\partial y} + \frac{\partial^2 z}{\partial x^2} \int X\, dx + \frac{\partial^2 z}{\partial y^2} \int Y\, dy \qquad (g)$$

If the membrane forces on the boundary of the shell are given, the respective boundary conditions can readily be expressed by means of Eqs. (e). If, in particular, the edge is connected with a vertical wall whose flexural rigidity is negligible or if the edge is free, then the edge forces normal to the elements ds of the boundary and proportional to $\partial^2 F/\partial s^2$ must vanish. Hence the variation of the stress function along such an edge must follow a linear law.

A Shell in the Form of an Elliptic Paraboloid. To illustrate the application of the method, let us take a shell in the form of an elliptic paraboloid (Fig. 232) with the middle surface

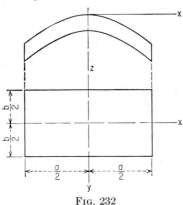

Fig. 232

$$z = \frac{x^2}{h_1} + \frac{y^2}{h_2} \qquad (h)$$

where h_1 and h_2 are positive constants. The sections $x =$ constant and $y =$ constant then yield two sets of parabolas, and the level curves are ellipses. Assuming solely a vertical load uniformly distributed over the ground plan of the shell and using Eqs. (f) and (g) we obtain

$$\frac{1}{h_2} \frac{\partial^2 F}{\partial x^2} + \frac{1}{h_1} \frac{\partial^2 F}{\partial y^2} = -\frac{p}{2} \qquad (i)$$

where $p = Z$ is the intensity of the load.

Let the shell be supported by four vertical walls $x = \pm a/2$, $y = \pm b/2$ in such a way that the reactive forces normal to the respective wall vanish along the boundary. Consequently, the boundary conditions for the function F are $\partial^2 F/\partial y^2 = 0$ on $x = \pm a/2$ and $\partial^2 F/\partial x^2 = 0$ on $y = \pm b/2$. Thus F may be a linear function in x and y on the boundary. Since terms linear in x or y have no effect on stresses [see Eqs. (e)], this is equivalent to the condition $F = 0$ on the whole boundary.

We satisfy Eq. (i) and make $F = 0$ on $y = \pm b/2$ by taking for F the expression

$$F = \frac{ph_1}{4}\left(\frac{b^2}{4} - y^2\right) + \sum_{n=1,3,5,\ldots}^{\infty} A_n \cosh\frac{n\pi x}{c} \cos\frac{n\pi y}{b} \qquad (j)$$

in which $c = b\sqrt{h_1/h_2}$. In order to fulfill the remaining condition $F = 0$ on $x = \pm a/2$, we first develop the algebraic term in expression (j) into the Fourier series

$$\frac{ph_1}{4}\left(\frac{b^2}{4} - y^2\right) = \frac{2ph_1 b^2}{\pi^3} \sum_{n=1,3,5,\ldots}^{\infty} \frac{1}{n}(-1)^{(n-1)/2} \cos\frac{n\pi y}{b} \qquad (k)$$

Substituting this in Eq. (j), making $x = \pm a/2$, and equating the result to zero we obtain for each $n = 1, 3, 5, \ldots$ the equation

$$\frac{2ph_1 b^2}{\pi^3 n^3}(-1)^{(n-1)/2} + A_n \cosh\frac{n\pi a}{2c} = 0 \qquad (l)$$

This yields the value of the coefficient A_n and leads to the final solution

$$F = \frac{ph_1}{4}\left[\frac{b^2}{4} - y^2 + \frac{8b^2}{\pi^3}\sum_{n=1,3,5,\ldots}^{\infty}(-1)^{(n+1)/2}\frac{1}{n^3}\frac{\cosh\dfrac{n\pi x}{c}}{\cosh\dfrac{n\pi a}{2c}}\cos\frac{n\pi y}{b}\right] \qquad (m)$$

To obtain the membrane forces we have only to differentiate this in accordance with the expressions (e) and to make use of the relations (c). The result is

$$N_x = -\frac{ph_2}{2}\sqrt{\frac{h_1^2 + 4x^2}{h_2^2 + 4y^2}}\left[1 + \frac{4}{\pi}\sum_{n=1,3,5,\ldots}^{\infty}(-1)^{(n+1)/2}\frac{1}{n}\frac{\cosh\dfrac{n\pi x}{c}}{\cosh\dfrac{n\pi a}{2c}}\cos\frac{n\pi y}{b}\right]$$

$$N_y = 2\frac{ph_1^2}{h_2}\sqrt{\frac{h_2^2 + 4y^2}{h_1^2 + 4x^2}}\sum_{n=1,3,5,\ldots}^{\infty}(-1)^{(n+1)/2}\frac{1}{n}\frac{\cosh\dfrac{n\pi x}{c}}{\cosh\dfrac{n\pi a}{2c}}\cos\frac{n\pi y}{b} \qquad (n)$$

$$N_{xy} = 2\frac{ph_1}{\pi}\sum_{n=1,3,5,\ldots}^{\infty}(-1)^{(n+1)/2}\frac{1}{n}\frac{\sinh\dfrac{n\pi x}{c}}{\cosh\dfrac{n\pi a}{2c}}\sin\frac{n\pi y}{b}$$

All series obtained above are convergent, the only exception being the last series, which diverges at the corners $x = \pm a/2$, $y = \pm b/2$. This fact is due to a specific property of the shell surface under consideration obtained by translation of a plane curve. The elements of such a surface are free of any twist, and for this reason the membrane forces N_{xy} fail to contribute anything to the transmission of the normal loading of the shell. As the forces N_x and N_y both vanish at the corner points of the shell, the shearing forces N_{xy} near these points have to stand alone for the transmission of the loading. Owing to the zero twist of the surface of the shells, this leads to an infinite increase of those shearing forces toward the corners of the shell. Practically, bending moments and transverse shearing forces will arise in the vicinity of the corners, should the edge conditions $N_x = 0$, $N_y = 0$ be fulfilled rigorously.

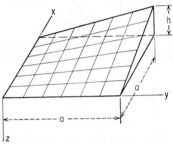

Fig. 233

A Shell in the Form of a Hyperbolic Paraboloid.[1] Another case where Pucher's method may be applied to advantage is a shell with a middle surface given by the equation

$$z = -\frac{xy}{c} \qquad (o)$$

in which $c = a^2/h$ (Fig. 233). Hence

$$\frac{\partial z}{\partial x} = -\frac{y}{c} \qquad \frac{\partial z}{\partial y} = -\frac{x}{c} \qquad \frac{\partial^2 z}{\partial x^2} = \frac{\partial^2 z}{\partial y^2} = 0 \qquad \frac{\partial^2 z}{\partial x \, \partial y} = -\frac{1}{c} \qquad (p)$$

Provided we have to deal with a vertical loading only, the differential equation (*f*) becomes

$$\frac{2}{c} \frac{\partial^2 F}{\partial x \, \partial y} = -Z \qquad (q)$$

which yields the result

$$N_{xy} = \frac{Zc}{2} \qquad (r)$$

Let us consider first a load of an intensity $Z = q$, uniformly distributed over the horizontal projection of a shell with edges free of normal forces. Then we have

$$N_{xy} = \frac{qc}{2} \qquad N_x = N_y = 0 \qquad (s)$$

Now consider the effect of the own weight of the shell equal to q_0 = constant per

[1] See F. Aimond, *Génie civil*, vol. 102, p. 179, 1933, and *Proc. Intern. Assoc. Bridge Structural Engrs.*, vol. 4, p. 1, 1936; also B. Laffaille, *Proc. Intern. Assoc. Bridge Structural Engrs.*, vol. 3, p. 295, 1935. Various cases of loading were discussed by K. G. Tester, *Ingr.-Arch.*, vol. 16, p. 39, 1947.

unit area of the surface. To this area corresponds an area

$$\cos \gamma = \frac{c}{\sqrt{x^2 + y^2 + c^2}}$$

of the horizontal projection of the shell. Hence

$$Z = \frac{q_0}{c} \sqrt{x^2 + y^2 + c^2} \tag{t}$$

and Eq. (r) yields

$$N_{xy} = \frac{q_0}{2} \sqrt{x^2 + y^2 + c^2}$$

Differentiating this with respect to y and then integrating the result with respect to x, or vice versa, both in accordance with Eqs. (e), we get

$$\tilde{N}_x = -\frac{q_0 y}{2} \log \frac{x + \sqrt{x^2 + y^2 + c^2}}{\sqrt{y^2 + c^2}}$$

$$\tilde{N}_y = -\frac{q_0 x}{2} \log \frac{y + \sqrt{x^2 + y^2 + c^2}}{\sqrt{x^2 + c^2}}$$

The true forces N_x and N_y are obtained from those expressions by means of Eqs. (c), in which the angles φ, θ are given by $\tan \varphi = -y/c$ and $\tan \theta = -x/c$.

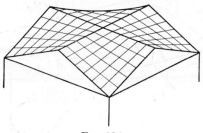

Fig. 234

Several shells of this kind may be combined to form a roof, such as shown in Fig. 234. It should be noted, however, that neither the dead load of the groin members, needed by such a roof, nor a partial loading—due, for instance, to snow—can be transmitted by the membrane forces alone; hence flexural stresses will necessarily arise.[1]

Of practical interest and worthy of mention are also the conoidal shells, which sometimes have been used in the design of cantilever roofs and dam walls.[2] Roof shells of this kind, however, with curved generatrices instead of straight ones, have also been used in structural applications.[3]

[1] See Flügge, *op. cit.*, p. 119, Flügge and Geyling, *op. cit.*, and F. A. Gerard, *Trans. Eng. Inst. Canada*, vol. 3, p. 32, 1959.

[2] The theory of the conoidal shell has been elaborated by E. Torroja, *Riv. ing.*, vol. 9, p. 29, 1941. See also M. Soare, *Bauingenieur*, vol. 33, p. 256, 1958, and Flügge, *op. cit.*, p. 127.

[3] See I. Doganoff, *Bautechnik*, vol. 34, p. 232, 1957.

CHAPTER 15

GENERAL THEORY OF CYLINDRICAL SHELLS

114. A Circular Cylindrical Shell Loaded Symmetrically with Respect to Its Axis. In practical applications we frequently encounter problems in which a circular cylindrical shell is submitted to the action of forces distributed symmetrically with respect to the axis of the cylinder. The stress distribution in cylindrical boilers submitted to the action of steam pressure, stresses in cylindrical containers having a vertical axis and submitted to internal liquid pressure, and stresses in circular pipes under uniform internal pressure are examples of such problems.

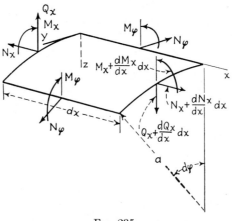

Fig. 235

To establish the equations required for the solution of these problems we consider an element, as shown in Figs. 228a and 235, and consider the equations of equilibrium. It can be concluded from symmetry that the membrane shearing forces $N_{x\varphi} = N_{\varphi x}$ vanish in this case and that forces N_φ are constant along the circumference. Regarding the transverse shearing forces, it can also be concluded from symmetry that only the forces Q_x do not vanish. Considering the moments acting on the element in Fig. 235, we also conclude from symmetry that the twisting moments $M_{x\varphi} = M_{\varphi x}$ vanish and that the bending moments M_φ are constant along the circumference. Under such conditions of symmetry

466

three of the six equations of equilibrium of the element are identically satisfied, and we have to consider only the remaining three equations, viz., those obtained by projecting the forces on the x and z axes and by taking the moment of the forces about the y axis. Assuming that the external forces consist only of a pressure normal to the surface, these three equations of equilibrium are

$$\frac{dN_x}{dx} a\, dx\, d\varphi = 0$$
$$\frac{dQ_x}{dx} a\, dx\, d\varphi + N_\varphi\, dx\, d\varphi + Za\, dx\, d\varphi = 0 \qquad (a)$$
$$\frac{dM_x}{dx} a\, dx\, d\varphi - Q_x a\, dx\, d\varphi = 0$$

The first one indicates that the forces N_x are constant,[1] and we take them equal to zero in our further discussion. If they are different from zero, the deformation and stress corresponding to such constant forces can be easily calculated and superposed on stresses and deformations produced by lateral load. The remaining two equations can be written in the following simplified form:

$$\frac{dQ_x}{dx} + \frac{1}{a} N_\varphi = -Z$$
$$\frac{dM_x}{dx} - Q_x = 0 \qquad (b)$$

These two equations contain three unknown quantities: N_φ, Q_x, and M_x. To solve the problem we must therefore consider the displacements of points in the middle surface of the shell.

From symmetry we conclude that the component v of the displacement in the circumferential direction vanishes. We thus have to consider only the components u and w in the x and z directions, respectively. The expressions for the strain components then become

$$\epsilon_x = \frac{du}{dx} \qquad \epsilon_\varphi = -\frac{w}{a} \qquad (c)$$

Hence, by applying Hooke's law, we obtain

$$N_x = \frac{Eh}{1-\nu^2}(\epsilon_x + \nu\epsilon_\varphi) = \frac{Eh}{1-\nu^2}\left(\frac{du}{dx} - \nu\frac{w}{a}\right) = 0$$
$$N_\varphi = \frac{Eh}{1-\nu^2}(\epsilon_\varphi + \nu\epsilon_x) = \frac{Eh}{1-\nu^2}\left(-\frac{w}{a} + \nu\frac{du}{dx}\right) \qquad (d)$$

From the first of these equations it follows that

$$\frac{du}{dx} = \nu\frac{w}{a}$$

---
[1] The effect of these forces on bending is neglected in this discussion.

and the second equation gives

$$N_\varphi = -\frac{Ehw}{a} \qquad (e)$$

Considering the bending moments, we conclude from symmetry that there is no change in curvature in the circumferential direction. The curvature in the x direction is equal to $-d^2w/dx^2$. Using the same equations as for plates, we then obtain

$$M_\varphi = \nu M_x$$
$$M_x = -D\frac{d^2w}{dx^2} \qquad (f)$$

where
$$D = \frac{Eh^3}{12(1-\nu^3)}$$

is the flexural rigidity of the shell.

Returning now to Eqs. (b) and eliminating Q_x from these equations, we obtain

$$\frac{d^2M_x}{dx^2} + \frac{1}{a}N_\varphi = -Z$$

from which, by using Eqs. (e) and (f), we obtain

$$\frac{d^2}{dx^2}\left(D\frac{d^2w}{dx^2}\right) + \frac{Eh}{a^2}w = Z \qquad (273)$$

All problems of symmetrical deformation of circular cylindrical shells thus reduce to the integration of Eq. (273).

The simplest application of this equation is obtained when the thickness of the shell is constant. Under such conditions Eq. (273) becomes

$$D\frac{d^4w}{dx^4} + \frac{Eh}{a^2}w = Z \qquad (274)$$

Using the notation

$$\beta^4 = \frac{Eh}{4a^2D} = \frac{3(1-\nu^2)}{a^2h^2} \qquad (275)$$

Eq. (274) can be represented in the simplified form

$$\frac{d^4w}{dx^4} + 4\beta^4 w = \frac{Z}{D} \qquad (276)$$

This is the same equation as is obtained for a prismatical bar with a flexural rigidity D, supported by a continuous elastic foundation and submitted to the action of a load of intensity Z.* The general solution of this equation is

$$w = e^{\beta x}(C_1 \cos \beta x + C_2 \sin \beta x)$$
$$+ e^{-\beta x}(C_3 \cos \beta x + C_4 \sin \beta x) + f(x) \qquad (277)$$

* See S. Timoshenko, "Strength of Materials," part II, 3d ed., p. 2, 1956.

GENERAL THEORY OF CYLINDRICAL SHELLS 469

in which $f(x)$ is a particular solution of Eq. (276), and $C_1, \ldots, C_4$ are the constants of integration which must be determined in each particular case from the conditions at the ends of the cylinder.

Take, as an example, a long circular pipe submitted to the action of bending moments M_0 and shearing forces Q_0, both uniformly distributed along the edge $x = 0$ (Fig. 236). In this case there is no pressure Z distributed over the surface of the shell, and $f(x) = 0$ in the general solution (277). Since the forces applied at the end $x = 0$ produce a local bending which dies out rapidly as the distance x from the loaded end increases, we conclude that the first term on the right-hand side of Eq. (277) must vanish.[1] Hence, $C_1 = C_2 = 0$, and we obtain

$$w = e^{-\beta x}(C_3 \cos \beta x + C_4 \sin \beta x) \qquad (g)$$

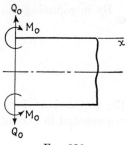

FIG. 236

The two constants C_3 and C_4 can now be determined from the conditions at the loaded end, which may be written

$$(M_x)_{x=0} = -D\left(\frac{d^2w}{dx^2}\right)_{x=0} = M_0$$
$$(Q_x)_{x=0} = \left(\frac{dM_x}{dx}\right)_{x=0} = -D\left(\frac{d^3w}{dx^3}\right)_{x=0} = Q_0 \qquad (h)$$

Substituting expression (g) for w, we obtain from these end conditions

$$C_3 = -\frac{1}{2\beta^3 D}(Q_0 + \beta M_0) \qquad C_4 = \frac{M_0}{2\beta^2 D} \qquad (i)$$

Thus the final expression for w is

$$w = \frac{e^{-\beta x}}{2\beta^3 D}[\beta M_0(\sin \beta x - \cos \beta x) - Q_0 \cos \beta x] \qquad (278)$$

The maximum deflection is obtained at the loaded end, where

$$(w)_{x=0} = -\frac{1}{2\beta^3 D}(\beta M_0 + Q_0) \qquad (279)$$

The negative sign for this deflection results from the fact that w is taken positive toward the axis of the cylinder. The slope at the loaded end is

---

[1] Observing the fact that the system of forces applied at the end of the pipe is a balanced one and that the length of the pipe may be increased at will, this follows also from the principle of Saint-Venant; see, for example, S. Timoshenko and J. N. Goodier, "Theory of Elasticity," 2d ed., p. 33, 1951.

obtained by differentiating expression (278). This gives

$$\left(\frac{dw}{dx}\right)_{x=0} = \frac{e^{-\beta x}}{2\beta^2 D}[2\beta M_0 \cos\beta x + Q_0(\cos\beta x + \sin\beta x)]_{x=0}$$

$$= \frac{1}{2\beta^2 D}(2\beta M_0 + Q_0) \quad (280)$$

By introducing the notation

$$\begin{aligned}\varphi(\beta x) &= e^{-\beta x}(\cos\beta x + \sin\beta x)\\ \psi(\beta x) &= e^{-\beta x}(\cos\beta x - \sin\beta x)\\ \theta(\beta x) &= e^{-\beta x}\cos\beta x\\ \zeta(\beta x) &= e^{-\beta x}\sin\beta x\end{aligned} \quad (281)$$

the expressions for deflection and its consecutive derivatives can be represented in the following simplified form:

$$\begin{aligned}w &= -\frac{1}{2\beta^3 D}[\beta M_0 \psi(\beta x) + Q_0 \theta(\beta x)]\\ \frac{dw}{dx} &= \frac{1}{2\beta^2 D}[2\beta M_0 \theta(\beta x) + Q_0 \varphi(\beta x)]\\ \frac{d^2 w}{dx^2} &= -\frac{1}{2\beta D}[2\beta M_0 \varphi(\beta x) + 2Q_0 \zeta(\beta x)]\\ \frac{d^3 w}{dx^3} &= \frac{1}{D}[2\beta M_0 \zeta(\beta x) - Q_0 \psi(\beta x)]\end{aligned} \quad (282)$$

The numerical values of the functions $\varphi(\beta x)$, $\psi(\beta x)$, $\theta(\beta x)$, and $\zeta(\beta x)$ are given in Table 84.[1] The functions $\varphi(\beta x)$ and $\psi(\beta x)$ are represented graphically in Fig. 237. It is seen from these curves and from Table 84

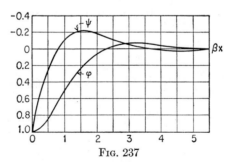

Fig. 237

that the functions defining the bending of the shell approach zero as the quantity βx becomes large. This indicates that the bending produced in the shell is of a local character, as was already mentioned at the beginning when the constants of integration were calculated.

If the moment M_x and the deflection w are found from expressions

[1] The figures in this table are taken from the book by H. Zimmermann, "Die Berechnung des Eisenbahnoberbaues," Berlin, 1888.

(282), the bending moment M_φ is obtained from the first of the equations (f), and the value of the force N_φ from Eq. (e). Thus all necessary information for calculating stresses in the shell can be found.

115. Particular Cases of Symmetrical Deformation of Circular Cylindrical Shells. *Bending of a Long Cylindrical Shell by a Load Uniformly Distributed along a Circular Section* (Fig. 238). If the load is far enough from the ends of the cylinder, solution (278) can be used for each half of

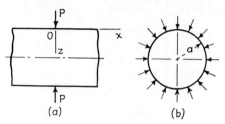

FIG. 238

the shell. From considerations of symmetry we conclude that the value of Q_0 in this case is $-P/2$. We thus obtain for the right-hand portion

$$w = \frac{e^{-\beta x}}{2\beta^3 D} \left[\beta M_0 (\sin \beta x - \cos \beta x) + \frac{P}{2} \cos \beta x \right] \quad (a)$$

where x is measured from the cross section at which the load is applied. To calculate the moment M_0 which appears in expression (a) we use expression (280), which gives the slope at $x = 0$. In our case this slope vanishes because of symmetry. Hence,

$$2\beta M_0 - \frac{P}{2} = 0$$

and we obtain

$$M_0 = \frac{P}{4\beta} \quad (b)$$

Substituting this value in expression (a), the deflection of the shell becomes

$$w = \frac{P e^{-\beta x}}{8\beta^3 D} (\sin \beta x + \cos \beta x) = \frac{P}{8\beta^3 D} \varphi(\beta x) \quad (283)$$

and by differentiation we find

$$\frac{dw}{dx} = -2\beta \frac{P}{8\beta^3 D} e^{-\beta x} \sin \beta x = -\frac{P}{4\beta^2 D} \zeta(\beta x)$$

$$\frac{d^2 w}{dx^2} = 2\beta^2 \frac{P}{8\beta^3 D} e^{-\beta x} (\sin \beta x - \cos \beta x) = -\frac{P}{4\beta D} \psi(\beta x) \quad (c)$$

$$\frac{d^3 w}{dx^3} = 4\beta^3 \frac{P}{8\beta^3 D} e^{-\beta x} \cos \beta x = \frac{P}{2D} \theta(\beta x)$$

TABLE 84. TABLE OF FUNCTIONS φ, ψ, θ, AND ζ

βx	φ	ψ	θ	ζ
0	1.0000	1.0000	1.0000	0
0.1	0.9907	0.8100	0.9003	0.0903
0.2	0.9651	0.6398	0.8024	0.1627
0.3	0.9267	0.4888	0.7077	0.2189
0.4	0.8784	0.3564	0.6174	0.2610
0.5	0.8231	0.2415	0.5323	0.2908
0.6	0.7628	0.1431	0.4530	0.3099
0.7	0.6997	0.0599	0.3798	0.3199
0.8	0.6354	−0.0093	0.3131	0.3223
0.9	0.5712	−0.0657	0.2527	0.3185
1.0	0.5083	−0.1108	0.1988	0.3096
1.1	0.4476	−0.1457	0.1510	0.2967
1.2	0.3899	−0.1716	0.1091	0.2807
1.3	0.3355	−0.1897	0.0729	0.2626
1.4	0.2849	−0.2011	0.0419	0.2430
1.5	0.2384	−0.2068	0.0158	0.2226
1.6	0.1959	−0.2077	−0.0059	0.2018
1.7	0.1576	−0.2047	−0.0235	0.1812
1.8	0.1234	−0.1985	−0.0376	0.1610
1.9	0.0932	−0.1899	−0.0484	0.1415
2.0	0.0667	−0.1794	−0.0563	0.1230
2.1	0.0439	−0.1675	−0.0618	0.1057
2.2	0.0244	−0.1548	−0.0652	0.0895
2.3	0.0080	−0.1416	−0.0668	0.0748
2.4	−0.0056	−0.1282	−0.0669	0.0613
2.5	−0.0166	−0.1149	−0.0658	0.0492
2.6	−0.0254	−0.1019	−0.0636	0.0383
2.7	−0.0320	−0.0895	−0.0608	0.0287
2.8	−0.0369	−0.0777	−0.0573	0.0204
2.9	−0.0403	−0.0666	−0.0534	0.0132
3.0	−0.0423	−0.0563	−0.0493	0.0071
3.1	−0.0431	−0.0469	−0.0450	0.0019
3.2	−0.0431	−0.0383	−0.0407	−0.0024
3.3	−0.0422	−0.0306	−0.0364	−0.0058
3.4	−0.0408	−0.0237	−0.0323	−0.0085
3.5	−0.0389	−0.0177	−0.0283	−0.0106
3.6	−0.0366	−0.0124	−0.0245	−0.0121
3.7	−0.0341	−0.0079	−0.0210	−0.0131
3.8	−0.0314	−0.0040	−0.0177	−0.0137
3.9	−0.0286	−0.0008	−0.0147	−0.0140

TABLE 84. TABLE OF FUNCTIONS φ, ψ, θ, AND ζ (Continued)

βx	φ	ψ	θ	ζ
4.0	−0.0258	0.0019	−0.0120	−0.0139
4.1	−0.0231	0.0040	−0.0095	−0.0136
4.2	−0.0204	0.0057	−0.0074	−0.0131
4.3	−0.0179	0.0070	−0.0054	−0.0125
4.4	−0.0155	0.0079	−0.0038	−0.0117
4.5	−0.0132	0.0085	−0.0023	−0.0108
4.6	−0.0111	0.0089	−0.0011	−0.0100
4.7	−0.0092	0.0090	0.0001	−0.0091
4.8	−0.0075	0.0089	0.0007	−0.0082
4.9	−0.0059	0.0087	0.0014	−0.0073
5.0	−0.0046	0.0084	0.0019	−0.0065
5.1	−0.0033	0.0080	0.0023	−0.0057
5.2	−0.0023	0.0075	0.0026	−0.0049
5.3	−0.0014	0.0069	0.0028	−0.0042
5.4	−0.0006	0.0064	0.0029	−0.0035
5.5	0.0000	0.0058	0.0029	−0.0029
5.6	0.0005	0.0052	0.0029	−0.0023
5.7	0.0010	0.0046	0.0028	−0.0018
5.8	0.0013	0.0041	0.0027	−0.0014
5.9	0.0015	0.0036	0.0026	−0.0010
6.0	0.0017	0.0031	0.0024	−0.0007
6.1	0.0018	0.0026	0.0022	−0.0004
6.2	0.0019	0.0022	0.0020	−0.0002
6.3	0.0019	0.0018	0.0018	+0.0001
6.4	0.0018	0.0015	0.0017	0.0003
6.5	0.0018	0.0012	0.0015	0.0004
6.6	0.0017	0.0009	0.0013	0.0005
6.7	0.0016	0.0006	0.0011	0.0006
6.8	0.0015	0.0004	0.0010	0.0006
6.9	0.0014	0.0002	0.0008	0.0006
7.0	0.0013	0.0001	0.0007	0.0006

Observing from Eqs. (*b*) and (*f*) of the preceding article that

$$M_x = -D\frac{d^2w}{dx^2} \qquad Q_x = -D\frac{d^3w}{dx^3}$$

we finally obtain the following expressions for the bending moment and shearing force:

$$M_x = \frac{P}{4\beta}\psi(\beta x) \qquad Q_x = -\frac{P}{2}\theta(\beta x) \qquad (284)$$

The results obtained are all graphically represented in Fig. 239. It is seen that the maximum deflection is under the load P and that its value as given by Eq. (283) is

$$w_{max} = \frac{P}{8\beta^3 D} = \frac{Pa^2\beta}{2Eh} \quad (285)$$

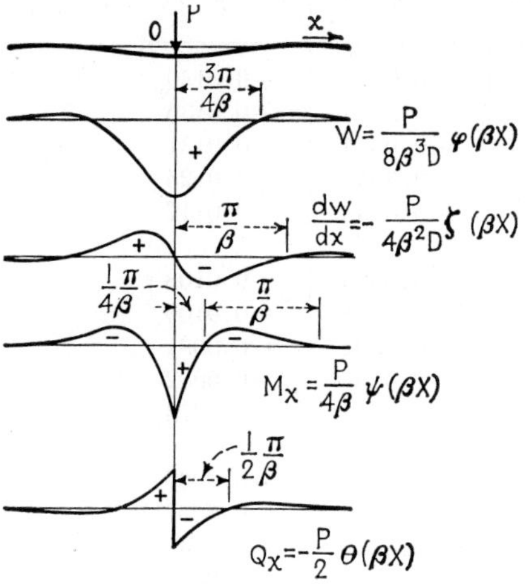

The maximum bending moment is also under the load and is determined from Eq. (284) as

$$M_{max} = \frac{P}{4\beta} \quad (286)$$

The maximum of the absolute value of the shearing force is evidently equal to $P/2$. The values of all these quantities at a certain distance from the load can be readily obtained by using Table 84. We see from this table and from Fig. 239 that all the quantities that determine the bending of the shell are small for $x > \pi/\beta$. This fact indicates that the bending is of a local character and that a shell of length $l = 2\pi/\beta$ loaded at the middle will have practically the same maximum deflection and the same maximum stress as a very long shell.

Fig. 239

Having the solution of the problem for the case in which a load is concentrated at a circular cross section, we can readily solve the problem of a load distributed along a certain length of the cylinder by applying the principle of superposition. As an example let us consider the case of a load of intensity q uniformly distributed along a length l of a cylinder (Fig. 240). Assuming that the load is at a considerable distance from the ends of the cylinder, we can use solution (283) to calculate the deflections.

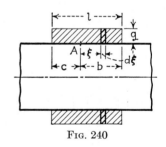

Fig. 240

The deflection at a point A produced by an elementary ring load of an intensity[1] $q\, d\xi$ at a distance ξ from A is obtained from expression (283) by substituting $q\, d\xi$ for P and ξ for x and is

$$\frac{q\, d\xi}{8\beta^3 D} e^{-\beta\xi}(\cos \beta\xi + \sin \beta\xi)$$

The deflection produced at A by the total load distributed over the

[1] $q\, d\xi$ is the load per unit length of circumference.

length l is then

$$w = \int_o^b \frac{q\,d\xi}{8\beta^3 D} e^{-\beta\xi}(\cos\beta\xi + \sin\beta\xi) + \int_o^c \frac{q\,d\xi}{8\beta^3 D} e^{-\beta\xi}(\cos\beta\xi + \sin\beta\xi)$$
$$= \frac{qa^2}{2Eh}(2 - e^{-\beta b}\cos\beta b - e^{-\beta c}\cos\beta c)$$

The bending moment at a point A can be calculated by similar application of the method of superposition.

Cylindrical Shell with a Uniform Internal Pressure (Fig. 241). If the edges of the shell are free, the internal pressure p produces only a hoop stress

$$\sigma_t = \frac{pa}{h}$$

and the radius of the cylinder increases by the amount

$$\delta = \frac{a\sigma_t}{E} = \frac{pa^2}{Eh} \qquad (d)$$

If the ends of the shell are built in, as shown in Fig. 241a, they cannot move out, and local bending occurs at the edges. If the length l of the

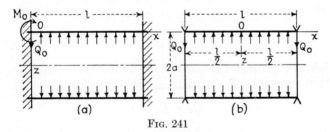

FIG. 241

shell is sufficiently large, we can use solution (278) to investigate this bending, the moment M_0 and the shearing force Q_0 being determined from the conditions that the deflection and the slope along the built-in edge $x = 0$ (Fig. 241a) vanish. According to these conditions, Eqs. (279) and (280) of the preceding article become

$$-\frac{1}{2\beta^3 D}(\beta M_0 + Q_0) = \delta$$

$$\frac{1}{2\beta^2 D}(2\beta M_0 + Q_0) = 0$$

where δ is given by Eq. (d).

Solving for M_0 and Q_0, we obtain

$$M_0 = 2\beta^2 D\delta = \frac{p}{2\beta^2} \qquad Q_0 = -4\beta^3 D\delta = -\frac{p}{\beta} \qquad (287)$$

We thus obtain a positive bending moment and a negative shearing force acting as shown in Fig. 241a. Substituting these values in expressions (282), the deflection and the bending moment at any distance from the end can be readily calculated using Table 84.

If, instead of built-in edges, we have simply supported edges as shown in Fig. 241b, the deflection and the bending moment M_x vanish along the edge $M_0 = 0$, and we obtain, by using Eq. (279),

$$Q_0 = -2\beta^3 D\delta$$

By substituting these values in solution (278) the deflection at any distance from the end can be calculated.

It was assumed in the preceding discussion that the length of the shell is large. If this is not the case, the bending at one end cannot be considered as independent of the conditions at the other end, and recourse must be had to the general solution (277), which contains four constants of integration. The particular solution of Eq. (276) for the case of uniform load $(Z = -p)$ is $-p/4\beta^4 D = -pa^2/Eh$. The general solution (277) can then be put in the following form by the introduction of hyperbolic functions in place of the exponential functions:

$$w = -\frac{pa^2}{Eh} + C_1 \sin \beta x \sinh \beta x + C_2 \sin \beta x \cosh \beta x$$
$$+ C_3 \cos \beta x \sinh \beta x + C_4 \cos \beta x \cosh \beta x \quad (e)$$

If the origin of coordinates is taken at the middle of the cylinder, as shown in Fig. 241b, expression (e) must be an even function of x. Hence

$$C_2 = C_3 = 0 \quad (f)$$

The constants C_1 and C_4 must now be selected so as to satisfy the conditions at the ends. If the ends are simply supported, the deflection and the bending moment M_x must vanish at the ends, and we obtain

$$(w)_{x=l/2} = 0 \quad \left(\frac{d^2w}{dx^2}\right)_{x=l/2} = 0 \quad (g)$$

Substituting expression (e) in these relations and remembering that $C_2 = C_3 = 0$, we find

$$-\frac{pa^2}{Eh} + C_1 \sin \alpha \sinh \alpha + C_4 \cos \alpha \cosh \alpha = 0 \quad (h)$$
$$C_1 \cos \alpha \cosh \alpha - C_4 \sin \alpha \sinh \alpha = 0$$

where, for the sake of simplicity,

$$\frac{\beta l}{2} = \alpha \quad (i)$$

From these equations we obtain

$$C_1 = \frac{pa^2}{Eh} \frac{\sin\alpha \sinh\alpha}{\sin^2\alpha \sinh^2\alpha + \cos^2\alpha \cosh^2\alpha} = \frac{pa^2}{Eh} \frac{2\sin\alpha \sinh\alpha}{\cos 2\alpha + \cosh 2\alpha}$$
$$C_4 = \frac{pa^2}{Eh} \frac{\cos\alpha \cosh\alpha}{\sin^2\alpha \sinh^2\alpha + \cos^2\alpha \cosh^2\alpha} = \frac{pa^2}{Eh} \frac{2\cos\alpha \cosh\alpha}{\cos 2\alpha + \cosh 2\alpha} \quad (j)$$

Substituting the values (j) and (f) of the constants in expression (e) and observing from expression (275) that

$$\frac{Eh}{a^2} = 4D\beta^4 = \frac{64\alpha^4 D}{l^4} \quad (k)$$

we obtain

$$w = -\frac{pl^4}{64 D \alpha^4}\left(1 - \frac{2\sin\alpha \sinh\alpha}{\cos 2\alpha + \cosh 2\alpha} \sin\beta x \sinh\beta x \right.$$
$$\left. - \frac{2\cos\alpha \cosh\alpha}{\cos 2\alpha + \cosh 2\alpha} \cos\beta x \cosh\beta x \right) \quad (l)$$

In each particular case, if the dimensions of the shell are known, the quantity α, which is dimensionless, can be calculated by means of notation (i) and Eq. (275). By substituting this value in expression (l) the deflection of the shell at any point can be found.

For the middle of the shell, substituting $x = 0$ in expression (l), we obtain

$$(w)_{x=0} = -\frac{pl^4}{64 D \alpha^4}\left(1 - \frac{2\cos\alpha \cosh\alpha}{\cos 2\alpha + \cosh 2\alpha}\right) \quad (m)$$

When the shell is long, α becomes large, the second term in the parentheses of expression (m) becomes small, and the deflection approaches the value (d) calculated for the case of free ends. This indicates that in the case of long shells the effect of the end supports upon the deflection at the middle is negligible. Taking another extreme case, viz., the case when α is very small, we can show by expanding the trigonometric and hyperbolic functions in power series that the expression in parentheses in Eq. (m) approaches the value $5\alpha^4/6$ and that the deflection (l) approaches that for a uniformly loaded and simply supported beam of length l and flexural rigidity D.

Differentiating expression (l) twice and multiplying it by D, the bending moment is found as

$$M_x = -D\frac{d^2w}{dx^2} = -\frac{pl^2}{4\alpha^2}\left(\frac{\sin\alpha \sinh\alpha}{\cos 2\alpha + \cosh 2\alpha} \cosh\beta x \cos\beta x \right.$$
$$\left. - \frac{\cos\alpha \cosh\alpha}{\cos 2\alpha + \cosh 2\alpha} \sin\beta x \sinh\beta x\right) \quad (n)$$

At the middle of the shell this moment is

$$(M_x)_{x=0} = -\frac{pl^2}{4\alpha^2} \frac{\sin\alpha \sinh\alpha}{\cos 2\alpha + \cosh 2\alpha} \qquad (o)$$

It is seen that for large values of α, that is, for long shells, this moment becomes negligibly small and the middle portion is, for all practical purposes, under the action of merely the hoop stresses pa/h.

The case of a cylinder with built-in edges (Fig. 241a) can be treated in a similar manner. Going directly to the final result,[1] we find that the bending moment M_0 acting along the built-in edge is

$$M_0 = \frac{p}{2\beta^2} \frac{\sinh 2\alpha - \sin 2\alpha}{\sinh 2\alpha + \sin 2\alpha} = \frac{p}{2\beta^2} \chi_2(2\alpha) \qquad (288)$$

where
$$\chi_2(2\alpha) = \frac{\sinh 2\alpha - \sin 2\alpha}{\sinh 2\alpha + \sin 2\alpha}$$

In the case of long shells, α is large, the factor $\chi_2(2\alpha)$ in expression (288) approaches unity, and the value of the moment approaches that given by the first of the expressions (287). For shorter shells the value of the factor $\chi_2(2\alpha)$ in (288) can be taken from Table 85.

TABLE 85

2α	$\chi_1(2\alpha)$	$\chi_2(2\alpha)$	$\chi_3(2\alpha)$
0.2	5.000	0.0068	0.100
0.4	2.502	0.0268	0.200
0.6	1.674	0.0601	0.300
0.8	1.267	0.1065	0.400
1.0	1.033	0.1670	0.500
1.2	0.890	0.2370	0.596
1.4	0.803	0.3170	0.689
1.6	0.755	0.4080	0.775
1.8	0.735	0.5050	0.855
2.0	0.738	0.6000	0.925
2.5	0.802	0.8220	1.045
3.0	0.893	0.9770	1.090
3.5	0.966	1.0500	1.085
4.0	1.005	1.0580	1.050
4.5	1.017	1.0400	1.027
5.0	1.017	1.0300	1.008

Cylindrical Shell Bent by Forces and Moments Distributed along the Edges. In the preceding section this problem was discussed assuming

[1] Both cases are discussed in detail by I. G. Boobnov in his "Theory of Structure of Ships," vol. 2, p. 368, St. Petersburg, 1913. Also included are numerical tables which simplify the calculations of moments and deflections.

GENERAL THEORY OF CYLINDRICAL SHELLS 479

that the shell is long and that each end can be treated independently. In the case of shorter shells both ends must be considered simultaneously by using solution (e) with four constants of integration. Proceeding as in the previous cases, the following results can be obtained. For the case of bending by uniformly distributed shearing forces Q_0 (Fig. 242a), the deflection and the slope at the ends are

$$(w)_{x=0, x=l} = -\frac{2Q_0\beta a^2}{Eh} \frac{\cosh 2\alpha + \cos 2\alpha}{\sinh 2\alpha + \sin 2\alpha} = -\frac{2Q_0\beta a^2}{Eh} \chi_1(2\alpha)$$
$$\left(\frac{dw}{dx}\right)_{x=0, x=l} = \pm \frac{2Q_0\beta^2 a^2}{Eh} \frac{\sinh 2\alpha - \sin 2\alpha}{\sinh 2\alpha + \sin 2\alpha} = \pm \frac{2Q_0\beta^2 a^2}{Eh} \chi_2(2\alpha) \quad (289)$$

In the case of bending by the moments M_0 (Fig. 242b), we obtain

$$(w)_{x=0, x=l} = -\frac{2M_0\beta^2 a^2}{Eh} \frac{\sinh 2\alpha - \sin 2\alpha}{\sinh 2\alpha + \sin 2\alpha} = -\frac{2M_0\beta^2 a^2}{Eh} \chi_2(2\alpha)$$
$$\left(\frac{dw}{dx}\right)_{x=0, x=l} = \pm \frac{4M_0\beta^3 a^2}{Eh} \frac{\cosh 2\alpha - \cos 2\alpha}{\sinh 2\alpha + \sin 2\alpha} = \pm \frac{4M_0\beta^3 a^2}{Eh} \chi_3(2\alpha) \quad (290)$$

In the case of long shells, the factors χ_1, χ_2, and χ_3 in expressions (289) and (290) are close to unity, and the results coincide with those given by

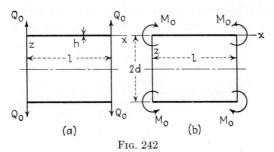

FIG. 242

expressions (279) and (280). To simplify the calculations for shorter shells, the values of functions χ_1, χ_2, and χ_3 are given in Table 85.

Using solutions (289) and (290), the stresses in a long pipe reinforced by equidistant rings (Fig. 243) and submitted to the action of uniform internal pressure p can be readily discussed.

Assume first that there are no rings. Then, under the action of internal pressure, hoop stresses $\sigma_t = pa/h$ will be produced, and the radius of the pipe will increase by the amount

$$\delta = \frac{pa^2}{Eh}$$

Now, taking the rings into consideration and assuming that they are absolutely rigid, we conclude that reactive forces will be produced between each ring and the pipe. The magnitude of the forces per unit length of the circumference of the tube will be denoted by P. The magnitude of P will now be determined from the condition that the forces P produce a deflection of the pipe under the ring equal to the expansion δ created by the internal pressure p. In calculating this deflection we observe that a portion of the tube between two adjacent rings may be considered as the shell shown in Fig. 242a and b. In this case $Q_0 = -\frac{1}{2}P$, and the magnitude of the bending moment M_0 under a ring is determined from the condition that $dw/dx = 0$ at that point. Hence from Eqs. (289) and (290) we find

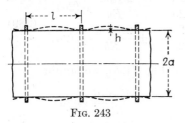

Fig. 243

$$-\frac{P\beta^2 a^2}{Eh}\chi_2(2\alpha) + \frac{4M_0\beta^3 a^2}{Eh}\chi_3(2\alpha) = 0$$

from which

$$M_0 = \frac{P\chi_2(2\alpha)}{4\beta\chi_3(2\alpha)} \qquad (p)$$

If the distance l between the rings is large,[1] the quantity

$$2\alpha = \beta l = \frac{l}{\sqrt{ah}}\sqrt[4]{3(1-\nu^2)}$$

is also large, the functions $\chi_2(2\alpha)$ and $\chi_3(2\alpha)$ approach unity, and the moment M_0 approaches the value (286). For calculating the force P entering in Eq. (p) the expressions for deflections as given in Eqs. (289) and (290) must be used. These expressions give

$$\frac{P\beta a^2}{Eh}\chi_1(2\alpha) - \frac{P\beta a^2}{2Eh}\frac{\chi_2^2(2\alpha)}{\chi_3(2\alpha)} = \delta = \frac{pa^2}{Eh}$$

or

$$P\beta\left[\chi_1(2\alpha) - \frac{1}{2}\frac{\chi_2^2(2\alpha)}{\chi_3(2\alpha)}\right] = \frac{\delta Eh}{a^2} = p \qquad (291)$$

For large values of 2α this reduces to

$$\frac{P\beta a^2}{2Eh} = \delta$$

which coincides with Eq. (285). When 2α is not large, the value of the reactive forces P is calculated from Eq. (291) by using Table 85. Solving Eq. (291) for P and substituting its expression in expression (p), we find

$$M_0 = \frac{p}{2\beta^2}\chi_2(2\alpha) \qquad (292)$$

This coincides with expression (288) previously obtained for a shell with built-in edges.

To take into account the extension of rings we observe that the reactive

[1] For $\nu = 0.3$, $2\alpha = 1.285l/\sqrt{ah}$.

forces P produce in the ring a tensile force Pa and that the corresponding increase of the inner radius of the ring is[1]

$$\delta_1 = \frac{Pa^2}{AE}$$

where A is the cross-sectional area of the ring. To take this extension into account we substitute $\delta - \delta_1$ for δ in Eq. (291) and obtain

$$P\beta \left[\chi_1(2\alpha) - \frac{1}{2} \frac{\chi_2^2(2\alpha)}{\chi_3(2\alpha)} \right] = p - \frac{Ph}{A} \qquad (293)$$

From this equation, P can be readily obtained by using Table 85, and the moment found by substituting $p - (Ph/A)$ for p in Eq. (292).

If the pressure p acts not only on the cylindrical shell but also on the ends, longitudinal forces

$$N_x = \frac{pa}{2}$$

are produced in the shell. The extension of the radius of the cylinder is then

$$\delta' = \frac{pa^2}{Eh} \left(1 - \frac{1}{2} \nu \right)$$

and the quantity $p(1 - \frac{1}{2}\nu)$ must be substituted for p in Eqs. (292) and (293).

Equations (293) and (291) can also be used in the case of external uniform pressure provided the compressive stresses in the ring and in the shell are far enough from the critical stresses at which buckling may occur.[2] This case is of practical importance in the design of submarines and has been discussed by several authors.[3]

116. Pressure Vessels. The method illustrated by the examples of the preceding article can also be applied in the analysis of stresses in cylindrical vessels submitted to the action of internal pressure.[4] In discussing the "membrane theory" it was repeatedly indicated that this theory fails to represent the true stresses in those portions of a shell close to the

---

[1] It is assumed that the cross-sectional dimensions of the ring are small in comparison with the radius a.

[2] Buckling of rings and cylindrical shells is discussed in S. Timoshenko, "Theory of Elastic Stability," 1936.

[3] See paper by K. von Sanden and K. Günther, "Werft und Reederei," vol. 1, 1920, pp. 163–168, 189–198, 216–221, and vol. 2, 1921, pp. 505–510.

[4] See also M. Esslinger, "Statische Berechnung von Kesselböden," Berlin, 1952; G. Salet and J. Barthelemy, *Bull. Assoc. Tech. Maritime Aeronaut.*, vol. 44, p. 505, 1945; J. L. Maulbetsch and M. Hetényi, *ASCE Design Data*, no. 1, 1944, and F. Schultz-Grunow, *Ingr.-Arch.*, vol. 4, p. 545, 1933; N. L. Svensson, *J. Appl. Mechanics*, vol. 25, p. 89, 1958.

edges, since the edge conditions usually cannot be completely satisfied by considering only membrane stresses. A similar condition in which the membrane theory is inadequate is found in cylindrical pressure vessels at the joints between the cylindrical portion and the ends of the vessel. At these joints the membrane stresses are usually accompanied by local bending stresses which are distributed symmetrically with respect to the axis of the cylinder. These local stresses can be calculated by using solution (278) of Art. 114.

Let us begin with the simple case of a cylindrical vessel with hemispherical ends (Fig. 244).[1] At a sufficient distance from the joints mn

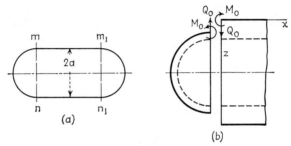

Fig. 244

and $m_1 n_1$ the membrane theory is accurate enough and gives for the cylindrical portion of radius a

$$N_x = \frac{pa}{2} \qquad N_t = pa \qquad (a)$$

where p denotes the internal pressure.

For the spherical ends this theory gives a uniform tensile force

$$N = \frac{pa}{2} \qquad (b)$$

The extension of the radius of the cylindrical shell under the action of the forces (a) is

$$\delta_1 = \frac{pa^2}{Eh}\left(1 - \frac{\nu}{2}\right) \qquad (c)$$

and the extension of the radius of the spherical ends is

$$\delta_2 = \frac{pa^2}{2Eh}(1 - \nu) \qquad (d)$$

Comparing expressions (c) and (d), it can be concluded that if we consider only membrane stresses we obtain a discontinuity at the joints as represented in Fig. 244b. This indicates that at the joint there must act

[1] This case was discussed by E. Meissner, *Schweiz. Bauztg.*, vol. 86, p. 1, 1925.

shearing forces Q_0 and bending moments M_0 uniformly distributed along the circumference and of such magnitudes as to eliminate this discontinuity. The stresses produced by these forces are sometimes called *discontinuity stresses*.

In calculating the quantities Q_0 and M_0 we assume that the bending is of a local character so that solution (278) can be applied with sufficient accuracy in discussing the bending of the cylindrical portion. The investigation of the bending of the spherical ends represents a more complicated problem which will be fully discussed in Chap. 16. Here we obtain an approximate solution of the problem by assuming that the bending is of importance only in the zone of the spherical shell close to the joint and that this zone can be treated as a portion of a long cylindrical shell[1] of radius a. If the thickness of the spherical and the cylindrical portion of the vessel is the same, the forces Q_0 produce equal rotations of the edges of both portions at the joint (Fig. 244b). This indicates that M_0 vanishes and that Q_0 alone is sufficient to eliminate the discontinuity. The magnitude of Q_0 is now determined from the condition that the sum of the numerical values of the deflections of the edges of the two parts must be equal to the difference $\delta_1 - \delta_2$ of the radial expansions furnished by the membrane theory. Using Eq. (279) for the deflections, we obtain

$$\frac{Q_0}{\beta^3 D} = \delta_1 - \delta_2 = \frac{pa^2}{2Eh}$$

from which, by using notation (275),

$$Q_0 = \frac{pa^2 \beta^3 D}{2Eh} = \frac{p}{8\beta} \tag{e}$$

Having obtained this value of the force Q_0, the deflection and the bending moment M_x can be calculated at any point by using formulas (282), which give[2]

$$w = \frac{Q_0}{2\beta^3 D}\, \theta(\beta x)$$

$$M_x = -D\frac{d^2 w}{dx^2} = -\frac{Q_0}{\beta}\, \zeta(\beta x)$$

Substituting expression (e) for Q_0 and expression (275) for β in the formula for M_x, we obtain

$$M_x = -\frac{ahp}{8\sqrt{3(1-\nu^2)}}\, \zeta(\beta x) \tag{f}$$

[1] E. Meissner, in the above-mentioned paper, showed that the error in the magnitude of the bending stresses as calculated from such an approximate solution is small for thin hemispherical shells and is smaller than 1 per cent if $a/h > 30$.

[2] Note that the direction of Q_0 in Fig. 244 is opposite to the direction in Fig. 236.

This moment attains its numerical maximum at the distance $x = \pi/4\beta$, at which point the derivative of the moment is zero, as can be seen from the fourth of the equations (282).

Combining the maximum bending stress produced by M_x with the membrane stress, we find

$$(\sigma_x)_{\max} = \frac{ap}{2h} + \frac{3}{4} \frac{ap}{h\sqrt{3(1-\nu^2)}} \zeta\left(\frac{\pi}{4}\right) = 1.293 \frac{ap}{2h} \qquad (g)$$

This stress which acts at the outer surface of the cylindrical shell is about 30 per cent larger than the membrane stress acting in the axial direction. In calculating stresses in the circumferential direction in addition to the membrane stress pa/h, the hoop stress caused by the deflection w as well as the bending stress produced by the moment $M_\varphi = \nu M_x$ must be considered. In this way we obtain at the outer surface of the cylindrical shell

$$\sigma_t = \frac{ap}{h} - \frac{Ew}{a} - \frac{6\nu}{h^2} M_x = \frac{ap}{h}\left[1 - \frac{1}{4}\theta(\beta x) + \frac{3\nu}{4\sqrt{3(1-\nu^2)}}\zeta(\beta x)\right]$$

Taking $\nu = 0.3$ and using Table 84, we find

$$(\sigma_t)_{\max} = 1.032 \frac{ap}{h} \qquad \text{at } \beta x = 1.85 \qquad (h)$$

Since the membrane stress is smaller in the ends than in the cylinder sides, the maximum stress in the spherical ends is always smaller than the calculated stress (h). Thus the latter stress is the determining factor in the design of the vessel.

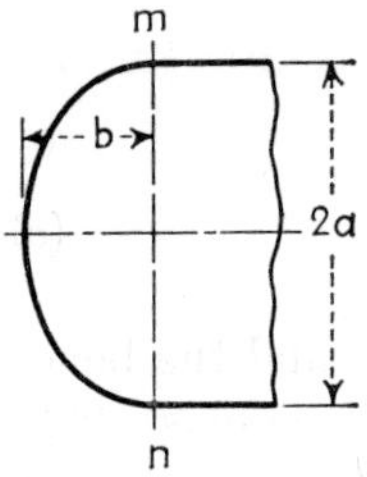

The same method of calculating discontinuity stresses can be applied in the case of ends having the form of an ellipsoid of revolution. The membrane stresses in this case are obtained from expressions (263) and (264) (see page 440). At the joint mn which represents the equator of the ellipsoid (Fig. 245), the stresses in the direction of the meridian and in the equatorial direction are, respectively,

Fig. 245

$$\sigma_\varphi = \frac{pa}{2h} \qquad \sigma_\theta = \frac{pa}{h}\left(1 - \frac{a^2}{2b^2}\right) \qquad (i)$$

The extension of the radius of the equator is

$$\delta_2' = \frac{a}{E}(\sigma_\theta - \nu\sigma_\varphi) = \frac{pa^2}{Eh}\left(1 - \frac{a^2}{2b^2} - \frac{\nu}{2}\right)$$

Substituting this quantity for δ_2 in the previous calculation of the shearing force Q_0, we find

$$\delta_1 - \delta_2' = \frac{pa^2}{Eh}\frac{a^2}{2b^2}$$

and, instead of Eq. (e), we obtain

$$Q_0 = \frac{p}{8\beta} \frac{a^2}{b^2}$$

It is seen that the shearing force Q_0 in the case of ellipsoidal ends is larger than in the case of hemispherical ends in the ratio a^2/b^2. The discontinuity stresses will evidently increase in the same proportion. For example, taking $a/b = 2$, we obtain, from expressions (g) and (h),

$$(\sigma_x)_{\max} = \frac{ap}{2h} + \frac{3ap}{h\sqrt{3(1-\nu^2)}} \zeta\left(\frac{\pi}{4}\right) = 2.172 \frac{ap}{2h}$$

$$(\sigma_t)_{\max} = 1.128 \frac{ap}{h}$$

Again, $(\sigma_t)_{\max}$ is the largest stress and is consequently the determining factor in design.[1]

117. Cylindrical Tanks with Uniform Wall Thickness. If a tank is submitted to the action of a liquid pressure, as shown in Fig. 246, the stresses in the wall can be analyzed by using Eq. (276). Substituting in this equation

$$Z = -\gamma(d - x) \qquad (a)$$

where γ is the weight per unit volume of the liquid, we obtain

$$\frac{d^4w}{dx^4} + 4\beta^4 w = -\frac{\gamma(d-x)}{D} \qquad (b)$$

A particular solution of this equation is

$$w_1 = -\frac{\gamma(d-x)}{4\beta^4 D} = -\frac{\gamma(d-x)a^2}{Eh} \qquad (c)$$

This expression represents the radial expansion of a cylindrical shell with free edges under the action of hoop stresses. Substituting expression (c) in place of $f(x)$ in expression (277), we obtain for the complete solution of Eq. (b)

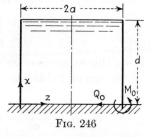

Fig. 246

$$w = e^{\beta x}(C_1 \cos \beta x + C_2 \sin \beta x) + e^{-\beta x}(C_3 \cos \beta x + C_4 \sin \beta x) - \frac{\gamma(d-x)a^2}{Eh}$$

In most practical cases the wall thickness h is small in comparison with both the radius a and the depth d of the tank, and we may consider the shell as infinitely long. The constants C_1 and C_2 are then equal to zero,

---

[1] More detail regarding stresses in boilers with ellipsoidal ends can be found in the book by Höhn, "Über die Festigkeit der gewölbten Böden und der Zylinderschale," Zürich, 1927. Also included are the results of experimental investigations of discontinuity stresses which are in a good agreement with the approximate solution. See also Schultz-Grunow, *loc. cit.*

and we obtain

$$w = e^{-\beta x}(C_3 \cos \beta x + C_4 \sin \beta x) - \frac{\gamma(d-x)a^2}{Eh} \qquad (d)$$

The constants C_3 and C_4 can now be obtained from the conditions at the bottom of the tank. Assuming that the lower edge of the wall is built into an absolutely rigid foundation, the boundary conditions are

$$(w)_{x=0} = C_3 - \frac{\gamma a^2 d}{Eh} = 0$$

$$\left(\frac{dw}{dx}\right)_{x=0} = \left[-\beta C_3 e^{-\beta x}(\cos \beta x + \sin \beta x)\right.$$

$$\left. + \beta C_4 e^{-\beta x}(\cos \beta x - \sin \beta x) + \frac{\gamma a^2}{Eh}\right]_{x=0} = \beta(C_4 - C_3) + \frac{\gamma a^2}{Eh} = 0$$

From these equations we obtain

$$C_3 = \frac{\gamma a^2 d}{Eh} \qquad C_4 = \frac{\gamma a^2}{Eh}\left(d - \frac{1}{\beta}\right)$$

Expression (d) then becomes

$$w = -\frac{\gamma a^2}{Eh}\left\{d - x - e^{-\beta x}\left[d \cos \beta x + \left(d - \frac{1}{\beta}\right)\sin \beta x\right]\right\}$$

from which, by using the notation of Eqs. (281), we obtain

$$w = -\frac{\gamma a^2 d}{Eh}\left[1 - \frac{x}{d} - \theta(\beta x) - \left(1 - \frac{1}{\beta d}\right)\zeta(\beta x)\right] \qquad (e)$$

From this expression the deflection at any point can be readily calculated by the use of Table 84. The force N_φ in the circumferential direction is then

$$N_\varphi = -\frac{Ehw}{a} = \gamma a d \left[1 - \frac{x}{d} - \theta(\beta x) - \left(1 - \frac{1}{\beta d}\right)\zeta(\beta x)\right] \qquad (f)$$

From the second derivative of expression (e) we obtain the bending moment

$$M_x = -D\frac{d^2 w}{dx^2} = \frac{2\beta^2 \gamma a^2 D d}{Eh}\left[-\zeta(\beta x) + \left(1 - \frac{1}{\beta d}\right)\theta(\beta x)\right]$$

$$= \frac{\gamma a d h}{\sqrt{12(1-\nu^2)}}\left[-\zeta(\beta x) + \left(1 - \frac{1}{\beta d}\right)\theta(\beta x)\right] \qquad (g)$$

Having expressions (f) and (g), the maximum stress at any point can readily be calculated in each particular case. The bending moment has its maximum value at the bottom, where it is equal to

$$(M_x)_{x=0} = M_0 = \left(1 - \frac{1}{\beta d}\right)\frac{\gamma a d h}{\sqrt{12(1-\nu^2)}} \qquad (h)$$

GENERAL THEORY OF CYLINDRICAL SHELLS

The same result can be obtained by using solutions (279) and (280) (pages 469, 470). Assuming that the lower edge of the shell is entirely free, we obtain from expression (c)

$$(w_1)_{x=0} = -\frac{\gamma a^2 d}{Eh} \qquad \left(\frac{dw_1}{dx}\right)_{x=0} = \frac{\gamma a^2}{Eh} \qquad (i)$$

To eliminate this displacement and rotation of the edge and thus satisfy the edge conditions at the bottom of the tank, a shearing force Q_0 and bending moment M_0 must be applied as indicated in Fig. 246. The magnitude of each of these quantities is obtained by equating expressions (279) and (280) to expressions (i) taken with reversed signs. This gives

$$-\frac{1}{2\beta^3 D}(\beta M_0 + Q_0) = +\frac{\gamma a^2 d}{Eh}$$

$$\frac{1}{2\beta^2 D}(2\beta M_0 + Q_0) = -\frac{\gamma a^2}{Eh}$$

From these equations we again obtain expression (h) for M_0, whereas for the shearing force we find[1]

$$Q_0 = -\frac{\gamma a d h}{\sqrt{12(1-\nu^2)}}\left(2\beta - \frac{1}{d}\right) \qquad (j)$$

Taking, as an example, $a = 30$ ft, $d = 26$ ft, $h = 14$ in., $\gamma = 0.03613$ lb per in.³, and $\nu = 0.25$, we find $\beta = 0.01824$ in.⁻¹ and $\beta d = 5.691$. For such a value of βd our assumption that the shell is infinitely long results in an accurate value for the moment and the shearing force, and we obtain from expressions (h) and (j)

$M_0 = 13,960$ in.-lb per in. $Q_0 = -563.6$ lb per in.

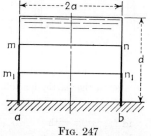

Fig. 247

In the construction of steel tanks, metallic sheets of several different thicknesses are very often used as shown in Fig. 247. Applying the particular solution (c) to each portion of uniform thickness, we find that the differences in thickness give rise to discontinuities in the displacement w_1 along the joints mn and m_1n_1.

These discontinuities, together with the displacements at the bottom ab, can be removed by applying moments and shearing forces. Assuming that the vertical dimension of each portion is sufficiently large to justify the application of the formulas for an infinitely large shell, we calculate the discontinuity moments and shearing forces as before by using Eqs. (279) and (280) and applying at each joint the two conditions that the adjacent portions of the shell have equal deflections and a common tangent. If the use of formulas (279) and (280) derived for an infinitely long shell cannot be justified, the general solution containing four constants of integration must be applied to each portion of the tank. The determination of the constants under such conditions becomes much more complicated, since the fact that each joint cannot be treated

---

[1] The negative sign indicates that Q_0 has the direction shown in Fig. 246 which is opposite to the direction used in Fig. 236 when deriving expressions (279) and (280).

independently necessitates the solution of a system of simultaneous equations. This problem can be solved by approximate methods.[1]

118. Cylindrical Tanks with Nonuniform Wall Thickness. In the case of tanks of nonuniform wall thickness the solution of the problem requires the integration of Eq. (273), considering the flexural rigidity D and the thickness h as no longer constant but as functions of x. We have thus to deal with a linear differential equation of fourth order with variable coefficients. As an example, let us consider the case when the thickness of the wall is a linear function of the coordinate x.[*] Taking the origin of the coordinates as shown in Fig. 248, we have for the thickness of the wall and for the flexural rigidity the expressions

$$h = \alpha x \qquad D = \frac{E\alpha^3}{12(1-\nu^2)} x^3 \qquad (a)$$

and Eq. (273) becomes

$$\frac{d^2}{dx^2}\left(x^3 \frac{d^2w}{dx^2}\right) + \frac{12(1-\nu^2)}{\alpha^2 a^2} xw = -\frac{12(1-\nu^2)\gamma(x-x_0)}{E\alpha^3} \qquad (b)$$

The particular solution of this equation is

$$w_1 = -\frac{\gamma a^2}{E\alpha} \frac{x-x_0}{x} \qquad (c)$$

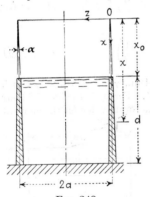

FIG. 248

This solution represents the radial expansion of a shell with free edges under the internal pressure $\gamma(x-x_0)$. As a result of the displacement (c) a certain amount of bending of the generatrices of the cylinder occurs. The corresponding bending moment is

$$M_x = -D \frac{d^2w_1}{dx^2} = -\frac{\gamma \alpha^2 a^2 x_0}{6(1-\nu^2)} \qquad (d)$$

This moment is independent of x and is in all practical cases of such small magnitude that its action can usually be neglected.

To obtain the complete solution of Eq. (b) we have to add to the particular solution (c) the solution of the homogeneous equation

$$\frac{d^2}{dx^2}\left(x^3 \frac{d^2w}{dx^2}\right) + \frac{12(1-\nu^2)}{\alpha^2 a^2} xw = 0$$

---

[1] An approximate method of solving this problem was given by C. Runge, *Z. Math. Physik*, vol. 51, p. 254, 1904. This method was applied by K. Girkmann in a design of a large welded tank; see *Stahlbau*, vol. 4, p. 25, 1931.

[*] H. Reissner, *Beton u. Eisen*, vol. 7, p. 150, 1908; see also W. Flügge, "Statik und Dynamik der Schalen," 2d ed., p. 167, Berlin, 1957. For tanks slightly deviating from the cylindrical form see K. Federhofer, *Österr. Bauzeitschrift*, vol. 6, p. 149, 1951; and for tanks with thickness varying in accordance with a quadratic law, see Federhofer, *Österr. Ingr.-Arch.*, vol. 6, p. 43, 1952. A parameter method, akin to that explained in Art. 40, has been used by H. Faure, *Proc. Ninth Congr. Appl. Mech. Brussels*, vol. 6, p. 297, 1957. Many data regarding the design of water tanks are found in W. S. Gray, "Reinforced Concrete Reservoirs and Tanks," London, 1954, and in V. Lewe, "Handbuch für Eisenbetonbau," vol. 9, Berlin, 1934.

which, upon division by x, can be also written

$$\frac{1}{x}\frac{d^2}{dx^2}\left(x^3\frac{d^2w}{dx^2}\right) + \frac{12(1-\nu^2)}{\alpha^2 a^2}w = 0 \qquad (e)$$

The solution of this equation of the fourth order can be reduced to that of two equations of the second order[1] if we observe that

$$\frac{1}{x}\frac{d^2}{dx^2}\left(x^3\frac{d^2w}{dx^2}\right) = \frac{1}{x}\frac{d}{dx}\left\{x^2\frac{d}{dx}\left[\frac{1}{x}\frac{d}{dx}\left(x^2\frac{dw}{dx}\right)\right]\right\}$$

For simplification we introduce the following symbols:

$$L(w) = \frac{1}{x}\frac{d}{dx}\left(x^2\frac{dw}{dx}\right) \qquad (f)$$

$$\rho^4 = \frac{12(1-\nu^2)}{\alpha^2 a^2} \qquad (g)$$

Equation (e) then becomes

$$L[L(w)] + \rho^4 w = 0 \qquad (h)$$

and can be rewritten in one of the two following forms:

$$\begin{array}{l}L[L(w) + i\rho^2 w] - i\rho^2[L(w) + i\rho^2 w] = 0 \\ L[L(w) - i\rho^2 w] + i\rho^2[L(w) - i\rho^2 w] = 0\end{array} \qquad (i)$$

where $i = \sqrt{-1}$.

We see that Eq. (h) is satisfied by the solutions of the second-order equations

$$L(w) + i\rho^2 w = 0 \qquad (j)$$
$$L(w) - i\rho^2 w = 0 \qquad (k)$$

Assuming that

$$w_1 = \varphi_1 + i\varphi_2 \qquad w_2 = \varphi_3 + i\varphi_4 \qquad (l)$$

are the two linearly independent solutions of Eq. (j), it can be seen that

$$w_3 = \varphi_1 - i\varphi_2 \quad \text{and} \quad w_4 = \varphi_3 - i\varphi_4 \qquad (m)$$

are the solutions of Eq. (k). All four solutions (l) and (m) together then represent the complete system of independent solutions of Eq. (h). By using the sums and the differences of solutions (l) and (m), the general solution of Eq. (h) can be represented in the following form:

$$w = C_1\varphi_1 + C_2\varphi_2 + C_3\varphi_3 + C_4\varphi_4 \qquad (n)$$

in which $C_1, \ldots, C_4$ are arbitrary constants. Thus the problem reduces to the determination of four functions $\varphi_1, \ldots, \varphi_4$, which can all be obtained if the complete solution of either Eq. (j) or Eq. (k) is known.

Taking Eq. (j) and substituting for $L(w)$ its meaning (f), we obtain

$$x\frac{d^2w}{dx^2} + 2\frac{dw}{dx} + i\rho^2 w = 0 \qquad (o)$$

---

[1] This reduction was shown by G. Kirchhoff, "Berliner Monatsberichte," p. 815, 1879; see also I. Todhunter and K. Pearson, "A History of the Theory of Elasticity," vol. 2, part 2, p. 92.

By introducing new variables

$$\eta = 2\rho \sqrt{ix} \qquad \zeta = w \sqrt{x} \qquad (p)$$

Eq. (o) becomes

$$\eta^2 \frac{d^2\zeta}{d\eta^2} + \eta \frac{d\zeta}{d\eta} + (\eta^2 - 1)\zeta = 0 \qquad (r)$$

We take as a solution of this equation the power series

$$\zeta_1 = a_0 + a_1\eta + a_2\eta^2 + \cdots \qquad (s)$$

Substituting this series in Eq. (r) and equating the coefficients of each power of η to zero, we obtain the following relation between the coefficients of series (s):

$$(n^2 - 1)a_n + a_{n-2} = 0 \qquad (t)$$

Applying this equation to the first two coefficients and taking $a_{-1} = a_{-2} = 0$, we find that $a_0 = 0$ and that a_1 can be taken equal to any arbitrary constant. Calculating the further coefficients by means of Eq. (t), we find that series (s) is

$$\zeta = C' \frac{\eta}{2} \left[1 - \frac{\eta^2}{2 \cdot 4} + \frac{\eta^4}{2 \cdot 4 \cdot 6} - \frac{\eta^6}{2 \cdot (4 \cdot 6)^2 \cdot 8} + \cdots \right] = C' J_1(\eta) \qquad (u)$$

where $J_1(\eta)$ is the Bessel function of the first kind and of the first order. For our further discussion it is advantageous to use the relation

$$J_1(\eta) = -\frac{d}{d\eta} \left[1 - \frac{\eta^2}{2^2} + \frac{\eta^4}{(2 \cdot 4)^2} - \frac{\eta^6}{(2 \cdot 4 \cdot 6)^2} + \cdots \right] = -\frac{dJ_0}{d\eta} \qquad (v)$$

in which the series in brackets, denoted by J_0, is the Bessel function of the first kind and of zero order. Substituting the expression $2\rho \sqrt{ix}$ for η [see notation (p)] in the series representing $J_0(\eta)$ and collecting the real and the imaginary terms, we obtain

$$J_0(\eta) = \psi_1(2\rho \sqrt{x}) + i\psi_2(2\rho \sqrt{x}) \qquad (w)$$

where

$$\psi_1(2\rho \sqrt{x}) = 1 - \frac{(2\rho \sqrt{x})^4}{(2 \cdot 4)^2} + \frac{(2\rho \sqrt{x})^8}{(2 \cdot 4 \cdot 6 \cdot 8)^2} - \cdots$$

$$\psi_2(2\rho \sqrt{x}) = -\frac{(2\rho \sqrt{x})^2}{2^2} + \frac{(2\rho \sqrt{x})^6}{(2 \cdot 4 \cdot 6)^2} - \frac{(2\rho \sqrt{x})^{10}}{(2 \cdot 4 \cdot 6 \cdot 8 \cdot 10)^2} + \cdots$$

(294)

The solution (u) then gives

$$\zeta_1 = -C'[\psi_1'(2\rho \sqrt{x}) + i\psi_2'(2\rho \sqrt{x})] \qquad (a')$$

where ψ_1' and ψ_2' denote the derivatives of the functions (294) with respect to the argument $2\rho \sqrt{x}$.

The second integral of Eq. (r) is of a more complicated form. Without derivation it can be represented in the form

$$\zeta_2 = C''[\psi_3'(2\rho \sqrt{x}) + i\psi_4'(2\rho \sqrt{x})] \qquad (b')$$

TABLE 86. TABLE OF THE $\psi(x)$ FUNCTIONS

x	$\psi_1(x)$	$\psi_2(x)$	$\dfrac{d\psi_1(x)}{dx}$	$\dfrac{d\psi_2(x)}{dx}$
0.00	+1.0000	0.0000	0.0000	0.0000
0.10	+1.0000	−0.0025	−0.0001	−0.0500
0.20	+1.0000	−0.0100	−0.0005	−0.1000
0.30	+0.9999	−0.0225	−0.0017	−0.1500
0.40	+0.9996	−0.0400	−0.0040	−0.2000
0.50	+0.9990	−0.0625	−0.0078	−0.2499
0.60	+0.9980	−0.0900	−0.0135	−0.2998
0.70	+0.9962	−0.1224	−0.0214	−0.3496
0.80	+0.9936	−0.1599	−0.0320	−0.3991
0.90	+0.9898	−0.2023	−0.0455	−0.4485
1.00	+0.9844	−0.2496	−0.0624	−0.4974
1.10	+0.9771	−0.3017	−0.0831	−0.5458
1.20	+0.9676	−0.3587	−0.1078	−0.5935
1.30	+0.9554	−0.4204	−0.1370	−0.6403
1.40	+0.9401	−0.4867	−0.1709	−0.6860
1.50	+0.9211	−0.5576	−0.2100	−0.7302
1.60	+0.8979	−0.6327	−0.2545	−0.7727
1.70	+0.8700	−0.7120	−0.3048	−0.8131
1.80	+0.8367	−0.7953	−0.3612	−0.8509
1.90	+0.7975	−0.8821	−0.4238	−0.8857
2.00	+0.7517	−0.9723	−0.4931	−0.9170
2.10	+0.6987	−1.0654	−0.5690	−0.9442
2.20	+0.6377	−1.1610	−0.6520	−0.9661
2.30	+0.5680	−1.2585	−0.7420	−0.9836
2.40	+0.4890	−1.3575	−0.8392	−0.9944
2.50	+0.4000	−1.4572	−0.9436	−0.9983
2.60	+0.3001	−1.5569	−1.0552	−0.9943
2.70	+0.1887	−1.6557	−1.1737	−0.9815
2.80	+0.0651	−1.7529	−1.2993	−0.9589
2.90	−0.0714	−1.8472	−1.4315	−0.9256
3.00	−0.2214	−1.9376	−1.5698	−0.8804
3.10	−0.3855	−2.0228	−1.7141	−0.8223
3.20	−0.5644	−2.1016	−1.8636	−0.7499
3.30	−0.7584	−2.1723	−2.0177	−0.6621
3.40	−0.9680	−2.2334	−2.1755	−0.5577
3.50	−1.1936	−2.2832	−2.3361	−0.4353
3.60	−1.4353	−2.3199	−2.4983	−0.2936
3.70	−1.6933	−2.3413	−2.6608	−0.1315
3.80	−1.9674	−2.3454	−2.8221	+0.0526
3.90	−2.2576	−2.3300	−2.9808	+0.2596

TABLE 86. TABLE OF THE $\psi(x)$ FUNCTIONS (Continued)

x	$\psi_1(x)$	$\psi_2(x)$	$\dfrac{d\psi_1(x)}{dx}$	$\dfrac{d\psi_2(x)}{dx}$
4.00	−2.5634	−2.2927	−3.1346	+0.4912
4.10	−2.8843	−2.2309	−3.2819	+0.7482
4.20	−3.2195	−2.1422	−3.4199	+1.0318
4.30	−3.5679	−2.0236	−3.5465	+1.3433
4.40	−3.9283	−1.8726	−3.6587	+1.6833
4.50	−4.2991	−1.6860	−3.7536	+2.0526
4.60	−4.6784	−1.4610	−3.8280	+2.4520
4.70	−5.0639	−1.1946	−3.8782	+2.8818
4.80	−5.4531	−0.8837	−3.9006	+3.3422
4.90	−5.8429	−0.5251	−3.8910	+3.8330
5.00	−6.2301	−0.1160	−3.8454	+4.3542
5.10	−6.6107	+0.3467	−3.7589	+4.9046
5.20	−6.9803	+0.8658	−3.6270	+5.4835
5.30	−7.3344	+1.4443	−3.4446	+6.0893
5.40	−7.6674	+2.0845	−3.2063	+6.7198
5.50	−7.9736	+2.7890	−2.9070	+7.3729
5.60	−8.2466	+3.5597	−2.5409	+8.0453
5.70	−8.4794	+4.3986	−2.1024	+8.7336
5.80	−8.6644	+5.3068	−1.5856	+9.4332
5.90	−8.7937	+6.2854	−0.9844	+10.1394
6.00	−8.8583	+7.3347	−0.2931	+10.3462

in which ψ_3' and ψ_4' are the derivatives with respect to the argument $2\rho\sqrt{x}$ of the following functions:

$$\psi_3(2\rho\sqrt{x}) = \frac{1}{2}\psi_1(2\rho\sqrt{x}) - \frac{2}{\pi}\left[R_1 + \log\frac{\beta 2\rho\sqrt{x}}{2}\psi_2(2\rho\sqrt{x})\right]$$

$$\psi_4(2\rho\sqrt{x}) = \frac{1}{2}\psi_2(2\rho\sqrt{x}) + \frac{2}{\pi}\left[R_2 + \log\frac{\beta 2\rho\sqrt{x}}{2}\psi_1(2\rho\sqrt{x})\right]$$

(295)

where

$$R_1 = \left(\frac{2\rho\sqrt{x}}{2}\right)^2 - \frac{S(3)}{(3\cdot 2)^2}\left(\frac{2\rho\sqrt{x}}{2}\right)^6 + \frac{S(5)}{(5\cdot 4\cdot 3\cdot 2)^2}\left(\frac{2\rho\sqrt{x}}{2}\right)^{10} - \cdots$$

$$R_2 = \frac{S(2)}{2^2}\left(\frac{2\rho\sqrt{x}}{2}\right)^4 - \frac{S(4)}{(4\cdot 3\cdot 2)^2}\left(\frac{2\rho\sqrt{x}}{2}\right)^8 + \frac{S(6)}{(6\cdot 5\cdot 4\cdot 3\cdot 2)^2}\left(\frac{2\rho\sqrt{x}}{2}\right)^{12} - \cdots$$

$$S(n) = 1 + \frac{1}{2} + \frac{1}{3} + \cdots + \frac{1}{n}$$

$$\log\beta = 0.57722$$

TABLE 86. TABLE OF THE $\psi(x)$ FUNCTIONS (Continued)

x	$\psi_3(x)$	$\psi_4(x)$	$\dfrac{d\psi_3(x)}{dx}$	$\dfrac{d\psi_4(x)}{dx}$
0.00	+0.5000	$-\infty$	0.0000	$+\infty$
0.10	+0.4946	−1.5409	−0.0929	+6.3413
0.20	+0.4826	−1.1034	−0.1419	+3.1340
0.30	+0.4667	−0.8513	−0.1746	+2.0498
0.40	+0.4480	−0.6765	−0.1970	+1.4974
0.50	+0.4275	−0.5449	−0.2121	+1.1585
0.60	+0.4058	−0.4412	−0.2216	+0.9273
0.70	+0.3834	−0.3574	−0.2268	+0.7582
0.80	+0.3606	−0.2883	−0.2286	+0.6286
0.90	+0.3377	−0.2308	−0.2276	+0.5258
1.00	+0.3151	−0.1825	−0.2243	+0.4422
1.10	+0.2929	−0.1419	−0.2193	+0.3730
1.20	+0.2713	−0.1076	−0.2129	+0.3149
1.30	+0.2504	−0.0786	−0.2054	+0.2656
1.40	+0.2302	−0.0542	−0.1971	+0.2235
1.50	+0.2110	−0.0337	−0.1882	+0.1873
1.60	+0.1926	−0.0166	−0.1788	+0.1560
1.70	+0.1752	−0.0023	−0.1692	+0.1290
1.80	+0.1588	+0.0094	−0.1594	+0.1056
1.90	+0.1433	+0.0189	−0.1496	+0.0854
2.00	+0.1289	+0.0265	−0.1399	+0.0679
2.10	+0.1153	+0.0325	−0.1304	+0.0527
2.20	+0.1026	+0.0371	−0.1210	+0.0397
2.30	+0.0911	+0.0405	−0.1120	+0.0285
2.40	+0.0804	+0.0429	−0.1032	+0.0189
2.50	+0.0705	+0.0444	−0.0948	+0.0109
2.60	+0.0614	+0.0451	−0.0868	+0.0039
2.70	+0.0531	+0.0452	−0.0791	−0.0018
2.80	+0.0455	+0.0447	−0.0719	−0.0066
2.90	+0.0387	+0.0439	−0.0650	−0.0105
3.00	+0.0326	+0.0427	−0.0586	−0.0137
3.10	+0.0270	+0.0412	−0.0526	−0.0161
3.20	+0.0220	+0.0394	−0.0469	−0.0180
3.30	+0.0176	+0.0376	−0.0417	−0.0194
3.40	+0.0137	+0.0356	−0.0369	−0.0204
3.50	+0.0102	+0.0335	−0.0325	−0.0210
3.60	+0.0072	+0.0314	−0.0284	−0.0213
3.70	+0.0045	+0.0293	−0.0246	−0.0213
3.80	+0.0022	+0.0271	−0.0212	−0.0210
3.90	+0.0003	+0.0251	−0.0180	−0.0206

TABLE 86. TABLE OF THE $\psi(x)$ FUNCTIONS (Continued)

x	$\psi_3(x)$	$\psi_4(x)$	$\dfrac{d\psi_3(x)}{dx}$	$\dfrac{d\psi_4(x)}{dx}$
4.00	−0.0014	+0.0230	−0.0152	−0.0200
4.10	−0.0028	+0.0211	−0.0127	−0.0193
4.20	−0.0039	+0.0192	−0.0104	−0.0185
4.30	−0.0049	+0.0174	−0.0083	−0.0177
4.40	−0.0056	+0.0156	−0.0065	−0.0168
4.50	−0.0062	+0.0140	−0.0049	−0.0158
4.60	−0.0066	+0.0125	−0.0035	−0.0148
4.70	−0.0069	+0.0110	−0.0023	−0.0138
4.80	−0.0071	+0.0097	−0.0012	−0.0129
4.90	−0.0071	+0.0085	−0.0003	−0.0119
5.00	−0.0071	+0.0073	+0.0005	−0.0109
5.10	−0.0070	+0.0063	+0.0012	−0.0100
5.20	−0.0069	+0.0053	+0.0017	−0.0091
5.30	−0.0067	+0.0044	+0.0022	−0.0083
5.40	−0.0065	+0.0037	+0.0025	−0.0075
5.50	−0.0062	+0.0029	+0.0028	−0.0067
5.60	−0.0059	+0.0023	+0.0030	−0.0060
5.70	−0.0056	+0.0017	+0.0032	−0.0053
5.80	−0.0053	+0.0012	+0.0033	−0.0047
5.90	−0.0049	+0.0008	+0.0033	−0.0041
6.00	−0.0046	+0.0004	+0.0033	−0.0036

Having solutions (a') and (b') of Eq. (r), we conclude that the general solution (n) of Eq. (e) is

$$w = \frac{\zeta}{\sqrt{x}} = \frac{1}{\sqrt{x}}[C_1\psi_1'(2\rho\sqrt{x}) + C_2\psi_2'(2\rho\sqrt{x}) + C_3\psi_3'(2\rho\sqrt{x})$$
$$+ C_4\psi_4'(2\rho\sqrt{x})] \quad (c')$$

Numerical values of the functions $\psi_1, \ldots, \psi_4$ and their first derivatives are given in Table 86.[1] A graphical representation of the functions $\psi_1', \ldots, \psi_4'$ is given in Fig. 249. It is seen that the values of these functions increase or decrease rapidly as the distance from the end increases. This indicates that in calculating the constants of integration in solution (c') we can very often proceed as we did with functions (281), i.e., by considering the cylinder as an infinitely long one and using at each edge only two of the four constants in solution (c').

[1] This table was calculated by F. Schleicher; see "Kreisplatten auf elastischer Unterlage," Berlin, 1926. The well-known Kelvin functions may be used in place of the functions ψ, to which they relate as follows: $\psi_1(x) = $ ber x, $\psi_2(x) = -$ bei x, $\psi_3(x) = -(2/\pi)$ kei x, $\psi_4 = -(2/\pi)$ ker x. For more accurate tables of the functions under consideration see p. 266.

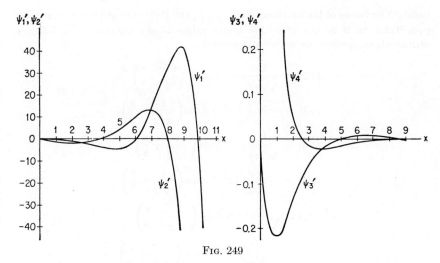

Fig. 249

In applying the general theory to particular cases, the calculation of the consecutive derivatives of w is simplified if we use the following relations:

$$\psi_1''(\xi) = \psi_2(\xi) - \frac{1}{\xi}\psi_1'(\xi)$$

$$\psi_2''(\xi) = -\psi_1(\xi) - \frac{1}{\xi}\psi_2'(\xi)$$

$$\psi_3''(\xi) = \psi_4(\xi) - \frac{1}{\xi}\psi_3'(\xi)$$

$$\psi_4''(\xi) = -\psi_3(\xi) - \frac{1}{\xi}\psi_4'(\xi)$$

(d')

where the symbol ξ is used in place of $2\rho\sqrt{x}$. From expression (c') we then obtain

$$N_\varphi = -\frac{Eh}{a}w = -\frac{E\alpha}{a}\sqrt{x}[C_1\psi_1'(\xi) + C_2\psi_2'(\xi) + C_3\psi_3'(\xi) + C_4\psi_4'(\xi)] \quad (e')$$

$$\frac{dw}{dx} = \frac{1}{2x\sqrt{x}}\{C_1[\xi\psi_2(\xi) - 2\psi_1'(\xi)] - C_2[\xi\psi_1(\xi) + 2\psi_2'(\xi)]$$
$$+ C_3[\xi\psi_4(\xi) - 2\psi_3'(\xi)] - C_4[\xi\psi_3(\xi) + 2\psi_4'(\xi)]\} \quad (f')$$

$$M_x = -D\frac{d^2w}{dx^2} = -\frac{E\alpha^3}{48(1-\nu^2)}\sqrt{x}\{C_1[(\xi)^2\psi_2'(\xi) - 4(\xi)\psi_2(\xi) + 8\psi_1'(\xi)]$$
$$- C_2[(\xi)^2\psi_1'(\xi) - 4(\xi)\psi_1(\xi) - 8\psi_2'(\xi)]$$
$$+ C_3[(\xi)^2\psi_4'(\xi) - 4(\xi)\psi_4(\xi) + 8\psi_3'(\xi)]$$
$$- C_4[(\xi)^2\psi_3'(\xi) - 4(\xi)\psi_3(\xi) - 8\psi_4'(\xi)]\} \quad (g')$$

$$Q_x = \frac{dM_x}{dx} = \frac{E\alpha^3\rho^2}{24(1-\nu^2)}\sqrt{x}\{C_1[\xi\psi_1(\xi) + 2\psi_2'(\xi)]$$
$$+ C_2[\xi\psi_2(\xi) - 2\psi_1'(\xi)] + C_3[\xi\psi_3(\xi) + 2\psi_4'(\xi)] + C_4[\xi\psi_4(\xi) - 2\psi_3'(\xi)]\} \quad (h')$$

By means of these formulas the deflections and the stresses can be calculated at any point, provided the constants $C_1, \ldots, C_4$ are determined from the edge conditions. The values of the functions $\psi_1, \ldots, \psi_4$ and their derivatives are to be taken from Table 86 if $2\rho \sqrt{x} \leq 6$. For larger values of the argument, the following asymptotic expressions are sufficiently accurate:

$$\psi_1(\xi) \approx \frac{1}{\sqrt{2\pi\xi}} e^{\xi/\sqrt{2}} \cos\left(\frac{\xi}{\sqrt{2}} - \frac{\pi}{8}\right)$$

$$\psi_2(\xi) \approx -\frac{1}{\sqrt{2\pi\xi}} e^{\xi/\sqrt{2}} \sin\left(\frac{\xi}{\sqrt{2}} - \frac{\pi}{8}\right)$$

$$\psi_1'(\xi) \approx \frac{1}{\sqrt{2\pi\xi}} e^{\xi/\sqrt{2}} \cos\left(\frac{\xi}{\sqrt{2}} + \frac{\pi}{8}\right)$$

$$\psi_2'(\xi) \approx -\frac{1}{\sqrt{2\pi\xi}} e^{\xi/\sqrt{2}} \sin\left(\frac{\xi}{\sqrt{2}} + \frac{\pi}{8}\right)$$

$$\psi_3(\xi) \approx \sqrt{\frac{2}{\pi\xi}} e^{-\xi/\sqrt{2}} \sin\left(\frac{\xi}{\sqrt{2}} + \frac{\pi}{8}\right) \quad (296)$$

$$\psi_4(\xi) \approx -\sqrt{\frac{2}{\pi\xi}} e^{-\xi/\sqrt{2}} \cos\left(\frac{\xi}{\sqrt{2}} + \frac{\pi}{8}\right)$$

$$\psi_3'(\xi) \approx -\sqrt{\frac{2}{\pi\xi}} e^{-\xi/\sqrt{2}} \sin\left(\frac{\xi}{\sqrt{2}} - \frac{\pi}{8}\right)$$

$$\psi_4'(\xi) \approx \sqrt{\frac{2}{\pi\xi}} e^{-\xi/\sqrt{2}} \cos\left(\frac{\xi}{\sqrt{2}} - \frac{\pi}{8}\right)$$

As an example, consider a cylindrical tank of the same general dimensions as that used in the preceding article (page 487), and assume that the thickness of the wall varies from 14 in. at the bottom to $3\frac{1}{2}$ in. at the top. In such a case the distance of the origin of the coordinates (Fig. 248) from the bottom of the tank is

$$d + x_0 = \tfrac{4}{3}d = 416 \text{ in.}$$

Hence, $(2\rho \sqrt{x})_{x=x_0+d} = 21.45$. For such a large value of the argument, the functions $\psi_1, \ldots, \psi_4$ and their first derivatives can be replaced by their asymptotic expressions (296). The deflection and the slope at the bottom of the tank corresponding to the particular solution (c) are

$$(w_1)_{x=x_0+d} = -\frac{\gamma a^2}{E\alpha} \frac{d}{d+x_0} \qquad \left(\frac{dw_1}{dx}\right)_{x=x_0+d} = -\frac{\gamma a^2}{E\alpha} \frac{x_0}{(x_0+d)^2} \qquad (i')$$

Considering the length of the cylindrical shell in the axial direction as very large, we take the constants C_3 and C_4 in solution (c') as equal to zero and determine the constants C_1 and C_2 so as to make the deflection and the slope at the bottom of the shell equal to zero. These requirements give us the two following equations:

$$\frac{1}{\sqrt{x}} [C_1 \psi_1'(2\rho \sqrt{x}) + C_2 \psi_2'(2\rho \sqrt{x})]_{x=x_0+d} = \frac{\gamma a^2}{E\alpha} \frac{d}{d+x_0}$$

$$\frac{1}{2x\sqrt{x}} \{C_1[2\rho \sqrt{x} \, \psi_2(2\rho \sqrt{x}) - 2\psi_1'(2\rho \sqrt{x})] - C_2[2\rho \sqrt{x} \, \psi_1(2\rho \sqrt{x}) \qquad (j')$$

$$+ 2\psi_2'(2\rho \sqrt{x})]\}_{x=x_0+d} = \frac{\gamma a^2}{E\alpha} \frac{x_0}{(d+x_0)^2}$$

Calculating the values of functions ψ_1, ψ_2 and their derivatives from the asymptotic formulas (296) and substituting the resulting values in Eqs. (j'), we obtain

$$C_1 = -269 \frac{\gamma a^2}{E\alpha} \frac{1}{\sqrt{d+x_0}} N$$

$$C_2 = -299 \frac{\gamma a^2}{E\alpha} \frac{1}{\sqrt{d+x_0}} N$$

where
$$N = (e^{-\xi/\sqrt{2}} \sqrt{2\pi\xi})_{\xi=21.45}$$

Substituting these values of the constants in expression (g') we find for the bending moment at the bottom

$$M_0 = 13{,}900 \text{ lb-in. per in.}$$

In the same manner, by using expression (h'), we find the magnitude of the shearing force at the bottom of the tank as

$$Q_0 = 527 \text{ lb per in.}$$

These results do not differ much from the values obtained earlier for a tank with uniform wall thickness (page 487).

119. Thermal Stresses in Cylindrical Shells. *Uniform Temperature Distribution.* If a cylindrical shell with free edges undergoes a uniform temperature change, no thermal stresses will be produced. But if the edges are supported or clamped, free expansion of the shell is prevented, and local bending stresses are set up at the edges. Knowing the thermal expansion of a shell when the edges are free, the values of the reactive moments and forces at the edges for any kind of symmetrical support can be readily obtained by using Eqs. (279) and (280), as was done in the cases shown in Fig. 241.

Temperature Gradient in the Radial Direction. Assume that t_1 and t_2 are the uniform temperatures of the cylindrical wall at the inside and the outside surfaces, respectively, and that the variation of the temperature through the thickness is linear. In such a case, at points at a large distance from the ends of the shell, there will be no bending, and the stresses can be calculated by using Eq. (51), which was derived for clamped plates (see page 50). Thus the stresses at the outer and the inner surfaces are

$$\sigma_x = \sigma_\varphi = \pm \frac{E\alpha(t_1 - t_2)}{2(1 - \nu)} \quad (a)$$

where the upper sign refers to the outer surface, indicating that a tensile stress will act on this surface if $t_1 > t_2$.

Near the ends there will usually be some bending of the shell, and the total thermal stresses will be obtained by superposing upon (a) such stresses as are necessary to satisfy the boundary conditions. Let us consider, as an example, the condition of free edges, in which case the stresses σ_x must vanish at the ends. In calculating the stresses and deformations in this case we observe that at the edge the stresses (a) result in uniformly distributed moments M_0 (Fig. 250a) of the amount

$$M_0 = -\frac{E\alpha(t_1 - t_2)h^2}{12(1 - \nu)} \quad (b)$$

To obtain a free edge, moments of the same magnitude but opposite in direction must be superposed (Fig. 250b). Hence the stresses at a free edge are obtained by superposing upon the stresses (a) the stresses produced by the moments $-M_0$ (Fig. 250b). These latter stresses can be readily calculated by using solution (278). From this solution it follows that

$$(M_x)_{x=0} = \frac{E\alpha(t_1 - t_2)h^2}{12(1 - \nu)} \quad (M_\varphi)_{x=0} = \nu(M_x)_{x=0} = \frac{\nu E\alpha(t_1 - t_2)h^2}{12(1 - \nu)} \quad (c)$$

$$(N_\varphi)_{x=0} = -\frac{Eh}{a}(w)_{x=0} = \frac{Eh}{a}\frac{M_0}{2\beta^2 D} = \frac{Eh\alpha(t_1 - t_2)}{2\sqrt{3}(1 - \nu)}\sqrt{1 - \nu^2} \quad (d)$$

It is seen that at the free edge the maximum thermal stress acts in the circumferential direction and is obtained by adding to the stress (a) the stresses produced by the moment M_φ and the force N_φ. Assuming that $t_1 > t_2$, we thus obtain

$$(\sigma_\varphi)_{\max} = \frac{E\alpha(t_1 - t_2)}{2(1 - \nu)}\left(1 - \nu + \frac{\sqrt{1 - \nu^2}}{\sqrt{3}}\right) \quad (e)$$

For $\nu = 0.3$ this stress is about 25 per cent greater than the stress (a) calculated at points at a large distance from the ends. We can therefore conclude that if a crack will occur in a brittle material such as glass due to a temperature difference $t_1 - t_2$, it will start at the edge and will proceed in the axial direction. In a similar manner the stresses can also be calculated in cases in which the edges are clamped or supported.[1]

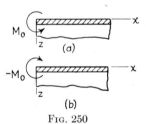

FIG. 250

Temperature Gradient in the Axial Direction. If the temperature is constant through the thickness of the wall but varies along the length of the cylinder, the problem can be easily reduced to the solution of Eq. (274).[2] Let $t = F(x)$ be the increase of the temperature of the shell from a certain uniform initial temperature. Assuming that the shell is divided into infinitely thin rings by planes perpendicular to the x axis and denoting the radius of the shell by a, the radial expansion of the rings due to the temperature change is $\alpha a F(x)$.

---

[1] Several examples of this kind are discussed in the paper by C. H. Kent, *Trans. ASME*, vol. 53, p. 167, 1931.

[2] See Timoshenko and Lessells, "Applied Elasticity," p. 146, 1925.

This expansion can be eliminated and the shell can be brought to its initial diameter by applying an external pressure of an intensity Z such that

$$\frac{a^2 Z}{Eh} = \alpha a F(x)$$

which gives

$$Z = \frac{Eh\alpha}{a} F(x) \qquad (f)$$

A load of this intensity entirely arrests the thermal expansion of the shell and produces in it only circumferential stresses having a magnitude

$$\sigma_\varphi = -\frac{aZ}{h} = -E\alpha F(x) \qquad (g)$$

To obtain the total thermal stresses, we must superpose on the stresses (g) the stresses that will be produced in the shell by a load of the intensity $-Z$. This latter load must be applied in order to make the lateral surface of the shell free from the external load given by Eq. (f). The stresses produced in the shell by the load $-Z$ are obtained by the integration of the differential equation (276), which in this case becomes

$$\frac{d^4 w}{dx^4} + 4\beta^4 w = -\frac{Eh\alpha}{Da} F(x) \qquad (h)$$

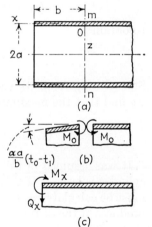

Fig. 251

As an example of the application of this equation let us consider a long cylinder, as shown in Fig. 251a, and assume that the part of the cylinder to the right of the cross section mn has a constant temperature t_0, whereas that to the left side has a temperature that decreases linearly to a temperature t_1 at the end $x = b$ according to the relation

$$t = t_0 - \frac{(t_0 - t_1)x}{b}$$

The temperature change at a point in this portion is thus

$$F(x) = t - t_0 = -\frac{(t_0 - t_1)x}{b} \qquad (i)$$

Substituting this expression for the temperature change in Eq. (h), we find that the particular solution of that equation is

$$w_1 = \frac{\alpha a}{b}(t_0 - t_1)x \qquad (j)$$

The displacement corresponding to this particular solution is shown in Fig. 251b, which indicates that there is at the section mn an angle of discontinuity of the magnitude

$$\frac{w_1}{x} = \frac{\alpha a}{b}(t_0 - t_1) \qquad (k)$$

To remove this discontinuity the moments M_0 must be applied. Since the stress σ_φ corresponding to the particular solution (j) cancels the stresses (g), we conclude that the stresses produced by the moments M_0 are the total thermal stresses resulting from the above-described decrease in temperature. If the distances of the cross section mn from the ends of the cylinder are large, the magnitude of the moment M_0 can be obtained at once from Eq. (280) by substituting

$$Q_0 = 0 \qquad \left(\frac{dw_1}{dx}\right)_{x=0} = -\frac{\alpha a}{2b}(t_0 - t_1)$$

to obtain[1]

$$M_0 = -\beta D \frac{\alpha a}{2b}(t_0 - t_1) \qquad (l)$$

Substituting for β its value from expression (275) and taking $\nu = 0.3$, we find that the maximum thermal stress is

$$(\sigma_x)_{\max} = \frac{6M_0}{h^2} = 0.353 \frac{E\alpha}{b}\sqrt{ah}(t_0 - t_1) \qquad (m)$$

It was assumed in this calculation that the length b to the end of the cylinder is large. If this is not the case, a correction to the moment (l) must be calculated as follows. In an infinitely long shell the moment M_0 produces at the distance $x = b$ a moment and a shearing force (Fig. 251c)[2] that are given by the general solution (282) as

$$M_x = -D\frac{d^2w}{dx^2} = M_0\varphi(\beta b)$$
$$Q_x = -D\frac{d^3w}{dx^3} = -2\beta M_0\zeta(\beta b) \qquad (n)$$

Since at the distance $x = b$ we have a free edge, it is necessary to apply there a moment and a force of the magnitude

$$-M_x = -M_0\varphi(\beta b) \qquad -Q_x = 2\beta M_0\zeta(\beta b) \qquad (o)$$

in order to eliminate the forces (n) (Fig. 251c).

---

[1] If $t_0 - t_1$ is positive, as was assumed in the derivation, M_0 is negative and thus has the direction shown in Fig. 251b.
[2] The directions M_x and Q_x shown in Fig. 251c are the positive directions if the x axis has the direction shown in Fig. 251a.

The moment produced by the forces (o) at the cross section mn gives the desired correction ΔM_0 which is to be applied to the moment (l). Its value can be obtained from the third of the equations (282) if we substitute in it $-M_0\varphi(\beta b)$ for M_0 and $-2\beta M_0 \zeta(\beta b)^*$ for Q_0. These substitutions give

$$\Delta M = -D\frac{d^2w}{dx^2} = -M_0[\varphi(\beta b)]^2 - 2M_0[\zeta(\beta b)]^2 \qquad (p)$$

As a numerical example, consider a cast-iron cylinder having the following dimensions: $a = 9\frac{11}{16}$ in., $h = 1\frac{3}{8}$ in., $b = 4\frac{1}{4}$ in.; $\alpha = 101 \cdot 10^{-7}$, $E = 14 \cdot 10^6$ psi,

$$t_0 - t_1 = 180°C$$

The formula (m) then gives

$$\sigma_{\max} = 7{,}720 \text{ psi} \qquad (q)$$

In calculating the correction (p), we have

$$\beta = \sqrt{\frac{3(1-\nu^2)}{a^2h^2}} = \frac{1}{2.84} \text{ (in.)}^{-1} \qquad \beta b = 1.50$$

and, from Table 84,

$$\varphi(\beta b) = 0.238 \qquad \zeta(\beta b) = 0.223$$

Hence, from Eq. (p),

$$\Delta M = -M_0(0.238^2 + 2 \cdot 0.223^2) = -0.156 M_0$$

This indicates that the above-calculated maximum stress (q) must be diminished by 15.6 per cent to obtain the correct maximum value of the thermal stress.

The method shown here for the calculation of thermal stresses in the case of a linear temperature gradient (i) can also be easily applied in cases in which $F(x)$ has other than a linear form.

120. Inextensional Deformation of a Circular Cylindrical Shell.[1]

If the ends of a thin circular cylindrical shell are free and the loading is not symmetrical with respect to the axis of the cylinder, the deformation consists principally in bending. In such cases the magnitude of deflection can be obtained with sufficient accuracy by neglecting entirely the strain in the middle surface of the shell. An example of such a loading condition is shown in Fig. 252. The shortening of the vertical diameter along which the forces P act can be found with good accuracy by considering only the bending of the shell and assuming that the middle surface is inextensible.

Let us first consider the limitations to which the components of displacement are subject if the deformation of a cylindrical shell is to be inextensional. Taking an element in the middle surface of the shell at a point O and directing the coordinate axes as shown in Fig. 253, we shall

---

* The opposite sign to that in expression (o) is used here, since Eqs. 282 are derived for the direction of the x axis opposite to that shown in Fig. 251a.

[1] The theory of inextensional deformations of shells is due to Lord Rayleigh, *Proc. London Math. Soc.*, vol. 13, 1881, and *Proc. Roy. Soc. (London)*, vol. 45, 1889.

502   THEORY OF PLATES AND SHELLS

denote by u, v, and w the components in the x, y, and z directions of the displacement of the point O. The strain in the x direction is then

$$\epsilon_x = \frac{\partial u}{\partial x} \tag{a}$$

In calculating the strain in the circumferential direction we use Eq. (a) (Art. 108, page 446). Thus,

$$\epsilon_\varphi = \frac{1}{a}\frac{\partial v}{\partial \varphi} - \frac{w}{a} \tag{b}$$

The shearing strain in the middle surface can be expressed by

$$\gamma_{x\varphi} = \frac{\partial u}{a\, \partial \varphi} + \frac{\partial v}{\partial x} \tag{c}$$

which is the same as in the case of small deflections of plates except that $a\, d\varphi$ takes the place of dy. The condition that the deformation is inexten-

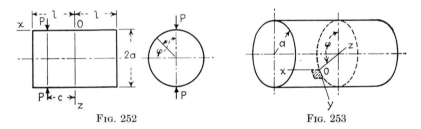

Fig. 252   Fig. 253

sional then requires that the three strain components in the middle surface must vanish; i.e.,

$$\frac{\partial u}{\partial x} = 0 \qquad \frac{1}{a}\frac{\partial v}{\partial \varphi} - \frac{w}{a} = 0 \qquad \frac{\partial u}{a\, \partial \varphi} + \frac{\partial v}{\partial x} = 0 \tag{d}$$

These requirements are satisfied if we take the displacements in the following form:

$$u_1 = 0$$
$$v_1 = a \sum_{n=1}^{\infty} (a_n \cos n\varphi - a'_n \sin n\varphi)$$
$$w_1 = -a \sum_{n=1}^{\infty} n(a_n \sin n\varphi + a'_n \cos n\varphi) \tag{e}$$

where a is the radius of the middle surface of the shell, φ the central angle, and a_n and a'_n constants that must be calculated for each particular case of loading. The displacements (e) represent the case in which all

GENERAL THEORY OF CYLINDRICAL SHELLS

cross sections of the shell deform identically. On these displacements we can superpose displacements two of which vary along the length of the cylinder and which are given by the following series:

$$u_2 = -a \sum_{n=1}^{\infty} \frac{1}{n} (b_n \sin n\varphi + b'_n \cos n\varphi)$$

$$v_2 = x \sum_{n=1}^{\infty} (b_n \cos n\varphi - b'_n \sin n\varphi) \qquad (f)$$

$$w_2 = -x \sum_{n=1}^{\infty} n(b_n \sin n\varphi + b'_n \cos n\varphi)$$

It can be readily proved by substitution in Eqs. (d) that these expressions also satisfy the conditions of inextensibility. Thus the general expressions for displacements in inextensional deformation of a cylindrical shell are

$$u = u_1 + u_2 \qquad v = v_1 + v_2 \qquad w = w_1 + w_2 \qquad (g)$$

In calculating the inextensional deformations of a cylindrical shell under the action of a given system of forces, it is advantageous to use the energy method. To establish the required expression for the strain energy of bending of the shell, we begin with the calculation of the changes of curvature of the middle surface of the shell. The change of curvature in the direction of the generatrix is equal to zero, since, as can be seen from expressions (e) and (f), the generatrices remain straight. The change of curvature of the circumference is obtained by comparing the curvature of an element mn of the circumference (Fig. 254) before deformation with that of the corresponding element m_1n_1 after deformation. Before deformation the curvature in the circumferential direction is

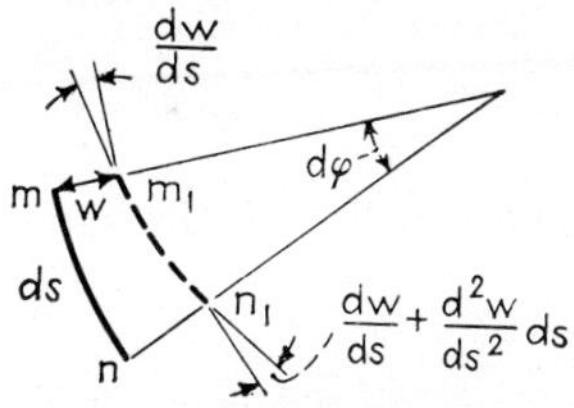

Fig. 254

$$\frac{\partial \varphi}{\partial s} = \frac{\partial \varphi}{a \, \partial \varphi} = \frac{1}{a}$$

The curvature of the element m_1n_1 after deformation is

$$\frac{\partial \varphi_1}{\partial s_1} \approx \frac{d\varphi + \frac{\partial^2 w}{\partial s^2} ds}{(a - w) \, d\varphi}$$

Hence the change in curvature is

$$\chi_\varphi = \frac{d\varphi + \frac{\partial^2 w}{\partial s^2} ds}{(a - w) d\varphi} - \frac{d\varphi}{a \, \partial\varphi} \approx \frac{1}{a^2}\left(w + \frac{\partial^2 w}{\partial \varphi^2}\right)$$

By using the second of the equations (d) we can also write

$$\chi_\varphi = \frac{1}{a^2}\left(\frac{\partial v}{\partial \varphi} + \frac{\partial^2 w}{\partial \varphi^2}\right) \tag{h}$$

The bending moment producing this change in curvature is

$$M_\varphi = -\frac{D}{a^2}\left(\frac{\partial v}{\partial \varphi} + \frac{\partial^2 w}{\partial \varphi^2}\right)$$

and the corresponding strain energy of bending per unit area can be calculated as in the discussion of plates (see page 46) and is equal to

$$\frac{D}{2a^4}\left(\frac{\partial v}{\partial \varphi} + \frac{\partial^2 w}{\partial \varphi^2}\right)^2 = \frac{D}{2a^4}\left(w + \frac{\partial^2 w}{\partial \varphi^2}\right)^2 \tag{i}$$

In addition to bending, there will be a twist of each element such as that shown at point O in Fig. 253. In calculating this twist we note that during deformation an element of a generatrix rotates[1] through an angle equal to $-\partial w/\partial x$ about the y axis and through an angle equal to $\partial v/\partial x$ about the z axis. Considering a similar element of a generatrix at a circumferential distance $a \, d\varphi$ from the first one, we see that its rotation about the y axis, as a result of the displacement w, is

$$-\frac{\partial w}{\partial x} - \frac{\partial^2 w}{\partial \varphi \, \partial x} d\varphi \tag{j}$$

The rotation of the same element in the plane tangent to the shell is

$$\frac{\partial v}{\partial x} + \frac{\partial \left(\frac{\partial v}{\partial x}\right)}{\partial \varphi} d\varphi$$

Because of the central angle $d\varphi$ between the two elements, the latter rotation has a component with respect to the y axis equal to[2]

$$-\frac{\partial v}{\partial x} d\varphi \tag{k}$$

From results (j) and (k) we conclude that the total angle of twist between the two elements under consideration is

$$-\chi_{x\varphi} a \, d\varphi = -\left(\frac{\partial^2 w}{\partial \varphi \, \partial x} + \frac{\partial v}{\partial x}\right) d\varphi$$

[1] In determining the sign of rotation the right-hand screw rule is used.
[2] A small quantity of second order is neglected in this expression.

and that the amount of strain energy per unit area due to twist is (see page 47)

$$\frac{D(1-\nu)}{a^2}\left(\frac{\partial^2 w}{\partial \varphi\, \partial x} + \frac{\partial v}{\partial x}\right)^2 \qquad (l)$$

Adding together expressions (i) and (l) and integrating over the surface of the shell, the total strain energy of a cylindrical shell undergoing an inextensional deformation is found to be

$$V = \frac{D}{2a^4}\iint\left[\left(\frac{\partial v}{\partial \varphi} + \frac{\partial^2 w}{\partial \varphi^2}\right)^2 + 2(1-\nu)a^2\left(\frac{\partial^2 w}{\partial \varphi\, \partial x} + \frac{\partial v}{\partial x}\right)^2\right]a\, d\varphi\, dx$$

Substituting for w and v their expressions (g) and integrating, we find for a cylinder of a length $2l$ (Fig. 252) the following expression for strain energy:

$$V = \pi Dl \sum_{n=2}^{\infty} \frac{(n^2-1)^2}{a^3}\left\{n^2\left[a^2(a_n^2 + a_n'^2) + \frac{1}{3}l^2(b_n^2 + b_n'^2)\right]\right.$$
$$\left. + 2(1-\nu)a^2(b_n^2 + b_n'^2)\right\} \qquad (297)$$

This expression does not contain a term with $n = 1$, since the corresponding displacements

$$v_1 = a(a_1 \cos \varphi - a_1' \sin \varphi)$$
$$w_1 = -a(a_1 \sin \varphi + a_1' \cos \varphi) \qquad (m)$$

represent the displacement of the circle in its plane as a rigid body. The vertical and horizontal components of this displacement are found by substituting $\varphi = \pi/2$ in expressions (m) to obtain

$$(v_1)_{\varphi=\pi/2} = -aa_1' \qquad (w_1)_{\varphi=\pi/2} = -aa_1$$

Such a displacement does not contribute to the strain energy.

The same conclusion can also be made regarding the displacements represented by the terms with $n = 1$ in expressions (f).

Let us now apply expression (297) for the strain energy to the calculation of the deformation produced in a cylindrical shell by two equal and opposite forces P acting along a diameter at a distance c from the middle[1] (Fig. 252). These forces produce work only during radial displacements w of their points of application, i.e., at the points $x = c$, $\varphi = 0$, and $\varphi = \pi$. Also, since the terms with coefficients a_n and b_n in the expressions for w_1 and w_2 [see Eqs. (e) and (f)] vanish at these points, only terms with coefficients a_n' and b_n' will enter in the expression for deformation. By using the

[1] The case of a cylindrical shell reinforced by elastic rings with two opposite forces acting along a diameter of every ring was discussed by R. S. Levy, J. Appl. Mechanics, vol. 15, p. 30, 1948.

principle of virtual displacements, the equations for calculating the coefficients a'_n and b'_n are found to be

$$\frac{\partial V}{\partial a'_n} \delta a'_n = -na\, \delta a'_n (1 + \cos n\pi) P$$

$$\frac{\partial V}{\partial b'_n} \delta b'_n = -nc\, \delta b'_n (1 + \cos n\pi) P$$

Substituting expression (297) for V, we obtain, for the case where n is an even number,

$$a'_n = -\frac{a^2 P}{n(n^2-1)^2 \pi D l}$$

$$b'_n = -\frac{nc P a^3}{(n^2-1)^2 \pi D l [\tfrac{1}{3} n^2 l^2 + 2(1-\nu) a^2]}$$

(n)

If n is an odd number, we obtain

$$a'_n = b'_n = 0 \qquad (o)$$

Hence in this case, from expressions (e) and (f),

$$u = \frac{Pa^3}{\pi Dl} \sum_{n=2,4,6,\ldots} \frac{ac \cos n\varphi}{(n^2-1)^2 [\tfrac{1}{3} n^2 l^2 + 2(1-\nu) a^2]}$$

$$v = \frac{Pa^3}{\pi Dl} \sum_{n=2,4,6,\ldots} \left\{ \frac{1}{n(n^2-1)^2} + \frac{ncx}{(n^2-1)^2 [\tfrac{1}{3} n^2 l^2 + 2(1-\nu) a^2]} \right\} \sin n\varphi \quad (p)$$

$$w = \frac{Pa^3}{\pi Dl} \sum_{n=2,4,6,\ldots} \left\{ \frac{1}{(n^2-1)^2} + \frac{n^2 cx}{(n^2-1)^2 [\tfrac{1}{3} n^2 l^2 + 2(1-\nu) a^2]} \right\} \cos n\varphi$$

If the forces P are applied at the middle, $c = 0$ and the shortening of the vertical diameter of the shell is

$$\delta = (w)_{\varphi=0} + (w)_{\varphi=\pi} = \frac{2Pa^3}{\pi Dl} \sum_{n=2,4,6,\ldots} \frac{1}{(n^2-1)^2} = 0.149 \frac{Pa^3}{2Dl} \qquad (q)$$

The increase in the horizontal diameter is

$$\delta_1 = -[(w)_{\varphi=\pi/2} + (w)_{\varphi=3\pi/2}] = \frac{2Pa^3}{\pi Dl} \sum_{n=2,4,6,\ldots} \frac{(-1)^{n/2+1}}{(n^2-1)^2} = 0.137 \frac{Pa^3}{2Dl} \qquad (r)$$

The change in length of any other diameter can also be readily calculated. The same calculations can also be made if c is different from zero, and the deflections vary with the distance x from the middle.

Solution (p) does not satisfy the conditions at the free edges of the shell, since it requires the distribution of moments $M_x = \nu M_\varphi$ to prevent any bending in meridional planes. This bending is, however, of a local character and does not substantially affect the deflections (q) and (r), which are in satisfactory agreement with experiments.

The method just described for analyzing the inextensional deformation of cylindrical shells can also be used in calculating the deformation of a portion of a cylindrical shell which is cut from a complete cylinder of radius a by two axial sections making an angle α with one another (Fig. 255). For example, taking for the displacements the series

$$u = -\frac{\alpha a}{\pi} \sum \frac{b_n}{n} \sin \frac{n\pi\varphi}{\alpha}$$

$$v = a \sum a_n \cos \frac{n\pi\varphi}{\alpha} + x \sum b_n \cos \frac{n\pi\varphi}{\alpha}$$

$$w = -\frac{\pi a}{\alpha} \sum n a_n \sin \frac{n\pi\varphi}{\alpha} - \frac{x\pi}{\alpha} \sum n b_n \sin \frac{n\pi\varphi}{\alpha}$$

we obtain an inextensional deformation of the shell such that the displacements u and w and also the bending moments M_φ vanish along the edges mn and m_1n_1. Such conditions are obtained if the shell is supported at points m, n, m_1, n_1 by bars directed radially and is loaded by a load P in the plane of symmetry. The deflection produced by this load can be found by applying the principle of virtual displacements.

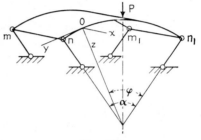

Fig. 255

121. General Case of Deformation of a Cylindrical Shell.[1]

To establish the differential equations for the displacements u, v, and w which define the deformation of a shell, we proceed as in the case of plates. We begin with the equations of equilibrium of an element cut out from the cylindrical shell by two adjacent axial sections and by two adjacent sections perpendicular to the axis of the cylinder (Fig. 253). The corresponding element of the middle surface of the shell after deformation is shown in Fig. 256a and b. In Fig. 256a the resultant forces and in Fig. 256b the

---

[1] A general theory of bending of thin shells has been developed by A. E. H. Love; see *Phil. Trans. Roy. Soc. (London)*, ser. A, p. 491, 1888; and his book "Elasticity," 4th ed., chap. 24, p. 515, 1927; see also H. Lamb, *Proc. London Math. Soc.*, vol. 21. For bending of cylindrical shells see also H. Reissner, *Z. angew. Math. Mech.*, vol. 13, p. 133, 1933; L. H. Donnell, *NACA Rept.* 479, 1933 (simplified theory); E. Torroja and J. Batanero, "Cubiertos laminares cilindros," Madrid, 1950; H. Parkus, *Österr. Ingr.-Arch.*, vol. 6, p. 30, 1951; W. Zerna, *Ingr.-Arch.*, vol. 20, p. 357, 1952; P. Csonka, *Acta Tech. Acad. Sci. Hung.*, vol. 6, p. 167, 1953. The effect of a concentrated load has been considered by A. Aas-Jakobsen, *Bauingenieur*, vol. 22, p. 343, 1941; by Y. N. Rabotnov, *Doklady Akad. Nauk S.S.S.R.*, vol. 3, 1946; and by V. Z. Vlasov, "A General Theory of Shells," Moscow, 1949. For cylindrical shells stiffened by ribs, see N. J. Hoff, *J. Appl. Mechanics*, vol. 11, p. 235, 1944; "H. Reissner Anniversary Volume," Ann Arbor, Mich., 1949; and W. Schnell, *Z. Flugwiss.*, vol. 3, p. 385, 1955. Anisotropic shells (together with a general theory) have been treated by W. Flügge, *Ingr.-Arch.*, vol. 3, p. 463, 1932; also by Vlasov, *op. cit.*, chaps. 11 and 12. For stress distribution around stiffened cutouts, see bibliography in L. S. D. Morley's paper, *Natl. Luchtvaarlab. Rappts.*, p. 362, Amsterdam, 1950. A theory of thick cylindrical shells is due to Z. Bazant, *Proc. Assoc. Bridge Structural Engrs.*, vol. 4, 1936.

resultant moments, discussed in Art. 104, are shown. Before deformation, the axes x, y, and z at any point O of the middle surface had the directions of the generatrix, the tangent to the circumference, and the normal to the middle surface of the shell, respectively. After deformation, which is assumed to be very small, these directions are slightly changed. We then take the z axis normal to the deformed middle surface, the x axis in the direction of a tangent to the generatrix, which may have become curved, and the y axis perpendicular to the xz plane. The

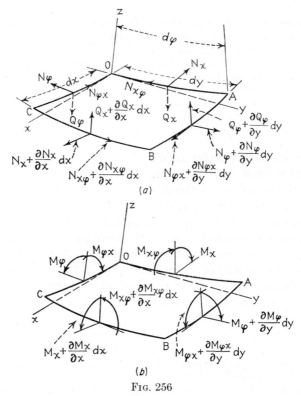

Fig. 256

directions of the resultant forces will also have been slightly changed accordingly, and these changes must be considered in writing the equations of equilibrium of the element $OABC$.

Let us begin by establishing formulas for the angular displacements of the sides BC and AB with reference to the sides OA and OC of the element, respectively. In these calculations we consider the displacements u, v, and w as very small, calculate the angular motions produced by each of these displacements, and obtain the resultant angular displacement by superposition. We begin with the rotation of the side BC with respect to the side OA. This rotation can be resolved into three component rotations with respect to the x, y, and z axes. The rotations of the sides OA and BC with respect to the x axis are caused by the displacements v and w. Since the displacements v represent motion of the sides OA and BC in the circumferential direction (Fig. 253), if a is the radius of the middle surface of the cylinder, the corresponding rotation of side OA about the x axis is v/a, and that of side BC is

$$\frac{1}{a}\left(v + \frac{\partial v}{\partial x}\,dx\right)$$

Thus, owing to the displacements v, the relative angular motion of BC with respect to OA about the x axis is

$$\frac{1}{a}\frac{\partial v}{\partial x}\,dx \tag{a}$$

Because of the displacements w, the side OA rotates about the x axis through the angle $\partial w/(a\,d\varphi)$, and the side BC through the angle

$$\frac{\partial w}{a\,\partial\varphi} + \frac{\partial}{\partial x}\left(\frac{\partial w}{a\,\partial\varphi}\right)dx$$

Thus, because of the displacements w, the relative angular displacement is

$$\frac{\partial}{\partial x}\left(\frac{\partial w}{a\,\partial\varphi}\right)dx \tag{b}$$

Summing up (a) and (b), the relative angular displacement about the x axis of side BC with respect to side OA is

$$\frac{1}{a}\left(\frac{\partial v}{\partial x} + \frac{\partial^2 w}{\partial x\,\partial\varphi}\right)dx \tag{c}$$

The rotation about the y axis of side BC with respect to side OA is caused by bending of the generatrices in axial planes and is equal to

$$-\frac{\partial^2 w}{\partial x^2}\,dx \tag{d}$$

The rotation about the z axis of side BC with respect to side OA is due to bending of the generatrices in tangential planes and is equal to

$$\frac{\partial^2 v}{\partial x^2}\,dx \tag{e}$$

The formulas (c), (d), and (e) thus give the three components of rotation of the side BC with respect to the side OA.

Let us now establish the corresponding formulas for the angular displacement of side AB with respect to side OC. Because of the curvature of the cylindrical shell, the initial angle between these lateral sides of the element $OABC$ is $d\varphi$. However, because of the displacements v and w this angle will be changed. The rotation of the lateral side OC about the x axis is

$$\frac{v}{a} + \frac{\partial w}{a\, d\varphi} \tag{f}$$

The corresponding rotation for the lateral side AB is

$$\frac{v}{a} + \frac{\partial w}{a\, d\varphi} + \frac{d}{d\varphi}\left(\frac{v}{a} + \frac{\partial w}{a\, d\varphi}\right) d\varphi$$

Hence, instead of the initial angle $d\varphi$, we must now use the expression

$$d\varphi + d\varphi\left(\frac{\partial v}{a\, \partial \varphi} + \frac{\partial^2 w}{a\, \partial \varphi^2}\right) \tag{g}$$

In calculating the angle of rotation about the y axis of side AB with respect to the side OC we use the expression for twist from the preceding article (see page 504); this gives the required angular displacement as

$$-\left(\frac{\partial^2 w}{\partial \varphi\, \partial x} + \frac{\partial v}{\partial x}\right) d\varphi \tag{h}$$

Rotation about the z axis of the side AB with respect to OC is caused by the displacements v and w. Because of the displacement v, the angle of rotation of side OC is $\partial v/\partial x$, and that of side AB is

$$\frac{\partial v}{\partial x} + \frac{\partial}{a\, \partial \varphi}\left(\frac{\partial v}{\partial x}\right) a\, d\varphi$$

so that the relative angular displacement is

$$\frac{\partial}{a\, \partial \varphi}\left(\frac{\partial v}{\partial x}\right) a\, d\varphi \tag{i}$$

Because of the displacement w, the side AB rotates in the axial plane by the angle $\partial w/\partial x$. The component of this rotation with respect to the z axis is

$$-\frac{\partial w}{\partial x} d\varphi \tag{j}$$

Summing up (i) and (j), the relative angular displacement about the z axis of side AB with respect to side OC is

$$\left(\frac{\partial^2 v}{\partial \varphi\, \partial x} - \frac{\partial w}{\partial x}\right) d\varphi \tag{k}$$

Having the foregoing formulas[1] for the angles, we may now obtain three equations of equilibrium of the element $OABC$ (Fig. 256) by projecting all forces on the x, y, and z axes. Beginning with those forces

---

[1] These formulas can be readily obtained for a cylindrical shell from the general formulas given by A. E. H. Love in his book "Elasticity," 4th ed., p. 523, 1927.

GENERAL THEORY OF CYLINDRICAL SHELLS 511

parallel to the resultant forces N_x and $N_{\varphi x}$ and projecting them on the x axis, we obtain

$$\frac{\partial N_x}{\partial x} dx\, a\, d\varphi \qquad \frac{\partial N_{\varphi x}}{\partial \varphi} d\varphi\, dx$$

Because of the angle of rotation represented by expression (k), the forces parallel to N_y give a component in the x direction equal to

$$-N_\varphi \left(\frac{\partial^2 v}{\partial \varphi\, \partial x} - \frac{\partial w}{\partial x} \right) d\varphi\, dx$$

Because of the rotation represented by expression (e), the forces parallel to $N_{x\varphi}$ give a component in the x direction equal to

$$-N_{x\varphi} \frac{\partial^2 v}{\partial x^2} dx\, a\, d\varphi$$

Finally, because of angles represented by expressions (d) and (h), the forces parallel to Q_x and Q_φ give components in the x direction equal to

$$-Q_x \frac{\partial^2 w}{\partial x^2} dx\, a\, d\varphi - Q_\varphi \left(\frac{\partial^2 w}{\partial \varphi\, \partial x} + \frac{\partial v}{\partial x} \right) d\varphi\, dx$$

Regarding the external forces acting on the element, we assume that there is only a normal pressure of intensity q, the projection of which on the x and y axes is zero.

Summing up all the projections calculated above, we obtain

$$\frac{\partial N_x}{\partial x} dx\, a\, d\varphi + \frac{\partial N_{\varphi x}}{\partial \varphi} d\varphi\, dx - N_\varphi \left(\frac{\partial^2 v}{\partial \varphi\, \partial x} - \frac{\partial w}{\partial x} \right) d\varphi\, dx$$

$$- N_{x\varphi} \frac{\partial^2 v}{\partial x^2} dx\, a\, d\varphi - Q_x \frac{\partial^2 w}{\partial x^2} dx\, a\, d\varphi - Q_\varphi \left(\frac{\partial^2 w}{\partial \varphi\, \partial x} + \frac{\partial v}{\partial x} \right) d\varphi\, dx = 0$$

In the same manner two other equations of equilibrium can be written. After simplification, all three equations can be put in the following form:

$$a \frac{\partial N_x}{\partial x} + \frac{\partial N_{\varphi x}}{\partial \varphi} - aQ_x \frac{\partial^2 w}{\partial x^2} - aN_{x\varphi} \frac{\partial^2 v}{\partial x^2}$$

$$- Q_\varphi \left(\frac{\partial v}{\partial x} + \frac{\partial^2 w}{\partial x\, \partial \varphi} \right) - N_\varphi \left(\frac{\partial^2 v}{\partial x\, \partial \varphi} - \frac{\partial w}{\partial x} \right) = 0$$

$$\frac{\partial N_\varphi}{\partial \varphi} + a \frac{\partial N_{x\varphi}}{\partial x} + aN_x \frac{\partial^2 v}{\partial x^2} - Q_x \left(\frac{\partial v}{\partial x} + \frac{\partial^2 w}{\partial x\, \partial \varphi} \right)$$

$$+ N_{\varphi x} \left(\frac{\partial^2 v}{\partial x\, \partial \varphi} - \frac{\partial w}{\partial x} \right) - Q_\varphi \left(1 + \frac{\partial v}{a\, \partial \varphi} + \frac{\partial^2 w}{a\, \partial \varphi^2} \right) = 0 \qquad (298)$$

$$a \frac{\partial Q_x}{\partial x} + \frac{\partial Q_\varphi}{\partial \varphi} + N_{x\varphi} \left(\frac{\partial v}{\partial x} + \frac{\partial^2 w}{\partial x\, \partial \varphi} \right) + aN_x \frac{\partial^2 w}{\partial x^2}$$

$$+ N_\varphi \left(1 + \frac{\partial v}{a\, \partial \varphi} + \frac{\partial^2 w}{a\, \partial \varphi^2} \right) + N_{\varphi x} \left(\frac{\partial v}{\partial x} + \frac{\partial^2 w}{\partial x\, \partial \varphi} \right) + qa = 0$$

512   THEORY OF PLATES AND SHELLS

Going now to the three equations of moments with respect to the x, y, and z axes (Fig. 256b) and again taking into consideration the small angular displacements of the sides BC and AB with respect to OA and OC, respectively, we obtain the following equations:

$$a\frac{\partial M_{x\varphi}}{\partial x} - \frac{\partial M_{\varphi}}{\partial \varphi} - aM_x\frac{\partial^2 v}{\partial x^2} - M_{\varphi x}\left(\frac{\partial^2 v}{\partial x\, \partial \varphi} - \frac{\partial w}{\partial x}\right) + aQ_{\varphi} = 0$$

$$\frac{\partial M_{\varphi x}}{\partial \varphi} + a\frac{\partial M_x}{\partial x} + aM_{x\varphi}\frac{\partial^2 v}{\partial x^2} - M_{\varphi}\left(\frac{\partial^2 v}{\partial x\, \partial \varphi} - \frac{\partial w}{\partial x}\right) - aQ_x = 0 \quad (299)$$

$$M_x\left(\frac{\partial v}{\partial x} + \frac{\partial^2 w}{\partial x\, \partial \varphi}\right) + aM_{x\varphi}\frac{\partial^2 w}{\partial x^2} + M_{\varphi x}\left(1 + \frac{\partial v}{a\, \partial \varphi} + \frac{\partial^2 w}{a\, \partial \varphi^2}\right)$$

$$- M_{\varphi}\left(\frac{\partial v}{\partial x} + \frac{\partial^2 w}{\partial x\, \partial \varphi}\right) + a(N_{x\varphi} - N_{\varphi x}) = 0$$

By using the first two of these equations[1] we can eliminate Q_x and Q_{φ} from Eqs. (298) and obtain in this way three equations containing the resultant forces N_x, N_{φ}, and $N_{x\varphi}$ and the moments M_x, M_{φ}, and $M_{x\varphi}$. By using formulas (253) and (254) of Art. 104, all these quantities can be expressed in terms of the three strain components ϵ_x, ϵ_{φ}, and $\gamma_{x\varphi}$ of the middle surface and the three curvature changes χ_x, χ_{φ}, and $\chi_{x\varphi}$. By using the results of the previous article, these latter quantities can be represented in terms of the displacements u, v, and w as follows:[2]

$$\epsilon_x = \frac{\partial u}{\partial x} \qquad \epsilon_{\varphi} = \frac{\partial v}{a\, \partial \varphi} - \frac{w}{a} \qquad \gamma_{x\varphi} = \frac{\partial u}{a\, \partial \varphi} + \frac{\partial v}{\partial x}$$

$$\chi_x = \frac{\partial^2 w}{\partial x^2} \qquad \chi_{\varphi} = \frac{1}{a^2}\left(\frac{\partial v}{\partial \varphi} + \frac{\partial^2 w}{\partial \varphi^2}\right) \qquad \chi_{x\varphi} = \frac{1}{a}\left(\frac{\partial v}{\partial x} + \frac{\partial^2 w}{\partial x\, \partial \varphi}\right) \quad (300)$$

Thus we finally obtain the three differential equations for the determination of the displacements u, v, and w.

In the derivation equations (298) and (299) the change of curvature of the element $OABC$ was taken into consideration. This procedure is necessary if the forces N_x, N_y, and N_{xy} are not small in comparison with their *critical* values, at which lateral buckling of the shell may occur.[3] If these forces are small, their effect on bending is negligible, and we can omit from Eqs. (298) and (299) all terms containing the products of the resultant forces or resultant moments with the derivatives of the small displacements u, v, and w. In such a case the three Eqs. (298) and the

[1] To satisfy the third of these equations the trapezoidal form of the sides of the element $OABC$ must be considered as mentioned in Art. 104. This question is discussed by W. Flügge, "Statik und Dynamik der Schalen," 2d ed., p. 148, Berlin, 1957.

[2] The same expressions for the change of curvature as in the preceding article are used, since the effect of strain in the middle surface on curvature is neglected.

[3] The problems of buckling of cylindrical shells are discussed in S. Timoshenko, "Theory of Elastic Stability," and will not be considered here.

first two equations of system (299) can be rewritten in the following simplified form:

$$a\frac{\partial N_x}{\partial x} + \frac{\partial N_{\varphi x}}{\partial \varphi} = 0$$

$$\frac{\partial N_\varphi}{\partial \varphi} + a\frac{\partial N_{x\varphi}}{\partial x} - Q_\varphi = 0$$

$$a\frac{\partial Q_x}{\partial x} + \frac{\partial Q_\varphi}{\partial \varphi} + N_\varphi + qa = 0 \qquad (301)$$

$$a\frac{\partial M_{x\varphi}}{\partial x} - \frac{\partial M_\varphi}{\partial \varphi} + aQ_\varphi = 0$$

$$\frac{\partial M_{\varphi x}}{\partial \varphi} + a\frac{\partial M_x}{\partial x} - aQ_x = 0$$

Eliminating the shearing forces Q_x and Q_φ, we finally obtain the three following equations:

$$a\frac{\partial N_x}{\partial x} + \frac{\partial N_{\varphi x}}{\partial \varphi} = 0$$

$$\frac{\partial N_\varphi}{\partial \varphi} + a\frac{\partial N_{x\varphi}}{\partial x} + \frac{\partial M_{x\varphi}}{\partial x} - \frac{1}{a}\frac{\partial M_\varphi}{\partial \varphi} = 0 \qquad (302)$$

$$N_\varphi + \frac{\partial^2 M_{\varphi x}}{\partial x\, \partial \varphi} + a\frac{\partial^2 M_x}{\partial x^2} - \frac{\partial^2 M_{x\varphi}}{\partial x\, \partial \varphi} + \frac{1}{a}\frac{\partial^2 M_\varphi}{\partial \varphi^2} + qa = 0$$

By using Eqs. (253), (254), and (300), all the quantities entering in these equations can be expressed by the displacements u, v, and w, and we obtain

$$\frac{\partial^2 u}{\partial x^2} + \frac{1-\nu}{2a^2}\frac{\partial^2 u}{\partial \varphi^2} + \frac{1+\nu}{2a}\frac{\partial^2 v}{\partial x\, \partial \varphi} - \frac{\nu}{a}\frac{\partial w}{\partial x} = 0$$

$$\frac{1+\nu}{2}\frac{\partial^2 u}{\partial x\, \partial \varphi} + a\frac{1-\nu}{2}\frac{\partial^2 v}{\partial x^2} + \frac{1}{a}\frac{\partial^2 v}{\partial \varphi^2} - \frac{1}{a}\frac{\partial w}{\partial \varphi}$$
$$+ \frac{h^2}{12a}\left(\frac{\partial^3 w}{\partial x^2\, \partial \varphi} + \frac{\partial^3 w}{a^2\, \partial \varphi^3}\right) + \frac{h^2}{12a}\left[(1-\nu)\frac{\partial^2 v}{\partial x^2} + \frac{\partial^2 v}{a^2\, \partial \varphi^2}\right] = 0 \qquad (303)$$

$$\nu\frac{\partial u}{\partial x} + \frac{\partial v}{a\, \partial \varphi} - \frac{w}{a} - \frac{h^2}{12}\left(a\frac{\partial^4 w}{\partial x^4} + \frac{2}{a}\frac{\partial^4 w}{\partial x^2\, \partial \varphi^2} + \frac{\partial^4 w}{a^3\, \partial \varphi^4}\right)$$
$$- \frac{h^2}{12}\left(\frac{2-\nu}{a}\frac{\partial^3 v}{\partial x^2\, \partial \varphi} + \frac{\partial^3 v}{a^3\, \partial \varphi^3}\right) = -\frac{aq(1-\nu^2)}{Eh}$$

More elaborate investigations show[1] that the last two terms on the left-hand side of the second of these equations and the last term on the left-hand side of the third equation are small quantities of the same order as those which we already disregarded by assuming a linear distribution of stress through the thickness of the shell and by neglecting the stretching of the middle surface of the shell (see page 431). In such a case it

---

[1] See Vlasov, *op. cit.*, p. 316, and, for more exact equations, p. 257.

will be logical to omit the above-mentioned terms and to use in analysis of thin cylindrical shells the following simplified system of equations:

$$\frac{\partial^2 u}{\partial x^2} + \frac{1-\nu}{2a^2}\frac{\partial^2 u}{\partial \varphi^2} + \frac{1+\nu}{2a}\frac{\partial^2 v}{\partial x\, \partial \varphi} - \frac{\nu}{a}\frac{\partial w}{\partial x} = 0$$

$$\frac{1+\nu}{2}\frac{\partial^2 u}{\partial x\, \partial \varphi} + a\frac{1-\nu}{2}\frac{\partial^2 v}{\partial x^2} + \frac{1}{a}\frac{\partial^2 v}{\partial \varphi^2} - \frac{1}{a}\frac{\partial w}{\partial \varphi} = 0 \qquad (304)$$

$$\nu\frac{\partial u}{\partial x} + \frac{\partial v}{a\, \partial \varphi} - \frac{w}{a} - \frac{h^2}{12}\left(a\frac{\partial^4 w}{\partial x^4} + \frac{2}{a}\frac{\partial^4 w}{\partial x^2\, \partial \varphi^2} + \frac{\partial^4 w}{a^3\, \partial \varphi^4}\right) = -\frac{aq(1-\nu^2)}{Eh}$$

Some simplified expressions for the stress resultants which are in accordance with the simplified relations (304) between the displacements of the shell will be given in Art. 125.

From the foregoing it is seen that the problem of a laterally loaded cylindrical shell reduces in each particular case to the solution of a system of three differential equations. Several applications of these equations will be shown in the next articles.

122. Cylindrical Shells with Supported Edges. Let us consider the case of a cylindrical shell supported at the ends and submitted to the

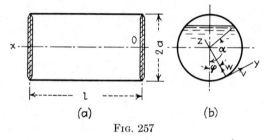

Fig. 257

pressure of an enclosed liquid as shown in Fig. 257.[1] The conditions at the supports and the conditions of symmetry of deformation will be satisfied if we take the components of displacement in the form of the following series:

$$u = \sum\sum A_{mn} \cos n\varphi \cos \frac{m\pi x}{l}$$

$$v = \sum\sum B_{mn} \sin n\varphi \sin \frac{m\pi x}{l} \qquad (a)$$

$$w = \sum\sum C_{mn} \cos n\varphi \sin \frac{m\pi x}{l}$$

in which l is the length of the cylinder and φ is the angle measured as shown in Fig. 257.[2]

[1] See S. Timoshenko, "Theory of Elasticity," vol. 2, p. 385, St. Petersburg, 1916 (Russian).

[2] By substituting expressions (a) in Eqs. (300) it can be shown that the tensile forces N_x and the moments M_x vanish at the ends; the shearing forces do not vanish, however, since $\gamma_{x\varphi}$ and $M_{x\varphi}$ are not zero at the ends.

The intensity of the load q is represented by the following expressions:

$$q = -\gamma a(\cos \varphi - \cos \alpha) \quad \text{when } \varphi < \alpha$$
$$q = 0 \quad \text{when } \varphi > \alpha \qquad (b)$$

in which γ is the specific weight of the liquid and the angle α defines the level of the liquid, as shown in Fig. 257b. The load q can be represented by the series

$$q = \sum\sum D_{mn} \cos n\varphi \sin \frac{m\pi x}{l} \qquad (c)$$

in which the coefficients D_{mn} can be readily calculated in the usual way from expressions (b). These coefficients are represented by the expression

$$D_{mn} = -\frac{8\gamma a}{mn\pi^2(n^2 - 1)} (\cos \alpha \sin n\alpha - n \cos n\alpha \sin \alpha) \qquad (d)$$

where $m = 1, 3, 5, \ldots$ and $n = 2, 3, 4, \ldots$

whereas

$$D_{m0} = -\frac{4\gamma a}{m\pi^2} (\sin \alpha - \alpha \cos \alpha) \qquad (e)$$

and

$$D_{m1} = -\frac{2\gamma a}{m\pi^2} (2\alpha - \sin 2\alpha) \qquad (f)$$

In the case of a cylindrical shell completely filled with liquid, we denote the pressure at the axis of the cylinder[1] by γd; then

$$q = -\gamma(d + a \cos \varphi) \qquad (g)$$

and we obtain, instead of expressions (d), (e), and (f),

$$D_{mn} = 0 \qquad D_{m0} = -\frac{4\gamma d}{m\pi} \qquad D_{m1} = -\frac{4\gamma a}{m\pi} \qquad (h)$$

To obtain the deformation of the shell we substitute expressions (a) and (c) in Eqs. (304). In this way we obtain for each pair of values of m and n a system of three linear equations from which the corresponding values of the coefficients A_{mn}, B_{mn}, and C_{mn} can be calculated.[2] Taking a particular case in which $d = a$, we find that for $n = 0$ and $m = 1, 3, 5, \ldots$ these equations are especially simple, and we obtain

$$B_{m0} = 0 \qquad C_{m0} = -\frac{m\pi}{\lambda \nu} A_{m0} = -\frac{\pi N}{3m\left[\lambda^2(1-\nu^2) + \frac{\eta^2}{3} m^4\pi^4\right]}$$

where

$$N = \frac{2\gamma a l^2 h}{\pi^2 D} \qquad \lambda = \frac{l}{a} \qquad \eta = \frac{h}{2l}$$

---

[1] In a closed cylindrical vessel this pressure can be larger than $a\gamma$.

[2] Such calculations have been made for several particular cases by I. A. Wojtaszak, *Phil. Mag.*, ser. 7, vol. 18, p. 1099, 1934; see also the paper by H. Reissner in *Z. angew. Math. Mech.*, vol. 13, p. 133, 1933.

For $n = 1$ the expressions for the coefficients are more complicated. To show how rapidly the coefficients diminish as m increases, we include in Table 87 the numerical values of the coefficients for a particular case in which $a = 50$ cm, $l = 25$ cm, $h = 7$ cm, $\nu = 0.3$, and $\alpha = \pi$.

Table 87. The Values of the Coefficients in Expressions (a)

m	$A_{m0}\dfrac{2 \cdot 10^3}{Nh}$	$C_{m0}\dfrac{2 \cdot 10^3}{Nh}$	$A_{m1}\dfrac{2 \cdot 10^3}{Nh}$	$B_{m1}\dfrac{2 \cdot 10^3}{Nh}$	$C_{m1}\dfrac{2 \cdot 10^3}{Nh}$
1	57.88	$-1{,}212.$	49.18	-66.26	$-1{,}183$
3	0.1073	-6.742	0.1051	-0.0432	$-\ \ \ 6.704$
5	0.00503	-0.526	0.00499	-0.00122	$-\ \ \ 0.525$

It is seen that the coefficients rapidly diminish as m increases. Hence, by limiting the number of coefficients to those given in the table, we shall obtain the deformation of the shell with satisfactory accuracy.

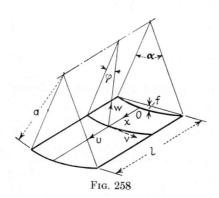

Fig. 258

123. Deflection of a Portion of a Cylindrical Shell.

The method used in the preceding article can also be applied to a portion of a cylindrical shell which is supported along the edges and submitted to the action of a uniformly distributed load q normal to the surface (Fig. 258).[1] We take the components of displacement in the form of the series

$$u = \sum\sum A_{mn} \sin\frac{n\pi\varphi}{\alpha} \cos\frac{m\pi x}{l}$$

$$v = \sum\sum B_{mn} \cos\frac{n\pi\varphi}{\alpha} \sin\frac{m\pi x}{l} \qquad (a)$$

$$w = \sum\sum C_{mn} \sin\frac{n\pi\varphi}{\alpha} \sin\frac{m\pi x}{l}$$

in which α is the central angle subtended by the shell and l is the length of the shell. It can be shown by substitution of expressions (a) in Eqs. (300) that in this way we shall satisfy the conditions at the boundary, which require that along the edges $\varphi = 0$ and $\varphi = \alpha$ the deflection w, the force N_φ, and the moment M_φ vanish and that along the edges $x = 0$ and $x = l$ the deflection w, the force N_x, and the moment M_x vanish.

[1] See Timoshenko, "Theory of Elasticity," vol. 2, p. 386, 1916.

The intensity of the normal load q can be represented by the series

$$q = \sum\sum D_{mn} \sin \frac{n\pi\varphi}{\alpha} \sin \frac{m\pi x}{l} \qquad (b)$$

Substituting series (a) and (b) in Eqs. (304), we obtain the following system of linear algebraic equations for calculating the coefficients A_{mn}, B_{mn}, and C_{mn}:

$$A_{mn}\pi\left[\left(\frac{am}{l}\right)^2 + \frac{(1-\nu)n^2}{2\alpha^2}\right] + B_{mn}\pi\frac{(1+\nu)amn}{2\alpha l} + C_{mn}\frac{\nu am}{l} = 0$$

$$A_{mn}\pi\frac{(1+\nu)amn}{2\alpha l} + B_{mn}\pi\left[\frac{(1-\nu)a^2m^2}{2l^2} + \frac{n^2}{\alpha^2}\right] + C_{mn}\frac{n}{\alpha} = 0 \qquad (c)$$

$$A_{mn}\nu\pi\frac{am}{l} + B_{mn}\frac{n\pi}{\alpha} + C_{mn}\left[1 + \frac{\pi^4 h^2}{12a^2}\left(\frac{a^2m^2}{l^2} + \frac{n^2}{\alpha^2}\right)^2\right] = D_{mn}\frac{a^2(1-\nu^2)}{Eh}$$

To illustrate the application of these equations let us consider the case of a uniformly distributed load[1] acting on a portion of a cylindrical shell having a small angle α and a small sag $f = a[1 - \cos(\alpha/2)]$. In this particular case expression (b) becomes

$$q = \sum_{1,3,5,\ldots}\sum_{1,3,5,\ldots} \frac{16q}{\pi^2 mn} \sin \frac{m\pi x}{l} \sin \frac{n\pi\varphi}{\alpha} \qquad (d)$$

and the coefficients D_{mn} are given by the expression

$$D_{mn} = \frac{16q}{mn\pi^2} \qquad (e)$$

Substituting these values in Eqs. (c), we can calculate the coefficients A_{mn}, B_{mn}, and C_{mn}. The calculations made for a particular case in which $\alpha a = l$ and for several values of the ratio f/h show that for small values of this ratio, series (a) are rapidly convergent and the first few terms give the displacements with satisfactory accuracy.

The calculations also show that the maximum values of the bending stresses produced by the moments M_x and M_φ diminish rapidly as f/h increases. The calculation of these stresses is very tedious in the case of larger values of f/h, since the series representing the moments become less rapidly convergent and a larger number of terms must be taken.

The method used in this article is similar to Navier's method of calculating bending of rectangular plates with simply supported edges. If only the rectilinear edges $\varphi = 0$ and $\varphi = \alpha$ of the shell in Fig. 258 are simply supported and the other two edges are built in or free, a solution similar to that of M. Lévy's method for the case of rectangular plates (see page 113) can be applied. We assume the following series for the components of displacement:

[1] The load is assumed to act toward the axis of the cylinder.

$$u = \sum U_m \sin \frac{m\pi\varphi}{\alpha}$$

$$v = \sum V_m \cos \frac{m\pi\varphi}{\alpha} \qquad (f)$$

$$w = \sum W_m \sin \frac{m\pi\varphi}{\alpha}$$

in which U_m, V_m, and W_m are functions of x only. Substituting these series in Eqs. (304), we obtain for U_m, V_m, and W_m three ordinary differential equations with constant coefficients. These equations can be integrated by using exponential functions. An analysis of this kind made for a closed cylindrical shell[1] shows that the solution is very involved and that results suitable for practical application can be obtained only by introducing simplifying assumptions. It could be shown that each set of the functions U_m, V_m, W_m contains eight constants of integration for each assumed value of m. Accordingly, four conditions on each edge $x = $ constant must be at our disposal. Let us formulate these conditions in the following three cases.

Built-in Edge. Usually such a support is considered as perfectly rigid, and the edge conditions then are

$$u = 0 \qquad v = 0 \qquad w = 0 \qquad \frac{\partial w}{\partial x} = 0 \qquad (g)$$

Should it happen, however, that the shell surface on the edge is free to move in the direction x, then the first of the foregoing conditions has to be replaced by the condition $N_x = 0$.

Simply Supported Edge. Such a hinged edge is not able to transmit a moment M_x needed to enforce the condition $\partial w/\partial x = 0$. Assuming also that there is no edge resistance in the direction x, we arrive at the boundary conditions

$$v = 0 \qquad w = 0 \qquad M_x = 0 \qquad N_x = 0 \qquad (h)$$

whereas the displacement u and the stress resultants $N_{x\varphi}$, $M_{x\varphi}$, and Q_x do not vanish on the edge.

The reactions of the simply supported edge (Fig. 259a) deserve brief consideration. The action of a twisting couple $M_{x\varphi} \, ds$, applied to an element $ABCD$ of the edge, is statically equivalent to the action of three forces shown in Fig. 259b. A variation of the radial forces $M_{x\varphi}$ along the edge yields, just as in the case of a plate (Fig. 50), an additional shearing force of the intensity $-\partial M_{x\varphi}/\partial s$, the total shearing force being (Fig. 259c)

$$T_x = Q_x - \frac{\partial M_{x\varphi}}{a \, \partial \varphi} \qquad (i)$$

The remaining component $M_{x\varphi} \, d\varphi$ (Fig. 259b) may be considered as a supplementary membrane force of the intensity $M_{x\varphi} \, d\varphi/ds = M_{x\varphi}/a$. Hence the resultant membrane force in the direction of the tangent to the edge becomes

$$S_x = N_{x\varphi} + \frac{M_{x\varphi}}{a} \qquad (j)$$

---

[1] See paper by K. Miesel, *Ingr.-Arch.*, vol. 1, p. 29, 1929. An application of the theory to the calculation of stress in the hull of a submarine is shown in this paper.

Free Edge. Letting all the stress resultants vanish on the edge, we find that the four conditions characterizing the free edge assume the form

$$N_x = 0 \qquad M_x = 0 \qquad S_x = 0 \qquad T_x = 0 \qquad (k)$$

where S_x and T_x are given by expressions (j) and (i), respectively.[1]

124. An Approximate Investigation of the Bending of Cylindrical Shells. From the discussion of the preceding article it may be concluded that the application of the general theory of bending of cylindrical shells in even the simplest cases results in very complicated calculations. To make the theory applicable to the solution of practical problems some further simplifications in this theory are necessary. In considering the membrane theory of cylindrical shells it was stated that this theory gives satisfactory results for portions of a shell at a considerable distance from the edges but that it is insufficient to satisfy all the conditions at the boundary. It is logical, therefore, to take the solution furnished by the membrane theory as a first

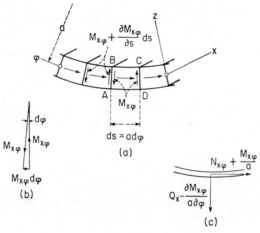

Fig. 259

approximation and use the more elaborate bending theory only to satisfy the conditions at the edges. In applying this latter theory, it must be assumed that no external load is distributed over the shell and that only forces and moments such as are necessary to satisfy the boundary conditions are applied along the edges. The bending produced by such forces can be investigated by using Eqs. (303) after placing the load q equal to zero in these equations.

In applications such as are encountered in structural engineering[2] the ends $x = 0$ and $x = l$ of the shell (Fig. 260) are usually supported in such a manner that the

---

[1] For a solution of the problem of bending based on L. H. Donnell's simplified differential equations see N. J. Hoff, *J. Appl. Mechanics*, vol. 21, p. 343, 1954; see also Art. 125 of this book.

[2] In recent times thin reinforced cylindrical shells of concrete have been successfully applied in structures as coverings for large halls. Descriptions of some of these structures can be found in the article by F. Dischinger, "Handbuch für Eisenbetonbau," 3d ed., vol. 12, Berlin, 1928; see also the paper by F. Dischinger and U. Finsterwalder in *Bauingenieur*, vol. 9, 1928, and references in Art. 126 of this book.

displacements v and w at the ends vanish. Experiments show that in such shells the bending in the axial planes is negligible, and we can assume $M_x = 0$ and $Q_x = 0$ in the equations of equilibrium (301). We can also neglect the twisting moment $M_{x\varphi}$. With these assumptions the system of Eqs. (301) can be considerably simplified, and

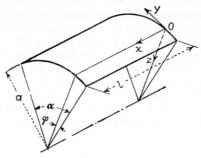

Fig. 260

the resultant forces and components of displacement can all be expressed in terms[1] of moment M_φ. From the fourth of the equations (301) we obtain

$$Q_\varphi = \frac{1}{a} \frac{\partial M_\varphi}{\partial \varphi} \qquad (a)$$

Substituting this in the third equation of the same system, we obtain, for $q = 0$,

$$N_\varphi = -\frac{\partial Q_\varphi}{\partial \varphi} = -\frac{1}{a} \frac{\partial^2 M_\varphi}{\partial \varphi^2} \qquad (b)$$

The second and the first of the equations (301) then give

$$\frac{\partial N_{x\varphi}}{\partial x} = \frac{1}{a}\left(Q_\varphi - \frac{\partial N_\varphi}{\partial \varphi}\right) = \frac{1}{a^2}\left(\frac{\partial M_\varphi}{\partial \varphi} + \frac{\partial^3 M_\varphi}{\partial \varphi^3}\right) \qquad (c)$$

$$\frac{\partial^2 N_x}{\partial x^2} = -\frac{1}{a}\frac{\partial^2 N_{x\varphi}}{\partial \varphi\, \partial x} = -\frac{1}{a^3}\left(\frac{\partial^2 M_\varphi}{\partial \varphi^2} + \frac{\partial^4 M_\varphi}{\partial \varphi^4}\right) \qquad (d)$$

The components of displacement can also be expressed in terms of M_φ and its derivatives. We begin with the known relations [see Eqs. (253) and (254)]

$$\epsilon_x = \frac{\partial u}{\partial x} = \frac{1}{Eh}(N_x - \nu N_\varphi)$$

$$\gamma_{x\varphi} = \frac{\partial u}{a\, \partial \varphi} + \frac{\partial v}{\partial x} = \frac{2(1+\nu)}{Eh} N_{x\varphi} \qquad (e)$$

$$\epsilon_\varphi = \frac{\partial v}{a\, \partial \varphi} - \frac{w}{a} = \frac{1}{Eh}(N_\varphi - \nu N_x)$$

[1] This approximate theory of bending of cylindrical shells was developed by U. Finsterwalder; see *Ingr.-Arch.*, vol. 4, p. 43, 1933.

GENERAL THEORY OF CYLINDRICAL SHELLS

From these equations we obtain

$$\frac{\partial u}{\partial x} = \frac{1}{Eh}(N_x - \nu N_\varphi)$$

$$\frac{\partial^2 v}{\partial x^2} = \frac{1}{Eh}\left[2(1+\nu)\frac{\partial N_{x\varphi}}{\partial x} - \frac{1}{a}\left(\frac{\partial N_x}{\partial \varphi} - \nu \frac{\partial N_\varphi}{\partial \varphi}\right)\right] \quad (f)$$

$$\frac{\partial^2 w}{\partial x^2} = \frac{1}{Eh}\left[a\left(\nu \frac{\partial^2 N_x}{\partial x^2} - \frac{\partial^2 N_\varphi}{\partial x^2}\right) + 2(1+\nu)\frac{\partial^2 N_{x\varphi}}{\partial x \partial \varphi} - \frac{1}{a}\left(\frac{\partial^2 N_x}{\partial \varphi^2} - \nu \frac{\partial^2 N_\varphi}{\partial \varphi^2}\right)\right]$$

Using these expressions together with Eqs. (b), (c), and (d) and with the expression for the bending moment

$$M_\varphi = -\frac{D}{a^2}\left(\frac{\partial v}{\partial \varphi} + \frac{\partial^2 w}{\partial \varphi^2}\right) \quad (g)$$

we finally obtain for the determination of M_φ the following differential equation of the eighth order:

$$\frac{\partial^8 M_\varphi}{\partial \varphi^8} + (2+\nu)a^2 \frac{\partial^8 M_\varphi}{\partial x^2 \partial \varphi^6} + 2\frac{\partial^6 M_\varphi}{\partial \varphi^6} + (1+2\nu)a^4 \frac{\partial^8 M_\varphi}{\partial x^4 \partial \varphi^4}$$

$$+ 2(2+\nu)a^2 \frac{\partial^6 M_\varphi}{\partial x^2 \partial \varphi^4} + \frac{\partial^4 M_\varphi}{\partial \varphi^4} + \nu a^6 \frac{\partial^8 M_\varphi}{\partial x^6 \partial \varphi^2} + (1+\nu)^2 a^4 \frac{\partial^6 M_\varphi}{\partial x^4 \partial \varphi^2}$$

$$+ (2+\nu)a^2 \frac{\partial^4 M_\varphi}{\partial x^2 \partial \varphi^2} + 12(1-\nu^2)\frac{a^6}{h^2}\frac{\partial^4 M_\varphi}{\partial x^4} = 0 \quad (h)$$

A particular solution of this equation is afforded by the expression

$$M_\varphi = A e^{\alpha \varphi} \sin \frac{m\pi x}{l} \quad (i)$$

Substituting it in Eq. (h) and using the notation

$$\frac{m\pi a}{l} = \lambda \quad (j)$$

the following algebraic equation for calculating α is obtained:

$$\alpha^8 + [2 - (2+\nu)\lambda^2]\alpha^6 + [(1+2\nu)\lambda^4 - 2(2+\nu)\lambda^2 + 1]\alpha^4$$

$$+ [-\nu\lambda^6 + (1+\nu)^2\lambda^4 - (2+\nu)\lambda^2]\alpha^2 + 12(1-\nu^2)\frac{a^2}{h^2}\lambda^4 = 0 \quad (k)$$

The eight roots of this equation can be put in the form

$$\alpha_{1,2,3,4} = \pm(\gamma_1 \pm i\beta_1) \qquad \alpha_{5,6,7,8} = \pm(\gamma_2 \pm i\beta_2) \quad (l)$$

Beginning with the edge $\varphi = 0$ and assuming that the moment M_φ rapidly diminishes as φ increases, we use only those four of the roots (l) which satisfy this requirement. Then combining the four corresponding solutions (i), we obtain

$$M_\varphi = [e^{-\gamma_1 \varphi}(C_1 \cos \beta_1\varphi + C_2 \sin \beta_1\varphi) + e^{-\gamma_2\varphi}(C_3 \cos \beta_2\varphi + C_4 \sin \beta_2\varphi)]\sin\frac{m\pi x}{l} \quad (m)$$

which gives for $\varphi = 0$

$$M_\varphi = (C_1 + C_3)\sin\frac{m\pi x}{l}$$

If instead of a single term (i) we take the trigonometric series

$$M_\varphi = \sum A_m e^{\alpha_m \varphi} \sin \frac{m\pi x}{l} \qquad (n)$$

any distribution of the bending moment M_φ along the edge $\varphi = 0$ can be obtained. Having an expression for M_φ, the resultant forces Q_φ, N_φ, and $N_{x\varphi}$ are obtained from Eqs. (a), (b), and (c).

If in some particular case the distributions of the moments $M\varphi$ and the resultant forces Q_φ, N_φ, and $N_{x\varphi}$ along the edge $\varphi = 0$ are given, we can represent these distributions by sine series. The values of the four coefficients in the terms containing $\sin (m\pi x/l)$ in these four series can then be used for the calculation of the four constants $C_1, \ldots, C_4$ in solution (m); and in this way the complete solution of the problem for the given force distribution can be obtained.

If the expressions for u, v, and w in terms of M_φ are obtained by using Eqs. (f), we can use the resulting expressions to solve the problem if the displacements, instead of the forces, are given along the edge $\varphi = 0$. Examples of such problems can be found in the previously mentioned paper by Finsterwalder,[1] who shows that the approximate method just described can be successfully applied in solving important structural problems.

125. The Use of a Strain and Stress Function. In the general case of bending of a cylindrical shell, for which the ratio l/a (Fig. 260) is not necessarily large, the effect of the couples M_x and M_{xy} cannot be disregarded. On the other hand, the simplified form [Eqs. (304)] of the relations between the displacements allows the introduction of a function[2] $F(x,\varphi)$ governing the state of strain and stress of the shell. Using the notation

$$c^2 = \frac{h^2}{12a^2} \qquad \xi = \frac{x}{a} \qquad \Delta = \frac{\partial^2}{\partial \xi^2} + \frac{\partial^2}{\partial \varphi^2} \qquad (a)$$

we can rewrite Eqs. (304) in the following form, including all three components X, Y, and Z of the external loading,

$$\frac{\partial^2 u}{\partial \xi^2} + \frac{1-\nu}{2}\frac{\partial^2 u}{\partial \varphi^2} + \frac{1+\nu}{2}\frac{\partial^2 v}{\partial \xi \partial \varphi} - \nu \frac{\partial w}{\partial \xi} = -\frac{(1-\nu^2)a^2}{Eh}X$$

$$\frac{1+\nu}{2}\frac{\partial^2 u}{\partial \xi \partial \varphi} + \frac{\partial^2 v}{\partial \varphi^2} + \frac{1-\nu}{2}\frac{\partial^2 v}{\partial \xi^2} - \frac{\partial w}{\partial \varphi} = -\frac{(1-\nu^2)a^2}{Eh}Y \qquad (305)$$

$$\nu \frac{\partial u}{\partial \xi} + \frac{\partial v}{\partial \varphi} - w - c^2 \Delta \Delta w = -\frac{(1-\nu^2)a^2}{Eh}Z$$

The set of these simultaneous equations can be reduced to a single differential equation by putting

$$u = \frac{\partial^3 F}{\partial \xi \partial \varphi^2} - \nu \frac{\partial^3 F}{\partial \xi^3} + u_0$$

$$v = -\frac{\partial^3 F}{\partial \varphi^3} - (2+\nu)\frac{\partial^3 F}{\partial \xi^2 \partial \varphi} + v_0 \qquad (306)$$

$$w = -\Delta \Delta F + w_0$$

---

[1] *Ibid.*
[2] Due to Vlasov, *op. cit.* Almost equivalent results, without the use of a stress function, were obtained by L. H. Donnell, *NACA Rept.* 479, 1933. See also N. J. Hoff, *J. Appl. Mechanics*, vol. 21, p. 343, 1954.

where u_0, v_0, w_0 are a system of particular solutions of the nonhomogeneous equations (305). As for the strain and stress function $F(\xi,\varphi)$, it must satisfy the differential equation

$$\Delta\Delta\Delta\Delta F + \frac{1-\nu^2}{c^2}\frac{\partial^4 F}{\partial \xi^4} = 0 \qquad (307)$$

which is equivalent to the group of Eqs. (305), if $X = Y = Z = 0$.* It can be shown that in this last case not only the function F but also all displacement and strain components, as well as all stress resultants of the shell, satisfy the differential equation (307).

For the elongations, the shearing strain, and the changes of the curvature of the middle surface of the shell, the expressions (300) still hold. The stress resultants may be represented either in terms of the displacements or directly through the function F. In accordance with the simplifications leading to Eqs. (304), the effect of the displacements u and v on the bending and twisting moments must be considered as negligible. Thus, with the notation

$$K = \frac{Eh}{1-\nu^2} \qquad D = \frac{Eh^3}{12(1-\nu^2)} \qquad (308)$$

the following expressions are obtained:

$$N_x = \frac{K}{a}\left[\frac{\partial u}{\partial \xi} + \nu\left(\frac{\partial v}{\partial \varphi} - w\right)\right] = \frac{Eh}{a}\frac{\partial^4 F}{\partial \xi^2\, \partial \varphi^2}$$

$$N_\varphi = \frac{K}{a}\left(\frac{\partial v}{\partial \varphi} - w + \nu\frac{\partial u}{\partial \xi}\right) = \frac{Eh}{a}\frac{\partial^4 F}{\partial \xi^4} \qquad (309)$$

$$N_{x\varphi} = \frac{K(1-\nu)}{2a}\left(\frac{\partial u}{\partial \varphi} + \frac{\partial v}{\partial \xi}\right) = -\frac{Eh}{a}\frac{\partial^4 F}{\partial \xi^3\, \partial \varphi}$$

$$M_x = -\frac{D}{a^2}\left(\frac{\partial^2 w}{\partial \xi^2} + \nu\frac{\partial^2 w}{\partial \varphi^2}\right) = \frac{D}{a^2}\left(\frac{\partial^2}{\partial \xi^2} + \nu\frac{\partial^2}{\partial \varphi^2}\right)\Delta\Delta F$$

$$M_\varphi = -\frac{D}{a^2}\left(\frac{\partial^2 w}{\partial \varphi^2} + \nu\frac{\partial^2 w}{\partial \xi^2}\right) = \frac{D}{a^2}\left(\frac{\partial^2}{\partial \varphi^2} + \nu\frac{\partial^2}{\partial \xi^2}\right)\Delta\Delta F \qquad (310)$$

$$M_{x\varphi} = -M_{\varphi x} = \frac{D(1-\nu)}{a^2}\frac{\partial^2 w}{\partial \xi\, \partial \varphi} = -\frac{D}{a^2}(1-\nu)\frac{\partial^2}{\partial \xi\, \partial \varphi}\Delta\Delta F$$

$$Q_x = -\frac{D}{a^3}\frac{\partial}{\partial \xi}\Delta w = \frac{D}{a^3}\frac{\partial}{\partial \xi}\Delta\Delta\Delta F$$

$$Q_\varphi = -\frac{D}{a^3}\frac{\partial}{\partial \varphi}\Delta w = \frac{D}{a^3}\frac{\partial}{\partial \varphi}\Delta\Delta\Delta F \qquad (311)$$

Representing the differential equation (307) in the form

$$(\Delta)^4 F + 4\gamma^4 \frac{\partial^4 F}{\partial \xi^4} = 0 \qquad (b)$$

where

$$\gamma = \sqrt[4]{\frac{3(1-\nu^2)a^2}{h^2}} \qquad (c)$$

---

* Further stress functions F_x, F_y, F_z were introduced by Vlasov, *op. cit.*, to represent the particular integral of Eqs. (305) if X, Y, or Z, respectively, is not zero.

we see that Eq. (307) is also equivalent to the group of four equations

$$\Delta F_n \pm \gamma(1 \pm i)\frac{\partial F_n}{\partial \xi} = 0 \qquad (d)$$

with $i = \sqrt{-1}$ and $n = 1, 2, 3, 4$. Putting, finally,

$$\begin{aligned}
F_1 &= e^{-\frac{1}{2}\gamma(1+i)\xi}\Phi_1 \\
F_2 &= e^{\frac{1}{2}\gamma(1+i)\xi}\Phi_2 \\
F_3 &= e^{-\frac{1}{2}\gamma(1-i)\xi}\Phi_3 \\
F_4 &= e^{\frac{1}{2}\gamma(1-i)\xi}\Phi_4
\end{aligned} \qquad (e)$$

for the four new functions Φ_n a set of four equations

$$\Delta \Phi_n + \mu_n i \Phi_n = 0 \qquad (f)$$

is obtained, in which for the constant μ_n we have to assume

$$\begin{aligned}
\mu_1 &= \mu_2 = -\frac{a}{2h}\sqrt{3(1-\nu^2)} \\
\mu_3 &= \mu_4 = \frac{a}{2h}\sqrt{3(1-\nu^2)}
\end{aligned} \qquad (g)$$

The form of each of the equations (f) is analogous to that of the equation of vibration of a membrane. In comparison with Eqs. (d), Eqs. (f) have the advantage of being invariant against a change of coordinates on the cylindrical surface of the shell.

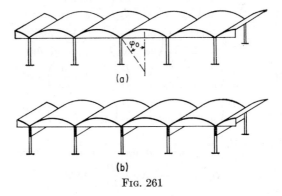

Fig. 261

126. Stress Analysis of Cylindrical Roof Shells.[1] Three typical roof layouts are shown in Figs. 261 and 265. The shells may be either continuous in the direction x or else supported only twice, say in the planes $x = 0$ and $x = l$. We shall confine ourselves to the latter case. We suppose the supporting structures to be rigid with

[1] See also "Design of Cylindrical Concrete Shell Roofs," *ASCE Manuals of Eng. Practice*, no. 31, 1952; J. E. Gibson and D. W. Cooper, "The Design of Cylindrical Shell Roofs," New York, 1954; R. S. Jenkins, "Theory and Design of Cylindrical Shell Structures," London, 1947; A. Aas-Jakobsen, "Die Berechnung der Zylindershalen," Berlin, 1958. Many data on design of roof shells and an interesting comparison of different methods of stress analysis may be found in *Proceedings of a Symposium on Concrete Shell Roof Construction*, Cement and Concrete Association, London, 1954.

GENERAL THEORY OF CYLINDRICAL SHELLS 525

respect to forces acting in their own planes, $x = $ constant, but as perfectly flexible with respect to transverse loading. In Fig. 261a the tension members at $\varphi = \varphi_0$ are flexible, whereas the shells shown in Figs. 261b and 265 are stiffened by beams of considerable rigidity, especially so in the vertical plane.

Any load distribution over the surface of the shell may be represented by the magnitude of its three components in the form of the series

$$X = \sum_{m=1}^{\infty} X_m(\varphi) \cos \frac{\lambda_m x}{a}$$

$$Y = \sum_{m=1}^{\infty} Y_m(\varphi) \sin \frac{\lambda_m x}{a} \tag{a}$$

$$Z = \sum_{m=1}^{\infty} Z_m(\varphi) \sin \frac{\lambda_m x}{a}$$

in which

$$\lambda_m = \frac{m\pi a}{l} \tag{b}$$

Likewise, let us represent the particular solutions u_0, v_0, w_0 in expressions (306) in the form

$$u_0 = \sum_{m=1}^{\infty} U_{om}(\varphi) \cos \frac{\lambda_m x}{a}$$

$$v_0 = \sum_{m=1}^{\infty} V_{om}(\varphi) \sin \frac{\lambda_m x}{a} \tag{c}$$

$$w_0 = \sum_{m=1}^{\infty} W_{om}(\varphi) \sin \frac{\lambda_m x}{a}$$

Expressions for the stress resultants N_x and M_x obtained from these series by means of Eqs. (309) and (310), in which $\xi = x/a$, show that the conditions (h) of Art. 123 for hinged edges are fulfilled at the supports $x = 0$ and $x = l$.

In order to obtain the general expressions for the displacements in the case

$$X = Y = Z = 0$$

we make use of the resolving function F (Art. 125) by taking it at first in the form

$$F_m = e^{\alpha \varphi} \sin \frac{\lambda_m x}{a} \tag{d}$$

Substitution of this expression in the differential equation (307) yields the following characteristic equation for α:

$$(\alpha^2 - \lambda_m^2)^4 + \frac{1 - \nu^2}{c^2} \lambda_m^4 = 0 \tag{e}$$

in which $c^2 = h^2/12a^2$. The eight roots of this equation can be represented in the form

$$\begin{aligned}
\alpha_1 &= \gamma_1 + i\beta_1 & \alpha_5 &= -\alpha_1 \\
\alpha_2 &= \gamma_1 - i\beta_1 & \alpha_6 &= -\alpha_2 \\
\alpha_3 &= \gamma_2 + i\beta_2 & \alpha_7 &= -\alpha_3 \\
\alpha_4 &= \gamma_2 - i\beta_2 & \alpha_8 &= -\alpha_4
\end{aligned} \qquad (f)$$

with real values of γ and β. Using the notation

$$\rho = \sqrt{\lambda_m} \sqrt[8]{\frac{1-\nu^2}{c^2}} \qquad \sigma = \frac{\lambda_m^2}{\rho^2} \qquad (g)$$

we obtain

$$\begin{aligned}
\gamma_1 &= \frac{\rho}{\sqrt[4]{8}} \sqrt{\sqrt{(1+\sigma\sqrt{2})^2+1}+1+\rho\sqrt{2}} \\
\gamma_2 &= \frac{\rho}{\sqrt[4]{8}} \sqrt{\sqrt{(1-\sigma\sqrt{2})^2+1}-(1-\rho\sqrt{2})} \\
\beta_1 &= \frac{1}{\gamma_1}\frac{\rho^2}{\sqrt{8}} \\
\beta_2 &= \frac{1}{\gamma_2}\frac{\rho^2}{\sqrt{8}}
\end{aligned} \qquad (h)$$

Returning to the series form of solution, we find that the general expression for the stress function becomes

$$F = \sum_{m=1}^{\infty} f_m(\varphi) \sin \frac{\lambda_m x}{a} \qquad (i)$$

where
$$f_m(\varphi) = C_{1m} e^{\alpha_1 \varphi} + C_{2m} e^{\alpha_2 \varphi} + \cdots + C_{8m} e^{\alpha_8 \varphi} \qquad (j)$$

and $C_{1m}, C_{2m}, \ldots$ are arbitrary constants.

We are able now to calculate the respective displacements by means of the relations (306). Adding to the result the solution (c), we arrive at the following expressions for the total displacements of the middle surface of the shell:

$$\begin{aligned}
u &= \sum_{m=1}^{\infty} (\lambda_m f_m'' + \nu\lambda_m^3 f_m + U_{om}) \cos \frac{\lambda_m x}{a} \\
v &= \sum_{m=1}^{\infty} [(2+\nu)\lambda_m^2 f_m' - f_m''' + V_{om}] \sin \frac{\lambda_m x}{a} \\
w &= \sum_{m=1}^{\infty} (2\lambda_m^2 f_m'' - f_m'''' - \lambda_m^4 f_m + W_{om}) \sin \frac{\lambda_m x}{a}
\end{aligned} \qquad (k)$$

where primes denote differentiation with respect to φ.

The strain and stress components now are obtained by means of expressions (300), (309), (310), and (311). In the most general case of load distribution four conditions

GENERAL THEORY OF CYLINDRICAL SHELLS 527

on each edge $\varphi = \pm\varphi_0$ are necessary and sufficient to calculate the constants C_{m1}, ..., C_{m8} associated with each integer $m = 1, 2, 3, \ldots$.

As an example, let us consider the case of a vertical load uniformly distributed over the surface of the shell. From page 460 we have

$$X = 0 \qquad Y = p \sin \varphi \qquad Z = p \cos \varphi \tag{l}$$

Hence the coefficients of the series (a) are defined by

$$X_m = \frac{2}{l} \int_0^l X \cos \frac{\lambda_m x}{a} \, dx = 0$$

$$Y_m = \frac{2}{l} \int_0^l Y \sin \frac{\lambda_m x}{a} \, dx = \frac{4p}{m\pi} \sin \varphi \quad (m)$$

$$Z_m = \frac{2}{l} \int_0^l Z \sin \frac{\lambda_m x}{a} \, dx = \frac{4p}{m\pi} \cos \varphi$$

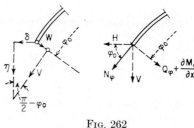

Fig. 262

in which $m = 1, 3, 5, \ldots$. An appropriate particular solution (c) is given by

$$U_{om} = A_{om} \cos \varphi \qquad V_{om} = B_{om} \sin \varphi \qquad W_{om} = C_{om} \cos \varphi \tag{n}$$

The coefficients A_{om}, B_{om}, and C_{om} are readily obtained by substitution of the expressions (c), (n), and (m) in Eqs. (305).

To satisfy the conditions of symmetry with respect to the meridian plane $\varphi = 0$, a suitable form of the function (j) is

$$f_m(\varphi) = A_{1m} \cos \beta_1\varphi \cosh \gamma_1\varphi + A_{2m} \sin \beta_1\varphi \sinh \gamma_1\varphi + A_{3m} \cos \beta_2\varphi \cosh \gamma_2\varphi$$
$$+ A_{4m} \sin \beta_2\varphi \sinh \gamma_2\varphi \tag{o}$$

in which β_1, β_2, γ_1, and γ_2 are defined by the expressions (h) and $m = 1, 3, 5, \ldots$.

In order to formulate the edge conditions on $\varphi = \pm\varphi_0$ in the simplest way, let us write the expressions for the vertical and horizontal components of the edge displacement and of the membrane forces on the edge as well (Fig. 262). We obtain

$$\eta = v \sin \varphi_0 + w \cos \varphi_0 \tag{p_1}$$
$$\delta = v \cos \varphi_0 - w \sin \varphi_0 \tag{p_2}$$

$$V = N_\varphi \sin \varphi_0 + \left(Q_\varphi + \frac{\partial M_{\varphi x}}{\partial x}\right) \cos \varphi_0 \tag{p_3}$$

$$H = N_\varphi \cos \varphi_0 - \left(Q_\varphi + \frac{\partial M_{\varphi x}}{\partial x}\right) \sin \varphi_0 \tag{p_4}$$

Finally, the rotation of the shell with respect to the edge line is expressed by

$$\chi = \frac{v}{a} + \frac{\partial w}{a \, \partial\varphi} \tag{p_5}$$

In all terms on the right-hand side of the foregoing expressions we have to put $\varphi = \varphi_0$. The following three kinds of edge conditions may be considered in particular.

Roof with Perfectly Flexible Tension Rods (Fig. 261a). Owing to many conr spans supposed to form the roof, the deformation of the roof can be consi symmetrical with respect to the vertical plane through an intermediate edr where the displacement δ and the rotation χ must vanish. Hence

$$v \cos \varphi_0 - w \sin \varphi_0 = 0 \tag{q_1}$$

$$v + \frac{\partial w}{\partial \varphi} = 0 \tag{q_2}$$

on $\varphi = \varphi_0$. Letting Q_0 be the weight of the tension rod per unit length, we have, by Eq. (p_3), a further condition

$$2V = Q_0 \tag{q_3}$$

in which Q_0, if constant, can be expanded in the series

$$Q_0 = \frac{4Q_0}{\pi} \sum_{m=1,3,5,\ldots}^{\infty} \frac{1}{m} \sin \frac{\lambda_m x}{a} \tag{p_6}$$

Finally, the elongation ϵ_x of the shell on the edge $\varphi = \varphi_0$ must be equal to the elongation of the tension member. If A_0 denotes the cross-sectional area of the latter and E_0 the corresponding Young modulus,[1] then we have, for $\varphi = \varphi_0$,

$$\frac{1}{E_0 A_0} \int_0^x 2N_{\varphi x}\, dx = \frac{\partial u}{\partial x} \tag{q_4}$$

in which the integral represents the tension force of the rod.

The further procedure is as follows. We calculate four coefficients $A_{1m}, \ldots, A_{4m}$ for each $m = 1, 3, 5, \ldots$ from the conditions (q_1), $\ldots$, (q_4). The stress function F is now defined by Eqs. (o) and (i), and the respective displacements are given by the expressions (306) or (k). Finally, we obtain the total stress resultants by means of expressions (309) to (311), starting from the known displacements, or, for the general part of the solution, also directly from the stress function F.

Roof over Many Spans, Stiffened by Beams (Fig. 261b). The conditions of symmetry

$$v \cos \varphi_0 - w \sin \varphi_0 = 0 \tag{r_1}$$

and

$$v + \frac{\partial w}{\partial \varphi} = 0 \tag{r_2}$$

on $\varphi = \varphi_0$ are the same as in the preceding case. To establish a third condition, let Q_0 be the given weight of the beam per unit length, h_0 its depth, $E_0 I_0$ the flexural rigidity of the beam in the vertical plane, and A_0 the cross-sectional area. Then the differential equation for the deflection η of the beam becomes

$$E_0 I_0 \frac{d^4 \eta}{dx^4} = Q_0 - 2V + 2 \frac{h_0}{2} \frac{\partial N_{\varphi x}}{\partial x} \tag{r_3}$$

the functions η, V, and Q_0 being given by the expressions (p_1), (p_3), and (p_6), respectively. The last term in Eq. (r_3) is due to the difference of level between the edge of the shell and the axis of the beam. As for the elongation ϵ_x of the top fibers of the beam, it depends not only on the tension force but also on the curvature of the beam. ...rving the effect of the curvature $d^2\eta/dx^2$, we obtain in place of Eq. (q_4) the

$$\frac{2}{E_0 A_0} \int_0^x N_{\varphi x}\, dx + \frac{h_0}{2} \frac{d^2 \eta}{dx^2} = \frac{\partial u}{\partial x} \tag{r_4}$$

[1] a tension member composed of two materials, say steel and concrete, ...-sectional area must be used.

GENERAL THEORY OF CYLINDRICAL SHELLS

The further procedure of analysis remains essentially the same as in the foregoing case.

The distribution of membrane forces and bending moments M_φ obtained[1] for the middle span of a roof, comprising three such spans in all, is shown in Fig. 263. In the direction x the span of the shell is $l = 134.5$ ft, the surface load is $p = 51.8$ psf, and the weight of the beam $Q_0 = 448$ lb per ft. Stress resultants obtained by means of the membrane theory alone are represented by broken lines.

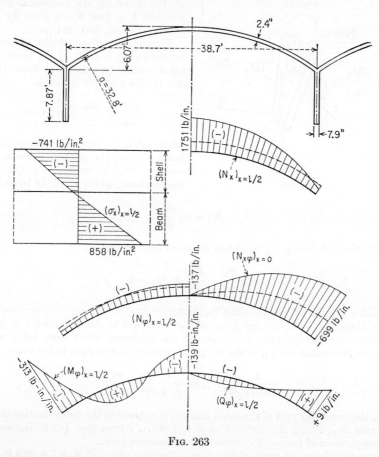

Fig. 263

One-span Roof, Stiffened by Beams (Fig. 265). In such a case we have to observe not only the deflection of the beam, given by the edge displacements η and δ, but the rotation of the beam χ as well (Fig. 264). The differential equation for the vertical deflection is, this time, of the form

$$E_0 I_0 \frac{d^4\eta}{dx^4} = Q_0 - V + \frac{h_0}{2}\frac{\partial N_{\varphi x}}{\partial x} \tag{s_1}$$

[1] By Finsterwalder, *loc. cit.*, using the method described in Art. 124; see also *Proc. Intern. Assoc. Bridge Structural Engrs.*, vol. 1, p. 127, 1932.

the notation being the same as in the previous case. The horizontal deflection is governed in like manner by the equation

$$E_0 I'_0 \frac{d^4}{dx^4}\left(\delta - \chi \frac{h_0}{2}\right) = -H \quad (s_2)$$

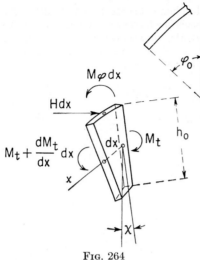

Fig. 264

in which $E_0 I'_0$ denotes the flexural rigidity of the beam in the horizontal plane, whereas δ, χ, and H are given by the expressions (p_2), (p_5), and (p_4).

The condition of equilibrium of couples acting on an element of the beam and taken about the axis of the beam (Fig. 264) yields a further equation

$$\frac{dM_t}{dx} - \frac{Hh_0}{2} + M_\varphi = 0 \quad (t)$$

where M_t is the torsional moment of the beam. Now, the relation between the moment M_t, the twist $\theta = \partial\chi/\partial x$, and the torsional rigidity C_0 of the beam is

$$M_t = C_0 \frac{d\chi}{dx} \quad (u)$$

Substituting this in Eq. (t), we obtain the third edge condition

$$C_0 \frac{d^2\chi}{dx^2} - \frac{Hh_0}{2} + M_\varphi = 0 \quad (s_3)$$

in which χ is given by the expression (p_5) and $\varphi = \varphi_0$.

The elongation ϵ_x of the top fibers of the beam due to the deflection δ may be neglected, the average value of ϵ_x through the thickness of the beam being zero. Therefore, the condition (r_4) of the foregoing case can be rewritten in the form

$$\frac{1}{E_0 A_0} \int_0^x N_{\varphi x}\, dx + \frac{h_0}{2} \frac{d^2\eta}{dx^2} = \frac{\partial u}{\partial x} \quad (s_4)$$

Again the remaining part of the stress analysis is reduced to the determination of the constants $A_{1m}, \ldots, A_{4m}$ for each $m = 1, 3, 5, \ldots$ from Eqs. (s_1) to (s_4) and to the computation of stresses by means of the respective series.

Figure 265 shows the stress distribution in the case of a shell with $l = 98.4$ ft and $\varphi_0 = 45°$. It is seen in particular that the distribution of the membrane stresses σ_x over the depth of the whole beam, composed by the shell and both stiffeners, is far from being linear. However, by introducing $\delta = 0$ as the edge condition instead of the condition (s_2), an almost linear stress diagram 2 could be obtained. If we suppose, in addition, that the rotation χ vanishes too, we arrive at a stress distribution given by curve 3.*

* For particulars of the calculation see K. Girkmann, "Flächentragwerke," 4th ed., p. 499, Springer-Verlag, Vienna, 1956. The diagrams of Figs. 265 and 263 are reproduced from that book by permission from the author and the publisher.

GENERAL THEORY OF CYLINDRICAL SHELLS 531

Various simplifications can be introduced into the rather tedious procedure of stress calculation described above.

Thus, if the ratio l/a is sufficiently large, the stress resultants M_x, Q_x, and $M_{x\varphi}$ can be disregarded, as explained in Art. 124. Again, the particular solution (c) may be replaced by a solution obtained directly by use of the membrane theory of cylindrical

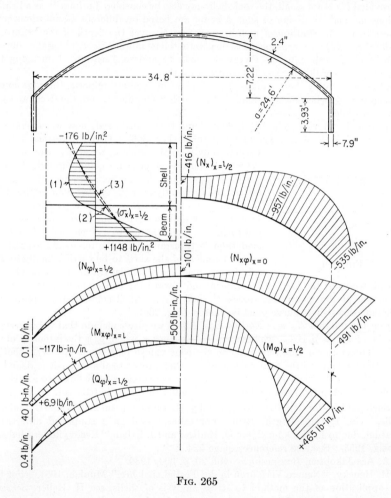

Fig. 265

shells (Art. 112). The corresponding displacements, needed for the formulation of the boundary conditions, could be obtained from Eqs. (309). The method considered in Art. 124 is simplified still more if from all derivatives with respect to φ needed to represent the strain and stress components, only those of the highest order are retained.[1]

On the other hand, the procedure of the stress computation can be greatly reduced by use of special tables for strain and stress components due to the action of the edge

[1] See H. Schorer, *Proc. ASCE*, vol. 61, p. 181, 1935.

forces on the cylindrical shell.[1] A method of iteration[2] and the method of finite differences[3] have also been used in stress analysis of shells.

If edge conditions on the supports $x = 0$, $x = l$ of the shell are other than those assumed on page 524, the stress disturbance arising from the supplementary edge forces would require special investigation.[4]

Provided l/a is not small, the roof shell may also be considered primarily as a beam.[5] Various methods of design of such a beam are based on different assumptions with respect to the distribution of membrane forces N_x over the depth of the beam. A possible procedure, for example, is to distribute the membrane forces along the contour of the shell according to the theory of elasticity and to distribute them along the generatrices according to the elementary beam theory.

In the case of very short roof shells continuous over many supports, the edge conditions on $\varphi = \pm\varphi_0$ become secondary, and a further simplification of the stress analysis proves possible.[6]

So far only circular cylindrical shells have been considered; now let us consider a cylindrical shell of any symmetrical form (Fig. 266). Given a vertical loading varying only with the angle φ, we always can obtain a cylindrical surface of pressure going through the generatrices A, C, and B. If, for instance, the load is distributed uniformly over the ground plan of the shell, the funicular curve ACB would be a parabola. Now suppose the middle surface of the shell to coincide with the surface of pressure due to a given load. The total load then is transmitted by the forces N_φ toward the edges A and B of the shell to be carried finally by the side beams over the whole length of the cylinder. If, instead, we want the load to be transmitted toward the end supports of the shell by the action of the membrane forces N_x and $N_{x\varphi}$, a shell contour overtopping the funicular (thrust-line) curve must be chosen (Fig. 266).

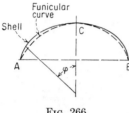

Fig. 266

From the relation $N_\varphi = -Za$ [see Eqs. (270)] we also conclude that for a vertical load, i.e., for $Z = p_v \cos \varphi$, we have $N_\varphi = -p_v a \cos \varphi$, where p_v is the intensity of the load. Therefore the ring forces N_φ on the edge vanish only when $\varphi_0 = \pi/2$, that is, when the tangents to the contour line of the shell are vertical at the edges A and B. This condition is satisfied by such contours as a semicircle, a semiellipse, or a cycloid,[7] which all overtop the pressure line due to a uniformly distributed load.

[1] Such tables (for $\nu = 0.2$) are given by H. Lundgren in his book "Cylindrical Shells," vol. 1, Copenhagen, 1949. For tables based on a simplified differential equation, due to L. H. Donnell, see D. Rüdiger and J. Urban, "Kreiszylinderschalen," Leipzig, 1955. See also references, page 524.

[2] A. Aas-Jakobsen, *Bauingenieur*, vol. 20, p. 394, 1939.

[3] H. Hencky, "Neuere Verfahren in der Festigkeitslehre," Munich, 1951. For the first application of the method to stress analysis of shells, see H. Keller, *Schweiz. Bauztg.*, p. 111, 1913. The relaxation method has been applied to stress analysis of shells by W. Flügge, "Federhofer-Girkmann-Festschrift," p. 17, Vienna, 1950.

[4] By application of Miesel's theory, *op. cit.*, or by an approximate method due to Finsterwalder, *op. cit.*

[5] This approach has especially been used by A. Aas-Jakobsen, *op. cit.*, p. 93.

[6] See B. Thürlimann, R. O. Bereuter, and B. G. Johnston, *Proc. First U.S. Natl. Congr. Appl. Mech.*, 1952, p. 347. For application of the photoelasticity method to a cylindrical shell (tunnel tube), see G. Sonntag, *Bauingenieur*, vol. 31, p. 408, 1956.

[7] For membrane stresses in shells of this kind see, for example, Girkmann, *op. cit.*, and A. Pflüger, "Elementare Schalenstatik," Berlin, 1957. The bending of semi-elliptical shells was considered by A. Aas-Jakobsen, *Génie civil*, p. 275, 1937. For other shapes of cylindrical roofs, see E. Wiedemann, *Ingr.-Arch.*, vol. 8, p. 301, 1937.

CHAPTER 16

SHELLS HAVING THE FORM OF A SURFACE OF REVOLUTION AND LOADED SYMMETRICALLY WITH RESPECT TO THEIR AXIS

127. Equations of Equilibrium. Let us consider the conditions of equilibrium of an element cut from a shell by two adjacent meridian planes and two sections perpendicular to the meridians (Fig. 267).[1] It can be concluded from the condition of symmetry that only normal stresses will act on the sides of the element lying in the meridian planes. The stresses can be reduced to the resultant force $N_\theta r_1 d\varphi$ and the resultant moment $M_\theta r_1 d\varphi$, N_θ and M_θ being independent of the angle θ which defines the position of the meridians. The side of the element perpendicular to the meridians which is defined by the angle φ (Fig. 267) is acted upon by normal stresses which result in the force $N_\varphi r_2 \sin\varphi\, d\theta$ and the moment $M_\varphi r_2 \sin\varphi\, d\theta$ and by shearing forces which reduce the force $Q_\varphi r_2 \sin\varphi\, d\theta$ normal to the shell. The external load acting upon the element can be resolved, as before, into two components $Y r_1 r_2 \sin\varphi\, d\varphi\, d\theta$ and $Z r_1 r_2 \sin\varphi\, d\varphi\, d\theta$ tangent to the meridians and normal to the shell, respectively. Assuming that the membrane forces N_θ and N_φ do not approach their critical values,[2] we neglect the change of curvature in deriving the equations of equilibrium and proceed as was shown in Art. 105. In Eq. (f) of that article, obtained by projecting the forces on the tangent to the meridian, the term $-Q_\varphi r_0$ must now be added to the left-hand side. Also, in Eq. (j), which was

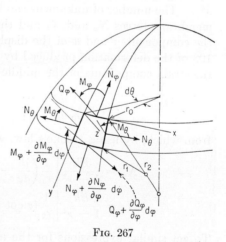

Fig. 267

---

[1] We use for radii of curvature and for angles the same notation as in Fig. 213.

[2] The question of buckling of spherical shells is discussed in S. Timoshenko, "Theory of Elastic Stability," p. 491, 1936.

obtained by projecting the forces on the normal to the shell, an additional term $d(Q_\varphi r_0)/d\varphi$ must be added to the left-hand side. The third equation is obtained by considering the equilibrium of the moments with respect to the tangent to the parallel circle of all the forces acting on the element. This gives[1]

$$\left(M_\varphi + \frac{dM_\varphi}{d\varphi} d\varphi\right)\left(r_0 + \frac{dr_0}{d\varphi} d\varphi\right) d\theta - M_\varphi r_0 \, d\theta - M_\theta r_1 \cos \varphi \, d\varphi \, d\theta$$
$$- Q_\varphi r_2 \sin \varphi \, r_1 \, d\varphi \, d\theta = 0$$

After simplification, this equation, together with the two equations of Art. 105, modified as explained above, gives us the following system of three equations of equilibrium:

$$\frac{d}{d\varphi}(N_\varphi r_0) - N_\theta r_1 \cos \varphi - r_0 Q_\varphi + r_0 r_1 Y = 0$$
$$N_\varphi r_0 + N_\theta r_1 \sin \varphi + \frac{d(Q_\varphi r_0)}{d\varphi} + Z r_1 r_0 = 0 \quad (312)$$
$$\frac{d}{d\varphi}(M_\varphi r_0) - M_\theta r_1 \cos \varphi - Q_\varphi r_1 r_0 = 0$$

In these three equations of equilibrium are five unknown quantities, three resultant forces N_φ, N_θ, and Q_φ and two resultant moments M_θ and M_φ. The number of unknowns can be reduced to three if we express the membrane forces N_φ and N_θ and the moments M_φ and M_θ in terms of the components v and w of the displacement. In the discussion in Art. 108 of the deformation produced by membrane stresses, we obtained for the strain components of the middle surface the expressions

$$\epsilon_\varphi = \frac{1}{r_1} \frac{dv}{d\varphi} - \frac{w}{r_1} \qquad \epsilon_\theta = \frac{v}{r_2} \cot \varphi - \frac{w}{r_2}$$

from which, by using Hooke's law, we obtain

$$N_\varphi = \frac{Eh}{1-\nu^2}\left[\frac{1}{r_1}\left(\frac{dv}{d\varphi} - w\right) + \frac{\nu}{r_2}(v \cot \varphi - w)\right]$$
$$N_\theta = \frac{Eh}{1-\nu^2}\left[\frac{1}{r_2}(v \cot \varphi - w) + \frac{\nu}{r_1}\left(\frac{dv}{d\varphi} - w\right)\right] \quad (313)$$

To get similar expressions for the moments M_φ and M_θ let us consider the changes of curvature of the shell element shown in Fig. 267. Considering the upper and the lower sides of that element, we see that the initial angle between these two sides is $d\varphi$. Because of the displacement v along the meridian, the upper side of the element rotates with

---

[1] In this derivation we observe that the angle between the planes in which the moments M_θ act is equal to $\cos \varphi \, d\theta$.

respect to the perpendicular to the meridian plane by the amount v/r_1. As a result of the displacement w, the same side further rotates about the same axis by the amount $dw/(r_1\,d\varphi)$. Hence the total rotation of the upper side of the element is

$$\frac{v}{r_1} + \frac{dw}{r_1\,d\varphi} \tag{a}$$

For the lower side of the element the rotation is

$$\frac{v}{r_1} + \frac{dw}{r_1\,d\varphi} + \frac{d}{d\varphi}\left(\frac{v}{r_1} + \frac{dw}{r_1\,d\varphi}\right)d\varphi$$

Hence the change of curvature of the meridian is[1]

$$\chi_\varphi = \frac{1}{r_1}\frac{d}{d\varphi}\left(\frac{v}{r_1} + \frac{dw}{r_1\,d\varphi}\right) \tag{b}$$

To find the change of curvature in the plane perpendicular to the meridian, we observe that because of symmetry of deformation each of the lateral sides of the shell element (Fig. 267) rotates in its meridian plane by an angle given by expression (a). Since the normal to the right lateral side of the element makes an angle $(\pi/2) - \cos\varphi\,d\theta$ with the tangent to the y axis, the rotation of the right side in its own plane has a component with respect to the y axis equal to

$$-\left(\frac{v}{r_1} + \frac{dw}{r_1\,d\varphi}\right)\cos\varphi\,d\theta$$

This results in a change of curvature

$$\chi_\theta = \left(\frac{v}{r_1} + \frac{dw}{r_1\,d\varphi}\right)\frac{\cos\varphi}{r_0} = \left(\frac{v}{r_1} + \frac{dw}{r_1\,d\varphi}\right)\frac{\cot\varphi}{r_2} \tag{c}$$

Using expressions (b) and (c), we then obtain

$$\begin{aligned}M_\varphi &= -D\left[\frac{1}{r_1}\frac{d}{d\varphi}\left(\frac{v}{r_1} + \frac{dw}{r_1\,d\varphi}\right) + \frac{\nu}{r_2}\left(\frac{v}{r_1} + \frac{dw}{r_1\,d\varphi}\right)\cot\varphi\right]\\ M_\theta &= -D\left[\left(\frac{v}{r_1} + \frac{dw}{r_1\,d\varphi}\right)\frac{\cot\varphi}{r_2} + \frac{\nu}{r_1}\frac{d}{d\varphi}\left(\frac{v}{r_1} + \frac{dw}{r_1\,d\varphi}\right)\right]\end{aligned} \tag{314}$$

Substituting expressions (313) and (314) into Eqs. (312), we obtain three equations with three unknown quantities v, w, and Q_φ. Discussion of these equations will be left to the next article.

We can also use expressions (314) to establish an important conclusion regarding the accuracy of the membrane theory discussed in Chap. 14. In Art. 108 the equations for calculating the displacements v and w were

---

[1] The strain of the middle surface is neglected, and the change in curvature is obtained by dividing the angular change by the length $r_1\,d\varphi$ of the arc.

established. By substituting the displacements given by these equations in expressions (314), the bending moments and bending stresses can be calculated. These stresses were neglected in the membrane theory. By comparing their magnitudes with those of the membrane stresses, a conclusion can be drawn regarding the accuracy of the membrane theory.

We take as a particular example a spherical shell under the action of its own weight (page 436). If the supports are as shown in Fig. 215a, the displacements as given by the membrane theory from Eqs. (*f*) and (*b*) (Art. 108) are

$$v = \frac{a^2 q(1+\nu)}{Eh}\left(\frac{1}{1+\cos\alpha} - \frac{1}{1+\cos\varphi} + \log\frac{1+\cos\varphi}{1+\cos\alpha}\right)\sin\varphi$$
$$w = v\cot\varphi - \frac{a^2 q}{Eh}\left(\frac{1+\nu}{1+\cos\varphi} - \cos\varphi\right) \tag{d}$$

Substituting these expressions into formulas (314) for the bending moments, we obtain

$$M_\theta = M_\varphi = \frac{qh^2}{12}\frac{2+\nu}{1-\nu}\cos\varphi \tag{e}$$

The corresponding bending stress at the surface of the shell is numerically equal to

$$\frac{q}{2}\frac{2+\nu}{1-\nu}\cos\varphi$$

Taking the ratio of this stress to the compressive stress σ given by the membrane theory [see Eqs. (257)], we find

$$\frac{q}{2}\frac{2+\nu}{1-\nu}\cos\varphi \bigg/ \frac{aq}{h(1+\cos\varphi)} = \frac{2+\nu}{2(1-\nu)}\frac{h}{a}(1+\cos\varphi)\cos\varphi$$

The maximum value of this ratio is found at the top of the shell where $\varphi = 0$ and has a magnitude, for $\nu = 0.3$, of

$$3.29\frac{h}{a} \tag{f}$$

It is seen that in the case of a thin shell the ratio (*f*) of bending stresses to membrane stresses is small, and the membrane theory gives satisfactory results provided that the conditions at the supports are such that the shell can freely expand, as shown in Fig. 215a. Substituting expression (*e*) for the bending moments in Eqs. (312), closer approximations for the membrane forces N_φ and N_θ can be obtained. These results will differ from solutions (257) only by small quantities having the ratio h^2/a^2 as a factor.

From this discussion it follows that in the calculation of the stresses in symmetrically loaded shells we can take as a first approximation the solution given by the membrane theory and calculate the corrections by means of Eqs. (312). Such corrected values of the stresses will be accurate enough if the edges of the shell are free to expand. If the edges are not free, additional forces must be so applied along the edge as to satisfy the boundary conditions. The calculation of the stresses produced by these latter forces will be discussed in the next article.

128. Reduction of the Equations of Equilibrium to Two Differential Equations of the Second Order. From the discussion of the preceding article, it is seen that by using expressions (313) and (314) we can obtain from Eqs. (312) three equations with the three unknowns v, w, and Q_φ. By using the third of these equations the shearing force Q_φ can be readily eliminated, and the three equations reduced to two equations with the unknowns v and w. The resulting equations were used by the first investigators of the bending of shells.[1] Considerable simplification of the equations can be obtained by introducing new variables.[2] As the first of the new variables we shall take the angle of rotation of a tangent to a meridian. Denoting this angle by V, we obtain from Eq. (a) of the preceding article

$$V = \frac{1}{r_1}\left(v + \frac{dw}{d\varphi}\right) \tag{a}$$

As the second variable we take the quantity

$$U = r_2 Q_\varphi \tag{b}$$

To simplify the transformation of the equations to the new variables we replace the first of the equations (312) by one similar to Eq. (255) (see page 435), which can be obtained by considering the equilibrium of the portion of the shell above the parallel circle defined by the angle φ (Fig. 267). Assuming that there is no load applied to the shell, this equation gives

$$2\pi r_0 N_\varphi \sin \varphi + 2\pi r_0 Q_\varphi \cos \varphi = 0$$

from which

$$N_\varphi = -Q_\varphi \cot \varphi = -\frac{1}{r_2} U \cot \varphi \tag{c}$$

Substituting in the second of the equations (312), we find, for $Z = 0$,

$$r_1 N_\theta \sin \varphi = -N_\varphi r_0 - \frac{d(Q_\varphi r_0)}{d\varphi}$$

---

[1] See A. Stodola, "Die Dampfturbinen," 4th ed., p. 597, 1910; H. Keller, *Mitt. Forschungsarb.*, vol. 124, 1912; E. Fankhauser, dissertation, Zurich, 1913.

[2] This method of analyzing stresses in shells was developed for the case of a spherical shell by H. Reissner, "Müller-Breslau-Festschrift," p. 181, Leipzig, 1912; it was generalized and applied to particular cases by E. Meissner, *Physik. Z.*, vol. 14, p. 343, 1913; and *Vierteljahrsschr. naturforsch. Ges. Zürich*, vol. 60, p. 23, 1915.

and, observing that $r_0 = r_2 \sin \varphi$, we obtain

$$N_\theta = -\frac{1}{r_1} \frac{d}{d\varphi}(Q_\varphi r_2) = -\frac{1}{r_1} \frac{dU}{d\varphi} \qquad (d)$$

Thus the membrane forces N_φ and N_θ are both represented in terms of the quantity U, which is, as we see from notation (b), dependent on the shearing force Q_φ.

To establish the first equation connecting V and U we use Eqs. (313), from which we readily obtain

$$\frac{dv}{d\varphi} - w = \frac{r_1}{Eh}(N_\varphi - \nu N_\theta) \qquad (e)$$

$$v \cot \varphi - w = \frac{r_2}{Eh}(N_\theta - \nu N_\varphi) \qquad (f)$$

Eliminating w from these equations, we find

$$\frac{dv}{d\varphi} - v \cot \varphi = \frac{1}{Eh}[(r_1 + \nu r_2)N_\varphi - (r_2 + \nu r_1)N_\theta] \qquad (g)$$

Differentiation of Eq. (f) gives[1]

$$\frac{dv}{d\varphi} \cot \varphi - \frac{v}{\sin^2 \varphi} - \frac{dw}{d\varphi} = \frac{d}{d\varphi}\left[\frac{r_2}{Eh}(N_\theta - \nu N_\varphi)\right] \qquad (h)$$

The derivative $dv/d\varphi$ can be readily eliminated from Eqs. (g) and (h) to obtain

$$v + \frac{dw}{d\varphi} = r_1 V = \frac{\cot \varphi}{Eh}[(r_1 + \nu r_2)N_\varphi - (r_2 + \nu r_1)N_\theta]$$
$$- \frac{d}{d\varphi}\left[\frac{r_2}{Eh}(N_\theta - \nu N_\varphi)\right]$$

Substituting expressions (c) and (d) for N_φ and N_θ, we finally obtain the following equation relating to U and V:

$$\frac{r_2}{r_1^2}\frac{d^2U}{d\varphi^2} + \frac{1}{r_1}\left[\frac{d}{d\varphi}\left(\frac{r_2}{r_1}\right) + \frac{r_2}{r_1}\cot \varphi - \frac{r_2}{r_1 h}\frac{dh}{d\varphi}\right]\frac{dU}{d\varphi}$$
$$- \frac{1}{r_1}\left[\frac{r_1}{r_2}\cot^2 \varphi - \nu - \frac{\nu}{h}\frac{dh}{d\varphi}\cot \varphi\right]U = EhV \qquad (315)$$

The second equation for U and V is obtained by substituting expressions (314) for M_φ and M_θ in the third of the equations (312) and using notations (a) and (b). In this way we find

[1] We consider a general case by assuming in this derivation that the thickness h of the shell is variable.

$$\frac{r_2}{r_1^2}\frac{d^2V}{d\varphi^2} + \frac{1}{r_1}\left[\frac{d}{d\varphi}\left(\frac{r_2}{r_1}\right) + \frac{r_2}{r_1}\cot\varphi + 3\frac{r_2}{r_1h}\frac{dh}{d\varphi}\right]\frac{dV}{d\varphi}$$
$$- \frac{1}{r_1}\left(\nu - \frac{3\nu\cot\varphi}{h}\frac{dh}{d\varphi} + \frac{r_1}{r_2}\cot^2\varphi\right)V = -\frac{U}{D} \quad (316)$$

Thus the problem of bending of a shell having the form of a surface of revolution by forces and moments uniformly distributed along the parallel circle representing the edge is reduced to the integration of the two Eqs. (315) and (316) of the second order.

If the thickness of the shell is constant, the terms containing $dh/d\varphi$ as a factor vanish, and the derivatives of the unknowns U and V in both equations have the same coefficients. By introducing the notation

$$L(\cdots) = \frac{r_2}{r_1^2}\frac{d^2(\cdots)}{d\varphi^2} + \frac{1}{r_1}\left[\frac{d}{d\varphi}\left(\frac{r_2}{r_1}\right) + \frac{r_2}{r_1}\cot\varphi\right]$$
$$\frac{d(\cdots)}{d\varphi} - \frac{r_1\cot^2\varphi}{r_2r_1}(\cdots) \quad (i)$$

the equations can be represented in the following simplified form:

$$L(U) + \frac{\nu}{r_1}U = EhV$$
$$L(V) - \frac{\nu}{r_1}V = -\frac{U}{D} \quad (317)$$

From this system of two simultaneous differential equations of the second order we readily obtain for each unknown an equation of the fourth order. To accomplish this we perform on the first of the equations (317) the operation indicated by the symbol $L(\cdots)$, which gives

$$LL(U) + \nu L\left(\frac{U}{r_1}\right) = EhL(V)$$

Substituting from the second of the equations (317),

$$L(V) = \frac{\nu}{r_1}V - \frac{U}{D} = \frac{\nu}{r_1Eh}\left[L(U) + \frac{\nu}{r_1}U\right] - \frac{U}{D}$$

we obtain

$$LL(U) + \nu L\left(\frac{U}{r_1}\right) - \frac{\nu}{r_1}L(U) - \frac{\nu^2}{r_1^2}U = -\frac{Eh}{D}U \quad (318)$$

In the same manner we also find the second equation

$$LL(V) - \nu L\left(\frac{V}{r_1}\right) + \frac{\nu}{r_1}L(V) - \frac{\nu^2}{r_1^2}V = -\frac{Eh}{D}V \quad (319)$$

If the radius of curvature r_1 is constant, as in the case of a spherical or a conical shell or in a ring shell such as is shown in Fig. 220, a further simplification of Eqs. (318) and (319) is possible. Since in this case

$$L\left(\frac{U}{r_1}\right) = \frac{1}{r_1} L(U)$$

by using the notation

$$\mu^4 = \frac{Eh}{D} - \frac{\nu^2}{r_1^2} \qquad (j)$$

both equations can be reduced to the form

$$LL(U) + \mu^4 U = 0 \qquad (320)$$

which can be written in one of the two following forms:

or
$$L[L(U) + i\mu^2 U] - i\mu^2[L(U) + i\mu^2 U] = 0$$
$$L[L(U) - i\mu^2 U] + i\mu^2[L(U) - i\mu^2 U] = 0$$

These equations indicate that the solutions of the second-order equations

$$L(U) \pm i\mu^2 U = 0 \qquad (321)$$

are also the solutions of Eq. (320). By proceeding as was explained in Art. 118, it can be shown that the complete solution of Eq. (320) can be obtained from the solution of one of the equations (321). The application of Eqs. (321) to particular cases will be discussed in the two following articles.

129. Spherical Shell of Constant Thickness. In the case of a spherical shell of constant thickness $r_1 = r_2 = a$, and the symbol (i) of the preceding article is

$$L(\cdots) = \frac{1}{a}\left[\frac{d^2}{d\varphi^2}(\cdots) + \cot\varphi \frac{d}{d\varphi}(\cdots) - \cot^2\varphi(\cdots)\right] \qquad (a)$$

Considering the quantity aQ_φ, instead of U, as one of the unknowns in the further discussion and introducing, instead of the constant μ, a new constant ρ defined by the equation

$$\rho^2 = \frac{a\mu^2}{2} = \sqrt{\frac{3a^2(1-\nu^2)}{h^2} - \frac{\nu^2}{4}} \qquad (b)$$

we can represent the first of the equations (321) in the following form:

$$\frac{d^2 Q_\varphi}{d\varphi^2} + \cot\varphi \frac{dQ_\varphi}{d\varphi} - \cot^2\varphi\, Q_\varphi + 2i\rho^2 Q_\varphi = 0 \qquad (322)$$

A further simplification is obtained by introducing the new variables[1]

$$x = \sin^2\varphi$$
$$z = \frac{Q_\varphi}{\sin\varphi} \qquad (c)$$

[1] This solution of the equation was given by Meissner, *op. cit.*

With these variables Eq. (322) becomes

$$x(x-1)\frac{d^2z}{dx^2} + \left(\frac{5}{2}x - 2\right)\frac{dz}{dx} + \frac{1-2i\rho^2}{4}z = 0 \qquad (d)$$

This equation belongs to a known type of differential equation of the second order which has the form

$$x(1-x)y'' + [\gamma - (\alpha + \beta + 1)x]y' - \alpha\beta y = 0 \qquad (e)$$

Equations (d) and (e) coincide if we put

$$\gamma = 2 \qquad \alpha = \frac{3 \pm \sqrt{5 + 8i\rho^2}}{4} \qquad \beta = \frac{3 \mp \sqrt{5 + 8i\rho^2}}{4} \qquad (f)$$

A solution of Eq. (e) can be taken in the form of a power series

$$y = A_0 + A_1 x + A_2 x^2 + A_3 x^3 + \cdots \qquad (g)$$

Substituting this series in Eq. (e) and equating the coefficients for each power of x to zero, we obtain the following relations between the coefficients:

$$A_1 = \frac{\alpha\beta}{1 \cdot \gamma} A_0 \qquad A_2 = \frac{(\alpha+1)(\beta+1)}{2(\gamma+1)} A_1$$

$$\cdots\cdots\cdots\cdots\cdots\cdots\cdots\cdots\cdots\cdots\cdots\cdots$$

$$A_n = A_{n-1}\frac{(\alpha+n-1)(\beta+n-1)}{n(\gamma+n-1)}$$

$$\cdots\cdots\cdots\cdots\cdots\cdots\cdots\cdots\cdots\cdots\cdots\cdots$$

With these relations series (g) becomes

$$y = A_0\left[1 + \frac{\alpha\beta}{1 \cdot \gamma}x + \frac{\alpha(\alpha+1)\beta(\beta+1)}{1 \cdot 2 \cdot \gamma(\gamma+1)}x^2 \right.$$
$$\left. + \frac{\alpha(\alpha+1)(\alpha+2)\beta(\beta+1)(\beta+2)}{1 \cdot 2 \cdot 3 \cdot \gamma(\gamma+1)(\gamma+2)}x^3 + \cdots\right] \qquad (h)$$

This is the so-called hypergeometrical series. It is convergent for all values of x less than unity and can be used to represent one of the integrals of Eq. (d). Substituting for α, β, and γ their values (f) and using the notation

$$\delta^2 = 5 + 8i\rho^2 = 5 + 4i\sqrt{\frac{12a^2(1-\nu^2)}{h^2} - \nu^2} \qquad (i)$$

we obtain as the solution of Eq. (d)

$$z_1 = A_0\left[1 + \frac{3^2 - \delta^2}{16 \cdot 1 \cdot 2}x + \frac{(3^2 - \delta^2)(7^2 - \delta^2)}{16^2 \cdot 1 \cdot 2 \cdot 2 \cdot 3}x^2 + \cdots\right] \qquad (j)$$

which contains one arbitrary constant A_0.

The derivation of the second integral of Eq. (d) is more complicated.[1] This integral can be written in the form

$$z_2 = z_1 \log x + \frac{1}{x} \varphi(x) \tag{k}$$

where $\varphi(x)$ is a power series that is convergent for $|x| < 1$. This second solution becomes infinite for $x = 0$, that is, at the top of the sphere (Fig. 267), and should not be considered in those cases in which there is no hole at the top of the sphere.

If we limit our investigation to these latter cases, we need consider only solution (j). Substituting for δ^2 its value (i) and dividing series (j) into its real and imaginary parts, we obtain

$$z_1 = S_1 + iS_2 \tag{l}$$

where S_1 and S_2 are power series that are convergent when $|x| < 1$. The corresponding solution of the first of the equations (321) is then

$$U_1 = az_1 \sin \varphi = I_1 + iI_2 \tag{m}$$

where I_1 and I_2 are two series readily obtained from the series S_1 and S_2.

The necessary integral of the second of the equations (321) can be represented by the same series I_1 and I_2 (see page 489). Thus, for the case of a spherical shell without a hole at the top, the general solution of the differential equation (320), which is of the fourth order, can be represented in the form

$$U = aQ_\varphi = AI_1 + BI_2 \tag{n}$$

where A and B are constants to be determined from the two conditions along the edge of the spherical shell.

Having expression (n) for U, we can readily find the second unknown V. We begin by substituting expression (m) in the first of the equations (321), which gives

$$L(I_1 + iI_2) = -i\mu^2(I_1 + iI_2)$$

Hence $\quad L(I_1) = \mu^2 I_2 \qquad L(I_2) = -\mu^2 I_1 \tag{o}$

Substituting expression (n) in the first of the equations (317) and applying expressions (o), we then obtain

$$EhaV = aL(U) + \nu U = (A\nu - Ba\mu^2)I_1 + (Aa\mu^2 + B\nu)I_2 \tag{p}$$

It is seen that the second unknown V is also represented by the series I_1 and I_2.

---

[1] Differential equations that are solved by hypergeometrical series are discussed in the book "Riemann-Weber, Die partiellen Differential-Gleichungen," vol. 2, pp. 1–29, 1901. See also E. Kamke, "Differentialgleichungen," vol. 1, 2d ed., p. 465, Leipzig, 1943.

Having the expressions for U and V, we can obtain all the forces, moments, and displacements. The forces N_φ and N_θ are found from Eqs. (c) and (d) of the preceding article. The bending moments M_φ and M_θ are obtained from Eqs. (314). Observing that in the case of a spherical shell $r_1 = r_2 = a$ and using notation (a), we obtain

$$M_\varphi = -\frac{D}{a}\left(\frac{dV}{d\varphi} + \nu \cot \varphi\, V\right)$$
$$M_\theta = -\frac{D}{a}\left(\nu \frac{dV}{d\varphi} + \cot \varphi\, V\right) \qquad (q)$$

In calculating the components v and w of displacement we use the expressions for the strain in the middle surface:

$$\epsilon_\varphi = \frac{1}{Eh}(N_\varphi - \nu N_\theta) \qquad \epsilon_\theta = \frac{1}{Eh}(N_\theta - \nu N_\varphi)$$

Substituting for N_φ and N_θ their expressions in U and V, we obtain expressions for ϵ_φ and ϵ_θ which can be used for calculating v and w as was explained in Art. 108.

In practical applications the displacement δ in the planes of the parallel circles is usually important. It can be obtained by projecting the components v and w on that plane. This gives (Fig. 267)

$$\delta = v \cos \varphi - w \sin \varphi$$

The expression for this displacement in terms of the functions U and V is readily obtained if we observe that δ represents the increase in the radius r_0 of the parallel circle. Thus

$$\delta = a \sin \varphi\, \epsilon_\theta = \frac{a \sin \varphi}{Eh}(N_\theta - \nu N_\varphi) = -\frac{\sin \varphi}{Eh}\left(\frac{dU}{d\varphi} - \nu U \cot \varphi\right) \qquad (r)$$

Thus all the quantities that define the bending of a spherical shell by forces and couples uniformly distributed along the edge can be represented in terms of the two series I_1 and I_2.

The ease with which practical application of this analysis can be made depends on the rapidity of convergence of the series I_1 and I_2. This convergence depends principally upon the magnitude of the quantity

$$\rho = \sqrt[4]{\frac{3a^2}{h^2}(1-\nu^2) - \frac{\nu^2}{4}} \qquad (s)$$

which, if ν^2 is neglected in comparison with unity, becomes

$$\rho \approx \sqrt[4]{3}\,\sqrt{\frac{a}{h}}$$

Calculations show[1] that for $\rho < 10$ the convergence of the series is satisfactory, and all necessary quantities can be found without much difficulty for various edge conditions.

As an example we shall take the case of a spherical shell submitted to the action of uniform normal pressure p (Fig. 268). The membrane stresses in this case are

$$\sigma_\varphi = \sigma_\theta = -\frac{pa}{2h} \quad (t)$$

and the corresponding membrane forces that keep the shell in equilibrium are

$$(N_\varphi)_{\varphi=\alpha} = -\frac{pa}{2} \quad (u)$$

By superposing on the membrane forces horizontal forces

$$H = \frac{pa}{2} \cos \alpha$$

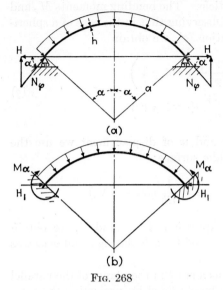

Fig. 268

uniformly distributed along the edge of the shell, we obtain the case, represented in Fig. 268a, in which the loaded shell is supported by vertical reactions of a horizontal plane. The stresses in this case are obtained by superposing on the membrane stresses (t) the stresses produced by the horizontal forces H. These latter stresses can be obtained by using the general solutions (n) and (p) and determining the constants A and B in these solutions so as to satisfy the boundary conditions

$$(N_\varphi)_{\varphi=\alpha} = H \cos \alpha \qquad (M_\varphi)_{\varphi=\alpha} = 0$$

The stresses obtained in this way for a particular case in which $a = 56.3$ in., $h = 2.36$ in., $\alpha = 39°$, $p = 284$ psi, and $\nu = 0.2$ are shown in Fig. 269.

By superposing on the membrane forces (u) the horizontal forces H_1 and bending moments M_α uniformly distributed along the edge, we can also obtain the case of a shell with built-in edges (Fig. 268b). The stresses in this case are obtained by superposing on the membrane stresses (t) the stresses produced in the shell by the forces H_1 and the moments M_α. These latter stresses are obtained as before from the general solutions (n) and (p), the constants A and B being so determined as to satisfy the boundary conditions

$$(\epsilon_\theta)_{\varphi=\alpha} = 0 \qquad (V)_{\varphi=\alpha} = 0$$

The total stresses obtained in this way for the previously cited numerical example are shown in Fig. 270.

From the calculation of the maximum compressive and maximum tensile stresses for various proportions of shells submitted to the action of a uniform normal pressure p, it was found[2] that the magnitude of these stresses depends principally on the

---

[1] Such calculations were made by L. Bolle, *Schweiz. Bauztg.*, vol. 66, p. 105, 1915.
[2] *Ibid.*

SHELLS FORMING SURFACE OF REVOLUTION 545

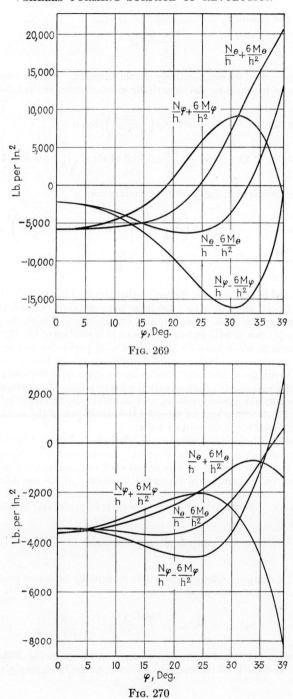

Fig. 269

Fig. 270

magnitude of the quantity

$$\frac{a}{h} \sin^2 \alpha$$

and can be represented by comparatively simple formulas. For the case represented in Fig. 268a these formulas for the numerically greatest stress are as follows:

For $\dfrac{a}{h} \sin^2 \alpha < 1.2$ $\qquad \sigma = -1.24p \left(\dfrac{a \sin \alpha}{h}\right)^2 \cos \alpha$

For $1.2 < \dfrac{a}{h} \sin^2 \alpha < 12$ $\qquad \sigma = \dfrac{ap}{2h} \left[\left(1.6 + 2.44 \sin \alpha \sqrt{\dfrac{a}{h}}\right) \cos \alpha - 1 \right]$

For the case represented in Fig. 268b the formulas are:

For $\dfrac{a \sin^2 \alpha}{h} < 3$ $\qquad \sigma = -p \left(\dfrac{a \sin \alpha}{h}\right)^2 \left[0.75 - 0.038 \left(\dfrac{a \sin \alpha}{h}\right)^2 \sin^2 \alpha \right]$

For $3 < \dfrac{a \sin^2 \alpha}{h} < 12$ $\qquad \sigma = -1.2 \dfrac{ap}{h}$

It was assumed in the foregoing discussion that the shell has no hole at the top. If there is such a hole, we must satisfy the boundary conditions on both the lower and the upper edges of the shell. This requires consideration of both the integrals (j) and (k) of Eq. (d) (see p. 541) and finally results in a solution of Eq. (320) which contains four constants which must be adjusted in each particular case so as to satisfy the boundary conditions on both edges. Calculations of this kind show[1] that, if the angle α is not small, the forces distributed along the upper edge have only a very small influence on the magnitude of stresses at the lower edge. Thus, since these latter stresses are usually the most important, we can obtain the necessary information for the design of a shell with a hole by using for the calculation of the maximum stresses the formulas derived for shells without holes.

The method of calculating stresses in spherical shells discussed in this article can also be applied in calculating thermal stresses. Assume that the temperatures at the outer and at the inner surfaces of a spherical shell are constant but that there is a linear variation of temperature in the radial direction. If t is the difference in the temperatures of the outer and inner surfaces, the bending of the shell produced by the temperature difference is entirely arrested by constant bending moments (see Art. 14):

$$M_\varphi = M_\theta = \frac{\alpha t D (1 + \nu)}{h} \qquad (v)$$

In the case of a complete sphere these moments actually exist and produce bending stresses the maximum values of which are

$$(\sigma_\varphi)_{\max} = (\sigma_\theta)_{\max} = \frac{6 \alpha t D (1 + \nu)}{h^3} = \frac{\alpha t E}{2(1 - \nu)} \qquad (w)$$

If we have only a portion of a sphere, supported as shown in Fig. 268a, the edge is free to rotate, and the total thermal stresses are obtained by superposing on stresses

[1] *Ibid.*

(w) the stresses that are produced in the shell by the moments

$$M_\alpha = -\frac{\alpha t D(1+\nu)}{h}$$

uniformly distributed along the edge. These latter stresses are obtained by using the method discussed in this article.[1] In the case shown in Fig. 268b the thermal stresses are given by formula (w), if the temperature of the middle surface always remains the same. Otherwise, on the stresses (w) must be superposed stresses produced by forces H and moments M_α which must be determined in each particular case so as to satisfy the boundary conditions.

130. Approximate Methods of Analyzing Stresses in Spherical Shells.

In the preceding article it has already been indicated that the application of the rigorous solution for the stresses in spherical shells depends on the rapidity of convergence of the series entering into the solution. The convergence becomes slower, and more and more terms of the series must be calculated, as the ratio a/h increases, i.e., as the thickness of the shell becomes smaller and smaller in comparison with its radius.[2] For such shells approximate methods of solution have been developed which give very good accuracy for large values of a/h.

One of the approximate methods for the solution of the problem is the method of asymptotic integration.[3] Starting with Eq. (320) and introducing, instead of the shearing force Q_φ, the quantity

$$z = Q_\varphi \sqrt{\sin \varphi} \qquad (a)$$

we obtain the equation

$$z^{\mathrm{IV}} + a_2 z^{\mathrm{II}} + a_1 z^{\mathrm{I}} + (\beta^4 + a_0)z = 0 \qquad (b)$$

in which

$$a_0 = -\frac{63}{16 \sin^4 \varphi} + \frac{9}{8 \sin^2 \varphi} + \frac{9}{16} \qquad a_1 = \frac{3 \cos \varphi}{\sin^3 \varphi}$$
$$a_2 = -\frac{3}{2 \sin^2 \varphi} + \frac{5}{2} \qquad 4\beta^4 = (1-\nu^2)\left(1 + \frac{12 a^2}{h^2}\right) \qquad (c)$$

It can be seen that for thin shells, in which a/h is a large number, the quantity $4\beta^4$ is very large in comparison with the coefficients a_0, a_1, and a_2, provided the angle φ is not small. Since in our further discussion we shall be interested in stresses near the edge where $\varphi = \alpha$ (Fig. 268) and

---

[1] Thermal stresses in shells have been discussed by G. Eichelberg, *Forschungsarb.*, no. 263, 1923. For shells of arbitrary thickness see also E. L. McDowell and E. Sternberg, *J. Appl. Mechanics*, vol. 24, p. 376, 1957.

[2] Calculations by J. E. Ekström in *Ing. Vetenskaps. Akad.*, vol. 121, Stockholm, 1933, show that for $a/h = 62.5$ it is necessary to consider not less than 18 terms of the series.

[3] See O. Blumenthal's paper in *Repts. Fifth Intern. Congr. Math.*, Cambridge, 1912; see also his paper in *Z. Math. Physik*, vol. 62, p. 343, 1914.

α is not small, we can neglect the terms with the coefficients a_0, a_1, and a_2 in Eq. (b). In this way we obtain the equation

$$z^{IV} + 4\beta^4 z = 0 \quad (d)$$

This equation is similar to Eq. (276), which we used in the investigation of the symmetrical deformation of circular cylindrical shells. Using the general solution of Eq. (d) together with notation (a), we obtain

$$Q_\varphi = \frac{1}{\sqrt{\sin \varphi}} [e^{\beta\varphi}(C_1 \cos \beta\varphi + C_2 \sin \beta\varphi) + e^{-\beta\varphi}(C_3 \cos \beta\varphi + C_4 \sin \beta\varphi)] \quad (e)$$

From the previous investigation of the bending of cylindrical shells we know that the bending stresses produced by forces uniformly distributed along the edge decrease rapidly as the distance from the edge increases. A similar condition also exists in the case of thin spherical shells. Observing that the first two terms in solution (e) decrease while the second two increase as the angle φ decreases, we conclude that in the case of a sphere without a hole at the top it is permissible to take only the first two terms in solution (e) and assume

$$Q_\varphi = \frac{e^{\beta\varphi}}{\sqrt{\sin \varphi}} (C_1 \cos \beta\varphi + C_2 \sin \beta\varphi) \quad (f)$$

Having this expression for Q_φ and using the relations (b), (c), and (d) of Art. 128 and the relations (p), (q), and (r) of Art. 129, all the quantities defining the bending of the shell can be calculated, and the constants C_1 and C_2 can be determined from the conditions at the edge. This method can be applied without any difficulty to particular cases and gives good accuracy for thin shells.[1]

Instead of working with the differential equation (320) of the fourth order, we can take, as a basis for an approximate investigation of the bending of a spherical shell, the two Eqs. (317).[2] In our case these equations can be written as follows:

$$\frac{d^2 Q_\varphi}{d\varphi^2} + \cot \varphi \frac{dQ_\varphi}{d\varphi} - (\cot^2 \varphi - \nu) Q_\varphi = EhV$$
$$\frac{d^2 V}{d\varphi^2} + \cot \varphi \frac{dV}{d\varphi} - (\cot^2 \varphi + \nu) V = -\frac{a^2 Q_\varphi}{D} \quad (g)$$

---

[1] An example of application of the method of asymptotic integration is given by S. Timoshenko; see *Bull. Soc. Eng. Tech.*, St. Petersburg, 1913. In the papers by Blumenthal, previously mentioned, means are given for the improvement of the approximate solution by the calculation of a further approximation.

[2] This method was proposed by J. W. Geckeler, *Forschungsarb.*, no. 276, Berlin, 1926, and also by I. Y. Staerman, *Bull. Polytech. Inst. Kiev*, 1924; for a generalization see Y. N. Rabotnov, *Doklady Akad. Nauk S.S.S.R.*, n.s., vol. 47, p. 329, 1945.

where Q_φ is the shearing force and V is the rotation of a tangent to a meridian as defined by Eq. (a) of Art. 128. In the case of very thin shells, if the angle φ is not small, the quantities Q_φ and V are damped out rapidly as the distance from the edge increases and have the same oscillatory character as has the function (f). Since β is large in the case of thin shells, the derivative of the function (f) is large in comparison with the function itself, and the second derivative is large in comparison with the first. This indicates that a satisfactory approximation can be obtained by neglecting the terms containing the functions Q_φ and V and their first derivatives in the left-hand side of Eqs. (g). In this way Eqs. (g) can be replaced by the following simplified system of equations:[1]

$$\frac{d^2 Q_\varphi}{d\varphi^2} = EhV$$
$$\frac{d^2 V}{d\varphi^2} = -\frac{a^2}{D} Q_\varphi \quad (h)$$

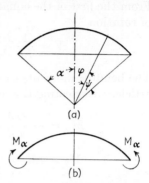

By eliminating V from these equations, we obtain

$$\frac{d^4 Q_\varphi}{d\varphi^4} + 4\lambda^4 Q_\varphi = 0 \quad (i)$$

where $\quad \lambda^4 = 3(1 - \nu^2)\left(\frac{a}{h}\right)^2 \quad (j)$

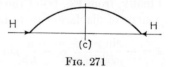

Fig. 271

The general solution of this equation is

$$Q_\varphi = C_1 e^{\lambda\varphi} \cos \lambda\varphi + C_2 e^{\lambda\varphi} \sin \lambda\varphi + C_3 e^{-\lambda\varphi} \cos \lambda\varphi + C_4 e^{-\lambda\varphi} \sin \lambda\varphi \quad (k)$$

Considering the case in which there is no hole at the top (Fig. 271a) and the shell is bent by forces and moments uniformly distributed along the edge, we need consider from the general solution (k) only the first two terms, which decrease as the angle φ decreases. Thus

$$Q_\varphi = C_1 e^{\lambda\varphi} \cos \lambda\varphi + C_2 e^{\lambda\varphi} \sin \lambda\varphi \quad (l)$$

The two constants C_1 and C_2 are to be determined in each particular case from the conditions at the edge ($\varphi = \alpha$). In discussing the edge conditions it is advantageous to introduce the angle $\psi = \alpha - \varphi$ (Fig. 271). Substituting $\alpha - \psi$ for φ in expression (l) and using the new constants

[1] This simplification of the problem is equivalent to the replacement of the portion of the shell near the edge by a tangent conical shell and application to this conical shell of the equation that was developed for a circular cylinder (Art. 114); see E. Meissner, "A. Stodola Festschrift," p. 406, Zürich, 1929.

C and γ, we can represent solution (l) in the form

$$Q_\varphi = Ce^{-\lambda\psi} \sin(\lambda\psi + \gamma) \tag{m}$$

Now, employing Eqs. (b), (c), and (d) of Art. 128, we find

$$N_\varphi = -Q_\varphi \cot\varphi = -\cot(\alpha - \psi)Ce^{-\lambda\psi}\sin(\lambda\psi + \gamma)$$
$$N_\theta = -\frac{dQ_\varphi}{d\varphi} = -\lambda\sqrt{2}\,Ce^{-\lambda\psi}\sin\left(\lambda\psi + \gamma - \frac{\pi}{4}\right) \tag{323}$$

From the first of the equations (h) we obtain the expression for the angle of rotation

$$V = \frac{1}{Eh}\frac{d^2Q_\varphi}{d\varphi^2} = -\frac{2\lambda^2}{Eh}Ce^{-\lambda\psi}\cos(\lambda\psi + \gamma) \tag{324}$$

The bending moments can be determined from Eqs. (q) of the preceding article. Neglecting the terms containing V in these equations, we find

$$M_\varphi = -\frac{D}{a}\frac{dV}{d\varphi} = \frac{a}{\lambda\sqrt{2}}Ce^{-\lambda\psi}\sin\left(\lambda\psi + \gamma + \frac{\pi}{4}\right)$$
$$M_\theta = \nu M_\varphi = \frac{a\nu}{\lambda\sqrt{2}}Ce^{-\lambda\psi}\sin\left(\lambda\psi + \gamma + \frac{\pi}{4}\right) \tag{325}$$

Finally, from Eq. (r) of the preceding article we find the horizontal component of displacement to be

$$\delta \approx -\frac{\sin\varphi}{Eh}\frac{dU}{d\varphi} = -\frac{a}{Eh}\sin(\alpha - \psi)\lambda\sqrt{2}\,Ce^{-\lambda\psi}\sin\left(\lambda\psi + \gamma - \frac{\pi}{4}\right) \tag{326}$$

With the aid of formulas (323) to (326) various particular cases can readily be treated.

Take as an example the case shown in Fig. 271b. The boundary conditions are

$$(M_\varphi)_{\varphi=\alpha} = M_\alpha \qquad (N_\varphi)_{\varphi=\alpha} = 0 \tag{n}$$

By substituting $\psi = 0$ in the first of the equations (323), it can be concluded that the second of the boundary conditions (n) is satisfied by taking the constant γ equal to zero. Substituting $\gamma = 0$ and $\psi = 0$ in the first of the equations (325), we find that to satisfy the first of the conditions (n) we must have

$$M_\alpha = \frac{a}{2\lambda}C$$

which gives

$$C = \frac{M_\alpha 2\lambda}{a}$$

Substituting values thus determined for the constants γ and C in expressions (324) and (326) and taking $\psi = 0$, we obtain the rotation and the horizontal displacement of the edge as follows:

$$(V)_{\psi=0} = -\frac{4\lambda^3 M_\alpha}{Eah} \qquad (\delta)_{\psi=0} = \frac{2\lambda^2 \sin\alpha}{Eh} M_\alpha \qquad (327)$$

In the case represented in Fig. 271c, the boundary conditions are

$$(M_\varphi)_{\varphi=\alpha} = 0 \qquad (N_\varphi)_{\varphi=\alpha} = -H\cos\alpha \qquad (o)$$

To satisfy the first of these conditions, we must take $\gamma = -\pi/4$. To satisfy the second boundary condition, we use the first of the equations (323) which gives

$$-H\cos\alpha = C\cot\alpha\sin\frac{\pi}{4}$$

from which we determine

$$C = -\frac{2H\sin\alpha}{\sqrt{2}}$$

Substituting the values of the constants γ and C in (324) and (326), we find

$$(V)_{\psi=0} = \frac{2\lambda^2 \sin\alpha}{Eh} H \qquad (\delta)_{\psi=0} = -\frac{2a\lambda\sin^2\alpha}{Eh} H \qquad (328)$$

It can be seen that the coefficient of M_α in the second of the formulas (327) is the same as the coefficient of H in the first of the formulas (328). This should follow at once from the reciprocity theorem.

Formulas (327) and (328) can readily be applied in solving particular problems Take as an example the case of a spherical shell with a built-in edge and submitted to the action of a uniform normal pressure p (Fig. 272a). Considering first the corresponding membrane problem (Fig. 272b), we find a uniform compression of the shell

$$N_\varphi = N_\theta = -\frac{pa}{2}$$

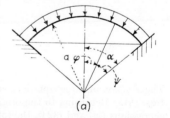

The edge of this shell experiences no rotation and undergoes a horizontal displacement

$$\delta = \frac{a\sin\alpha}{Eh}(N_\theta - \nu N_\varphi) = -\frac{pa^2(1-\nu)}{2Eh}\sin\alpha \qquad (p)$$

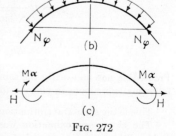

Fig. 272

To obtain the solution of the given problem we superpose on the membrane forces of Fig. 272b forces and moments uniformly distributed along the edge as in Fig. 272c. These forces and moments are of such magnitude that the corresponding horizontal displacement is equal and opposite to the displacement (p), and the corresponding rotation of the edge is equal to zero.

In this way, by using formulas (327) and (328), we obtain the following equations for the determination of M_α and H:

$$-\frac{4\lambda^3}{Eah} M_\alpha + \frac{2\lambda^2 \sin \alpha}{Eh} H = 0$$

$$\frac{2\lambda^2 \sin \alpha}{Eh} M_\alpha - \frac{2a\lambda \sin^2 \alpha}{Eh} H = \frac{pa^2(1-\nu)}{2Eh} \sin \alpha$$

from which

$$M_\alpha = -\frac{pa^2(1-\nu)}{4\lambda^2} = -\frac{pah}{4}\sqrt{\frac{1-\nu}{3(1+\nu)}}$$

$$H = \frac{2\lambda}{a \sin \alpha} M_\alpha = -\frac{pa(1-\nu)}{2\lambda \sin \alpha}$$

(q)

The negative signs indicate that M_α and H have directions opposite to those shown in Fig. 271.

The approximate equations (h) were obtained by neglecting the unknown functions Q_φ and V and their first derivatives in the exact equations (g). A better approximation is obtained if we introduce the new variables[1]

Substituting

$$Q_1 = Q_\varphi \sqrt{\sin \varphi} \qquad V_1 = V \sqrt{\sin \varphi}$$

$$Q_\varphi = \frac{Q_1}{\sqrt{\sin \varphi}} \qquad V = \frac{V_1}{\sqrt{\sin \varphi}}$$

in Eqs. (g), we find that the terms containing the first derivatives of Q_1 and V_1 vanish. Hence, to obtain a simplified system of equations similar to Eqs. (h), we have to neglect only the terms containing the quantities Q_1 and V_1 in comparison with the terms containing the second derivatives of the same quantities. This gives

$$\frac{d^2Q_1}{d\varphi^2} = EhV_1$$

$$\frac{d^2V_1}{d\varphi^2} = -\frac{a^2}{D} Q_1$$

The solution of these equations can be obtained in the same manner as in the case of Eqs. (h). Returning to the original variables Q_φ and V, we then obtain, instead of expressions (m) and (324), the following solutions:[2]

$$Q_\varphi = C \frac{e^{-\lambda\psi}}{\sqrt{\sin(\alpha-\psi)}} \sin(\lambda\psi + \gamma)$$

$$V = -\frac{2\lambda^2}{Eh} C \frac{e^{-\lambda\psi}}{\sqrt{\sin(\alpha-\psi)}} \cos(\lambda\psi + \gamma)$$

(329)

Proceeding now in exactly the same way as in our previous discussion, we obtain the following expressions in place of formulas (323), (325), and (326):

[1] This is the same transformation as was used by O. Blumenthal; see Eq. (a), p. 547.
[2] The closer approximation was obtained by M. Hetényi, *Publs. Intern. Assoc. Bridge Structural Engrs.*, vol. 5, p. 173, 1938; the numerical example used in the further discussion is taken from this paper.

$$N_\varphi = -\cot(\alpha - \psi) C \frac{e^{-\lambda\psi}}{\sqrt{\sin(\alpha - \psi)}} \sin(\lambda\psi + \gamma)$$

$$N_\theta = C \frac{\lambda e^{-\lambda\psi}}{2\sqrt{\sin(\alpha - \psi)}} [2\cos(\lambda\psi + \gamma) - (k_1 + k_2)\sin(\lambda\psi + \gamma)]$$

$$M_\varphi = \frac{a}{2\lambda} C \frac{e^{-\lambda\psi}}{\sqrt{\sin(\alpha - \psi)}} [k_1 \cos(\lambda\psi + \gamma) + \sin(\lambda\psi + \gamma)] \quad (330)$$

$$M_\theta = \frac{a}{4\nu\lambda} C \frac{e^{-\lambda\psi}}{\sqrt{\sin(\alpha - \psi)}} \{[(1 + \nu^2)(k_1 + k_2) - 2k_2]\cos(\lambda\psi + \gamma) + 2\nu^2 \sin(\lambda\psi + \gamma)\}$$

$$\delta = \frac{a \sin(\alpha - \psi)}{Eh} C \frac{\lambda e^{-\lambda\psi}}{\sqrt{\sin(\alpha - \psi)}} [\cos(\lambda\psi + \gamma) - k_2 \sin(\lambda\psi + \gamma)]$$

where
$$k_1 = 1 - \frac{1 - 2\nu}{2\lambda} \cot(\alpha - \psi)$$

$$k_2 = 1 - \frac{1 + 2\nu}{2\lambda} \cot(\alpha - \psi)$$

Applying formulas (330) to the particular cases previously discussed and represented in Fig. 271b and c, we obtain, instead of formulas (327) and (328), the following better approximations:

$$(V)_{\psi=0} = -\frac{4\lambda^3 M_\alpha}{Eahk_1} \quad (\delta)_{\psi=0} = \frac{2\lambda^2 \sin\alpha}{Ehk_1} M_\alpha \quad (331)$$

$$(V)_{\psi=0} = \frac{2\lambda^2 \sin\alpha}{Ehk_1} H \quad (\delta)_{\psi=0} = -\frac{\lambda a \sin^2\alpha}{Eh}\left(k_2 + \frac{1}{k_1}\right) H \quad (332)$$

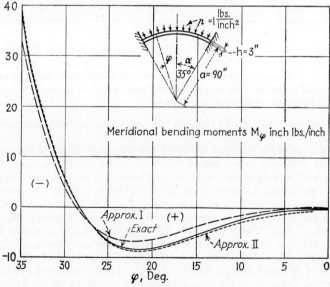

Fig. 273

By applying these formulas to the particular case shown in Fig. 272a, second approximations for the reactive moments M_α and reactive forces H are readily obtained.

To compare the first and second approximations with the exact solution, we shall consider a numerical example in which $a = 90$ in., $h = 3$ in., $\alpha = 35°$, $p = 1$ psi, and $\nu = \frac{1}{6}$. The first and second approximations for M_φ have been calculated by using the first of the equations (325) and the third of equations (330) and are represented by the broken lines in Fig. 273. For comparison the exact solution[1] has also been calculated by using the series of the preceding article. This exact solution is represented by the full line in Fig. 273. In Fig. 274 the force N_θ as calculated for the same numerical

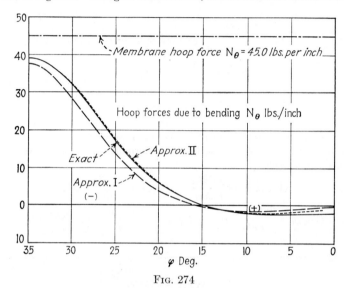

Fig. 274

example is shown. From these two figures it can be concluded that the second approximation has very satisfactory accuracy. Observing that in our example the ratio a/h is only 30 and the angle $\alpha = 35°$ is comparatively small, it can be concluded that the second approximation can be applied with sufficient accuracy in most cases encountered in present structural practice.[2]

[1] It was necessary to take 10 terms in the series to obtain sufficient accuracy in this case.

[2] In the case in which the angle α is small and the solution (329) is not sufficiently accurate, the shell may be considered "shallow" and treated accordingly (see Art. 132). Application of the equations of finite differences to the same problem has been made by P. Pasternak, Z. angew. Math. Mech., vol. 6, p. 1, 1926. The case of non-isotropic shells is considered by E. Steuermann, Z. angew. Math. Mech., vol. 5, p. 1, 1925. One particular case of a spherical shell of variable thickness is discussed by M. F. Spotts, J. Appl. Mechanics (Trans. ASME), vol. 61, 1939, and also by F. Tölke, Ingr.-Arch., vol. 9, p. 282, 1938. For the effect of concentrated loads, see F. Martin, Ingr.-Arch., vol. 17, p. 107, 1949, and Art. 132. The problem of nonsymmetrical deformation of spherical shells is considered by A. Havers, Ingr.-Arch., vol. 6, p. 282, 1935. Further discussion of the same problem in connection with the stress analysis of a spherical dome supported by columns is given by A. Aas-Jakobsen, Ingr.-Arch., vol. 8, p. 275, 1937.

131. Spherical Shells with an Edge Ring. In order to reduce the effect of the thrust of a dome in its action upon the supporting structure, an edge ring (Figs. 275a and 276a) is sometimes used. The vertical deflection of this ring, supported either continuously or in a number of points, may be neglected in the following analysis.

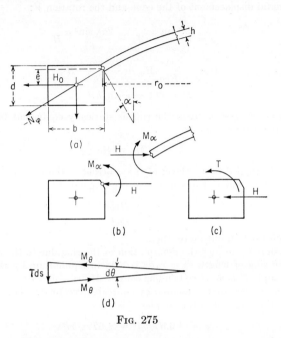

Fig. 275

Let us consider the conditions on the edge $\varphi = \alpha$ of the dome carrying some distributed, symmetrical load. The membrane forces N_φ, N_θ due to this load would produce, according to Eq. (r) (page 543) an increase of the radius $r_0 = a \sin \alpha$ equal to

$$\delta_0 = \frac{r_0}{Eh}(N_\theta - \nu N_\varphi)_{\varphi=\alpha} \qquad (a)$$

This displacement will be accompanied by a rotation of the edge tangent

$$V_0 = \frac{1}{a}\left(v + \frac{dw}{d\varphi}\right)_{\varphi=\alpha} = \frac{1}{Eh}\left[\cot\varphi(1+\nu)(N_\varphi - N_\theta) - \frac{d}{d\varphi}(N_\theta - \nu N_\varphi)\right]_{\varphi=\alpha} \qquad (b)$$

according to results obtained on page 538, and by a thrust

$$H_0 = -\cos\alpha(N_\varphi)_{\varphi=\alpha} \qquad (c)$$

The corresponding tension force in the ring is $H_0 r_0$, and the elongation is $\epsilon_\theta = \dfrac{H_0 r_0}{Ebd}$, where E denotes Young's modulus of the material of the ring. The increase of the radius r_0 due to the action of H_0 will be

$$\delta_1 = \epsilon_\theta r_0 = \frac{H_0 r_0^2}{Ebd} \qquad (d)$$

In order to bring the edge deformation of the shell in accordance with the deformation of the ring, let us apply along the circumference of both the edge and the ring uniformly distributed couples of an intensity M_α and radial forces of an intensity H (Fig. 275b). Using the results (327) and (328), we obtain the following expressions for the horizontal displacement of the edge and the rotation V:

$$\delta = \frac{2\lambda^2 \sin \alpha}{Eh} M_\alpha - \frac{2a\lambda \sin^2 \alpha}{Eh} H$$

$$V = -\frac{4\lambda^3}{Eha} M_\alpha + \frac{2\lambda^2 \sin \alpha}{Eh} H \qquad (e)$$

where

$$\lambda^4 = 3(1 - \nu^2)(a/h)^2$$

The action of M_α and H upon the ring is statically equivalent to the combined action of the overturning couples

$$T = M_\alpha + He \qquad (f)$$

and of forces H applied on the level of the centroid of the ring section (Fig. 275c). These latter cause a radial displacement of the ring equal to

$$\delta_2 = \frac{Hr_0^2}{Ebd} \qquad (g)$$

as follows from Eq. (d), but no rotation.

It still remains to consider the deformation of the ring due to the couples T. An element of the ring of length $ds = r_0 \, d\theta$ is held in equilibrium by the action of an overturning couple $T \, ds$ and two bending couples $M_\theta = T \, ds/d\theta = Tr_0$ (see Fig. 275d, where all three couples are represented by equivalent vectors). Thus, the maximum hoop stress in the ring due to the couples T is

$$\sigma = \pm 6M_\theta/bd^2 = \pm 6Tr_0/bd^2$$

The corresponding unit elongation of the top and bottom fibers of the ring is seen to be $\epsilon = \pm 6Tr_0/Ebd^2$, respectively. Hence the rotation of the transverse section of the ring becomes

$$V_2 = \frac{2r_0}{d} |\epsilon| = \frac{12Tr_0^2}{Ebd^3} = \frac{12r_0^2}{Ebd^3}(M_\alpha + He) \qquad (h)$$

where $|\epsilon|$ denotes the absolute value of the largest unit elongation.

Now, the total horizontal displacement of the shell edge must be equal to that of the ring, and the same holds for the rotation. This yields the following relations:

$$\delta_0 + \delta = \delta_1 + \delta_2 + V_2 e \qquad (i)$$
$$V_0 + V = V_2 \qquad (j)$$

in which the term $V_2 e$ represents the effect of the rotation on the radial displacement of the ring at the level of the edge of the shell. After substitution of the expressions (a) to (h) for the displacement and the rotation in (i) and (j), we obtain two linear equations for the unknown values of M_α and H. These values also define the constants of integration of the approximate solution, as shown in Art. 130. The total stress resultants and deflections of the shell can be found then by combining the effect of membrane forces with the effect of bending, this latter being expressed, for example, by Eqs. (323), (324), and (325).

As an illustrative example, let us consider a spherical dome (Fig. 276a) with $a = 76.6$ ft, $\alpha = 40°$, $r_0 = 49.2$ ft, $h = 2.36$ in., and the cross-sectional dimensions of the

SHELLS FORMING SURFACE OF REVOLUTION

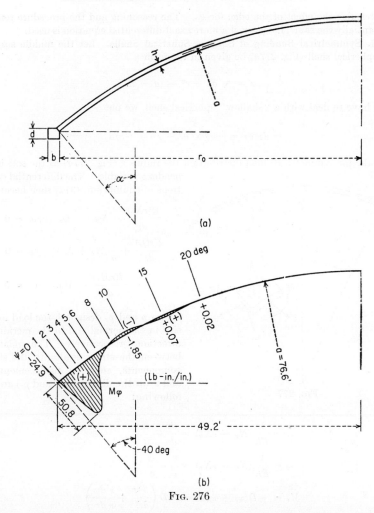

Fig. 276

ring $b = 1.97$ ft, $d = 1.64$ ft, and $e \approx d/2$; the modulus E is the same for the shell and the ring, and the constant ν is assumed equal to zero. The dome is submitted to the action of its own weight $q = 41$ psf of the surface of the dome. The membrane forces due to this load are given by Eqs. (257), and the procedure of computation indicated above leads to the following values of the edge forces:[1]

$$M_\alpha = -24.84 \text{ lb-in. per in.}$$
$$H = -8.95 \text{ lb per in.}$$

The corresponding values of bending moment M_φ are shown in Fig. 276b.

In the foregoing the simplified differential equation (i), Art. 130, has been employed

[1] The details of computation may be found in K. Girkmann, "Flächentragwerke," 4th ed., p. 442, Springer-Verlag, Vienna, 1956. The diagram Fig. 276b is reproduced here by courtesy of Professor K. Girkmann and the Springer-Verlag, Vienna.

to determine the effect of the edge forces. The reasoning and the procedure remain substantially the same, however, if a more exact differential equation is used.

132. Symmetrical Bending of Shallow Spherical Shells. Let the middle surface of a spherical shell (Fig. 277a) be given in the form

$$z = \sqrt{a^2 - r^2} - (a - z_0) \tag{a}$$

If we have to deal with a "shallow" spherical shell, we put

$$dz/dr = -r/\sqrt{a^2 - r^2} \approx -r/a$$

and take for symmetrical[1] load distribution the radius r (Fig. 277a) as the sole independent variable. The differential equations of equilibrium (312) then become

$$\frac{d(rN_r)}{dr} - N_\theta - \frac{r}{a} Q_r + rp_r = 0 \tag{b}$$

$$\frac{d(rQ_r)}{dr} + \frac{r}{a}(N_r + N_\theta) + rp = 0 \tag{c}$$

$$\frac{d(rM_r)}{dr} - M_\theta - rQ_r = 0 \tag{d}$$

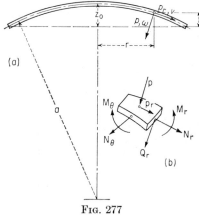

Fig. 277

where p and p_r designate the load intensity in the normal and in the meridional direction, respectively. The relations between the stress resultants, the strain components, and the displacements w and v (in the directions p and p_r) are the following:

$$\epsilon_r = \frac{1}{Eh}(N_r - \nu N_\theta) = \frac{dv}{dr} - \frac{w}{a}$$

$$\epsilon_\theta = \frac{1}{Eh}(N_\theta - \nu N_r) = \frac{v}{r} - \frac{w}{a} \tag{e}$$

$$M_r = -D(\chi_r + \nu\chi_\theta) = -D\left(\frac{d^2w}{dr^2} + \frac{\nu}{r}\frac{dw}{dr}\right)$$

$$M_\theta = -D(\chi_\theta + \nu\chi_r) = -D\left(\frac{1}{r}\frac{dw}{dr} + \nu\frac{d^2w}{dr^2}\right) \tag{f}$$

in which

$$D = \frac{Eh^3}{12(1-\nu^2)} \tag{g}$$

Now we take the fact into account that the effect of transverse shear Q_r on membrane forces in Eq. (b) can be neglected in the case of a shallow shell. Assuming, furthermore, that the load term p_r is derivable from a load potential Ω, so that $p_r = -d\Omega/dr$, we satisfy Eq. (b) by setting

---

[1] The general theory of shallow spherical shells, due to E. Reissner, is free from this limitation; see *J. Math. and Phys.*, vol. 25, p. 80, 1946; vol. 25, p. 279, 1947.

SHELLS FORMING SURFACE OF REVOLUTION

$$N_r = \frac{1}{r}\frac{dF}{dr} + \Omega$$
$$N_\theta = \frac{d^2F}{dr^2} + \Omega \tag{h}$$

where F is a stress function. It is easy to verify also that the relations (e) between the strain and the displacement components correspond to the equation of compatibility:

$$\frac{1}{r^2}\frac{d}{dr}\left(r^2\frac{d\epsilon_\theta}{dr}\right) - \frac{1}{r}\frac{d\epsilon_r}{dr} + \frac{1}{a}\Delta w = 0 \tag{i}$$

in which $\Delta = d^2/dr^2 + (1/r)(d/dr)$.

Combining Eqs. (e) and (i), we arrive at the following fundamental equation for F and w:

$$\Delta\Delta F + \frac{Eh}{a}\Delta w = -(1 - \nu)\Delta\Omega \tag{j}$$

In order to obtain a second fundamental relation between the same functions, we substitute Q_r from (d) in Eq. (c). We obtain

$$\frac{d}{dr}\left[\frac{d(rM_r)}{dr} - M_\theta\right] + \frac{r}{a}(N_r + N_\theta) + rp = 0 \tag{k}$$

Using now the expressions (f) and (h) in combination with Eq. (k), we find

$$\Delta\Delta w - \frac{1}{Da}\Delta F = \frac{p}{D} + \frac{2\Omega}{Da} \tag{l}$$

Finally, let us write the expressions for the vertical shearing force Q_v and the horizontal displacement δ, both of which may be used in formulating the edge conditions of the shell. We obtain

$$Q_v = Q_r + \frac{r}{a}N_r \qquad \delta = v - \frac{r}{a}w \tag{m}$$

in which the expression for the transverse force

$$Q_r = -D\frac{d}{dr}(\Delta w) \tag{n}$$

is of the same form as in the theory of plates.

In the case $p = \Omega = 0$, the integration of the simultaneous equations (j) and (l) can be carried out by multiplying Eq. (j) by a factor $-\lambda$ and adding the result to Eq. (l). This yields

$$\Delta\Delta(w - \lambda F) - \lambda(Eh/a)\Delta(w + F/\lambda h\, DE) = 0 \tag{o}$$

From (o) we obtain an equation for a single function $w - \lambda F$ by putting $\lambda = -1/\lambda h\, DE$; that is,

$$\lambda = \frac{i}{Eh^2}\sqrt{12(1 - \nu^2)} \tag{p}$$

where $i = \sqrt{-1}$. Let us also introduce a characteristic length l defined by the relation $\lambda Eh/a = i/l^2$, so that

$$l = \frac{\sqrt{ah}}{\sqrt[4]{12(1-\nu^2)}} \qquad (q)$$

The differential equation (o) then assumes the form

$$\Delta\Delta(w - \lambda F) - \frac{i}{l^2}\Delta(w - \lambda F) = 0 \qquad (r)$$

Next, setting

$$w - \lambda F = \Phi + \Psi \qquad (s)$$

we obtain Φ and Ψ as the general solution of the equations

$$\Delta\Phi = 0 \qquad \Delta\Psi - \left(\frac{\sqrt{i}}{l}\right)^2 \Psi = 0 \qquad (t)$$

The respective solutions are of the form

$$\Phi = A_1 + A_2 \log x \qquad (u)$$
$$\Psi = A_3[\psi_1(x) + i\psi_2(x)] + A_4[\psi_4(x) + i\psi_3(x)] \qquad (v)$$

where

$$x = \frac{r}{l} \qquad (w)$$

A_n are arbitrary complex constants and $\psi_1(x), \ldots, \psi_4(x)$ are functions defined on page 490 and tabulated in Table 86. Using the solutions (u) and (v) and a set of real constants C_n and separating in Eq. (s) real and imaginary parts after substitution, we can obtain the following general expressions for the normal deflection w and the stress function F:[1]

$$w = C_1\psi_1(x) + C_2\psi_2(x) + C_3\psi_3(x) + C_4\psi_4(x) + C_5 \qquad (x)$$

$$F = -\frac{Eh^2}{\sqrt{12(1-\nu^2)}}[-C_1\psi_2(x) + C_2\psi_1(x) - C_3\psi_4(x) + C_4\psi_3(x) + C_6 \log x] \qquad (y)$$

To illustrate the use of the foregoing results, let us consider a shallow shell with a very large radius subjected to a point load P at the apex $r = 0$.

In such a case we have to satisfy the obvious condition

$$Q_v = -\frac{P}{2\pi r} = -\frac{P}{2\pi l x} \qquad (z)$$

while w, dw/dr, N_r, and N_θ must be finite at $r = 0$, and w, M_r, and M_θ must vanish for $r = \infty$. Using the first of the expressions (m) to satisfy Eq. (z), we obtain

$$C_6 = \frac{Pa}{2\pi} \frac{\sqrt{12(1-\nu^2)}}{Eh^2}$$

and for the other constants we get the values

[1] It can be shown that a term $C_7 \log x$ must be omitted in expression (x), while a constant term C_8 can be suppressed as immaterial in expression (y).

$$C_4 = 0 \qquad C_3 = \frac{\pi}{2} C_6$$
$$C_1 = C_2 = C_5 = 0$$

Accordingly, the final results are

$$w = \frac{\sqrt{3(1-\nu^2)}}{2} \frac{Pa}{Eh^2} \psi_3(x)$$

$$F = \frac{Pa}{4} \left[\psi_4(x) - \frac{2}{\pi} \log x \right]$$

Since $\psi_3(0) = 0.5$ we obtain for the deflection of the shell at the point of the application of the load the value

$$w_0 = \frac{\sqrt{3(1-\nu^2)}}{4} \frac{Pa}{Eh^2}$$

The distribution of the membrane stresses $\sigma_r = N_r/h$ and $\sigma_\theta = N_\theta/h$ and that of the bending stresses $\sigma'_r = \mp 6M_r/h^2$ and $\sigma'_\theta = \mp 6M_\theta/h^2$ on the upper surface of

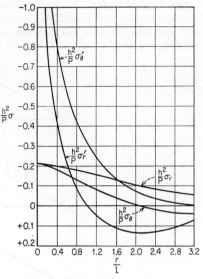

Fig. 278

the shell (for which the upper signs must be taken) are shown in Fig. 278.

When the central load P is uniformly distributed over a circular area of a small radius c, the following results hold at the center of the loaded area $r = 0$:

$$w_0 = \frac{\sqrt{12(1-\nu^2)}}{\pi} \frac{Pa}{Eh^2} \left[\frac{1}{\mu^2} - \frac{\pi}{2\mu} \psi'_4(\mu) \right]$$

$$\sigma_r = \sigma_\theta = -\frac{\sqrt{12(1-\nu^2)}}{2\pi} \frac{P}{h^2} \left[\frac{1}{\mu^2} - \frac{\pi}{2\mu} \psi'_4(\mu) \right]$$

$$\sigma'_r = \sigma'_\theta = \pm \frac{3(1+\nu)}{2} \frac{P}{h^2} \frac{\psi_3(\mu)}{\mu}$$

where
$$\mu = \frac{c}{l} = \sqrt[4]{12(1-\nu^2)} \frac{c}{\sqrt{ah}}$$

Since the expressions (x) and (y) contain six arbitrary constants in all, any symmetrical conditions at the center and on the outer edge of the shell could be fulfilled.

It should be noted also that, as far as bending is concerned, a shallow spherical shell behaves somewhat like a plate on an elastic foundation. This time the characteristic length is given by Eq. (q) instead of expression (a), page 260, which we had in the case of the plate. Thus, when l as defined by Eq. (q) is small compared with the radius of the edge, this is equivalent to the case of a plate on a very rigid foundation. The deflections and the bending moments at the center of such a shell are affected very little by the respective conditions on the outer edge, which only govern the state of the edge zone of the shell.[1]

---

[1] For inextensional deformations of shallow elastic shells see M. W. Johnson and E. Reissner, *J. Math. and Phys.*, vol. 34, p. 335, 1956; singular solutions were considered by W. Flügge and D. A. Conrad, *Stanford Univ. Tech. Rept.* 101, 1956. Some of the

133. Conical Shells. To apply the general equations of Art. 128 to the particular case of a conical shell (Fig. 279a), we introduce in place of the variable φ a new variable y which defines the distance from the apex of the cone. The length of an infinitesimal element of a meridian is now dy, instead of $r_1\, d\varphi$ as was previously used. As a

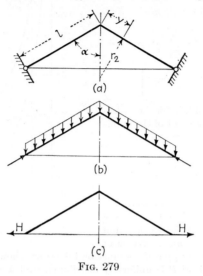

Fig. 279

result of such changes in the variables, the following transformations of the derivatives with respect to φ are necessary:

$$\frac{d}{d\varphi} = r_1 \frac{d}{dy} \qquad \frac{d^2}{d\varphi^2} = \frac{d}{d\varphi}\left(r_1 \frac{d}{dy}\right) = r_1^2 \frac{d^2}{dy^2} + \frac{dr_1}{d\varphi}\frac{d}{dy}$$

With these transformations, the symbol (i) in Art. 128 becomes

$$L(\cdots) = r_2 \frac{d^2(\cdots)}{dy^2} + \left(\frac{dr_2}{dy} + \frac{r_2}{r_1}\cot\varphi\right)\frac{d(\cdots)}{dy} - \frac{1}{r_2}\cot^2\varphi(\cdots) \qquad (a)$$

Observing that for a cone the angle φ is constant and using notation α for $\pi/2 - \varphi$ (Fig. 279), we obtain

$$r_2 = y\tan\alpha \qquad \frac{dr_2}{dy} = \tan\alpha$$

---

previous results were already given by J. W. Geckeler, *Ingr.-Arch.*, vol. 1, p. 255, 1930. General differential equations for curved plates (shallow shells) were established by K. Marguerre, *Proc. Fifth Intern. Congr. Appl. Mech.*, 1938, p. 93. For bending of shallow shells of translation, see G. Ae. Oravas, *Österr. Ingr.-Arch.*, vol. 11, p. 264, 1957, and for nonlinear bending of shallow spherical shells, R. M. Simons, *J. Math. and Phys.*, vol. 35, p. 164, 1956. For bending of shallow helicoidal shells see E. Reissner, *J. Appl. Mechanics*, vol. 22, p. 31, 1955. Helicoidal shells were also considered by R. Malcor, *Travaux*, vol. 32, p. 605, December, 1948, and by L. Solomon, *Priklad. Mat. Mekhan.*, vol. 18, p. 43, 1954. For shallow shells see V. Z. Vlasov, "A General Theory of Shells," Moscow, 1949.

Substituting the expressions into (a) and putting $r_1 = \infty$, the symbol $L(\cdots)$ becomes

$$L(\cdots) = \tan\alpha \left[y \frac{d^2(\cdots)}{dy^2} + \frac{d(\cdots)}{dy} - \frac{1}{y}(\cdots) \right]$$

Equations (321) of Art. 128 are then

$$\tan\alpha \left(y \frac{d^2 U}{dy^2} + \frac{dU}{dy} - \frac{U}{y} \right) \pm i\mu^2 U = 0$$

or, with $U = r_2 Q_\varphi = y \tan\alpha \, Q_y$,*

$$y \frac{d^2(yQ_y)}{dy^2} + \frac{d(yQ_y)}{dy} - Q_y \pm \frac{i\mu^2 y Q_y}{\tan\alpha} = 0$$

Using the notation (j) of Art. 128 and introducing the new notation

$$\lambda^4 = \frac{\mu^4}{\tan^2\alpha} = \frac{Eh}{D}\cot^2\alpha = \frac{12(1-\nu^2)}{h^2}\cot^2\alpha \qquad (b)$$

we finally obtain

$$y \frac{d^2(yQ_y)}{dy^2} + \frac{d(yQ_y)}{dy} - Q_y \pm i\lambda^2 y Q_y = 0 \qquad (c)$$

Considering the first of these equations, we transform it into the known Bessel equation by introducing, instead of y, a new variable

$$\eta = 2\lambda \sqrt{i} \sqrt{y} \qquad (d)$$

which gives

$$\frac{d^2(yQ_y)}{d\eta^2} + \frac{1}{\eta}\frac{d(yQ_y)}{d\eta} + \left(1 - \frac{4}{\eta^2}\right)(yQ_y) = 0 \qquad (e)$$

A similar equation has already been discussed in the treatment of a cylindrical shell of nonuniform thickness (Art. 118). The functions $\psi_1, \ldots, \psi_4$ which were introduced at that time and whose numerical values are given in Table 86 can also be applied in this case. The general solution for yQ_y which satisfies both of the equations (c) can then be represented in the following form:[1]

$$yQ_y = C_1\left[\psi_1(\xi) + \frac{2}{\xi}\psi_2'(\xi)\right] + C_2\left[\psi_2(\xi) - \frac{2}{\xi}\psi_1'(\xi)\right]$$
$$+ C_3\left[\psi_3(\xi) + \frac{2}{\xi}\psi_4'(\xi)\right] + C_4\left[\psi_4(\xi) - \frac{2}{\xi}\psi_3'(\xi)\right] \qquad (f)$$

where $\xi = 2\lambda\sqrt{y}$, and the primes denote derivatives with respect to ξ. From our previous discussion and from the values of Table 86 we know that the functions ψ_1

* The subscript y is used instead of φ in the further discussion of conical shells.
[1] A comprehensive discussion of conical shells is given in F. Dubois' doctoral dissertation "Über die Festigkeit der Kegelschale," Zürich, 1917; this paper also contains a series of numerical examples with curves illustrating the stress distribution in conical shells having various angles at the apex. The case of an arbitrary loading has been considered by N. J. Hoff, *J. Appl. Mechanics*, vol. 22, p. 557, 1955, and thermal stresses by J. H. Huth, *J. Aeronaut. Sci.*, vol. 20, p. 613, 1953.

and ψ_2 and their derivatives ψ'_1 and ψ'_2 have an oscillatory character such that the oscillations are damped out rapidly as the distance y decreases. These functions should be used in investigating the bending of a conical shell produced by forces and moments distributed uniformly along the edge $y = l$. The functions ψ_3 and ψ_4 with their derivatives also have an oscillatory character, but their oscillations increase as the distance y decreases. Hence the third and fourth terms in solution (f), which contain these functions and their derivatives, should be omitted if we are dealing with a complete cone. The two constants C_1 and C_2, which then remain, will be determined in each particular case from the boundary conditions along the edge $y = l$.

In the case of a truncated conical shell there will be an upper and a lower edge, and all four constants $C_1, \ldots, C_4$ in the general solution (f) must be considered to satisfy all the conditions at the two edges. Calculations show that for thin shells such as are commonly used in engineering and for angles α which are not close to $\pi/2$, the forces and moments applied at one edge have only a small effect on the stresses and displacements at the other edge.[1] This fact simplifies the problem, since we can use a solution with only two constants. We use the terms of the integral (f) with the constants C_1 and C_2 when dealing with the lower edge of the shell and the terms with constants C_3 and C_4 when considering the conditions at the upper edge.

To calculate these constants in each particular case we need the expressions for the angle of rotation V, for the forces N_y and N_θ, and for the moments M_y and M_θ. From Eqs. (c) and (d) of Art. 128 we have

$$N_y = -Q_y \tan \alpha$$
$$N_\theta = -\frac{dU}{dy} = -\frac{d(r_2 Q_y)}{dy} = -\frac{d(yQ_y)}{dy} \tan \alpha \qquad (g)$$

From the first of the equations (317) we obtain the rotation

$$V = \frac{1}{Eh} L(U) = \frac{\tan^2 \alpha}{Eh} \left[y \frac{d^2(yQ_y)}{dy^2} + \frac{d(yQ_y)}{dy} - Q_y \right] \qquad (h)$$

The bending moments as found from Eqs. (314) are

$$M_y = -D \left(\frac{dV}{dy} + \frac{\nu}{y} V \right)$$
$$M_\theta = -D \left(\frac{V}{y} + \nu \frac{dV}{dy} \right) \qquad (i)$$

By substituting $y \tan \alpha$ for a in Eq. (r) of Art. 129 we find

$$\delta = \frac{y \sin \alpha \tan \alpha}{Eh} \left[-\frac{d(yQ_y)}{dy} + \nu Q_y \right] \qquad (j)$$

Thus all the quantities that define the bending of a conical shell are expressed in terms of the shearing force Q_y, which is given by the general solution (f). The functions $\psi_1, \ldots, \psi_4$ and their first derivatives are given in Table 86 for $\xi < 6$. For larger values of ξ the asymptotic expressions (296) (page 496) of these functions can be used with sufficient accuracy.

[1] For $\alpha \approx 84°$, F. Dubois found that the stress distribution in a truncated conical shell has the same character as that in a circular plate with a hole at the center. This indicates that for such angles the forces and the moments applied at both edges must be considered simultaneously.

As an example we take the case represented in Fig. 279a. We assume that the shell is loaded only by its weight and that the edge $(y = l)$ of the shell can rotate freely but cannot move laterally. Considering first the corresponding membrane problem (Fig. 279b), we find

$$N_\theta = -qy \sin \alpha \tan \alpha$$
$$N_y = -\frac{qy}{2 \cos \alpha} \tag{k}$$

where q is the weight per unit area of the shell. As a result of these forces there will be a circumferential compression of the shell along the edge of the amount

$$\epsilon_\theta = \frac{1}{Eh}(N_\theta - \nu N_y) = -\frac{ql}{2 \cos \alpha\, Eh}(2 \sin^2 \alpha - \nu) \tag{l}$$

To satisfy the boundary conditions of the actual problem (Fig. 279a) we must superpose on the membrane stresses given by Eqs. (k) the stresses produced in the shell by horizontal forces H (Fig. 279c) the magnitude of which is determined so as to eliminate the compression (l). To solve this latter problem we use the first two terms of solution (f) and take

$$yQ_y = C_1\left[\psi_1(\xi) + \frac{2}{\xi}\psi_2'(\xi)\right] + C_2\left[\psi_2(\xi) - \frac{2}{\xi}\psi_1'(\xi)\right] \tag{m}$$

The constants C_1 and C_2 will now be determined from the boundary conditions

$$(M_y)_{\xi=2\lambda\sqrt{l}} = 0 \qquad (\delta)_{\xi=2\lambda\sqrt{l}} = -\epsilon_\theta l \sin \alpha = \frac{ql^2 \tan \alpha}{2Eh}(2 \sin^2 \alpha - \nu) \tag{n}$$

in which expressions (i) and (j) must be substituted for M_y and δ. After the introduction of expression (m) for yQ_y, expressions (i) and (j) become

$$M_y = \frac{2}{\xi^2}\left\{C_1\left[-\xi\psi_2'(\xi) + 2(1-\nu)\psi_2(\xi) - \frac{4(1-\nu)}{\xi}\psi_1'(\xi)\right]\right.$$
$$\left.+ C_2\left[\xi\psi_1'(\xi) - 2(1-\nu)\psi_1(\xi) - \frac{4(1-\nu)}{\xi}\psi_2'(\xi)\right]\right\} \tag{o}$$

$$\delta = \frac{y \sin \alpha}{Eh}(N_\theta - \nu N_y) = -\frac{\sin \alpha \tan \alpha}{2Eh}\left\{C_1\left[\xi\psi_1'(\xi) - 2\psi_1(\xi) - \frac{4}{\xi}\psi_2'(\xi)\right]\right.$$
$$\left.+ C_2\left[\xi\psi_2'(\xi) - 2\psi_2(\xi) + \frac{4}{\xi}\psi_1'(\xi)\right]\right\}$$
$$+ \frac{\nu \sin \alpha \tan \alpha}{Eh}\left\{C_1\left[\psi_1(\xi) + \frac{2}{\xi}\psi_2'(\xi)\right] + C_2\left[\psi_2(\xi) - \frac{2}{\xi}\psi_1'(\xi)\right]\right\} \tag{p}$$

Substituting $2\lambda\sqrt{l}$ for ξ in expressions (o) and (p) and using Table 86 or expressions (296), we obtain the left-hand sides of Eqs. (n). We can then calculate C_1 and C_2 from these equations if the load q and the dimensions of the shell are given. Calculations show that for shells of the proportions usually applied in engineering practice the quantity ξ is larger than 6, and the asymptotical expressions (296) for the functions entering in Eqs. (o) and (p) can be used. An approximate solution for conical shells, similar to that given in the preceding article for spherical shells, can also readily be developed.

The case of a conical shell the thickness of which is proportional to the distance y from the apex can also be rigorously treated. The solution is simpler than that for the case of uniform thickness.[1]

134. General Case of Shells Having the Form of a Surface of Revolution.

The general method of solution of thin-shell problems as developed in Art. 128 can also be applied to ring shells such as shown in Fig. 220. In this way the deformation of a ring such as shown in Fig. 280a can be discussed.[2] Combining several rings of this kind, the problem of compression of corrugated pipes such as shown in Fig. 280b can be treated.[3] Combining several conical shells, we obtain a corrugated pipe as shown in Fig. 280c. The compression of such a pipe can be investigated by using the solution developed for conical shells in the previous article. The method of Art. 128 is also applicable to more general surfaces of revolution provided that the thickness of the wall varies in a specific manner, that the general equations (315) and (316) obtain the forms (317).[4] The solution of these equations, provided it can be obtained, is usually of a complicated nature and cannot readily be applied in solving practical problems.

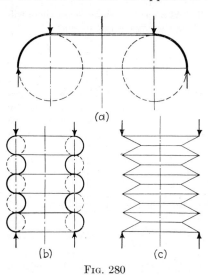

Fig. 280

[1] Meissner, *Vierteljahrsschr. naturforsch. Ges. Zürich*, vol. 60, p. 23, 1915; see also E. Honegger, "Festigkeitsberechnung von Kegelschalen mit linear veränderlicher Wandstärke," doctoral thesis, Zürich, 1919. For the case of an arbitrary loading see H. Nollau, *Z. angew. Math. Mech.*, vol. 24, p. 10, 1944.

[2] Problems of this kind are rigorously treated in the paper by H. Wissler, "Festigkeitsberechnung von Ringflächenschalen," doctoral thesis, Zürich, 1916. For toroidal shells see also R. A. Clark, *J. Math. and Phys.*, vol. 29, p. 146, 1950; for those with an elliptical cross section, see R. A. Clark, T. I. Gilroy, and E. Reissner, *J. Appl. Mechanics*, vol. 19, p. 37, 1952. Short axisymmetrical shells under edge loading have been considered by G. Horvay, C. Linkous, and J. S. Born, *J. Appl. Mechanics*, vol. 23, p. 68, 1956. For calculation of annular, conical, and spherical shells in combination with a flat bottom, see G. Horvay and I. M. Clausen, *J. Appl. Mechanics*, vol. 22, p. 25, 1955.

[3] Such corrugated pipes were considered by K. Stange, *Ingr.-Arch.*, vol. 2, p. 47, 1931. R. A. Clark and E. Reissner have considered some corrugated pipes as "nearly cylindrical shells"; see *J. Appl. Mechanics*, vol. 23, p. 59, 1956. For the theory of such shells see also E. F. Burmistrov, *Priklad. Mat. Mekhan.*, vol. 13, p. 401, 1949.

[4] See Meissner paper, *loc. cit.*

At the same time, all the existing solutions indicate that, for thin shells for which the angle φ is not small, the stresses produced by forces and moments uniformly distributed along the edge are of a local character and die out rapidly as the distance from the edge increases. This fact suggests the use in more general cases of the same kind of approximate solutions as were discussed in the case of spherical shells. Starting with the general equations (315) and (316) (page 538), we neglect on the left-hand sides of these equations the functions U and V and their first derivatives in comparison with the second derivatives.[1] This results in the following simplified system of equations:

$$\frac{r_2}{r_1^2}\frac{d^2U}{d\varphi^2} = EhV$$
$$\frac{r_2}{r_1^2}\frac{d^2V}{d\varphi^2} = -\frac{U}{D} \quad (a)$$

Differentiating the first of these equations twice, we obtain

$$\frac{d^2}{d\varphi^2}\left(\frac{r_2}{r_1^2}\frac{d^2U}{d\varphi^2}\right) = \frac{d^2}{d\varphi^2}(EhV) \quad (b)$$

If after differentiation we again retain on each side only one term containing the derivative of the highest order of the functions U and V, we obtain

$$\frac{r_2}{r_1^2}\frac{d^4U}{d\varphi^4} = Eh\frac{d^2V}{d\varphi^2} = -\frac{Ehr_1^2}{r_2}\frac{U}{D} \quad (c)$$

After the introduction of the notation

$$\lambda^4 = \frac{1}{4}\frac{Ehr_1^2}{r_2D} = 3(1-\nu^2)\frac{r_1^4}{r_2^2h^2} \quad (d)$$

---

[1] This method of obtaining an approximate solution in a general case is due to J. W. Geckeler, *Forschungsarb.*, no. 276, p. 21, Berlin, 1926. An extension of Blumenthal's method of asymptotic integration on the general case of shells in form of a surface of revolution was given by E. Steuermann, *Proc. Third Intern. Congr. Appl. Mech.*, vol. 2, p. 60, 1930. For the method of asymptotic integration see also F. B. Hildebrand, *Proc. Symposia Appl. Math.*, vol. 3, p. 53, 1950. For the general theory of shells and the limits of its application see F. B. Hildebrand, E. Reissner, and G. B. Thomas, *NACA Tech. Note* 1833, 1949; W. Zerna, *Ingr.-Arch.*, vol. 17, p. 149, 1949; A. E. Green and W. Zerna, *Quart. Mech. Appl. Math.*, vol. 3, p. 9, 1950; H. Parkus, *Österr. Ingr.-Arch.*, vol. 4, p. 160, 1950; J. K. Knowles and E. Reissner, *J. Math. and Phys.*, vol. 35, p. 351, 1957; H. Neuber, *Z. angew. Math. Mech.*, vol. 29, p. 97, 1949. The effect of transverse shear deformation on shells of revolution has been considered by P. M. Naghdi, *Quart. Appl. Math.*, vol. 15, p. 41, 1957. Advances in the nonlinear theory of shells are especially due to N. A. Alumyae, K. Z. Galimov, and K. M. Mushtari; see bibliography in A. S. Volmir, "Flexible Plates and Shells," Moscow, 1956. See also Z. Parszewski, *Proc. Ninth Intern. Congr. Appl. Mech.*, vol. 6, p. 280, Brussels, 1957; G. Schwarze, *Ingr.-Arch.*, vol. 25, p. 278, 1957.

Eq. (c) becomes

$$\frac{d^4U}{d\varphi^4} + 4\lambda^4 U = 0 \qquad (e)$$

This is of the same form as Eq. (i) in Art. 130, which was obtained for spherical shells. The difference between the two equations consists only in the fact that the factor λ, given by expression (d), is no longer constant in the general case but varies with the angle φ. Since the function U dies out rapidly as the distance from the edge increases, we can obtain a satisfactory approximate solution of Eq. (e) by replacing λ by a certain constant average value. The approximate solution previously obtained for a sphere can then be directly applied here.

To obtain a more satisfactory result the shell can be divided by parallel circles into several zones for each of which a certain constant average value of λ is used. Beginning with the first zone at the edge of the shell, the two constants of the general solution (329) are obtained from the conditions at the edge in the same manner as was illustrated for a spherical shell. Then all quantities defining the deformations and stresses in this zone are obtained from Eqs. (330). The values of these quantities at the end of the first zone give the initial values of the same quantities for the second zone. Thus, after changing the numerical value of λ for the second zone, we can continue the calculations by again using the general solution (329).[1]

If the factor λ can be represented by the expression

$$\lambda = \frac{a}{b + \varphi}$$

in which a and b are constants, a rigorous solution of Eq. (e) can be obtained.[2] However, since Eq. (e) is only an approximate relation, such a rigorous solution apparently has little advantage over the previously described approximate calculation.[3]

---

[1] An application of this method to the calculation of stresses in full heads of pressure vessels is given in the paper by W. M. Coates, *Trans. ASME*, vol. 52, p. 117, 1930.

[2] See Geckeler, *op. cit.* An application of this solution to the calculation of stresses in a steep-sided dome is given by W. Flügge; see "Statik und Dynamik der Schalen," 2d ed., p. 194, Berlin, 1957. Shells with varying thickness were also considered by C. N. DeSilva and P. M. Naghdi, *Quart. Appl. Math.*, vol. 15, p. 169, 1957.

[3] For bibliography regarding shells, see also the books of W. Flügge, *op. cit.;* K. Girkmann, *op. cit.;* and R. L'Hermite, "Resistance des matériaux théorique et expérimentale," Paris, 1954. The theory of prismatic and pyramidal shells is considered in the above-mentioned books, and also by J. Born, "Faltwerke," Stuttgart, 1954. For bibliography in the field of roof shells see especially A. Aas-Jakobsen, *op. cit.*, and *Proc. Symposium on Concrete Shell Roof Construction,* Cement and Concrete Association, London, 1954.

NAME INDEX

Aas-Jakobsen, A., 456, 507, 524, 532, 554, 568
Adams, E. H., 367
Aggarwala, B. D., 341
Aimond, F., 464
Alblas, J. B., 210, 322, 338
Allen, J. W., 319
Alumyae, N. A., 567
Andrä, W., 362
Ashwell, D. G., 418
Atanasoff, J. V., 205, 350

Badaruddin, S., 281
Baron, F. M., 328
Barta, J., 205
Barthelemy, J., 481
Bassali, W. A., 293
Batanero, J., 507
Bay, H., 185
Bazant, Z., 507
Beer, R., 364
Behlendorff, E., 437
Bereuter, R. O., 532
Berger, E. R., 197, 345, 350
Berger, H. M., 424
Bergsträsser, M., 143
Beyer, K., 61
Biezeno, C. B., 67, 350
Birkhoff, G. D., 104
Black, L. D., 276
Blokh, V. I., 253
Blumenthal, O., 547, 548, 552, 567
Bogunović, V., 208, 211
Bolle, L., 166, 544, 546
Boobnov, I. G., 6, 197, 208, 209, 478
Born, J., 568
Born, J. S., 566
Borowicka, H., 281
Boussinesq, J., 84, 166, 280, 304
Bowen, G., 363
Brandes, G., 319
Bryan, G. H., 311
Burmistrov, E. F., 566
Byerly, W. E., 165

Ćalisev, K. A., 218
Caminade, R., 258
Carrier, G. F., 298, 377
Castigliano, A., 168
Chabloz, E., 304, 307
Chang, C. C., 381
Chenea, P. F., 307
Chien Wei-Zang, 421
Clark, R. A., 566
Clausen, I. M., 566
Clebsch, A., 67, 72, 82, 102, 284, 290, 364, 380
Coan, J. M., 425
Coates, W. M., 568
Collatz, L., 399
Conrad, D. A., 561
Conway, H. D., 207, 298, 303, 306, 381
Cooper, D. W., 524
Courant, R., 335, 341, 362
Csonka, P., 507
Czerny, F., 185, 244

Dantu, M., 276, 362, 363
Day, C. L., 445
Dean, W. R., 298, 442
De Silva, C. N., 568
Deverall, L. I., 298, 340, 341
Dimitrov, N., 219
Dischinger, F., 449, 519
Dixon, A. C., 298
Doganoff, I., 465
Donnell, L. H., 507, 519, 522, 532
Drucker, D. C., 362
Dubois, F., 563, 564

Eck, B., 404
Egger, H., 408
Eichelberg, G., 547
Eichinger, A., 258
Ekström, J. E., 547
Elgood, W. N., 277
El-Hashimy, M., 322, 329
Emde, F., 176

Emperger, F. von, 449
Esslinger, M., 481
Estanave, E., 113, 124
Evans, T. H., 202

Fankhauser, E., 537
Favre, H., 177, 304, 307, 319, 362, 488
Federhofer, K., 197, 403, 408, 415, 488
Feodossiev, V. I., 402
Finsterwalder, U., 519, 520, 529, 532
Fletcher, H. J., 273
Flügge, W., 285, 290, 437, 443, 453, 456, 461, 465, 488, 507, 512, 532, 561, 568
Föppl, A., 260, 290, 419, 423
Föppl, L., 303, 419
Forchheimer, P., 436
Frederick, D., 281
Frey, K., 251
Friedrichs, K. O., 404
Fung, Y. C., 418
Funk, P., 197

Galerkin, B. G., 104, 124, 133, 211, 219, 234, 246, 295, 312, 313, 316, 347
Galimov, K. Z., 567
Garabedian, C. A., 104
Geckeler, J. W., 436, 548, 562, 567, 568
Gerard, F. A., 465
Germain, S., 82
Geyling, F., 461, 465
Gibson, J. E., 524
Gilg, B., 177, 179
Gilroy, T. I., 566
Girkmann, K., 197, 210, 236, 308, 364, 389, 449, 488, 530, 532, 557, 568
Girtler, R., 316
Gittleman, W., 307
Gladwell, G. M. L., 293
Goland, M., 323
Goldenveiser, A. L., 437
Goodier, J. N., 44, 69, 73, 92–94, 98, 280, 321, 362, 382, 417, 469
Gorbounov-Posadov, M. I., 281
Goriupp, K., 208, 213
Göttlicher, H., 316
Goursat, E., 340
Grabert, G., 363
Grammel, R., 67, 303, 350
Gran Olsson, R., 173, 176, 299, 303, 306, 311
Gray, W. S., 488
Green, A. E., 168, 567
Grein, K., 255
Grinberg, G. A., 197
Günther, K., 481

Guyon, Y., 371
Guzmán, A. M., 319

Haas, T., 258
Habel, A., 281
Haberland, G., 362
Hageman, J. G., 258
Hajnal-Konyi, K., 292
Halbritter, F., 277
Happel, H., 277, 313
Hassé, H. R., 298
Hausrath, A. H., 408
Havers, A., 554
Hearmon, R. F. S., 367
Hencky, H., 173, 197, 404, 419, 532
Hersey, M. D., 404
Hertz, H., 260, 269, 280
Hetényi, M., 481, 552
Higdon, R. A., 104
Hilbert, D., 335, 341
Hildebrand, F. B., 567
Hoeland, G., 329
Hoff, N. J., 507, 519, 522, 563
Hofmann, R., 308, 310
Hogg, A. H. A., 280, 281
Höhn, E., 485
Holgate, S., 377
Holl, D. L., 104, 210, 218, 278
Holzer, H., 299
Homberg, H., 375
Honegger, E., 566
Hoppmann, W. H., 368
Horvay, G., 566
Huang, M. K., 207, 298
Huber, M. T., 156, 366, 372, 376
Huffington, N. J., 368
Huth, J. H., 563

Ishkova, A. G., 281

Jahnke, E., 176
Janes, W. C., 281
Jaramillo, T. J., 210, 336
Jenkins, R. S., 524
Jensen, V. P., 218, 236, 319, 360
Johnson, M. W., 561
Johnston, B. G., 532
Jung, H., 104, 281, 336, 340

Kaiser, R., 419, 426
Kantorovich, L. V., 348
Kármán, Th. von, 418
Keller, H., 532, 537

NAME INDEX

Kelvin, Lord, 45, 49, 84, 88, 265, 494
Kent, C. H., 498
Kirchhoff, G., 83, 88, 489
Knowles, J. K., 567
Koepcke, W., 224
Koiter, W. T., 210, 338
König, H., 443
Koppe, E., 104, 418
Koyalovich, B. M., 197
Krettner, J., 319
Krieger, R., 362
Krilov, A. N., 399
Kromm, A., 173
Kudriavtzev, N. V., 298

Laffaille, B., 464
Lagrange, J. L., 82
Lamb, H., 507
Landwehr, R., 363
Lardy, P., 319
Laws, B. C., 197
Lechnitzky, S. G., 364, 367, 368, 377
Lee, G. H., 362
Leibenson, L. S., 104
Leitz, H., 197
Leonhardt, F., 362
Lepper, H. A., 143
Lessells, J. M., 498
Lévy, M., 113, 166, 196, 336, 517
Levy, R. S., 505
Levy, S., 404, 425
Lewe, V., 251, 253, 254, 277, 488
L'Hermite, R., 258, 568
Liebman, H., 362
Linkous, C., 566
Lobo, G., 61
Lockwood Taylor, J., 381
Lösch, F., 116
Lourye, A. I., 293, 341
Love, A. E. H., 72, 102, 197, 311, 507, 510
Luisoni, C. J., 319
Lundgren, H., 532

McCormick, F. J., 281
McDowell, E. L., 547
MacGregor, C. W., 210
MacNeal, R. H., 363
McPherson, A., 404
Magness, L. S., 368
Magnus, W., 146
Malcor, R., 562
Mansfield, E. H., 418
Marcus, H., 92, 123, 205, 220, 236, 253, 254, 360
Marguerre, K., 104, 280, 281, 562

Mariotte, E., 111
Martin, F., 457, 554
Massonnet, C., 371
Maulbetsch, J. L., 96, 162, 481
Meissner, E., 482, 483, 537, 540, 549, 566
Melan, E., 162, 290
Mesnager, A., 362
Michell, J. H., 72, 102, 293
Miesel, K., 518, 532
Moore, R. L., 205
Morcovin, V., 377
Morley, L. S. D., 507
Morse, R. F., 381
Müggenburg, H., 298
Muhs, H., 278
Müller, E., 218
Müller, W., 276, 290
Murphy, G., 277
Mushtari, K. M., 567
Muskhelishvili, N. I., 293, 341
Muster, D. F., 298

Nádai, A., 45, 69, 114, 145, 156, 197, 220, 221, 246, 250, 257, 293, 295, 316, 327, 341, 350, 391, 402, 415
Naghdi, P. M., 281, 307, 567, 568
Naruoka, M., 371
Nash, W. A., 67, 210
Nasitta, K., 293
Navier, C. L., 108, 113, 276
Neuber, H., 567
Nevel, D. E., 276
Newmark, N. M., 143, 206, 207, 236, 245, 415
Nielsen, N. J., 254, 360
Nollau, H., 566
Nowacki, W., 162, 375

Oberhettinger, F., 146
Odley, E. G., 203
Ōhasi, Y., 376
Ohlig, R., 104, 218
Olsen, H., 375
Oravas, G. Ae., 562

Panov, D. Y., 402
Papkovitch, P. F., 114
Parkus, H., 162, 507, 567
Parszewski, Z., 567
Paschoud, J., 307
Pasternak, P., 554
Pearson, K., 72, 82, 83, 489
Perry, C. L., 313
Pettersson, O., 392

Pflüger, A., 369, 436, 532
Pichler, O., 299
Pickett, G., 205, 281
Pigeaud, G., 141
Poisson, S. D., 51, 84
Pozzati, P., 253
Pucher, A., 192, 329, 449, 453, 461

Rabotnov, Y. N., 507, 548, 567
Ramberg, W., 404
Raville, M. E., 281
Rayleigh, Lord, 501
Reinitzhuber, F., 375
Reismann, H., 391
Reissner, E., 166, 168, 173, 281, 293, 319, 322, 341, 396, 415, 449, 558, 561, 562, 566, 567
Reissner, H., 289, 292, 449, 488, 507, 515, 537
Ritz, W., 345, 347
Roark, R. J., 289, 293
Rock, D. H., 197, 351
Roš, M., 258
Rowley, J. C., 281
Rüdiger, D., 532
Runge, C., 443, 488

Sadowsky, M. A., 298
Saint Venant, B. de, 67, 72, 82, 83, 102, 364, 380
Saito, A., 298
Salet, G., 481
Sanden, K. von, 481
Sattler, K., 371
Savin, G. N., 321, 324, 341, 377
Schade, H. A., 372
Schaefer, H., 363
Schäfer, M., 168
Schambeck, H., 319
Schleicher, F., 266, 494
Schmerber, L., 319
Schmidt, H., 290
Schnell, W., 507
Schorer, H., 531
Schultz-Grunow, F., 341, 481, 485
Schultze, E., 278
Schwarze, G., 567
Schwieger, H., 362
Sekiya, T., 298
Sen, B., 313
Sengupta, H. M., 313
Seydel, E., 367
Shaw, F. S., 298, 362
Siess, C. P., 206, 207, 245, 319
Simmons, J. C., 97

Simons, R. M., 562
Smotrov, A., 214, 315
Sneddon, I. N., 104, 336
Soare, M., 465
Sokolnikoff, E. S., 104
Sokolnikoff, I. S., 104
Solomon, L., 562
Sonntag, G., 532
Southwell, R. V., 362
Spotts, M. F., 554
Staerman, I. Y., 548
Stange, K., 404, 566
Sternberg, E., 547
Steuermann, E., 554, 567
Stevenson, A. C., 104, 341
Stiles, W. B., 207
Stippes, M., 408, 425
Stodola, A., 537
Sturm, R. G., 205
Svensson, N. L., 481
Syed Yusuff, 421
Szabo, I., 281

Tait, P. G., 45, 49, 84
Terzaghi, K., 259
Tester, K. G., 464
Thoma, D., 458
Thomas, G. B., 567
Thorne, C. J., 205, 220, 273, 298, 340, 350
Thürlimann, B., 532
Timoshenko, S., 8, 24, 27, 28, 36, 44, 69, 73, 76, 92–94, 98, 113, 115, 131, 190, 197, 280, 308, 318, 362, 382, 389, 390, 393, 395, 396, 398, 400, 417, 431, 468, 469, 481, 498, 512, 514, 516, 533, 548
Todhunter, I., 72, 82, 83, 489
Tölke, F., 257, 341, 554
Torroja, E., 465, 507
Trenks, K., 369
Truesdell, C., 449
Tungl, E., 197

Uflyand, Y. S., 206, 338
Urban, J., 532
Uspensky, J. V., 318

Van der Eb, W. J., 213
Vint, J., 277
Vlasov, V. Z., 457, 507, 513, 522, 523, 562
Volmir, A. S., 404, 415, 567

Wahl, A. M., 61
Watson, G. N., 265

Way, S., 6, 404, 407, 421
Wegner, U., 185
Weil, N. A., 415
Weinmeister, J., 375
Weinstein, A., 197, 351
Westergaard, H. M., 245, 273, 275, 328
Wiedemann, E., 449, 532
Wieghardt, K., 363
Wissler, H., 566
Witrick, W. H., 418
Woinowsky-Krieger, S., 69, 70, 72, 93, 104, 113, 158, 162, 220, 236, 251, 258, 280, 298, 313, 338, 371, 375
Wojtaszak, I. A., 197, 515
Wolf, F., 362

Wood, M. R., 362
Wotruba, K., 363

Yeh Kai-Yuan, 421
Yi-Yuan Yu, 323, 341
Yonezawa, H., 371
Young, D., 203, 205, 206
Young, D. H., 113

Zandman, F., 362
Zerna, W., 449, 507, 567
Zimmermann, H., 470
Zurmühl, R., 362

SUBJECT INDEX

Anisotropic plates, 364
  bending of, strain energy in, 377
  circular, 376
  elliptic, 376
  rectangular, 371
  rigidities of, flexural, 366
Approximate investigation of bending, of continuous plates, 236
  of cylindrical shells, 519
  of shells having form of surface of revolution, 567
  of spherical shells, 547
Asymptotic integration of equation for bending of spherical shells, 547
Average curvature, 35

Bending of plates, to anticlastic surface, 37, 44
  bipolar coordinates in, 290, 298
  characteristic functions in, 334
  under combined action of lateral load and forces in middle plane of plate, 378
  cylindrical (see Cylindrical bending)
  to cylindrical surface, 4
  to developable surface, 47, 418
  elongations due to, 38, 384
  having initial curvature, 27, 393
  influence surfaces in, 328, 329, 332
  by lateral load, 79
  methods in, approximate, 325
  by moments distributed along edges, 37, 180
  photoelasticity in, 362
  plane-stress analogy with, 363
  polar coordinates in, 282
  pure (see Pure bending of plates)
  rigorous theory of, 98
  singularities in, 325
  to spherical surface, 43
  virtual-displacements principle in, 342, 387
  (See also Approximate investigation of bending; Strain energy in bending; specific types of plates)

Bipolar coordinates, use in bending of plates, 290, 298
Boundary conditions, for built-in edges, 83, 171
  for curvilinear boundary, 87
  for cylindrical roof shells, 518, 527
  for elastically supported edges, 86
  for free edges, 83, 171
  Kirchhoff's derivation of, 88
  for simply supported edges, 83, 171
  in using finite differences, 361
Buckling (critical) load, 389, 392
Built-in edge, boundary conditions for, 83
  (See also Clamped edges)

Cantilever plate, 210, 327, 336
  reaction under partial load, 334
Characteristic functions in bending of plates, 334
Circular hole, in circular plate, 17, 58, 303
  in infinitely extended plate, 319
  in square plate, 322
  stress concentration around, 321
Circular inclusion, 323
Circular plates, circular hole in, 58, 61, 303
  with clamped edges, 55, 68, 290
  under combined action of lateral load and forces in middle plane of plate, 391
  corrugated, 404
  critical load for, 392
  deflections of, large, 396
  differential equation for, 53, 54, 283
  on elastic foundation, 259
  under linearly varying load, 285
  loaded, at center, 67
    eccentrically, 290
    symmetrically, 53
    uniformly, 54
  supported at several points, 293
  symmetrical bending of, 51
  theory of bending, corrections to, 70, 72
  of variable thickness, 298, 304

Clamped edges, boundary conditions for, 83, 171
  expressed in finite differences, 361
  circular plates with, 55, 68, 290
  rectangular plates with, 197
Columns, equidistant, plates supported by rows of, 245
Combined action of lateral load and forces in middle plane of plate, 378
  circular plates under, 391
  rectangular plates under, 380, 387
  strain energy in case of, 383
Complex variable method applied in bending, of anisotropic plates, 377
  of isotropic plates, 340
Concentrated load, cantilever plate under, 210, 327, 336
  centrally applied, circular plate under, 67
  cylindrical shell under, 505
  eccentrically applied, circular plate under, 290
  footing slab under, 221, 307
  local stresses under, 69
  on plate on elastic foundation, 263, 267, 275, 280
  rectangular plate under, with clamped edges, 203
    of infinite length, 144
    with simply supported edges, 111, 141
Conical dome, wind pressure on, 451
Conical shells, bending of, 562
  membrane stresses in, 439, 451
Conoidal shell, 465
Constant strength, shells of, 442
Continuous rectangular plates, 229
  approximate design of, 236
  (*see* Large deflections)
  supported by rows of columns, 245
  with two edges simply supported, 229
Corrugated circular plate, 404
Corrugated pipes under axial compression, 566
Corrugated plates, flexural rigidity of, 367
Critical load, for circular plate, 392
  for rectangular plate with supported edges, 389
Curvature, average, 35
  initial, bending of plates having, 27, 393
  measured with reflected light, 363
  principal, 36
  of slightly bent plates, 33
Curved plates, initially, bending of, 27, 393
Curvilinear boundary, conditions for, 87

Cylindrical aeolotropy in plates, 377
Cylindrical bending, of bottom plates in hull of ship, 21
  of plates, 4
  with built-in edges, 13
  differential equation for, 4
  on elastic foundation, 30
  with elastically built-in edges, 17
  with simply supported edges, 6
Cylindrical roof shells, 460, 519, 524
  boundary conditions for, 518, 527
Cylindrical shells, bending of, approximate investigation of, 519
  bent by forces distributed along edges, 478
  deflection of, general equation for, 514
  deflection of uniformly loaded portion of, 516
  deformation of, inextensional, 501
  under hydrostatic pressure, supported at ends, 514
  reinforced by rings, 479
  stress and strain function in investigation of, 522
  symmetrically loaded, 466
  theory of, general, 466
    membrane, 457
  thermal stresses in, 497
  under uniform internal pressure, 475
Cylindrical tanks, with nonuniform wall thickness, 488
  with uniform wall thickness, 485

Deflection, of circular plates, 51, 285
  of elliptical plates, 310, 312
  of laterally loaded plates, 79
  differential equation for, 82
  large (*see* Large deflections)
  limitations regarding, 47, 72
  small, 79
  of plates under combined lateral loading and forces in middle plane, 378
  of portion of cylindrical shell, 516
  of rectangular plates with simply supported edges, 105
    under concentrated load, 111, 141
    due to temperature gradient, 162
    under hydrostatic load, 124
    of infinite length, 4, 149
    partially loaded, 135
    under sinusoidal load, 105
    under triangular load, 130
    uniformly loaded, 109, 113
    of variable thickness, 173
    with various edge conditions, 180
    (*See also* Rectangular plates)

SUBJECT INDEX

Deflection, strain energy method in calculating, 342, 400, 412
  of triangular plates simply supported, bent by moments uniformly distributed along boundary, 94
  under concentrated load, 314, 316
  due to temperature gradient, 96
  uniformly loaded, 313, 317
Deformation of shells without bending, 429
Developable surface, bending of plate to, 47, 418
Diaphragms under uniform pressure, conical, 562
  spherical, 544
Differential equation, for bending of plates, anisotropic, 365
  under combined lateral loads and loading in middle plane of plate, 379
  to cylindrical surface, 4
  with large deflections, 398, 417, 418
  under lateral loads, 82
  for bending of spherical shells, 540
  for deflection of membranes, 419
  for symmetrical bending of cylindrical shells, 468
Discontinuity stresses, in ellipsoidal boiler ends, 484
  in pressure vessels, 483
Displacements in symmetrically loaded shells, 445
Dome, conical, 451
  spherical (see Spherical dome)

Elastic inclusion, 323
Elastic properties of plywood, 367
Elastic solid, semi-infinite, plate resting on, 278
Ellipsoidal ends of boiler, 484
Ellipsoidal shells, 440
Elliptic functions, use in theory of plates, 341
Elliptic paraboloid, shell in form of, 462
Elliptic plates, uniformly loaded, with clamped edges, 310
  with simply supported edges, 312
Elongations due to bending of plates, 38, 384
Energy method, applied in bending, of cylindrical shells, 505
  of plates, 342, 347
  in calculating large deflections, 412, 419
Exact theory of plates, 98
Experimental methods, 362

Finite-difference equation, for deflection of rectangular plates, 351
  for large deflections, 398, 419
  operators used in formulating, 360
  for skewed plates, 357
Flat slabs, 245
  circular, 292
  rectangular, 245
    in form of strip, 255
    over many panels, 245
    over nine panels, 253
    reversed, 276
    rigid connection with column, effect of, 257
Flexural rigidity, of plates, 5
  of anisotropic material, 365
  corrugated, 367
  of shells, 432
Free edge, boundary conditions for, 83, 171
  expressed in finite differences, 361

Green's function in bending of plates, 112, 328, 334
Gridwork system, bending of, 369

Helicoidal shells, 562
Hole (see Circular hole)
Hull of ships, bending of bottom plates of, 21
Hyperbolic paraboloid, shell in form of, 464

Images, method of, 156, 225, 314
Inclusion, elastic, 323
  rigid, in plate, 323
Inextensional deformation, of cylindrical shells, 501
  of plates with large deflections, 418
Infinite length, plate of, 4, 149
Influence surfaces in bending of plates, 328
  example of use, for circular plates, 329
  for continuous plates, 332
  for rectangular plates, 329
Initial curvature, bending of plates with, 27, 393

Large deflections, 396
  approximate formulas for, 400, 410, 416
  calculation of, strain-energy method in, 400, 412

Large deflections, of circular plates, 396
  under concentrated load, 412
  under edge moments, 396
  uniformly loaded, 400, 404, 408
  differential equations for, 398, 417, 418
  of rectangular plates, uniformly loaded, 421, 425
    clamped, 421
    simply supported, 425
Lateral vibration of plates, 334
Limitations on application of customary theory, 47, 72, 165
Local stresses under concentrated load, 69
Long rectangular plates, 4, 149
  bending of, to cylindrical surface, 4
  with built-in edges, 13
  under concentrated load, 144
  on elastic foundation, 30
  with elastically built-in edges, 17
  with simply supported edges, 6
  small initial cylindrical curvature in, 27

Membrane equation, application in bending of plates, 92, 351
Membrane forces in shells, 433
  expressed in terms of displacements, 523, 534
  use of stress function in calculating, 461
Membrane theory of shells, 429
  cylindrical, 457
  in form of surface of revolution, 433
Membranes, circular, deflection of, 403
  differential equation for deflection of, 419
  square, deflection of, 420
Methods, approximate, in bending of plates, 325
Middle plane of plate, 33
Middle surface, of plate, 33
  of shell, 429
Mohr's circle for determination, of curvatures, 36
  of moments, 40, 359
Moments, bending and twisting, of shells, 430
  expressed in terms of displacements, 523, 535
  determined by Mohr's circle, 40, 359
  of plates, relation with curvature, 81, 283
  twisting, 39, 41, 81

Navier solution, for portion of cylindrical shell, 516
  for simply supported plates, 108, 111
Neutral surface, 38

Nonlinear problems in bending of plates, 308
Nonsymmetrically loaded shells, 447

Orthotropic plates, 364

Photoelasticity in bending of plates, application of, 362
"Photostress" method, 362
Plane-stress analogy with bending of plates, 363
Plates (see specific types of plates)
Plywood, elastic properties of, 367
Poisson's ratio, effect on stresses in plates, 97
  numerical value of, 97
Polar coordinates, in bending of plates, 282
  bending and twisting moments expressed in, 283
  differential equation for deflection in, 53, 54, 283
  for large deflections in, 418
  strain energy expressed in, 345, 346
Polygonal plates, 93, 341
Pressure vessels, 481
  discontinuity stresses in, 483
Principal curvature, 33, 36
  planes of, 36
Pure bending of plates, anticlastic surface in, 44
  limitation of deflection in, 47
  particular cases of, 42
  relation between bending moments and curvature in, 39
  slope and curvature in, 33
  strain energy of, 46

Reactions at boundary of plates, relation with deflection, 84
  of simply supported rectangular plates, under hydrostatic load, 128, 132
    under triangular load, 134
    under uniform load, 120
Rectangular plates, anisotropic, 371
  with clamped edges, 197
  under concentrated load, 111, 141, 144, 203
  continuous, 229, 236, 245
  deflection calculation, by energy method, 342, 347
    by finite difference method, 351
    by method of reversion, 349
  (See also Deflection)

SUBJECT INDEX

Rectangular plates, under hydrostatic pressure, 124
  of infinite length, 4, 149
  long (*see* Long rectangular plates)
  partially loaded, 135
  semi-infinite, 221, 225
  simply supported, 105
    under uniform load, 109, 113
    under sinusoidal load, 105
    under triangular load, 130
    under uniform load, 109, 113, 240
  of variable thickness, 173
  with various edge conditions, 180
    all edges built-in, 197, 245
    all edges elastically supported or free and resting on corner points, 218
    all edges simply supported, 105, 240
    three edges built-in, one edge free, 211
      one edge simply supported, 205, 244
    three edges simply supported, one edge built-in, 192, 241
    two adjacent edges simply supported, other edges built-in, 207, 243
    two opposite edges simply supported, one edge free, fourth edge built-in or simply supported, 208
      two others built-in, 185, 242
      two others free or supported elastically, 214
Reflected light, measuring of curvatures with, 363
Relaxation method, 362
Reversion method, 349
Rigid inclusion in plate, 323
Rigidity, flexural, of plates, 5
  anisotropic, 365
  of shells, 432
Rigorous theory of plates, 98
Ring, reinforcing, of spherical dome, 555
Ring-shaped plates, 58, 303
Roof shells, cylindrical, 460, 519, 524

Sector, plates in form of, 295
Semicircular plate, clamped, 298
  simply supported, 295
Semi-infinite rectangular plates, 221
  under concentrated load, 225
  under uniform load, 221
Shallow spherical shells, 558
Shear (*see* Transverse shear)
Shearing forces, of cylindrical shells expressed in terms of displacements, 523

Shearing forces, of plates, relation with deflection, 82, 284
Shearing strain in plate, 41
Shearing stress in plate, 41
Shells, conical, 439, 451, 562
  conoidal, 465
  of constant strength, 442
  cylindrical (*see* Cylindrical shells)
  deflections of, strain energy method of calculating, 505
  deformation of, without bending, 429
  ellipsoidal, 440
  flexural rigidity of, 432
  in form, of elliptic paraboloid, 462
    of hyperbolic paraboloid, 464
    of surface of revolution (*see* Surface of revolution)
    of torus, 441, 566
  nonsymmetrically loaded, 447
  spherical, 436
    wind pressure on, 449
  symmetrically loaded (*see* Symmetrically loaded shells)
Simply supported edges, boundary conditions for, 83, 171
  circular plates with, 56, 68
  rectangular plates with, 105
Singularities in bending of plates, 325
Skewed plates, 318, 357
Spherical dome, under action of its weight, 436
  bending of, approximate analysis, 547
  bending stress calculation for, example, 554
  with edge ring, 555
  membrane forces in, 436
  shallow, 558
  supported at isolated points, 453
  under wind pressure, 449
Strain energy in bending, of anisotropic plates, 377
  of isotropic plates, 88
    expressed in polar coordinates, 345, 346
    for large deflections, 400, 412
    in pure bending, 47
Strain energy method in calculating deflections, of plates, large, 400, 412
  small, 342
  of shells, 505
Stress function, in calculating membrane forces of shells, 461
  in general theory of cylindrical shells, 522
  in resolving equations for large deflections, 413, 417

Stresses in plate, normal, 42
   Poisson's ratio effect on, 97
   shearing, 41, 42
   (*See also* Thermal stresses)
Successive approximation in calculating bending stresses in shells, 552
Surface of revolution, shells having form of, 433, 533
   bending stresses in, 566
   symmetrically loaded, 433, 533
      displacements in, 445
         equations for determining membrane forces in, 434
         particular cases of, 436
Symmetrically loaded shells, 433, 533
   spherical, 436, 540, 547
Synclastic surface, 37

Tanks, of constant strength, 443
   cylindrical, with nonuniform wall thickness, 488
      with uniform wall thickness, 485
   spherical, 437
Thermal stresses, in cylindrical shells, 497
   in plates with clamped edges, 49
   in simply supported rectangular plates, 162
   in spherical shells, 546
   in triangular plates, 95
Thick plates, 69, 72, 98
Toroidal shells, 441, 566
Transforms, use in theory of plates, 336
Transverse shear, effect of, on deflections of plates, 72, 165
   on stresses around hole, 322

Triangular load, rectangular plates under, 130
Triangular plates, clamped in all or two sides, 315
   equilateral, simply supported, 313
      bending, by concentrated load, 314
         by edge moments, 94
         by uniform load, 313
      thermal stresses in, 95
   in form of isosceles right triangle, 316
Twist of surface, 35
Twisting moment, 39
   in terms of deflection, 41, 81

Uniform load, plates under, circular, 54
      rectangular, clamped, 197
         continuous, 229, 236
         simply supported, 109, 113
   portion of cylindrical shell under, 516
   on spherical shell, 544

Variable thickness, plates of, circular, 298, 305
   rectangular, 173
   ring-shaped, 303
Vibration, lateral, of plates, 334
Virtual displacements, application of principle in bending, of plates, 342, 387
   of shells, 505

Wedge-shaped plates, 337
Wind pressure on dome, conical, 451
   spherical, 449